1979

SHOCK

CONTRIBUTORS

IULIU ŞUTEU, M.D., Ph.D.
Professor of Surgery, Bucharest University School of Medicine,
Head of the Division of Surgery, The Emergency Clinical Hospital, Bucharest;
General, the Army Medical Corps,
formerly Surgeon-in-Chief of the Central Military Hospital, Bucharest.

TRAIAN BĂNDILĂ, M.D., Sc.D.
Associate Professor of Anaesthesiology,
Head of the Department of Anaesthesia and Intensive Therapy,
The Central Military Hospital, Bucharest;
Colonel, the Army Medical Corps.

ATANASIE CAFRIŢĂ, M.D., Sc.D.
Senior anaesthesiologist,
Department of Anaesthesia and Intensive Therapy,
The Central Military Hospital, Bucharest;
Colonel, the Army Medical Corps.

ALEXANDRU I. BUCUR, M.D., Sc.D.
Surgeon-in-Charge, The Central Military Hospital, Bucharest;
Captain, the Army Medical Corps;
Assistant Instructor, Bucharest University School of Medicine,
The Emergency Clinical Hospital, Bucharest

VASILE CÂNDEA, M.D., Sc.D.
Surgeon-in-Charge, The Central Military Hospital, Bucharest;
Head of the Division of Cardiovascular Surgery, Fundeni Clinical Hospital;
Lieut. Colonel, the Army Medical Corps.

SHOCK

PATHOLOGY
METABOLISM
SHOCK CELL
TREATMENT

IULIU ŞUTEU, M.D, Ph.D.
TRAIAN BĂNDILĂ, M.D, Sc.D.
ATANASIE CAFRIŢĂ, M.D, Sc.D.
ALEXANDRU I. BUCUR, M.D, Sc.D.
VASILE CÂNDEA, M.D, Sc.D.

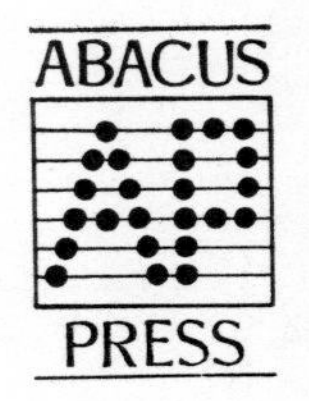

TUNBRIDGE WELLS, KENT

Revised, up-dated English version of the Romanian edition of ŞOCUL, published by Editura Militară, Bucharest, in 1973.

ABACUS PRESS,
Abacus House, Speldhurst Road, Tunbridge Wells, Kent, England.

British Library Cataloguing in Publication Data

Shock: pathology, metabolism, shock cell, treatment.
Bibl. — Index.
ISBN 0-85626-091-6

1. Şuteu, Iuliu 2. Sturza, Ioana 3. Bucur, Alexandru I
4. Dimitriu, Dorin 5. Boothe, S Whittingham

617'.21 RB150.S5

Shock—Collected works
Cellular therapy—Collected works

Printed in Romania

PUBLISHER'S NOTE

This book was originally published in Romania in 1973 under the title Șocul, *with a Foreword by Mr Theodore Burghele, the President of the Academy of the Socialist Republic of Romania. The authors have been working as a team, under the guidance of the Professor of Surgery of the Bucharest University School of Medicine, Iuliu Șuteu, for a number of years and this volume is a summary of their discoveries and the conclusions they have reached on the intricate problem of shock states. For the English edition, the original text has been completely revised and up-dated and additional illustrations have been provided.*

English version of the revised Romanian work
ŞOCUL by I. ŞUTEU, T. BĂNDILĂ, A. CAFRIŢĂ, A. I. BUCUR, V. CÂNDEA
published by EDITURA MILITARĂ, Bucharest, 1973

Translated by IOANA STURZA
Translation Editor S. WHITTINGHAM BOOTHE
Scientific reviewer Prof. ION TEODORESCU-EXARCU, M. D., Ph. D.

Original drawings by ALEXANDRU I. BUCUR, M. D.
Draftsman: LUCIAN TOMA
Jacket Design: EUGEN STOIAN

Editor: Colonel DORIN DIMITRIU, engineer, Head of Scientific Department, Editura Militară.
Lay-out: ANDREI DUŢĂ

CONTENTS

Preface XV

Abbreviations XX

Chapter 1. GENERAL INTRODUCTION
(Iuliu Şuteu)

1.1. Semantics 1
1.2. History 6
1.3. Ecology 13
1.4. Nosology 17
1.4.1. Definitions 17
1.4.2. The present notion of shock 19
1.4.3. Delimitations 22
1.5. Systematics 27
1.5.1. Aetiopathogenic classifications 28
1.5.2. Clinical classifications 36
1.5.3. Evolutionary classifications 38

Selected bibliography 48

Chapter 2. EXPERIMENTAL SHOCK
(Iuliu Şuteu, Atanasie Cafriţă, Traian Băndilă, Alexandru I. Bucur, Vasile Cândea)

2.1. Experimental versus clinical shock 51
2.2. The themes of personal investigations 52
2.3. Rectocolic administration of a complex antishock solution 53
2.4. Hyperthermic shock induced by controlled hyperthermia 55
2.5. Study of the regulation of the splanchnic circulation in traumatic shock 58
2.6. Haemodynamic study of traumatic shock and the therapeutical value of ganglioplegics 60

2.7. Study of the microcirculation in experimental shock with the ^{51}Cr radioisotope 63
2.8. The efficiency of some pharmacodynamic agents in controlling the disturbances of the microcirculation in experimental shock (study carried out with ^{51}Cr radioisotope) 66
2.9. Visceral dynamics of the parameters of the acid-base balance in traumatic shock 75
2.10. Experimental study of the shock cell 83
2.11. Experimental study of the lymphatic circulation in shock 88

Selected bibliography 93

Chapter 3. GENERAL PATHOLOGICAL PHYSIOLOGY
(Iuliu Şuteu, Alexandru I. Bucur)

3.1. The 'human' reaction to shock 95
3.2. Pathophysiological theories 97
3.2.1. Isolated theories 97
3.2.2. Typical pathophysiological moments 99
3.3. Topographic and dynamic stages 100
3.3.1. Topographic stages 101
3.3.2. General pathophysiological scheme 102
3.3.3. Dynamic stages 104
3.4. Fundamental pathophysiological links 106
3.4.1. Importance of the volume of the circulating fluid lost 106
3.4.2. Gradient of circulating fluid lost 107
3.4.3. The importance of catecholamines 109
3.4.4. The importance of perfusion of the microcirculation 110
3.5. Irreversibility and other controversies 112
3.5.1. Catecholamines and irreversibility 112
3.5.2. Fluid deficit and pressor medication 112
3.5.3. Irreversibility and the interval up to its onset 113
3.6. Cybernetic elements 114
3.6.1. Automatic regulation system 114
3.6.2. Automatic biological system 116
3.6.3. Cybernetic scheme of shock 116
3.7. Genetic system 118
3.7.1. Gene and operon 118
3.7.2. Genetic code and homeostasis 120
3.8. Regulation levels 121
3.8.1. Genetic regulation 121
3.8.2. Metabolic regulation 121

3.8.3. Hormonal regulation 123
3.8.4. Neuronal regulation of the carotid glomus 124

Selected bibliography 125

Chapter 4. THE NEURONAL AND ENDOCRINE SYSTEM
(Alexandru I. Bucur, Atanasie Cafriță)

4.1. Complex neuronal systems 129
4.2. Cholinergic and adrenergic space 132
4.2.1. Acetylcholine 135
4.2.2. Catecholamines 135
4.3. The particular properties of neuronal perfusion 138
4.4. The particular properties of neuronal metabolism .. 139
4.5. The criteria of cerebral death 141
4.6. The oscillations of neuronal integration 142
4.7. Neurohormonal structures 144
4.7.1. Hypothalamus 145
4.7.2. The pituitary body 147
4.8. The hormonal systems in shock 148
4.8.1. Hormonal mechanisms; cyclic AMP 149
4.8.2. The cybernetics of the endocrine structures .. 150
4.8.3. Hypothalamo-hypophyseal hormones 151
4.8.4. Thyroid and pancreatic hormones 153
4.8.5. Steroid hormones 154
4.8.6. Renin and erythropoietin 158
4.9. Complex hormonal regulation 159
4.9.1. Blood rheodynamics 159
4.9.2. Volaemia and electrolytaemia; portal osmostat 160
4.9.3. Metabolism regulation 161
4.10. Acute hormonal dysfunctions 162
4.10.1. States of shock due to hormonal deficit 162
4.10.2. States of shock due to hormonal excess 165
4.11. Regulation of the immune system 167
4.11.1. General mechanisms 167
4.11.2. Thymus 168

Selected bibliography 170

Chapter 5. VASCULAR DYNAMICS
(Iuliu Șuteu, Atanasie Cafriță, Alexandru I. Bucur)

5.1. General data 175
5.1.1. Haemodynamostat 177
5.1.2. Immediate vascular reactions 178
5.1.3. Oscillations of the haemodynamostat in shock 180

5.2. The heart in shock 183
5.2.1. Intrinsic factors 183
5.2.2. Metabolism of the heart in shock 184
5.2.3. Regulation of cardiac performance 185
5.2.4. Impairment of cardiac performance 187
5.3. Regional flow 189
5.3.1. Flow-tissular perfusion relationship 190
5.3.2. Features of regional flow 193
5.3.3. The significance of arterial pressure 194
5.4. Venous return 195
5.4.1. The significance of the central venous pressure 195
5.4.2. Characteristics of portal circulation 197
5.5. The microcirculation 198
5.5.1. The sphincteral apparatus and the receptors........ 200
5.5.2. Extrinsic and intrinsic regulation 201
5.5.3. Regional morphofunctional features 204
5.5.4. The microcirculation and the stages of shock 206

Selected bibliography 207

Chapter 6. INTERSTITIAL COMPARTMENT
(Vasile Cândea, Alexandru I. Bucur)

6.1. Collagen 213
6.2. Interstitial hydroelectrolytic oscillations 213
6.3. Interstitial cells 215
6.4. Glycolysis and transcapillary transport 216
6.5. The reticuloendothelial system 217
6.5.1. Phagocytosis 217
6.5.2. Antibody genesis 218
6.6. Interstitial vasoactive substances 219
6.6.1. Histamine 219
6.6.2. Endogenous heparin 220
6.6.3. Serotonin 221
6.6.4. Prostaglandins 222
6.7. Metabolic oscillations of the interstitium 222
6.8. The lymphatic circulation 224
6.8.1. Lymphatic rheodynamics 224
6.8.2. Lymphatic transport in shock 227
6.8.3. Lymphatic flow oscillations in shock........ 228

Selected bibliography 231

Chapter 7. ELEMENTS OF HAEMORHEOLOGY
(Iuliu Şuteu, Atanasie Cafriţă, Alexandru I. Bucur)

7.1. Plasma 236

7.1.1. The functions of the intravascular fluid...... 236
7.1.2. Intercompartmental biological barriers 240
7.1.3. The biology of water 243
7.1.4. Macromolecular systems 243
7.1.5. Osmosis and polyelectrolytes 244

7.2. Blood cells 246
7.2.1. The red blood cell 246
7.2.2. The thrombocyte 249

7.3. The blood 250
7.3.1. Viscosity 250
7.3.2. The laws of intravascular flow 252
7.3.3. Closing pressure 254

7.4. The coagulolytic system 255
7.4.1. The sludge phenomenon 258
7.4.2. Disseminated intravascular coagulation 260
7.4.3. Platelet agglutination and lipid embolism 268
7.4.4. The plasmin system 270
7.4.5. The kinin system 273

Selected bibliography 276

Chapter 8. METABOLISM
(Atanasie Cafriţă, Alexandru I. Bucur)

8.1. Metabolic priorities in states of shock 281
8.2. Energy yielding function 282
8.2.1. Cellular metabolic typology 283
8.2.2. Oxigen and carbon dioxide cycle 285
8.2.3. Proton cycle; acidosis and alkalosis 287
8.2.4. The redox systems and adenylates 291
8.2.5. Electrolytes 294
8.2.6. Vitamins and enzymes 297

8.3. Metabolic lines 298
8.3.1. Organic fuels 298
8.3.2. The muscles 299
8.3.3. The liver cell and the glycostat 300
8.3.4. Adipocyte and the lipostat 303
8.3.5. The nitrate functions 305
8.3.6. *L*-tyrosine metabolism 306
8.3.7. Regulation of the enzymatic chains 307

Selected bibliography 310

Chapter 9. THE SHOCK CELL
(Iuliu Şuteu, Atanasie Cafriţă, Alexandru I. Bucur)

9.1. Points of view 313
9.2. Cellular integration of the macrosystem 314

9.3. Intracellular regulations 316
9.3.1. Enzymatic crossroads 316
9.3.2. Enzymatic dynamics of the shock cell 318
9.4. Cellular organelles 322
9.4.1. The nucleus and the genetic code............ 325
9.4.2. The membrane system.................... 325
9.4.3. The energy-producing apparatus — the mitochondria 326
9.4.4. The digestive apparatus — the lysosomes 327
9.5. Structural aspect of the shock cell 328
9.5.1. Histology of the shock cell................ 329
9.5.2. Infrastructure of the shock cell 329
9.6. Biophysics of the shock cell 330
9.6.1. Elements of irreversible thermodynamics 330
9.6.2. Thermodynamic oscillations in the shock cell.. 334
9.6.3. Therapeutical inflexion points 337
9.6.4. The reversible-irreversible borderline at cellular level 340
Selected bibliography 341

Chapter 10. THE VISCERA IN SHOCK
(Iuliu Șuteu)
10.1. The gastrointestinal tract 346
10.1.1. The stomach in shock 346
10.1.2. The intestine in shock 350
Selected bibliography 353
10.2. The subordinate digestive glands 354
10.2.1. The liver in shock 354
Selected bibliography 357
10.2.2. The pancreas in shock 358
Selected bibliography 360
10.3. The kidney in shock *(Atanasie Cafriță, Traian Băndilă)* 360
Selected bibliography 366
10.4. The lung in shock *(Alexandru I. Bucur)* 367
Selected bibliography 373

Chapter 11. ELEMENTS OF THERAPY
(Iuliu Șuteu, Traian Băndilă, Atanasie Cafriță)
11.1. Management of shock-inducing lesions 376
11.1.1. Notions of organization, management and treatment 378
11.1.2. The role of surgery 379

11.2. Emergency treatment 382
11.2.1. Cardiorespiratory resuscitation 382
11.2.2. Artificial ventilation and oxygen therapy...... 386
11.3. Blood volume substitutes 388
11.3.1. The blood 391
11.3.2. Plasma substitutes 394
11.3.3. Crystalloid solutions 396
11.4. Rheodynamic medication 398
11.4.1. Medication of the α- and β-receptors 398
11.4.2. Cardiotonics and diuretics 402
11.4.3. Coagulolytic equilibrium medication.......... 403
11.5. Cellular re-equilibration 405
11.5.1. Metabolic medication 405
11.5.2. Corticosteroid therapy.............................. 406
Selected bibliography 407

Chapter 12. TYPES OF SHOCK
(Iuliu Şuteu, Atanasie Cafriţă)

12.1. Haemorrhagic shock.............................. 413
12.2. Traumatic shock and its variants 418
12.2.1. General 418
12.2.2. Traumatic shock from firearm wounds 420
12.2.3. The crush syndrome 421
12.2.4. Operative and anaesthetic shock 422
12.2.5. Blast shock 424
12.3. Bacteriemic shock.............................. 425
12.4. Burn shock.............................. 432
12.5. Anaphylactic shock 434
12.6. Cardiogenic shock.............................. 437
Selected bibliography 440

Authors index 443

Between pages 328 and 329 there are 16 colour figures and 25 monochrome electron microphotographs.

PREFACE

Shock has been recognized as a clinical syndrome for almost 2000 years, but today it is still a controversial phenomenon in the medical world. In the historical and geographical universe man is an infinitesimal entity subject to the action of a complexity of endo- and exogenous factors — the 'environmental pressure'. Contemporary life contains a great variety of harmful factors, of varied physical, chemical and/or biological origin that are the potential generators of *critical* situations, of true *states of shock*, from the viewpoint of medical pathology.

Conditioned by the outstanding technical progress in all fields, these deleterious factors increase in numbers and intensity at an unprecedented rate; their action, however, may be avoided, limited, prevented or removed. Although vast programmes of prevention have been drawn up, the world statistics communicated at the first International Symposium on accidents causing shock, held in Washington in 1970, gave impressive figures referring to trauma. Today it is known that one of ten inhabitants of the earth suffers yearly an injury that will necessitate several days hospitalisation, that two-thirds of the cases are youths, and that the care of a casualty with multiple injuries — in shock — demands a surgical intervention that often exceeds the complexity of cardiac surgery and transplant surgery; nevertheless, the organisation of centres of investigation and treatment of cases of severe trauma and shock is only in its initial stage in many countries. Among the present ecologic accidents is the 'trauma epidemic' which, unfortunately, occupies an important place.

Resulting from the multiple conflicts between human beings and the harmful factors of their environment, the world morbidity rate for trauma is greater than the morbidity rate for cardiovascular diseases, and four times that for cancer.

Apart from patients suffering from severe shock with acute functional imbalance, many are brought to operation or to intensive care units suffering from chronic diseases which result in severe functional and morphologic disturbances, and which justify the present notion of 'chronic shock'.

As a rule, in the clinic, a diagnosis of shock seems to be comparatively simple: cold or moist skin, arterial blood pressure below 80 mmHg, cardiac index less than 2500 ml/min/m^2, urinary output less than 25 ml/hour and arterial standard bicarbonate below 2 mEq. But these criteria may become rigid and, therefore, useless if not followed up functionally. In point of fact, understanding of the complex phenomenology of shock implies a deep-rooted knowledge of the pathophysiology of shock and its fluctuating course.

When shock was conceived as exclusively haemodynamic, systemic arterial pressure served as a guide for the clinical stages and the therapeutic approach; later, when shock was understood at rheologic level, the criteria for assessing the stage of shock became the lowering of the effective circulating blood volume and perfusion of the microcirculation. The notion of 'shock viscus' was coined (shock liver, kidney, intestine, lung, etc.). In 1935, Alexis Carrel stated that 'Man is the hardiest of all animals, . . . however, our organs are fragile. They are damaged by the slightest shock. They disintegrate as soon as blood circulation stops'.

The next stage in our understanding of the phenomenology of shock 'descended' to the intimacy of the cellular enzymatic chains; for a long time lactic acidosis came to be considered the most reliable prognostic marker of the severity of shock.

Today we have gone beyond this stage. The pathobiochemical and pathobiophysical events in the course of shock are analysed at molecular and atomic level, where the very logic of the cybernetic and thermodynamic organization of the human body is actually disturbed. Therefore, considering the cell as the lesional site where the main endogenous antishock struggle takes place in the macrosystem, in writing this book we considered that the notion of *shock cell* offers the best basis for studying the disturbances produced by shock down to molecular level, particularly as Science in the last decade has furnished an avalanche of data useful for our understanding of the biologic occurrences. The thermodynamic approach to shock was possible by separate consideration and then integration in the behaviour of the energy subsystems in the shock cell. The characteristics of the shock cell also allow for systematization of the governing and control relationships between the cellular space and the extracellular space.

Most of the studies of the last few years deal with the cellular metabolic response to shock, and monitoring of the biochemical parameters most reliable for assessing the severity of shock. Monitoring has resulted in determination of flow parameters whose variations permit the institution of a prompt replacement of the fluid lost. It has become obvious that the hydroelectrolytic reaction to shock does not consist only in a simple loss of intravascular effective circulating fluid; studies have likewise been carried out on the role of the interstitium and lymphatics in shock, on crystalloids that are now being substituted for colloids for making up volume losses, especially in the first phases of shock. The first adverse responses to dextrans, mannitol, diuretics, etc. are recorded, and the much discussed electrolytic solutions have finally proved extremely useful, permitting re-equilibration of the interstitial space and decrease in the number of perfusions and, therefore, of the untoward post-transfusional reactions.

The reactions of the human body to shock are still imperfect and incomplete, a true shock-inducing agent always being a surprisingly intense pathogenic stimulus to which our homeostatic systems have not yet become fully adapted, and are not yet 'phylogenetically' aware. Therefore, the immediate endogenic measures the body takes, known at present as the *reactional syndrome*, are restricted by the inherent sacrifices this demands. Otherwise stated, the human body is at first panicstricken and tries to save itself by reacting over-hastily and, as a result, a certain disorder ensues which may sometimes in itself be fatal.

Prolonged selective vasoconstriction may give rise to irreparable damage in vital areas, among which the lung, today considered to be in man the most sensitive target organ of hypotension, hypovolaemia and septic states. On the other hand, enzymatic hyperglycolysis endeavours to supply energy to the shock cell, loading it, however, with protons up to the dangerous level of hydrolase autodigestion.

Almost all the occurrences with an immediate adaptive trend that appear in shock are of disproportionate intensity and duration, impeding the therapy that has at times to attenuate or stimulate these same occurrences, depending on the state of shock.

The alpha effect of catecholamines is at first useful both haemodynamically and metabolically (proven by selective blocking with bretilium tossilate of norepinephrine release which has never improved but, on the contrary, aggravated shock), but an excess of alpha stimulation is harmful if it persists in certain areas. Thus, α-lytic drugs have gained favour in the therapy of shock but they do not yet occupy an important place because excessive dilatation of the microcirculation, which they bring about, may be dangerous and difficult to control.

In order to emphasize the uncertainty of the therapeutical measures taken today it is sufficient to recall that some authors recommend β-stimulating drugs (especially in cardiogenic and bacteraemic shock) and others use β-blocking agents (especially for the pulmonary protection they offer). The favourable action of catecholamine precursors in the late phases of shock is likewise well known, while α-stimulants are administered with great care in bacteraemic shock because of their antiphlogistic effect.

Although of bewildering aetiologies, the numerous types of shock encountered in the clinic have a common pathophysiological centre whose segments fall into a fundamental and characteristic biological pattern.

It is almost unanimously recognized today that shock has a multiphase, oscillating development, but the moment of transition from one phase to another is difficult to establish clinically. These phases are, however, of great practical importance since they at times impose paradoxical changes in the treatment applied which must follow closely in the wake of the progressive waves of shock and exercise a continuous intracellular action sustaining the enzymatic and membraneous apparatus.

The oscillating course of shock is also unpredictable, differing entirely from the usual systemic reaction of the after-effects of an injury; hence, in its dynamics, shock often reaches the borderline of irreversibility over which it will pass in the absence of adequate treatment.

It is not yet known for certain, especially at clinical level, which are the limits of the reversible-irreversible; it appears certain, however, that this point is decided at the level of the *shock cell*, long before the terminal clinical signs appear. In the intimacy of cellular metabolism certain transformations occur a long while before they become clinically apparent; therefore the clinician must be permanently aware of the state of shock at enzymatic level and that of the functional reflexes of nucleic acids, so that the intracellular introduction of each therapeutical gesture must be borne in mind. The clinic thus acquires a deeper perspective, without losing its fascinating facet, an insight into the intracellular source of life, the old term of 'clinical picture' taking on three dimensions and becoming a 'clinical relief'.

The surgeon of today is almost always aware of the pathophysiological axis of shock, and faces, together with the anaesthetist-resuscitator, problems of tissular hypoperfusion and disturbances in the homeostasis of the human body at every step.

'The modern surgeon must be acquainted with surgical pathophysiology', asserts Prof. Th. Burghele, recalling at the same time the classical condition announced by René Leriche: '. . . for the sake of quality the surgeon of today must rise from the rank of anatomist to the dignity of physiologist'.

The chapters deal with different aspects in the study of shock and may therefore seem to be unbalanced. The themes contained in previous well-known works may only be outlined or merely mentioned in some chapters, or dealt with in detail in others and then taken up again in further chapters where the subject of shock was deemed to be insufficiently treated or needing to be brought up to date.

The reader will not find in this book precise instructions but only principles concerning the investigation and therapy of each type of shock. The great variety of clinical aspects calls for individual consideration of each particular case; hence we have avoided any attempt at a schematical representation.

This is not a book on clinical practice and treatment of shock but a discussion on the pathophysiology of shock, based on the biological perspective of this phenomenon and with additional new data on biochemical, hormonal, genetic, cybernetic, rheodynamic and thermodynamic aspects with an attempt at bringing into relief some of the facets that until recently have been neglected.

To those who will perhaps regret that certain facts or ideas have not been sufficiently emphasized we answer by quoting Montesquieu: 'When treating a subject it is not necessary to exhaust it; it is sometimes sufficient to draw attention to it'.

Long-term investigation of a complex biological phenomenon such as shock calls for team work and presupposes the cooperation of numerous specialists. The authors of this volume undertook the role of initiating the experimental models and of promoting their implementation; they also made the decision to combine experimental findings with clinical experience and, especially, to take on the responsibility of gathering together suitable material about shock, which by its very nature is composed of diverse elements.

The studies were carried out in Bucharest in the Department of Experimental Medicine of the Central Military Hospital (Dr T. Giurgiu) and the Department of Experimental Nuclear Medicine of the Fundeni Clinical Hospital (Dr O. Cavulea). Histologic examinations were performed and interpreted in the Laboratory of Pathologic Morphology of the Central Military Hospital (Dr I. Strîmbeanu). The biochemical constants and plasma enzymes were determined in the Laboratory of Biochemistry of the same hospital (pharmacist Dr C. Apreotesei) and haematologic studies in the Center of Haematology of the Armed Forces (Dr C. Iercan and Dr R. Perlea) and the Laboratory of Haematology of the Central Military Hospital (Dr C. Satmari).

We wish to express our gratitude to all who have participated and contributed so generously in the multiple themes required in the study of shock; we likewise acknowledge our grateful thanks to doctors D. Singer, P. Ionescu, D. Bărboi, T. Safta, A. Verdeș and a numerous group of enthusiasts without whose practical aid and advice our endeavours would have been inadequate.

To our colleagues who by their work and skill down the years have made of our team a closely knit working unit, we owe more than a debt, for this volume belongs to them, too, to an equal extent.

Our most cordial thanks are due to Prof. Ion Teodorescu-Exarcu for his outstanding help, comments and advice.

We cannot say how deeply indebted we are and how much we value the work of our colleagues in the field of medicine, whatever their specialty, who meet at the patients bedsides to allay their daily suffering by understanding, self-denial and professional integrity.

The editors of the Military Publishing House have always been solicitous and have constantly shown a keen interest in medical publications; thanks to them transforming a rough manuscript into a published book became a pleasant work of collaboration.

THE AUTHORS

Bucharest, December, 1976.

ABBREVIATIONS

AB	actual bicarbonate
ACTH	adrenocorticotrophic hormone
ADH	antidiuretic hormone
ADP	adenosine diphosphoric acid
AMP	adenosine monophosphoric acid
ARAS	ascending reticular activating system
ARF	acute renal failure
ATP	adenosine triphosphoric acid
ATP-ase	adenosine triphosphatase
AV shunt	arteriolo-venular shunt
BB	buffer bases
BE	base excess
BP	blood pressure
CI	cardiac index
cAMP	cyclic AMP (3', 5', AMP)
Cit a, b, c, etc.	cytochrome a, b, c, etc.
CITOX	cytochrome-oxidase
CNS	central nervous system
CO	cardiac output
CoA	coenzyme A
CO_2 total	total carbon dioxide
COMT	catechol-ortho-methyl-transferase
CRF	corticotrophic releasing factor (*see* CRH)
CRH	corticotrophic releasing hormone
C_2SC_0A	active acetate
CSF	cerebrospinal fluid
CVP	central venous pressure
DHAP	dihydroxyacetone phosphate
DIC	disseminated intravascular coagulation
DNA	deoxyribonucleic acid
DOCA	deoxy-11-corticosterone acetate
1,3-DPG	1,3-diphosphoglycerate
1,2-DPG	1,2-diphosphoglycerate
ECBV	effective circulating blood volume
ECG	electrocardiogram
EM	Embden-Meyerhof cycle
EMK	Embden-Meyerhof-Krebs cycle
FA	acid phosphorylase
FAC	acid phosphatase
FAD	flavinadeninedinucleotide
FFA	free fatty acids
FFAC	free fatty acids cycle (Lynen)
FSH	follicular stimulating hormone
FSRH	follicular stimulating releasing hormone
FT	total phosphorylase
FWC	free water clearance
F-6-P	fructose-6-phosphate
F-1,6-DP(FDP)	fructose-1,6-diphosphate
GABA	gamma amino-butyric acid
GFR	glomerular filtration rate
GH (*see* STH)	growth hormone
GOT	glutamic oxalacetic transaminase
GPT	glutamic pyruvic transaminase
GRH	growth releasing hormone
GSH	hydrogenated glutathione

GSSG	dehydrogenated glutathione
G-1-P	glucose-1-phosphate (Cori ester)
G-6-P	glucose-6-phosphate (Robison ester)
G-6-Pase (G-6-PA)	glucose-6-phosphatase
G-6-PDH	glucose-6-phosphate-dehydrogenase
Hb	haemoglobin
HbO_2	oxyhaemoglobin
HES	hydroxyethyl starch
Ht	haematocrite
ICP	intracranial pressure
IDH	isocitric-dehydrogenase
IgM (IgG, IgE, etc.)	immunoglobulin M, G, E, etc.
ICSH	islet cell stimulating hormone
K_i	intracellular potassium
K_e	extracellular potassium
LDH	lactic-dehydrogenase
LH	luteinizing hormone
LPEP	leucinaminopeptidase
LRH	luteinizing releasing hormone
LTH	luteotrophic hormone
MAO	monoamino-oxidase
MDF	myocardial depressor factor
MDH	malic-dehydrogenase
NAD	nicotinamide adenine dinucleotide
NADP	nicotinamide adenine dinucleotide phosphate
Na_i	intracellular sodium
Na_e	extracellular sodium
OAA	oxaloacetic acid
OP	oxidative phosphorylation
OPC	oxidative phosphorylation chain
PAOR	postaggressive oscillating reaction
PASR	postaggressive systemic reaction
PIOR	postinjury oscillating reaction
PISS	postinjury systemic syndrome
PEP	phosphoenolpyruvate
PFK	phosphofructokinase
PGDH	phosphoglyceraldehyde-dehydrogenase
PVP	polyvinyl pyrrolidon
PCO_2	partial carbon dioxide pressure
P_aCO_2	partial (arterial) carbon-dioxide pressure
P_aO_2	partial (arterial) oxygen pressure
P_vO_2	partial (venous) oxygen pressure
P_vCO_2	partial (venous) carbon dioxide pressure
P_AO_2	partial (alveolar) oxygen pressure
P_ACO_2	partial (alveolar) carbon dioxide pressure
pH_a	actual pH
RBC	red blood cell
RBF	renal blood flow
RES	reticuloendothelial system
RF(RH)	releasing factor (releasing hormone)
RI	respiratory index
RNA	ribonucleic acid
SB	standard bicarbonate
SDH	succinic-dehydrogenase
SFEZA	sulphatase
STH *(see* GH)	somatotrophic hormone
T_3	triiodothyronine
T_4	thyroxine
THAM (TRIS)	trihydroxymethylamino-methane
TPR	total peripheral resistance
TRF	thyrotropic releasing factor
TRH	thyrotropic releasing hormone
TSH	thyrostimulating hormone
UO	urinary output
VDM	vasodilatator material
VEM	vasoexcitator material

'... The Elder Pliny used to say that no book was so bad but that some part of it might be profitable ...'

PLINY THE YOUNGER, 61 A.D.

1
GENERAL INTRODUCTION

Shock, although not a scientific term, is a useful one.

WALTER, F. BALLINGER, 1968

1.1 SEMANTICS
1.2 HISTORY
1.3 ECOLOGY
1.4 NOSOLOGY
 1.4.1 Definitions
 1.4.2 The present notion of shock
 1.4.3 Delimitations
1.5 SYSTEMATICS
 1.5.1 Aetiopathogenic classifications
 1.5.2 Clinical classifications
 1.5.3 Evolutionary classifications

1.1 SEMANTICS

From the very beginning the term *shock* has signified a clinical state almost always characterised by hypotension and a particular overall functional exhaustion of the body.

As long ago as 1952 Ronald W. Raven stated that: 'a complex phenomenon such as shock cannot be limited by, or contained within, a definition and any grouping of words is incapable of defining it' [71].

None of the definitions of shock given up to the present time has managed to render its entire meaning. This is the primary argument given by those who propose to give up the term shock.

But the viability of this medical notion has unquestionable biological roots, that guarantee it an established place in clinical medicine. Therefore, in order to maintain the rich intrinsic content of the concept of shock it is better to abandon all attempts at a definition, because the latter will always be poor in characterising elements of the concept. Deloyers' remark is also well known: 'shock is more readily recognised than described and more readily described than defined' [91].

The notion of shock immediately brings up the image of a critical state of the human organism produced by an additional and harmful energy factor.

Following the action of such a *shock-inducing factor* the body is either destroyed or manages to re-equilibrate its vital functions and to survive. These two possibilities are conditioned by numerous variables that characterise the terms of the reaction between *the shock-inducing factor* and *the shocked organism*, to which may be added the therapeutical elements in clinical medicine.

To call shock 'a violent switch of life' [99] appears more literary than useful from the practical point of view but it is nonetheless true that to use the term *shock* as an independent nosological diagnosis is exaggerated and also inaccurate.

For instance, a diagnosis of 'septic shock from septicaemia with *Bacillus proteus*' and 'haemorrhagic shock from rupture of the oesophageal varices' appears correct and precise at first sight, but it contains a syntactic and also a physiopathologic inversion. In itself, shock is not a diagnosis; it is a biological state of the utmost gravity, not only the consequence of a cause that can be diagnosed precisely; it is the severe evolution of a pathological cause, which may however also evolve without shock. A correctly formulated diagnosis would be: 'septicaemia due to *B. proteus* with bacterial shock' and 'upper digestive haemorrhage (from portal hypertension) with haemorrhagic shock'.

The first advantage of such a diagnosis is that of underlining the cause of shock which represents moreover the *principal therapeutic objective*.

A portal hypertension with upper digestive haemorrhage consecutive to the rupture of the œsophageal varices has certain particularities compared, for instance, with haemorrhagic shock from splenic rupture or traumatic transection of an artery.

A second advantage is the possibility of including the stage of shock in the diagnosis (*see* Section 1.5.3); this is of particular importance for the choice and volume of the therapeutical approach. (For instance: 'upper digestive haemorrhage with haemorrhagic shock in a late stage').

The third, and not the least, advantage is that of conceiving the concept of shock as an *evolutive modality* — with a physiopathologic involvement of a certain intensity and duration — within a nosologic entity clearly defined in clinical practice.

Shock cannot be considered only as a symptom, a disease or a syndrome; it cannot be separated from the lesional nosologic entities known in the clinic.

Therefore *shock is only a pathologic manifestation of systemic amplitude and a certain duration*, that appears from the moment in which the primary lesions exceed (in intensity and/or duration of their action) what might be called the 'level of shock'. Above this level haemorrhage induces haemorrhagic shock, septicaemia becomes bacteriemic shock, osteomuscular injury generates traumatic shock, etc.

The concept of shock as an overall biological manifestation is also based upon experimental data that confirm the superposability of the interpretations — at least along general lines— throughout the scale of animals. It is one of the reasons for which the results of experimental investigations may be partly extrapolated for clinical use. But this extrapolation may also be dangerous and often inapplicable in clinical practice (*see* Chapter 2)

The idea that shock cannot be defined — since it is a fundamental biological entity — has often been expressed, as well as the resemblance of shock to an essential pathological condition of the inflammatory, allergic or carcinogenic type. The many similarities between the microcirculation and cellular phenomena in shock and those in an inflammatory reaction (Teodorescu-Exarcu, 1970) are striking [95]; a likeness also exists between the immune phenomena of shock and the allergic reaction, in general, as well as between cellular enzymatic functional and structural disorganisation in shock and that in the neoplastic cell (Şuteu, 1972) [92]. The existence of 'common patterns' gives rise, when the inflammation and allergy exceed the character of local phenomena, to general reactions transformed into the well known forms of shock: bacteriemic and anaphylactic, respectively (even if the latter is still subject to controversy). Also known are the coagulolytic disturbances that accompany certain neoplasias and develop into disseminated intravascular coagulation with consumption coagulopathy and fibrinolysis (phenomena often encountered in shock), as well as the hyperglycolysis of the 'shock cell' with a real Crabtree effect initially described as specific to the cancer cell.

We are therefore faced with 'certain fundamental response modalities' of the human body to the variety of perturbations aiming against its existence and which, in fact, reflect the biological perspective of the phenomenon of shock.

For man, shock may be considered as an unexpected ontogenetic stress of great amplitude, and for the species a factor of phylogenetic modelling (Baglioni, 1966; Monod, 1970; Şuteu and Bucur, 1974).

Survival of the human organism in a critical state of shock-inducing intensity only takes place at the cost of a rapid and brutal adaptation of its standards of genetic adjustment, which in shock are stressed beyond the phylogenetically admitted regimen. Therefore, spontaneous and/or therapeutical re-equilibration after shock implies a series of episomic changes (somations) or even operonal changes — 'acute mutations' (Dobzhansky, 1968), that sharply stress the elasticity of the metabolic pathways of the system (*see* Section 3.7).

It should not be forgotten that the term 'shock' was used at first in cases of severe bleeding caused by war wounds, frequently followed by death after a symptomatology that has become characteristic for the 'state of shock'. Nobody used 'shock' for the phenomena that result in a meningoencephalitis, an acute intoxication, asphyxia, etc.

The existence of an *interval of time*, necessary for a global mobilisation of the energogenetic mechanisms of the macrosystem is important in the use of the concept of shock.

Another condition for using the notion of shock is the *anatomical and functional integrity* (at any rate at the beginning) of the central neuroendocrine system, the 'dispatcher' able to generalise within the organism a reaction of the magnitude of shock. As a rule, head and brain injuries and primary comas are not accompanied by the appearence of shock, but a shock may be accompanied by coma (*see* Chapter 4). On the other hand, prior exhaustion of the neuroendocrine (for instance in cerebral lesions) and of the cellular enzymatic chains (for instance in acute intoxications) brings about a weak general response and death may occur before the onset of the actual shock status. This situation is met with in campaign conditions, in wartime, when soldiers are physically exhausted and underfed, or in peacetime, in aged, needy patients, suffering from irreversible chronic diseases.

The third condition is the existence of a *pathologic haemodynamic phenomenon* that implicitly affects cellular nutrition. In fact, all the reasons for which the *effective circulating blood volume* (ECBV) does not competently fulfil its multiple roles with regard to the tissues, will result in the various known states of shock.

Although exacting and urgent, this conditional triad of shock which has now become classical — interval of time, initial integrity of the neuroglandular system and a haemodynamic phenomenon of pathologic fluid redistribution — was also found to be present in other conditions that were not unanimously considered as generating typical shock (burns accompanied by combustion shock, a septicaemia followed by bacterial shock or myocardial infarct with cardiogenic shock).

Today, to the distress of the amateurs of rules and classifications, there are over a hundred pathologic circumstances that may give rise to shock [15, 20, 33, 39, 42, 78, 81, 87, 101], but all of them have 'a common pathogenic spinal column' upon which distinct traits are inserted, conferring particularities upon each type of shock. Briefly, in all the types of shock, whatever the initiating mechanism, the 'final common pathway' appears to be *inadequate tissue perfusion* (Hardaway, 1966) [40], which is promptly reflected in an *inadequate cellular metabolism* (Şuteu, 1973) [92].

More useful than a restricted definition is an understanding of the evolution of the concept of shock. Initially, shock received its haemodynamic imprint from its connection with a vascular or cardiac wound, and it will probably never escape from that meaning unless replaced by a new term. It is necessary to point out from the very beginning that it is not the capacity of the vascular tree that counts but its fluid content which — bathing the entire cellular 'federation' of the macro-organism — represents the actual *informational canal* of integration of the whole system. This is the reason why circulatory phenomena can never be separated from the concept of shock since without them it would have no meaning (*see* Chapters 5 and 7).

On the battlefield, the casualties with crushed limbs added to the significance of shock another integration theory, that of the coordinating role of the nervous, or better named neuroglandular, system.

It was then observed that systemic circulatory phenomena (determined in the clinic by the peripheral pulse and arterial pressure) are indirect, and not always reliable, indicators of tissular perfusion for which the ECBV in the microcirculation is in fact responsible. The terms *regional flow* and *capillary perfusion* reflect the actual state more precisely because they depend directly upon the ECBV. It becomes very clear that *hypotension is not synonymous with shock* and that all too often in the past the blood pressure has been treated rather than the patient. Tissue perfusion and cardiac output are now realised to be much more important than blood pressure *per se*.

The role of the heart in shock was also clarified. It is true that reduced coronary perfusion in shock leads to early myocardial depression and death. Animals that had myocardial protection from ischemia did not demonstrate immediate dramatic change but slower deterioration of the whole myocardial function; these late effects of shock on the heart are probably induced by the composition of the blood arising from the periphery, in other words 'effects of ischemic body on nonischemic heart' (Mazor, 1973).

As soon as the idea of shock as *inadequate capillary perfusion* (Lillehei,

1964; Hardaway, 1966; Ballinger, 1968) became accepted, it was understood that the true victim of all the acute material changes taking place in the shocked body is *the cell*, whose distress decides upon the reversibility or irreversibility of the shock.

For instance the degree of lactic acidaemia is proportional to the degree of inadequacy in the metabolic cellular pathways. This 'lesional cellular scene' may give to shock the term *acute global cellular dyshomeostasis* (Şuteu, 1973). Today, the redoubts of the intracellular organelles are being conquered one by one and it is also known how they can be helped when suffocated under conditions of shock (*see* Chapters 8 and 11). Biochemical and biophysical 'dissection' of the cellular phenomenology has given us an insight into the thermodynamical aspect of shock, i.e. that of stirring up and then annulling the source of positive entropy (*see* Section 9.6), since the life of a human body is sustained by permanent maintenance of this source of positive entropy.

Uncertainty in the clinical and physiopathological interpretation of the dynamic phenomena that characterise shock still persists: it is not yet known sufficiently to what extent, and when, the redistribution of the blood with centralisation of the circulation, massive increase of the catecholamines and cortisol concentrations in the humors, the appearance of endogenous vasoactive proteases, the triggering of fibrinolysis, etc., are helpful or harmful. Therefore, a deeper, more comprehensive and mobile understanding of the phenomena precipitated in the shocked organism, with their antagonist and agonist tendencies, is necessary; only thus will it be possible to influence them therapeutically to the benefit of the organism. For Dillon (1972) shock remains 'an equation with multiple unknown factors, whose solution necessitates a stochastic analysis of all the parameters of homeostasis'.

Moreover, we must not overlook the fact that the parameters of the intravascular sector that we measure as a rule only reflect indirectly and not always too accurately the perturbations of the intravascular compartment. In order to observe these perturbations direct study of the cell and interstitium throughout the duration of shock becomes absolutely necessary (*see* Chapters 6 and 9).

The *shock cell* is a notion that has clearly imposed itself today in the minds of clinicians and represents a theme and direct tool of investigation, as well as a target for modern therapy [92]. Study of the cell in shock is a means of broadening our understanding of the phenomenology of shock. It likewise reveals the pathophysiological characteristics of the cellular populations that belong to various tissues and organs and which participate to different extents to the suffering of the shocked macrosystem in its entirety. From here the terms *shock liver cell*, *shock erythrocyte*, *shock endotheliocyte*, etc. have been derived. Knowledge of the degree of the effect on the cellular systems in terms of their intravisceral topographic spread, will allow us to increase the efficiency of therapy, since detection of the onset of insufficiency at the level of the 'shock organ' is tardy (*see* Chapter 10). Although, the terms *shocked kidney*, *liver*, *intestine*, or *lung*, are still 'modern terms', in fact they signify a late and very pronounced visceral failure, with a poor evolution.

Shock is a 'mode of response' of the human body to environmental aggression, a mode that is phylogenetically codified in the neuroglandular systems of command and control. The response is coordinated by circulatory and metabolic routes, according to a dynamics whose oscillations are initially corrective but which may very rapidly become harmful in themselves. It is known that almost all vital response phenomena to stimuli of any kind, take place according to a fundamental *pattern*, called the *primordial biological pattern* (Grigg, 1967) [37].

In man the *pattern* expresses the psychical effect produced by an exterior action perceived globally [23]. It sometimes happens that we can only partly express the complex significance of a phenomenon which we grasp by intuition as a whole; this is a pattern that is sometimes indescribable, but which may be thought of and understood owing to the very specificity of its intuitional 'design' (Couffignal, 1966).

Graphically, the biological *primordial pattern* is represented by a curve with unequal slopes: a rapid ascent reaching a peak and a smoother descent tending asymptotically to infinity (Figure 1.1). Muscular jerks, twitching of the nervous system, etc., may be represented graphically in this way. Integration of this curve results in the well known sigmoid that symbolises most biological phenomena (saturation, hormonal levels, enzymatic kinetics, *p*H balance, etc.). The initial curve X may also have a flat shape X' if the energy activity (on the ordinate) is slower. The difference between curves X and X' may be compared to that between a sprinter and a walker (Grigg, 1967), between an acute and a chronic disease, between a hyperergic (violent) form of shock and an attenuated response to injury, etc.

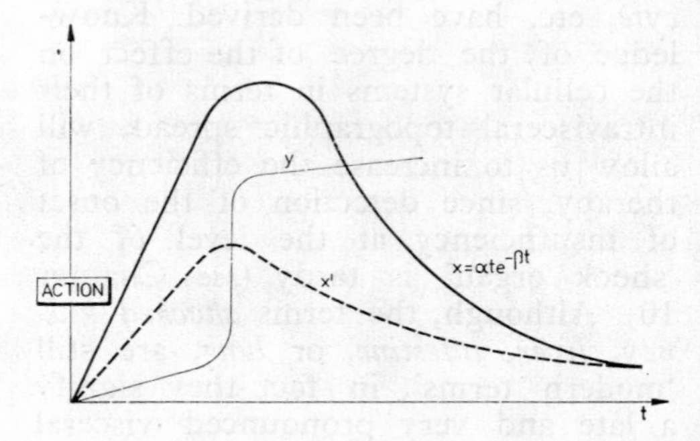

Figure 1.1. Graphical representation of the 'fundamental vital response' function (modified after Grigg). The asymmetrical response, X, may sometimes be slow, X', or an integral of sigmoid, Y.

These curves actually include the entire phenomenology of shock, with its dynamics as understood today down to the intimacy of the cellular systems. If the pattern of the fundamental biological response — considered as an isolated phenomenon — is represented by the asymmetrical curve described, the *evolutive pattern* takes on a *sinusoidal* shape that means mixed positive and negative 'biological pulses' (*see* Section 1.5.3.). Almost all the chapters of this book attempt to plot the pathophysiological events of shock along the *sinusoid*, which appears to express the biological oscillations of homeostasis more reliably

1.2 HISTORY

Some 2000 years ago *Celsus* left us an admirable description of the clinical picture of acute haemorrhage: 'when the heart is wounded, much blood issues, the pulse fades away, the colour of the skin becomes extremely pallid, a cold and malodorous sweat suddenly covers the body, the extremities turn cold and death quickly follows'.

1230. The word shock is used for the first time as a verb in the sense of 'to hit'. In the medieval tournaments this term of English *(to shock)* or Flemish origin *(shocken)*, suggested the blow of the lance on the breastplate.

1523. 'Shock' used as a noun to imply a strong, brutal blow.

1640. The figurative meaning of the word appears, i.e. to perturb, to upset, etc.

1743. The French surgeon *Le Dran* calls shock 'the state of the organism after penetration of a bullet or shell'. In this year the word 'shock' itself first appeared in English medical literature in a translation of Le Dran's work: *Reflections Drawn from Experiences with Gunshot Wounds.*

1795. Shock was for a time represented in the English language by the word 'collapse' from the Latin 'conlapsus'; this connotation persists in the descriptive German word 'Zusammensenkung'. These are the roots of a lot of semantic controversies. It is considered that *James Latta* introduced the term of shock in medicine in the sense used today.

Larrey gives numerous descriptions of the clinical pictures of the severely wounded in Napoleon's wars.

1825. *Travers* uses the term of *traumatic shock* to describe the stage of 'prostration with excitation' or the 'erratic form' in which the patient is agitated, suffocates and asks for water.

1827. *Guthrie* in his treatise on wounds caused by vapours uses the term of shock which he had already employed in his textbook (published in 1815) on wounds of the extremities produced by firearms.

Pirogov described 'traumatic stupor' which corresponds to the first stage of shock: 'with inert hands and feet, the wounded patient is numb, he does not move, cry out, groan or complain, participates in nothing, wants nothing, the body is cold, the skin pale, cadaveric, his sight fixed and hazy, the pulse filiform, hardly perceptible; when questioned he does not answer or only murmurs to himself'.

Golz, following experiments on frogs, demonstrated that shock, by vasomotor reflexes, affects not only the heart, as believed until then, but also the whole circulatory system.

1848. *François Frank* produces shock-like circulatory manifestations by lesion of the trigeminal nerve.

1856. One year after the description of *Thomas Addison*'s disease, *Brown-Sequard* demonstrated experimentally that life without the adrenals is not possible; he also drew attention to the disorders of the capillary circulation in shock, which he called the 'inhibition of exchanges'. He may therefore be considered as the pioneer who introduced capillary and cellular metabolism in the picture of shock.

1867. *Morris* writes a monograph: *Practical treatise on shock after operations and injuries.*

Fischer demonstrates experimentally that dilatation of the deep veins, especially of the portal system, occurs in shock.

Bloodgood calls the accumulation of blood in the venous bed an *intravascular haemorrhage.*

1879. *Mapothei* shows that in shock the fundamental alteration is vasoconstriction and not arteriolar dilatation.

1883. In the *International Encyclopaedia of Surgery* the chapter on the pathologic physiology of shock concludes: 'to sum up shock is an example of reflex paralysis in the strict sense of the word, a suspending of the reflexes affecting all the functions of the nervous system and not limited only to the heart and blood vessels'. Apart from recommending the administration of opium, strychnine, belladonna and digitalis, *Richardson* advocates the intravenous or *per os* use of ammonium for maintaining the fluidity of the blood and counteracting coagulation, a phenomenon considered as characteristic of the advanced stages of shock.

1896. Lister asserts that anesthesia and the newly discovered anesthetic substances will annihilate not only pain but also shock.

1901. In the *American Treatise of Surgery, Park* makes a distinction between shock and collapse, describing the following evolutive variants: *torpid shock, erectile shock* and *delayed shock* and recommends saline solutions as being as efficient as blood.

1899—1903. Crile, the renowned surgeon of Cleveland, adds to the vasomotor inhibitory reflexes of shock the inhibitory reflexes for the heart, lungs, thermoregulation and trophicity. He considers shock as a state of 'vasomotor collapse', therefore a vasodilatation due to the lack of peripheral response to sympathetic stimulation [25]. This great experimenter suggested the existence of a visceral system (including the brain, thyroid, adrenals, liver and muscles) able to transform potential energy into kinetic energy, and asserted that there are as many forms of shock as there are causes which can perturb this multivisceral system, and that the only fluid medium which can fulfil all the functions of the human blood is the transfused human blood.

1905. Malerne takes up again the vasoconstrictive theory of shock initiated by *Mapothei.*

1909. Henderson upholds that the fundamental disorder in shock is venodilatation with diminished venous return to the heart, which might be due to hyperventilation; he thus propounded the theory of 'acapnia'. He observed that the blood pressure falls in spite of the hyperactivity of the vasomotor nervous system and arteriolo-constriction [41].

1912. Trendelenburg observed in the cat the discharge of large amounts of adrenaline in the suprarenal vein after experimental haemorrhage.

1914. Mann noted in the dog with hemorrhage an increase in peripheral resistance due to vasoconstriction.

1917. Bainbridge and *Trevan* show that diminution of the blood volume is the basic pathological element in shock.

1918. Dale and *Richards* demonstrate that histamine produces disturbances of the shock type but fail to find this substance in the traumatic lesional focus [26].

1914—1920. Cannon and *Bayliss,* physiologists, and then the surgeons *Quénu* and *Delbet* develop the theory of toxic resorption in shock. The existence of toxemia starting from disintegration and histolysis of the devitalised tissues lends support to the idea of an early surgical intervention for removal of the toxic focus. A logical explanation is given to shock produced following the removal of a tourniquet and importance is attributed to the liver as the key organ in the centre of pathophysiological circuits. For *Quénu* shock is a 'traumatic toxemia'. *Quénu* and *Duval* first describe the three *H*s of clinical shock: *hypotension, hypothermia, hypoesthesia.*

1919. Cannon offers the concept of 'isovolaemic shock', i.e. without the external loss of fluids [19].

Richet describes anaphylactic shock and was criticised by those who admitted only two entities: *traumatic shock and operative shock.*

1923. Douglas introduces adrenaline in the treatment of shock in order to control vascular collapse.

1928. The natural hormone of the adrenals is extracted and *Swingle* and

Pfiffmer demonstrate that this hormone — *natural cortine* — may restore life both experimentally and clinically.

1930. Blalock and *Phemister* are among the first to show that the late administration of fluids to patients in shock has no longer any therapeutical effect. This *late shock* which no longer responds to any treatment was called *irreversible shock. Blalock* defines shock as a discrepancy between container and content and appraises at more than 35% the amount of circulating fluid retained in the zone of the injured focus [9]. This explains the onset of traumatic shock also without the release of toxins (the most suspect being histamine).

1923—1937. Cannon [19] elucidates the phenomena of blood redistribution produced by catecholamines, the brain and heart being the organs preferentially supplied. He indicates morphine for blocking the vasoconstrictor effect of pain.

1937. Dragstedt coins the term 'peptone shock'.

Moon's monograph appears on the phenomena of capillary dilatation and permeabilisation in shock. *Moon* draws up a pathophysiological scheme valid for all forms of shock. Injuries, burns, microbial or tissular toxins, etc. act upon the capillary endothelium and trigger a vicious circle: capillary atonia — hyperpermeability — plasmexodia — diminution of the circulating blood mass — hemoconcentration — decrease of the cardiac output — reduction of the peripheral circulations — tissular hypoxia — acidosis — capillary atonia. This scheme led to a most important therapeutical step: *filling of the vascular space with fluids. Moon* also reintroduced the notions of isovolemic and hypovolemic shock.

1940. Scudder emphasised the role of hemoconcentration and of hyperpotassemia in the evolution of shock and drew up the principles of the treatment of these disorders.

1941. Bywaters described acute renal insufficiency in the crush injury syndrome [17].

1941. Freeman demonstrated that shock after prolonged perfusion with adrenaline was similar to haemorrhagic or traumatic shock. Thus the term *adrenaline shock* was created, as well as the logical support for the use of adrenolytics in shock.

1941. Grant and *Reeves* published in England the Proceedings of a symposium dedicated to: *Observations of the General Effects of Injury in Man.*

1942. Wiggers and *Werle* underline the role of left heart failure in the progress of shock. *Wiggers* developed a model of experimental haemorrhagic shock still used today.

1944. Westerfield reports on shock caused by kallikreine.

1948. Ahlquist described α- and β-adrenergic tissular peripheral receptors.

1949. Hench and *Kendall* communicate the results obtained following the administration of cortisone in inflammatory diseases, marking the beginning of the well known period of the therapeutical use of glucocorticoid hormones.

1950. Selye defines shock as the sum of all general unspecific phenomena caused by sudden exposure to stimuli for which the organism is not adapted either quantitatively or qualitatively [79].

1950. Wiggers publishes his monumental work on shock.

1951. Knisely explains the phenomenon of the intravascular cellular agglutination and coins the term 'sludge'.

1953. At the Scientific Medical Conference of the Ministry of Health in Romania shock was defined as an *acute corticovisceral syndrome (Țurai).* Although the role of the nervous system was outlined very soon this idea degenerated into a useless absolutism. The practical notions of *compensated* and *decompensated* shock were introduced and the bases of a treatment for the prevention of decompensation were established. Immediately, *Diacenko* and the entire group of research workers of the Kirov Medico-Military Academy in Leningrad rose against the corticovisceral theory of shock and against the terms of compensated and decompensated shock, since this implies overlooking the causal organic lesions whose treatment wrongly becomes of secondary importance.

1954. Ricker proposes as a 'local shock' mechanism the triad: nerve—vessel—tissue, the basis of which is the axon reflex.

1955. Lillehei develops the very useful theory of *stagnating anoxia* that appears owing to venular contraction, a stage in the evolution of shock towards irreversibility. He also proposes pharmacological blocking of the α-receptors for early release of the microcirculation [52].

1955. Fine, Lillehei and *Weil* describe the major role in shock of the intestinal bacteria [33].

1955. Spink and his group in Minnesota initiate investigations on bacterial shock which rapidly becomes a unanimously recognised entity.

1955. Rusher showed that in man the capillary bed contains normally 5-7% of the circulating blood volume, which represents in fact only one-third of the active capillary bed, alternately supplied by local self-regulating processes. In shock, during acidosis, by releasing the entire capillary network, 10-15% of the effective circulating volume may be sequestered.

1955—1960. Zweifach, Chambers and *Shorr* studied the microcirculation under normal conditions and in shock, and together with *Mazur* described the 'vasoexcitatory material' (renin-like) and the 'vasodilatatory material' (produced by the liver in the late phase of shock).

1956. Knisely, Eliot and *Gelin* report on their investigations on the rheology of the blood in shock [46].

1958. Natof and *Sadove* substitute 'cardiovascular collapse' for the term of shock, once more starting semantic controversies that still continue.

1958. Huckabee [42] published the first general work on metabolism in anaerobiosis.

1960. Kulowski's monograph '*Crash Injuries*' was edited by Charles C. Thomas Publisher in the U.S.A. [48].

1960. Werle [102] wrote a detailed study on polypeptides and their importance in tissular perfusion disturbances.

1960. The works of *Sutherland* [93] then appeared concerning the role of adenosine 3-5-phosphate as mediator of the action of catecholamines and other hormones *(cyclic 3′5′-AMP).*

1961. Simeone upholds that the clinical picture of shock results from a plural etiology manifested, however, by a common pathological route: *inadequate perfusion of the organs* [87].

1962. A vast literature concerning all the phenomena of shock is published in '*Shock — Pathogeny and Therapy*', edited by the CIBA Symposium, Academic Press, New York.

1962. Moore opens the stage of investigations on the metabolism of shock, dividing the evolution of shock into three phases: catabolic, intermediary and anabolic. His monograph is now in the ninth edition [60].

1962. Lewis and *Hollander* assume that the precapillary sphincters, bathed in acid or stagnant blood, lose their tonus.

1962. Cope and *Litwin* [22] publish the first works to emphasise the importance of the lymphatic circulation in haemorrhagic shock.

1963. Burghele publishes, in Romania and Paris (Masson), his monograph '*The Kidney in Shock*' [13].

1963. Palmeiro proposes splanchnicectomy in shock.

1963. Zweifach publishes an index of all the works on shock that appeared between 1950 and 1962 [103].

1963. 'The Chemistry of Trauma', a monograph, is published by *Rhoads* and *Howard* [72].

1964. The volume '*Shock*' is published in Boston and in it *Hershey* emphasises the tissular metabolic elements of shock [42].

1964. Hayasaka and *Howard* publish a monographic study on shock including both experimental and clinical data.

1965. Weil demonstrates that large doses of cortisone lower peripheral arteriolar resistance in shock.

1965. Mills and *Moyer* publish a large volume dealing with the pathogeny and treatment of shock, presented at the 12th Hahnemann Symposium [58].

1965. An important Symposium on Shock is held in Boston, where *Hardaway* expounds the problem of microcoagulation, *Litwin* that of blood viscosity, *Moore* the terminal mechanisms that end in death, *Cahill* involvement of the lung, *Shoemaker* the principles of pathological physiology in shock, etc.

1965. Hunt's book on '*Accident Prevention and Life Saving*' is published in England by Livingstone Ltd. [43].

1965. Thal and *Wilson* [98] sum up the facts known to date on the pathological physiology of shock.

1965. Cahill [18] begins his first investigations on the importance of the lung in shock.

1966. Hardaway publishes his fundamental work on intravascular disseminated coagulation, suggesting the treatment of shock with fibrinolytics and heparin [40]. The same subject is dealt with in another book by McKay.

1966. Mandache publishes in Romania a study on experimental and clinical investigations concerning the pathological physiology of circulation and immunity in shock [53].

1960—1966. Laborit and his school [49, 50] develop theoretically and therapeutically the interpretation of shock as a 'disharmonic postaggressive oscillating reaction', also emphasising the therapeutic importance in shock of chlorpromazine and pharmacological blocking of the ascending reticular activator system. A vast programme of research was carried out on intracellular enzymatic anomalies and on the totality of postaggressive reactions. In 1964 Laborit published '*The Reaction of the Organism to Aggression in Shock*' and in 1965 '*Metabolic Regulation*'.

1967. The *Lister Centenary* held in England in April consisted almost entirely of problems linked to the maintenance of homeostasis in conditions of major metabolic, respiratory and circulatory disturbances.

1967. Weil and *Shubin* publish the '*Diagnosis and Treatment of Shock*' with a marked surgical and clinical character [101].

1967. Shoemaker [81] publishes his monograph '*Shock; Chemistry, Physiology and Therapy*'.

1968. In Romania *Chiotan* and *Cristea* in '*Shock — Pathologic Physiology — Clinics — Treatment*' bring up to date the knowledge accumulated concerning our understanding and treatment of this phenomenon [20].

1968. Shepro and *Fulton* wrote the book *'Microcirculation as related to shock'*.

1968. Teodorescu-Exarcu publishes in Romania *'The Science of General Surgical Aggression'* which rapidly becomes a basic textbook for any surgeon who wishes to understand the imperatives of pathophysiological thinking in surgery [95]. In *1974* the same author publishes another book on the same subject *'Biochemical Pathology'* [96].

1968. 'The Management of Trauma' by *Ballinger*, *Rutherford* and *Zuidema* is edited by Saunders [5].

1969. Dogru launches the paradox according to which the irreversibility of shock is due to the disappearance of certain substances useful to the organism that no longer circulate systemically (are blocked), rather than to the appearance of certain harmful substances.

1969. The International *'Symposium on Metabolic Alterations in Shock'* was held in Freiburg.

1970. Porter and *Knight* realised a splendid work about *'Energy metabolism in trauma'*.

1970. The University of Illinois Press published the book entitled *'Corticosteroids in the Treatment of Shock'* by *Shumer* and *Nyhus*.

1970. The 6th European Conference on Microcirculation and Shock was held in Aalborg.

1971. An International Symposium in Wiesbaden was dedicated to *'Protease Inhibition in Shock Therapy'*.

1971. A Symposium on the *'Fundamental Mechanism of Shock'*, held in Oklahoma City, symbolises some of the turning points in our approach to shock and shock therapy.

1971. Hershey and *Del Guercio* dedicated a little monograph to *'Septic Shock in Man'*.

1971. A new monograph edited by *A. P. Thal* together with a group of contributors is published in Chicago: *'Shock. A Physiologic Basis for Treatment'*. O. H. Wangensteen in the preface and W. C. Shoemaker in the presentation emphasise the importance of understanding shock at the biochemical level.

1972. Forscher, *Lillehei* and *Stubbs* wrote a monograph about *'Shock in Low and High-Flow States'* which was published by the Excerpta Medica Foundation. It is an interesting point of view about the classification of shock states.

1972. Ingram [44] reviews in an ample study the latest results obtained in investigations on complement and its role in poststress conditions.

1972. In London appears *'Proceedings of the Conference on Shock'* under the editorship of *Ledingham* and *McAllister*.

1973. Shires, *Carrico* and *Canizaro* offered their monumental work about *'Shock'* edited by Saunders.

1973. The programme for the first Spring Meeting in New York City in Hilton Hotels included many symposia and panel discussions about shock and trauma.

1973. Schwartz, *Lillehei*, *Shires*, *Spencer* and *Storer* [78] develop the theoretical fundamentals of our present day understanding of the pathophysiology and treatment of shock in their volume on *'Principles of Surgery'*.

1976. Messmer and *Sunder-Plassmann* wrote a splendid chapter on shock in their monograph *'Surgical Pathophysiology'*.

At present there are well known centres of investigations of the phenomenology of shock, and intensive therapy divisions in all major clinics in which the 'critical patients' in shock are monitored and managed.

1.3 ECOLOGY

The terminology of the conflicts waged by the human body with the elements of its 'ecological niche' has not been standardised.

The human body in its relationship with the environment, as well as in the optimal registers of its internal medium, finds numerous 'surprises' lying in wait for it: *stress, aggression, shock, trauma*. There is an overlapping of these terms but also differences that do not allow us to abandon any of them. This demands, however, a correct use of each term, especially as each may be used in turn to describe the evolutionary aspects of the same case.

A first notion, very much used at present, is that of *stress*.

This notion — which quickly passed beyond medical frontiers — means in fact a 'tensional state', a progressive exhaustion of energy. The tension between the organism and its environment may develop insidiously, starting with adaptation and passing to stress of the autonomic endocrine system (the hypothalamo-hypophyseo-adrenocortical axis). This is the so-called *repeat chronic stress*, that is a long term tension with two effects: energy spoliation of the neuroendocrine axis and psychical exhaustion due to the constant emotional effort of adaptation (Selye, 1950) [79].

But stress may also be *acute*, in which case the violence of the stimuli causes a brutal discharge of this axis. One of the possible consequences of such a 'corticoid torrent' may also be expressed by the well known 'stress ulcer' encountered in the evolution of hemorrhagic, traumatic or burn shock (*see* Section 10.1.1).

Stress must be considered as a very general term suggesting an imbalance between organism and environment, with obligate elements of emotional-motivational exhaustion and tension of the endocrine axis. *Therefore the shock status appears as a particular case of stress, with an accentuated vascular and metabolic content.*

Another unspecific term, with reference to which shock is likewise a particular case, is that of biological *aggression*. The human body functions as an open energy system, involving constant exchange with the environment. When the quality or intensity of the energy exchanges take on an aggressive character that exceeds the adaptive reactivity of the organism, the physiological and/or anatomical integrity of the system in its entirety is perturbed. If the exchanges between organism and environment usually have equivalent values, *aggression* appears to be a violent *attack* that produces a *lesion*. Initially localised, the lesion bears the imprint of the specificity of the etiological agent; but its systemic reflection loses its initial characteristics and becomes unspecific. The local and the systemic postaggressional alterations form the *lesional syndrome*.

If the violence of aggression does not immediately result in death, the body has a respite to mobilise complex systemic reactions of compensation — which are also unspecific — that form the *reactional syndrome* (Figure 1.2).

From the merging of elements of the lesional and reactional syndrome stems the manifestation, which varies in intensity, of the *postaggressive disease*, which obviously also includes more restricted situations of the 'post-traumatic disease', another unspecific term used by some authors [1, 2, 8, 29, 67, 98, 99, 101].

The term *trauma* encompasses a large range of insults to the body. It

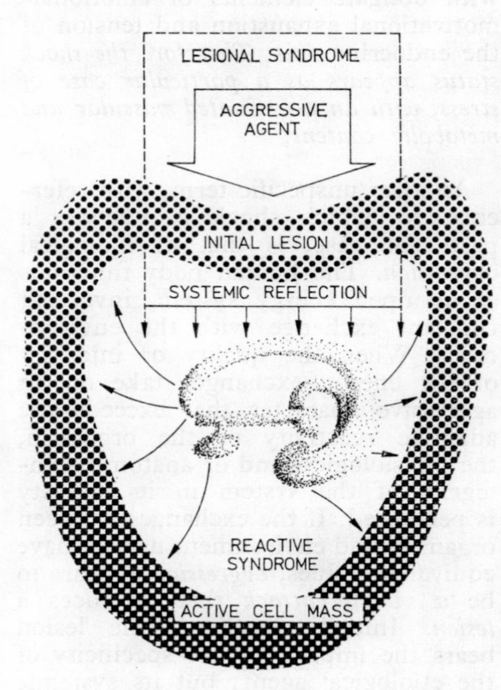

Figure 1.2. The systemic reflection of the *initial specific lesion* is analysed in the central neuro-endocrine circuits and generalised in the organism in the form of a *rectional unspecific syndrome*

is natural to consider *trauma* primarily as the result of the body striking, or being struck by, some object; but one must also consider injuries inflicted by chemical, electrical or thermal insult and those caused by changes in environmetal pressure or gravitational force (Ballinger and Zuidema, 1968) [5].

Some injuries have diffuse systemic effects; others mainly involve one or two organ systems. It is not possible to incorporate such an array of variables into a simple description of the effects of trauma on the body.

Local response to trauma is the wound, which has its complex phases of healing.

The 'local wound' takes various forms: crushing, laceration, a penetrating wound, burn, etc.

Some systemic conditions may have an adverse influence on wound healing: cortisone therapy, anemia, diabetes, uremia, heart failure, scurvy and severe hypoproteinemia. But more remarkable, however, is how little systemic conditions interfere with wound healing.

The ability to heal a wound even while the body is in a continued state of catabolism ('the primacy of the wound') is still one of the great wonders and mysteries of the body (Rutherford, 1968).

The systemic effects of injury are commonly described in terms of fluid and electrolyte shifts, endocrine activity, nitrogen balance and weight change. All these systemic changes are called *the postinjury systemic syndrome* (Brooks, 1967; Sevitt, 1970) [12, 80].

Although the postaggressive pathophysiological alterations that appear in the organism are the same, they have been described in different ways: *postaggressive oscillating reaction* (PAOR) [48, 49], *postaggressive systemic reaction* (PASR) [94, 95, 96], *postinjury systemic syndrome* (PISS) [80], *postaggressive stages* [54, 55], or, more simply, *response to trauma* [60, 61].

Within the broad nosological framework of postaggressive states a number of particular reactions are also included: hemostatic, inflammatory, hypercoagulating and fibrinolytic, immunologic, febrile. They coexist in the complex *shockogenic postaggressive systemic reaction*, which by its amplitude and intensity includes them all, at least to a certain extent. Moreover, shock is characterised by a multicellular metabolic energy strain. Thus, the shock status is classed as an intense and prolonged PASR or PISS (Teodorescu-Exarcu, 1975) [96].

The terms of *stress, injury, aggression* and *shock,* are related and always represent conflicting states between two aleatory variables: the reactivity of the organism and the quantitative and qualitative parameters of the lesional agents (Figure 1.3).

Reactivity is the capacity of the organism to respond to the action of certain environmental agents. This 'response' proves to be specific, proper to each separate individual and depends upon his constitution or 'terrain'. The constitution represents the totality of the factors that confer upon an organism the quality of giving a particular response to environmental stress; it is the cumulative reactivity and morphofunctional individuality of the individual [69, 94]. Study of the constitution (the terrain or soil) — edaphology — has established the existence at hierarchical levels (organism, organ, tissue, cell, organelle) of certain stable (inherited), labile (acquired) and temporary (aleatory) elements.

The following factors form part of the constitution:

(i) *genetic factors,* with their severe, codified self-control but with an appreciable rate of spontaneous mutations. Their expression in phenotypical characters results in biochemical, plastic, functional and immunological particularities;

(ii) *peristatic factors,* which include alimentation, the pre-existing or concomitant pathology, the geographical and social microclimate;

(iii) *chronobiological factors* that establish two nodal points — puberty and climacterium, i.e. the moment of flow and ebb of the sexual hormones;

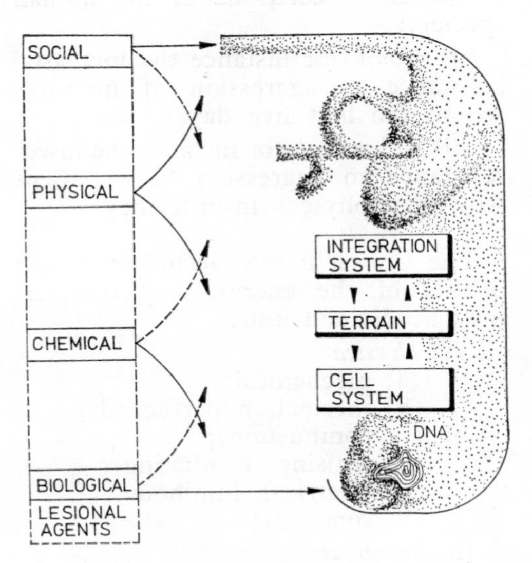

Figure 1.3. The constitution of an organism (the 'terrain') results from the function of the entire cellular system, coordinated by the neuroglandular integration system; according to their nature and intensity the lesional agents may attack, from the beginning, one or both systems.

(iv) *sex-linked factors* confer upon females a better protection against aggression (owing perhaps to the double representation of the somatic material of the *X* gonosome).

On responding to aggression, the constitution expresses an adequate reactivity, normoergy (orthoergy) or a denatured reactivity, pathergy (dysergy), that leads to qualitative disturbances (allergy, anaphylaxis) or global quantitative disturbances (hyperergy, hypoergy or anergy).

Shock may be considered as a dysergy of a hyperergic type.

In their entirety, the factors that make up reactivity contain hereditary or acquired features, pertaining to the:

(i) *species* (such as the increased resistence to cortisone of the human species);

(ii) *group* (for instance the increased resistance to aggression of neonates within the first five days);

(iii) *individual* (for instance the lower resistance to aggression of a patient with hypophyseal insufficiency).

The *lesional agents*, according to the quality of the energy they transfer, may be divided into:

- (i) *physical:*
  - (a) mechanical;
  - (b) thermal (hyperthermia, combustion);
  - (c) ionising (irradiating);
  - (d) electrical, luminous, sonorous;
- (ii) *chemical:*
  - (a) acids, bases, salts;
  - (b) anesthetics;
  - (c) overdose drugs (acetylcholine, epinephrine and norepinephrine, histamine, morphine);
- (iii) *biological:*
  - (a) bacteria;
  - (b) viruses;
- (iv) *psychological:* overwork, emotions;
- (v) *endogenous perturbation of homeostasis.*

The first four groups belong to the external biotic or abiotic environment, and the last group includes metabolic aggressions (hypoxia, acidosis, dysionia, etc.), which likewise reflect a change in the outer medium.

According to their intensity, the lesional agents may be *destructive* (explosions, electrocution, etc.) or *deviating;* only the latter category can bring about a postlesional state with or without shock.

In terms of the degree of alteration of the cellular biologic parameters, Moore (1968) divides the aggressors into three large categories (Figure 1.4). This is a classification of appositional type, whose single criterion is the harmfulness of certain lesional situations. Shock is classed in group 3 where the intensity of the cellular lesions are of maximum amplitude. Not all aggressive agents generate a state of shock; hence, the term of *shock-inducing etiological agent implies the reaching of an obligate lesional intensity threshold* (Simeone, 1963; Dunphy, 1971; Sevitt, 1970; Shires, 1970; Shoemaker, 1970) [29, 80, 82, 83, 87].

A local lesion (a wound of the soft tissues, with or without important vascular lesions, inflammation, etc.) may have a slight effect on global homeostasis and remission may occur by *circumscribed stages.* If, however, the local lesion exceeds in its amplitude a certain limit, called the *shock level,* it will affect homeostasis of the entire organism. The shock thresholds have been established for certain situations: burns over more than 15% of the body surface; haemorrhage more than 25% of the circulating blood volume (reduction of the blood volume to below 2000 ml/m² body surface, or a plasma loss of over 800-900 ml).

The lesions produced by exogenous kinetic energies form the large group of mechanical injuries, which have their etiological roots in two large categories of shock: haemorrhagic and traumatic, that often coexist.

In the clinic, traumatic pathology may be encountered in various forms:

(i) *local tissular alterations* (wounds of the soft parts, limited burns, etc.);

(ii) *loco-regional lesions* (wounds with visceral lesions, fractures);

(iii) *shock-generating lesions*, of the previous types but of increased intensity;

It is of course evident that the clinical meaning of the term shock has only a literary relationship to the common meaning of the expression 'emotional shock' in which a violent

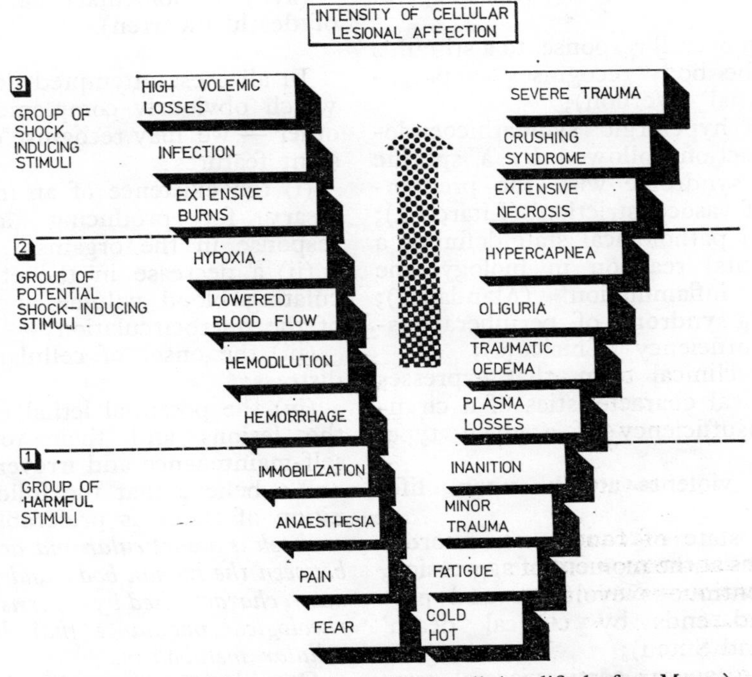

Figure 1.4. Hierarchy of the harmful stimuli (modified after Moore).

(iv) *local and distal, immediate or late complications* of the foregoing forms.

1.4 NOSOLOGY

1.4.1 DEFINITIONS

An essential definition risks incompleteness and an exhaustive definition may be misleading.

There have been many attempts to define shock in precise terms; they usually failed to encompass its many facets or were too complex to be generally adapted.

psychological alteration appears, without however bringing about organic these turbances.

A fairly haphazard list of several attempts at defining shock seems to be more useful than to take one of them as absolute. Thus, shock has been defined as:

(i) 'inadequate capillary perfusion' (Hardaway);

(ii) 'a polyetiologic syndrome characterised by a decrease of tissular blood flow below a certain critical level necessary for the normal display of the obligate metabolic processes' (Fine and Gelin);

(iii) 'A disorderly, postaggressive oscillating reaction, pathological by its intensity and duration' (Laborit);

(iv) 'an alarm reaction appearing after sudden exposure to a stimulus to which the body is not adapted' (Selye);

(v) 'an overall response to a stimulus which the body recognises as potentially lethal' (Vernon);

(vi) 'a hyperergic sympathicoendocrine reaction followed by a specific vascular syndrome with the predominance of vasoconstriction' (Litarczek);

(vii) 'a pathological state defining a fundamental reaction in biology, the same as inflammation' (Mandache);

(viii) 'a syndrome of peripheral vascular insufficiency' (Chiricuță);

(ix) 'a clinical term that expresses the physical characteristics of a circulatory insufficiency of a given type' (Weil);

(x) 'a violent attack upon life' (Dillon);

(xi) 'a state of functional disorder that begins at the moment of aggression, has a continuous evolution and progress and ends by clinical death' (Mareș and Șuteu);

(xii) 'an acute corticovisceral syndrome' (Țurai);

(xiii) 'an unspecific reaction of the mechanisms of adaptation to marked alteration of the organism-environment balance' (Chiotan and Cristea);

(xiv) 'a syndrome due to the action upon the organism of an aggressive factor, that develops following decrease in tissular blood perfusion below a critical level and above a certain duration' (Brînzeu);

(xv) 'a syndrome characterised by an acute reduction in the nutritive blood supply to vital tissues, associated with a disproportion between oxygen supply and demand, and inadequate elimination of the acid metabolites from the tissues' (Messmer and Sunder-Plassmann in *Surgical Pathophysiology*, New York, 1974).

(xvi) 'a rude unhinging of the machinery of life' (Gross);

(xvii) 'a momentary pause in the act of death' (Warren).

In all these attempted definitions — which obviously complement one another — we may recognise certain constant features:

(i) the existence of an interval necessary for producing a systemic response in the organism;

(ii) a decrease in the effective circulating blood volume and perfusion of the microcirculation;

(iii) the onset of cellular metabolic distress;

(iv) the potential lethal character of the lesions and their tendency to self-maintenance and irreversibility.

We believe that the following definition of shock is preferable:

shock is a particular and acute conflict, between the human body and its environment, characterised by a persistent haemorheological imbalance that deeply alters cellular metabolism.

Considering that the intracellular alterations appear to be the most pathognomonic of the state of shock and that they are the consequence of changes in the vascular space, shock might be defined as *the brutal projection in the intracellular space of an imbalance of the extracellular space.* This underlines the fact that primary suffocation of the cell only, as in hydrogen cyanide intoxication, does not represent a state of shock, and on the other hand that shock appears only when interstitial and vascular fluid perturbations bring about cellular distress throughout the entire body; the cellular boundary would thus establish also a limit and a condition to shock.

A cybernetic definition of shock is also possible if it is considered as *a complex and potentially irreversible disorder of the governing and control systems.* Biologically, shock may be defined as an *ontogenetic cellular stress of great amplitude*, genetically as *a violent strain upon the adaptive elasticity of the genome*, and thermodynamically as a *perturbation of the source of positive entropy of the macrosystem.*

Although the attempts at defining this controversial term of shock are incomplete and imperfect, the great practical utility of this term is evident and it will certainly be maintained in the clinic for a long time. The evocative force of this short suggestive notion — similar to a genetic code triplet — is sufficient guarantee.

1.4.2 THE PRESENT NOTION OF SHOCK

If we pass over the details and only follow up the evolution of the content of the notion of shock from the viewpoint of the great theories, several stages may be discerned [11, 16, 56, 66, 74, 84, 103].

The first is the *descriptive-clinical* stage from Celsus up to Larrey and Pirogov.

The second stage, starting at the beginning of the century, is characterised by a tendency to explain shock on the basis of the theory of nervous connections, consisting in reflex inhibition of vasomotricity provoked reflexly by the shock agent (Golz). To this Crile adds a 'reflex inhibition of the lungs, heart, thermoregulation, etc.'. During this period, although shock was mistakenly considered as a 'generalised paralysis of the vascular nervous tone', several findings, still considered to be true today, were made: splanchnic venodilatation (Fischer), which Bloodgood compares with an *intravascular haemorrhage;* inhibition of the capillarocellular exchanges (Brown-Sequard) and the idea of controlling the viscosity of the blood (Richardson).

The following stage, which may be called that of the *toxic theory of shock* was initiated by Cannon and Bayliss and theorised by Quénu and Delbet. The disaggregated damaged tissues, the products of bacterial histolysis and microbial toxins are considered as essential sources of aggressiveness in shock. This toxic theory, of such value in crystallising the notion of shock, also helped to explain the pathological physiology of shock caused by crushing and compression. The investigations carried out with a view to removing from the traumatic focus the 'toxic substances' led to the discovery of histamine (Dale) and the large family of tissular oligopeptides. The main criticism brought to the toxic theory is the lack of specificity of the isolated substances (Kobald, 1964).

Another, predominantly pathophysiological stage, elucidated the idea of the loss of fluid in shock, in which Moon's scheme is the essential guide: irrespective of its etiology, shock consists of a deficit in the circulating blood volume. This set the basis for the substitution of fluids (Figure 1.5).

In the following stage, dominated by Pavlov's School of physiology, shock was considered as a reaction of the adaptation system of the organism to environmental aggression, a reaction in which the central nervous system (CNS), and especially the cortex, plays a fundamental role; thus, the harmful impulses cause cortico-subcortical disorders that bring about disturbances in all the functional systems of the organism. This theory supplied the elements of a unity in the pathological physiology of shock, introduced the notions of compensated and decompensated shock and the idea of the prevention of decompensation.

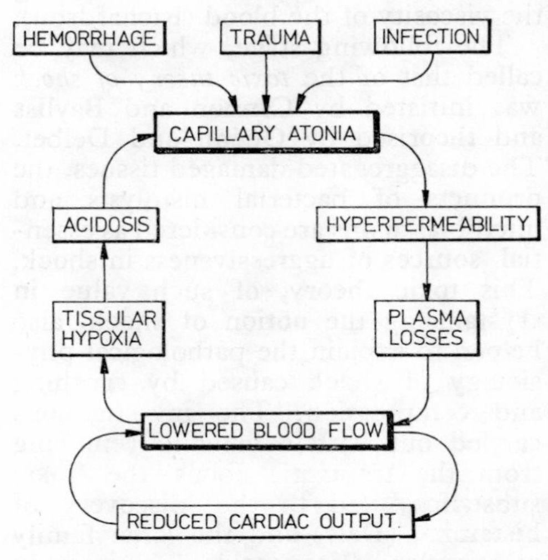

Figure 1.5. The pathophysiological scheme of shock (modified after Moon).

But this theory, held to be absolute, inevitably gave rise to criticisms of which the following may be mentioned:

(i) a purely functional interpretation of compensation and decompensation that ignores the causal organic lesions and favours excessive resuscitation, in many instances without treating the cause (hemorrhage, peritonitis, infection, etc.);

(ii) exaggerated role of the cortex, as being inevitably damaged, at least functionally, in shock; it is known that patients with head injuries do not develop shock and Pavlov himself emphasised the great morphofunctional conservation potential of the cortex in difficult situations. In response to this exaggerated role of the cortex the term of 'local shock' was proposed (Ricker), based upon the axon reflex, as opposed to the great syndrome of 'general shock'.

The contemporary stage is characterised by the tendency to introduce cybernetic elements in the approach to shock, clarifying certain aspects of its pathophysiological scheme. The reticular substance, the diencephalolimbic system and cortex enter together in reverberating circuits that then direct, hierarchically, on autocontrol loops, the endocrine, circulation and intracellular organelle systems.

The unitary 'neuro-endocrine-vasculo-cellular' concept has proved today to be the closest to the reality of shock [12, 21, 31, 35, 45, 53, 58, 88].

Shock affects all the cells, lowering the energy process to a level sufficient for a short time to preserve life, if not the totality of the functions [99]. At present, the investigations carried out are designed to establish the clinical, pathophysiological and therapeutical characteristics of each shock-inducing etiology and, on the other hand, to determine the evolutive markers of the state of shock in general. Today it is unanimously admitted that cellular perfusion is more reliably reflected by the *p*H and lactacidemia than by blood pressure values [32, 51, 65, 68, 66].

Actually, measurement of the blood pressure — such a simple and useful gesture in practice — may lead in the case of shock to errors rather than to precision. In order to frame the intensity and stage of shock, a multivariable matrix of values, forcibly reflecting the microcirculation and cellular metabolism, must be investigated. Siegel, combining nine parameters considered as reference points in the physiological space, found in septic shock that patients with blood pressure (BP) values within normal limits had a severe evolutionary pattern of shock [86].

Our present understanding of shock has undergone essential mutations. In the former concept of shock, according to which the collapse of the BP due to vasodilatation is the fundamental pathological link, the administration of fluids and vasoconstrictor substances, the Trendelenburg position and heat were considered essential. The new concept that views the low tissular perfusion with vasoconstriction in the microcirculation as the foremost disturbance, contraindicates vasoconstrictors and peripheral heating, since they produce an untoward withdrawal of blood from the circulation.

A notion that has appeared lately in the classification of shock by stages is that of *irreversible shock*. At a certain degree of disorganisation in cellular regulation a transition towards irrreversibility occurs, but no objective, repeatable criteria exist that make it possible in practice to guess the very moment at which transition from reversible to irreversible shock occurs (*see* Sections 3.5. and 9.6). Up to the present the term of irreversible shock is unanimously accepted for experimental shock. Functional derangement of the cell membranes, toxemia, exaggerated metabolic defect with denaturation of the nucleic acids and especially labilisation of the lysosomes are only some of the elements that bring about cellular distress [6, 7, 10, 20, 50, 54, 76, 91]. This is a further argument in favour of the study centred upon the *shock cell.*

In this stage of our knowledge, the discouraging meaning of the term of 'irreversible' shock appears to foreclose its use in clinical medicine as it implies the uselessness of therapy (Brînzeu, 1969, proposes that it should only be used as a 'posthumous diagnosis') [11]. But the terms 'refractory shock' and 'irreversible shock' are used in the clinic and in the medical literature not in order to designate an inevitable aggravation, but with the idea of drawing attention to the necessity of mobilising all therapeutical means in order to convert irreversible into the reversible. Demonstrative in this connection is Davis's well known anatomopathological communication: at the necropsy of 169 patients who died from shock diagnosed premortem as 'irreversible', he found in 41% of the cases major lesions that could still have been treated surgically (massive haemorrhage, peritonitis, pulmonary embolism, etc.); when irreversible shock appeared in previously operated patients, lesions that could still be treated surgically were detected at necropsy in 72% of the cases [28]. Surgery thus seems to be a deshocking act of the first importance, an emergency that must not be invalidated by the frequently mistaken label of 'irreversible shock'. In other words surgery must be considered as explorative and curative even in the most severe, and apparently hopeless cases.

The term 'irreversible shock' is therefore present in all the new classifications of shock per clinical stages and may be accepted, since today it has a broader meaning, that of signifying the irreversibility of the progressive self aggravating endogenous phenomenology of shock, in a given biological terrain, *in the absence of any therapeutical correction* (*see* Section 1.5.3). It is obvious that far from suggesting the uselessness of treatment, the stage of irreversible shock emphasises the necessity of highly efficient therapeutical measures [34, 38, 70, 86, 104].

But in clinical practice to talk about *irreversible shock* in the case of a patient would be tantamount to saying that all treatment should be abandoned. Such a concept is dangerous since in human shock the standard laboratory data never can predict with certainty the further course of the critically ill. For this reason, the term 'irreversible shock' theoretically remains correct but near the patient's bed the states of shock which do not respond to the methods of treatment now available in hospital are better only named 'refractory shock'.

1.4.3 DELIMITATIONS

As compared to the general notions of *injury*, *trauma*, *aggression*, and *stress*, *shock* appears as a distinct term since it signifies an overall haemodynamic and metabolic disorder, that appears at a given lesional-aggressive threshold and shows a tendency to an oscillating and, in most instances, aggravating self-maintenance. The essential condition for such a state of shock to occur is disruption of the entire cell population of the organism, mediated by the vascular element, lasting some time and tending to autoaggravation.

It is likewise necessary to dissociate from shock certain terms with a broader pathophysiological meaning, such as: *response to trauma* (Moore), *adaptation syndrome* (Selye), *postaggresive (post-injury) oscillating reaction* (Laborit), *postaggressive systemic reaction* (Teodorescu-Exarcu) or *post traumatic syndrome* and *postaggresive states* (Mareș-Șuteu), all expressing states of conflict of the organism with the factors of its ecological niche. Merging of the *lesional syndrome* elements with those of the *reactive syndrome* bring into relief the main pathophysiological and clinical features of these states, including that of shock [92].

Normally, after a stage of reception and initial imbalance of the macrosystem there follows a mobilisation of the neuroendocrine-metabolic systems of regulation, re-equilibration being obtained by the sequence of stages that resemble graphically a sinusoid, characteristic of any biological process (*see* Section 1.5.3). As a rule, these oscillating conflicting circumstances are *spontaneously reversible* and, being congruous, end by maintaining life and the adaptation to the new environmental situation.

There are however certain particular situations, when, by exceeding the biological variance admitted by homeostasis, the violence of the lesional stimuli cause a systemic reaction so brutal that it affects the synchronism of the regulation feedback loops and impresses upon the process an *oscillating*, *spontaneously irreversible*, *incongruous character*; these are the *shock states*. Bearing in mind the marked tendency to the lack of adaptation in shock-inducing conflicts it is admitted that spontaneous survival is only possible at the expense of rapid functional mutations, which a sufficiently elastic genetic apparatus alone can achieve [5, 63, 79, 90]. To save the individuals of his species man uses exogenous infusion of therapeutical elements which, to be efficient, demand dynamic understanding of the organism in shock.

It is now necessary to give a brief list of the semantic controversies that gravitate around the concept of shock. The excessive preoccupation of some authors to discuss whether shock is or is not a *symptom*, *disease* or *syndrome* seems useless; these medical notions are too narrow to include a phenomenon of evident biological complexity. Therefore the definitions of genetic type, as well as the broad pathophysiological theories are more applicable and efficient in practice. It seems more useful to deal only with the term of *shock* as such and evaluate its viability, especially as in 1955 Moore proposed to give it up, as it is an obsolete, illogical notion, since it includes cases of great clinical and therapeutical variety. Although this proposal still persists today, the term *shock* has been maintained and has enriched its content, simplifying its meanings and offering practical therapeutical models that come closer and closer to pathophysiological reality. The word *shock* imposes itself by its plastic value, suggesting disorders of great amplitude, and by the reflection of a clinical state of progressive gravity, as demanding urgent therapeutical measures.

In 1965 Moore again criticised the term *shock*, considering that it trivialised and hid the true causes of death, which were in his opinion: obliterating pneumonia, infections with a positive hemoculture, acute renal insufficiency, massive haemorrhage and direct injury of the vital organs. Moore's saying 'there are as many types of shock as there are ways of dying' is well known.

As shock appears to be recognised in almost all causes of death at least two amendments should be added:

(i) If we want indeed to maintain the term shock it is necessary to limit it to the pathological states that appear following a brutal action (of certain endo- or exogenous factors) upon a previously equilibrated organism, eliminating the terminal states or supra-acute deaths.

(ii) In itself shock is not a *cause* of death; actually, the same cause that triggered the phenomena of a state of shock is also the final cause of death. This emphasises the fact that shock is one of the *evolutive modalities* of an organism towards death, *one of the most heroic struggles of the human body against death.*

The most frequent terms with respect to which shock state has to be defined are:

Syncope, which from the viewpoint of pathological physiology consists of a temporary loss of consciousness with retrograde amnesia due to an insufficient flow of blood to the brain. As a matter of fact syncope has extracerebral causes that lead to the loss of consciousness and must be differentiated from the numerous intracerebral causes that may reduce the local blood flow (thrombosis, etc.) [85]. Cardiac rate disturbances and intracardiac obstruction, pulmonary embolism and irritation of the carotid sinus are some of the causes of syncope.

In shock, numerous causes that may produce syncopes are encountered (especially 'the state of onset syncope'), but total loss of consciousness (temporary as in syncope, or prolonged as in coma) does not characterise shock. Although between syncope (determined by vagal predominance) and shock (characterised by sympathetic predominance) there are more differences than similarities; the terms 'neurogenic shock' or 'primary shock' have been used for syncope, and proved as useless as they are inaccurate. This inadvertent use of terms may also be due to the uncertainty of the term 'syncope' which is a complex clinical syndrome that signifies not only loss of consciousness but also arrest of the circulation and respiration with collapse of the blood pressure and pulse. The pathologists some time ago differentiated two syncope variants: neurogenic and cardio-vasculo-respiratory.

Today it appears preferable to use the simple terms *respiratory arrest* or *cardiac arrest* and to give up the terms '*cardiac syncope*' and '*respiratory syncope*'. It seems indicated to maintain syncope only for its predominantly neurogenic manifestations and only for the sake of immediate clinical simplicity (because the diagnosis of its etiology is at any rate necessary).

Lypothymia is a symptom and consists in a fugacious and incomplete loss of consciousness and a diminution of the vascular tone. Plethysmographic studies have shown that a fall in BP is due to a sudden dilatation of the blood vessels in the lower extremities. This rapid fall in blood pressure is immediately followed by bradycardia, cold sweating, pallor of the skin, loss of consciousness. The patient shortly recovers without treatment; bradycardia differentiates this so-called 'vagovasal attack' from shock in which tachycardia usually appears.

Coma, a well defined clinical entity, consists in partial or complete loss of consciousness, with initial maintenance of the vegetative functions and diminution or disappearance of the relation functions [100].

Apart from the primary lesions of the brain coma may also appear in metabolical disturbances of cerebral energy produced by hypoxia (hypoxemic or/and haemodynamic), endotoxins, electrolytic imbalance, acidosis, etc., causes frequently encountered also in shock, especially in the final stages.

Therefore, shock is characterised by maintained vigilance of the corticosubcortical formations (with particular dynamics, *see* Section 4.6) whereas coma implies suspension of the mechanisms maintaining the state of vigilance and sustaining the vegetative functions at the level of regulation of the isolated organ. In a late stage, shock may also be accompanied by coma; as Weil says 'shock starts with anxiety and ends with coma' [101].

Collapse is a vagal reaction of short duration which is not followed by organ damage as a result of oxygen deficiency. In 1901 Park defined collapse as a 'pathologic situation with intrinsic causes' in contrast to shock which develops after 'an act of physical violence from the outside'.

In general, shock may be said to symbolise a *blow* and collapse a *breakdown*. Indeed, the term 'collapse' is used in the sense of a global yielding of the forces of the organism caused in principle by circulatory failure. However, it must not be forgotten that collapse means a sudden narrowing of a cavity or lumen up to eventual approximation of the walls. This may occur in the pulmonary alveoles or cerebral ventricles. Only for the vascular tree has it been agreed to consider collapse both as a sharp fall in the content *(haematogen vascular collapse)* and as a sudden dilatation of the vascular bed *(vasomotor collapse)*. In turn vasomotor collapse may be brought about by nervous inhibition *(vasovagal collapse)* or by a lack of reactivity *(exhaustion collapse)* (Table 1.1).

Exhaustion and decompensated collapse also correspond to the late stage of shock, taking on the particular form of *decompensated hyperergic collapse* owing to hypertonia of the postcapillarovenular sphincters (Litarczek, 1974) [52].

Collapse and shock are distinct clinical forms, with pathophysiological roots and therapeutical principles that are almost always contrary, and hence may sometimes coexist. If collapse only defines a pure hemodynamic disturbance, consisting of a *discrepancy* between the content and the container (appearing as a rule after vasodilatation), shock defines a hemodynamic disturbance based upon vasoconstriction and bringing about a volaemic diplacement that inevitably generates a dysmetabolic cellular stage.

The saying: 'collapse is treated with vasopressors and shock chiefly with vasodilators' has become classical [53].

But the hazy points reappear as soon as we try to understand these notions under their clinical dynamic aspect.

Table 1.1

THE DIFFERENTIAL DIAGNOSIS OF COLLAPSE VARIANTS ACCORDING TO THE PATHOLOGICAL SITUATIONS IN WHICH THEY APPEAR

Collapse variants and situations in which they appear	**Vasovagal collapse** (carotid sinus syndrome)	**Areactive-exhaustion collapse** (primary lesional syndrome, anaphylaxis, intoxications)	**Hyperergic collapse** (decompensation in the course of shock)	**Hematogenic collapse** (acute hypovolaemic situations)
Inotropism		—	+—	+—
Chronotropism	—	—	+	+
Cardiac output		—	+—	+—
Venous tonus		—	+	+
Arteriolar tonus	— —	—	+—	+—
Hypotonia *MCR*	+ +	+		
Hypertonia *MCR*			++	+
BP	—	—	+—	—
CVP	—	—+	+—	——
ECBV			——	—
Venous return		—	+—	—
Cellular perfusion			——	—
Skin temperature	+	+—	—	——

Considered as a dissociation between the circulating fluid and the vascular trunk, collapse is a fugacious clinical period since there immediately follows either a *cardiac arrest* when the circulatory collapse is the clinical manifestation of a deficit of the central pump, and then we prefer to use the latter term, or the collapse in itself represents a *pathogenic onset coefficient* (for instance in massive haemorrhage), when the persistence of tissular hypoperfusion will add, with time, cellular distress and a state of shock. It is likewise possible that the deficit of the central pump should not result in an irreversible cardiac arrest but should bring about tissular hypoperfusion which, when maintained for some time may generate the so-called *cardiogenic shock*, an assumption which at present meets with a general consensus.

The reality of these phenomena has shown that collapse may exist without shock, as well as shock without collapse episodes. When do the latter appear in the developmental stages of shock? First of all in the onset — during that rapid period that has so many names — collapse may be one of the pathogenic coefficients generating shock, but not the only one. Collapse may also appear in the late stages of shock when the vascular tonigenic mechanisms break down (this atonia refers to the micro- rather than to the macrocirculation). For the relationship between shock and collapse see also the paragraph on the derivatives of the notion of shock.

The 'anesthesiological' separation of collapse into hot and cold (Grant and Reeve, 1941) emphasises a particularly interesting fact: in the systemic circulation there is hypotension, whereas the microcirculation may be closed *(cold collapse)* or dilated *(hot collapse)*. This dissociation also appears in certain clinical classifications of shock *(see* Section 1.5.2).

Acute peripheral circulatory failure is a term very close in meaning to that of circulatory collapse, but in contrast to the latter it may also be included in the broad clinical classification of all circulatory failures, stressing the peripheral perfusion deficit. It is perhaps the clinical term that should be preferred to that of collapse, for the episodes of decrease in the ECBV that develop in the course of shock. The etiopathogenic classification of acute peripheral circulatory failure (cardiogenic, hypovolaemic, vasoplegic) only includes the pathological situations that may generate shock, but not shock itself. *Acute peripheral circulatory failure only represents the haemodynamic component of shock and is directly dependent on ECVB values.*

Some of the 'derivatives' of the notion of shock must also be discussed:

(i) *Psychical shock,* which it would be better to renounce deliberately, as it gives us the complex meaning of the notion of psychological shock but nothing in addition to the term *psychical stress* (Selye).

(ii) *Chronic shock* — the term includes the general elements of shock: *hypovolaemia*, generally due to dehydration, *hypoxia* caused by anemia, generalised *electrolytic disturbances* and progressive cellular *metabolic disorganisation*. Therefore the clinical reality of chronic shock has to be accepted as a form of evolution of the state of shock, the possible exaggeration of the parameter, time only enriching its content (Burghele).

(iii) *Secondary shock* — a term sometimes used to differentiate classical shock from *primary shock*, whose onset is marked by collapse. It is a point of view which starts from a clinical fact that can be interpreted and introduces a term that is not felicitous [67]. Let it be assumed that a state of collapse opens the clinical scene of a shock, but this does not justify the synonym collapse — primary shock (shock in being), since many collapses do not generate shock.

It should also be mentioned that in many instances syncope is mistakenly taken for 'primary shock'. Neither is the term 'secondary shock' given to shock, when not preceded by collapse, adequate.

Immediately after impact of the shock agent there is a short period of breakdown of the vital functions, with *hypotension* as a rule. This transitory period bears many names: *initial lesional stage, primary lesional syndrome, immediate depression or vasomotor collapse* (Selye and Laborit). If this transitory period is considered as a *collapse generating shock* and the pathophysiological deduction is reached that the only possible relationship is from collapse to shock, then certain inaccuracies may creep in. First, elements of collapse alone are to be found in the initial stage of shock and even these may be absent; secondly, in the dynamics of shock collapse may appear also later; in this case it is not merely a hypotension but also a severe atonia of the micro- and macrocirculation with hypertonia of the postcapillarovenular sphincters *(decompensated hyperergic collapse)*. Neither is the opposite filiation from shock to collapse true, since collapse only appears as a non-compulsory manifestation in the evolution of shock. Collapse must be understood only as a *circulatory aspect* of advanced shock, as a pronounced vascular pathophysiological link (Mandache).

Today our understanding of the pathological physiology of tissular hypoperfusion allows for a solution of the semantic controversies between the terms of shock and collapse, which it is better to dissociate than to join together. The clinical elements of collapse that appear against a background of shock at the onset and/or evolution of the latter may be termed more exactly *acute hypovolaemia, rapid decrease in ECBV, acute peripheral circulatory failure,* etc. In this way, the old and obsolete clinical notion of collapse can be avoided.

(iv) *Shock and contra-shock* are the two phases by which Selye characterised the dynamics of the 'alarm reaction' to aggression (*see* Section 1.5.3).

(v) *Local shock* although it may have all the pathophysiological elements of the picture of shock, an essential one is missing: the generalised, systemic element of the state of shock, and hence, this term can only be regarded as a figure of speech.

1.5 SYSTEMATICS

There are numerous etiopathogenic, clinical and staging classifications and almost a hundred types of shock have been described. Their list, not their classification, sometimes in alphabetical order, can be studied in several recent publications [13, 16, 20, 39, 42, 52, 60, 81, 101].

This is a list of some forms of shock (most of these terms are wrongly used) commonly referred to in the literature:

Acetylcholinic shock
Allergic shock
Anaphylactic shock
Anaphylactoid shock
Anesthetic shock
Bacterial shock
Bacteriemic shock
Burn shock
Caloric shock
Cardiogenic shock
Colloidal shock
Combustion shock
Crushing shock
Endotoxin shock
Epinephrinic shock
Haematogenic shock
Haemorrhagic shock
Heterotransfusion shock
Histamine shock
Hypovolaemic shock
Hyperthermic shock
Iatrogenic shock
Medicamentous shock
Neurogenic shock
Obstetrical shock
Oligemic shock
Operative shock
Peptone shock
Peritoneal shock
Postoperative shock
Postpartum shock
Protein shock
Septic shock
Seric shock
Spinal shock
Surgical shock
Tourniquet shock
Toxin shock
Traumatic shock
Tumbling shock
Vasogenic shock
Vasovagal shock

Some types of shock may be criticised from the viewpoint of their significance or are superimposed on other types: the shock of burned patients is the same as combustion shock, and operative, postoperative, anesthetic, obstetrical, surgical, postpartum, iatrogenic or medicamentous shock may eliminate one another in a more careful analysis. Hence, we have considered an exhaustive list of the types of shock communicated in the literature as of lesser interest and shall only discuss the fundamental forms of shock.

There are many more real pathological situations in the clinic that may take on the evolutive modality of a state of shock than there are names given to this state. It is of primary importance to recognise and treat in time these pathological causes that may generate shock [37, 86]. To consider the cause of a pathological situation, accompanied by shock, as a shock entity is exaggerated and may lead to confusion. Even if 'shocks': anaphylactic, anaphylactoid, allergic, histaminic, peptonic, protein, seric, acetylcholinic, colloidal, plasmatic, etc. have certain particularities (almost always established only by experiment) their mechanism of production is based, in general, upon a disturbance in vasomotor regulation with closely similar clinical aspects.

The stages of shock are sometimes mistaken [20] for types of shock: compensated, decompensated, irreversible, immediate, delayed, latent, etc. Similarly, doubtful and not very clear are the terms: emotional, epigastric, psychological, pure, 'typical' torpid shock, etc.

For the sake of orientation in this labyrinth of a terminology that is still being used without a sufficiently critical approach, it should be recalled that any type of shock is characterised by a *vascular phenomenon of readjustment and a cellular phenomenon of metabolic effort*. If the uniformity of the circulatory phenomena is not always complied with, the cellular disturbances are constant and give a unity to all forms of shock.

For a general consensus to exist concerning the classification of shock there ought to be a single criterion (aetiopathogenic or clinical), reliably reflecting the state of shock. As our insufficient knowledge of the pathological physiology of shock — as a biological phenomenon — does not yet permit such a 'conventional approach', the existing classifications consider as absolute one of the pathological coefficients that appears to be more important at a given moment in our understanding of the phenomenology of shock.

The evolution of our knowledge of shock is also reflected in the classifications that have been drawn up with time: the causes that might induce acute circulatory failure were initially classified, then the causes that modify systemic volaemia and finally those disturbing tissular perfusion. An attempt at an aetiopathogenic classification was made according to the quality of the field of receptors receiving the impact of the shock-inducing stimulus (Ţurai), classification that holds the germs of a cybernetical classification concerning the possibilities of the onset of shock.

All these classifications will be discussed in detail in the following paragraphs, but what must be emphasised once again is the importance of the perturbations affecting the cells, that are the beneficiaries of all the exchanges of substances and energy between the organism in its entirety and the environment. Therefore, considering that all the causes which may provoke a decrease in tissular perfusion suppress in the first instance the oxygen supply to the cell, we believe that *hypoxia* is the pathogenic coefficient closest to the distress of the *shock cell* and at the same time the closest to reality, although not sufficient. Next to the acute oxygen deficit, the constant humoral reaction in the state of shock triggered by *hypovolaemia* must be constantly borne in mind.

A classification today may also start from a consideration of shock as a brutal disruption of the cybernetic mechanisms of regulation of homeostasis. The informational circuits may be disturbed, either separately or combined, in the reception, relay, command and performance sections, bringing about the onset of all the different particular types of shock.

1.5.1 AETIOPATHOGENIC CLASSIFICATIONS

If the aetiological classification does not prove useful, determination of the shock-inducing agent and location of the primary lesions is of unquestionable practical therapeutical importance. Therefore a classification of the causal type, such as that in table 1.2, may be of particular interest.

Details concerning the principal aetio-pathogenic situations generating shock offer a general outlook on many clinical situations.

The loss of circulating fluid certainly sums up most of the causes of shock and those more frequently encountered. The loss may be *real* (haemorrhage, plasma losses, vomiting, diarrhœa, perspiration) in which case there is a depletion of the total hydric capital and therefore, a *circulating hypovolaemia with global hypovolaemia*. But the loss may also be *relative*, by pathological redistribution between the various spaces (stagnation of the fluid in the serosa, in the gastrointestinal tract, in the focus of an open fracture, in the damaged soft tissues, in the interstitial

Table 1.2 THE CAUSAL CLASSIFICATION OF SHOCK

Pathophysiological modalities	Shock-inducing causes		Anatomoclinical aspects
Sharp fall in the circulating volume	haemorrhage	external	arterial wound
		internal	rupture of the parenchyma extrauterine pregnancy
		exteriorised	bleeding ulcer esophagogastric varices
	plasmarrhage	external	burns
		internal	peritonitis, obstruction
	water loss (acute dehydration)		vomiting, diarrhœa digestive fistulas
Deficiencies of the cardiac pump	intrinsic organofunctional lesions		infarct, rhythm disturbances
	extrinsic lesions		cardiac tamponade
Pulmonary deficiencies	pleural		pneumothorax, haemothorax
	parenchymatous		massive atelectasis, embolism
Infections	locoregional		peritonitis, mediastinitis, extensive phlegmon
	generalised		septicaemia
Thromboembolism	thrombosis		vena cava, vena porta
	embolism		pulmonary, aortoiliac
Allergic accidents	medicamentous non-medicamentous		multiple clinical situations
Acute endocrine deficiencies or hyperfunction	acute insufficiency		adrenal, hypophyseal
	acute necrosis		pneumococci, anticoagulants (Sheehan syndrome)
	acute hyperfunction		pheochromocytoma thyrotoxicosis

space, or even in the microcirculation and/or the large veins) and therefore, a *circulating hypovolaemia with overall normovolaemia*. The term of circulating hypovolaemia might be replaced more exactly by that of low effective circulating blood volume. Similarly, although the use of 'normovolaemic shock' or 'isovolaemic shock' (signifying the global hydric volume) corresponds to a pathophysiological reality it may sometimes give rise to errors of interpretation. What counts in shock is the value of the ECBV, and a deficiency of this parameter is common to all states of shock considered classically as hypovolaemic or normovolaemic. The term 'normovolaemic shock' gives the impression that nothing unusual has happened in the fluid capital of the organism; actually, intercompartmental changes, pathological by their brutality and intensity, are manifested immediately by a decrease in ECBV. It may thus be asserted that the decrease in ECBV is always the effect of a *dysvolaemia* with intravascular hypovolemia, that may develop against the background of a global hypo or even normovolaemia. If the actual fluid losses lower the ECBV directly the other shock-inducing causes may result indirectly in the same disturbance by intercompartmental redistribution. All the 'great causes' of shock lower directly or indirectly the ECBV.

The inefficiency of the central pump lowers the cardiac output and therefore the ECBV. Bacteriaemia retains the fluid in the microcirculation (by a direct action on the smooth muscular fibre) and in the serosa and interstitial spaces by a retention of the inflammatory type. Hypersensitivity produced by a certain (anaphylactic) or alleged allergic mechanism (anaphylactoid) causes massive intravascular retention of the fluid; a similar effect is that of the pharmacochemical mechanisms of neurovasomotor disconnection. Any cause of major obstruction of the vascular tree may certainly play the

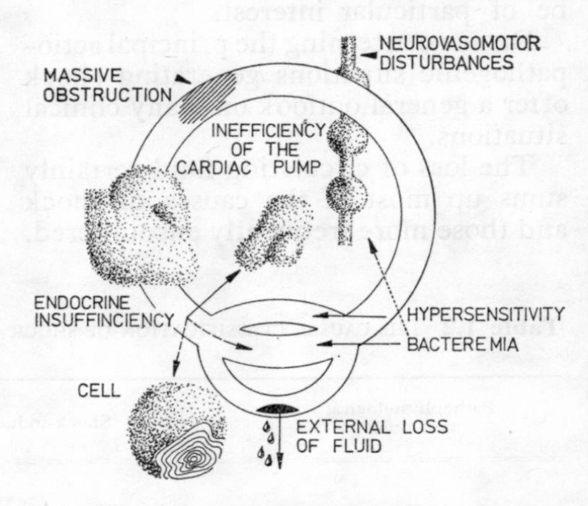

Figure 1.6. The causes of the states of shock listed in Table 1.2. Note that the endocrine deficits may also affect directly the cells.

same role of interrupting the fluid circuit as an inefficient heart. Finally, as may be seen from Figure 1.6 the endocrine causes of shock are associated from the beginning to vasomotor and metabolic deficiences.

According to the *quality* of the lost fluid the following pathophysiological 'variants' may be encountered:

(i) loss of whole blood (arterial, venous or/and capillary) as in all traumatic or/and operative wounds;

(ii) loss of red blood cells as in massive haemolysis;

(iii) predominant outer loss of plasma as in burns, or interior loss as in accentuated serous exudates;

(iv) loss of water and especially of electrolytes, as in diarrhœa, vomiting, fistulas or interstitial sequestra with œdema up to anasarca;

(v) predominant loss of water in severe polypnea or after tracheostomy.

Worthy of note is the pathogenic classification of Ţurai (1968) in terms of the field of receptors upon which the causal agent acts:

(i) shock due to trauma of the exteroceptors (for instance, burns);

(ii) shock due to trauma of the proprioceptor field (for instance traumatic shock of the patient with fractures);

(iii) shock due to trauma of the interoceptor field (for instance, transfusion shock, anaphylactic shock, bacterial shock);

(iv) complex shock, corresponding to the clinical reality in most cases, the initial receptivity being combined [97].

Although criticised because it only deals with a transient moment in the evolution of shock, this classification contains up-to-date ideas of the *surface* and *quality* of the receptor field.

It is likewise Ţurai who drew up a pathophysiological classification of the mechanisms lowering tissular perfusion including three categories (slightly modified):

(i) *Central deficit:*
- (a) myocardial lesion (infarct, trauma, myocarditis);
- (b) intrinsic ventricular filling disturbances (ventricular and supraventricular tachyarhythmias);
- (c) extrinsic ventricular filling disturbances (haemopericardium, haemopneumothorax, pericarditis);
- (d) ventricular emptying disturbances (massive pulmonary embolism).

(ii) *Volaemic deficit:*
- (a) haemorrhage (external, internal, exteriorised);
- (b) plasma losses (burns, obstruction, pancreatitis);
- (c) dehydration (vomiting, diarrhœa).

(iii) *Vasomotor tonus deficit:*
- (a) spinal anesthesia and therapy with ganglioplegics;
- (b) iatrogenic functional exclusion of the adrenals;
- (c) anaphylaxis;
- (d) intoxication with neurovasculotropic substances.

Another classification that must be mentioned is the English one adopted at Lister's centenary (1967) in which the terms of hypo- and normovolaemic shock are encountered, as well as the metabolic causes of shock (acidosis) and head injuries. The greatest asset of this classification is its symmetry but at the expense of certain inaccuracies. According to this classification shock may be divided into:

Shock
- hypovolaemic
  - haemorrhage
  - dehydration
  - plasma losses
- normovolaemic
  - cardiogenic
  - toxic
  - allergic
  - metabolic (acidosis)
  - neurogenic

Most pathophysiological classifications deal only with the vascular aspect, that is the stage of blood oxygen transport to the tissues. The essential pathophysiological criterion is considered to be *reduction of the effective circulating blood volume*, which may be brought about by:

(i) *absolute fluid losses* (haemorrhage, plasma protein losses, water and electrolyte losses);

(ii) *relative fluid losses* due to increase of the vascular bed or stocking in the 'third compartment' (anaphylaxis, obstruction, peritonitis), with temporary retention of stagnant endovascular, interstitial or digestive and serous fluid;

(iii) *disorders of the cardiopulmonary pump* (Figure 1.7).

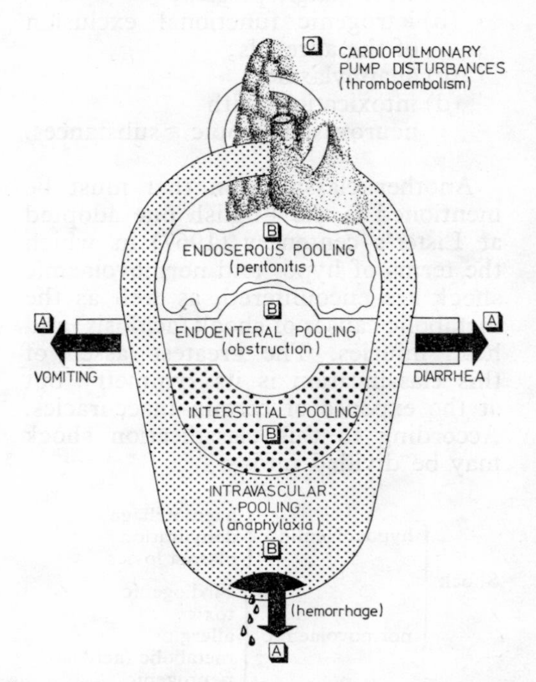

Figure 1.7. Decrease of the ECBV due to the actual loss of fluids (A); by pooling of the fluids (B); or disorders of the cardiopulmonary pump (C).

Almost all the classical etiopathogenic classifications of shock differentiate three categories (Darrow):

(i) *hypovolaemic* or *haematogenic* (Harison in 1939) also called *oligohaemic* by Fishberg in 1940;

(ii) *vasogenic*, also incorrectly called vasoplegic and/or normovolaemic;

(iii) *cardiogenic*.

To these three possibilities we must add a fourth, *mixed shock*, which actually reflects the most frequent situations. In its evolution any shock becomes mixed by the involvement of new pathogenic coefficients.

Perhaps the classification closest to the pathological physiology of the state of shock is that of Teodorescu-Exarcu who considers that there are two main causes of shock which lower the ECBV: a central cause (cardiogenic shock) and a peripheral cause (non-cardiogenic shock):

Cardiogenic shock (the initial defect is central, with primary decrease of the cardiac output and increase in the CVP)	(i) the heart does not empty: primary myocardial lesion, primary or secondary energodynamic lesion; (ii) the heart does not fill; tamponade or tachycardia more than 180 beats/min.
Non-cardiogenic shock (the initial defect is in the microcirculation; secondary decrease of the cardiac output; low CVP).	(i) hypovolaemia due to exteriorised loss of fluid; (ii) hypovolaemia due to intravascular retention (intravascular haemorrhage) or intercompartmental extravascular retention.

If the essential pathophysiological axis of shock is taken into consideration: dysvolaemia → decrease of the effective cellular perfusion volume → hypoxia → disturbed metabolism → acidosis → cellular destruction, then a pathophysiological classification can be drawn up according to the mode of onset of dysvolaemia (Table 1.3).

This classification illustrates the fact that the absolute external loss of fluid and/or intercompartmental changes result in the same major pathogenic coefficient: the decrease of the ECBV. This example of the situations in isolated pathophysiological circumstances joins together the slow and acute lesional modes of onset, which however almost always act in a complex way representing, either alone or in association, the causes of shock. The sequence of these pathophysiological

Table 1.3 THE PATHOPHYSIOLOGICAL CLASSIFICATION OF SHOCK IN TERMS OF THE MODE OF ONSET OF DYSVOLAEMIA

DYSVOLAEMIA		Decrease of effective circulating blood volume: Lesion	Type of shock
true circulating hypovolaemia	haemorrhage	vascular wound	haemorrhagic
	erythrorrhage	haemolysis	heterotransfusion
	plasma losses	burns	burn
	protein losses	vomiting	intestinal obstruction
	electrolyte losses	diarrhœa	peritoneal
	water losses	perspiration	caloric
	loss of lymph	lesion of the thoracic duct and burns	traumatic burn
relative circulating hypovolaemia	lesions of the cardiac pump	infarct, tamponade, myocardosis	cardiogenic
	lesions of the nervous vasomotor systems	medicamentous intoxications	acetylcholinic epinephrinic
	locoregional or disseminated intravascular obstruction	thromboembolism disseminated intravascular coagulation	traumatic bacteriemic
	extravascular obstruction	tourniquet, crushed tissues	tourniquet crushing
	extravascular retention of fluids ('The third compartment')	obstruction of the digestive tract	intestinal obstruction pancreatitis
		serous retention	peritoneal

situations in the fundamental forms of shock occurs as in Figure 1.8.

On following this pathophysiological sequence an attempt can be made to systematise the causes that interfere with the transport of oxygen from the air to the mitochondrial system, and which thus generate the metabolic imbalance specific of shock. Oxygen has two barriers to cross — an *alveolar-capillary* and a *blood-tissular barrier* — and three segments: air, blood and intracellular pathways (Figure 1.9).

All the causes that disturb the blood routes of oxygen transport result in *transport hypoxia*. These causes may act slowly (chronic anemia hypoxia) or brutally (acute posthaemorrhagic anemia and stagnant haemodynamic hypoxia). These rapid disruptions of the oxygen transport segment generate states of shock.

If it is admitted that the beneficiary of the entire system of functional efficiency of the organism is the *cell*, and the cybernetic mechanisms of organisation and control are represented by the nerve fibres and circulating blood, then shock may be considered as a *violent disorder of the systems of intercellular regulation*. These disorders that generate shock may affect the information circuits at:

(i) the level of the receptors (particularly traumatic shock, burn shock, etc.);

(ii) the level of the transport (especially haemorrhagic shock);

(iii) the level of the command (more frequently in bacteriemic shock, anesthetic shock);

(iv) the level of the execution (especially in anaphylactic, bacteriemic shock, etc.).

The examples are hazardous and certainly open to criticism; however, they are only designed to suggest the dominant mechanism. In any form of shock all the regulation systems are obviously perturbed simultaneously or successively, for instance, haemorrhagic shock affects the interoceptors (reception), lowers the ECBV (transport), brings about ischaemia of the neuronal centres (command) and collapse of the microcirculation, blocking the cells or the tissues (execution).

Figures 1.8 and 1.9 show that each cause disturbing the supply of fluid and/or oxygen to the cells may generate a state of shock. At the molecular level, the circulation of matter — in particulate or energy form — is the essence of existence of any organism. For the circulation of matter and energy to be efficient and the benefit maximum, it is necessary to have a *system of transport*, a *pump system* and especially a *regulating system* guaranteeing an optimal biological maintenance of the macrosystem. Disturbances that appear in any of these systems may bring about

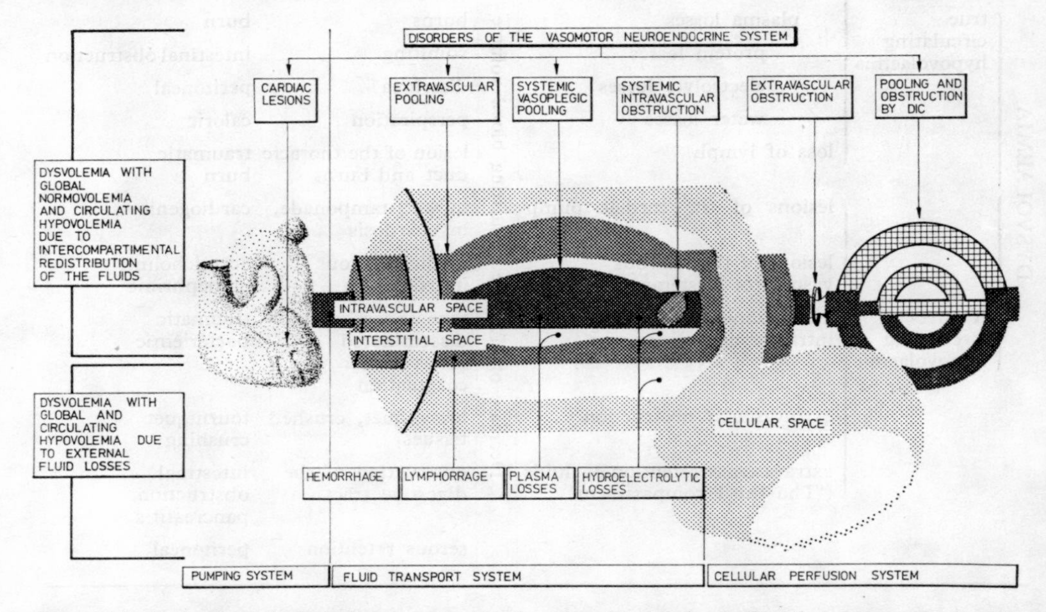

Figure 1.8. Pathophysiological situations that lead to the state of shock. Blood rheodynamic disturbances may occur in the pumping system, transport system, cellular perfusion and neurohormonal regulation.

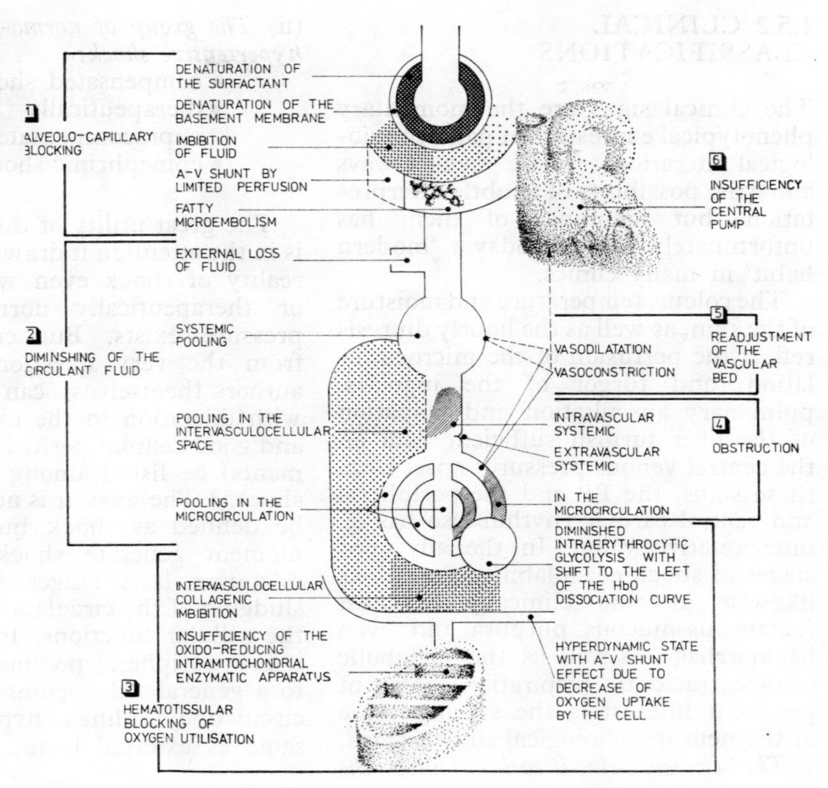

Figure 1.9. The causes that may prevent the transport of oxygen from the alveole to the mitochondrial respiratory chain.

a state of shock, i.e. what must in general be understood as a critical state, which questions the continuing existence of the individual as an organised system. Therefore, the heuristic value of the classification of the causes of shock proposed by Blalock in 1930 gives it a perennial substance: *haematogenic shock* (the fluid vehicle is affected), *vasogenic shock* (the cause of distress is the tubular system of transport), *cardiogenic shock* (deficiency of the central pump) and *neurogenic shock* (disturbed regulation) (Figure 1.10).

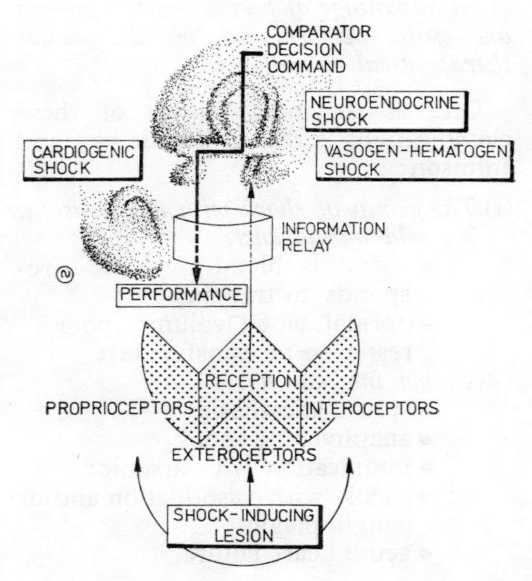

Figure 1.10. Cybernetic modes of onset of the state of shock.

1.5.2 CLINICAL CLASSIFICATIONS

The clinical signs are the momentary phenotypical expression of pathophysiological alterations. These signs always allow the possibility of a subtle interpretation, but ignorance of them has unfortunately become today a 'modern habit' in many clinics.

The colour, temperature and moisture of the skin, as well as the hourly diuresis reflect the perfusion of the microcirculation, and turgor of the jugulars, pulmonary auscultation and palpation of the liver furnish sufficient data on the central venous pressure; apart from these signs, the BP and the peripheral and central cardiac rhythm also has an interpretative value. In the advanced stages of shock coagulability alterations likewise become clinically manifest (cutaneous-mucous purpura and even haemorrhage, as well as the metabolic changes (acidotic respiration). Also of pertinent interest is the staging value of the neuropsychological clinical signs.

The clinical classifications have the great advantage of permitting the prompt and safe performance of the major therapeutical procedures.

One of the best known of these classifications is that of Hardaway and Johnson:

(i) *The group of shock with hypotension:*

(a) *cold integuments:*

- reduced blood volume (responds to transfusion);
- normal blood volume (poor response to transfusion);

(b) *hot integuments:*

- spinal anesthesia;
- anaphylactic shock;
- intoxication with arsenic;
- shock with vasodilation and/or ganglioplegics;
- acute heart failure.

(ii) *The group of normo- and hypertensive shock:*

- compensated shock;
- therapeutically supracompensated shock;
- epinephrinic shock.

The great utility of this classification is in the attention it draws to the clinical reality of shock even when a normal or therapeutically normalised blood pressure exists. But criticism arises from the very question put by the authors themselves: can the situations with dilatation in the microcirculation and good cellular perfusion (hot integuments) be listed among the 'states of shock'? The answer is no. They cannot be defined as shock but may at any moment generate shock if arteriolodilatation lasts longer and results in sludging of the circulation, jeopardising the cellular functions. In point of fact it is a peripheral pooling of fluids due to a generalised opening of the microcirculation, leading to hypovolaemia, the same as external losses.

Starting from the way in which circulating hypovoleamia is exteriorised the following triad of the clinical classification may be useful:

The onset of circulating hypovolaemia (decrease of the ECBV) may occur by:

(i) an actual loss of fluids (pale, moist, cold skin);

(ii) arteriolocapillary stagnation (red, moist, hot skin);

(iii) capillarovenous stagnation (cold, dry, cyanotic skin).

A classification of particular therapeutical interest is that of J. J. Byrne [16] (table 1.4).

Table 1.5 gives the practical clinical criteria and therapeutical response in the more frequent types of shock.

Table 1.4 Clinical classification of the states of shock
(Modified after Byrne)

Clinical sign	Type of shock	Therapeutical action
Cold	hypovolaemic shock (haemorrhagic, traumatic, burn, operative)	administration of fluids vasodilators
Hot	vasoplegic shock (anaphylactic, heterotransfusional, bacteriemic)	vasoconstrictive corticoids antibiotics (sometimes)
Congestive	cardiogenic shock of pulmonary embolism and acute heart failure	bleeding cardiotonics
With visceral distention	shock of obstruction, peritonitis and acute gastric dilatation	aspiration electrolytic re-equilibration administration of fluids surgical intervention

Table 1.5 Some of the clinical and therapeutical criteria in the differential diagnosis between the fundamental types of shock

		Type of shock	BP	CVP	Jugular turgor	Skin	Thirst	Transfusion effects	Vasodilators	Vasoconstrictors	Cardiotonics	Cortison	Surgery
			Clinical criteria					Therapeutical criteria					
Dysvoleamia	True global hypovolaemia	Haemorrhagic shock	↓	↓	−	cold	+	++	+	−	±	±	+
		traumatic shock	↓	↓	−	cold	+	+	+	−	±	±	+
		burn shock	↓	↓	−	cold	+	+	+	−	±	±	±
	Relative global hypovolaemia	bacterial shock	↓	↓	−	cold	±	±	++	−	±	+	
		cardiogenic shock	↓	↑	+	cya-nosed	±	bleeding performed	±	±	+		
		anaphylactic shock	↓	↓	−	hot	±	±		+	±	+	
		shock due to: obstruction peritonitis pancreatitis	↓	↓	−	cold	±	±	±	−	±	+	+
	Real hyper-volaemia	supracompen-sated shock	↑	↑	+	moist hot	disgust for water	bleeding performed	+	−	+		

1.5.3 EVOLUTIONARY CLASSIFICATIONS

The classification per stages of evolution, the most useful from the practical point of view, is also the least clear with regard to the terminology. The fact that no objective criteria (clinical or laboratory) have yet been found to characterise one or other stage of shock accounts for the great number of classifications per stages proposed, and their relative nosological correspondence. It is especially the moments of transition between the stages — the points of inflexion — that evade even the most attentive therapeutical follow up.

There are stages in the general response to aggression, explained by an *oscillating reaction*, designated and interpreted differently by various authors: adaptation syndrome (Selye); postaggressive oscillating reaction (Laborit); postaggressive systemic reaction (Teodorescu-Exarcu); postinjury systemic syndrome (Sevitt); response to trauma (Moore) (*see* Section 1.4.3).

The evolutive modes of response to aggression may take on the following aspects:

(i) the lesion causes an imbalance that does not exceed the homeostatic standards admitted phylogenetically; a harmonic postaggressive oscillating reaction (postaggressive systemic reaction) appears by means of which the organism buffers the local and systemic effects of the lesion and manages to obtain a re-equilibration (Figure 1.11*A*);

(ii) the lesion causes an imbalance beyond the limits of usual homeostasis and a large dysharmonic postaggressive oscillating reaction (postaggressive systemic reaction) with two possible evolutions, depending upon the intensity of the lesion, time of action of the aggressive agent and the body's reactive

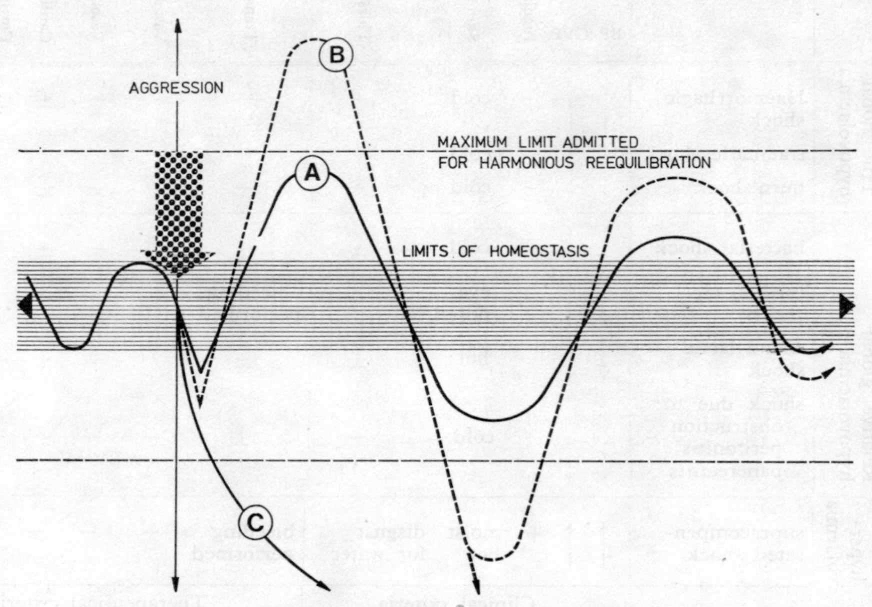

Figure 1.11. Evolutionary modes of response to aggression.

capacity: the imbalance continues up to exitus or towards re-equilibration by the energy and substantial efforts of the organism (endogenous, natural or spontaneous re-equilibration) and/or therapy (artificial or exogenous equilibration) (Figure 1.11*B*);

(iii) the lesion is massively destructive, causing a violent imbalance that does not permit the onset of a postaggressive oscillating reaction (postaggressive systemic reaction) and rapidly ends in death (Figure 1.11*C*). The first evolutive category includes the true postaggressive oscillating and systemic reactions which do not generate critical states, as appear in the second (shock states) and third category (sudden death).

The states of shock may be considered as dysharmonic modes of evolution of postaggressive oscillating or systemic reactions.

The postaggressive (postinjury) oscillating reaction (PAOR) is an unspecific defence reaction and homeostatic re-equilibration. Irrespective of the nature of the initial aggression, the organism mobilises similar neuroendocrino-tissular servomechanisms since it always has in view the same goal: to achieve maximum efficiency in the shortest interval and minimal energy expense, general re-equilibration and removal of the local consequences of the initial aggression. The uniformity of the reactive system of the PAOR justifies the aphorism that 'the organism is not polyglot' (Laborit).

Any PAOR begins by a moment of *immediate depression*, triggered by the aggresive agent *(primary lesional syndrome)*. There then follows a sequence in three phases — *catabolic, inversion and anabolic* — in the course of which complex metabolic and haemodynamic alterations take place. Spontaneous reversible PAOR is prolonged, the anabolic phase gradually recovering the plastic and energy reserves consumed in the preceding phases.

In its evolution PAOR brings about pathophysiological fluxes, often contradictory, that permanently contain the germs of a dysharmonic transformation , a veritable 'decompensation', followed by a state of shock. Therefore PAOR must be considered as a 'boomerang' [48] that must be controlled and arrested in time not to destroy the organism which launched it with the best of intentions.

The PAOR, pathologic by its duration (in sympathicotonic individuals, prolonged aggression, untimely therapy) or intensity (initial arrest without the typical onset of PAOR, or without managing the inversion phase, the catabolic phase remaining predominant) may result in the appearance of a state of shock.

The postaggressive systemic reaction (PASR) *or postinjury systemic syndrome* (PISS). The notion of *systemic reaction* clearly emphasises the general character of the response to aggression that brings about '*functional, biochemical, biophysical and enzymatic disturbances in all the tissues . . . , modifies the reactivity of the organism . . . , and has as substrate deep functional transformations of the endocrinovegetative systems of regulation*' (Teodorescu-Exarcu). The extinction of PASR (PISS) develops in four phases, superimposed on the same sinusoid as PAOR:

(i) *the catabolic phase (adrenergicocorticoid)* appears soon after the aggression and consists of functional stress governed by the cathecolamines ('the defence hormones') and catabolising hormones ('energogenetic hormones');

(ii) *the transition phase* resides in diminution of the events of the first phase *(corticoid decline)* and is closely conditioned by the cessation or removal of the aggressive factors;

(iii) *the anabolic phase* is the phase of acetylcholine and anabolising hormonal predominance in which the muscular force recovers up to complete metabolic equilibration;

(iv) *the phase of gain in body weight* continues the preceding phase by virtue of a positive loop of the oscillating reaction, with a favourable nitrate balance.

The response to trauma. Within the general framework of the response to any type of injury (including the operative one) the phases of PASR are perfectly superposable over those described by Moore. *Hormonal dynamics* stands at the basis of Moore's stages;

(i) *the adrenocorticoid phase* in which there is an increase in the ADH, thyroxin and the ACTH-cortisol complex, and of 17-hydroxysteroids in the plasma and urine;

(ii) *the phase of corticoid withdrawal*, during which the hormonal concentrations of the preceding phase decrease (the 'turning point');

(iii) *the phase of spontaneous anabolism* (early recovery) in which there is an equilibration of the catabolising and anabolising tendencies;

(iv) *the phase of lipid gain* (late recovery) when insulin and 17-ketosteroids (dehydroandrosterone-testosterone) become predominant.

These attempts at systematisation of the postaggressive disturbances are due to older observations concerning the evolution in *antagonic waves* of systemic pathophysiological alterations. Table 1.6 superposes the stages mentioned in the preceding paragraphs especially in order to emphasise the duration of this fundamental *sinusoidal reaction of re-equilibration* of the organism: more rapid phases at first (consumption) followed by prolonged phases (repair) up to return to the steady state.

The reactive system mobilised by the human body is insured along these stages by the following temporary pathophysiological sequence:

(i) *immediate imbalance* due to the acute changes produced by aggression furnishes the *global lesional information* with respect to which the body must act;

(ii) *the neurovegetative reaction*, the immediate intervention whose prompt-

Table 1.6 SUPERPOSAL IN TIME OF THE STAGES OF POSTINJURY STATES

Unspecific postinjury reaction	Phase I	Phase II	Phase III	Phase IV
PAOR	catabolic	inversion	anabolic protein and lipid	
Response to trauma	adrenocorticoid	corticoid withdrawal	spontaneous anabolism	gain in body weight
PASR (PISS)	catabolic (aggressive)	transition (return)	spontaneous anabolism (muscular force)	gain in body weight
Duration	1–3 days	2–4 days	8–14 days	months

ness insures regional readjustment of the fluids and mobilisation of the most labile cellular metabolic lines (for instance epinephrine accentuates glycogenolysis, etc.);

(iii) *the endocrinohumoral reaction*, slower, more diffuse, but more persistent and economical than the preceding one, insures metabolic readaptation of all the enzymatic lines;

(iv) *the intracellular genotypical reaction of adaptation* (the 'reasoner' and effector of all the information received due to the sequence of the three previous stages); it decides by controlled operon stress the predominance of metabolic enzymatic systems, regulating them in keeping with the critical state of the organism. This critical state is also reflected and relayed directly or by mediation to the entire cell population (Figure 1.12).

The pathophysiological axis of the oscillating stages of shock has also been described by Mareș and Șuteu in 1956, within the group of *post-aggresive stages:* a phase of contact of the aggresive agent with the body, in which local lesional and functional alterations appear; a phase of neuroreflex functional disturbances, with vasomotor and trophic disturbances and an evolutive tendency, and a phase of local and distal dystrophic phenomena [54].

Selye schematised the *adaptation syndrome*, differentiating a 'shock phase' with hyperfunction of the neuroendocrine axis and a phase of 'countershock' generated by the hormones released in the first phase. It is a very broad point of view but of obvious value due to its generality. These two phases together represent *the alarm reaction of the adaptation syndrome*, which may be continued either by a *resistant phase* or an *exhaustion phase*, according to the 'strength' which the aggressive factor had to cope with [79].

Owing to its rhythmicity, the phenomenology of shock is of the PAOR,

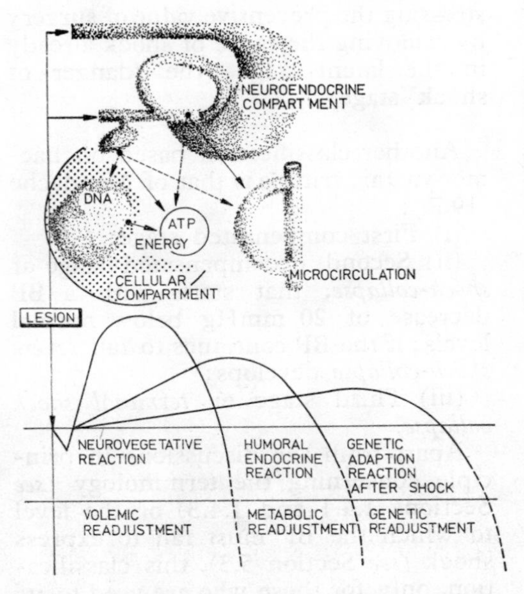

Figure 1.12. The sequence of the waves of neuroendocrino-genetic reaction re-equilibration.

PASR, PISS type, etc. But the sinusoid particularities of the evolution of shock are both quantitative (the amplitude of the loops is out of proportion with the spontaneous re-equilibration possibilities of the organism) and qualitative (the shape of these loops is often strongly asymmetrical, deviating from the coordinates of a harmonically buffered reaction).

Brînzeu (1969) proposes the following classification, applicable in the clinic:

(i) *latent shock*, that lasts as long as the cause generating shock, without the institution of shock itself;

(ii) *declared shock with circulatory compensation;*
(iii) *declared shock with circulatory decompensation;*
(iv) *shock in the terminal stage.*

This classification has the merit of stressing the preventive value of surgery by removing the cause of shock already in the latent stage (the 'danger of shock' stage).

Another classification based on haemodynamic criteria is that of Mandache (1966):
(i) First compensated stage;
(ii) Second decompensated stage or *shock-collapse*, that starts with a BP decrease of 20 mmHg below normal levels; if the BP continues to fall *severe shock-collapse* develops;
(iii) Third stage or *terminal shock collapse*.

Apart from the discussions in principle concerning the terminology *(see* Sections 1.4.1 and 1.4.3) or the level to which the BP must fall to express shock *(see* Section 5.3), this classification, only for those who are used to its dynamics, may be considered of practical utility in most types of shock [53].

A mixed pathophysiological and clinical classification per stages, of great prognostic and therapeutical interest, is that of Saegesser, modified by us (table 1.7).

Just as any classification that tries to landmark the dynamics of shock — always contradictory — this listing of stages has overlooked certain realities that have not escaped criticism: actually the microcirculation begins to suffer already in the vasoconstriction stage, and in the last stage apart from atonia the microcirculation is blocked by microthromboses, etc.

This classification per stages certainly brings into relief several points of great value: it offers an outline of the essential clinical signs (colour of the skin, vigil, sensation of thirst, respiration, temperature, etc.), and especially draws attention to two moments in which the peripheral pulse and BP cross each other, moments which mark the haemodynamic borderlines between the three stages (from relative bradycardia with normal BP in the first stage, to tachycardia with a weak peripheral pulse and decrease of the BP, and the last stage when the pulse becomes imperceptible and the BP falls towards zero.

The classifications per stages discussed in the foregoing paragraphs have a common trait, that of being limited to the vascular element and clinical expression of shock. Modern classifications start from stages of the PAOR, PASR or PISS type and also include important evolutive aspects of the metabolic oscillations.

A well known classification per stages is that proposed by Petrov and Postnicov:

(i) *the stage of compensated shock, of excitation or erectile;*

(ii) *the stage of decompensated shock, of inhibition or torpid;*

(iii) *the terminal stage of shock, preagonic.*

At first the terms of compensated and decompensated only had a circulatory significance, but subsequently metabolic aspects were also included. Chiotan and Cristea (1968) proposed:

(i) first stage, with a tendency to compensation;

(ii) second stage, haemodynamic and metabolic reactive decompensation, with a natural tendency towards recovery;

(iii) third stage, recovery and convalescence.

Table 1.7 CLASSIFICATION OF THE STAGES OF SHOCK
(Modified after Saegesser)

	Stages of shock		
Clinical sings	Vasoconstriction	Vasodilatation	Vascular atonia
Colour of skin	pale	cyanosis	grey cyanosis
Consciousness	obnubilation	obnubilation	coma
Pupil diameter	miosis	mydriasis	fix mydriasis
BP	normal	low	low to zero
Peripheral pulse	bradycardia	tachycardia	rare and weak pulse
Peripheral veins	normal	collapsed	distended
Respiration	superficial, normal rhythm	superficial, accelerated rhythm	superficial, accelerated rhythm
Thirst	moderate	very great	even absent
Skin hypothermia	present	diminished	accentuated
Diuresis	decreases	disappears	d'sappears

A classification with a more pronounced metabolic character is that of Davis (1967), who divides shock into:

(i) shock in the anoxic phase (lasting 1-3 days);

(ii) shock in the catabolic phase (lasting 2-3 weeks);

(iii) shock in the anabolic phase (lasting up to 10 weeks).

The terms 'compensated shock' and 'decompensated shock' are worth discussing separately, not so much due to their rich historical background, but because of the influence they still have on our 'way of thinking' about therapeutical principles. These terms were accepted in Romania by Țurai [97], Mareș [54], and Chiricuță [21].

Compensated shock is the label for the stage with normal BP or mild hypotension, with an hourly urinary output of 40 ml, for which a treatment with fluids is sufficient. In this stage *the prevention of decompensation must be held in view.*

Decompensated shock is the label of the stage with progressive hypotension, hourly urinary output of less than 40 ml and the onset of cellular, metabolic and then morphological irreversible lesions. Apart from fluids, a medication acting upon metabolism and the microcirculation must be administered.

The biphasic evolution of shock, of the compensated—decompensated type, has been long ago compared to the concept of reversible-irreversible, among others by Mareș and Șuteu, who, subdividing the stage of decompensated shock, differentiated:

(i) *compensated shock* (of cortical excitation);

(ii) *reversible decompensated shock* (of cortical inhibition);

(iii) *irreversible decompensated shock* (terminal).

In spite of their practical value the terms of compensated and decompensated shock have become — in our present understanding of the phenomenology of shock — both inaccurate and dangerous. Starting from the dynamics of the activity of the CNS (which has undergone recent changes in interpretation) these terms will always remain linked to BP values, whose unreliability is today generally admitted *(see* Sections 3.5 and 5.1). Therefore in compensated shock with low BP vasoconstrictors were recommended, whereas the adequate medication is of the α-blocking type, associated with massive rehydration. Even if some authors have adopted the term decompensated shock to the alterations of the microcirculation, this does not save the situation: on the one hand, impairment of the microcirculation begins already in the first stage (the so-called compensation by vasoconstriction which in fact is an 'overcompensation') and on the other, the worn-out term of 'compensated-decompensated shock' used in the clinic no longer manages to explain the dynamics of the cellular metabolic perturbation complexes. However substitution of the term by reversible-irreversible shock also meets with certain difficulties *(see* Section 1.4.3). This always happens when a pathophysiological classification per stages tries to penetrate within the limits of the clinical stages. An argument in favour of the classification of stages on the compensated-decompensated theme is that shock in the irreversible stage corresponds preferably to experimental shock. It has been asserted that the term decompensated shock used in the human clinic sounds less pessimistic than irreversible shock. Some authors uphold that the natural evolution of the decompensated stage develops towards reconversion to a normal state or to exitus, which only appears as an accident in the oscillating course of shock. However, one should not overlook the fact that shock is not a simple PAOR or PASR, *but a different quantitative and qualitative variant*, whose *fundamental evolutionary peculiarity is this very autoaggravating progress;* only a sufficient elasticity of the genome and/or therapeutical intervention, constantly synchronised with the stage of shock, can re-equilibrate the organism. Death is not a simple 'accident' but the untoward end that has to be avoided from the very beginning. *The tendency towards irreversibility exists from the onset of the state of shock, and therefore the notion of compensated shock is in fact deceptive (see* Section 9.6.3). Moreover, the modern classifications per stages (see the following paragraph) do not deny, but accentuate the necessity and possible re-equilibration of shock considered clinically in the irreversible stage.

The classification gaining ground today derives from the formula so sharply criticised at first: *reversible shock* (spontaneously or/and therapeutically) and *irreversible shock* (spontaneously and then therapeutically). Lillehei was among the first to explain the pathophysiological basis of these stages: *ischemic hypoxia* of the microcirculation is reversible, and *stagnant hypoxia* is potentially irreversible [52].

The transition toward irreversibility implies the *moment in the steady state dynamics when the regulating systems speed deleteriously and the organism as a whole is potentially irrecoverable although biologically still extant* [73, 97, 104]. In this case only increased therapy will gain ground (*see* Sections 3.5, 4.5. and 9.6). In the clinic it is difficult to establish the very moment of transition towards irreversibility, since it is decided at the infraclinical level. Therefore, clinical irreversibility may be in many instances a fallacy, persistence of the primary lesion generating shock being ignored. The diagnosis and efficient treatment of the lesion maintaining shock and causing its irreversibility is often abandoned in favour of a complex therapy of the unspecific, systemic syndrome. For this reason the terms 'compensated-decompensated shock' should be abandoned since they draw too much attention only to the general pathophysiological reaction. And for the same reason the diagnosis of the stage of irreversible shock may be pathophysiologically real but in clinical practice the term is used only to imply a maximum increase in the intensity of the treatment; actually near the bedside of a critically ill patient we only discuss the 'refractory' stage of shock.

We consider the classification of Hardaway as the closest to our present knowledge of the pathological physiology of shock and shall describe it, slightly modified, in order to render still more clearly the parallelism between the dynamics of the microcirculation and of the cellular disturbances (Figure 1.13):

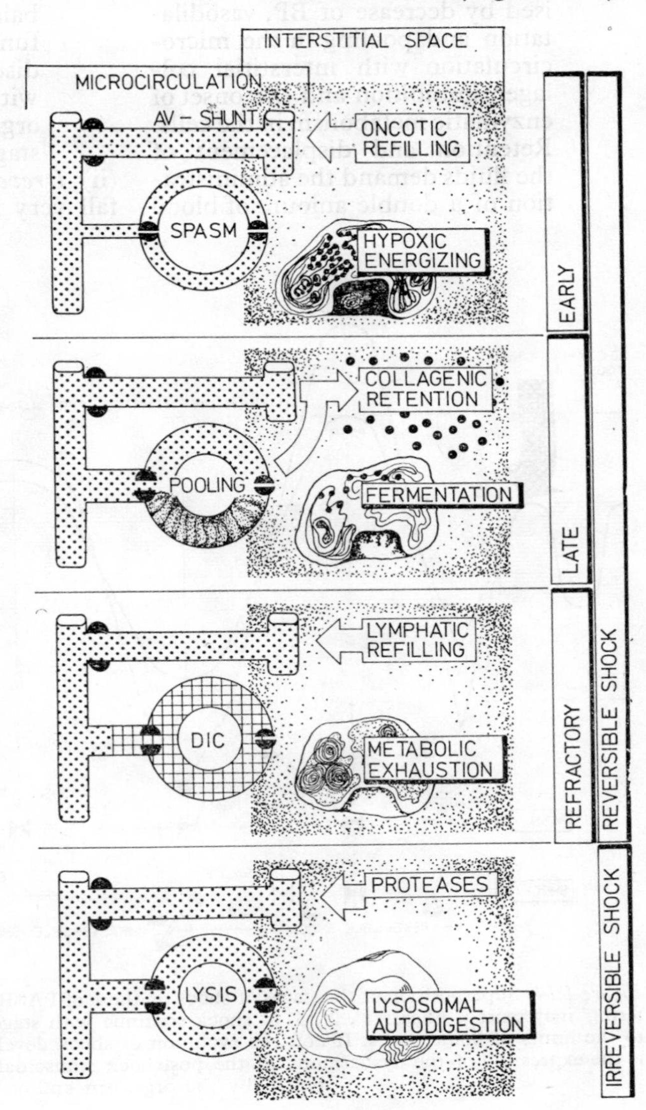

Figure 1.13. The stages of the state of shock with the main events in the microcirculation, interstitial space and cellular space.

(i) *Reversible shock*, with three evolutionary stages:

(a) *early reversible shock* with normal BP, vasoconstriction in the microcirculation (pale integuments, oliguria) and cellular hypoxia;

(b) *late reversible shock*, characterised by decrease of BP, vasodilatation and pooling in the microcirculation with interstitial collagen imbibition and the onset of enzymatic metabolism in the cells. Retention and displacement of the fluids demand the administration of a double amount of blood or substitutes as compared with the losses calculated;

(c) *refractory reversible shock* with aleatory BP, blood cell sludging and disseminated intravascular coagulation (DIC) in the microcirculation, as well as oscillating disturbances of the coagulolytic balance; the cell caters to acid functions that are systematically discharged; as a rule, infections with Gram-negative germs and organ deficiences develop in this stage.

(ii) *Irreversible shock:* the BP may fall very low and the microcirculation

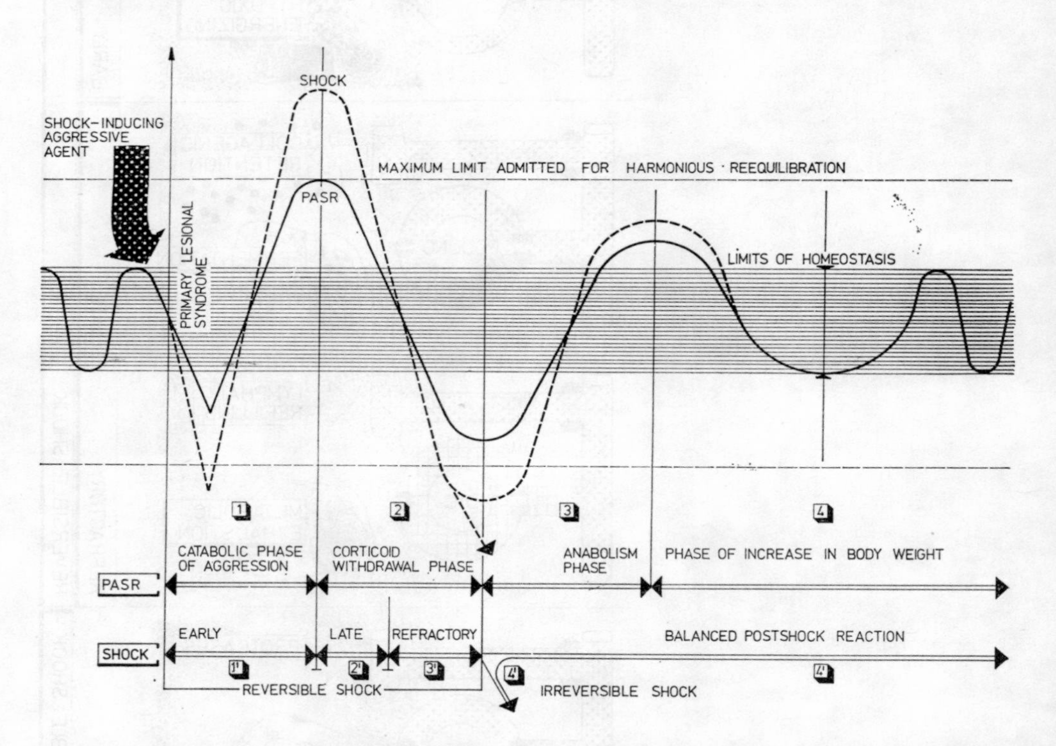

Figure 1.14. Superposition of the stages of shock on those of PASR (PISS). The possibility of a 'buffering' may exist if phases 1′, 2′, 3′ of shock continue with stages 3 and 4 of PASR up to return to the limits of homeostasis. If not, the evolution of shock develops towards irreversibility (4′) and is the expression of the dysharmony of the postshock sinusoidal reaction that could not be equilibrated by the organism and/or therapy.

and blood flow enter upon autoaggravating circuits giving rise to irreparable systemic damage; in the cell, acidosis activates lysosomal hydrolases that start cellular digestion, followed by their destruction, disorganisation and death. Spread of the necrosis zones and plasmatic generalization of the hydrolases spell death to the entire organism.

This evolutive classification of shock that we have been using for the last ten years in our daily practice and which has allowed us to apply a dynamic, target therapy, has been more efficient than the older concept of compensated-decompensated shock. Table 1.8 gives comparisons between this classification and some of the other classifications of shock per stages, and figure 1.14 gives the correspondence of the phases of shock and the phases that characterise the postaggressive harmonious oscillating reactions.

Table 1.8 SUPERPOSAL OF SOME OF THE CLASSIFICATIONS OF SHOCK PER STAGES

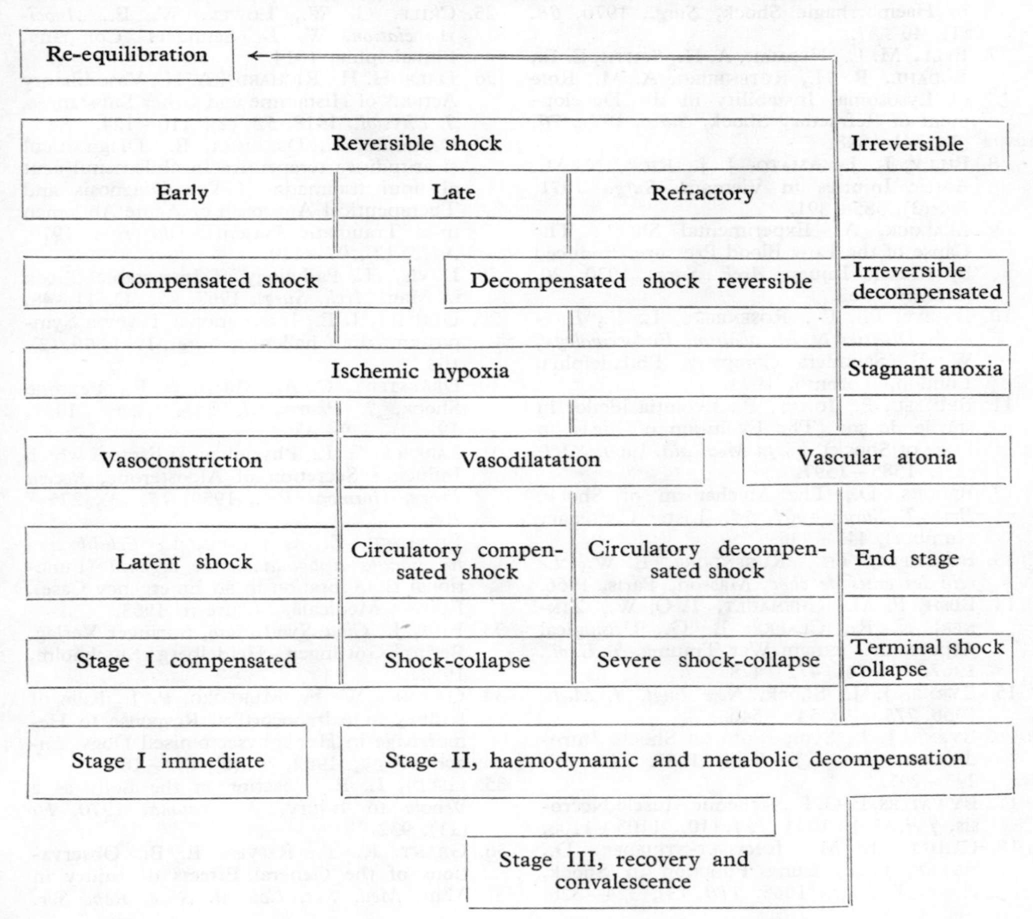

SELECTED BIBLIOGRAPHY

1. ALBERT, C. A., Le collapsus circulatoire: physiopathologie et traitement, *Laval Méd.*, 1970, *41*, (11), 1117–1125.
2. ALDMAN, B., Road Accidents in Sweden, *J. Trauma*, 1970, *10*, (11), 921–924.
3. ANDREASEN, N. J. C.; NORRIS, A. S.; HARFORD, C. E., Incidence of Long-Term Psychiatric Complications in Severely Burned Adults, *Ann. Surg.*, 1973, *174*, (5), 785–793.
4. BAGLIONI, C., *Molecular Evolution in Man*, John Hopkins Press, Baltimore, 1966.
5. BALLINGER, W. F., RUTHERFORD, R. B., ZUIDEMA, G. D., *The Management of Trauma*, W. B. Saunders Company, Philadelphia and London, 1968.
6. BAUE, A. E., SAYEED, M. M., Alterations in the Functional Capacity of Mitochondria in Haemorrhagic Shock, *Surg.*, 1970, *68*, (1), 40–47.
7. BELL, M. L., HERMAN, A. H., SMITH, E. E., EGDAHL, R. H., RUTENBURG, A. M., Role of Lysosomal Instability in the Development of Refractory Shock, *Surg.*, 1971, *70*, (3), 341–348.
8. BILLY, L. J., AMATO, J. J., RICK, N. M., Aortic Injuries in Vietnam, *Surg.*, 1971, *70*, (3), 385–391.
9. BLALOCK, A., Experimental Shock: The Cause of the Low Blood Pressure Produced by Muscle Injury. *Arch. Surg.*, 1930, *20*, (6), 959–962.
10. BONDY, Ph. K., ROSENBERG, L. E., *Duncan's Diseases of Metabolism, Endocrinology*, W. B. Saunders Company, Philadelphia, London, Toronto, 1974.
11. BRÎNZEU, P., IGNAT, P., Evoluţia ideilor în stările de şoc (The Evolution of Ideas on State of Shock), *Viaţa Medicală*, 1969, *XVI*, (23), 1585–1597.
12. BROOKS, D., The Mechanism of Shock, *Brit. J. Surg.*, 1967, *54*, (Lister Centenary Number), 441–446.
13. BURGHELE, TH., RUGENDORF, E. W., *Le rein des états de choc*, Masson, Paris, 1966.
14. BUSH, F. M., CHENAULT, T. O. W., ZINNER, N. R., CLARKE, B. G., Urological Aspects in Vietnam War Trauma, *J. Urol.*, 1967, 101 (4), 472–478.
15. BYRNE, J. J., Shock, *New Engl. J. Med.*, 1966, *275*, (10), 543–546.
16. BYRNE, J. J., Symposium on Shock. Introduction, *Amer. J. Surg.* 1965, *110*, (3), 293–297.
17. BYWATERS, E. G. L., Ischemic Muscle Necrosis, *J.A.M.A.*, 1944, *124*, (10), 1103–1108.
18. CAHILL, J. M., JONASSET-STRIEDER, D., BYRNE, J. J., Lung Function in Shock, *Amer. J. Surg.*, 1965, *110*, (3), 324–328.
19. CANNON, W. B., *Traumatic Shock*, D. Appleton Co., New York, 1923.
20. CHIOTAN, N., CRISTEA, I., *Şocul (Shock)*, Editura Medicală, Bucureşti, 1968.
21. CHIRICUŢĂ, I., ROMAN, L., Dinamica microcirculaţiei în şocul compensat şi decompensat (Dynamics of the Microcirculation in Compensated and Decompensated Shock), *Chirurgia*, 1971, *XX*, (4), 289–302.
22. COPE, O., LITWIN, S. B., Contribution of Lymphatic System to Replenishment of Plasma Volume Following Hemorrhage, *Ann. Surg.*, 1962, *156*, (6), 655–661.
23. COUFFIGNAL, L., *La cibernétique*, Presses Univ. de France, Paris, 1966.
24. COWLEY, R. A., ATTAR, S., LA BROSSE, E., Some Significant Biochemical Parameters Found in 300 Shock Patients, *J. Trauma*, 1969, *9*, (11), 926–938.
25. CRILE, G. W., LOWER, W. E., *Anoci-Association*, W. B. Saunders Company, Philadelphia, 1914.
26. DALE, H. H., RICHARDS, A. N., Vasodilatory Actions of Histamine and Other Substances, *J. Physiol.*, 1918, *52*, (2), 110–124.
27. DANICICO, I., DRĂGHICI, B., Diagnosticul şi atitudinea terapeutică în abdomenul acut al unui traumatizat (The Diagnosis and Therapeutical Approach of Acute Abdomen in a Traumatic Patient), *Chirurgia*, 1970, *XIX*, (2), 97–110.
28. DAVIS, H., Pathology of Intractable Shock in Man, *Arch. Surg.*, 1967, *95*, (1), 44–48.
29. DUNPHY, J. E., International Trauma Symposium: the Challenge, *Surg.*, 1971, *69*, (2), 162.
30. DRAGSTEDT, C. A., MEAD, F. B., Peptone Shock, *J. Pharmacol. Exp. Ther.*, 1937, *59*, (3), 429–436.
31. FARRELL, G. L., Physiological Factors which Influence Secretion of Aldosterone, *Recent Progr. Hormone Res.*, 1959, *15*, (3), 275–286.
32. FILIPESCU, Z., CURELARU, I., *Echilibrarea funcţională a bolnavilor în urgenţă* (Functional Equilibration in an Emergency Case), Editura Medicală, Bucureşti, 1963.
33. FINE, J., *Ciba Symposium*, Springer Verlag, Berlin - Göttingen - Heidelberg - Stockholm, 1962.
34. GANONG, W. F., MURLORO, P. J., Role of Kidney in Adrenocortical Response to Hemorrhage in Hypophysectomised Dogs, *Endocrinology*, 1962, *70*, (2), 182–188.
35. GELIN, L. E., Reaction of the Body as a Whole to Injury, *J. Trauma*, 1970, *10* (11), 932–939.
36. GRANT, R. T., REEVES, E. B., Observations of the General Effects of Injury in Man, *Med. Res. Council, Spec. Rep. Ser.*

No. 277, London, His Majesty's Stationery Office, 1941.
37. GRIGG, E. R. N., *Biologic Relativity*, Amaranth Books, Chicago, 1967.
38. GUYTON, A. C., *Function of the Human Body*, W. B. Saunders Company, Philadelphia-London-Toronto, 1969.
39. HARDAWAY, R. M., *Syndromes of Disseminated Intravascular Coagulation with Special Reference to Shock and Haemorrhage*, Charles C. Thomas, Springfield-Illinois, 1966.
40. HARDAWAY, R. M., Clinical Management of Shock, *Milit. Med.*, 1969, *134*, (9), 643–650.
41. HENDERSON, J., Acapnia and Shock, *Amer. J. Physiol.*, 1909, 24, (1), 66–74.
42. HUCKABEE, N. E., The Relationship of Pyruvate and Lactate During Anaerobic Metabolism, *J. Clin. Invest.*, 1958, *37*, (3), 264–272.
43. HUNT, J. H., *Accident Prevention and Life Saving*, Livingstone Ltd., Edinburgh and London, 1965.
44. INGRAM, D. G., *Biological Activities of Complement*, S. Karger Edit., Basel, München, Paris, London, New York, Sydney, 1972.
45. JUDGE, R.D., ZUIDEMA, G.D., *Physical Diagnosis*, Little Brown Comp., Boston, 1963.
46. KNISELY, M. H., ELIOT, T. S., BLOCH, H. E., Sludged Blood in Traumatic Shock, *Arch. Surg.*, 1945, *51*, (2), 220–228.
47. KOBALD, E. E., LOBELL, R., KATZ, W., THAL, A. P., Chemical Mediators Released by Endotoxin, *Surg. Gynec. Obstet.*, 1964, *118*, (5), 807–812.
48. KULOWSKI, J., *Crash Injuries*, Charles C. Thomas Publisher, Springfield, Illinois, 1960.
49. LABORIT, H., *Réaction organique à l'agression et choc*, Masson, Paris, 1955.
50. LABORIT, H., BARON, C., Effets de la monosemicarbazone de la bêta-naphtoquinone dans le choc hémorragique expérimental, *Agressol.*, 1971, *12*, *(1)*, 25–30.
51. LILLEHEI, R. C., LONGERBEAM, J. L., BLOCH, J. H., MANAX, W. G., The Nature of Irreversible Shock: Experimental and Clinical Observations, *Ann. Surg.*, 1964, *160* (4), 682–710.
52. LITARCZEK, G., Medicaţia vasoactivă şi cardiotonică în tratamentul tulburărilor circulatorii în şoc (Vasoactive Medication and Cardiotonics in the Treatment of Circulatory Disturbances in Shock), *Revista Sanitară Militară*, 1974, *LXXVI*, (5), 427–434.
53. MANDACHE, F., *Fiziopatologia circulaţiei şi imunitatea în şoc* (Physiopathology of the Circulation and the Immunity in Shock), Editura Academiei Republicii Socialiste România, Bucureşti, 1966.
54. MAREŞ, E., ŞUTEU, I., ATANASESCU S., *Şocul traumatic în chirurgia de campanie* (Traumatic Shock in Camp Surgery), Editura Militară, Bucureşti, 1956.
55. MAREŞ, E.; ŞUTEU, I., BĂNDILĂ, T., CÎNDEA, V., ILIE, A., Date noi în fiziopatologia şi tratamentul şocului (New Data in the Pathophysiology and Treatment of Shock), *Revista Sanitară Militară*, 1963, *LIX*, (6), 887–895.
56. MAYS, E. T., Complex Penetrating Hepatic Wounds, *Ann. Surg.*, 1971, *173*, (3), 421–428.
57. MAZOR, A., ROGEL, S., Cardiac Aspects of Shock: Effects of Ischemic Body on Nonischemic Heart, *Ann. Surg.*, 1973, *178*, (2), 128–133.
58. MILLS, L. C., MOYER, J. H., *Shock and Hypotension; Pathogenesis and Treatment*, the Twelfth Hahnemann Symposium, Grune and Stratton, New York, 1965.
59. MONOD, J., *Le hasard et la nécessité*, Edit. du Seuil, Paris, 1970.
60. MOORE, F. D., *Metabolic Care of the Surgical Patient*, W. B. Saunders Company, Philadelphia, 1968.
61. MOORE, F. D., Terminal Mechanisms in Human Injury, *Amer. J. Surg.*, 1965, *110*, (3), 317–323.
62. MORARU, I., *Medicina legală* (Forinsic Medicine), Bucureşti, 1967.
63. MOULIAS, R., MULLER-BERAT, C. N., Immunogénétique et susceptibilité à l'infection. Nouveaux aperçus sur la notion de 'terrain' en pathologie infectieuse, *Presse Méd.*, 1970, *70*, (5), 225–230.
64. MOYER, C. A., MARGRAF, H. W., MONAFO, W. W., Burn Shock and Extravascular Sodium Deficiency. Treatment with Ringer's Solution with Lactate, *Arch. Surg.*, 1965, *90*, (6), 799–805.
65. NAHAS, G. G., in *Shock and Hypotension*, Grune and Stratton, New York, 1965.
66. OLLEDART, R., MANSBERGER, A., R., The Effect of Hypovolemic Shock on Bacterial Defense, *Amer. J. Surg.*, 1965, *110*, (2), 302–311.
67. PLACA, A., CRIVDA, S., Choc, collapsus, collapsus prolongé, état similaire, *Presse Méd.*, 1965, *73*, (19), 1097–1101.
68. POCIDALO, J. J., *Réanimation et choc*, Flammarion, Paris, 1968.
69. POSTELNICU, D., SĂHLEANU, V., Problema constituţiei în medicina modernă (Constitutional Problems in Modern Medicine), *Viaţa Medicală*, 1970, *XVII*, (10), 435–440.
70. RANDALL, H. T., HARDY, J. D., MOORE, F. D., *Manual of Preoperative and Postoperative Care*, W. B. Saunders Company, Philadelphia and London, 1968.

71. RAVEN, R. W., *Surgical Care*, Butterworth and Comp. Ltd., London, 1952.
72. RHOADS, J. E., HOWARD, J. M., *The Chemistry of Trauma*, Charles C. Thomas Publ., Springfield, 1963.
73. RICH, N. M., HUGHES, C. W., Vietnam Vascular Registry, Preliminary Report, *Surg.*, 1969, *65*, (2), 218–224.
74. ROBBINS, S. L., *Pathology*, W. B. Saunders Company, Philadelphia, 1967.
75. RUTHERFORD, R. B., WEST, R. L., HARDAWAY, R. M., Coagulation Changes During Experimental Hemorrhagic Shock, *Ann. Surg.*, 1966, *164*, (2), 203–212.
76. SCHLOERB, P. R., Shock and Metabolism, *Surg. Gynec. Obst.*, 1969, *128*, (2), 315–319.
77 SCHUMER, W., Evolution of the Modern Therapy of Shock: Science versus Empiricism, *Surg. Clin. N. Amer.*, 1971, *51*, (1), 3–13.
78. SCHWARTZ, S. I., LILLEHEI, R. C., SHIRES, G. T., SPENCER, F. C., STORER, E. H., *Principles of Surgery*, McGraw-Hill Book Comp., 1973.
79. SELYE, H., *The Physiology and Pathology of Exposure to Stress*, Acta Inc., Montreal, 1950.
80. SEVITT, S., Reflections on Some Problems in the Pathology of Trauma, *J. Trauma*, 1970, *10*, (11), 962–973.
81. SHOEMAKER, W. C., *Shock ; Chemistry, Physiology and Therapy*, Charles C. Thomas, Springfield, Illinois, 1967.
82. SHOEMAKER, W. C., NOHR, A., PRINTEN, K. J., Use of Sequential Physiologic Measurements for Evolution and Therapy of Uncomplicated Septic Shock, *Surg. Gynec. Obst.*, 1970, *131*, (2), 245–254.
83. SHIRES, G. T., Initial Management of the Severely Injured Patient, *J. Amer. Med. Assoc.*, 1970, *213*, (11), 1873–1880.
84. SHIRES, G. T., CARRICO, C. J., COHN, D., The Role of Extracellular Fluid in Shock. In *Shock*, Hershey, S. G. (editor), Boston, Little Brown Comp., 1964.
85. SHILLINGFORD, J. P., Syncope, *Ann. J. Cardiol.*, 1970, *26*, (6) 609–612.
86. SIEGEL, J. H., GOLDWYN, R. M., FRIEDMAN, H. P., Pattern and Process in the Evolution of Human Septic Shock, *Surg.*, 1971, *70*, (2), 232–245.
87. SIMEONE, F. A., Shock, Trauma and the Surgeon, *Ann. Surg.*, 1963, 158 (6), 759–762.
88. STAHL, T. H., Pressure-Flow Factors in the Renal Excretory Response to Hemorrhage, *Surg. Forum*, 1965, *16*, (1), 3–12.
89. STEPHENSON, H. E., Jr., *Cardiac Arrest and Resuscitation*, The C. V. Mosby Company, Saint Louis, 1974.
90. STUGREN, I., *Evoluţionismul în secolul 20*, (Evolutionism in the 20th Century), Editura politică, Bucureşti, 1969.
91. ŞUTEU, I., CAFRIŢĂ, A., BUCUR, AL., Concepţii actuale în fiziopatologia şi tratamentul şocului (Present Concepts in the Pathophysiology and Treatment of Shock), *Revista Sanitară Militară*, 1971, *LXXIV*, (1), 9–22.
92. ŞUTEU, I., CAFRIŢĂ, A., BUCUR, AL., Aspecte infracelulare în şocul experimental (Infracellular Aspects in Experimental Shock), *Revista Sanitară Militară*, 1972, *LXXV*, (3), 357–366.
93. SUTHERLAND, E. W., RALL, T. W., The Relationship of Adenosine 3-5-phosphate and Phosphorylase to the Action of Catecholamines and Other Hormones, *Pharmacol. Rev.*, 1960, *12*, (3), 265–272.
94. TEODORESCU-EXARCU, I., *Reacţiile organismului la agresiune* în vol. *Studiul terenului în chirurgie* (Reactions of the Organism to Stress in: Study of the Patient's Constitution in Surgery) (sub red. Acad. Th. Burghele), Editura Medicală, Bucureşti, 1965, 60–72.
95. TEODORESCU-EXARCU, I., *Agresologie chirurgicală* (Surgical Stress), Editura Medicală, Bucureşti, 1968.
96. TEODORESCU-EXARCU, I., *Patologie Biochimică* (Biochemical Pathology), Editura Medicală, Bucureşti, 1974.
97. ŢURAI, I., Patogenia şi tratamentul şocului (The Pathogeny and Treatment of Shock), *Chirurgia*, 1960, *IX*, (4), 205–216.
98. THAL, A. P., WILSON, R. F., Shock; in *Current Problems in Surgery*. (A series of monthly clinical monographs). Chicago, Year Book Medical Publishers, Inc., September, 1965.
99. VERNON, S., Shock, A Design for Survival, *J. Abdom. Surg.*, 1970, *12*, (1), 1–2.
100. VOICULESCU, V., *Stările comatoase* (Comatous States), Editura Medicală, Bucureşti, 1968.
101. WEIL, M. H., SCHUBIN, H., *Diagnosis and Treatment of Shock*, Williams and Wilkins Comp., Baltimore, Maryland, 1967.
102. WERLE, E., *Kallikrein, Kallidin and Related Substances, Polypeptides which Affect Smooth Muscle and Blood Vessels*, New York, Pergamon Press, 1960.
103. ZWEIFACH, B. W., *Annotated Bibliography on Shock 1950–1962*, Publ. National Res. Council, Washington, 1963.
104. * * * *Abdominal and Genito-Urinary Injuries*, Military Surgical Manuals National Research Council, W. B. Saunders Company, Philadelphia and London, 1945.
105. * * * *Early Care of the Injured Patient*, by the Committee on Trauma American College of Surgeons, W.B. Saunders Comp., Philadelphia-London-Toronto, 1972.

2 EXPERIMENTAL SHOCK

The differences among the various animals and man appear to be in the susceptibility of the various organs to the haemodynamic disturbances of shock.

RICHARD, C. LILLEHEI, 1964

2.1 Experimental versus clinical shock

2.2 The themes of personal investigations

2.3 Rectocolic administration of a complex antishock solution

2.4 Hyperthermic shock induced by controlled hyperthermia

2.5 Study of the regulation of the splanchnic circulation in traumatic shock

2.6 Haemodynamic study of traumatic shock and the therapeutical value of ganglioplegics

2.7 Study of the microcirculation in experimental shock with the ^{51}Cr radioisotope

2.8 The efficiency of some pharmacodynamic agents in controlling the disturbances of the microcirculation in experimental shock (study carried out with ^{51}Cr radioisotope)

2.9 Visceral dynamics of the parameters of the acid-base balance in traumatic shock

2.10 Experimental study of the shock cell

2.11 Experimental study of the lymphatic circulation in shock

2.1 EXPERIMENTAL VERSUS CLINICAL SHOCK

Interpretation of the results obtained in experimental studies of shock and extrapolation of the observations to man should be carried out with reserve and a keen discrimination.

Two arguments are however always true:

(i) The haemodynamic and metabolic behaviour of the animal differs from that of humans in direct proportion to the rungs on the scale of the species; only experiments carried out on sub-primates and primates are in any way comparable with the phases of shock in man [6, 11, 16];

(ii) The models of traumatic shock (Noble-Collip) and haemorrhagic shock (Wiggers-Johnson, Fine, Reitman, etc.) have been strongly criticised recently [25, 30, 31, 41], because shock almost always occurs in men who are not under anesthesia, and bleeding in the human clinical cases never occurs with the swiftness of the gradients achieved experimentally in animals; arterial pressure in the clinic does not possess 'periods of constancy' that can be maintained by Wiggers' balloon for standardisation of the stages of shock; moreover, in the clinic not all cases of shock are lethal, as they are in most acute experiments.

It is extremely difficult, if not impossible, to control shock studies in humans. Although clinical shock and trauma study units have been developed in recent years, the results are limited.

Many different opinions exist concerning the same problem in experimental shock. The animals commonly used in shock experiments differ as regards their organ systems which represent the 'weak link' in response to shock. For example, the hepatic vein sphincters of the dog are very sensitive to *p*H changes, adrenergic stimuli and vasoactive substances, but these sphincters are less developed and appear to be of lesser significance in primates.

Contraction of the liver and spleen in the dog and cat supplies 30% of the blood volume of the total ECBV, whereas in humans this blood volume is obtained from the skin.

Similarly, the histological picture of acute endotoxemia varies considerably in terms of the species; in the rat a massive accumulation of leukocytes is observed in the lung; in the rabbit, focal necrosis of the liver, and in the dog, submucosal haemorrhage throughout the intestinal tract. These appear to be the major pathologic events.

Today, the classical models of experimental shock (haemorrhagic, traumatic and endotoxin) have been modified to be as similar as possible to the human clinical cases.

By causing graded bleeding (50% blood losses in 2-3 hours) and by the injection of live *E. coli* the experimental shock models are much closer to human shock [11, 16]. Until recently, the protective action of splanchnic denervation and of antiadrenergic agents in endotoxin shock seemed to have been established; more recently, however, it was observed that this protection no longer exists following the injection of live *E. coli* (as occurs in the human clinic) [17, 31].

It should not be forgotten that the informational value of the shock-inducing agent is globally perceived by the human organism, and that its response takes place in keeping with its own regulation systems, conferring upon it all its reactivity characteristics, i.e. the personal note of human shock in general and of each shocked subject in particular.

2.2 THE THEMES OF PERSONAL INVESTIGATIONS

Experimental investigations on shock were initiated in the Central Military Hospital of Bucarest in 1950-1952, instituting a well-established tradition [1, 3, 4, 7, 12, 13, 23, 24, 26, 37, 38].

The research work carried out by the authors of this volume lasted more than ten years during which experimental shock was explored, especially traumatic, haemorrhagic and thermal shock, pathological states which are closer to surgical practice and intensive therapy and, obviously, to situations arising on the battle field.

Experimental models for bacteriemic, burn or other types of shock were not attempted, but we have had ample clinical experience in this connection. The results were reported in earlier works. The experimental models tested with time reflect, in their entirety, the evolution of our knowledge of the phenomenology of shock. In almost each instance we tried to find the grounds of therapeutical criteria.

Starting from the observed fact of splanchnic vasoconstriction and the features of portal circulation in shock, our experiments began in 1952-1956 with research work on the value of rectocolic administration of procaine and enteral oxygen therapy [21, 35].

With the first models we tried to release the spasms of the suprahepatic veins in the late stages of shock by introducing *Norartrinal (Noradrenaline, Levophed)* into the celiac trunk through a catheter [34]. Local, controlled administration of pressor amines in the late phases of shock, when the endogenous deposits are depleted, is an older concept for us and has been taken up again by many investigators. The contemporary endeavours of Laborit's French school to find the best therapeutic method for rebuilding the catecholamine deposits in irreversible shock are well known.

We initiated investigations on thermal shock and devised an original apparatus for the induction of hyperthermia, by means of which we studied a series of constants and, particularly, the coagulolytic balance. Hyperthermia is not only a shock-inducing agent specific to our modern times but, when controlled, may represent a very useful therapeutic method [33].

The study of the rheology of the microcirculation was one of our main subjects of research; using radioisotopes with erythrocytic fixation, we were able to reveal visceral pooling of the blood fluid, and especially the therapeutic value of certain drugs, subsequently introduced in current clinical practice *(Levomepromasine, Hydergin,* etc.). The experience gained in the use of ganglioplegics in shock allowed us to use them under controlled, individualised administration in the clinic [7, 35, 36].

During the last few years our attention has been drawn to the study of metabolic disturbances, which in the last instance always appear to decide the fate of the shocked organism. The *shock cell* has become a subject of fundamental research work. Metabolic disturbances were studied either indirectly, by measuring *p*H variations in the essential visceral pedicles, or directly by investigating the intracellular histoenzymological lesions and the structural leptonic ones [4, 39]. In this case, too, we tried to check the efficiency of some drugs upon the membrane and enzymatic cellular systems *(Levomepromasine* and cortisone appeared to be the most useful).

Three main themes were developed in our recently experiments on shock:

(i) study of the lymphatic circulation territorialised at the organ level;

(ii) the influence of hypoxia and hyperthermia on the genetic immune parameters, with a view to the conservation and transplantation of the isolated organ;

(iii) study of the functional and structural alterations in the *shock cell* and determination of the best drugs for protecting the cellular subsystems.

These experimental data were processed statistically by methods applicable to the biological phenomena studied. Considering the lots as Bernoulli samples, the *t (Student)* test was applied for calculating the difference between two series of variations. At smaller frequencies, of less than 30, but not smaller than 5, Pearson's χ^2 test was used. For rare events *Poisson's* distribution and the *Wald-Serra* sequential calculus were introduced [10, 21, 32].

2.3 RECTOCOLIC ADMINISTRATION OF A COMPLEX ANTISHOCK SOLUTION

This study is of historical importance for the research workers of the Central

Military Hospital, being the first experimental shock model used in 1950 with practical implications of unquestionable value. The experiments started from the known fact that after parenteral administration procaine produces a certain degree of general analgesia. Bigelow and Harrison (1944) noted an increase in the cutaneous threshold of pain after intravenous administration of procaine to human subjects. However, the mechanism by means of which this tertiary amine of aminobenzoic acid ester achieves its general effects is still obscure. Its standard chemical formula is $C_{13}H_{20}N_2O_2 \cdot HCl$ (4-aminobenzoyldiaethylaminoaethanol hydrochloride) and its structural formula is:

$$\left[H_2N{-}C_6H_4{-}C(=O){-}O{-}CH_2CH_2N^{\oplus}H(C_2H_5)_2 \right] Cl^{\ominus}$$

Systemic analgesic effects are obtained with high doses (up to 800 mg), last at most 10-20 minutes and rapidly diminish within the following 60 minutes [44].

It appears reasonable to explain procaine analgesia by the central action of the substance or that of one of its hydrolysis products, most probably diaethylaminoaethanol [8].

Material and method. The formula of a complex antishock solution was established in doses valid also for man: procaine (1 g), morphine (0.02 g), luminal (0.10 g), alcohol (10 g), glucose (5 g) and distilled water (100 ml); in the experiments (dogs) the doses were administered in terms of body weight. The study included 55 dogs; in 45, curarised but not anesthetised, mixed traumatic shock was induced by crushing the front and back leg on one side with a wooden hammer, associated with traction of the mesenteries and trituration of 1 m intestine for 10 minutes.

Ten dogs, not submitted to shock, were used for establishing whether procaine is destroyed or not in the colon and for determining the rate of its portal absorption. To this end a 25 cm descending colon segment with intact mesentery, was isolated and cannulated, introducing 1% procaine (100 ml), without pressure. Procaine (colorimetric dose) was found in the blood after 7-15 min and persisted for 60-90 min. Similar results were obtained in humans: 7-10 min after rectal introduction of 1% procaine (100 ml) it was found in the blood and persisted for 90-120 min. The inactivation capacity of the blood (by procainesterase) is of 100 mg (i.e. 10 ml 1%) per minute.

Therefore, intrarectocolic administration of procaine offers the means of a prolonged perfusion without any risk of overdosage.

The 45 shocked dogs were separated into six lots:

(i) lot 1 (10 dogs) developed a severe state of shock followed by death after a mean interval of 1 h 44 min (Figure 2.1);

(ii) lot 2 (8 dogs) which received 10% chloral hydrate (2 ml per kg body weight) developed shock and died on average after 3 h 2 min;

(iii) lot 3 (10 dogs) received a 1% procaine intravenous perfusion (20 drops/min.); the average survival was of 4 h 14 min;

(iv) lot 4 (10 dogs) received chloral hydrate narcosis (the same as lot 2) and intravenous perfusion with procaine, the duration of survival being of 7 h 51 min;

(v) lot 5 (9 dogs) were administered a rectal enema (60-80 ml) with 1% procaine; survival 3 h 52 min;

(vi) lot 6 (8 dogs) received a rectal enema containing a complex antishock solution, meant to enhance the quality of chloral hydrate by using luminal and morphine; survival was 6 h 47 min.

The results show that the antishock action of procaine administered systemically is significant (comparison of the lots, two by two, always gave $\chi^2 > 9, p < 0.01$).

Subsequent investigations lend support to these findings (*see* Section 2.8). In spite of its contradictory action (sometimes sympatholytic, sometimes parasympatholytic) procaine has an essential *normotonic effect*, by amphomimetic stabilisation of the nervous control relays. Follow up in the six lots of the general picture of shock (pressure variations, respiration, hematocrit, plasma density, general behaviour, etc.) showed it to be more

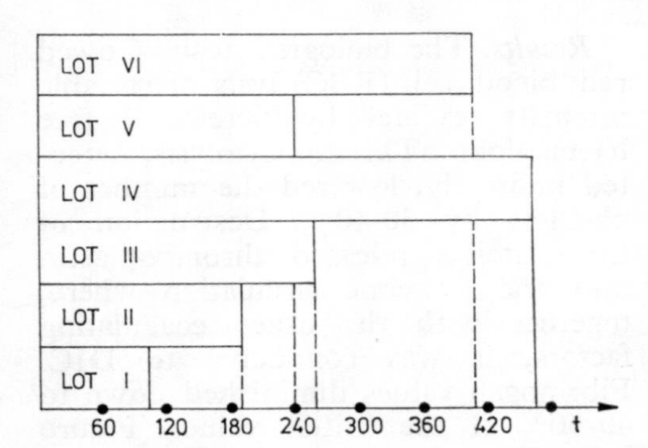

Figure 2.1. Survival in the animals of six experimental lots.

attenuated following the use of procaine alone, and especially when potentiated by other substances with a known stabilising action on the CNS (Figure 2.2). Confirmation of the value of

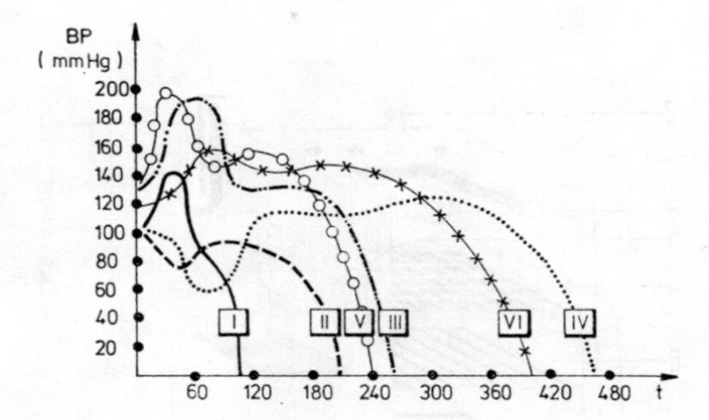

Figure 2.2. Mean blood pressure curves. Note the 'benign' aspect of the curve in lot VI (very close to that of lot IV).

rectocolic administration *(rectocolon procainisation)* thus offers the advantage of a simple unspecific method of antishock protection, that can readily be applied in the case of large numbers of casualities, with severe progressive states of shock.

2.4 HYPERTHERMIC SHOCK INDUCED BY CONTROLLED HYPERTHERMIA

The choice of this procedure was determined by theoretical considerations, checked experimentally, and according to which the bacterial flora and viruses, as well as the neoplasic cell are destroyed at temperatures of 42-43°C. To annul the balance between thermogenesis and thermolysis we used the pharmacodynamic action of certain drugs with a selective action on the ascending reticular activating system (ARAS), the site of the thermoregulating centres. By blocking these centres rectal temperature was increased to 44-45°C.

Material and method: an apparatus devised by us (I. Şuteu, T. Giurgiu, P. Ionescu, A. Cafriţă) was used — the

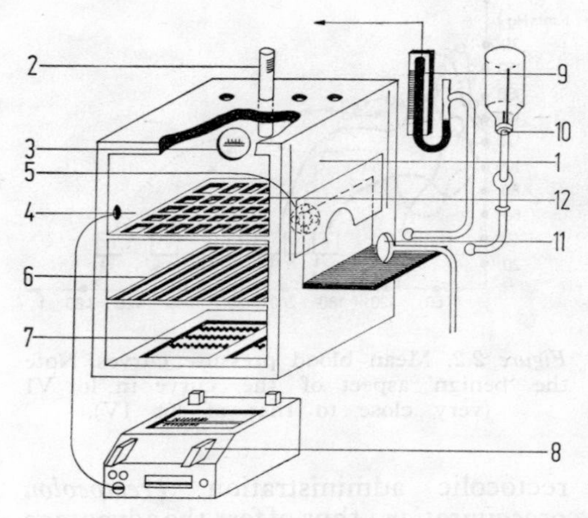

Figure 2.3. Schematical representation of the pyrostat:
1 — box of the apparatus; *2* — thermometer; *3* — thermostat; *4* — rectal electrode; *5* — ventilator; *6* — asbestos screen; *7* — heat source; *8* — thermocouple; *9* — Ludwig manometer; *10* — perfuser; *11* — oxygen mask; *12* — asbestos hammock.

pyrostat, whose microclimate may be raised to a temperature of 100-110 °C by means of two dry-heat electric generators, with a ventilation system for homogenisation. The apparatus was adapted to an electric thermocouple for recording aesophageal and rectal temperatures (Figure 2.3). The study was conducted on 20 dogs, 40 rabbits and 32 rats. Before being introduced into the thermostat each animal received a dose of phenothiazine *(Nozinan)* or butyrphenol *(dihydrobenzperidol)*; the carotid was then catheterised with polyethylene tubes for graphical recording of the BP and the jugular vein for withdrawing blood samples at certain intervals for biochemical and haematological determinations and testing the coagulolytic balance (thromboelastogram). Blood was collected before introducing the animal in the pyrostat. At 43 °C the animal was taken out of the pyrostat and cooled. The rectocolic temperature continued to increase for ten minutes up to 44-44.7 °C and even 47 °C (Figure 2.4).

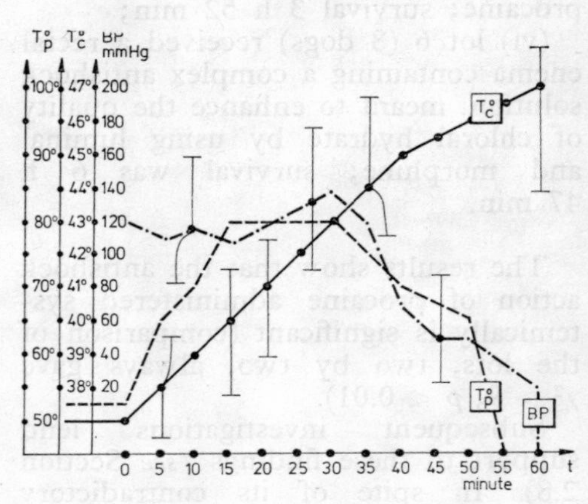

Figure 2.4. Mean intracolon temperature (T^o_c) and intracarotid arterial pressure (BP) curves with reference to pyrostat temperatures (T^o_p) in the onset of hyperthermal shock and after 60 minutes.

Results. The biological tests showed red blood cell (RBC) lysis of variable intensity revealed by increase in free haemoglobin. Thrombocytolysis, detected indirectly, lowered the number of platelets by 30-40%. Destruction of the platelets released thromboplastin into the systemic circulation where, together with the other coagulating factors, it was conducive to DIC. Fibrinogen values diminished down to 40-50% of the initial values (Figure 2.5). The Quick, Howell and Lee-White time showed oscillating values

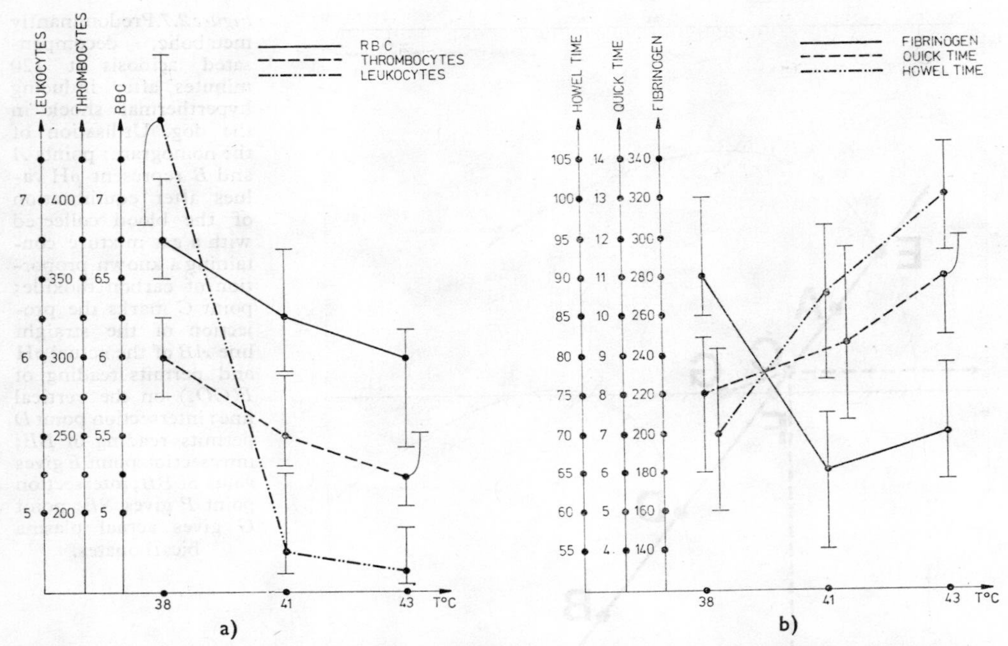

Figure 2.5. Mean RBC, thrombocyte and leukocyte count (*a*), Quick time, Howel time and fibrinogen recorded at 38°, 41° and 43 °C (*b*).

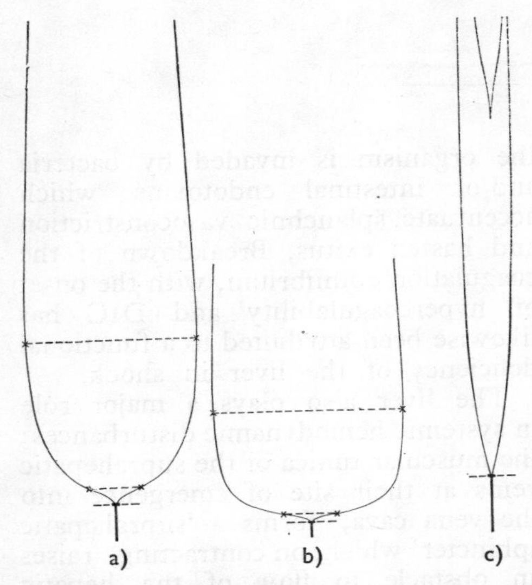

Figure 2.6. Thromboelastogram aspects in the onset of hyperthermal shock (*a*, normal), at 41 °C (*b*, hypercoagulability) and at 43 °C (*c*, hypocoagulability).

which compared with the thromboelastogram aspects revealed the evolution of hypercoagulability from the initial stage up to hypercoagulability with evident fibrinolysis aspects (Figure 2.6).

The acid-base balance, determined with the *Astrup* microequipment, was great altered. In the initial phase the presence of ventilatory alkalosis was found and, after 120 min, metabolic acidosis, at first compensated then decompensated (Figure 2.7).

The results obtained confirm the fact that thermal disintegration of the blood platelets, with the release of thromboplastin and activation of the plasma factors, triggers DIC. The mechanisms of coagulolytic balance (thrombin-antithrombin, plasmin-antiplasmin) are exceeded under hyperthermia conditions, and consumption coagulopathy and fibrinolysis develop.

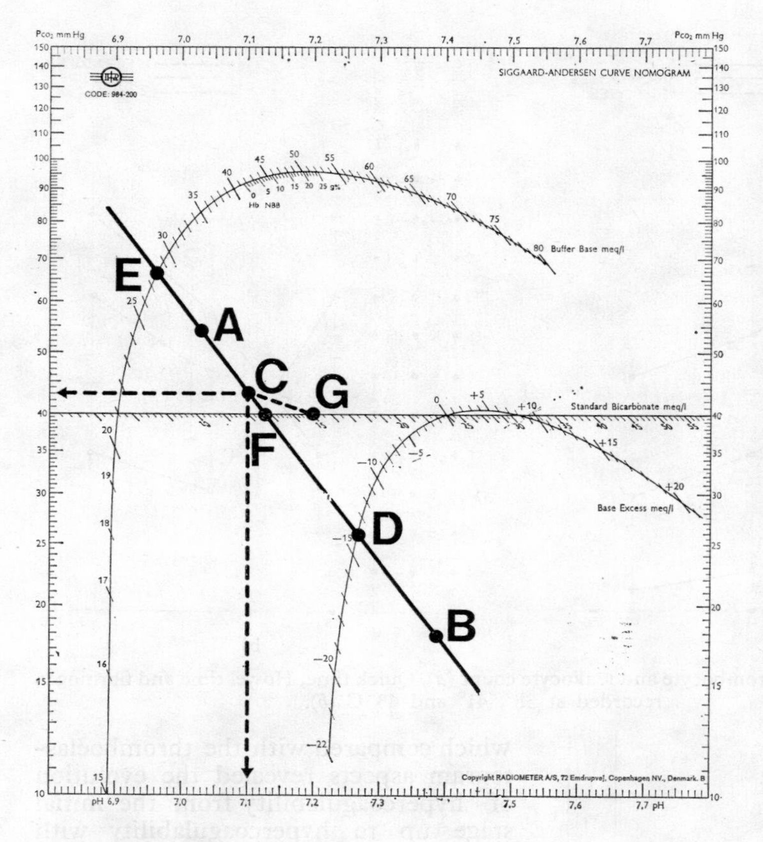

Figure 2.7. Predominantly metabolic, decompensated acidosis at 120 minutes after inducing hyperthermal shock in the dog. Utilisation of the nomogram: points *A* and *B* represent *p*H values after equilibration of the blood collected with a gas mixture containing a known proportion of carbon dioxide; point *C* marks the projection of the straight line *AB* of the actual *p*H and permits reading of $P(CO_2)$ on the vertical line; intersection point *D* permits reading of *BE*; intersection point *E* gives value of *BB*; intersection point *F* gives *SB*; point *G* gives actual plasma bicarbonates.

2.5 STUDY OF THE REGULATION OF THE SPLANCHNIC CIRCULATION IN TRAUMATIC SHOCK

Zweifach and Shorr [42] postulated that in shock disturbances of the hepatorenal regional circulation develop due to the successive appearance in and disappearance from the circulation of a 'vasodilator factor'. This factor was also incriminated by Fine [12] in the irreversibility of shock. The hypoxic liver no longer exercises its role of barrier in the path of bacteria and toxins; the hepatic production of the physiological bacteriolytic (properdin) diminishes and, in consequence, the organism is invaded by bacteria and/or intestinal endotoxins which accentuate splanchnic vasoconstriction and hasten exitus. Breakdown of the coagulation equilibrium, with the onset of hypercoagulability and DIC has likewise been attributed to a functional deficiency of the liver in shock.

The liver also plays a major role in systemic hemodynamic disturbances; the muscular tunica of the suprahepatic veins at their site of emergence into the vena cava, forms a 'suprahepatic sphincter' which, on contracting, raises an obstacle to flow of the hepatic blood. The liver, therefore, is a reservoir in which the blood may be stored or ejected into the general circulation. The amount of blood that can be rendered to the systemic circulation is equivalent to 30-55% of the weight of the liver. The functional suprahepatic sphincter is known to contract following the administration of histamine and to relax after pressor amines. However, in the initial stages of shock, in spite of the existing hypercatecholaminemia, the suprahepatic veins contract in spasms due to the lack of reactivity to the pressor amines (probably also induced by Fine's endotoxin factor). There is still debate about this 'suprahepatic sphincter' and the interpretation of its behaviour in shock.

It seems, however, that if in the *late stages* of shock, these functional suprahepatic sphincters could be bathed in pressor amines their spasm would be lysed. Today, it has been established that dilatation of the posthepatic functional sphincter is brought about by the effect of catecholamines (*see* Section 5.4.2).

Material and method. The experiment was carried out on three lots of twelve dogs each.

Shock was produced by cominuted fractures of the leg, muscular attrition and haemorrhage. After cannulation of the carotid the animals were curarised in order to immobilise them during the trauma and afterwards. The BP and respiration were recorded by kinography immediately after the trauma, and the drug administered when the systemic BP fell to about 50 mm Hg.

The first lot received 0.03 mg noradrenaline kg body weight by the intravenous route, continuing to measure the BP values. In the second lot the femoral artery was exposed and an ureteral catheter introduced up to the aorta above the emergence of the celiac trunk, through which the same amount of noradrenaline was administered. In the third lot noradrenaline was introduced in the aorta after the administration of a ganglioplegic. BP and respiration were followed-up until the animals died or for 3-5 hours after the administration of the drug.

Results. In the first lot after repeated administration of the drug by the intravenous route, the BP gradually fell until death ensued 3 hours after shock.

The second lot received noradrenaline in the aorta at the moment in which the BP fell to 50 mmHg. After the first administration an increase of up to 75-80% of the initial values was obtained and lasted for about 2—4 hours after which there was a slow decrease. The animals were sacrificed on average after 4 hours when the BP oscillated between 60-80 mmHg (Figure 2.8).

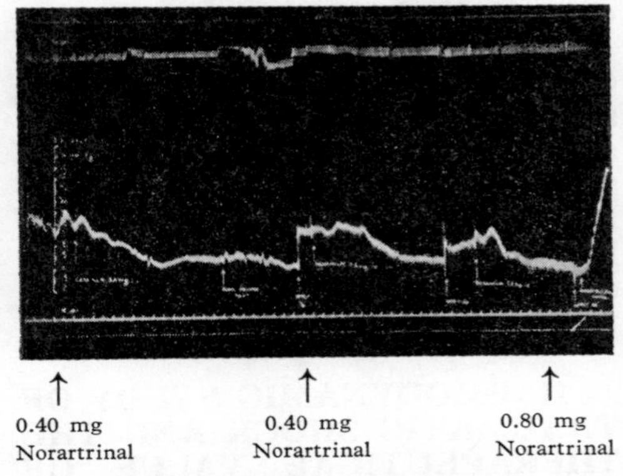

↑ 0.40 mg Norartrinal	↑ 0.40 mg Norartrinal	↑ 0.80 mg Norartrinal

Figure 2.8. Arterial pressure kinogram in a dog of lot II.

To the third lot, ganglioplegics (*Hexamethonium*, 0.5 mg/kg body weight) were administered after shock and when the BP fell to 30 mmHg noradrenaline was introduced in the aorta, obtaining a sharp, spectacular increase (Figure 2.9). Application of the t test to the BP values obtained in lots 1 and 3 gave a significance of $p < 0.001$.

Similar results were also obtained in the clinic in case of shock when

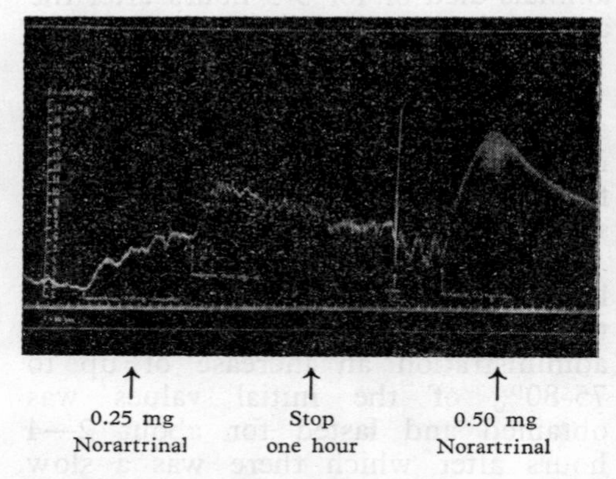

Figure 2.9. Arterial pressure kinogram in a dog of lot III.

BP values were remedied with α and β stimulants after the administration of a ganglioplegic.

Noradrenaline appears to relay spasm of the suprahepatic veins (β effect) and thus releases into the circulation all the blood stagnating in the portal system; there results an increase in BP of 25-30% maintained for 2-3 hours, also due to increase of the cardiac output.

2.6 HAEMODYNAMIC STUDY OF TRAUMATIC SHOCK AND THE THERAPEUTICAL VALUE OF GANGLIOPLEGICS

It is known that return of the blood initially withdrawn for producing shock arrests the evolution towards irreversibility, when done after a 'critical interval' (in the late phase of shock). A vasoactive medication must be obligately associated in this stage in order to mobilise the pool in the microcirculation. Starting from Fine's assertion that shock is a severe interruption in the distribution of the blood mass [12], which in most cases stagnates in the splanchnic area, our experiments were designed to overcome the dams of the intrahepatic and intestinal microcirculation and remit to the systemic circulation the stagnant blood of these areas.

In the first experimental stage *hexamethonium bromide* was used as the ganglioblocking agent; consequent to ganglionic blocking of the sympathetic vasomotor impulses, vasodilatation occurs in the microcirculation, followed by opening up of the pre- and post-capillary sphincteral dams.

Material and method. The experiment was carried out on 25 dogs; mixed traumatic shock was produced by crushing the hind legs from the knees down and traction of the mesentery.

The 25 dogs were separated into three lots: lot 1 (6 dogs) controls; lot 2 (9 dogs) to which *hexamethonium* was administered 15 minutes after the injury, and lot 3 which received *hexamethonium* when the BP fell to half the initial value. *Hexamethonium* was administered in 0.5 mg/kg body weight doses. After the effect of the drug had passed and the BP showed a tendency to increase, 0.25 mg *hexamethonium*/kg body weight was readministered. The acute experiment lasted 3-5 hours, after which the animal was sacrificed and fragments of the liver, lungs, kidneys, spleen, intestine and mesentery were collected for histologic examination.

Results and interpretation. Shock, manifested clinically by a fall in the BP below 80 mmHg, developed after different intervals in terms of the individual reactivity of the animals. After the injury, the control animals showed a brief increase in BP values, consequent to the marked discharge

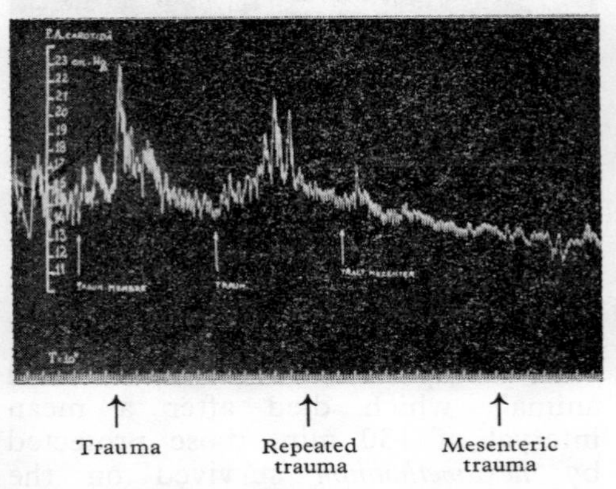

↑ Trauma ↑ Repeated trauma ↑ Mesenteric trauma

Figure 2.10. Arterial pressure kinogram in a control dog.

of catecholamines (α effect). After 10-15 minutes the BP gradually fell for 2-3 hours up to death (Figure 2.10).

In lot 2 in which *hexamethonium* was administered immediately after producing shock, the BP fell to minimal values, then after 15 min increased to a mean value of 60 mmHg (Figure 2.11). In two animals (Figure

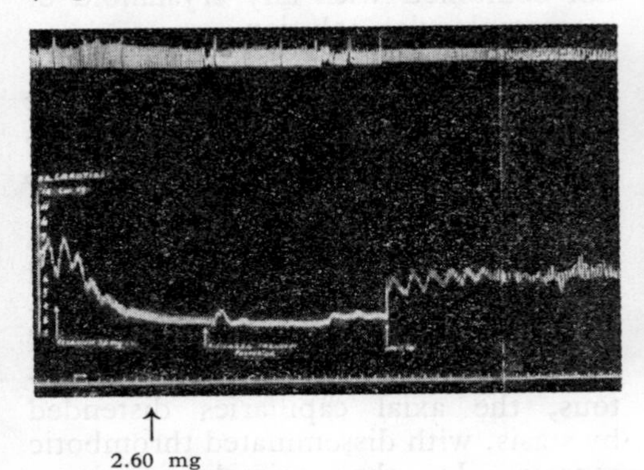

↑ 2.60 mg Hexamethonium

Figure 2.11. Arterial pressure kinogram in a dog of lot II.

2.12), BP was 80 mmHg after 70 min, and after twice repeating the administration of *hexamethonium* (3 mg doses) it rose to 110 mmHg within 10 min.

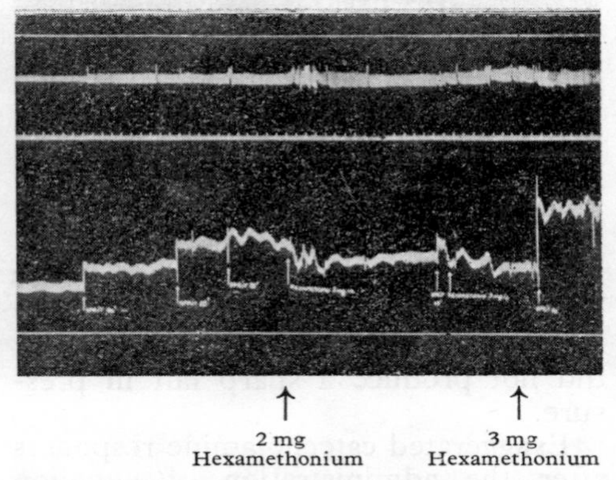

↑ 2 mg Hexamethonium ↑ 3 mg Hexamethonium

Figure 2.12. Arterial pressure kinogram in a dog of lot II (repeated hexamethonium doses).

In lot 3 *hexamethonium* was administered when the BP values were about 60-70 mmHg. There was an immediate decrease to 30-40 mmHg, which lasted longer than in the preceding lot (Figure 2.13). At 120 minutes (Figure 2.14) the BP was maintained at acceptable values but showed no tendency to return to normal values as in lot 2. However, the readministration of *hexamethonium*

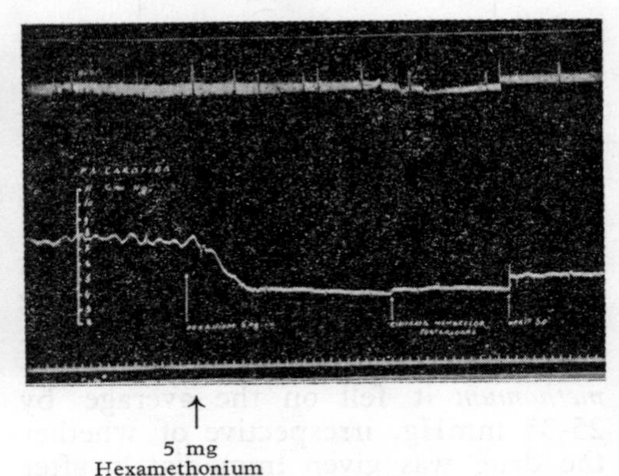

↑ 5 mg Hexamethonium

Figure. 2.13. Arterial pressure kinogram in a dog of lot III.

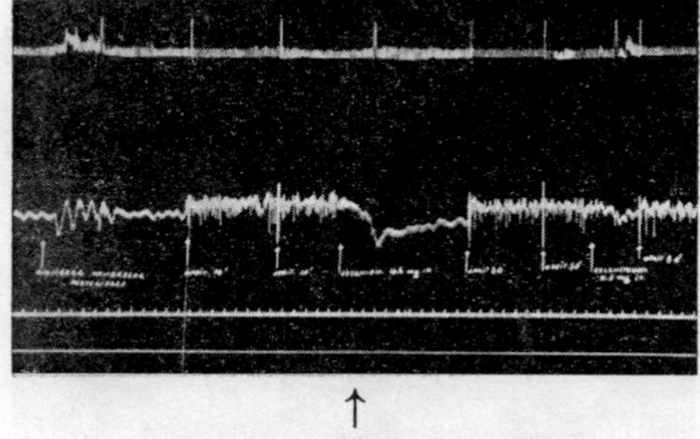

Figure 2.14. Arterial pressure kinogram in a dog of lot III (repeated hexamethonium doses).

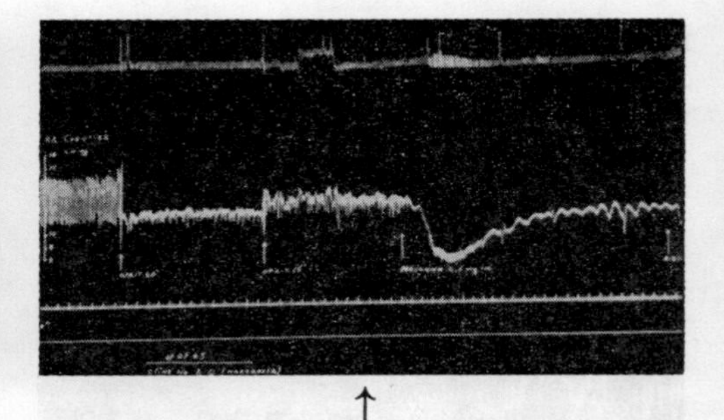

Figure 2.16. Readministration of hexamethonium after Norartrinal.

did not produce a sharp fall in pressure.

Exaggerated catecholamine responses after the administration of ganglion blocking agents have been noted. In all the cases in which the BP fell to

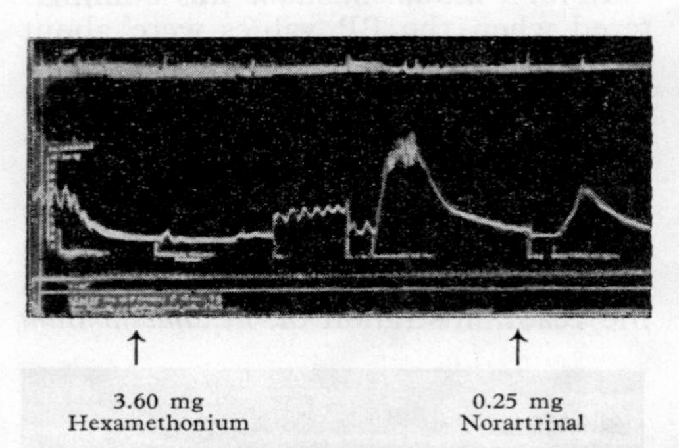

Figure 2.15. Norartrinal (Noradrenaline-Levophed) administered at 140 minutes after the onset of shock.

20 mmHg noradrenaline brought about a rise in the BP to 140-170 mmHg 140 min after the onset of shock. Following the administration of *hexamethonium* it fell on the average by 25-35 mmHg, irrespective of whether the drug was given immediately after shock or when the BP were around 70 mmHg (Figure 2.16).

In comparison with the control animals which died after a mean interval of 130 min, those protected by *hexamethonium* survived on the average 140 min more ($p < 0.001$). Three dogs were sacrificed 4 hours after shock when they still had a BP of about 60 mmHg. In contrast to the controls which were in a state of extreme agitation, those which had received *hexamethonium* were quiet, did not react to external stimulants, breathed regularly and had a cardiac rate that did not exceed 10% of the initial values. It should be emphasised that the ganglion blocking agent was not combined with any crystalloid or macromolecular solution.

The histologic examination (Figure 2.17) revealed in the lungs and kidneys of the controls intense blood stasis, with vascular dilatation and disseminated thromboses. The liver exhibited enormous dilatation of the sinusal capillaries, thromboses of the centrolobular and portal veins. In the small intestine the villi were dilated, œdematous, the axial capillaries distended by stasis, with disseminated thrombotic aspects. In the animals receiving *hexamethonium* the predominant lesion was blood stasis, minor dystrophic

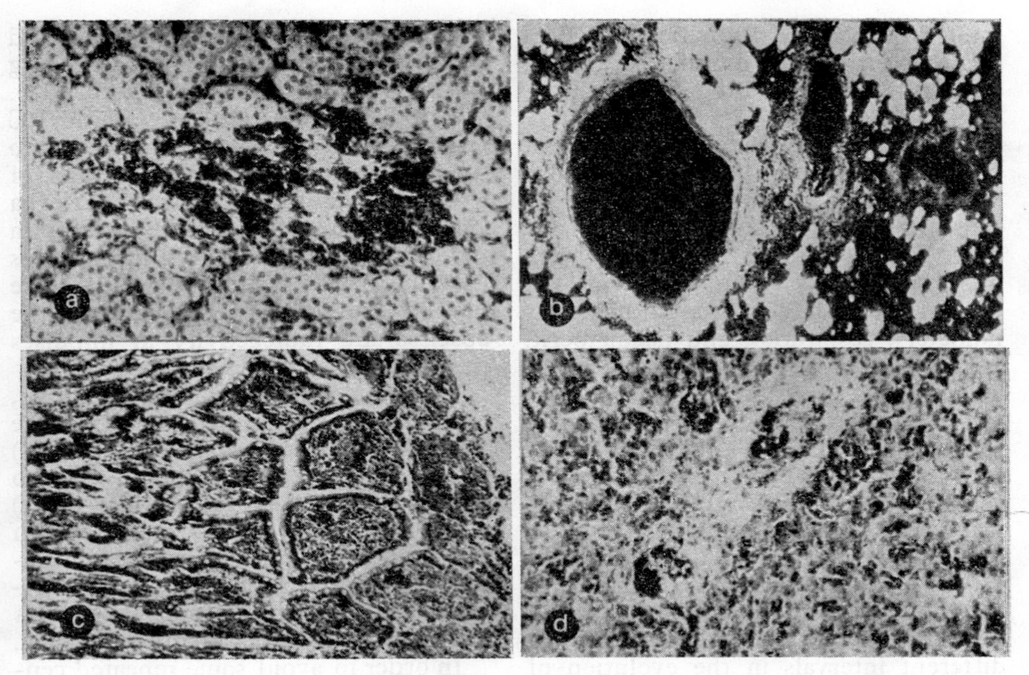

Figure 2.17:
a — disseminated glomerulotubular thromboses *(HE,* ×80); *b* — massive thromboses of the pulmonary parenchyma *(HE,* ×160); *c* — intracapillary disseminated coagulation and stasis in the intestinal villi *(HE,* ×80); *d* — interhepatocytic sinusoid dilatations, with centrolobular microthromboses and stasis *(HE,* ×80).

lesions in the liver, of the turbid intumescence type, but not a picture of vascular thrombosis.

Conclusions. In the animals protected by *hexamethonium* the course of shock was simpler and slower and the BP, in contrast to the controls (with descending curves down to exitus), remained level, even showing a slight tendency to rise. In some of the animals 3 hours after shock the BP values were close to the initial ones.

We assume that maintenance of the life of the animals protected by *hexamethonium* is due to the absence of vasoconstriction and the prevention of splanchnic pooling in the microcirculation.

2.7 STUDY OF THE MICROCIRCULATION IN EXPERIMENTAL SHOCK WITH THE ^{51}Cr RADIOISOTOPE

Starting from the fact that vasoconstriction is the dominant link in the pathogenic chain of shock, due to the consequent pooling of the blood in the microcirculation, we tried to demonstrate the presence of this *blood pool* in the microcirculation by radioisotopic methods. Pooling, demonstrated indirectly by measuring the blood flow at various levels by means of electromagnetic flowmeters, is the 'forerunner' of intracellular organelle distress. Our results demonstrated the visceral formation of a pool running parallel to diminution on the circulating blood volume.

Material and method. The experiment included 37 dogs. Traumatic shock was produced without anesthesia by crushing the hind legs (from the knee down). The classical parameters were monitored: BP (cannulation of the carotid, recording with a Ludwig manometer), respiration and haematocrit. In addition, distribution of the radioactive blood mass and blood volume were determined with radioactive sodium chromate (^{51}Cr), whose half life is 28 days. For distribution of the blood mass a Vakutronic apparatus was used, with a scintillation counter, previously screened with lead plates (10 mm thick), with an 8 mm slit at the bottom. The counter was applied over the same anatomical region at different intervals in the evolution of shock: hepatic area (at its densest point); intestinal area (halfway between the xyphoid process and pubis);

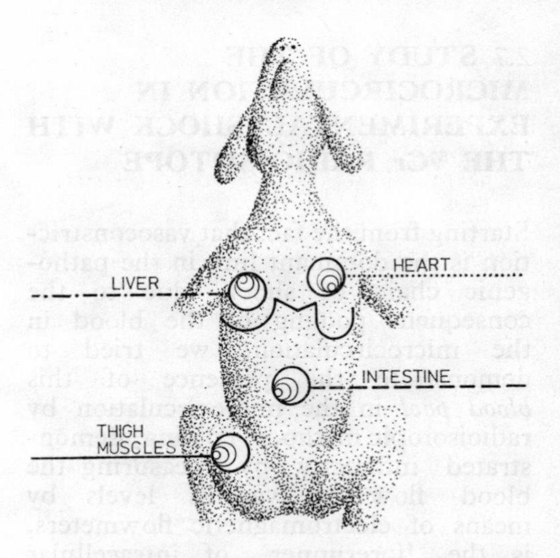

Figure 2.18. Regions scanned by the scintillation counter (see Figures 2.19, 2.20 and 2.23–2.31).

thigh muscles (anterior area); and heart (fourth intercostal space along the anterior axillary line) (Figure 2.18).

The RBC were labelled with 35 μC ^{51}Cr/kg body weight introduced into the systemic circulation by means of a catheter in the vena cava. After a 15 min interval for homogenisation radioactivity was measured in the areas described. The values recorded were plotted, the time in minutes on the abscissa and radioactivity in impulses per minute on the ordinate.

The total circulating blood volume, erythrocytic and plasmatic blood volume were determined in the blood (30 ml) collected from the jugular vein, to which ^{51}Cr (30 μC/kg body weight) was added. The blood was kept in the thermostat for 20 min, then distributed into five test tubes and centrifuged for 15 min; normal saline was substituted for the plasma supernate.

In order to avoid some repeated centrifugations the hexavalent chromium was reduced to trivalent chromium with ascorbic acid; 5 ml of the total 30 ml labelled blood was injected through the cava cannula. After 15 min, during which the labelled RBC mixed with circulating blood, 1 ml blood was collected for counting the impulses and 1 ml for haematocrit. After causing shock and in terms of BP values, a further ml was collected for determining residual radioactivity. The radioactive blood (5 ml) was again administered. After another 10 min blood samples were collected once more. This was repeated until death of the animal, the blood volume was calculated by dispersion of the mean values, calculating the standard deviation [28, 31].

Measurement of regional radioactivity and withdrawal of blood samples for determining the circulating blood volume were done concomitantly.

Each dog was his own control, i.e. the measurements taken before shock; 4-5 hours after shock (or after death) fragments were collected from the liver, small intestine, large intestine, spleen, kidneys and lungs for histologic examination.

Results and discussions. The dynamics of radioactivity in the course of shock showed a constant increase in the number of pulses at the level of the abdominal viscera, the increase ranging between 14 and 141% above the initial values recorded before shock ($p < 0.001$). The count rate at the level of the heart showed a decrease of 30-60% of the initial values ($p < 0.001$) (Figure 2.19).

Radioactivity gradually augmented immediately after shock running parallel to the evolution of the latter (fall of the BP). Increase in radioactivity in the viscera and muscles and its decrease in the heart became more accentuated in the first 1-2 hours after shock.

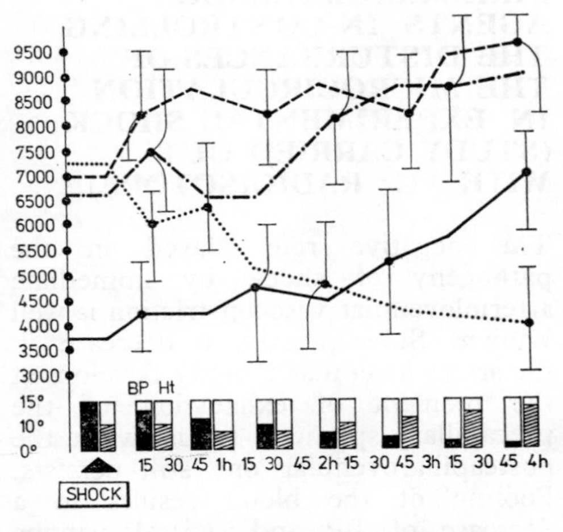

Figure 2.19. Mean pulse counts recorded in the four regions of the body in the entire group of animals.

The mean values (Figure 2.20) of the entire lot indicated an increase in radioactivity in the intestinal area of 25% after the first hour, of 48% after two hours and after two more

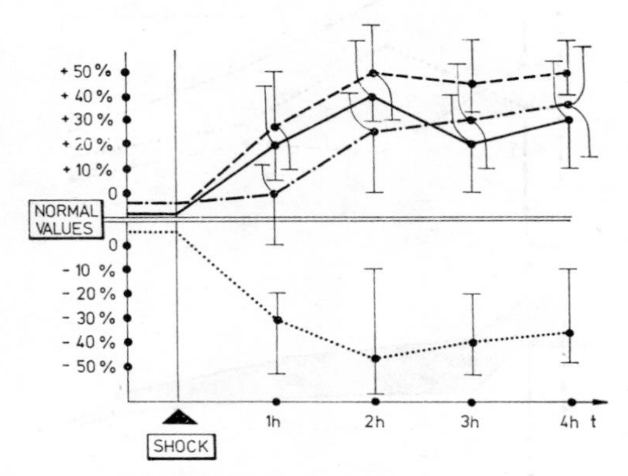

Figure 2.20. Procentual variation of the pulse count for the entire group of animals.

hours up to 50% of the initial values ($p < 0.001$).

In the liver, one hour after shock the increase was of only 5%, after 2 hours of 23% and up to 37% after 4 hours ($p < 0.01$).

Radioactivity in the thigh muscles increased by 20% after 1 hour, by 41% after two hours, then ceased, a slight diminution even being recorded ($p < 0.001$).

In the heart, after one hour radioactivity decreased by 35% and reached a low at two hours (50%), then gradually augmented ($p < 0.001$).

Determination of the circulating blood volume, erythrocytic and plasmatic volume, corroborated with the haematocrit, reflecting the gradual diminution of the circulating blood volume (Figure 2.21). One hour after shock the total blood volume fell on average by 50% of the initial value, producing an inversion of the haematocrit ratio in favour of the RBC ratio which in irreversible shock reaches up to 80% ($p < 0.001$).

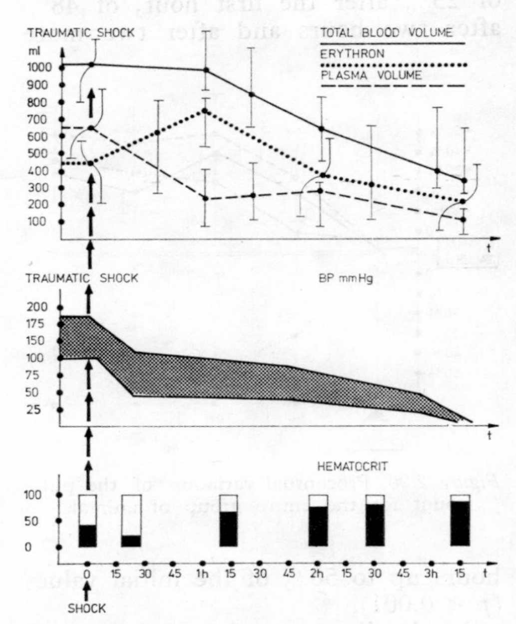

Figure 2.21. Variations of the mean blood volume, arterial pressure and hematocrit in the entire group of animals.

Table 2.1 gives the global blood volumes, erythrocytic and plasma blood volumes recorded in six of the shocked dogs. The total circulating blood volume falls in the mean to about 40% of the initial value after 3 hours, erythrocytic volume to about 50% and plasma volume to 20-25%, corresponding to a haematocrit of 75%.

Regional radioactivity revealed the defect in the distribution of the blood mass, a defect due to asymmetrical vasoconstriction induced by the immediate postagressive reaction syndrome.

The fact that within the first 60 min the count rate at the level of the liver is only 25% of the increase observed in the viscera at umbilical level shows that in the initial period the liver participates very little in sequestering the blood. Subsequently, after the first hour of shock, the liver accumulates more blood than the viscera in the umbilical region, because the efficiency of the catecholamines decreases in acid medium, releasing the splanchnic reservoir.

After the first two hours of shock, the dissociation of vascular motility in the microcirculation creates the conditions of a blood sequester, reflected in the sharp increase in the count rate per minute; in the following two hours it is less marked both in the liver and intestines, as well as in the thigh muscles, with a slight concomitant increase of radioactivity in the heart. This may be accounted for by exhaustion of the venular tone and postcapillarovenular sphincter, followed by refilling of the systemic circulation with part of the blood pooled in the microcirculation.

2.8 THE EFFICIENCY OF SOME PHARMACODYNAMIC AGENTS IN CONTROLLING THE DISTURBANCES OF THE MICROCIRCULATION IN EXPERIMENTAL SHOCK (STUDY CARRIED OUT WITH ^{51}Cr RADIOISOTOPE)

The negative role played in the pathogeny of shock by immediate arteriolovenular vasoconstriction is well known. Subsequently, a decrease in the arteriolovenular tonicity develops at the moment of exhaustion of the precapillary sphincter tonus, while the postcapillarovenular one still persists. Pooling of the blood results in a decrease of BP and central venous pressure (CVP). Blood pooling leads to the taking up of blood from the

Table 2.1 EVALUATION OF THE BLOOD VOLUMES WITH ^{51}Cr

Weight of dog (kg)	Blood volume before shock (ml)				Blood volume after shock (ml)											
					60 min				120 min				180 min			
	Total volume	Erythrocytic volume	Plasma volume	Ht (%)	Total volume	Erythrocytic volume	Plasma volume	Ht (%)	Total volume	Erythrocytic volume	Plasma volume	HT (%)	Total volume	Erythrocytic volume	Plasma volume	Ht (%)
10.500	1100 ±3.16	473 ±2.64	627 ±3.46	43	1028 ±4.12	776 ±3.16	250 ±3.46	65	675 ±2.44	368 ±2.64	285 ±2.23	60	404 ±2.82	263 ±2.82	141 ±2.23	75
6.200	573 ±4.12	217 ±3.60	356 ±3.31	38	500 ±3.31	187 ±2.23	313 ±2.44	39	236 ±3.16	135 ±2.82	101 ±3.31	61	80 ±2.44	56 ±2.23	24 ±2.69	81
8.400	790 ±2.23	373 ±3.16	417 ±2.82	39	573 ±4.12	313 ±3.16	260 ±2.64	61	431 ±2.23	291 ±3.31	140 ±2.82	70	183 ±2.82	129 ±2.23	54 ±2.44	83
9.500	1000 ±4.15	413 ±1.62	587 ±3.14	42	940 ±3.18	675 ±2.85	265 ±4.12	61	615 ±2.84	320 ±3.24	295 ±3.18	52	420 ±3.14	230 ±2.42	190 ±2.14	55
7.100	725 ±3.15	310 ±2.40	415 ±1.82	42	600 ±2.42	280 ±4.18	320 ±2.14	47	360 ±2.14	210 ±2.54	150 ±3.08	60	200 ±4.20	130 ±1.86	70 ±2.18	65
9.200	922 ±2.13	406 ±2.90	516 ±1.82	44	816 ±3.80	480 ±2.14	336 ±4.16	59	520 ±1.76	362 ±2.25	158 ±2.64	69	336 ±2.16	236 ±1.74	100 ±3.34	70

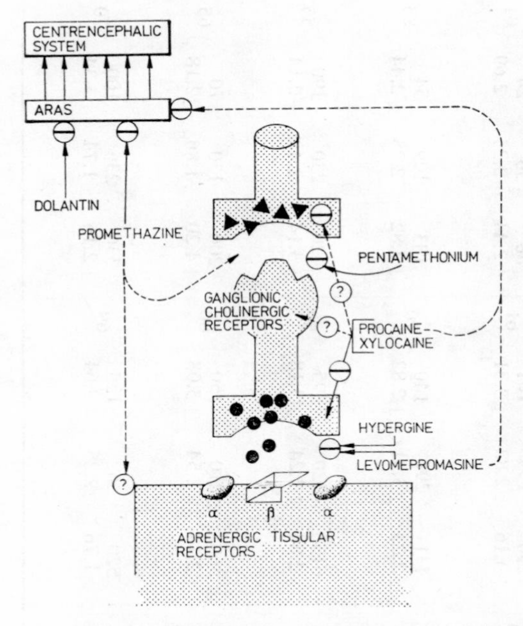

Figure 2.22. The probable site of action of the drugs used in the experiment.

reservoir (in Wiggers' classical experiment).

Attenuation of the initial hypersympathicotonia may slow down or even arrest the evolution of shock towards irreversibility. The pharmacodynamic agents that lower the sympathetic tonus may also act either upon the sympathetic ganglia or upon the endings of the postganglionic fibres or adrenergic receptors of the neuromuscular junctions of the effector organs (Figure 2.22).

Material and method. The efficiency of certain pharmacodynamic sympatholytic and adrenolytic agents in preventing or controlling vasoconstriction was studied by using the preceding experimental model described.

In the second stage of the experiment the 64 dogs were separated into seven lots, for testing the antishock efficiency of the following drugs: *Levomepromasine*

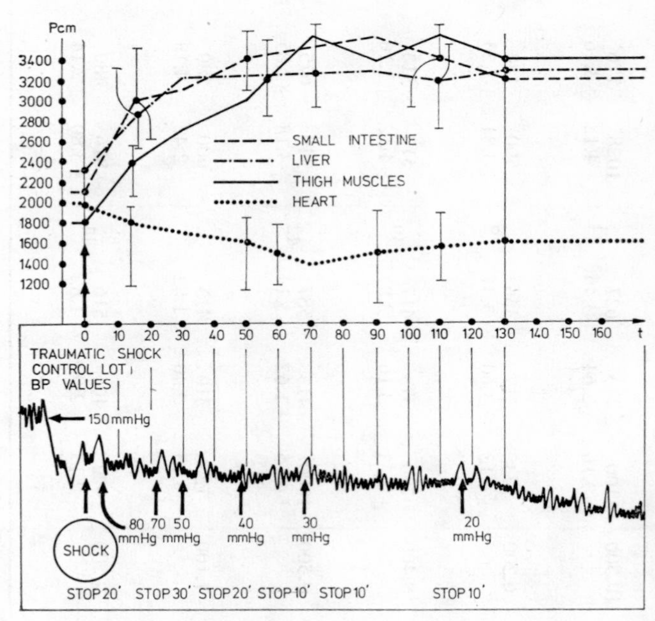

Figure 2.23. Mean values of the pulse count per minute recorded in the four visceral areas (upper graph); curves plotted versus arterial blood pressure (below). The symbols are the same as in figures 2.18, 2.19, and 2.20.

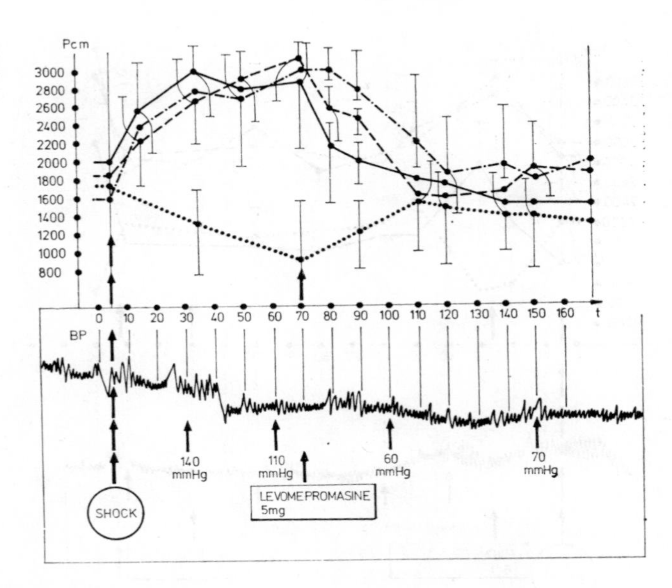

Figure 2.24. Mean pulse count per minute after the administration of Levomepromasine (Nozinan).

(10 dogs); *pentamethonium* (8 dogs); *Hydergin* (8 dogs); *promethazine* (8 dogs); *pethidine* (8 dogs); *procaine* (6 dogs); and *Xylocaine* (6 dogs). The control lot included ten animals.

Results. After shock, in the controls, the BP curve gradually fell and the pulse count proportionally increased in the liver, intestines and thigh muscles, but diminished in the heart (Figure 2.23). Determination of the blood volume with ^{51}Cr pointed to a progressive decrease in the ECBV.

The following results were obtained in the lots treated with sympatho-adrenolytics:

Levomepromasine (Nozinan): its peripheral effect on the microcirculation may be listed in the category of α-blocking drugs. It is administered in 0.5 mg/kg body weight amounts at the moment in which the BP reaches 110-100 mmHg. As may be seen from Figure 2.24 before administration a massive accumulation of RBC takes place in the liver, intestine and thigh muscles and a proportional decrease in the heart; following the administration of the drug and its α-lytic action upon the microcirculation, the pooled blood is returned to the systemic microcirculation although the BP still remains very low (60-70 mmHg). Thirty to forty minutes after the administration of *Nozinan* the pulse count rate returns close to regular, normal values; the BP curve shows a slight ascension and after 60-80 min may reach 90-110 mmHg. The administration of a pressor amine (0.05 mg noradrenalin/kg body weight) two hours after *Nozinan* produces a further increase in the BP, running parallel to a marked increase in the count rate (Figure 2.25).

Pentamethonium (Pendiomide). Ganglioplegics also suppress the adrenomedullary phase of the postaggressive reaction, thus avoiding the abrupt discharge of catecholamines. Pendiomide is an elective blocking agent of the postsynaptic cholinergic receptors.

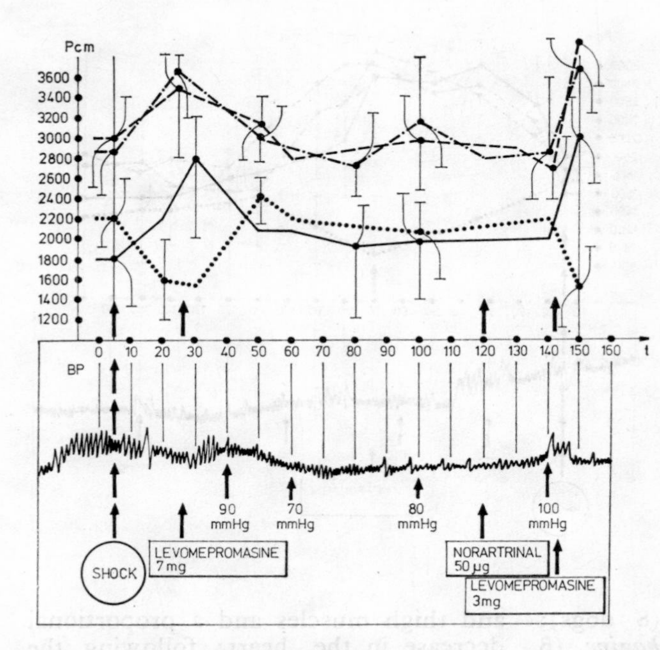

Figure 2.25. Mean pulse count per minute after the administration of Nozinan and Norartrinal.

The drug was administered after shock (2 mg/kg body weight) at the moment when BP values had fallen to 100-110 mmHg (Figure 2.26). After 8-10 min the BP fell down to 40% of the initial values; in the splanchnic region the pulse count gradually fell after 20-30 min almost to initial values and stayed level for 2 hours. The BP curve showed a tendency to

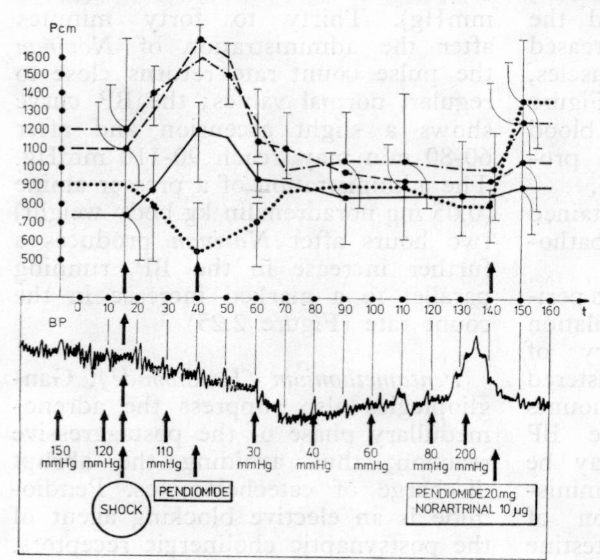

Figure 2.26. Mean pulse count per minute after the administration of pentamethonium (Pendiomide).

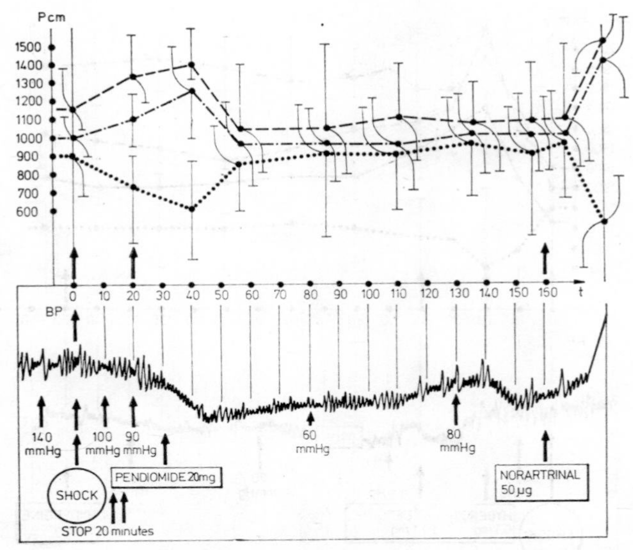

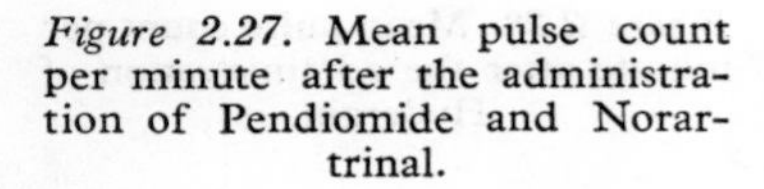
Figure 2.27. Mean pulse count per minute after the administration of Pendiomide and Norartrinal.

a progressively increase, attaining 90-110 mmHg. During all this period the count rate was within normal limits. The administration of noradrenaline (Figure 2.27) produced an increase in the BP up to 200 mmHg, without a proportional increase in the number of pulses, reflecting an increase in the efficiency of the cardiac work under the action of the drug (positive inotropic effect). It is worthy of note that 2 hours after shock the administration of noradrenaline only produced in the control dogs a slight increase in the BP and a considerable augmentation of the pulse count rate in splanchnic projection, followed within 6-10 min by exhaustion and death.

Hydergin is an α-blocking agent with a maximum effect at splanchnic level (excepting the liver where congestion develops); 0.1 mg doses/10 kg body weight where administered when the BP was 90 mmHg, at 35 min after producing shock (Figure 2.28). The BP fell to 30 mmHg and then gradually increased in the course of 2 hours to 100 mmHg. The pulse count showed a slow decrease at the level of the intestines and liver, returning to initial values after 80 min, but not in the liver.

Promethazine (Phenergan) was administered in 2 mg doses/kg body weight (Figure 2.29). After inducing the shock, the BP slowly fell and the pulse count increased considerably, but administration of the drug did not produce significant variations, which were similar to those in the control animals.

Pethidine (Dolantin) interrupts or blocks transmission of the pulses to the thalamocortical and ARAS systems. The drug (2 mg/kg body weight) was administered slowly by intravenous route, 30 min after shock, when the BP was about 110 mmHg. After 10 min it fell to 60 mmHg, then followed a slow ascensional curve. The pulse count rate was not modified by the

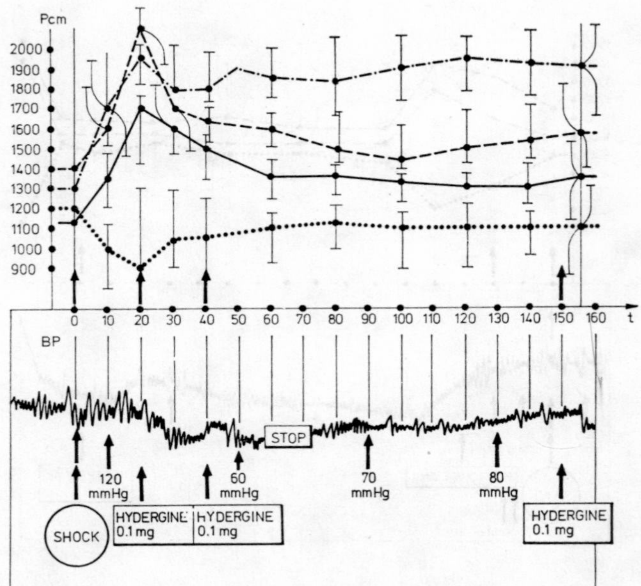

Figure 2.28. Mean pulse count per minute after the administration of Hydergin.

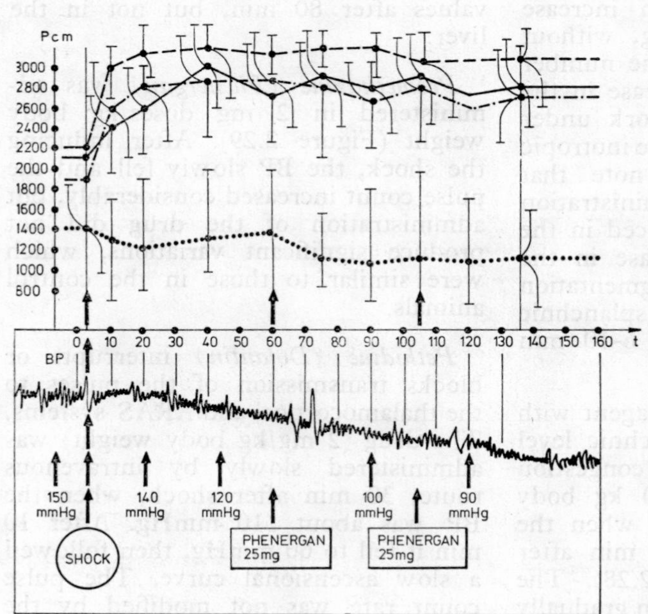

Figure 2.29. Mean pulse count per minute after the administration of promethasine (Phenergan).

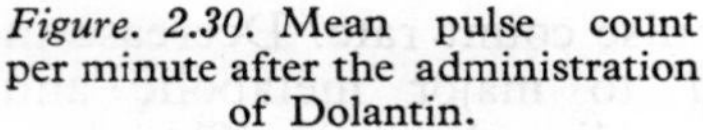
Figure. 2.30. Mean pulse count per minute after the administration of Dolantin.

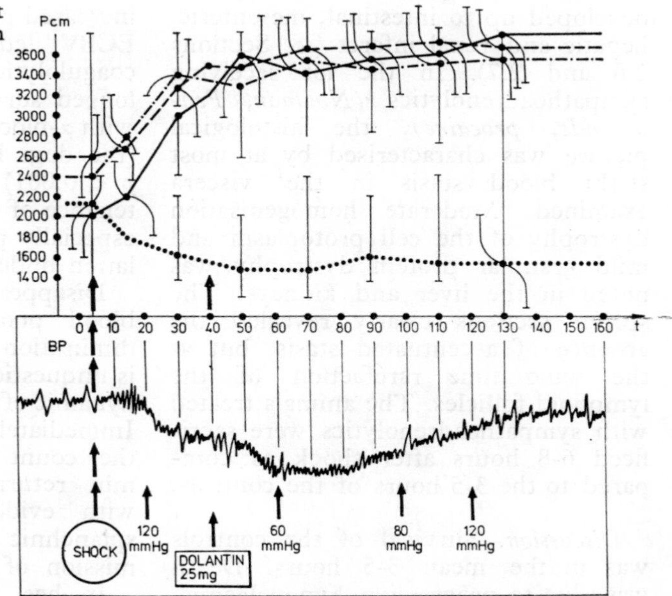

drug, being similar to that in the controls (Figure 2.30).

Procaine. The effect of procaine is manifested at the synaptic terminations of the preganglionic axon; it depresses transmission to certain central synapses, especially the association neurons and certain multisynaptic pathways in the spinal chord and brain stem. The antishock potential of procaine is due to the complexity of its pharmacodynamic action, which has not yet been fully elucidated (see Section 2.3). Procaine was administered in 1% solution, 10 mg/kg body weight, by means of a catheter introduced in the cava vein, at the moment in which the BP fell to 60-70 mmHg (Figure 2.31). Slow injection of the drug did not accentuate hypotension. BP values remained constant for 40 min, then gradually increased. Reinjection of procaine increased the BP to 110 mmHg. There was an abrupt increase in the scan count rate by more than 80% of the initial values, which returned, however to normal 10 min after the administration of procaine, oscillating around the initial values. At the level of the heart the count rate fell after procaine then returned to normal values.

Xylin (Xylocaine, lignocaine), administered in doses of 10 mg/kg body weight did not modify the BP or the pulse count rate. A double dose did not bring about an evident decrease of the count rate in the abdominal viscera but caused cardiorespiratory arrest. After resuscitation the dog recovered, with a normal BP and pulse count close to its initial value.

Histological findings. Cross-sections of the various organs from the control animals (untreated traumatic shock) showed a wide range in the intensity of the pathological process, which developed up to intestinal, mesenteric, hepatic and renal infarct (*see* Sections 2.6 and 2.7). In the lot receiving sympathoadrenolytics *(Nozinan, Pendiomide, procaine)*, the histological picture was characterised by at most slight blood stasis in the viscera examined. Moderate homogenisation dystrophy of the cell protoplasm and mild granular protein dystrophy was noted in the liver and kidneys. The spleen sections clearly revealed the absence of accentuated stasis, but at the same time rarefaction of the lymphoid follicles. The animals treated with sympathoadrenolytics were sacrificed 6-8 hours after shock as compared to the 3-5 hours of the controls.

Discussion. Survival of the controls was in the mean 3-5 hours. Death was due to progressive hypovolaemia, illustrated by fall of the BP to zero and the accumulation of blood in the splanchnic area, demonstrated by the increased pulse count rate. Decrease in ECBV led to major metabolic and coagulolytic disturbances. The prolonged survival of the animals treated with α-blocking agents or ganglioplegics (the dogs lived 6-8 hours after shock, $p < 0.001$) may be attributed to maintenance of the systemic circulation and especially perfusion of the microcirculation under almost normal conditions.

Disappearance of the splanchnic blood pools (illustrated by marked diminution in the pulse count rate) is unquestionably due to the pharmacodynamic effect of the α-blocking drugs. Immediately after their administration the count rate fell, then after 60-90 min returned to its initial values, with evident disappearance of the splanchnic pool and simultaneous remission of the ECBV and BP.

It has been postulated that the pharmacodynamic effect in shock of sympathoadrenolytics of the neuro- and ganglioplegic group is due especially

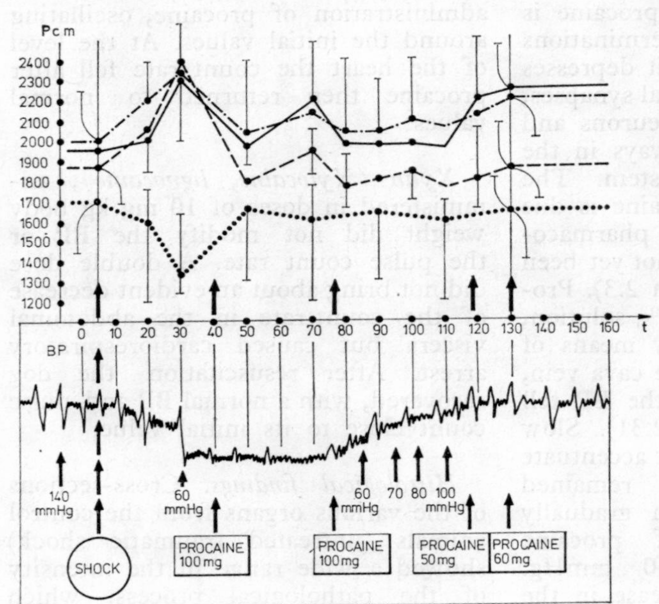

Figure 2.31. **Mean pulse count per minute after the administration of Procaine.**

to their role as electron suppliers. Actually, all pharmacodynamic agents with obvious antishock properties contain reducing chemical elements. Hypertonic glucose, neuro- and ganglioplegics, *Rheomacrodex*, heparin and the α-blocking agents are all electron donor substances. In the experimental model used no other medication was associated (perfusions with glucose, dextrans, hydrocortisone) just in order to emphasise the selective pharmacodynamic effect of the sympathoadrenergic drugs used.

Conclusions. *Levomepromasine* has remarkable antishock properties when administered by the intravenous route. Release of the blood by annulling of the sympathoadrenergic vasoconstrictor effect delays the onset of the irreversible phase of shock.

Pentamethonium gives very good results. The tendency of the BP to increase together with decrease in the pulse count at splanchnic level confirms the unquestionable antishock efficiency of this ganglioplegic.

Hydergin has an obvious effect at intestinal level but in the liver produces congestion with blood sequestration.

Promethasine. The effect is doubtful: within the limit of the parameters investigated no antishock effect could be made evident.

Pethidine (Dolantin) did not improve the circulation and the pulse count was superposable on that of the controls.

Procaine. The increase in BP after the slow administration of procaine and the disappearance of splanchnic vasoconstriction attest to the antishock value of this substance (*see* Section 2.3).

Xylocaine has antishock properties only when administered in toxic doses.

2.9 VISCERAL DYNAMICS OF THE PARAMETERS OF THE ACID-BASE BALANCE IN TRAUMATIC SHOCK

The study had in view the determination of the changes occurring in the acid-base balance in the venous blood of the visceral pedicles (portal system, suprahepatic veins and renal veins) and the correlation between the changes of these parameters in the arterial and venous blood of the systemic circulation (femoral artery and vein), as well as the correlation between the classical parameters of shock and those of the acid-base balance.

Material and method. The experiments were carried out on 15 dogs, in five of which the organic acids and partial carbon dioxide pressure were determined apart from the parameters of the acid-base balance. The shock model was similar to that described in Sections 2.6 and 2.7.

For the acid-base balance blood samples were collected at three different moments in the evolution of shock, in 1 ml syringes, previously heparinised. The initial sample was withdrawn immediately after laparotomy and was considered the control sample for each animal. The second sample was taken when the BP reached 60-70 mmHg, the CVP 3 cm of water and diuresis had stopped. The third was collected when the BP was 40 mmHg and CVP 1-2 cm of water.

Blood was collected from the femoral artery and vein, suprahepatic veins, mesenteric vein, splenic vein, renal vein (Figure 2.32).

The following acid-base balance parameters were determined:

pH_a, actual pH;

SB, standard bicarbonate;

Table 2.2 MEAN PARAMETERS OF THE ACID-BASE BALANCE DETERMINED IN THREE SAMPLES COLLECTED IN THE COURSE OF EXPERIMENTALLY-INDUCED SHOCK IN DOGS

	BP mmHg	Arterial blood	Femoral vein	Supra-hepatic vein	Mesenteric vein	Splenic vein	Renal vein
*pH*a	130	7.278	7.213	7.221	7.221	7.208	7.262
	70	7.240	7.164	7.173	7.161	7.153	7.214
	40	7.082	6.962	6.990	7.003	6.992	6.983
SB (mEq)	130	19.223	16.896	17.300	17.415	15.685	18.750
	70	15.915	14.500	16.308	14.431	15.531	16.085
	40	13.054	11.654	11.285	12.200	11.169	11.692
BB (mEq)	130	45.446	42.000	43.769	39.038	42.523	43.735
	70	40.900	35.692	38.154	35.938	36.346	40.362
	40	32.258	28.485	30.346	28.577	28.323	29.223
BE (mEq)	130	−6.962	−8.923	−9.708	−8.500	−9.662	−7.285
	70	−11.135	−12.769	−12.331	−13.446	−12.985	−11.385
	40	−16.423	−17.817	−18.315	−18.085	−19.562	−17.754
P(CO_2) (mmHg)	130	46.462	52.692	51.846	50.269	50.231	48.462
	70	45.577	56.962	51.923	51.192	53.385	50.692
	40	54.423	63.192	53.923	55.154	55.154	63.077
AB (mEq)	130	20.935	20.996	20.738	19.223	19.523	20.423
	70	17.708	17.469	17.238	16.392	17.254	18.154
	40	14.823	14.046	13.404	13.269	12.954	14.277
CO_2 total (mEq)	130	22.019	22.008	22.224	20.578	21.443	21.762
	70	18.732	18.424	18.405	17.240	18.155	19.182
	40	44.123	15.209	14.120	14.368	14.184	15.427

BB, buffer bases;
BE, base excess;
$P(CO)_2$, partial carbon dioxide pressure;
AB, actual bicarbonate;
CO_2 total, total carbon dioxide.

The acid-base balance constants were determined according to the nomogram of Andersen and Engel (Figure 2.33).

Lactic acid (Barker-Sumerson method) and pyruvic acid (Friedman method) were determined in two

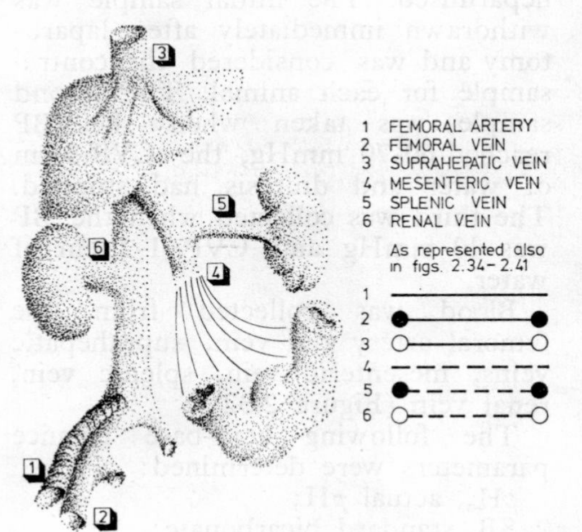

Figure 2.32. Site of blood withdrawal (symbols are also used in Figures 2.34–2.41).

Figure 2.33. Mixed acidosis (for symbols see Figure 2.7).

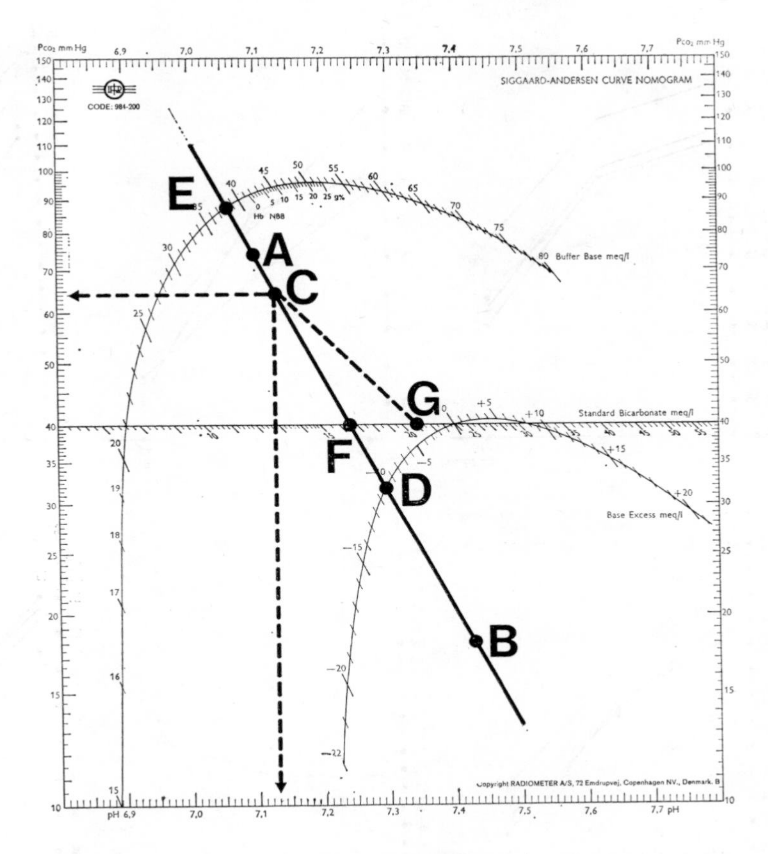

systemic venous and arterial blood samples at the beginning of the experiment (controls) and after an interval of 60-90 min when the BP remained at 30-40 mmHg. The plasma ionogram was determined in two systemic venous blood samples, collected under the conditions mentioned.

For determination of the partial oxygen pressure the blood was collected from the same arteries and veins as for the other determinations.

The mean acid-base balance values (pH_a, SB, BB, BE, $P(CO)_2$, AB and total CO_2) determined in the visceral pedicles are indicated in table 2.2 and Figure 2.34.

The results obtained may be summed up as follows:

(i) the mean volaemia constants, lactic acid, pyruvic acid and plasma sodium values are given in Figure 2.35;

(ii) the mean serum ionogram values refer only to sodium (Figure 2.35); potassium and calcium did not present significant changes;

(iii) $P(O_2)$ did not show significant oscillations in the blood samples studied.

Correlations between the pH_a values of the suprahepatic, mesenteric and splenic venous blood, on the one hand, and lactic acid and haematocrit of the femoral venous blood, on the other, are shown in Figure 2.36.

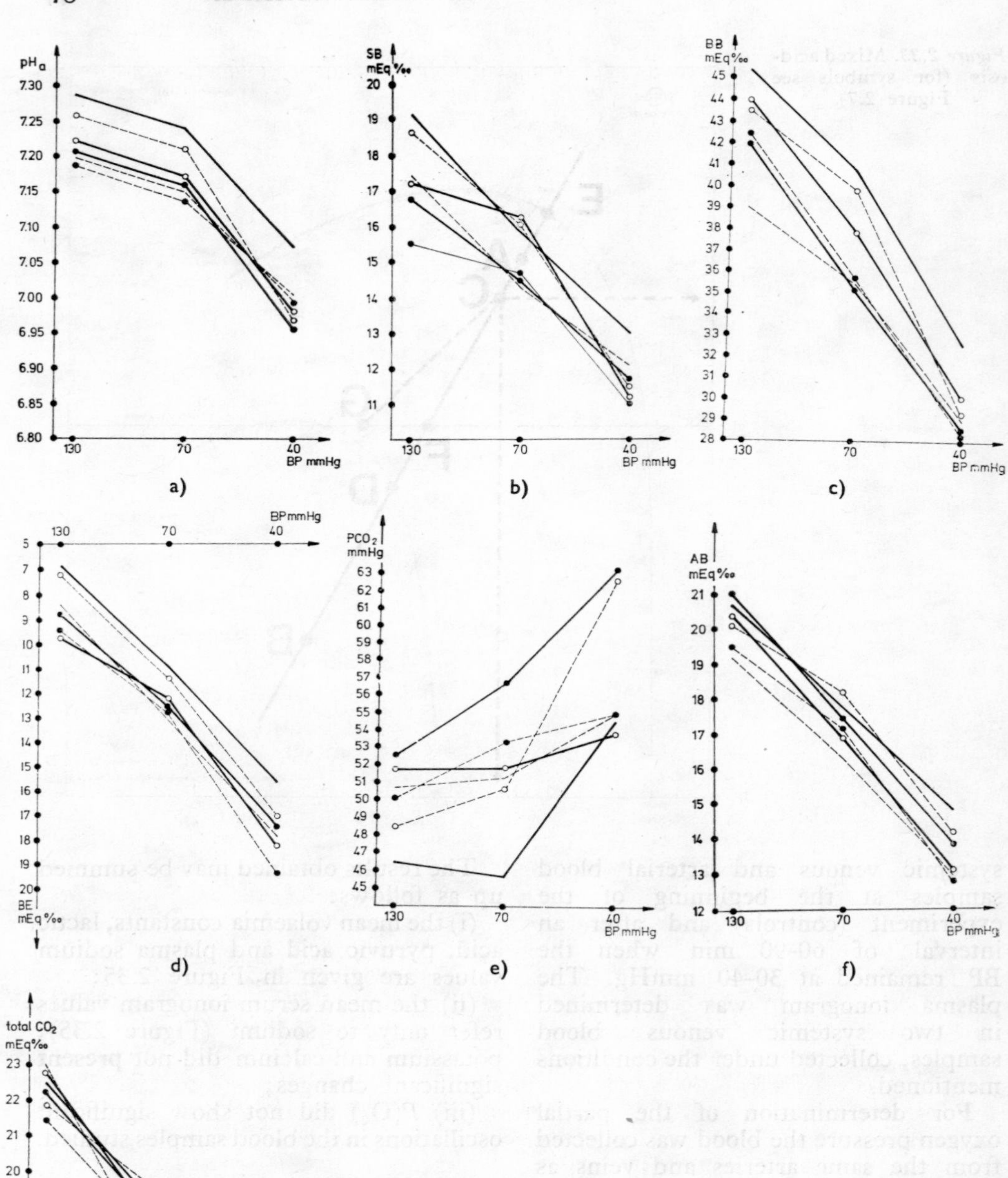

Figure 2.34. Mean values of the acid-base balance parameters determined in the blood of the visceral vascular pedicles, recorded at different moments in the evolution of shock (arterial blood pressure on the abscissa): *a*–pH_a; *b*–*SB*; *c*–*BB*; *d*–*BE*; *e*–$P(CO_2)$; *f*–*AB*; *g*–CO_2 total.

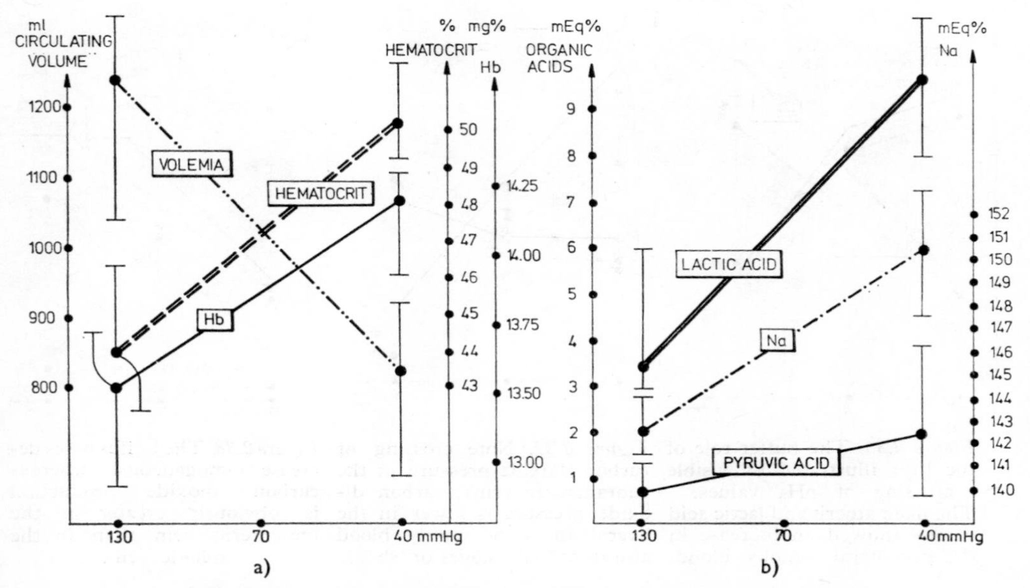

Figure 2.35. Some constants determined in the course of shock:
a—mean volaemia constants; *b*—biochemical values determined in the course of shock.

Correlations of the mean $P(CO_2)$ and BB values in the suprahepatic, mesenteric and splenic venous blood are plotted in Figure 2.37.

Correlations of the mean $P(CO_2)$ and BB values in the mesenteric and splenic veins, on the one hand, and lactic acid and haematocrit in the femoral vein, on the other, are shown in Figure 2.38.

Correlations between the mean $P(CO_2)$ and BB values in the suprahepatic veins and lactic acid and haematocrit in the femoral vein are indicated in Figure 2.39.

The correlations between the various parameters of the acid-base balance, haematocrit and lactic acid in the femoral artery and vein and renal vein are shown in the following figures:

(i) Figure 2.40, correlations between mean pH_a values in the femoral artery, femoral and renal veins and mean lactic acid and haematocrit values in the peripheral venous blood;

(ii) Figure 2.41, correlations between the mean $P(CO_2)$ and BB values in the femoral artery, femoral and renal vein and the mean lactic acid and haematocrit values in the peripheral venous blood.

Discussion. Absolute and mean values of all the parameters studied pointed to a state of mixed, predominantly *metabolic acidosis* in the blood of all the areas explored. The metabolic factor weighs more heavily in the balance than the respiratory factor in producing acidosis.

The pH_a *value* remained significantly high — as was to be expected — in the arterial blood, and low in the femoral vein. *It is worthy of note, however, that the pH values of the renal vein blood were initially very close to the arterial pH values and*

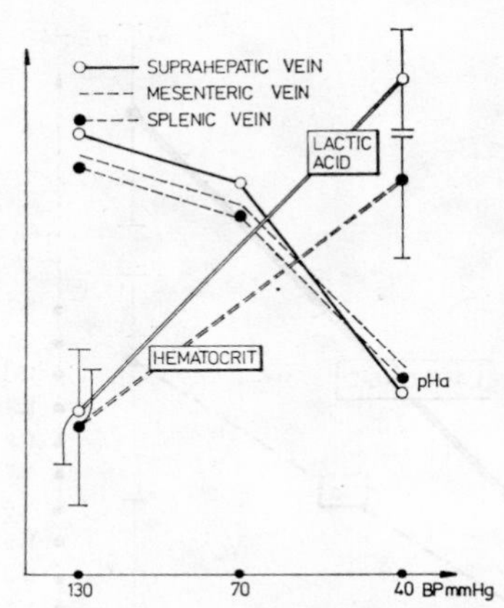

Figure 2.36. The buffer role of the liver illustrated by visible crossing of pH_a values. The haematocrit and lactic acid values showed an increase in the peripheral venous blood.

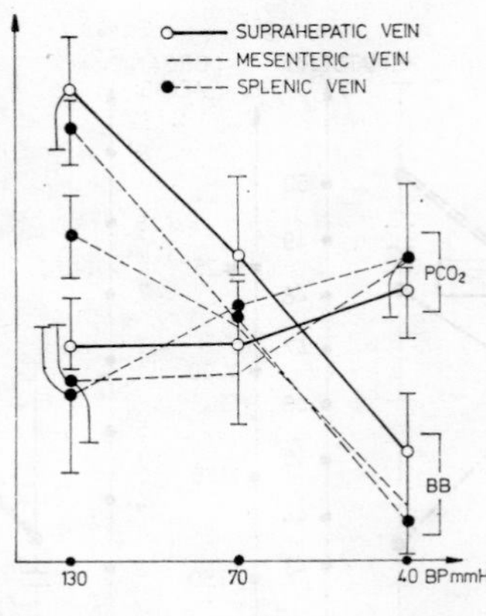

Figure 2.37. Note 'crossing' of carbon dioxide pressure in the suprahepatic veins; carbon dioxide pressure is lower in the liver than in the supply blood also in the late stages of shock.

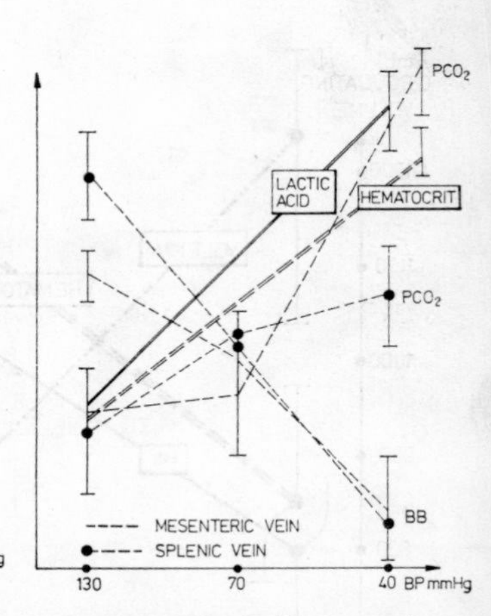

Figure 2.38. The buffer bases decrease homogeneously whereas carbon dioxide production is obviously greater in the mesenteric vein than in the splenic vein.

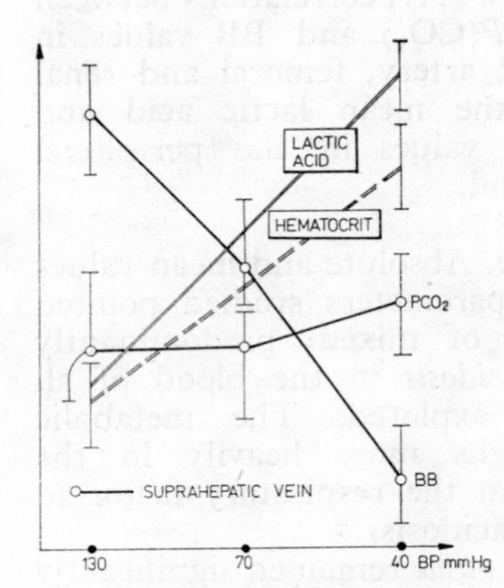

Figure 2.39. A sharp fall in buffer bases as compared to the slow production of carbon dioxide. The discrepancy between the decrease of *BB* and the elevation of lactacidemia is significant.

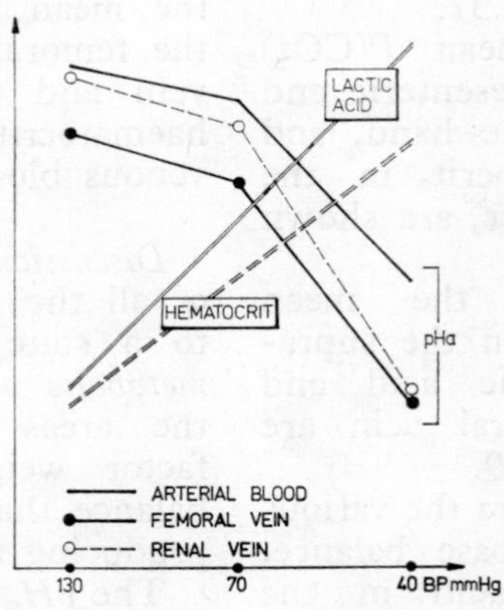

Figure 2.40. pH_a in the renal veins starts from values close to systemic arterial pH_a and finally reaches systemic venous pH_a values.

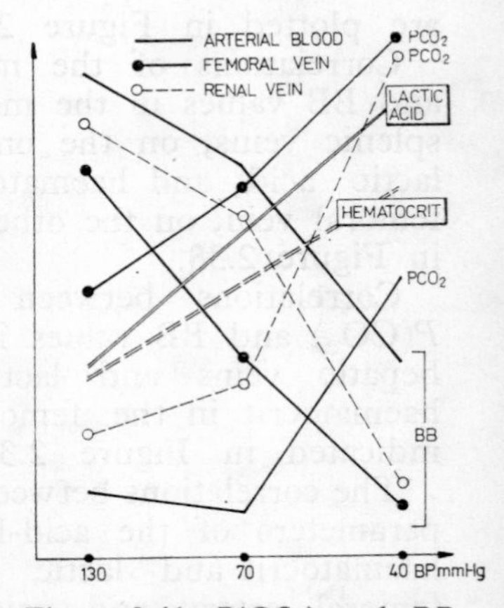

Figure 2.41. $P(CO_2)$ and *BB* values in the renal vein are close to systemic venous values after 'closure' of the renal circulation below *BP* values of 70 mmHg.

significantly differed from the values found in the other veins studied ($p < 0.001$). This confirms once again the role of the kidney in maintaining acid-base homeostasis, a role that ceases with the disappearance of diuresis. At BP values below 60-70 mmHg, the renal vein *p*H differs from that of the arterial blood, significantly approaching the values found in the femoral vein ($p < 0.001$).

Plasma bicarbonates decrease with aggravation of the state of shock and fall to 11-12 mEq% before death. At BP of 60-70 mmHg standard bicarbonates in the arterial blood and renal and suprahepatic veins present similar values which are however greater by 2-3 mEq than in the blood of the peripheral and portal system. This may be accounted for by the retention of bicarbonate ions in the kidney and by the role of the liver cell in lactic acid metabolism.

Buffer bases have similar dynamics probably due to the changes in the concentration of bicarbonate ions they contain.

The dynamics of excess bases runs parallel to that of the other metabolic parameters (SB and BB) of the acid-base balance, probably due to the same causes.

Partial carbon dioxide pressure increased in all the blood samples with aggravation of the shock.

$P(CO_2)$ in the arterial blood is identical or slightly lower than the initial values up to BP values of 70 mmHg, attributable to hyperventilation of the animals in this period and a satisfactory pulmonary perfusion. $P(CO_2)$ then increases in all the venous blood samples, except for the renal vein blood in which it is by 7-8 mmHg lower than in the peripheral venous blood, probably due to the use of carbon dioxide for the synthesis of bicarbonate in the tubular cells. At BP values of 30-40 mmHg, $P(CO_2)$ increases in the renal vein blood, being almost identical to that in the peripheral venous blood. These values are by 10-11 mmHg higher than in the blood samples from the mesenteric, splenic and suprahepatic circulation. This might be accounted for by arrest of renewed bicarbonate synthesis in the kidney at BP values below 60-70 mmHg and by the metabolism of the renal tissue in shock, far more intense than that of the other organs studied.

Actual bicarbonates and total carbon dioxide decreased in all the blood samples with accentuation and lengthening of the state of shock. No statistically significant differences were observed between the various samples.

The dynamics of the actual and mean lactic acid and pyruvic acid values showed an increase in the peripheral venous blood.

The *actual ionogram values* pointed to a significant increase in sodium, accounted for by the haemoconcentration characteristic of traumatic shock. The results were not conclusive as regards the changes in potassium and calcium, probably due to the more accentuated degree of metabolic inertia of these ions.

$P(CO_2)$, $CO_{2\ total}$, pH_a correlations. The progressive decrease in pH_a and $CO_{2\ total}$ was accompanied by an increase in $P(CO_2)$ in the final stage of shock. The highest divergence between these values was found in the arterial blood and the closest values in the peripheral venous blood. Similar values were found in the renal vein and arterial blood, and intermediate values in the suprahepatic and portal system.

$P(CO_2)$, BE and pH_a correlations were the same as shown in the preceding paragraph.

Lactic acid, haematocrit, pH_a correlations in the splenic, mesenteric and suprahepatic venous blood revealed a progressive increase in lactic acid and haematocrit in the peripheral blood, accompanied by progressive decrease of the *p*H in the suprahepatic, mesenteric and splenic veins. There is a maximum divergence in the final stage of shock that may be explained by hypoxia of the organs considered and by haemoconcentration (relative polyglobulia).

$P(CO_2)$ and BB *correlations* in the mesenteric and splenic venous blood with lactic acid values in the peripheral blood demonstrated an increase of $P(CO_2)$ running parallel to a constant decrease of the buffer bases, changes that became more marked with the increase in lactic acidosis and haemoconcentration.

$P(CO_2)$ and BB *correlations in the portal blood* showed a progressive increase of the former and decrease of the latter. The degree of metabolic acidosis (decrease of BB) and respiratory acidosis (increased $P(CO_2)$) was less accentuated in the suprahepatic venous blood, *confirming the role of the liver in correcting the acid-base imbalance.* In the suprahepatic veins there was a net decrease in total bases, corresponding to their similarly significant increase in the peripheral blood in the final stage of shock (owing to the failure of the liver to metabolise lactic acid in this stage). The decrease of total bases may likewise be correlated with increase in the haematocrit. Increase in $P(CO_2)$ in the suprahepatic venous blood is incomparably less important in the production of suprahepatic blood acidosis than the decrease in BB.

Study of the correlation of pH_a *in the femoral artery, femoral and renal veins and haematocrit in the peripheral blood*

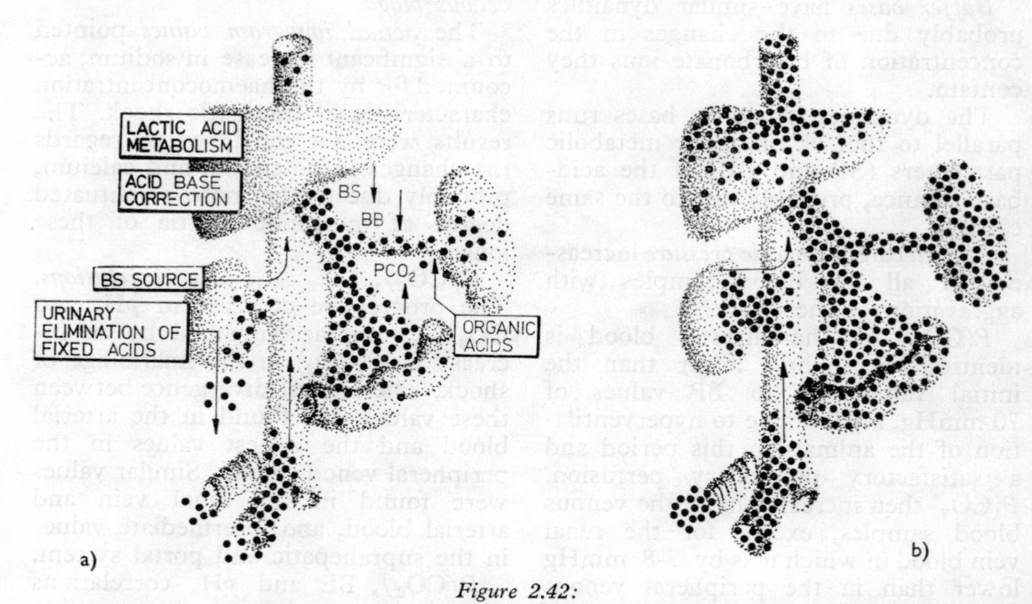

Figure 2.42:
a — initial role of the liver and kidney in maintaining the acid-base balance; *b* — after interrupting the renal irrigation this role is lost. The *p*H of the blood leaving the liver in the late stage of shock is similar to that of the portal blood and femoral vein blood.

showed a constant decrease of the *p*H with increase in lactic acidosis and haematocrit. The most accentuated changes were observed in the peripheral venous blood and to a lesser extent in the femoral artery.

Changes in the pH of the renal venous blood ran close to those in the arterial blood in the initial phase of shock and to those in the venous blood in the late phases.

$P(CO_2)$ and BB correlations in the femoral artery and vein and lactic acid and haematocrit in the peripheral venous blood revealed a permanent increase in $P(CO_2)$ and a decrease in total bases with increase in lactic acidosis and hematocrit. The most marked changes occurred in the femoral venous blood and the least important in the femoral arterial blood. $P(CO_2)$ values in the renal blood in the initial phase of shock were very close to those in the venous blood in the terminal phase. The BB showed similar fluctuations. These findings demonstrate once again the role of the kidney in correcting the acid-base imbalance, a role that is only possible in the initial phases of shock (Figure 2.42).

2.10 EXPERIMENTAL STUDY OF THE SHOCK CELL

Shock may be considered as a generalised reflection of the intracellular enzymatic processes and structural alterations of the subcellular organelles. Rapid alterations in the parameters of the *extracellular space* are received by the *cellular space* through several categories of information: decrease in plasma $P(O_2)$, *p*H variations, osmolarity and resistivity changes, increase in plasma glycoprotein and sterol hormones, increase of endo- and/or exogenous toxins, of hetero- and/or autoantibodies, etc.

In shock, part of the information belongs to the lesional syndrome, part to the reaction syndrome, the latter containing messages integrated by neuroendocrine pathways for a differentiated struggle of the entire organism. The cell is faced with an intense stress, being obliged to readjust its functional subsystems with an abruptness that often generates chaos. In most cases its endogenous effort is insufficient and exogenous therapy must be brought into action. An opportune, correct therapy implies, however, a deep insight into the complex events taking place in the *shock cell*. The cells of the organism do not participate to an equal extent in the energetic effort applied to annihilation of the disorders caused by shock and just as certainly do not act separately. There are cellular territories with an alert genetic and enzymatic 'equipment', ready for an immediate intervention, that ensures either the *cybernetics* or the *energetics* of the reactional syndrome, but there also exist reserve cellular territories, more difficult to mobilise and zero in front of an acute situation, such as shock. Moreover, the cells do not act by random agglomeration but by an ordered hierarchy governed by regulation systems that maintain life of the organism as a whole. Therefore the term 'shock cell' is used conventionally to facilitate our understanding of the intimate happenings.

The experiments had in view:

(i) to follow up functional enzymatic and electron-optical physical alterations in the course of mixed shock;

(ii) to establish the criteria for finding the limits of reversible-irreversible at cell level;

(iii) to check the value of drugs stabilising the enzymatic and membrane subsystems at the cellular space levels.

Material and method. Acute mixed hemorrhagic and traumatic shock was produced by the Wiggers method, modified by us *(see* preceding Sections). After cannulation of the carotid and jugular the abdominal cavity was opened by a longitudinal midline incision. The BP, CVP and skin temperature were recorded continuously, and plasma *p*H and lactacidemia at given intervals. Hemorrhage was produced up to BP values of 70 mmHg, and was maintained constant for 60 min by traction of the mesenteries for 5 min every 10 min (Figure 2.43). A new bleeding brought BP values down to 40 mmHg, likewise maintained by traction of the mesenteries at the same rate. Two hours after the onset, trauma was continued until shock developed up to death.

Cortisone (used as 6-methylprednisolon hemisuccinate-*Urbason)* was administered to eight of the 15 dogs in megadoses of 30 mg/kg body weight, administered by the intravenous route in two equal parts at 10 and 30 min after the first bleeding.

These eight dogs were considered to be under enzymatic and membraneous protection at the cellular level.

Before the onset shock, at one and two hours afterwards, and even three in case of longer survival, the following samples were collected:

From the portal vein and right atrium 12 ml whole blood for the biochemical determination of acid and alkaline phosphatases, amylase, lactic dehydrogenase (LDH) and the transaminases. The

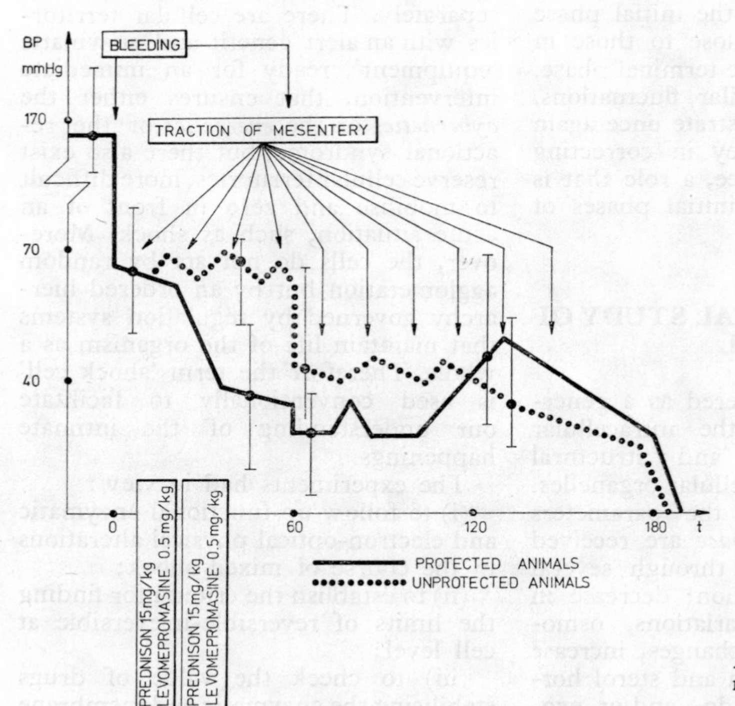

Figure 2.43. Experimental model used for study of the 'shock cell'.

portal and suprahepatic plasma enzymes were determined by the following methods: Bodansky for phosphatases, Wohlgemuth for amylase, spectrophotometry in U.V. light for LDH, Reitman-Frenkel (modified by Paget) for transaminases [5, 9].

Tissue specimens were collected from the kidney, intestine and muscles, and each was divided into three fragments:

(i) the first fragment was fixed in glutaraldehyde and postfixed in osmic acid (Pallade technique), included in Vestopal after Luft's technique modified by Petrovici *et al.* [21, 27, 43]; after sectioning in the ultramicrotome, mounting and suprastaining, visualization of the grids and photographing was done with an Opton electron microscope at different magnifications: ×1500, ×4000 ×8000, ×12 000, × ×16 000 and ×30 000;

(ii) the second fragment was frozen with dry ice, sectioned in the cryostat and incubated in media containing the following enzymes substrates: acid phosphorylase (FA); total phosphorylase (FT); glucose-6-phosphatase (G-6-PA); glucose-6-phosphate-dehydrogenase (G-6-PDH); phosphoglyceraldehyde-dehydrogenase (PGDH); lactic-dehydrogenase (LDH); isocitric-dehydrogenase (IDH); succindehydrogenase (SDH); malic-dehydrogenase (MDH); cytochrome-oxidase (CITOX); ATP-ase; diaphorases (NAD and NADP); sulphatase (SFZA); leucinaminopeptidase (LPEP); and acid phosphatase (FAC) [9, 43];

(iii) the third fragment was cut in two: one part was fixed in 10% formaldehyde and the other in Lillie fixator. Staining was carried out with: hematoxylin-eosin, trichrome Masson, PAS, Herovici, Kurnick, Sudan and Best carmine [15, 19, 20, 28].

Results. The samples collected before inducing shock were considered the controls, those taken after one hour considered in reversible shock, and after two hours in irreversible shock.

In unprotected dogs plasma enzymatic determinations revealed a net increase in hydrolases (acid phosphatase and amylase) and a slow increase in the protected animals. On the other hand LDH, GOT and GPT increased in both lots; the overall increase of plasma enzymes was due especially to the increase in the portal prehepatic areas (Figure 2.44).

In the non-protected dogs, evidence was found of glycolytic hyperactivity that was gradually exhausted, permanent hypoactivity of the Krebs cycle, oxidation-phosphorylation activity in the first phase, activity of the pentose shunt in the final stage and hydrolase hyperactivity in both stages. In the protected dogs, enzymatic activity was stable in both stages along all the metabolic lines (Table 2.3).

Electron microscopy showed particular lesions in the reversible stage: vacuolisation and ballooning of the mitochondria, but also intact or hypertrophic ones, clusters of lysosomes, nuclear heterochromatin margination and depletion of glycogen granules in the liver. In the irreversible phase opacification and homogenisation of the mitochondrial matrix, compact or voided mitochondria and confluence of monster lysosomes was noted in addition.

In the protected animals the infracellular lesions were less conspicuous (*see* Section 9.5.2).

Histologically, depletion of glycogen in the liver and kidneys, vascular microthromboses, lipoprotein dystrophy up to necrosis were noted in the irreversible phase of shock (*see* Section 9.5.1).

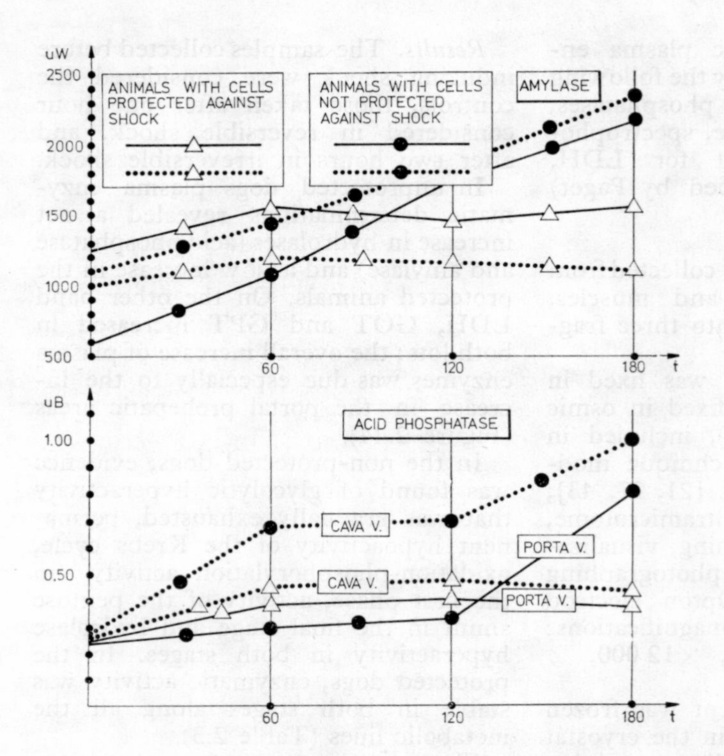

Figure 2.44. Biochemical determination of hydrolases (amylase and acid phosphatase) in the dynamics of shock in animals with 'protected' and 'unprotected' cells. Note increased portal supply of hydrolases in the 'unprotected' dogs.

Discussion. Under adverse hypoxic conditions the *shock cell* tries to produce sufficient energy to survive, but not implicitly to maintain the complexity of its functions in their entirety; hence for the sacrifices made by some cells there is no remission.

In the stage of shock considered reversible, phosphorylase hyperactivity with glycogen depletion is recorded, as well as activation of glycolysis enzymes, inhibition of the pentose shunt, a Krebs cycle that cannot cover demands (being restricted by the minimal O_2 supply) and a respiration chain where NAD and cytochromoxidase endeavour to furnish ATP energy. In the irreversible stage of glycolysis, which bore the brunt of the first stage, this chain is also completely exhausted.

The accumulated acid functions break down the membranes of the lysosomal sacs with the release of hydrolases that had already increased in the reversible stage of shock.

The electron microscope reveals the disappearance of the granular reticulum and the presence of tubulovesicular plaques of smooth reticulum. The mitochondria swell and become clear or hypertrophic, the lysosomes accumulate in large amounts, glycogen granules disappear from the liver, margination of the nuclear chromatin develops; all these lesions appear to be reversible. Disorganisation and vacuolisation of the reticulum follows, the mitochondria are transformed into amorphous masses, the lysosomes become huge and confluent, encompassing the mitochondria, microbodies, etc., the nuclear mem-

Table 2.3 MEAN QUALITATIVE OF ENZYMATIC ACTIVITY

Group	Time	Tissue	FA	FT	G-6-PDH	PGDH	LDH	SDH	IDH	MDH	G-6-PA	CITOX	ATP-ase	NAD	NADP	SFZA	LPEP	FAC
7 unprotected dogs	Time I Onset of shock	Liver cell	±	+	±	+	+	++	+	+	±	+	+	+	+	±	±	±
		hepatic lobule	−	+	±	+	+	+	+	+	+	+	±	+	+	±	−	+
	Time II 1 h after shock	Liver cell	+++	++	++	+++	+++	+	+	±	+	++	++	++	+	+++	++	++
		hepatic lobule	++	+	++	++	+++	+	+	±	+	+	+	+	+	++	+++	+++
		renal cortex	+++	++	+++	+++	+++	±	+	±	++	++	++	++	+	++	+++	+++
		renal medulla	++	++	++	++	+++	±	±	−	+	±	+	++	+	++	++	++
	Time III 2 h after shock	liver cell	±	+	±	+	+	+	+	−	+++	±	−	±	±	+++	+++	++
		hepatic lobule	−	−	−	+	+	±	+	+	++	+	±	+	+	++	+++	++
		renal cortex	±	+	+	−	+	+	±±	±+	++	±	±	±	±	+++	++	++
		renal medulla	±	±	±	−	+	+	+	±	+++	±	−	±	±	++	++	+++
8 protected dogs	Time I Onset of shock	Liver cell	+	+	±	+	++	++	+	+	±	+	+	+	++	±	±	±
		hepatic lobule	+	+	±	+	+	++	+	+	±	+	+	+	+	−	±	±
	Time II 1 h after shock	Liver cell	+	+	+	+	++	+	+	±	++	±	±	+	+	+	++	+
		hepatic lobule	+	+	+	+	++	+	+	+	+	+	±	+	+	+	+	+
		renal cortex	±	+	+	+	++	+	+	±	+	+	+	+	+	±	+	±
		renal medulla	+	+	+	+	+	+	+	±	+	±	+	+	+	+	+	±
	Time III 2 h after shock	Liver cell	±	+	±	+	+	±	±	±	+	+	±	+	+	+	±	+
		hepatic lobule	±	+	±	+	+	+	±	±	+	±	±	+	+	±	+	+
		renal cortex	±	+	±	+	+	+	+	+	+	±	+	+	+	+	+	+
		renal medulla	±	±	+	+	+	+	+	+	+	+	+	+	+	+	++	++

branes are ragged; all these lesions are irreversible.

Therapy of the shock cell must be founded upon a deep understanding of its enzymatic and infrastructural alterations in the course of shock. The shocked cell will be sustained by protection of its biological membrane, of its reducing functions, and re-equilibration of its metabolic circuits. In the course of this study it was found that large cortisone and levomepromasine doses offer enzymatic stability in glycolysis, the Krebs cycle, pentoses and oxidation chain and the absence of hydrolase hyperactivity. Worthy of note was also the lack of severe lesions on the electron microscopic images. Stability of the membrane was revealed by plasma enzymatic determinations: amylase and acid phosphatase increased to a greater extent in the unprotected dogs and did not increase in those with an intact lysosomal membrane; on the other hand, LDH, GOT and GPT, which act topochemically on the reticulum increased to the same extent, regardless of protection.

Conclusions. In the shock cell, the genetic apparatus attempts a rapid adaptation by arresting synthesis and overstraining the energy systems. As the Krebs cycle is limited by hypoxia glycolytic hyperfunction develops with an increase in acidity, corrosion of the lysosomal sacs and discharge of hydrolases, at first intracellularly then into the systemic circulation. The infracellular lesions affect especially the lysosomes, mitochondria and endoplasmatic reticulum and can obviously be correlated with functional cellular alterations. The reversible-irreversible limit in shock is certainly decided at cellular level, perhaps with enzymatic exhaustion of glycolysis and loss of selectivity of the lipoprotein membranes. Drugs stabilising the membranes and promoting metabolic reduction manage to equilibrate the cell. They will form part of the therapy, together with the medication acting on the microcirculation and with volemia substitutes.

2.11 EXPERIMENTAL STUDY OF THE LYMPHATIC CIRCULATION IN SHOCK

The experiment was carried out on 30 dogs, weighing 10-14 kg. A mixed traumatic and haemorrhagic shock was produced, according to the technique described in the previous sections.

Lymph samples were collected by catheterisation of the thoracic duct, carotid, right atrium and mesenteric vein (Figure 2.45), at three different moments in the course of shock, established according to haemodynamic criteria, the same as in the previous experiments (*see* Section 2.10).

Exposure and catheterisation of the thoracic duct by left paramedian cervicotomy was carried out 30 min after the injection of Evans blue in the foot-pad. After incision of the pectoral muscle the thoracic duct was exposed between Pirogoff's venous confluent and the trachea. The thoracic duct empties either through a single arch or through a 'delta', a more difficult variant (Figure 2.46).

In this experiment we had in view:

(a) determination of the lymphatic output;
(b) variation of the acid-base balance parameters;
(c) variations in protein concentrations and electrophoretic fractions;
(d) variations in total lipid and cholesterol concentrations;

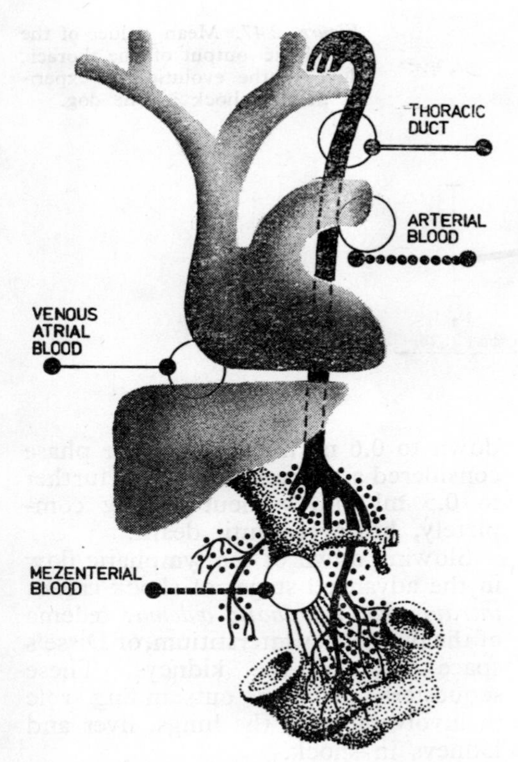

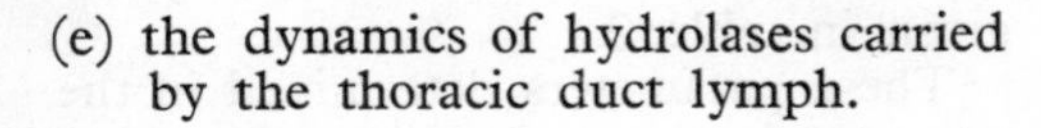

Figure 2.45. Site of collection of the samples used in the experiment (symbols used in Figures 2.49 and 2.50).

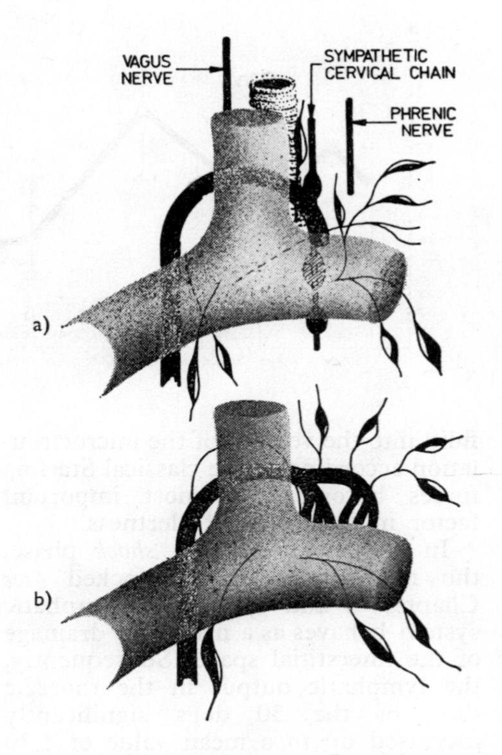

Figure 2.46. Anatomy of the arch of the thoracic lymphatic trunk: *a* — single arch; *b* — 'delta' arch.

(e) the dynamics of hydrolases carried by the thoracic duct lymph.

Results.

(a) Very interesting findings were made on following up the lymphatic output. The thoracic duct lymph in the dog is derived from the intra-abdominal viscera, in a proportion of more than 90%, its composition and amounts reflecting the metabolic events of this anatomical space. The lymphatic parameters, next to those of the portal vein reflect splanchnic metabolism more reliably than the arterial or venous parameters of the general circulation.

The lymphatic output behaves like a general sinusoid of the state of shock proving once again 'the wisdom of the body' when faced with difficult situations.

In the phase of *early reversible shock* a rapid increase is recorded in the lymphatic output, probably due to the set of receptors on the parietal smooth muscle cells; it may be assumed that constriction (not spasm) of the lymphatic ducts takes place in the pool of catecholamines discharged at the beginning of shock. From the onset output of 0.90 ml/min it increases to 1.10 ml/min for 15-20 min (Figure 2.47). The output then falls to 0.80 ml/min, as penetration of the interstitial

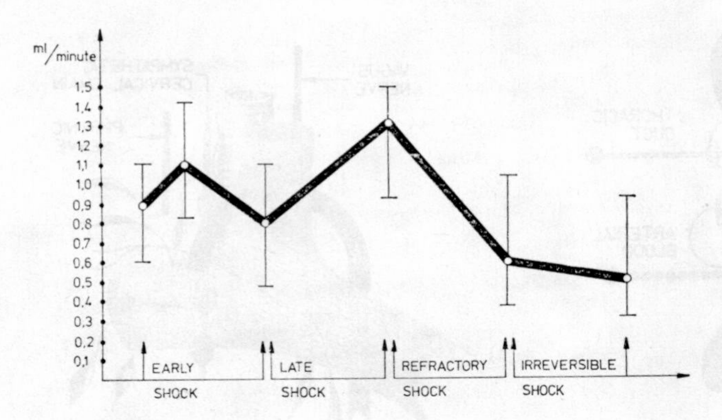

Figure 2.47. Mean values of the lymphatic output of the thoracic duct in the evolution of experimental shock in the dog.

fluid into the venules of the microcirculation according to the classical Starling forces becomes the most important factor in this stage of alertness.

In *the late reversible shock* phase, the microcirculation is blocked *(see* Chapters 5 and 7) and the lymphatic system behaves as a necessary drainage of the interstitial space. Subsequently, the lymphatic output in the thoracic duct of the 30 dogs significantly increased up to a mean value of 1.30 ml/min, i.e. within an hour 78 ml and in 24 hours 1872 ml lymph returned to the circulation of the system.

In man, accepting the output as 2.5 ml/min, the lymphatic return to the circulation will be of 150 ml per hour and 3600 ml per day. This may be regarded as a true 'plasma self perfusion' which the human body makes use of when the venous return is lowered following the microcirculation *(see* Section 6.8). In this stage the venulo-lymphatic shunts open and the pores of the lymphatic capillaries become wider.

In the phase of *refractory reversible shock* the microcirculation network is blocked by DIC and the interstitium is dominated by the collagenic sponge that sequesters water and electrolytes. In this stage the lymphatic flow slows down to 0.6 ml/min, and in the phase considered as *irreversible,* it falls further to 0.5 ml/min, without ceasing completely, however, until death.

Slowing down of the lymphatic flow in the advanced stages of shock causes *intravisceral lymphatic œdema:* œdema of the pulmonary interstitium, of Disse's space and of the kidneys. These sequesters play an outstanding role in involvement of the lungs, liver and kidneys in shock.

(b) The parameters of the acid-base balance, determined as in the experiment described in Sections 2.9, are given in table 2.4.

These parameters determined in the thoracic duct lymph reflect the onset of the predominantly metabolic acidosis, specific of shock and run parallel to the values found in the systemic arterial and venous blood: *in all the stages of shock the lymph is more alkaline than the venous blood, the lymphatic discharge playing the part of a true 'buffer' in the acidotic drama of shock* (Figure 2.48).

(c) The concentration of total proteins in the thoracic duct lymph steadily increased in the course of experimental shock from 3.9 g % at the onset, to 4.5 g % in the late

Table 2.4

MEAN PARAMETERS OF THE ACID-BASE BALANCE DETERMINED IN THREE SAMPLES COLLECTED IN THE COURSE OF EXPERIMENTAL SHOCK IN THE DOG

	PA mm Hg	Lymph	Mesenteric vein	Right atrium	Femoral artery
	130	7.521	7.23	7.310	7.320
pH_a	60	7.491	7.179	7.160	7.170
	40	7.410	6.821	6.700	7.060
	130	22	16.500	18.250	17.500
SB	60	16.5	12.100	15.500	14.200
(mEq)	40	11.5	7.300	8.250	7.950
BB	130	61	40.500	49.500	46.000
(mEq)	60	40.5	31.100	36.000	35.250
	40	20	21.500	23.500	22.000
BE	130	−2	−9.780	−7.900	−9.000
(mEq)	60	−9	−13.500	−13.800	−16.500
	40	−15	−19.900	−22.000	−22.000
$P(CO_2)$	130	10	46.112	39.100	35.500
(mm	60	10	49.192	43.500	39.250
Hg)	40	12	53.154	46.250	43.120
AB	130	0	15.500	19.500	16.000
(mEq)	60	12.5	9.250	15.000	12.520
	40	6.2	5.600	6.000	5.800
CO_2 total	130	0	18.500	19.600	17.125
(mEq)	60	12.5	12.350	26.200	15.200
	40	6.2	6.300	7.100	6.000

phase of shock and 4.9 g % in the refractory stage.

Albumins, 'the mobile protein mass' discharged into the microcirculation significantly increased, bringing about a rise in osmotic pressure with a plasma-expander effect. It is a known fact that within the first four hours after haemorrhage the lymph brings into the circulation half the amount of proteins lost in the course of bleeding. These proteins are 'extracted' from the interstitial space which will then be depleted also of its fluid, its osmotic pressure falling.

In the dynamics of shock the following globulins were separated electrophoretically:

(i) α_1-globulins decrease in the lymph but increase in the portal blood;

(ii) α_2-globulins increase in the portal and the arterial blood and in the lymph only during the later reversible stage of shock, probably to the detriment of haptoglobulins that appear to play an undetermined role in postaggressive states;

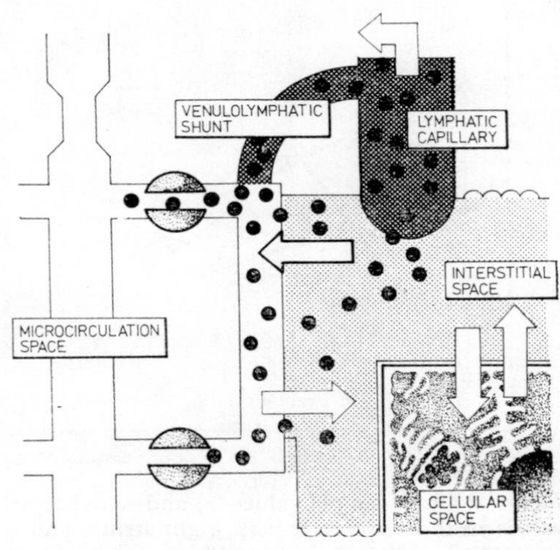

Figure 2.48. Microanatomy of the venulo-lymphatic shunt.

(iii) β-globulins decrease in the course of shock in the lymph, the portal circulation and the greater circulation;

(iv) γ-globulins significantly increase in the lymph of the thoracic duct, especially IgG.

(d) No changes were noted in total lipid levels in the portal vein and systemic circulation; in the lymph however the initial value was at the onset 958 mg %, then fell to 658 mg % in the late phase of shock and increased to 962 mg % in the refractory stage.

These data reflect a good function of the liver cell in the first phases of shock when it can still use the energy supplied by lipids, and the accumulation of lipids in the phases when 'shock liver' develops.

(e) Release of the hydrolases from the lysosomes of each cell is equivalent to self-destruction of the cells by autodigestion, hence the term of 'suicide cell sacs' or 'grave diggers of the cell' given to lysosomes. Progressive acidosis in the late stages of shock reaches the level at which these lysosomal sacs discharge their content, flooding the fluids of the body with hydrolases.

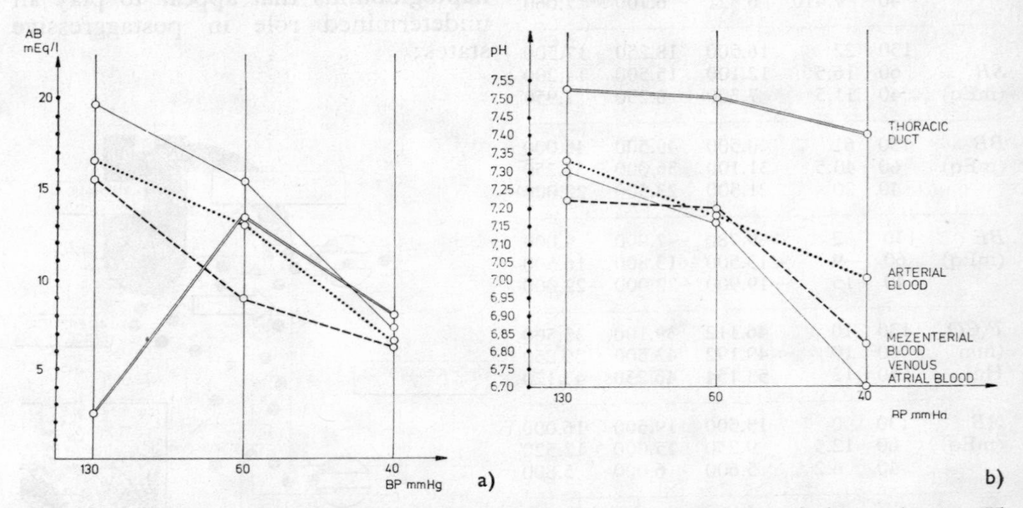

Figure 2.49. Mean *p*H values (a) and actual bicarbonate (b) in the thoracic lymph duct and mean *p*H values in the mesenteric vein, right atrium and arterial blood. Note the alkalinising tendency of the lymph especially in the first phases of shock.

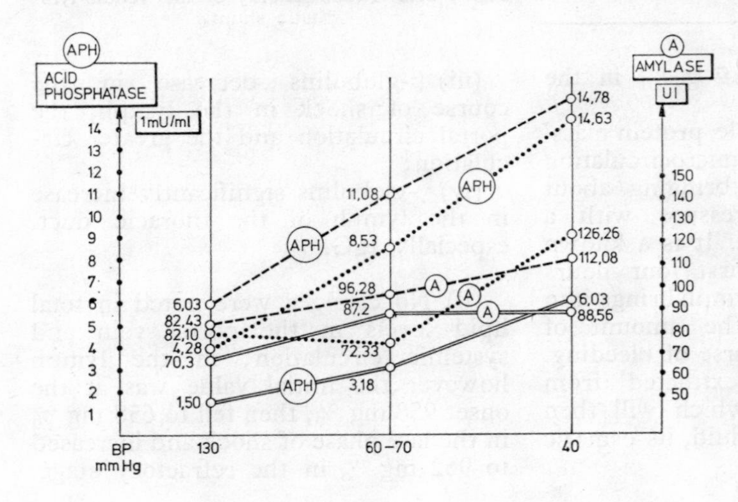

Figure 2.50. Acid phosphatase and amylase values in the thoracic lymph duct, mesenteric vein and arterial blood in the dynamics of shock.

In our experiment acid phosphatase and amylase were determined in the thoracic duct (*see* Figure 2.49). It is obvious that in the final stages of shock the lymph is loaded with hydrolases, derived especially from the intestinal area, in far greater titres than in the portal or systemic blood.

SELECTED BIBLIOGRAPHY

1. AUGUSTIN, A., VASSERSTRÖM, V., GIURGIU, T., GRIGORESCU, C., MARINESCU, I., IONESCU, P., SAFTA, T., Cercetări experimentale şi clinice privind valoarea radiorenogramei în şocul hemoragic (Experimental and Clinical Study Concerning the Value of the Radiorenogram in Hemorrhagic Shock), *Revista sanitară militară*, 1965, *LXI*, (Special no.), 71–73.
2. BASSIN, R., VLADECK, B. C., KIM, S. I., SHOEMAKER, W. C., Comparison of Hemodynamic Responses of Two Experimental Shock Models with Clinical Hemorrhage, *Surg.*, 1971, *69* (5), 722–729.
3. BĂNDILĂ, T., CAFRIŢĂ, A., TOMA, T., SURDULESCU, S., Reanimarea pe timpul transportului (Resuscitation During Transport), *Revista sanitară militară*, 1965, *LXI*, (special no.), 78–81.
4. BĂNDILĂ, T., Echilibrul aciobazic în şocul traumatic (Acid-base Balance in Traumatic Shock), Doctor's Thesis, I.M.F. Bucharest, 1972.
5. BERGMEYER, H. V., *Methods of Enzymatic Analysis*, Academic Press, New York-London, 1963.
6. BUCKBERG, G., COHN, J., DARLING, C., Escherichia Coli Bacteriemic Shock in Conscious Baboons, *Ann. Surg.*, 1971, *173*, (I), 122–130.
7. CAFRIŢĂ, A., BĂNDILĂ, T., GIURGIU, T., STRÎMBEANU, I., SAFTA, T., SURDULESCU, S., PLEŞCA, M., Gangliopleglcele în şoc, (Ganglioplegics in Shock), *Revista sanitară militară*, 1965, *LXI*, (special no.), 84.
8. CONN, E. E., STUMPF, P. K., *Outlines of Biochemistry*, Springer-Verlag, Berlin-Heidelberg-New York, 1963.
9. DICULESCU, I., A Histochemical Analysis of Dehydrogenase Variety in the Different Types of Muscular Tissues, *J. Histochem. Cytochem.*, 1964, *28*, (12), 145–150.
10. DUHAMEL, J., ROQUES, J. C., Deux techniques de statistique simples comportant un minimum de calculs, utilisables en biologie et en médecine, *Biol. Méd.*, 1971, *60*, (2), 177–203.
11. EHRLICH, F. E., An Experimental Shock Model Simulating Clinical Haemorrhagic Shock, *Surg. Gynec. Obst.*, 1969, *129*, (6), 1173–1180.
12. FINE, J., *Ciba Symposium Shock*, Springer, Stockholm - Berlin - Göttingen - Heidelberg, 1962.
13. GIURGIU, T., IONESCU, P., SAFTA, T., Cercetări asupra proprietăţilor imunobiologice ale homo şi heterotransplantelor liofilizate, (Investigations on the Immunobiologic Properties of Lyophilized Homo and Heterotransplants), *Revista sanitară militară*, 1965, *LXI*, (special no.), 237–242.
14. GIURGIU, T., IONESCU, P., CAFRIŢĂ, A., Hemodiluţie cu soluţii coloidale în şocul hemoragic — cercetări experimentale, (Haemodilution with Colloidal Solutions in Haemorrhagic Shock – Experimental Investigations), *Revista sanitară militară*, 1972, *LXXV*, (3), 347–356.
15. GURR, E., *Methods of Analytical Histology and Histochemistry*, Leonard Hill Books Ltd., London-New York, 1958.
16. HAYASAKA, H., HOWARD, J. M., *Septic Shock. Experimental and Clinical Studies*, Charles C. Thomas Publ., Springfield, Illinois, 1964.
17. HINSHOW, L. B., Mechanism of Decreased Venous Return. Subhuman Primate-Administered Endotoxin, *Arch. Surg.*, 1970, *100*, (5), 600–606.
18. LILLEHEI, R. C., MCLEAN, L. D., The Intestinal Factor in Irreversible Endotoxin Shock, *Ann. Surg.*, 1958, *148*, (4), 515–520.
19. LILLIE, R. D., *Histopathologic Technique and Practical Histochemistry*, The Blakistone Comp., New York-Toronto, 1954.
20. LISON, L., *Histochimie et cytochimie animales. Principes et méthodes*, 2-ème édition, Gauthier-Villars, Paris, 1953.
21. LUFT, J. H., Improvements in Epoxy Resin Embedding Method, *J. Biophys. Biochem. Cytol.*, 1961, *14* (9), 409–414.
22. MAREŞ, E., ŞUTEU, I., GIURGIU, T., BACIU, D., CAVULEA, O., Cercetări privind oxigenoterapia enterală în şocul experimental (Investigations on Enteral Oxygenotherapy in Experimental Shock), *Revista sanitară militară*, 1965, *LXI*, (special no.), 32–37.

23. MAREŞ, E., ŞUTEU, I., STRÎMBEANU, I., VĂIDEANU, C., GIURGIU, T. DĂNCIULOIU, A., Influenţa diferitelor procedee de contenţie asupra consolidării homotransplantelor osoase liofilizate (The Influence of Different Contention Procedures on the Consolidation of Lyophilized Bone Homotransplants), *Revista sanitară militară*, 1965, *LXI* (special no.), 206–215.
24. NICULESCU, G., STRÎMBEANU, I., CHINŢA, G., FILIP, I., Homogrefa de piele în tratamentul arsurilor întinse (Skin Homograft in the Treatment of Extensive Burns), *Revista sanitară militară*, 1965, *LXI*, (special no.), 196–197.
25. NOBLE, R. L., COLLIP, J. B., Quantitative Method for Production of Experimental Traumatic Shock without Haemorrhage in Unanaesthetized Animals, *Quart. J. Exp. Physiol.*, 1942, *31*, (1), 187–199.
26. OANCEA, T., ATANASESCU, S., POPESCU, P., BACHMAN, M., GIURGIU T., IONESCU, P., SAFTA, T., VOICU, G., SINGER, D., Cercetări experimentale privind eficienţa unor măsuri terapeutice în şocul prin suflu, (Experimental Investigations on the Efficacy of Some Therapeutical Measures in Blast Shock), *Revista sanitară militară*, 1965, *LXI*, (special no.), 58–62.
27. PALADE, G., A Study of Fixation for Electron Microscopy, *J. Exp. Med.*, 1952, *95*, (2), 285–297.
28. PEARCE, E. A. G., *Histochemistry Theoretical and Applied*, Churchill Ltd. London, 1954.
29. POSTELNICU, T., TĂUTU, P., *Metode matematice în medicină şi biologie*, (Mathematical Methods in Medicine and Biology), Editura tehnică, Bucharest, 1971.
30. RETY, N. A., COUVES, C. M., Effects of Slow Infusion of Endotoxin in the Dog, *Canad. J. Surg.*, 1969, *12*, (4), 493–496.
31. RHOADS, J. E., HOMARD, J. M., *The Chemistry of Trauma*, Springfield, Charles C. Thomas Publ., 1963.
32. STEINBACH, M., *Prelucrarea statistică în medicină şi biologie*, (Statistical Processing in Medicine and Biology), Ed. Academiei R. S. România, Bucharest, 1961.
33. ŞUTEU, I., GIURGIU, T., IONESCU, P., SAFTA, T., BĂNDILĂ, T., Unele aspecte fiziopatologice şi terapeutice ale şocului hipertermic experimental (Some Pathophysiologic and Therapeutical Aspects in Experimental Hyperthermic Shock), *Revista sanitară militară*, 1965, *LXI*, (special no.) 48 51.
34. ŞUTEU, I., CAFRIŢĂ, A., BĂNDILĂ, T., GIURGIU, T., STRÎMBEANU, I., IONESCU, P., VERDEŞ, A., Valoarea aminelor presoare în reglarea circulaţiei splanhnice în şoc. (The Value of Pressor Amines in Regulation of the Splanchnic Circulation in Shock), *Revista sanitară militară*, 1965, *LXI*, (special no.), 54–57.
35. ŞUTEU, I., Contribuţii experimentale asupra combaterii şocului la răniţi pe cîmpul de luptă (Experimental Contributions to Shock Control in the Wounded on the Battle Field), *Revista sanitară militară*, 1958, *LXI*, (4), 569–581.
36. ŞUTEU, I., CAFRIŢĂ, A., Studiul experimental al evoluţiei şi tratamentului şocului traumatic cu ajutorul radioizotopului ^{51}Cr, (Experimental Study of the Evolution and Treatment of Traumatic Shock with ^{51}Cr), *Revista sanitară militară*, 1967, *LXIII*, (4), 575–585.
37. ŞUTEU, I., CAFRIŢĂ, A., Eficienţa comparată a unor agenţi farmacodinamici în combaterea tulburărilor hemodinamice din şocul traumatic experimental, studiate cu radioizotopul ^{51}Cr, (Comparative Efficacy of Some Pharmacodynamic Agents in the Control of Hemodynamic Disturbances in Experimental Traumatic Shock, Studied with ^{51}Cr), *Revista sanitară militară*, 1968, *LXIV*, (6), 821–832.
38. ŞUTEU, I., GIURGIU, T., IONESCO, P., SAFTA, T., L'hyperthermie induite controlée. Etude expérimentale, *J. Chir.*, *(Paris)*, 1968, *95*, (5–6), 669–674.
39. ŞUTEU, I., CAFRIŢĂ, A., BUCUR, I. A., Aspecte infracelulare în şocul experimental, (Infracellular Aspects in Experimental Shock), *Revista sanitară militară*, 1972, *LXXV*, (3), 357–366.
40. WIGGERS, C. J., OPDYKE, D. F., JOHNSON, J. R., Portal Pressure Gradients, under Experimental Conditions, Including Haemorrhagic Shock, *Amer. J. Physiol.*, *146*, (2), 192–196.
41. WIGGERS, C. J., *The Physiology of Shock*, Commonwealth Fund., New York, 1950.
42. ZWEIFACH, B. W., Microcirculation, *Sc. Amer.*, 1959, *200*, (1) 56–64.
43. * * * *Carleton's Histological Technique*, revised and rewritten by DRURY, R. A.B., and WALLINGTON, E. A.. Oxord Univ. Press, New York-Toronto 1966.
44. * * * *Dictionnaire Vidal*, 50e édition, O.V.P., Paris 1974.

3 GENERAL PATHOLOGICAL PHYSIOLOGY

We should consider Nature's reaction as first aid until the doctor arrives.

ALBERT SCHWEITZER

3.1 The 'human' reaction to shock
3.2 Pathophysiological theories
3.2.1 Isolated theories
3.2.2 Typical pathophysiological moments
3.3 Topographic and dynamic stages
3.3.1 Topographic stages
3.3.2 General pathophysiological scheme
3.3.3 Dynamic stages
3.4 Fundamental pathophysiological links
3.4.1 Importance of the volume of the circulating fluid lost
3.4.2 Gradient of circulating fluid lost
3.4.3 The importance of catecholamines
3.4.4 The importance of perfusion of the microcirculation
3.5 Irreversibility and other controversies
3.5.1 Catecholamines and irreversibility
3.5.2 Fluid deficit and pressor medication
3.5.3 Irreversibility and the interval up to its onset
3.6 Cybernetic elements
3.6.1 Automatic regulation system
3.6.2 Automatic biological system
3.6.3 Cybernetic scheme of shock
3.7 Genetic system
3.7.1 Gene and operon
3.7.2 Genetic code and homeostasis
3.8 Regulation levels
3.8.1 Genetic regulation
3.8.2 Metabolic regulation
3.8.3 Hormonal regulation
3.8.4 Neuronal regulation of the carotid glomus

3.1 THE 'HUMAN' REACTION TO SHOCK

After the shock-inducing conflict the human organism generates a rapid inhibitory reaction: *immediate imbalance* [71] or *immediate shock* [32, 65]. This reaction is characterised by severe restriction of the vital energodynamic functions and is due to the promptness with which the *parasympathetic* — that is the oldest phylogenetic defence system — tries to protect the organism; in man this protection is, however, inadequate, representing an attempt to circumvent a firm response corresponding to the severity of the lesional agent, and may be taken for true 'cowardice'. The body is, notwithstanding, soon convinced that it must fight in order to survive and immediately mobilises the *orthosympathetic* system, an energising catabolic system with a rapid efficiency. But even in man this system is still imperfect, its activity is temporary and is accompanied by sacrifices that very soon become dangerous. As shock may be described as 'a rude unhinging of the machinery of life' [75] it is normal that the systemic effects on homeostasis should appear to be slightly chaotic.

In an accident there is both a localised lesion (the wound) and complex reations in the whole body (the systemic effects of injury). Man's system of regulation and control receives information of the event, at first by nervous pathways then, slowly, by chemical messengers relayed by humoral pathways. A response is organised, which is also twofold — nervous and humoral — to save life, and it is addressed to each cellular group of the body. The systemic effects of injury combined with the complex general response of the body generate the state of shock. We must never forget that most of the phenomena of shock are in fact homeostatic attempts on the part of the organism to save itself: severe prolonged vasoconstriction, selective redistribution of fluids and anaerobiosis.

Whatever the initial lesional etiological factor is, as soon as it exceeds the 'shock threshold' a particular generalised reaction is triggered urgently demanding that the entire organism, whose existence is endangered, should join the struggle.

Generalisation takes place at different speeds through the fundamental performance pathways:

(i) the nervous system of relation (predominantly acetylcholine effector space);

(ii) the autonomic nervous system (acetylcholine and catecholamine effector space);

(iii) the endocrine system (hormonal effector space);

(iv) the metabolic system (enzymatic effector space);

(v) the genetic system (operonal effector space).

Between these systems there is a well known structural, functional and also pharmacological merging together, whose informational connections are assured by the coding rules of the genetic system contained in each cell unit.

Faced with aggression, the animal organism may respond by fleeing or by fighting, the dominant role being played by the nervous system of relation [32]. But to aggressions of shock-inducing intensity the human body responds by an *interiorised struggle*, consisting of a violent effort at adaptation of the endogenous energy systems, the dominant role being played by the autonomic nervous system, and by the endocrine and metabolic systems. Biogenic amines, polypeptide and sterol hormones organise these 'human interiorised struggles' i.e. shock [70].

The intensity and swiftness of the general response differs according to the visceral region: hypoxia and the wave of morphofunctional cellular lesions that accompany it avoid for some time the brain, heart and also the lungs but severely affect the intestine, kidneys and lung from the beginning. This is true for man; for other species the affected area varies.

These responses have deep roots in the pattern of innervation (the splanchnic as major efferent sympathetic pathway for the intestinal microcirculation); of the histoarchitectonic type (fragility of the gas and fluid phase balance in the lung); of the angioarchitectonic type (renal, hepatic lobes, etc.); or of the enzymatic chains (within the preferential metabolic limits in the brain, myocardium and endocrine glands) [15, 16, 30, 40, 62].

The inadequate cellular perfusion is first reflected in the tissues with an intense metabolism, supplied by the postcapillary system *(portal system* and *rete admirabilis)*. Hence, the liver, renal tubules and pituitary — the key viscera sustaining the organism in shock — are unusually sensitive in man [5, 45, 75].

Circumscribing homeostasis within the region of certain privileged organs occurs due to a shift in the programme, learnt and verified by phylogenetic experience and selection. The catecholamines are deciphered by the receptors that have a given spatial arrangement, and the entire hormonal torrent is received by the intramembrane, intranuclear and/or enzymatic receptors. This reception is understood more complexely at present, including *c*AMP, as a 'second messenger' in man. This behaviour is engraved in the genetic code and constantly becomes more perfect under our very eyes. *The response of the human system to aggression* will certainly be in the future better balanced and more economical, more efficient and 'cybernetic', without the numerous contradictions and imperfections of today.

Not all the organs of immediate vital importance are protected to the same extent; the brain and the heart are sheltered but not the lung which has become today the critical zone of shock [41, 50]. Another obvious proof that the 'wisdom of the organism' is still incomplete is the damage caused to the kidney, intestine, pancreas and liver, which the organism cannot always protect; for this 'voluntary sacrifice' made in the course of shock, the human body has to pay very soon a price that it is not always able to honour.

3.2 PATHOPHYSIOLOGICAL THEORIES

Numerous explanations have been given to the phenomenology of shock. Almost all have left us at least one pathophysiological link that has been verified *in time* by practice and which forms part of the common scheme of our understanding of shock today.

3.2.1 ISOLATED THEORIES

It has been asserted that there are as many theories concerning the pathological physiology of shock as there are types of shock described. Worthy of note among these theories are the following:

(i) *The cardiac failure theory* unquestionably valid in cardiogenic shock; but the role of decrease of the cardiac output has been extended as a major factor of importance also to other noncardiogenic types of shock. Although the adversaries of this theory emphasise the fact that *decrease in the cardiac output is only the consequence of peripheral rheodynamic disturbances*, it is unanimously admitted today that the myocardium must be sustained in all kinds of shock.

(ii) *The endotoxin theory*. In short, any hypovolemia with hypoxia blocks the reticulo-endothelial system (RES) which can no longer detoxify what it receives, especially from the intestinal flora; the toxins accumulated in the blood then generate rheodynamic and visceral disturbances. Although antibiotics do not confer an evident protection in any type of shock the theory has been largely accepted.

(iii) *The histaminic theory* was sustained for some time until it was found that histamine is rapidly inactivated in the blood and antihistamines do not give satisfactory results in the clinic. Recently, this theory has been taken up again, showing that histamine release is very low in animals previously adapted to experimental shock.

(iv) *The vasospasm theory* maintains that the fundamental factor in shock is catecholamine vasoconstriction. This is a widely accepted theory which regardless of the various initial causes explains the inadequate tissular perfusion in a significant proportion of the body.

(v) *The disseminated intravascular coagulation theory* considers that thrombotic blocking of the microcirculation is the essential event in shock.

(vi) *The metabolic theory* stresses the importance of enzymatic and cellular membrane involvement; this may likewise be criticised if it is recalled that cellular impairment is the consequence of inadequate perfusion of the microcirculation. However, even if shock does not begin at the cellular level (in the classical and current view of the concept of shock) it *is always and from the beginning projected in the intracellular space.*

(vii) *The nervist and endocrine theories* have prepared the way to our present understanding of the phenomenology by the cybernetic theory of shock.

The fact that none of the theories could fully explain shock is due to its etiopathogenic and clinical polymorphism; but a merging of these theories makes it possible to foresee a unitary concept.

The nervist theory postulates the neuroautonomic reaction in the response to aggression; the neurohumoral theory has proved the actual increase in catecholamines; the vasomotor theory has established initial splanchnic vasoconstriction as a constantly present pathogenic coefficient; the toxic theory suggested the role of intestinal toxins and proteases as factors of irreversibility; the theory of hydroelectrolytic stocking in the traumatised zones underlined one of the important causes of hypovolaemia; the metabolic theory decided upon the prognostic marker role of lactic acidaemia, etc.

The knowledge accumulated during the last two decades has enriched the content of each of these theories, from which a new theory has arisen which might be called the *neuro-endocrino-vasculo-metabolic theory* [13, 43, 61, 70, 72, 80]. Comparatively recent data concerning the physiology of the reticular substance and limbic

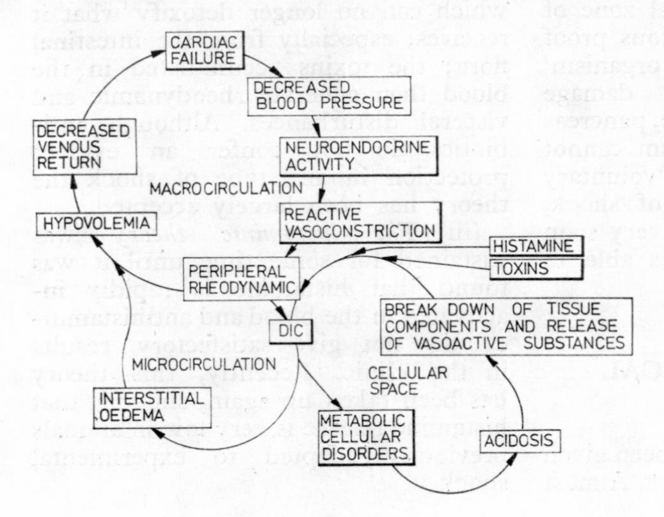

Figure 3.1. General scheme of the intricate circulatory and tissue parameters that various 'theories' point out as being the most important.

system (nervist theory), the role of α and β receptors in the dynamics of the microcirculation (vasomotor theory), the role of the 'bundle' of 'depressor tissular hormones' and hydrolases (toxic theory), rheologic disturbances (stocking theory), have set the basis of the unitary theory of the pathologic physiology of shock. Our knowledge in the fields of immunity, energetics, cybernetics and pharmacotherapy acquired during the last few years have elucidated many obscure points. However, the etiology of each state of shock may bring in many instances a personal note that favours one pathogenic factor to the detriment of another [36].

In the course of shock rheodynamic and metabolic parameters are of primary interest; but most of them tend to be brought back to normal values by means of counter regulatory mechanisms.

To the left of the circles in Figure 3.1 are the circulatory parameters reflecting the macro- and the micro-circulation. The parameters on the right, which one might call tissular parameters, are far more difficult to measure under clinical conditions; they are, notwithstanding, those which, especially during the later stages of shock, influence its final course so deeply.

It is obviously a mistake to put too much emphasis on only one biological system and to forget that this system may only be a fiddler in an orchestra.

The same factors act in various systems, some of which are aggravating (e.g. lysis and kinin systems) and some of which counter-balance (e.g. clotting and lysis systems) the action of the others at different moments. Another interaction is that of the reduced microcirculation which may be due to vasoconstriction, vaso-occlusion by thrombus formation or vasocompression by interstitial œdema.

At the other extreme too little is known about the alteration of the subcellular biochemical mechanisms.

Although the complex event that we call shock has not a single common pathway, there are several pathophysiological steps that may be described in a general chapter; but it is useless to try and fit one shoe to all feet.

3.2.2 TYPICAL PATHOPHYSIOLOGICAL MOMENTS

Questions perturb all those who wish to find the common traits of this complex phenomenon: can there be anything common among over one hundred types of shock described in the medical literature? Their etiological polymorphism is discouraging but it should be recalled that 'the human organism is not polyglot' (Laborit), its fundamental pathophysiological mechanisms speak the same language. There is a general response reaction, qualitatively common to any type of conflict between the human organism and its environment. Congruous synchronisation of all the links of the systemic response may sometimes re-establish the situation starting from a postaggressive (postinjury) oscillating systemic reaction.

In cases of conflicts generating shock the intensity of the lesional agents force and disorganise the adaptive links, modifying the response reaction qualitatively, rendering it unsynchronised, dysharmonic and evolutionarily deleterious to the macrosystem.

There are several *pathophysiological moments* encountered in most kinds of shock, moments that are common and knit together in a unitary concept:

(i) *The real or relative ECBV deficit* is always accompanied by a primary or secondary decrease in the cardiac output and increase in peripheral resistance; the catecholamine factor encircles these elements with an auto-aggravating haemodynamic ring.

(ii) The main stimulators of catecholamine discharge: *hypovolaemia*, *hypoxia*, *lactic acidosis*, etc. are the constant pathogenic coefficients of the states of shock (when not corrected they maintain the release of catecholamines in excess).

(iii) *Rheodynamic disturbance* of the microcirculation represents the major impairment in states of shock since it affects the energy and substance supply to the cells.

(iv) *Cellular hypoxia* shifts the enzymatic chains towards fermentation, with in consequence insufficient energy to cater for the excessive demands imposed upon the macrosystem, and a massive accumulation of metabolic detritus, resulting in acidosis. In hypoxia, the cells accumulate protons: then the extravascular vasoactive amines (histamine and serotonine) activate the plasma vasoactive kinins; hypoxia, acidosis, hypercatecholaminemia and hypercortisolemia potentiate the toxic effects, of maximum gravity especially for the myocardium [64].

(v) *Progressive acidosis*, attaining the critical point at which the cell life ceases, produces necrosis, focal at first, then confluent and finally generalised.

(vi) *Hypotension* as a symptom is of secondary importance; apparently 'compensated' shock, according to the BP criterion, may be accompanied by inefficient cellular perfusion. The old term 'hypertensive shock' masks severe cellular impairment, due to excessive vasoconstriction. A low BP is preferable if cellular perfusion is sufficient. The 'actual compensation' must be followed up at intracellular level.

(vii) *Cellular involvement*. Irrespective of the cause and evolutionary patterns of the states of shock, the cell is soon affected, involving the DNA strands of the intranuclear code, the cytoplasmatic enzymatic chains and, finally, the membranes. Breakdown of their architecture is equivalent to the irreversible disorganisation of the cell.

3.3 TOPOGRAPHIC AND DYNAMIC STAGES

The pathological physiology of shock has focused the attention of many generations of clinicians and researchers, but the treatment has always remained empirical. If in the last two decades a rational treatment has gained ground it is certainly due to pathological physiology that is better understood and that it has passed beyond the threshold of the cellular maze.

Circulatory alterations are awlays impressive, dominating the clinical picture of shock from the onset up to re-equilibration or death. On the other hand, intracellular disturbances are at first not reflected clinically, or perhaps today's interpretation of semeiology does not attribute to them what is actually theirs; finally however cellular impairment also becomes clinically manifest, mediated by circulatory disturbances. Hence, the pathophysiology of shock is dominated in the clinic by circulatory alterations, but from the simple notion of hypotension (or the confused term of collapse) it has reached a deeper understanding of the alterations in the microcirculation space, whose clinical expression is at present well known. *The intracellular micropathophysiology* of shock offers a therapeutical possibility of increased efficiency.

Shock-inducing factors brutally modify the circulation of fluids in the organism. Surgery and anesthesia represent a deliberate 'violation of homeostasis' (Byrne) which may sometimes produce shock. Volaemic isotopic determinations performed after major operations for cancer, as well as after cholecystectomy or vagotomy with pyloroplasty are well known: the red blood cell count may decrease by 15%, plasma by 24% and the total extracellular fluid up to 15% [76, 77].

3.3.1 TOPOGRAPHIC STAGES

According to the intensity and to the duration and topographic space affected, any state of shock passes through two major stages, artificially separated from one another:

(i) extracellular stage (fluid disturbances);

(ii) intracellular stage (metabolic perturbances).

The extracellular stage includes the totality of the alterations in neuroendocrine regulation and the circulation of systemic fluids, in the macro- and microcirculation, and also in the lymphatic and interstitial circulation.

The intracellular stage rapidly follows the preceding stage with an avalanche of informational messengers and points of insertion up to the intimacy of the polynucleotidic chain of the genetic code; this stage includes all the changes in intracellular regulation of the enzymatic chains.

The organism in its effort tries to prevent the disturbances of the extracellular space from affecting the phylogenetically more sensitive vital parameters of the intracellular space. Hence, generalisation of the state of shock to the active cellular mass of the entire macrosystem offers two advantages:

(i) energy is required by all the cellular subsystems for neutralisation of the shock-inducing energy;

(ii) dispersion and dilution of the perturbation of the extracellular space parameters that are more readily corrected by the totality of the cells.

The cellular space appears to be the main lesional scene where victory or defeat will take place. Until lately, the cell was considered to suffer only in the late phases of shock and therapeutical emphasis was only put upon haemodynamics. The experimental and clinical studies of the last few years have shown that *lactic acidosis*, the expression of early cellular metabolic sacrifices, appears before any sign of decrease in tissular perfusion (for instance hypotension and a fall in urinary output). It has likewise been observed experimentally, by monitoring, that the transmembrane potential decreases and the selective ionic pumps are blocked immediately after impact of the shock-inducing agent with the macrosystem. Acid material, lysosomal hydrolases, polypeptides, adenylates, potassium, endotoxins rapidly flood the internal medium, aggravating the membrane lesions and altering the fluidity of the blood. Anti-DNA antibodies have been isolated shortly after the onset of shock, their titre pointing to the severity of the shock and their presence reflecting a *rapid autoimmune attack upon the genetic matrix* of protein synthesis and of the system of cellular coordination [7, 53].

As in any schematic representation, the pathophysiological stages common to any type of shock are but artificial snapshots of a phenomenon well known for its evolutive plasticity:

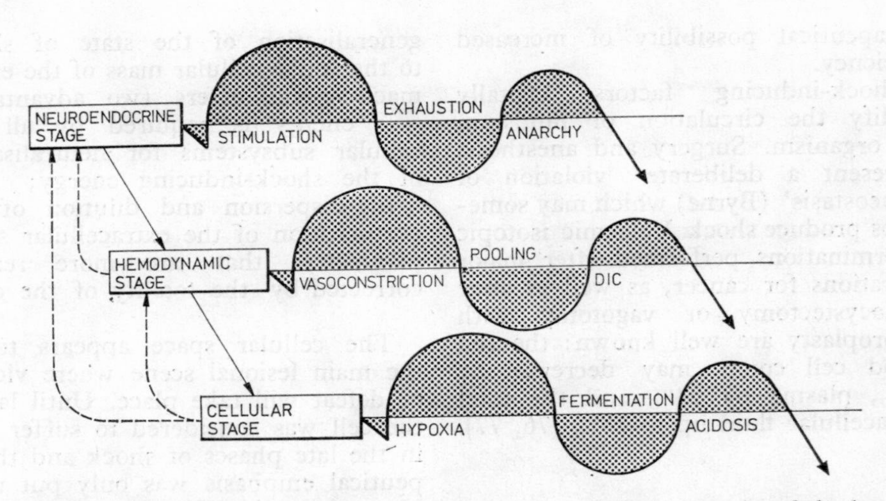

Figure 3.2. The linkage of the three different phases (*stages*) in the intensity of shock.

(i) *The stage triggering the neuroendocrine reaction* includes:
  (a) the substage of lesional reception;
  (b) the substage of the central integration circuits;
  (c) the substage of neurohormonal efference.

(ii) *The haemodynamic stage* includes:
  (a) the systemic macrocirculation substage;
  (b) the microcirculation substage;
  (c) the interstitial lymphatic substage.

(iii) *The cellular stage* is divided into:
  (a) the metabolic effort substage;
  (b) the metabolic exhaustion substage;
  (c) the substage of irreversible structural lesions.

There is a certain hierarchy, an interconditioning and simultaneous translation in time of these stages (Figure 3.2).

In its evolution each substage passes through a phase of reversible functional and structural lesional alterations, which may then become irreversible. The intensity of the structural lesions increases from biochemical to leptonic and histologic level, finally taking on an anatomoclinical macroscopic relief. Each of these stages and substages forms, in spite of the odds against continuity, the subjects of chapters 4, 5, 6, 7, 8 and 9.

3.3.2 GENERAL PATHOPHYSIOLOGICAL SCHEME

For the sake of simplification the pathophysiological state of shock may be schematically represented as shown in Figure 3.3: the shock inducing factors offer acute lesional information that is rapidly relayed by specific neuronal pathways to the ascending lemnisci and especially by unspecific pathways (ARAS) up to the cortico-diencephalohypophyseal dispatcher system; here the generalising order is released, whose peripheral expression is a brutal sustained increase in plasma epinephrine, norepinephrine and cortisol concentrations. These chemical messengers direct the redistribution of intracellular fluids and enzymatic chains. Due to the obstruction of the microcirculation with pooled, hypercoagulable blood (especially in the lungs, intestine, kidneys, liver, skin, etc.) the decrease in the ECBV becomes more accentuated (or is initiated in some instances). Impairment of the microcirculation involves both vascular alterations (spasms, dilatation and atonia with dysynergism of the pre- and post-capillary sphincters) and alterations of the blood fluid (pooling, sludging, disseminated intravascular coagulation, fibrinolysis). The consequence is immediate cellular distress, since hypoxia brings about from the very beginning sugar fermentation with acidosis and opening of the lysosomal sacs which, by releasing the hydrolase digestive enzymes, disorganise and destroy the cell.

Along general lines, the pathophysiological axis described is the common route of any type of shock but not necessarily followed step by step. At the more important crossroads there are numerous collaterals that may join the common route, but there is also the possibility of leaping over some of the stages. Traumatic, thermal and haemorrhagic shock seem to comply best with the entire axis, but there are many types of shock that start upon the common pathway from one of its pathogenic links. Thus, cardiogenic shock begins by a deficit of the central pump, anaphylactic shock by an 'attack' of the vascular tree, 'adrenalinic shock' by vasoconstriction of the microcirculation and cellular metabolic exhaustion and crushing shock by toxemic flooding, etc.

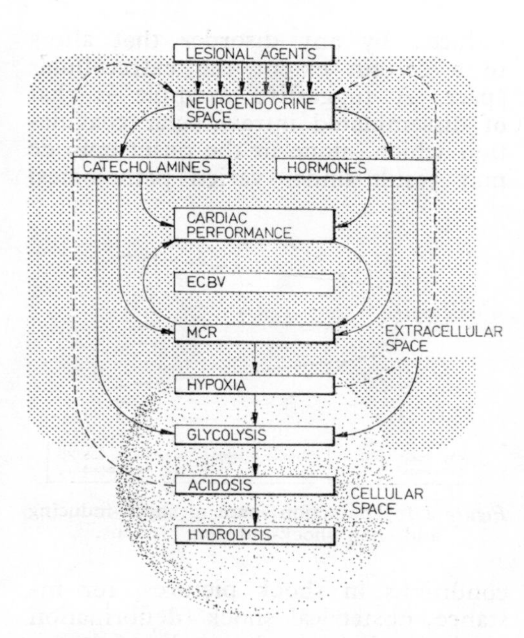

Figure 3.3. General scheme of the pathophysiological axis of shock.

Whatever the sequence of the events, the conditions of the shock status have to be complied with: *fluid disturbances in the extracellular space, reflected in the cellular space, resulting in an energy depletion of great intensity*. Hence, at least the latter stages of the common pathways of shock must of necessity be gone through: *rheodynamic impairment of the microcirculation and intracellular enzymatic affection.*

Therefore the lesions that involve from the very onset only the neuroendocrine system or regulation or only the cellular level of metabolic effort cannot generate a state of shock (Figure 3.4).

It is known that as a rule head injuries and enzymatic cellular intoxications do not generate states of shock; on the other hand shock may be induced by any disorder that alters to a certain extent the extracellular space. A suggestive example is that of disseminated intravascular coagulation which, alone in the onset episode may result under varied pathological

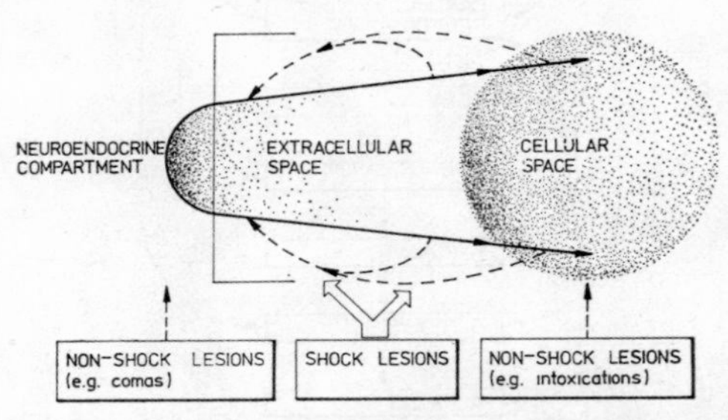

Figure 3.4. The action space of shock-inducing and non shock-inducing lesions.

conditions in shock pictures, for instance, obstetrical shock (defibrination syndrome in uteroplacental and hypophyseal apoplexy), shock in acute haemorrhagic-necrotic pancreatitis, shock in acute viral diseases and acute pneumococcal infections (Waterhouse-Friderichsen syndrome), shock in retroperitoneal haematomas, etc.

Bacteriemic shock has been assumed to attack the cell initially and is considered as one of the most interiorised types of shock. Some earlier experimental studies demonstrated the protective effect of splanchnic denervation and of antiadrenergic agents in traumatic and endotoxin shock; more recent investigations, however, based upon the same model but using live *E. coli* in LD_{80}, showed that although splanchnic vasoconstriction did not appear, the cell was not protected (decrease in the *p*H and arteriovenous oxygen gradient) and the mortality rate and intestinal lesions were the same. This proves the existence of a direct cellular attack in situations similar to those occurring in man [31].

3.3.3 DYNAMIC STAGES

After the schematical outline of the topographical pathophysiology of the phenomenology of shock it is now pertinent to discuss its dynamics, in the course of which we shall meet with the sinusoidal curve once again.

The shock-inducing information received at the periphery is transmitted by specific and unspecific centropetal pathways (somato-autonomic-humoral) to the diencephalocortical level where, due to the brutality of the stimuli (inadequate, nociceptive) a supraliminal inhibition of *temporary protection* develops (against the usual background of resting physiological occlusion).

There then follows an interval that has been designated as: immediate shock (Selye), immediate depression, vegetative collapse, primary lesional syndrome (Laborit) [35, 36], immediate postaggressive imbalance (Teodorescu-Exarcu) [71]. Whatever the term used, this transient interval is characterised by a decrease in the functional tonus of the neurones of the corticoautonomic centres, a state of blocking reflected by the well known group of '*H*s': cutaneous hypoesthesia, osteotendinous hyper-reflectivity, hypothermia, hypovolaemia, hypotension, hypoxia, hypoglycemia, hypochloremia, etc. Reflex functional inhibition of the liver or kidneys may take place, as well as alterations in the permeability and active exchanges through the membranes, etc.; this initial homeostatic storm has a reflex mechanism and also probably a humoral one (acetylcholinic, histaminic, etc.).

These savage alterations give rise to new pathogenic coefficients or 'stimuli of the second order' that amplify the initial ones (for instance hypovolaemia and hypotension caused by intravascular pooling and loss of the vasomotor tonus, may be added to the initial hypovolaemia caused by the actual loss of blood, plasma, etc.). In bacteriemic shock it has been assumed that initial hypotension, characteristic of immediate depression, develops due to the action of the bacterial lipopolysaccharide in about two minutes, mediated by the complement system (*see* Section 4.11).

Detected by the interoceptor fields (endovascular and intratissular) 'stimuli of the second order' are added syntropically to the stimuli of the initial shock-inducing lesion (that continues to emit) and reach the same diencephalocortical space. In the meantime, this physiological occlusion is rapidly 'washed away' by hypertonia of the ARAS, which triggers violent cortical dyssynchronisation of the arousal type. The state of alert cortical awareness, specific of early shock (the former 'erectile' stage) determines hyperactivity of the centrencephalic system, mutually maintained by reverberation. In this centrencephalic system, the posterolateral hypothalamus (orthosympathetic) is a kind of common gate of entry of all the activating stimuli received by the thalamus, limbic formations and isocortical areas, whose integration will govern hypertonia of the *catecholamine ergotropic system* and *hypophysoadrenal catabolic hormonal system*. The well known systemic hemodynamic redistribution also takes place in the microcirculation and the cells have to initiate an energising metabolic effort. The ascending slope of the first positive sinusoid loop reaches a peak that marks the limit of the energy resources of the organism; this is also the limit of the *early reversible shock (see* Section 1.5.3 and Figures 1.14 and 35).

On the descending asymmetrical slope of the first positive loop the organism continues to be in full effort, showing however a tendency to a slow decrease. This is the *phase of late reversible shock*, a longer and more frequently encountered stage in the clinic that necessitates skilled therapy since it contains the germs of decline.

The following phase begins when the sinusoid crosses the line of physiological balance and passes into the negative loop: the phase of *refractory reversible shock*. Reversibility is still possible if the maximum depression reached by the sinusoid is compatible with life, at a level at which the organism still has time to mobilise the energy mechanisms of return to a physiological balance.

Re-entry within the physiological limits admitted takes place after several buffering oscillations, when reversibility is maintained.

However, an irreversible evolution is just as possible, depression of the negative loop continuing to levels incompatible with life *(irreversible shock)*. The oscillating dynamics of the cortico-centrencephalic system pass in the meantime from the state of alert (in early shock) to active internal metabolic inhibition (in the late shock stage) and then to passive external inhibition (in the refractory stage of shock) and anarchy, with the final appearance of neuronal structural lesions (in irreversible shock) *(see* Section 4.6).

This oscillating cortical dynamics in shock is also complied with at all the other hierarchically assembled levels: subcortical centres and endocrine, haemodynamic, metabolic systems, etc. (Figure 3.5).

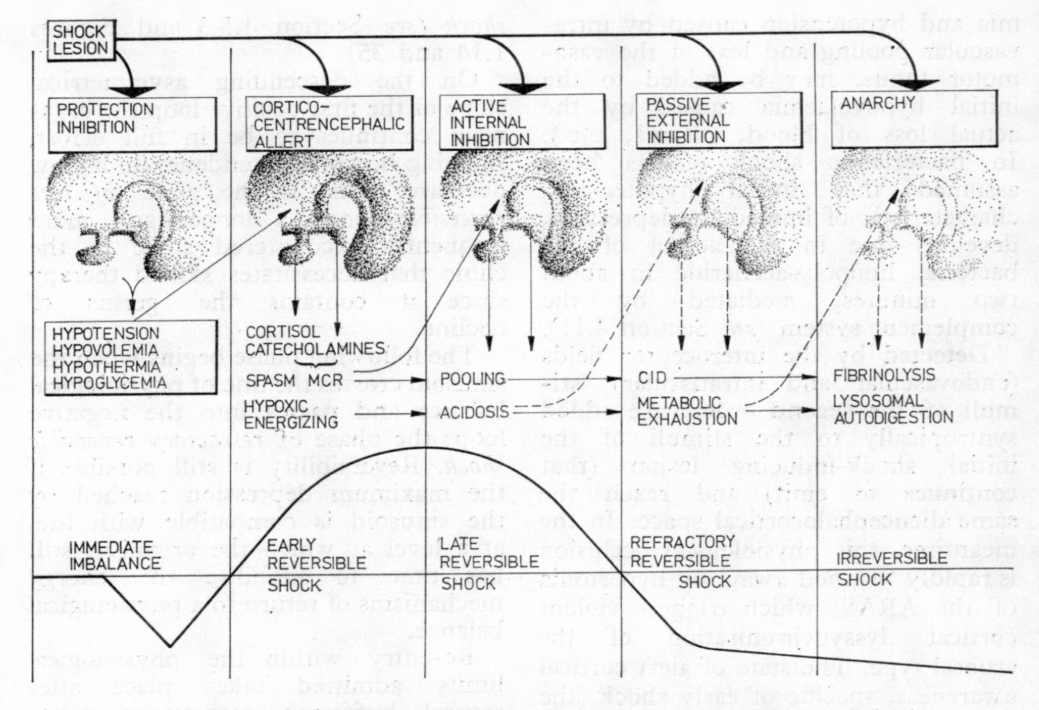

Figure 3.5. Vertical hierarchy and horizontal staging of the dynamics of shock.

3.4 FUNDAMENTAL PATHOPHYSIOLOGICAL LINKS

The evolutive aspects of shock depend upon the following parameters:

(i) volume of lost fluid (in absolute values);

(ii) structure of lost fluid (amount of proteins, electrolytes, etc.);

(iii) the rate of fluid depletion (gradient and duration of hypovolaemia).

3.4.1 IMPORTANCE OF THE VOLUME OF THE CIRCULATING FLUID LOST

An actual or relative fall in the ECBV is important in its absolute value (table 3.1 and *see* also table 7.3).

It is true that the exterior losses (vascular haemorrhage, haematemesis and/or melena, refractory vomiting, profuse diarrhœa, accentuated perspiration, plasma depletion in burns, etc.) are evident but not always readily appraisable quantitatively. When fluid accumulates in the serous spaces (haemoperitoneum, haemopericardium, pleurisy, etc.) the losses are still more difficult to determine. It is obviously, even harder to assess the fluid losses following upon the primary redistribution of fluid between the various compartments (exudates, obstruction, massive thromboembolism, pancreatitis, anaphylaxis, etc.). Not even today can a correct answer always be given to Cannon's question put in 1923: 'where is the blood that leaves the circulation?'

Table 3.1 THE ABSOLUTE AMOUNT OF BLOOD FLUID LOST REFLECTED BY THE INTENSITY OF SOME PARAMETERS

	ECBV losses			
	< 10%	10—20 %	20—30 %	> 30%
Cardiac output	↓	↓↓	↓↓↓	↓↓↓
Peripheral resistance	↑	↑↑	↑↑↑	↑↑
Venous return	—	↓	↓↓	↓↓↓
CVP and BP	—	↓	↓↓	↓↓↓
Tachycardia	—	++	+++	+++
Tissular irrigation	↓	↓↓	↓↓↓	↓↓↓
Hypoxia	—	±	++	+++
Acidosis	—	—	+	+++
Urinary flow	—	↓	↓↓	↓↓↓
Skin temperature	↓	↓↓	↓↓↓	↓↓↓

An external fluid loss is rapidly followed by refilling at the expense of the extravascular fluid and may be successfully compensated by a therapeutical exogenous supply.

A fall in the ECBV due to a primary redistribution between the compartments has a less optimistic prognosis. On the one hand, it starts by anarchy of the natural mechanisms of spontaneous compensation and, on the other, therapeutical completion of the lost volume runs almost always counter to the pathological dynamics of the endogenous displacement of fluids, often resulting in global overloading.

External haemorrhage is the prototype of acute, isotonic reduction of the intravascular volume. There follows fluid donation of the interstitial space which, in milder hemorrhage, may bring the blood volume back to normal within 20-24 hours.

Burns, injuries, peritonitis, pancreatitis, obstruction are however situations in which the transcapillary flow of fluids takes place inversely: the interstitial space becomes hypertrophic due to parasitic pooling of water, electrolytes and even proteins, drawn from the plasma, when the selective permeability of the capillary is affected.

3.4.2 GRADIENT OF CIRCULATING FLUID LOST

The features of the haemodynamic and metabolic response, common to all states of shock, are closely related to the rate of decrease of the ECBV. Between a rapid and a slow haemorrhage there are qualitative differences in the principal parameters of homeostasis, even if the absolute loss of ECBV is the same.

In the clinic, a conventional criterion of *four hours* may be established: if the loss of ECBV, followed by a state of shock occurs in less than four hours the volume loss is considered rapid; if this lasts more than four hours then volume loss is considered slow and prolonged.

According to the rate of fluid depletion account has always to be taken, on applying the treatment in the clinic, of the following pathophysiological differences:

(i) Even if the total volume depletion is smaller in a rapid than in a slow loss, in the former the ECBV and especially the index of the central circulating blood volume fall much lower as the compensatory redistributions do not have time to intervene. A slow loss, even if more important quantitatively, is always better supported (because it has time to influence the extravascular space). The central circulating volume is reduced by about two-thirds of the total blood volume following rapid depletion and only by one-half in the case of slow depletion [77].

(ii) In a rapid volume loss, the cardiac output, cardiac index, blood pressure and central venous pressure fall more rapidly and seriously and return more slowly to normal values; only the BP due to intense vasoconstriction exceeds in the stage of compensation the values obtained after a slow volume depletion. The intensity of vasoconstriction (therefore of the discharge of catecholamines) is directly proportional to the gradient of volume depletion.

(iii) The sharp fall in cardiac output and stroke volume output following rapid volume losses probably expresses acute reduction of the venous return (which actually accounts for the decrease in the central blood volume). No close relationships exist between the circulating central blood volume, ECBV and CVP; at first the ECBV decreases, then the CVP increases, probably by neurogenic mechanisms of the venous tonus. When the ECBV returns to normal and the cardiac output increases, the venous return certainly increases and the CVP may be high but decreases in most cases (the right ventricle and pulmonary circulation no longer obstruct the venous return).

(iv) The stroke volume of the left ventricle increases, especially after improvement of the systole (intrinsic and extrinsic positive inotropic effects) in slow depletion, whereas in rapid losses there is a temporary increase due to acceleration of the rate of flow, i.e. worsening of the diastole; the left ventricle is far more affected in abrupt fluid losses.

(v) In rapid volume depletion, running parallel to an accentuated decrease of the available peripheral oxygen, the arterial *p*H decreases, the $P_a(CO_2)$ increases and there is a marked elevation in tissular oxygen extraction, differences in arteriolo-venular oxygen, and tissular oxygen consumption. Arterial saturation with oxygen, the total arterial content in oxygen and $P_a(CO_2)$ decrease at the same rate, regardless of the rate of volume loss.

(vi) Blood rheological disturbances always appear very rapidly, which explains the blocking of tissular perfusion by collapse of the microcirculation and cellular affection even in supra-acute volume depletion. In an experiment on dogs the withdrawal of 60 ml blood/kg body weight resulted in exsanguination in 7-8 min, the arterial BP attaining 0 within 5 minutes. At BP values of 40 mmHg sludging appeared in the retinal arterioles; the $P_a(O_2)$ was not modified but $P_a(CO_2)$ fell due to hyperventilation [76].

(vii) Metabolic alterations depend upon the circulation of gases, the lung being the critical organ; the lung in shock suffers in direct proportion to the duration of the onset and maintenance of hypovolaemia. In rapid losses the $P_a(O_2)$ remains almost normal, whereas in slow volume losses it decreases (parallel to reduction of haemoglobin concentration).

(viii) The *available* peripheral oxygen decreases by one-half in both types of volume losses; in rapid volume losses, the low cardiac output may account for the diminished oxygen release to the tissues, but in slow losses when the cardiac output is hardly affected diminution in the release of oxygen to the tissues is obviously due to an intraerythrocytic blocking of oxygen (*see* Section 7.2.1).

(ix) Rapid volume losses result in accentuated acidosis; it has also been assumed that hypoxia and acidosis are only harmful to the heart at very low values ($p\text{H} = 7.1$), so that in this case, too, the erythrocytic mechanisms of oxygen release can be incriminated instead of the decrease in cardiac output.

Rapid hypovolemia leads to accentuated acidosis owing to a decrease in perfusion and the efficient amount of oxygen utilised by the tissues. In slow volaemic losses in which accentuated trachypnea appears, the mechanisms of hyperventilation stimulation are set in motion and $P_a(CO_2)$ decreases still further. Slow hypovolaemia upsets haemodynamics and tissular perfusion to a lesser extent and, although the amount of oxygen utilised by the tissues is small, pulmonary mechanisms of carbon dioxide elimination interfere and may even cause gas alkalosis instead of metabolic acidosis.

Conclusions. The rate of volume depletion imprints a particular pattern upon the hemorheodynamic and metabolic alterations, whose distinct features demand a differentiation of the therapeutical approach (Walters, 1969).

In abrupt volemic losses the first to suffer is tissular perfusion (haemorheodynamics), whereas in slower losses metabolic and pulmonary impairment are more manifest. Metabolic affection in abrupt volume losses is of the uncompensated acidotic type. Apart from haemorheodynamic compensation, slow volume losses also allow for metabolic compensation (in the course of which the pulmonary mechanisms may even develop gas alkalosis).

In both modes of volemic depletion the fundamental pathological axis is blocking of the oxygen and carbon dioxide transport system at the lung and red blood cell level (*see* Section 7.2.1).

3.4.3 THE IMPORTANCE OF CATECHOLAMINES

Well established in the pathogenesis of shock appears to be the flow of catecholamines which marks the beginning of the reaction syndrome. It is the most rapid phenomenon that affects the smooth muscle cells of the microcirculation and intracellular energy-producing enzymatic chains [66]. This leap in the concentration of catecholamines was demonstrated objectively by plasma and urinary determinations in laboratory animals and the human clinic. The figures reported vary because various, non-standard methods were used for the determinations. The increase in catecholamines ranges from 30% in haemorrhagic shock [39], or a 10 to 35-fold increase in thermal shock, three hours after the accident [14], up to a 100 to150-fold increase of epinephrine and a 50 to 100-fold increase of norepinephrine after burns [78]. *It is assumed that in shock catecholamines may show a 30 up to 300-fold increase as compared to initial levels* [67, 68].

The major stimuli of hypercatecholemia are *hypovolaemia* and *hypoxia.* A *hypovolaemia* of only 10% of the intravascular blood volume is a powerful stimulant of the adrenomedulla, a fact that has to be borne in mind since hypovolemia up to 10% is not clinically manifest. Pallor and contraction of the peripheral veins in the immediate postoperative stage points to a discharge of catecholamines and the normal BP values recorded may be misleading.

Lactic acid likewise seems to stimulate the adrenomedulla [78]; therefore acidosis accentuates the central release of catecholamines although peripherally it attenuates their effect.

Hypoxia causes rapid stimulation whereas increased carbon dioxide concentrations are followed by a slow but constant increase in catecholamines [20].

Steroids potentiate the vascular response of catecholamines [28] with which they interact by complex metabolic pathways *(see* Chapter 8).

Anticholinesterases accentuate the peripheral catecholamine response, an effect also produced by serotonin [42, 55].

The catecholamines are 'recognised' by their specific receptors following their histofunctional spread *(see* Section 5.5.1), thus directing their inter- and intravisceral redistribution (for instance intrarenal, intrahepatic, etc.).

This rapid adaptation reaction involves serious sacrifices: rheologic disturbances immediately develop, with an increase in viscosity and aggravation of disseminated intravascular coagulation. Cellular metabolic failure will be expressed by progressive acidosis up to the protonic concentration favouring hydrolytic digestion incompatible with maintenance of the cellular morphological integrity and functions.

Like the adrenergic receptors, the cholinergic receptors appear to show a varied functional behaviour. For instance, the cholinergic receptors of the contractile cells in the myocardium are richer in SH groups than those of the nodal tissue [2, 60, 63], and hence atropine blocks the cardiac rhythm more than the inotropism.

3.4.4 THE IMPORTANCE OF PERFUSION OF THE MICROCIRCULATION

The dynamic alterations at the level of the receptors of the microcirculation represent the essential link in shock. This is the site of attachment of catecholamines, the *myoneural junction* or *adrenoeffector end-plate*. Hypercatecholaemia is maintained in shock by numerous pathogenic coefficients (hypovolaemia not corrected, hypoxia, etc.) that overlap and act beyond the limit at which hypercatecholaemia becomes deleterious, starting one of the fundamental self-promoting vicious circles of shock (Figure 3.6). The lack of

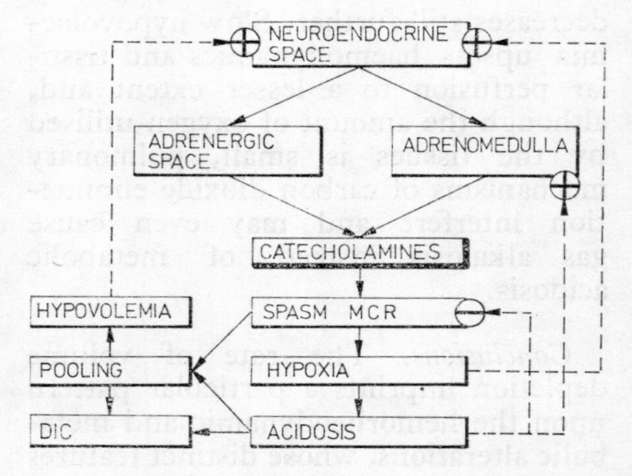

Figure 3.6. Automaintenance of catecholamine release.

justification of a treatment with pressor amines (formerly upheld for redressing the systemic BP) appears only too clearly, since they stand at the basis of the very phenomena that must be corrected. Today it has been unanimously admitted that 'it is not essential for the pressure values to be optimal but it is essential for the tissues to be perfused by an optimal blood output per unit tissue volume' (Litarczek, 1974).

The haemodynamic disturbances produced by catecholamines may be studied at three levels: the cardiac pump, systemic vascular tree (arterial and venous) and microcirculation *(see* Chapter 5). Pathogenical cohesion is realised by blocking the microcircula-

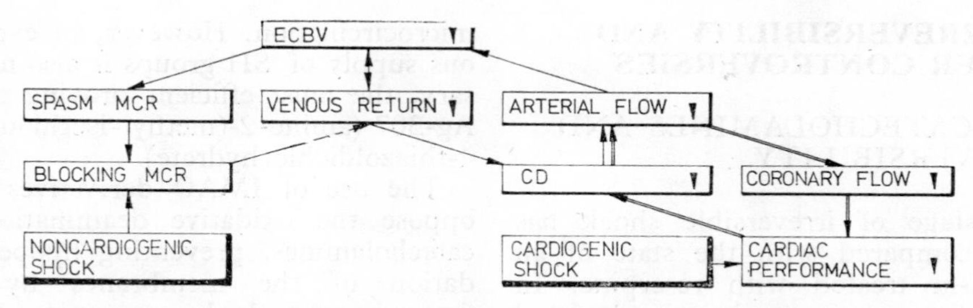

Figure 3.7. Autoaggravating haemodynamic loops in cardiogenic and non-cardiogenic shock.

tion (self-maintained peripheral blocking: spasm → acidosis → DIC → spasm) that results in an insufficient venous return, thus bringing to a close the main vicious circle of haemodynamics. At a given moment the heart is also taken up in an aggravating haemodynamic vicious circle, well known in general pathological physiology, that leads to secondary heart failure, also explaining the mechanism of cardiogenic shock (Figures 3.7 and 3.8).

The increase in peripheral resistance is the essential cause of inefficient cellular perfusion, but there is however a particular form of shock in which systemic peripheral resistance is not affected: in pulmonary embolism, an effector heart with an initially good venous return produces hypertension in the pulmonary circulation where the lesion consists in fact also of an increased resistance, but in this instance of the pulmonary circulation. The heart rapidly fails, the venous return decreases and the ECBV, which initially showed no disturbances, likewise decreases.

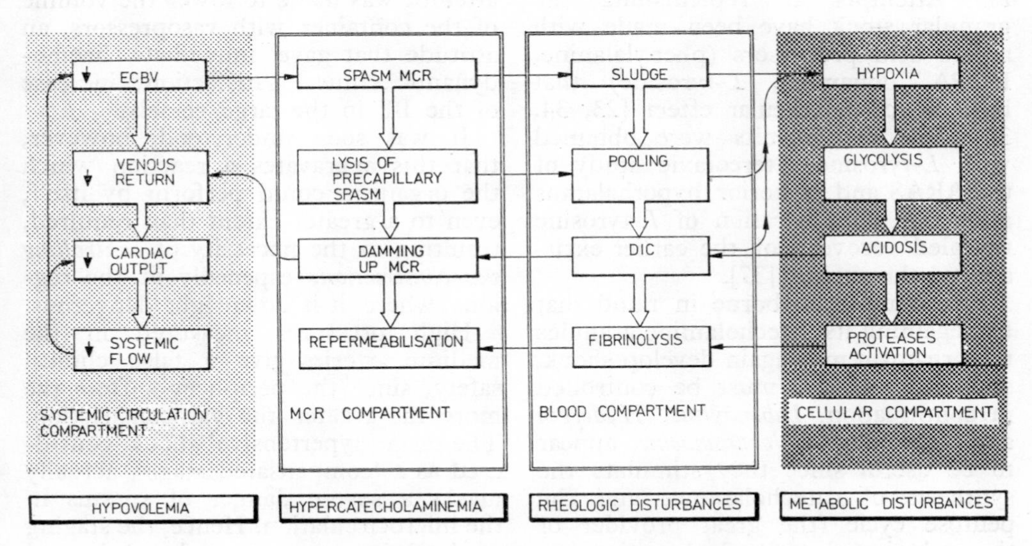

Figure 3.8. The interrelationship of some haemorheodynamic and metabolic systems in shock.

3.5 IRREVERSIBILITY AND OTHER CONTROVERSIES

3.5.1 CATECHOLAMINES AND IRREVERSIBILITY

The stage of irreversible shock has been compared with the state of an organism treated with reserpine. In this phase the organism is depleted of its own catecholamine granules. Therapeutically undesirable up to this phase they would now be more than necessary especially for their metabolic energising role. However, they cannot be administered as such even in this phase for two reasons: first, exogenous catecholamines do not incite a response from the receptors in this stage of shock (the endogenous catecholamines alone appear to be 'recognised') and secondly, because they stand at 'the origin of the pathological state that must be treated' [37].

The only possibility is activation of the *endogenous synthesis of catecholamines.* Attempts at replenishing the granular stock have been made with amino acid precursors (phenylalanine, DOPA, dopamine, *L*-tyrosine) that have no cardiovascular effect [23, 34, 35]. The best results were obtained with *L*-tyrosine. Stereotaxic study of the ARAS and posterior hypothalamus after the administration of *L*-tyrosine revealed recovery of the earlier excitability thresholds [37].

But it must be borne in mind that by regaining its catecholamine granules the organism may again develop shock, so that the latter must be controlled in continuation. *Dibenzyline*, *Hydergin* and especially *Chlorpromasine* appear to be useful since they stimulate the synthesis of catecholamines and the pentose cycle (the great provider of the reducing functions) that also has lytic effects on the spasms of the microcirculation. However, an exogenous supply of SH groups is also necessary, the most efficient proving to be Ag-307 (amino-2-(methyl-isothioureil)-4-thiazoldichlorhydrate).

The use of IMAO derivatives that oppose the oxidative deamination of catecholamines, preventing lipoperoxidation of the membranes by the formation of hydrogen peroxide is likewise necessary *(see* Section 8.3.6).

3.5.2 FLUID DEFICIT AND PRESSOR MEDICATION

As the fluid system is conceived as 'an elastic container with a pump and resistance' (Hardaway, 1967), any discrepancy between the content and the container must be corrected. In general, the container is refilled with fluids, but in very few cases does this alone solve the critical situation. It is also necessary to act upon the walls of the container. Initially, an attempt was made to lower the volume of the container with vasopressors, an attitude that gave immediate haemodynamic clinical satisfaction: increase of the BP in the large vessels.

It was soon understood, however, that this aggravates a reaction which the organism could perform by itself, even to a greater extent than required, resulting in the necessity of *controlling vasoconstriction*, especially in the regions where it is downright dangerous.

High pressures measured in the medium arteries give a false clinical safety, since the periphery suffers the more the greater the vasoconstriction. The term 'hypertensive' shock, considered as a 'compensated stage', actually masks a severe danger of spasms in the microcirculation. Hence, the staging of shock into compensated and decompensated, a predominantly clinical terminology based on not very certain criteria of evolution has been abandoned (*see* Section 1.5.3). *Haemodynamic compensation does not reflect affection of the tissues, where the prognosis of shock is actually decided.*

Fluid depletion has to be carefully investigated in order to be correctly interpreted. This sometimes requires investigations with ^{51}Cr for the red blood cells, ^{125}I-serum albumin for plasma, ^{22}Na for the extracellular space and ^{3}H for total water; the results of the determinations will allow for correct evaluation of the total losses and fluid displacement.

As the interstitium appears to absorb an appreciable hydroionic volume (Dillon and Fulton, 1973), the concept of an obligatory therapeutic supply of crystalloids was taken up again and sustained as a 'novelty' by Shires (1974). Other authors however, likewise using modern methods, did not find such marked fluid deficiencies in the interstitium (Gutelius, 1973). Persistence of the dilemma: crystalloids or colloids may perhaps be solved in practice (by a mixed administration), but the question still remains and produces more mishaps in the clinic; and especially the moot question is: *when* colloids and *when* crystalloids? (*see* Section 6.2; all the authors noted in this paragraph are in the bibliography of Chapter 6).

3.5.3 IRREVERSIBILITY AND THE INTERVAL UP TO ITS ONSET

Many clinical, infraclinical and even thermodynamic criteria have been looked for in order to define the borderline between reversibility and irreversibility in shock.

The clinical criteria evidently appear to be more practical, but this reversible-irreversible borderline, governed by energy-inducing mechanisms maintaining the vital phenomenon are manifested clinically in a confused way, at least in our present understanding. The line of demarcation between refractory shock and irreversible shock is not sharp; clinically, it should never be assumed to have taken place until death supervenes. The road to our understanding of the course of events from the cellular substrate to the clinical scene is still studded by unknown barriers. Several criteria have been suggested: irreversibility begins at BP values below 20 mmHg, less than 3 g % haemoglobin, less than 10% haematocrit. Clinically irreversibility may also be considered to set in when the rate and intensity of the pulse reach a terminal decrease ('the second bradycardia'), also accompanied by a complete fall in BP. Practically, it can also be sustained that irreversibility begins when the efficiency of a correct treatment can no longer be controlled.

Other 'theories' on the causes of irreversibility have also been developed. Irreversibility has been assumed to be generated by excessive vasoconstriction due to α-stimulation of resistence in the microcirculation, which then accentuates hypovolaemia and hypercatecholaemia; paradoxically, irreversibility due to excessive β-stimulation has likewise been hypothesised [39, 44, 68, 80]. The major element of irreversibility has also been assumed to be phleboconstriction. Hypovolaemia, not corrected in time, is likewise implicated (in general in experimental shock, if volemia is corrected within less than two hours the animal is saved; volaemia recovered after six hours is no longer efficient).

The progress of acidosis is a reliable marker of the severity of shock (acidosis must be understood as reflecting metabolic impairment indirectly, since simple therapeutical alkalinisation of the internal medium is insufficient if the metabolic causes of acidosis are not corrected). Of late, many clinicians have correlated, in intensive care units, the severity of shock with intra-arterial lactate levels; arterial lactate concentrations of more than 120 mg % is believed to signify the irreversible metabolic stage of shock.

It has likewise been asserted that irreversibility is determined for any type of shock, by intestinal permeability to endotoxins [13]. Some uphold that irreversibility begins with deterioration of the sinusal mechanism maintaining the BP [21]. Metabolically, irreversibility has been explained by blocking the entry of pyruvic acid, as active acetate, in the EMK cycle [64].

The most modern theory of irreversibility is that which considers the moment of intracellular rupture of the lysosomal sacs as being decisive for cellular disorganisation and disintegration [11, 18, 33, 37, 38, 69]. Solubilisation and activation of lysosomal hydrolases in connection with hypoxia and cellular acidosis were first noted by Beufay and De Duve in 1957 [11]. The circulating levels of acid phosphatase, β-glucuronidase and cathepsins were found to be increased in various types of shock both experimentally and clinically. The onset of this 'anarchic digestion' actually marks the decease within an organ of a cellular group with maximum acidosis; the phenomenon as such (cellular hydrolase autodigestion) does not necessarily imply a fatal prognosis for the whole body, which may be saved but at the cost of irreversible cicatrices. The pancreatic and hepatic lysosomes seem to be the most sensitive to acidosis; two hours after occlusion of the superior mezenteric artery they become labile, eliminating the dangerous hydrolases [19].

Some research workers are still sceptical with regard to the importance of lysosomal hydrolases in shock. Schumer in typical experimental haemorrhagic shock did not find any 'harmful hydrolase charge' in the plasma or in the liver cells 2, 4, 6 and 24 hours after the onset of shock [64]. Although the validity of the lysosomal theory has been questioned it remains an important event that plays a role in cellular disorders.

It cannot be questioned, however, that if the onset of the shock state takes place as a rule in the extracellular fluid transport system, *the moment of decline is decided intracellularly*.

A knowledge of this moment, so useful to clinicians, shows us the necessity of furthering our understanding of the 'shock cell citadel' *(see Chapter 9)*.

3.6 CYBERNETIC ELEMENTS

3.6.1 AUTOMATIC REGULATION SYSTEM

In the human body certain regulation processes take place that are controlled by complex dynamic automatic systems, made up of a web of elementary control elements, with a cyclic character, based on the general scheme of any closed-circuit automatic system (with an inverse reaction) (Figure 3.9).

The control organ transmits signals through direct connecting pathways to *the controlled organ*; the latter will perform an action modifying its status prior to the signal received. This event is signalled back to the control organ, informing it about the new conditions (and the control organ will issue new orders in compliance with the newly created conditions.)

Actually, the human body works with automatic control systems of the category of *automatic follow up systems* (which also include the simpler systems of automatic regulation and the systems of programmed automatic regulation).

An automatic follow up system presupposes that the magnitude of references received by the control organ permanently varies (in terms of the set of sources of supplementary information in the macrosystem), as well as the status magnitude of the controlled object. Moreover, the automatic follow up biological systems, as well as most of the technical ones have in their structure non-linear functional elements, with a high rate of possible perturbations. Therefore, the biological automatic systems have a high degree of indetermination, i.e. a high informational entropy.

Control processes involve the reception, processing and transmission of *information.* Any message that communicates an event is considered as information; the latter implies the fusion of a physical support and semantics [8].

The *support* of information is the physical phenomenon associated with semantics (the same information may be obtained by cable or telephone).

The *semantics* of information is its intrinsic significance, sense, pattern, meaning.

Equivalent informations have identical semantics and different supports (the same message transmitted by axons through Hermann currents cross the synaptic space through acetylcholine quanta at a certain frequency).

The transformation of an information involves a change only in the support, whereas a mutation implies a change in semantics.

Information is transmitted on the basis of a *code* (the law of transformation of semantic symbols). Therefore the code is a rule (convention) of correspondence of certain symbols belonging to different systems of communication (Couffignal, 1966.)

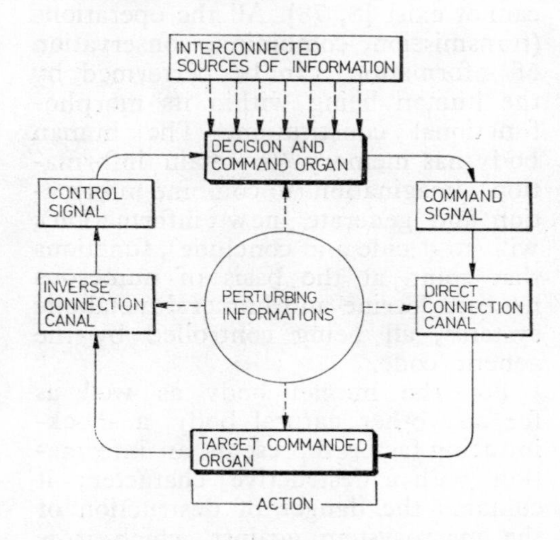

Figure 3.9. General scheme of a closed circuit automatic system.

The sense of circulation of an information is assured by a material carrier — *the signal,* that contains the semantically coded information which it is able to transmit. Thus, the signal has a strictly semantic informational aspect, that can be converted into action, and an energy aspect, of material support, that can be measured, conserved, etc. The transmission of signals necessitates an *emitter*, a *receptor* and a *spatial canal* from the start to the point of arrival of the information contained in the signals.

The emitter and receptor must be tuned to each other (to know the same code).

3.6.2 AUTOMATIC BIOLOGICAL SYSTEM

Apart from the exchange of substance and energy the body also constantly exchanges information with its environment. Information thus represents a particular type of relationship between material processes, without which it cannot exist [8, 78]. All the operations (transmission, combining, conservation of information) can be performed by the human being within its morpho-functional constitution. The human body has memory (to retain information), imagination (to combine information and generate new information), will (to decide and conclude), functions that stand at the basis of numerous neuroendocrine and humoral metabolic systems, all being controlled by the genetic code.

For the human body as well as for any other natural body, a shock-inducing factor represents an information with a destructive character; it contains the danger of destruction of the macrosystem against which it is directed [8, 52, 79].

The field of action of the human body, conceived as a cybernetic system is onto- and phylogenetically dimensioned in time and space. Its scope is twofold: to realise a biologically competent individual and to ensure its perpetuation [27]. The endogenous source of information of the human body is the genetic code. This is included, in all placental mammals, in a polynucleotide chain 'weighing' 7×10^{-9} mg DNA [56]. In man, this chain breaks down in the course of cellular division always into 46 fragments, called chromosomes.

Fourty-four of these chromosomes, called autosomes, are responsible for the somatic structure and function and two, called gonosomes, for sexualisation. That is the autosomes ensure the ontogenesis of the individual and the gonosomes the phylogenesis of the species [27].

The DNA polynucleotide chain is organised on the principle of a code contained throughout the entire cellularity of the organism [4, 29, 49, 72]. This 'tagging' is expressed from the moment of amphimixia up to final exhaustion of the genetic message; it is the 'duration of genetic extinction' [17], that takes on the same form as the universally valid mode of evolution in time of any vital phenomenon (the primordial pattern) [22]. Therefore slower mobilisation of the genetic mechanisms in the aged explains in principle some very well known facts: the torpid evolution of cancer in the aged but also the rich morbid spectrum of old age; the increased incidence of chromosomal anomalies due to the loss of genetic material, established by cytogenetic methods [9]; the severe evolution of shock in old people in whom mobilisation of the energy response is slow [70].

3.6.3 CYBERNETIC SCHEME OF SHOCK

Let us consider the human organism and its environment and assimilate their mutual physical actions to an *automatic regulation system*, which is analysed only qualitatively, as the quantitative aspects in living beings are still obscure.

In the particular case of an *aggressive action* of the environment upon the human being, a primary lesion will be produced either on the outer surface of the body (crushing, burns, etc.) or from the very beginning an interior lesion (infection, intoxication, etc.). If the primary lesion develops a general response sufficient to displace the parameters of homeostasis beyond their physiologically admitted amplitude, then a general state of shock appears.

The initial domain of the aggressive action is limited to a restricted region of the body and to a certain duration, its target being to produce a change, brutal in the case of shock, in the homeostasis of the macrosystem. The aggressive agent with the value of a shock-inducing factor, causing a lesional alteration, is also the *source of information* for a given receptor field. The initial receptors are also the *translators*, operating transformation of the support of the information but maintaining its 'shock-inducing' semantics. They may supply at most a spatial coding, that may particularise within a limited interval the state of shock according to the receptor field predominantly affected (extero-, proprio- or interoceptors) (Figure 3.10).

Moreover, throughout its entire transmission along informational channels, the 'shock-inducing information' will undergo in its numerous neuroglandular tissular relays only equivalent transformations, with a change in the energy support but bearing the same initial

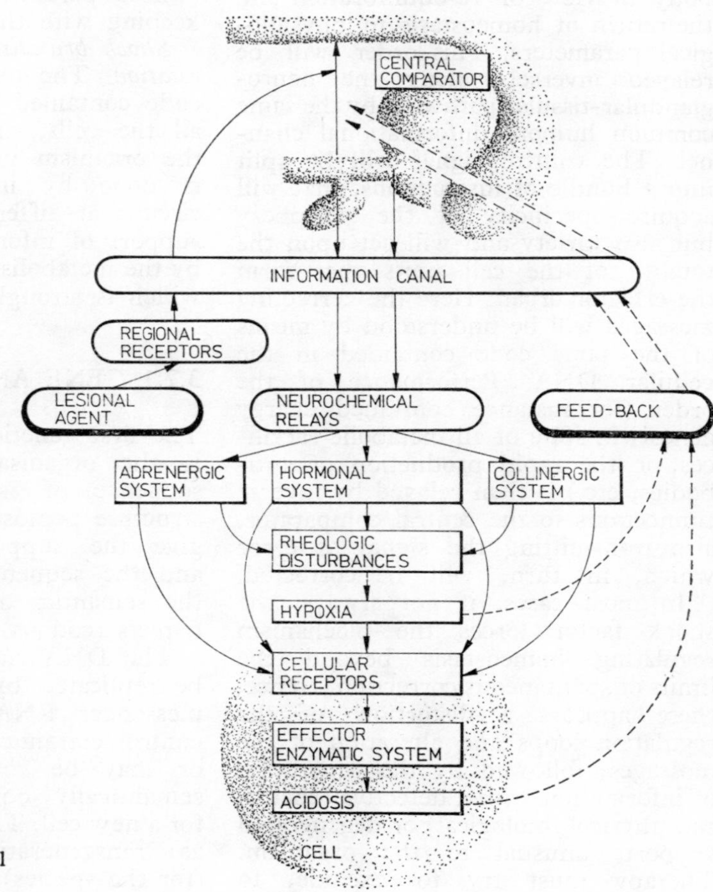

Figure 3.10. Cybernetic scheme of the pathological physiology of shock.

lesional significance for which the whole cellularity of the organism will be alerted. The equivalent transformations are assured by the presence of certain attuned devices (decoding and recoding) of the transducers contained in the relays.

The shock-inducing information thus reaches *the central comparator*, that works with spatial standards of regulation, formed phylo- and ontogenetically and inserted in the genetic code. Here decisions are taken, and all orders elaborated for correcting the lesional syndrome by mobilisation of the whole body in view of re-equilibration and the return of homeostasis to physiological parameters. This order will be relayed inversely along the neuro-glandular-tissular pathways by the same common humoral informational channel. The control signal will be split into a bundle of instructions, that will acquire specificity at the periphery but also variety and will act upon the totality of the cell units that form the effector organ. Here the correcting messages will be understood by means of the same code contained in the cellular DNA. Performance of the order (for instance continued energy genesis in spite of all metabolic sacrifices, or the rapid production of antibodies, etc.) is then relayed by inverse connections to the central comparator, also transmitting the signal of error which, in turn, will be corrected.

In most cases, if not always, the shock factor forces the mechanism regulating homeostasis beyond the limits of spontaneous correction. Hence, there appears a distortion of the regulation loops and alteration of the messages, followed by the generation of information with a defective (chemical, physical, biological) or degenerated support, unusual to the organism. Therapy must try to exclude, to diminish or to reconvert this information (*see* Chapter 9 and Section 11.5).

3.7 GENETIC SYSTEM

The findings made during the last two decades have shown that the zygote contains in his genetic code the entire programme of development of the organism; ontogenesis will develop this programme, build and consolidate it phylogenetically and express it in terms of ambient conditions (whose parameters are in general in keeping with the programme).

Shock brutalises this ecological interrelation. The total amount of genetic code contained to an equal extent in all the cells, directs the function of the organism in its entirety. Control is done by intricate informational circuits at different levels, the energy support of information being ensured by the metabolism of the macrosystem, which is strongly disturbed in shock.

3.7.1 GENE AND OPERON

The first genetic informational circuit is the organisation of DNA in a sequence of nucleotides, in whose structure pentose and phosphoric acid give the support and the energy, and the sequence of aminated bases the semantics of a code arranged in triplets (codons) [10, 54].

The DNA chain within a cell may be replicated by segments in shorter messenger RNA chains (which will control extranuclear protein synthesis) or may be totally replicated in a semantically equivalent DNA chain for a new cell. The genetic code operates as a transgeneration information channel (for the species) or may be commuted through an intracellular informational channel (for the individual) (Figure 3.11).

The genetic code may transmit unchanged information or with small errors, e. g. mutations. These may be gross (lethal), moderate or even insignificant, favourable or unfavourable to the species [25]. The rate of appearance of mutations is imposed by a concrete ecological situation, naturally supplying a permanent background of spontaneous mutations, from among which the laws of natural selection choose the 'good' from the 'bad' mutations (Ohno) [56].

Shock situations may therefore be considered for the human body as states of conflict generating an avalanche of 'forced' mutations [1]. Only organisms possessing a great plasticity will be able to mobilise a rapid functional adaptation. In defective or aged organisms adaptability is restricted.

To be efficient — and therefore to operate cybernetically — the code needs a regulation system. Its functional unit is *the gene* which is postulated to be made up of several hundred nucleotides.

The gene is considered as the basic genetic unit and has several important properties for shock situations.

Although very stable, the gene may undergo occasional, sharp mutations (it may be considered a mutational unit, a *muton*). Biochemical mutation exceeds, however, in finesse genetic mutation, since it may affect a single nucleotidic triplet or a single nucleotide of a triplet.

Shock is a typical example of a discharge of mutations, of variable extent which will readjust the cellular enzymatic and membrane systems.

The gene has a 'temporal dimension' (Gedda, 1967) in terms of the energy of its informational support *(ergon)*; the interval of time in which the informational semantics contained in its code is expressed naturally and realised by the gene bears the name of *chronon*; this term actually expresses

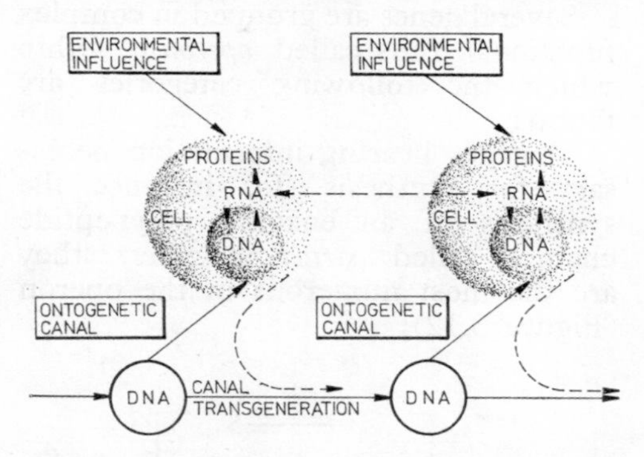

Figure 3.11. Philogenetical and ontogenetical commuting system of the pro-species and pro-individual genetic code.

the specific extinction time of a gene, from the moment it begins acting to the moment of its exhaustion [17]; shock forces the setting into action of a huge amount of genes deposited in the 'reserve' operons of the genome.

The gene is capable of endo-reduplication, a phenomenon that stands at the basis of the evolution and development of the genome [3]. In shock this is the way in which an increased 'avalanche' of the metabolic flow is realised by glycolytic pathways.

The cybernetics of intracellular expression of the genetic code uses two fundamental mechanisms: inactivation and genic compensation, by means of which accommodation and rapid adaptation of the genome can be obtained in states of shock. Transition from a state of potential information to a state of active information takes place in terms of the metabolic requirements of each cell in its functional moments; the messages of the genes are not permanently expressed intracellularly, but are commuted and disconnected according to a certain temporal sequence and the metabolic requirements of the cell.

Several genes are grouped in complex functional units called *operons*, within which the following categories are found:

(i) genes bearing information necessary for synthesis (for instance the synthesis of an enzyme polypeptide chain), called *structure genes;* they are the most numerous in the operon (Figure 3.12);

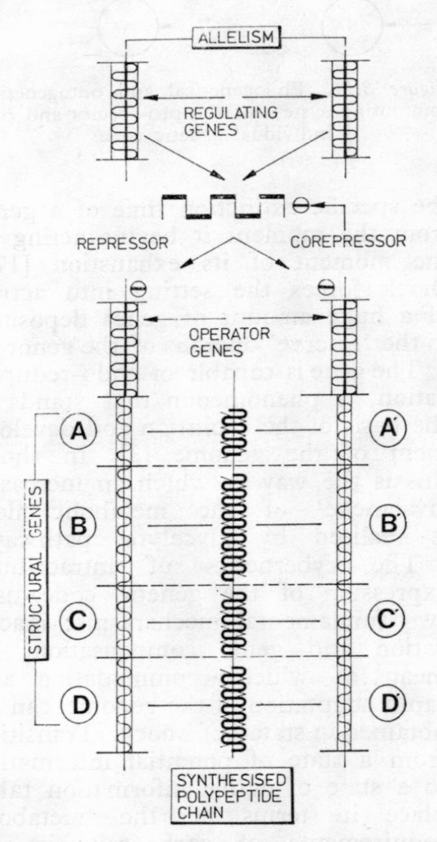

Figure 3.12. Double safety operonal regulation system.

(ii) *regulation genes*, i.e. *the regulating gene* emitting a signal *(repressor)* and *the operating gene* that receives the signal and stops the sequence of syntheses contained within the structural genes. The hypothesis of a repressor has been recently confirmed; repressors have been isolated and it was established that they are structurally proteins and not nucleic acids [29, 48].

The operon appears as an automatic system, supplied with regulating genes (masters) and structural genes (slaves), able to control its synthesis output and functioning time. Therefore control of the synthesis of an enzyme by the corresponding operon takes place only at the moment in which the substrate specific of the enzyme appears and which behaves like a corepressor, temporarily annulling the repressor and deblocking the operator gene. Utilisation of the substrate will deinhibit the repressor and leave the operon at rest.

3.7.2 GENETIC CODE AND HOMEOSTASIS

The regulation system of the operon is manifested in the intracellular space whereas the second informational circuit that uses the genetic code is realised in the extracellular space, by means of a complex neuroendocrinotissular cybernetic system formed of a suite of devices and having a vast peripheral receptor: the cells of the organism. All the links of the system are assembled genetically by the omnipresence of the same amount and quality of genetic code. Thus, the central dispatchers are in tune with a receptive periphery, able to decipher the information received and emit the retroactive signal [26, 27].

Within all the structures of the body there is a vast circulation of messages with various semantics; they are sent out, transported, received, understood and performed at great distances in the organism; the code of these messages and their action programme is retained by the genetic code operons.

The genetic system controls the entire intracellular morphofunctional placement. The metabolic elements have, on the one hand, an *intracellular circulation* and, on the other, an *extracellular circulation*, forming a vast informational network that bathes the entire cellularity, rendering it functionally harmonious. In the extracellular circuit messengers are relayed whose specificity decreases from the protein ones to the sterols, amines, monoses, electrolytes and gases.

The human organism is usually required to equilibrate a constant *physiological imbalance* with the envirlinment — characteristic of its very ofe as an open system – whereas the shocked human organism is faced with a *violent pathological imbalance* which will stretch to a maximum all the hierarchical loops of regulation. The living organism appears as a system organised in view of the efficient performance of a programme under certain environmental conditions (Milcu and Ionescu, 1968). Shock alteration of the environmental conditions will be counter to the efficiency of the programme.

The system of cellular regulation may also be perturbed directly, irradiation being a well known example. A veritable 'biochemical shock' occurs, in which the 'metabolic reflexes' of the nucleic acids are affected [12]. In any kind of shock, if the classical meaning of shock is accepted, it begins by upsetting the extracellular fluid space, and these disturbances are projected from the very beginning upon the genetic system of intracellular regulation.

Correlation between the disappearance of the chromatin corpuscle (equivalent to chromatin activation) and various stress situations (hunger, irradiation, burns, etc.) was suggested some time ago [6]. The early appearance of anti-DNA autoantibodies, before evidence of acidosis or hydrolases [7, 53] is further proof of important stereochemical alteration of the nucleoproteins.

3.8 **REGULATION LEVELS**

For a limited lesion the local reparatory processes are sufficient but in shock the entire organism has to be mobilised for re-equilibration, involving the following regulation and control systems:

(i) genetic (operon);
(ii) metabolic (enzyme-substrate relationship);
(iii) endocrinal (hormonal);
(iv) nervous (electrochemical).

3.8.1 GENETIC REGULATION

In shock, genetic regulation initially intensifies the output of all the operons but especially of those that are responsible for the control enzymes of the metabolic energy-producing, catabolising lines (*see* Figure 3.12 and Section 9.1).

3.8.2 METABOLIC REGULATION

Enzymatic regulation is based on an immediate or distal inverse connection. Figure 3.13 gives a schematic representation of the possible feedback relations between two dependent variables, A and B, and an independent variable X. In the case of a positive feedback, if X increases, A also increases which brings about an increase in B, which in turn

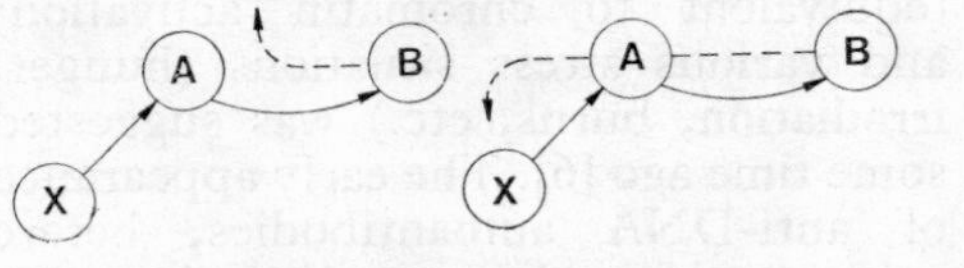

Figure 3.13. Feedback relation: *a* – positive; *b* – negative, between two dependent variables A and B and an independent variable X.

accentuates the increase of A. A non-economical *runaway* phenomenon takes place, as in thrombin autocatalysis, hyperglycolysis, acceleration of the corticosubcortical neuronal circuits, etc. In the case of a negative feedback, increase of B arrests the increase of A, therefore an economic equilibrium phenomenon takes place, as in most of the physiological phenomena of the organism.

In any chemical metabolic reaction the concentration of the end product acts by negative feedback upon the enzyme that controls the reaction. The end product brings about steric alteration of the enzyme, fixing itself upon the latter at a locus different from that of its specific activity. This spatial neoconfiguration (allosteric) blocks the enzyme until the reaction product (that has acted as inhibitor) is utilised in the system (Figure 3.14).

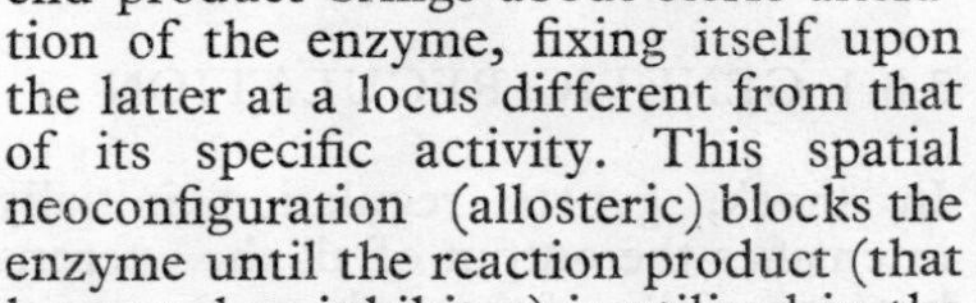

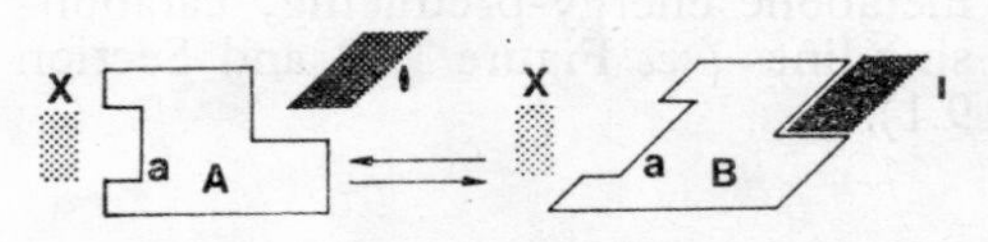

Figure 3.14. Allosteric regulation: the enzyme in its active form '*A*' may act with substrate '*X*' by its active locus, *a*. If at another locus an inhibitor '*I*' is temporarily fixed, then the latter will modify the enzyme sterically and in its new form '*B*' it will no longer be able to use its active locus, *a*.

A suggestive example is that of the role of the enzyme synthesising citrate (citratesynthase) which decides upon the turnover rate of the EMK cycle (Figure 3.15). Citratesynthase

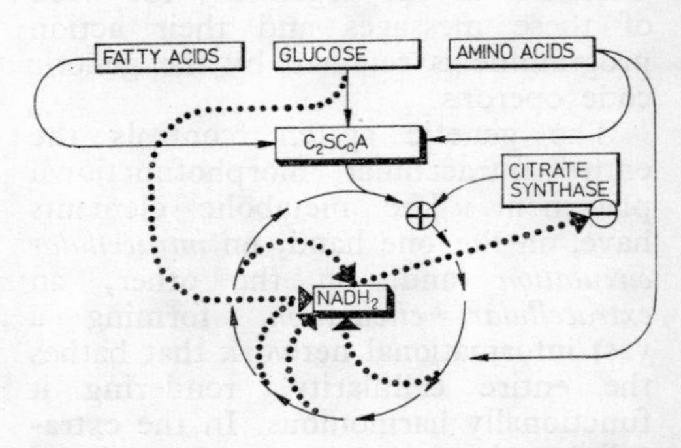

Figure 3.15. One of the fundamental metabolic arrests in shock: blocking of the Krebs cycle by excess $NADH_2$ that cannot be dehydrogenised in the oxidation chain.

condenses oxalacetate with the acetyl group, offered by active acetate (C_2SCoA) for the formation of citrate. In this reaction (and then throughout the whole of the EMK cycle) a reduction takes place, the hydrogen atoms being fixed on the reduced nicotinamide adenine dinucleotide ($NADH_2$) and leading to the oxidation-phosphorylation chain where adenosine triphosphate (ATP) is produced. $NADH_2$ appears to be the 'key' of the transformation of nutrients into energy. Studies on *E. coli* have demonstrated its regulating role on citratesynthase: excess $NADH_2$ (which is always present in shock, the oxidation chain not functioning in the absence of oxygen) inhibits the formation of citratesynthase, and excess adenosine monophosphate (AMP) (almost absent in shock) accelerates it. In the bacteria the diameter of citratesynthase increased from 100 to 120 Å in the presence of $NADH_2$, which probably 'swells' the enzyme (allosterically). The addition of AMP results in a return to the initial volume and activity of the enzyme.

3.8.3 HORMONAL REGULATION

Hormonal regulation is based upon exceptionally efficient cybernetic mechanisms, acting at different levels, as may be seen from the generalising scheme which Milcu calls the *endocrinon* [46, 47] (*see* Section 4.8).

At present, the hormonal regulation has been enriched with a new intracellular step: the action of *c*AMP.

Adenosine 3′,5′-monophosphate cyclic (*c*AMP) is a regulator of cell functions and mediates the physiological responses of cells to many hormones (Sutherland, 1960)

A hormone is transported from the cells of origin to the cells of a target organ by body fluids. The hormone acts as a *first messenger* by altering the levels of *c*AMP in the target cell.

The hormone does this by interacting with and regulating the activity of *adenylcyclase*, which is a membrane bound enzymatic system in the target cell; the enzyme catalyses the conversion of intracellular ATP to *c*AMP (Figure 3.16).

The *intracellular* level of *c*AMP acts as a *second messenger* charging the target cell to perform a specific function.

There is no specificity for *c*AMP *per se* but only for *c*AMP (whose level is connected with the hormone) combined with the responding enzymatic systems within the cell. For instance, catecholamines act *via c*AMP within the liver cell (to perform phosphorylase activation), within the fat cell (to perform lipolysis), within the heart cell (to perform inotropic increase) and within the uterus cell (to perform relaxation).

A significant decrease of *c*AMP, especially in the liver cells, was observed in the shocked animals; this event reflected a cellular deficit in ATP or a fundamental cellular defect within the membrane system.

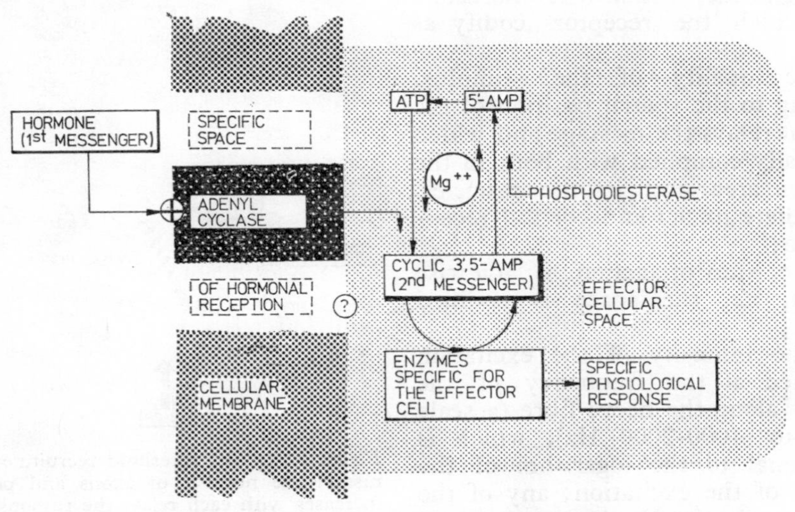

Figure 3.16. Second messenger system involving adenylcyclase.

3.8.4 NEURONAL REGULATION OF THE CAROTID GLOMUS

Neurochemical regulation is characterised by its swiftness and precision. It is based upon the transmission of information by frequency modulated currents along the neuronal structures (somatic and autonomic) and by the specificity of the colinergic and adrenergic receptors *(see* Section 4.2).

The nervous system is an aggregate of neurones melded trophically with the neuroglia and connected to one another to aid of the processing and transmission of information. Information is relayed to the neuronal transport system and processed by the receptors (tele-, extero-, proprio-, interoceptors), which are also the biological translators, transforming, by means of the 'operator', an excitation physicochemical magnitude into fundamental signals of the 'all or none' type (impulses with a modulated frequency). The signals bear to the centres an informational message that announces not only the existence of the physicochemical excitation magnitude, but also its characteristics, which the receptors codify as follows:

(i) the quality of the stimulus, according to the spatial position of the fibre that relays the message and which is sterically complied with both in the reception area and in the fascicles, intermediate nuclei and cortical reception area;

(ii) the site of action is likewise specified, i.e. the space is codified spatially;

(iii) the intensity of the excitation is codified at the frequency of the impulses along the nerve fibre (a scale of up to 1000-2000 Hz), which is proportional to the logarithm of the intensity of the excitation; any of the receptors stimulated above a certain limit — as is the case of shock agents — reaches a maximum threshold of the possible frequency modulation, beyond which the sensation of pain appears;

(iv) the duration of the action of the stimulus is codified by the interval at which the frequency modulated signals succeed one another along the afferent fibre; almost always, even when the action of the stimulus persists, there is a period of accommodation.

In shock, the lesional stimulus (for instance multiple trauma) apart from its intensity, which is inadequate from the very beginning, offers with time new local pathological coefficients that aggravate the nociceptive afference (after the initial pain appear new stimuli caused by œdema and acidosis of the destroyed tissues, or an untimely mobilisation or infection with its entire local suite).

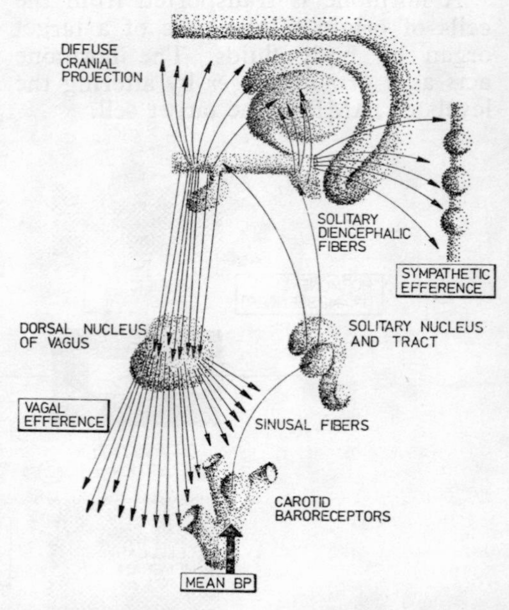

Figure 3.17. The threshold recruitment mechanisms: the number of axons and perikaryons increases with each relay, the response reaction thus becoming general.

In order to understand the role of afference a suggestive example can be given of one of the first centripetal systems of information, which is of particular importance in shock: the carotid sinus system (Figure 3.17). In the carotid glomus the mean blood pressure values are monitored, and a branch of the glossopharyngeal nerve relays the information to the medulla oblongata to the tractus solitarius nucleus from where the diffuse cranial projection starts (solitario-reticulo-thalamo-hypothalamo-cortical) [51].

The sinusal baroreceptors modulate low (high) pressures into low (high) frequencies. This is not however the only possible means of modulation; at a given pressure threshold only a certain number of sinusal fibres are stimulated and if the pressure threshold increases the number of stimulated fibres also increases. This is called *threshold recruitment*. At each overlying relay the number of stimulated fibres increases on the principle of a pyramidal coefficient, that depends upon the initial frequency (which in turn depends upon the carotid pressure value), frequency maintained along the entire cone of corticopetal fibres, that form a true spatial standard [59]. Physiological pressure values (optimal) give a progressive recruitment (optimal) based upon a pyramidal coefficient (optimal) that forms the functional tonus of an *optimal space standard*, well determined phylogenetically.

The standards are so intricate that part of the spatial circulation standard may also serve for respiration. These standards compare the frequencies above or below the optimal zone, modify the pyramidal coefficient and tend to shift the physiological space standard, which will trigger correction. Along the evolutionary phyllum disturbing influences have broadened the optimal zones of the neural spatial standards, but the humoral ones have still remained very narrow and are therefore always very sensitive and prompt in their correcting intervention.

SELECTED BIBLIOGRAPHY

1. ANFINSEN, C. B., *The Molecular Basis of Evolution*, John Wiley and Sons, New York, 1959.
2. ASKERKHANOV, R. P. L., ALIEV, O. M. G., The Characteristic of Acetylcholine Metabolism in Accordance with the Operative Trauma and the Anesthetic Used in Various Pathological Conditions, *Vestn. Khir. Grekov*, 1970, *105*, (10), 97–100.
3. BAGLIONI, C., *Molecular Evolution in Man*, in *Proceedings of the Third International Congress of Human Genetics*, Chicago, Sept. 5–10, The John Hopkins Press, Baltimore, 1966.
4. BEADLE, G. W., Biochemical genetics, *Chem. Rev.*, 1945, (37), 15–19.
5. BROOKS, D. K., The Mechanism of Shock, *Brit. J. Surg.*, 1967, *54*, Lister Centenary Number, 441–446.
6. CARATZALI, A., NACHTIGAL, S., *Nuclear Sex Appendages of Leucocytes*, in *Selected Topics on Genital Anomalies and Related Subjects*, Charles C. Thomas, Springfield, Illinois, 1971.
7. CLEUDINNEN, B. G. A Study of Antibodies to *DNA* in the Sera of Shocked Patients, *Ann. Surg.*, 1969, *170*, (6), 1021–1024.
8. COUFFIGNAL, L., *La cibernétique*, Presses Universitaires de France, Paris, 1966.
9. COURT-BROWN, W. M., *Human Population Cytogenetics*, North-Holland Research Monographs, vol. V, Amsterdam, 1967.
10. CRICK, F. H. C., The Genetic Code, *Sci. Amer.*, 1962, *207*, (1), 66–74.
11. DEDUVE, C., Le rôle des lysosomes en pathologie cellulaire, *Triangle*, 1970, *9*, (6), 200–208.

12. FILIPOVICH, I. V., KOSHCHEENKO, N. N., ROMANTZEV, E. F., The Mechanism of Biochemical Shock. I, The Correlation between the Accumulation of some Thiol Radioprotectos in Rat Tissues and Biochemical Changes Induced by Them, *Biochem. Pharmacol.*, 1970, *19*, (9), 2533–2540.
13. FINE, J., *Ciba Symposium Shock*, Springer, Berlin - Göttingen - Heidelberg - Stockholm, 1962.
14. FREY, R., AHNEFELD, F. W., Recherches sur le taux de amines catéchiques postopératoires et post-traumatiques, *Anest. Analg. Réanim.*, 1965, *22*, (3), 471–474.
15. FULLEE, R. W., Differences in the Regulation of Tyrosine Aminotransferase in Brain and Liver, *J. Neurochem.*, 1970, *17*, (4), 539–543.
16. FUXE, K., HOKFELT, T., Histochemical Fluorescence Detection of Changes in Central Monoamine Neurones Provoked by Drugs Acting on the *CNS*, *Triangle*, 1971, *10*, (3), 73–84.
17. GEDDA, L., COSA, D., BRENCI, G., Chronon and the Problem of Anticipation, *Acta Genet. Med. et Gemell.*, 1967, *XVI*, (3), 217–228.
18. GERGELY, M., HORPACSY, G., BARANKAY, T., HEZSAI L., A verzeses hypotensia fokanak és időtartamanak hatása eber, illetve altattot kutyakra, *Kisérl. orvos-tud.*, 1970, *22*, (5), 488–495.
19. GLENN, T. M., LEFER, A. M., Role of Lysosomes in the Pathogenesis of Splanchnic Ischemia Shock in Cats, *Circulat. Res.*, 1970, *27*, (5), 783–797.
20. GOLDMAN, R. H., HARRISON, D. C., The effects of Hypoxia and Hypercarbia on Myocardial Catecholamines, *J. Pharmacol. Exp.*, 1970, *174*, (2), 307–314.
21. GOOTMAN, P. M., COHEN, M. I., Efferent Splanchnic Activity and Systemic Arterial Pressure, *Amer. J. Physiol.*, 1970, *219*, (4), 897–903.
22. GRIGG, E. R.M., *Biologic Relativity*, Amaranth Books, Chicago, 1967.
23. GUENTER, C. A., HINSHAW, L. B., Hemodynamic and Respiratory Effects of Dopamine on Septic Shock in the Monkey, *Amer. J. Physiol.*, 1970, *219*, (2), 335–339.
24. HEMS, D. A., BROSNAN, J. T., Effects of Ischaemia on Content of Metabolism in Rat Liver and Kidney "in vivo", *Biochem. J.*, 1970, *120*, (1), 105–111.
25. HSIA, D. Y-Y., *Inborn Errors of Metabolism, Clinical Aspects*, Year Book Medical Publishers Inc., Chicago, 1966.
26. IONESCU, B., MAXIMILIAN, C., DUMITRACHE, C., Fenomene cibernetice în endocrinologie (Cybernetic Phenomena in Endocrinology), *Progresele științei*, 1969, *5*, (2), 59–68.
27. IONESCU, B., DUMITRACHE, C., BUCUR, A., Interpretare cibernetică a efectului informației sexualizante în disgenezia 47, XXY (Cybernetic Interpretation of the Effect of Sexualizing Information in Dysgenesis), *Studii și Cercetări de Endocrinologie*, 1971, *22*, (4), 247–255.
28. IVERSEN, L. L., SALT, P. J., Inhibition of Catecholamine Uptake by Steroids in the Isolated Rat Heart, *Brit. J. Pharmacol.*, 1970, *40*, (3), 528–533.
29. JACOB, F., MONOD, J., Genetic Regulatory Mechanisms in the Synthesis of Protein, *J. Molec. Biol.*, 1961, *3*, (3), 318–322.
30. JUDGE, R. D., ZUIDEMA, G. D., *Physical Diagnosis*, Little Brown Comp., Boston, 1963.
31. KUX, M., HOLMES, D. D., HINSHAW, L. B., MASION, W. H., Effects of Injection of Live E. coli Organisms on Dogs after Denervation of the Abdominal Viscera, *Surg.*, 1971, *69*, (3), 392–398.
32. LABORIT, H., *Réaction organique à l'agression et choc*, Masson et Cie., Paris, 1955.
33. LABORIT, H., BARON, C., WEBER, R., Essai de mise en évidence expérimentale du rôle des lysosomes dans l'évolution du choc hémorragique irréversible, *Agressol.*, 1967, *8*, (1), 23–33.
34. LABORIT, H., BARON, C., WEBER, R., Traitement du choc hémorragique expérimental dit 'irréversible'. Rôle des groupes SH et de la restauration des réserves intracorticulaires en catécholamines, I. Vue d'ensemble, *Agressol.*, 1969, *10*, 189–198.
35. LABORIT, H., BARON, C., WEBER, B., Traitement du choc hémorragique expérimental dit 'irréversible'. Rôle des groupes SH et de la restauration des réserves intracorticulaires en catécholamines, II. Etude expérimentale, *Agressol.*, 1969, *10*, (3), 199–204.
36. LABORIT, H., BARON, C., WEBER, B., Traitement du choc hémorragique expérimental dit 'irréversible'. Rôle des groupes SH et de la restauration des réserves intracorticulaires en catécholamines. III. Etude stéréotaxique des stimulations cérébrales, *Agressol.*, 1969, *10*, (3), 205–215.
37. LABORIT, H., LONDON, A., Rôle des lysosomes dans les lésions de l'infarctus expérimental du myocarde et dans l'action protectrice des inhibiteurs de la monoamine oxydase, *Agressol.*, 1969, *10*, (4), 303–308.
38. LARCAN, A., STOLTZ, J. F., *Microcirculation et Hémorhéologie*, Masson, Paris, 1970.
39. LILLEHEI, R. C., LONGERBEAM, J. K., BLOCH, J. H., MANAX, W. G., The Nature of Irreversible Shock, *Ann. Surg.*, 1964, *160*, (4), 682–710.

40. LUNDSGAARD-HANSEN, P., SCHILT, W.; HEITMAN, L., OROZ, M.; BUCHLER, A., LEMEUNIER, A., Influence of the Agonal Period on the Postmortem Metabolic State of the Heart: A Problem in Cardiac Preservation, *Ann. Surg.*, 1971, *174*, (5), 744–754.
41. MCLAUGHLIN, J. S., Physiologic Consideration of Hipoxemia in Shock and Trauma, *Ann. Surg.*, 1971, *173*, (5), 667–679.
42. MALIK, K. U., Potentiation by Anticholinesterases of the Response of Rat Mesenteric Arteries to Sympathetic Postganglionic Nerve Stimulation, *Circulat. Res.*, 1970, *27*, (5), 647–655.
43. MANDACHE, F., *Fiziopatologia circulației și imunitatea în șoc* (Pathologic Physiology of the Circulation and Immunity in Shock), Editura Academiei Republicii Socialiste România, București, 1966.
44. MAREȘ, E., ȘUTEU, I., ATANASESCU, S., *Chirurgia de campanie* (Camp Surgery), Editura Militară, București, 1956.
45. MAZOR, A., ROGEL, S., Cardiac Aspects of Shock, *Ann. Surg.*, 1973, *178*, (2), 128–133.
46. MILCU, ȘT. M., IONESCU, B., Mecanisme cibernetice în sistemul endocrin (Cybernetic Mechanism in the Endocrine System), *St. cerc. Endocrinol.*, 1967, *18*, (5), 385–392.
47. MILCU, ȘT. M., IONESCU, B., The 'Escort Phenomenon' in the Endocrine Pathology, *Revue Roumaine d'Endocrinologie*, 1968, *5*, (2), 147–150.
48. MONOD, J., *Le hasard et la nécessité*, Edit· du Seuil, Paris, 1970.
49. MORARU, I., ANTOHI, ST., *Introducere în genetica moleculară*, (Introduction to Molecular Genetics), Editura Medicală, București, 1966.
50. MOORE, F. D., Terminal Mechanisms in Human Injury, *Amer. J. Surg.*, 1965, *110*, (2), 317.
51. NEISTADT, A., SCHWARTZ, S. I., Effects of Electrical Stimulation of the Carotid Sinus Nerve in Reversal of Experimentally Induced Hypertension, *Surg.*, 1967, *61*, (6), 923–931.
52. NICOLAU, E., *Introducere în cibernetică*, (Introduction to Cybernetics), Editura Tehnică, București, 1963.
53. NICK, W. V., DNA Antibodies in Primate Sera during Haemorrhagic and Endotoxic Shock, *Amer. J. Surg.*, 1968, *115*, (6), 769–773.
54. NIRENBERG, M. W., The genetic code, *Sci. Amer.*, 1963, *208*, (1), 80–87.
55. NISHINO, K., IRIKURA, T.; TAKAYANAGI, I., Mode of Action of 5-Hydroxytryptamine on Isolated Rat Vas Deferents, *Nature*, 1970, *228*, (5271), 564–565.
56. OHNO, S., *Sex Chromosomes and Sex-linked Genes*, Springer-Verlag, Berlin-Heidelberg-New York, 1967.
57. PRIANO, L. L., WILSON, R. D., TRABER, D. L., Cardiorespiratory Alterations in Unanesthezized Dogs due to Gram-negative Bacterial Endotoxin, *Amer., J. Physiol.*, 1971, *220*, (3), 705–711.
58. REHDER, E., ENQUIST, I. F., Species Differences in Response to Cortisone in Wounded Animals, *Arch. Surg.*, 1967, *94*, (1), 74–78.
59. REPCIUC, E., CONSTANTINESCU, P., FILOTTI, I., *Modelajul homeostaziei* (The Modelling of Homeostasis). Communication at the Symposium on Cybernetics, Brașov, 1964.
60. ROBERTS, C. N., KONIOVIC, J., Differences in the Chronotropic and Inotropic Inhibitors and Certain Blocking Agents, *J. Pharmacol. Exp. Ther.*, 1969, *169*, (1), 109–119.
61. RUTHERFORD, R. B., WEST, R. L., HARDAWAY, R. M., Coagulation Changes During Experimental Hemorrhagic Shock, *Ann. Surg.*, 1966, *164*, (2), 203–212.
62. SĂHLEANU, V., Geneza și semnificația hiperplaziilor glandelor endocrine (The Genesis and Significance of Hyperplasisa of the Endocrine Glands). *St. cerc. Endocrinol.*, 1966, *17* (4), 309.
63. SCHLID, H. O., Récepteurs et classification des médicaments, *Triangle*, 1969, *9*, (4), 132–137.
64. SCHUMER, W., NYHUS, L. M., Corticosteroid Effect and Biochemical Parameters of Human Oligemic Shock, *Arch. Surg.*, 1970, *100*, (4), 405–408.
65. SELYE, H., *The Physiology and Pathology of Exposure to Stress*, Acta Inc. Medical Publisher, Montreal, 1950.
66. SENDELBECK, L. R., YATES, F. E., Adrenal Cortical and Medullary Hormones in Recovery of Tissues from Local Injury, *Amer. J. Physiol.*, 1970, *219*, (4), 845–853.
67. SHOEMAKER, W. C., Pathophysiologic Mecanism in Shock and their Therapeutic Implication, *Amer. J. Surg.*, 1965, *110*, (3), 337–341.
68. SHOEMAKER, W. C., *Shock: Chemistry, Physiology and Therapy*, Charles C. Thomas Edit., Springfield-Illinois, 1967.
69. SUTHERLAND, E. W., RALL, T. W., The Relationship of Adenosine 3-5-Phosphate and Phosphorylase to the Action of Catecholamines and Other Hormones, *Pharmacol. Rev.*, 1960, *12*, (3), 265–272.
70. ȘUTEU, I., CAFRIȚĂ, A., BUCUR, I. A., Concepții actuale în fiziopatologia și tratamentul șocului (Present Concepts in the Pathophysiology and Treatment of Shock),

Revista Sanitară Militară, 1972, *LXXV*, (3), 267–276.
71. Teodorescu-Exarcu, I., *Patologie biochimică*, Editura Medicală, Bucureşti, 1974.
72. Thal, A. P., Sardesai, V. M., Shock and the Circulating Polypeptides, *Amer. J. Surg.*, 1965, *110*, (3), 308–312.
73. Turpin, R., Lejeune, J., *Les chromosomes humains*, Gauthier-Villars, Paris, 1965.
74. Ţurai, I., Patogenia şi tratamentul şocului (The Pathogeny and Treatment of Shock), *Chirurgia*, 1960, *IX*, (4), 205–216.
75. Vernon, S., Shock. A Design for Survival, *J. Abdom. Surg.*, 1970, *12*, (1), 1 – 2.
76. Walters, G., Circulatory Patterns in Clinical Shock, *Postgrad. Med. J.*, 1969, *45*, (526), 497–502.
77. Weil, M. H., Schubin, H., *Diagnosis and Tratment of Shock*, Williams and Wilkins Co., Baltimore, Maryland, 1967.
78. Westphal, U., *Steroid-Protein Interactions*, Springer-Verlag, Heidelberg-New York, 1971.
79. Wiener, N., Schade, P. I., *Progress in Biocybernetics*, Elsevier Publ. Comp., Amsterdam-London-New York, 1965.
80. Wiggers, C. J., *The Physiology of Shock*, Commonwealth Fund., New York, 1950.

4

THE NEURONAL AND ENDOCRINE SYSTEM

The search for truth is in one way hard and in another easy. For it is evident that no one can master it fully, nor miss it wholly. But each adds a little to our knowledge of nature. And from all the facts assembled, there arises a certain grandeur.

ARISTOTLE

4.1 Complex neuronal systems
4.2 Cholinergic and adrenergic space
4.2.1 Acetylcholine
4.2.2 Catecholamines
4.3 The particular properties of neuronal perfusion
4.4 The particular properties of neuronal metabolism
4.5 The criteria of cerebral death
4.6 The oscillations of neuronal integration
4.7 Neurohormonal structures
4.7.1 Hypothalamus
4.7.2 The pituitary body
4.8 The hormonal systems in shock
4.8.1 Hormonal mechanisms; cyclic AMP
4.8.2 The cybernetics of the endocrine structures
4.8.3 Hypothalamo-hypophyseal hormones
4.8.4 Thyroid and pancreatic hormones
4.8.5 Steroid hormones
4.8.6 Renin and erythropoietin
4.9 Complex hormonal regulation
4.9.1 Blood rheodynamics
4.9.2 Volaemia and electrolytaemia; portal osmostat
4.9.3 Metabolism regulation
4.10 Acute hormonal dysfunctions
4.10.1 States of shock due to hormona deficit
4.10.2 States of shock due to hormonal excess
4.11 Regulation of the immune system
4.11.1 General mechanisms
4.11.2 Thymus

4.1 COMPLEX NEURONAL SYSTEMS

The central nervous system (CNS) is organised functionally as an ensemble of neuronal circuits of the feedback type, arranged vertically, horizontally and circularly (Figure 4.1) [37, 65, 73]. In shock, the following systems are at first directly involved in informational integration of the CNS regulation:

(1) *The lemniscus system* containing the ascending specific sensitivosensory pathways (spinothalamic band, Reil's band, trigeminal band and lateral band).

In shock, pain stimuli take the usual mixed, specific and then unspecific, afferent pathways. By collaterals of the specific pathways *(via* the spinothalamic and Reil bands) the pain stimuli reach the ARAS which they activate. But, the specific afference of pain that rapidly reaches the cortex only gives rise to a brief reflex activity, since only 20% of the centripetal lemniscal fibres reach the cortex through the specific transthalamic pathways; the rest 80% of their ascending informational flow ends in the reticular substance at different levels [95]. A slower, unspecific, but far more complex afference, develops reaching the cortex diffusely and generating alert reactions to the initial painful stimuli. The reticular substance acts as an important station representing the core of all the recurrent connexions of the loops in Figure 4.1. The reticular area extends ventrally to the hypothalamohypophyseal region, upwards to the diencephalolimbic area and downward to the spinal tonigenic mechanisms (Figure 4.2).

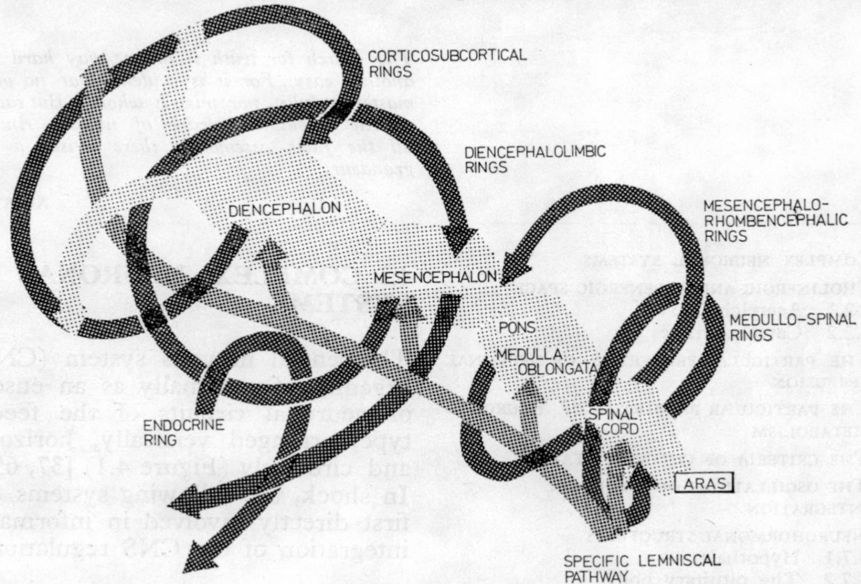

Figure 4.1. The principal neuronal integration systems. The endocrine loop is directly involved with the limbic loops.

(2) *The ascending reticular activating system* (ARAS) formed of a neuronal network situated between the bulbar decussation of the pyramids and unspecific thalamoseptal nuclei; it is continued caudally with the medullary reticular substance and cranially with the diffuse thalamic system, that insures cortical projection. It contains more than 100 nuclei of various size and a web of connexions including the specific neurofasciculate systems (Eccles, 1957; Jenkins, 1969; Ungar, 1963).

(3) *The limbic system*, with phylogenetically old structures, in which the history of the struggle with the environment is well preserved, has an essential function in the organisation and elaboration of the neuroendocrine responses. It is therefore natural that the limbic area, in its modern context, should play a primordial role in the mechanisms of adaptation and control of the general behaviour of the organism in shock.

The phylogenetic evolution of the behaviour to aggression expresses the ratio of the neocortex to the limbic system: even if the numerator has developed reaching well known proportions in man the denominator has also substantially progressed so that the ratio has been maintained at constant values [54, 69]. It is very clear today that in human beings the limbic system suffered not an involutional, but an evolutional process (Karli) [69]. The cortical segments, the limbic system

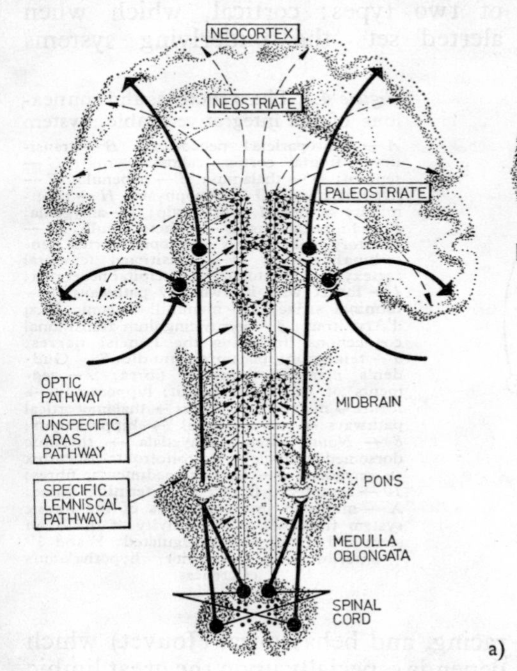

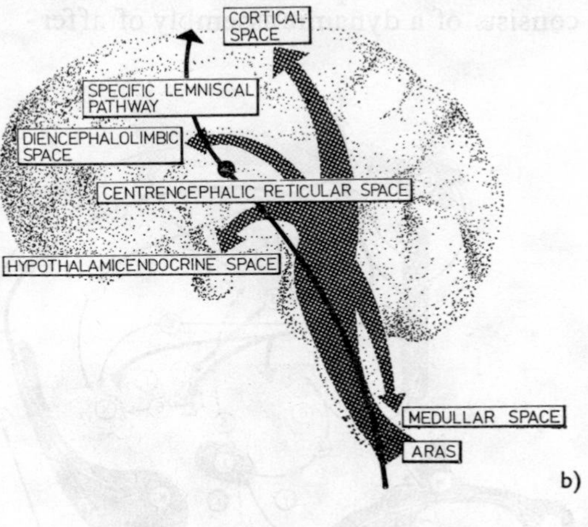

Figure 4.2:

a — vertical-transverse projection of the specific pathways (lemniscus) and their connection with the reticular space. In the inset the centrencephalic reticular system; *b* — the efferences of the centrencephalic reticular space towards the unspecific thalamus and hypothalamus, limbic areas, cortical and medullar spaces.

(that is 'awake' even when the neocortex is 'asleep') and its multiple rings (that anchor it to all the diencephalomesencephalic structures) permit its interference at any crucial moment in the life of the individual: in wakefulness (sensory afference and general activity), reactivity, memory, emotions, in feeding and sexual behaviour and, obviously in its response to aggression, etc. The misleading, restricted interpretation of a 'limbic olfactory apparatus' (corresponding at present only to the rhinencephalon) can no longer be admitted. Moreover, anosmic animals have a well developed olfactory system. The interpretation of limbic structures as vestigial remains must also be given up; suggestive in this connection is the fornix, the broad efferent pathway of the hippocampus, which in man numbers four times as many fibres as in the monkey, thus forming a more important morphological pathway than the pyramidal bundle [95]. The notion of limbic system has undergone during the last decade 'a quantitative and qualitative mutation along anatomic lines and physiologic competence' [69]. Of particular importance are the connexions of the limbic integration system (Figure 4.3).

Motivation of the global behaviour of the organism confronted by shock agents has a double character:

(i) a *quantitative* character that represents the intensity of the response to the shock stimulus;
(ii) a *qualitative* character, indicating the direction of the response and tending as a rule to save the macrosystem, but in an alternating, contradictory way.

The state of 'post-shock motivation' consists of a dynamic assembly of afferences and reafferences recorded by the receptors, then followed by confrontation with the memorised information (in this case overwhelmed) and the elaboration of a multivectorial behaviour (disorganised).

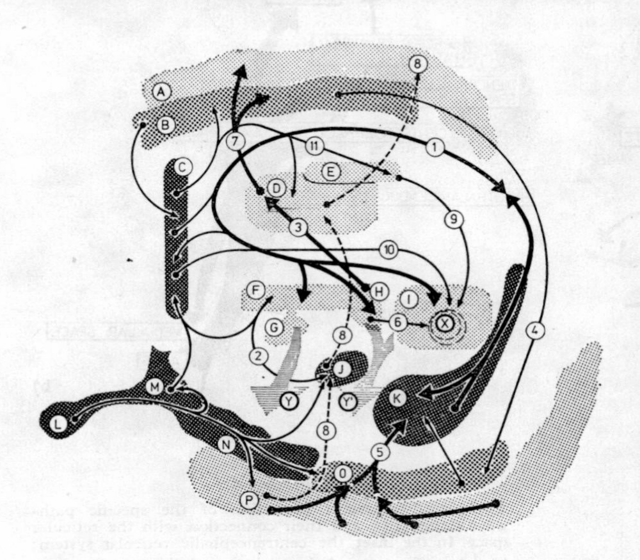

Figure 4.3. The structures and connexions of the integrative limbic system *A* — frontoparietal neocortex; *B* — transitional frontal cortex (callosal gyrus); *C* — septum; *D* — thalamus; *E* — habenula; *F* — hypothalamus; *G* — hypophysis; *H* — mammillary bodies; *I* — midbrain; *J* —amygdala; *K* — hippocampus; *L* — olfactory bulb; *M* — olfactory tubercle; *N* — allopaleocortex (entorhinal gyrus); *O* — transitional temporal cortex; *P* — frontotemporooccipital neocortex; *1* — fornix with its two-way pathways; *2* — terminal striae; *3* — mammillothalamic Vicq d'Azyr tract; *4* — gyrus cingulum transitional connections (including the Lancisi nerves; *5* — temporohippocampus bundle; *6* — Gudden's mammillotegmental fibres; *7* — segments of the *Papez* circuit: hippocampus → fornix → mammillary bodies → thalamocortical pathways → callosal gyrus → hippocampus; *8* — *Nauta* circuit: amygdala → thalamic dorsomedian nucleus → orbitofrontal isocortex → amygdala; *9* — habenulopeduncular fibres; *10* — septotectal fibres; *11* — terminal striae; *X* — midbrain projection area of the limbic system from where the activity of the brain stem and spinal cord is regulated; *Y* and *Y'* — anterior and posterior hypothalamus efferences.

In the CNS there are neuronal systems with a specific sensitivity for clearly defined variations of the inner medium, and unspecific activator neuronal systems that perform a particular modulation (passing for instance from a 'metabolic level' to an 'impulse level' of the action potentials). These unspecific systems act especially by limbic circuits and in the centrencephalic space. Centripetal shock messages are subject to a modulator control, initially in the reticular formation of the mesencephalon. Here, fluctuations of the state of cortical vigilance has repercussions upon the afferent pathways which they stimulate in excess by a positive feedback runaway phenomenon.

The state of vigilance is constantly of two types: cortical, which when alerted sets the underlying systems racing, and behavioural (Jouvet) which depends especially upon the great limbic system, i.e. upon the main integration space of the complex humoral response to shock. This response, elaborated and modulated in the limbic rings is then directed by the ergotropic hypothalamus along the sympathomedulloadrenergic pathways and by the anterotuberal hypothalamus along the hypophysocorticosterol pathways (Figure 4.3).

4.2 CHOLINERGIC AND ADRENERGIC SPACE

In the first phases of shock the neuronal ensemble of the CNS, well supplied with oxygen and glucose, organises the *reaction syndrome*, a complex of 'neuro-endocrine instructions' that will produce a rapid increase in the titre of the neurochemical and ergotropic hormonal messages [111].

In the CNS space most synapses use acetylcholine quanta as messengers; the peripheral cholinergic space that contains all the neuroneuronal synapses and *parasympathetic neuroeffector synapses* is very vast (Figure 4.4). Catecholamines are used only in the *orthosympathetic neuroeffector synapses.* The spatial distribution of the two categories of neuromessengers has remarkable effects: although topographically the adrenergic sector is more restricted than the cholinergic one it has approximately the same peripheral effectiveness surface; the adrenergic sector having proximal acetylcholinic relays is vassal to the cholinergic system (Euler) [49].

The adrenergic effector area is the actual domain of the endogenous struggle against shock. Along the entire adrenergic neuroeffector relay system hypertonia develops, implying an increase in the concentration of catecholamines released into the neurotissular space and also of acetylcholine concentration in the clefts of the neuroneuronal relay synapses.

The peripheral effect of the neurochemical messengers is mediated by the receptors. The tissular receptor is the stereochemical area situated on the surface or in the intimacy of a cell to which the messenger attaches itself. The messenger communicates his order which is read and performed according to a code.

For catecholamines there are at least two types of receptors in the tissues: α and β [3, 4]; reception of the catecholamines by the α receptors (norepinephrine is taken up especially by these receptors) triggers the effects shown in table 4.1. If the order is received by receptors (which take up especially epinephrine) the effects are the opposite or altogether different. In general α receptors are associated

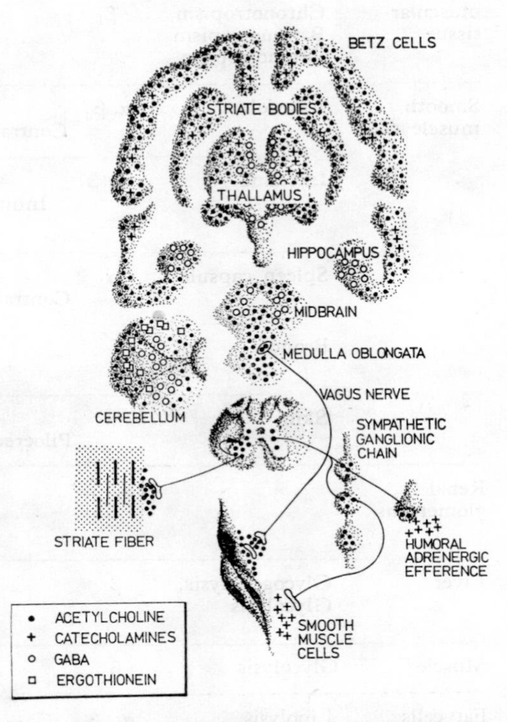

Figure 4.4. The (central and peripheral) cholinergic space is far greater than the adrenergic space. The effector surface is however equilibrated.

with excitation functions (except the intestine) and the β receptors with conservative, inhibitory functions (except the heart). The organism discharges into the systemic circulation a variable amount of catecholamines and the effects obtained differ according to the different equipment of the tissues with receptors suitable for the functions of the respective tissues *(see* Section 5.5.1).

Table 4.1 THE PHYSIOLOGICAL EFFECTS OF CATECHOLAMINES

		Type of receptor	Epinephrine		Norepinephrine	
			α effect	β effect	α effect	β effect
Cardiac muscular tissue	Inotropism	β_1		+++		±
	Chronotropism			++		+
	Bathmotropism			+		+
	Dromotropism			+		−
Smooth muscle cell	Arteriole	α, β_2	± Contractility	+++ Relaxation	+++ Contractility	± Relaxation
	Intestine	α, β	+ Inhibition of motility	++	++ Inhibition of motility	+
	Spleen capsule	α	+ Contractility		+ Contractility	
	Bronchi	β		+++ Relaxation		+ Relaxation
	Skin	α	+ Piloerection		+ Piloerection	
Renal glomerulus		α, β		+ Increased diuresis	++ Decreased diuresis	+ Increased diuresis
Liver	Glycogenolysis, Glycolysis	β		+++		+
Muscle	Glycolysis	β		+		+
Fat cells	Lipolysis	α, β	++	+	++	+

The new drugs with an elective action (excitatory or inhibitory) that act upon the α and β receptors respectively favour channeling of the regional effects of catecholamines. Determination of the dissociation constants of certain β blocking adrenergic agents in various organs of the rabbit suggest the existence of several kinds of visceral β receptors [22]. The sensitivity of α receptors likewise differs: the precapillary sphincters respond to norepinephrine dilutions of 1:10 000 000 and the metaarteriole to a 1:4 000 000 dilution. The receptors appear to be recruited according to their sensitivity in terms of catecholamine concentrations. In general the β receptors are sensitive to catecholamines before the α receptors and thus small doses of endogenous catecholamines solve the conservative resting requirements. When the concentration of catecholamines released increases, the α receptors are also alerted with increased effector consequences. This classical pharmacodynamic paradox of the small catecholamine doses is well known. Between the activity of the α receptors and the amount of endogenous catecholamines liberated there is an inverse relationship: blocking of the α receptors stimulates the release of endogenous catecholamines in continuation like a local feedback. 'Modulation' of vasoconstrictor response of the microcirculation by liberation into the active tissues of some humoral factors with β-like qualities, improves tissular perfusion [11, 16, 18, 36, 72, 97].

4.2.1 ACETYLCHOLINE

The cholinergic space and cholinergic receptors are well known [52, 114]. In shock there is a metabolically directed increase in the synthesis of acetylated choline (*see* Section 8.3.6). A decrease in the concentration of acetylcholinesterase has likewise been reported, which might initially accelerate the ganglionic relay mechanisms of the postlesional reaction syndrome.

4.2.2 CATECHOLAMINES

The enzymatic battery for the synthesis and breakdown of catecholamines is to be found in the brain, in ectopic chromaffin cells or organised cells in the adrenomedulla and tissular adrenoeffector junctions. Between these zones there exists a balance and studies with ^{14}C-thyrosine have demonstrated that removal of the adrenal medulla was followed by a four-fold increase in the synthesis of catecholamines in the tissues [96].

The polyhedrical chromaffin cells, distributed in the form of anastomotic cords in the adrenal medulla respond to acetylcholine stimulation of the splanchnic nerves (medullary segment T_{10}-L_1). Insulin hypoglycemia stimulates especially the epinephrine producing cells, and reserpine and α-lytics those cells synthetising norepinephrine. The catecholamine content in the glandular tissue is of 2-4 mg/g tissue, of which 20-30% is norepinephrine.

The biosynthesis of catecholamines starts from phenylalanine and is controlled by thyrosine hydroxylation and transamination. Catecholamines are rapidly inactivated (only 4% appear as such in the urine), their catabolism being a combination of orthomethylation (80%) and oxidative deamination (20%) (Figure 4.5). The half-life of catecholamines is very short: in the course of 2-3 complete systemic circuits they are taken up by the visceral receptors or inactivated [10, 18, 106, 112].

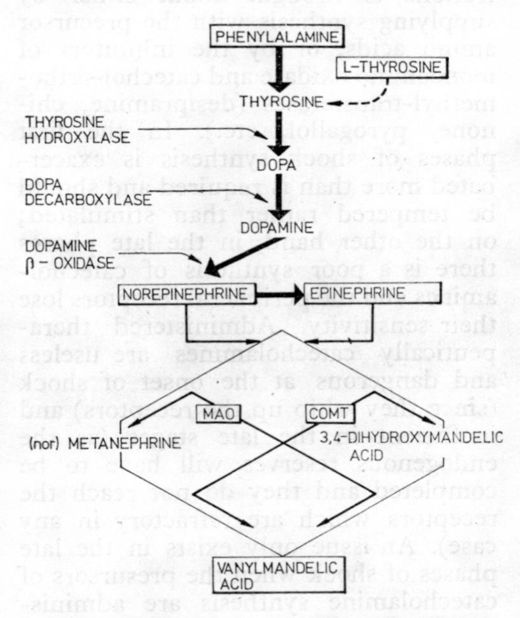

Figure 4.5. The synthesis and degradation of catecholamines. Orthomethylation is ensured by catecholorthomethyltransferase (COMT) and oxidative deamination by monoaminooxidase (MAO).

In shock, catecholamines are liberated either owing to neurogenic stimuli and the action of other biogenic amines, or by the direct effect of hypovolemia, hypoxia, lactacidemia and sterols, or the administration of α-lytic and hypotensive drugs (*see* Section 3.4.3) [31, 32].

The biogenic amines have common mechanisms of action: histamine and serotonin in small doses have catecholamine-like effects, especially when catechol-orthomethyl-transferase is inhibited; the effect is obtained either directly or indirectly, stimulating the synthesis of catecholamines in the adrenal medulla [23, 45, 49] (*see* Section 4.10.2). *The adrenergic system is the most rapid implement of the reaction syndrome in shock* (Figure 4.6).

An increase in catecholamine concentrations is brought about either by supplying synthesis with the precursor amino acids, or by the inhibitors of monoamino oxidase and catechol-orthomethyl-transferases (desipramine, chinone, pyrogallol, etc.). In the first phases of shock synthesis is exacerbated more than is required and should be tempered rather than stimulated; on the other hand, in the late phases there is a poor synthesis of catecholamines and the peripheral receptors lose their sensitivity. Administered therapeutically catecholamines are useless and dangerous at the onset of shock (since they whip up the receptors) and inefficient in the late stages (as the endogenous reserves will have to be completed and they do not reach the receptors which are refractory in any case). An issue only exists in the late phases of shock when the presursors of catecholamine synthesis are administered, wihch will stimulate endogenous synthesis, capable of refilling the depleted stores and reawakening peripheral reception [77].

The administration of tyrosine may build up again the α-constrictor effect of catecholamines on the receptors of the microcirculation. This effect is membranous, catecholamines favouring the release of calcium from the reticulum pouches at the disposal of myosin-ATP-ase. Tyrosine is assumed to serve as a metabolic precursor of the EMK cycle, elective for the neural structures and managing a 'target' resuscitation of the latter [78, 79]. Moreover, it replenishes the cholinergic stock of acetylcholine and catecholamine granular deposits (which will also activate the phosphorylase and lipase energy-yielding systems), a desirable effect in the phase of shock considered refractory or even irreversible; at the same time, the vasoconstrictor effect, that always generates shock, will be buffered with α-lytics (*see* Section 8.3.6).

The fact that the nerve impulse releases acetylcholine and catecholamine quanta in the peripheral synaptic spaces has been precisely established, whereas the chemical messenger of the synapses that integrates the state of shock in the CNS is far less understood [19, 21, 27, 93, 122, 126]. The following substances have been assumed to play this role:

Acetylcholine which has a broad area of activity also in the CNS (Figure 4.4).

Catecholamines, *serotonin* and the *P. substance*, differ structurally but have in general the same topographic distribution in the CNS as synaptic chemical mediators [55, 58, 103, 107, 115, 124].

These three substances might only have a modulating role in synaptic transmission in the CNS, since they are also to be found in large amounts in the neuroglia [109]. Catecholamine concentrations increase in the CSF during operations on the CNS [73]. Corticoids lower the intracerebral concentrations of serotonin and catecholamines probably by inhibiton of triptophan metabolism [15].

γ-*aminobutyric* acid (GABA) has a depressive effect on neuronal structures [2, 8, 56, 104, 105]; it is produced by

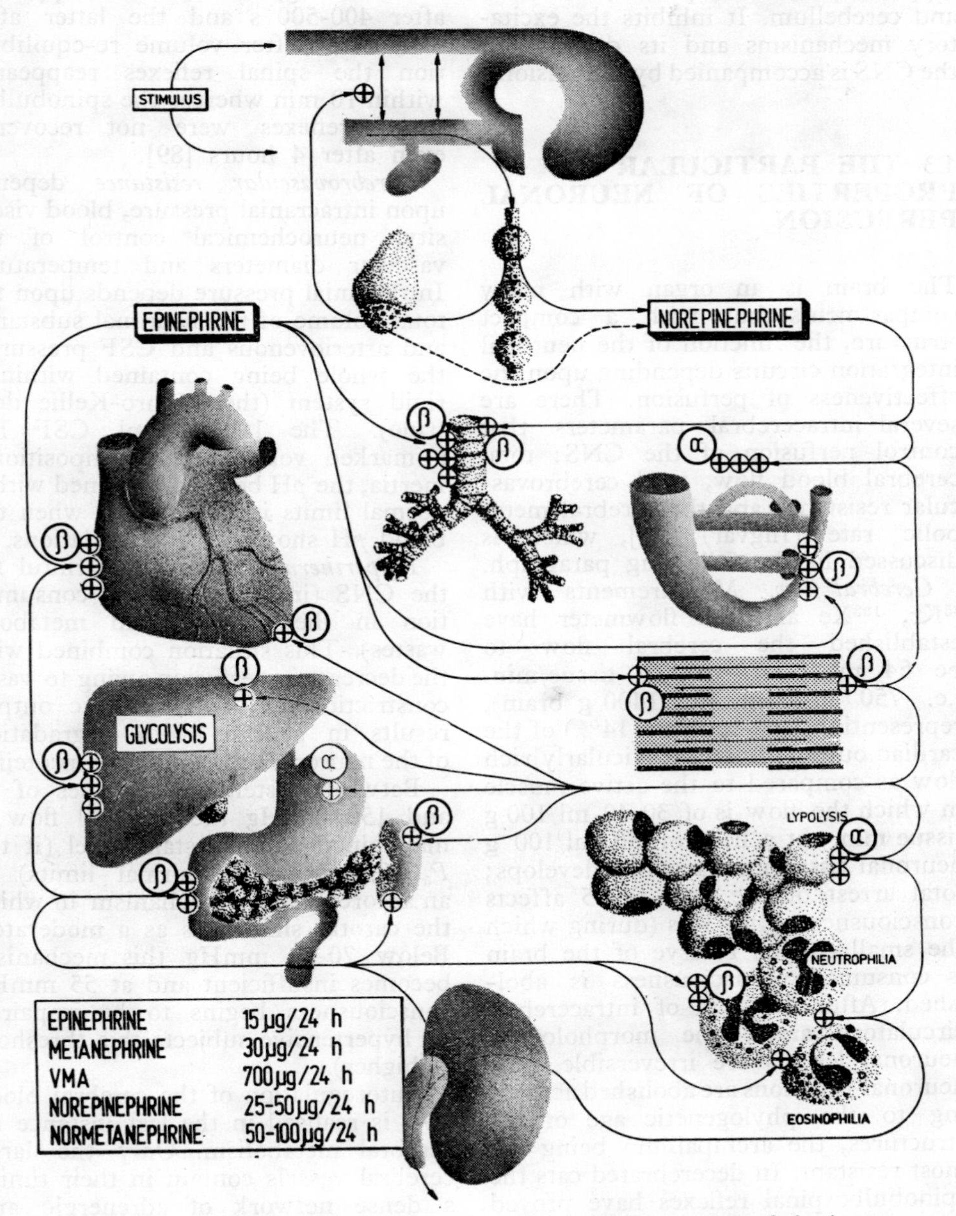

Figure 4.6. The principal physiological actions of endogenous catecholamines.

glutamic acid decarboxylation and is to be found in high concentrations in the hippocampus, diencephalon, midbrain and cerebellum. It inhibits the excitatory mechanisms and its decrease in the CNS is accompanied by convulsions.

4.3 THE PARTICULAR PROPERTIES OF NEURONAL PERFUSION

The brain is an organ with many compartments but with a compact structure, the function of the neuronal integration circuits depending upon the effectiveness of perfusion. There are several intracerebral parameters that control perfusion of the CNS: total cerebral blood flow, total cerebrovascular resistance and the cerebral metabolic rate (Ingvar) [63], which is discussed in the following paragraph.

Cerebral flow. Measurements with ^{85}Kr, ^{133}Xe and the flowmeter have established the cerebral flow to be 54 ml (+ 12)/100 g tissue/min, i.e. 750 ml/min. (for 1400 g brain), representing one-seventh (14%) of the cardiac output. It is a particularly rich flow as compared to the active muscle in which the flow is of 30-40 ml/100 g tissue/min. At a flow of 30 ml/100 g neuronal tissue/min syncope develops; total arrest of the flow for 5 affects consciousness; after 10 s (during which the small oxygen reserve of the brain is consumed) consciousness is abolished. After 4-5 min of intracerebral circulatory arrest the morphological neuronal lesions are irreversible. The neuronal functions are abolished according to the phylogenetic age of the structures, the archipallium being the most resistant. In decerebrated cats the spinobulbospinal reflexes have proved more vulnerable to hypoxia than the segmentary and medullary polysynaptic reflexes. On maintaining the BP at 30-40 mmHg, the former disappeared after 400-500 s and the latter after 550-800 s. After volume re-equilibration the spinal reflexes reappeared within 10 min whereas the spinobulbospinal reflexes were not recovered even after 4 hours [89].

Cerebrovascular resistance depends upon intracranial pressure, blood viscosity, neurochemical control of the vascular diameters and temperature. Intracranial pressure depends upon the total volume of the neuronal substance and arteriovenous and CSF pressures, the whole being contained within a rigid system (the Monro-Kellie doctrine). The 100-200 ml CSF has a marked volume and compositional inertia, the *p*H being maintained within normal limits in shock even when the blood *p*H shows marked variations.

Hyperthermia is always harmful for the CNS (increases oxygen consumption in the neurons and metabolic wastes). This situation combined with the decrease in perfusion owing to vasoconstriction and/or low cardiac output results in acidosis with degradation of the neuronal enzymes and proteins.

Between systemic BP values of 60 and 150 mmHg the cerebral flow is maintained at a constant level (if the $P_a(CO_2)$ is within normal limits) by an autoregulation mechanism in which the carotid sinus acts as a moderator. Below 70-60 mmHg this mechanism becomes insufficient and at 55 mmHg consciousness begins to be impaired (in hypertensive subjects this threshold is higher).

Autoregulation of the cerebral blood flow is realised in the last instance by cerebral metabolism. Only the large cerebral vessels contain in their tunica a dense network of adrenergic and cholinergic filaments; in the arterioles and parenchymatous microcirculation there is no such network, thus protecting the neuronal structures against the fluctuations of the neurogenic vascular tonus [63]. In these areas chemical regulation appears to be of first importance. The two essential metabolic waste products H^+ and carbon dioxide increase the local flow and are thus removed.

The inhalation of 5 up to 7% carbon dioxide increases the cerebral flow by 75% (by vasodilatation). A decrease in P_aCO_2 (by hyperventilation) lowers the cerebral flow, the lactate production of the nervous structures tending to correct gas alkalosis [94]. Carbon dioxide concentration is independent of P_aO_2 or pH; decrease in the pH is a much weaker cerebral vasodilator. Hypoxia may however also increase the cerebral flow; at a normal P_aO_2 the shifts in P_aCO_2 are predominant; if the latter become exaggerated the low P_aO_2 values take over the control. P_aCO_2 over 20% causes convulsions [10]. Of late, studies have been carried out on electrolytic regulation of the cerebral flow; however the role of sodium and calcium concentrations have not yet been fully established [34, 74].

4.4 THE PARTICULAR PROPERTIES OF NEURONAL METABOLISM

In order to understand the effects of cerebral hypoxia in states of shock it is useful to recall the particular properties of neuronal energy metabolism.

Of the substances that can be 'burnt' glucose alone is permitted to pass the selection of the lipoprotein membrane of the blood-brain barrier. Hence, bicarbonate, administered systemically does not correct cerebral acidosis [84]. The barrier is however readily crossed by gases, oxygen and carbon dioxide entering but also rapidly leaving the CNS. *Consequently, the nervous structures have no available reserves of carbohydrates and oxygen.* The brain depends directly upon fluid and oxygen supply, i.e. the circulation of fluids and the metabolism of the entire organism.

The oxidation of glucose with the production of carbon dioxide and water is the fundamental reaction that maintains neuronal vitality, more than in any other tissue. The sequence of catabolic reactions for obtaining energy from carbohydrates (Figure 4.7) may be separated into three stages:

(i) glycolysis (EM cycle) that takes place in eleven steps, each directed by a distinct enzyme;

(ii) the citric acid cycle (Krebs). The ten steps of this 'metabolic heart' release hydrogen ions and carbon dioxide gas;

(iii) the third stage is a chain of six dehydrogenation steps in which the electrons of the hydrogen pairs are transported and given up to oxygen, essential for the normal termination of all the reactions of the three stages, its absence arresting the steps of the entire system.

In the absence of oxygen glucose remains at the lactate stage, producing only 10% of the total energy (*see* Chapter 8).

Although the brain represents 2.2% of the human body weight it utilises 65% of the total glucose under physical resting conditions.

The regional cerebral blood flow, established isotopically, shows that only 20% is distributed to the white matter; the neurons (the grey matter) that represent 20% of the cerebral cellular population utilise 80% of the amount of glucose and oxygen offered by perfusion of the CNS.

The other neuronal constituents (fatty acids, lipids, amino acids) are likewise permanently metabolised and synthetised in the course of their metabolic turnover [62]. Their catabolic cycles are included in the Krebs cycle. This does not however contribute, under normal conditions, to the production of energy. There are, notwithstanding, two pathological situations, common to shock, in which neuronal catabolism is altered:

(i) *Hypoxia*, acute, due to any cause, produces in man within 4-8 min irreversible destruction of the neuronal functions (at 37 °C). (Actually, 4-5 min anoxia after cerebral flow pooling induces cortical morphological alterations). Arrest of the circulation is therefore more severe than perfusion with hypoxic blood [68]. Other organs are far more resistant to the absence of oxygen: the skeletal muscles may function in hypoxia for a long interval by hypoxic blood [68]. Other organs are far more resistant to the absence of oxygen: the skeletal muscles may function in hypoxia for a long interval by

rapid adaptation to anaerobic metabolism. In the neonate, the neurons in anaerobiosis may function somewhat better, succindehydrogenase being less active than lactic dehydrogenase [60].

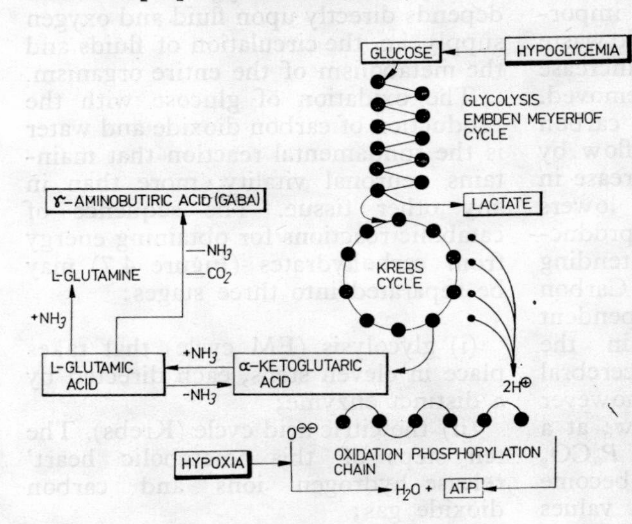

Figure 4.7. The particularities of neuronal metabolism. Hypoglycemia stops the entire energy yielding system, and hypoxia arrests retrogradely the stages of oxidative-phosphorylation and of the Krebs cycle, up to lactate that accumulates in glycolysis steps. As a rule when the Krebs cycle functions, α-ketoglutaric acid can fix two ammonia molecules forming glutamine. When it can fix a single ammonia molecule there is an accumulation of GABA and when the Krebs cycle also slows down ammonia accumulates.

(ii) *Hypoglycemia* is deleterious to the CNS. In 90 min the small cerebral reserves of carbohydrates are utilised and the neurons start to catabolise their own lipids and proteins; as they have no lipidoprotein reserves in the energy pathways the neurons discharge structural lipoproteins and nucleic acids resulting from self-destruction; in one hour's hypoglycemia they consume 50% of their own neuronal structures (as in insulin coma). Therefore, severe hypoxia and hypoglycemia do not arrest the production of neuronal energy but destroy the intraneuronal structures, consuming them. In other words energy is produced through a 'pathological' pathway.

The group of *glutamic*, *aspartic* and *γ-aminobutyric acids* is important for neuronal metabolism, representing 75% of the free amino acids contained in the neuron. The source of these acids is however also glucose: for instance glutamic acid is produced by the Krebs cycle from α-ketoglutaric acid. Glutamic acid has then two possibilities:

(i) the formation of glutamine, that takes up the neuronal excess of ammonium. In shock, ammonium genesis increases, reflecting the degradation of neuronal structural proteins; ammonia will withdraw α-ketoglutaric acid from the Krebs cycle, to form glutamine (Weil-Malherbe cycle). This auto-deintoxication process stops however with the end of the Krebs cycle;

(ii) the formation of γ-aminobutyric acid by decarboxylation specific of the CNS neurons; γ-aminobutyric acid may then serve as a metabolic substrate, as a synaptic messenger with an inhibitory effect or as a metabolic feedback component. Its intracerebral accumulation in shock is, now at any rate, considered undesirable [8, 56, 105].

4.5 THE CRITERIA OF CEREBRAL DEATH

Owing to the preferential irrigation and to metabolic peculiarities that do not permit deficiences in the oxygen and glucose supply, the nervous substance is extremely sensitive. The rapidity and irreversibility of morphological neuronal lesions, as well as their incompatibility with survival — at least organised survival — of the entire macrosystem, makes the *CNS the decisive organ in establishing the actual death of the individual.*

Respiratory and cardiac stop cannot be implicitly identified with death.

The moment of biological death (real death) is considered equal to organic death of the brain.

In practice, several possibilities are encountered:

(i) the circulation may be taken up again 4-5 min after cardiac stop;

(ii) the brain has been damaged directly, but by resuscitation and active treatment respiration and circulation can be maintained for days and weeks; in such cases although the individual has ceased to exist as an organism with an organised brain, physically he can be considered as an object 'that can still be killed' [5].

(iii) the circulation is taken up again after cardiac stop but the period of reanimation of the CNS has been exceeded (exceeded coma).

In the latter two situations the question arises of how long cardiorespiratory resuscitation must be continued.

Proof of *cerebral death* must be found.

Death that ends a chronic disease, with deterioration of the vital organs does not necessitate proof of cerebral death.

In the two situations mentioned when a certain doubt exists the patient has to be followed up for another 12 hours, without recording the cerebral angiogram but with continued isoelectric EEG. The first or repeated cardiac stop is considered the moment of death. *If cerebral angiography is repeated then 30 minutes observation and resuscitation are sufficient* [5, 53, 65, 117].

In the clinic the situation may be either very simple or extremely complex. The interval of safety, at least for the EEG, which has been discussed by various national and international societies and medical specialities, is 3 to 48 hours. Today the general trend in practice is to consider a double carotid angiogram as the most reliable criterion of cerebral death.

As a rule, cerebral death is established by the surgeon together with the anaesthetist and neurologist and involves three large groups of criteria.

(i) The clinical criteria of cerebral death described by Mollaret and Goulon in 1959.

(ii) The EEG criteria. The tracing must be isoelectric not only flat since the latter may still reveal a low voltage activity. Although not a sign of organic death an isoelectric EEG tracing represents a sign of functional death lending support to a diagnosis of cerebral death.

In practice it is considered that bilateral mydriasis, respiratory arrest, areflexia and an inexisting EEG tracing are sufficient to establish cerebral death [53].

(iii) The angiographic criteria offer elements of certainty. The death of an organ can only be asserted when it has been proven that its perfusion has stopped for an interval longer than the specific survival time of the respective organ. For the brain it is sufficient to prove objectively by seriography that there is a total circulatory arrest, both carotid and vertebral, of 4-5 min at normal temperature and 5-20 min in hypothermia. It may be stated that the circulation ceases whenever the intracranial pressure exceeds the systemic BP. In these conditions the BP/intracranial pressure (ICP) ratio is equal to or smaller than 1:

$$\frac{BP}{ICP} \leqslant 1$$

Hence, the classical angiographic method establishes death of the brain by the lack of visualisation of the blood vessels, which may however be 'crushed' by raised intracranial pressure. At present, a pressure cerebral angiographic method has been described which, in carefully selected cases, can record the cerebral angiogram even in the presence of raised intracerebral pressure [5].

4.6 THE OSCILLATIONS OF NEURONAL INTEGRATION

The term 'neurogenic' shock is inadequate and confusing [117]. It reflects situations in which the rapid development of a peripheral fluid pool rapidly lowers the venous return, the cardiac output and hence the cerebral flow, affecting consciousness. The unspecific EEG alterations are the same for hypoxia as for hypoglycemia, etc. This condition may also develop following upon the pain of a myocardial infarct and here, too, one might include the states of shock after relaxation of the venous areas following the evacuation of fluids (ascites, pleurisy, etc.) and all 'reflex shocks' (pleural, peritoneal, etc.), formerly called 'precardiac' by some authors.

In all these situations a decrease occurs in the cerebral blood flow or, more precisely, reduction of the neuronal metabolic efficiency. But these events do not imply an actual state of shock. Therefore in none of the situations mentioned does the term of 'neurogenic shock' actually correspond to the pathophysiological reality. A decrease in the cerebral flow appears in the different stages of any kind of shock and represents a stage in cerebral involvement ('shock brain') and not the complex phenomenon of shock itself. When a deficit in cerebral perfusion occurs suddenly and is transient, the clinical terms of syncope, fainting, etc. seem more adequate although not sufficiently clear (*see* Section 1.4.3). It is advisable to describe these clinical states more precisely in the aetiologic diagnosis.

In shock the central neuronal structures follow the same oscillating dynamics which they impose upon the entire macrosystem *(see* Section 3.3.3).

Before alerting the cortex, the ARAS is in turn activated by specific neuronal collateral pathways, by humoral chains, as well as by the reverberating circuits from the cerebellum and cerebral cortex (Figure 4.8).

The nervous substrate in the organisation of the state of alertness is the centrencephalic system, that includes the mesencephalic reticular substance, ergotropic hypothalamus, specific thalamic nucleic and limbic circuits. The bundle of shock information initially skirts the neocortex (physiological occlusion maintained by the ARAS offers this temporary protection) and causes brutal blocking of the centrencephalic formations. Clinically, this moment corresponds to a state of immediate depression that may end by cardiorespiratory stop if the shock is severe and the balance of the organism unstable.

If the organism resists, the bundle of shock information borne by the ARAS will break through the physiological block and 'hit' the cortex where it will induce desynchronisation of the alert type. Metabolically, this will be reflected by an excessive utilisation of oxygen, glucose, creatinphosphate and ATP, and peripherally by a discharge of acetylcholine, histamine and catecholamines. Clinically, a conscious sensation of pain develops, then gradually fades.

The cortex alerts the subcortical centers (limbic and striated bodies, reverberating circuits); in turn they will realert the cortex by positive feedback, by runaway. The ARAS fully participates in this phenomenon of deteriorating avalanche (reticulo-cortico-reticular), which may already lead in this phase to fatigue of the CNS following irreversible structural neuronal lesions (Mori and Simeone, 1970) [89].

To these may be added the humoral reverberations that flood the ARAS in

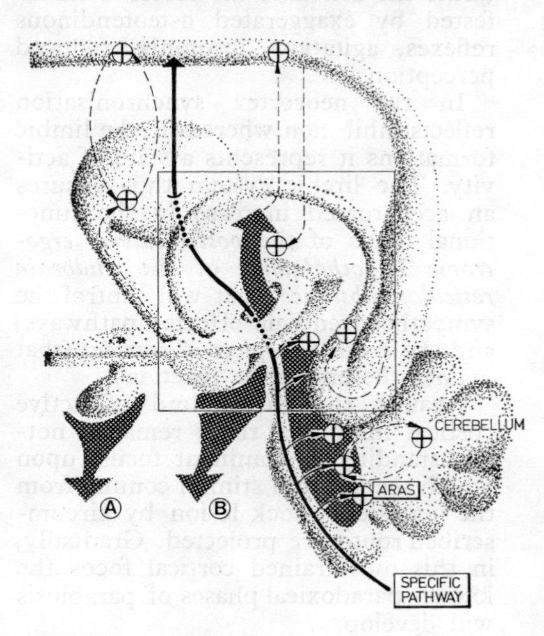

Figure 4.8. Neuronal dynamics in the first phases of shock. Note the positive feedback (runaway), the centrencephalic space (in the inset) and the two large efferences of the *reaction syndrome* in shock: endocrine, (*A*) and adrenergic neuronal (*B*).

high concentrations. Of the substances that alert the ARAS by humoral pathways (acetylcholine, epinephrine, ADH, ACTH, histamine, hypoglycemia, serotonin, low *p*H, increase in $P_a(CO_2)$ and decrease in $P_a(O_2)$) the most certain in shock is ADH *(see* Section 4.8.3).

When the neuronal and neuroglial metabolic endeavour is able to establish an *active internal protection inhibition*, the cortex will be protected. The subcortical centres remain in their previous state of alert, which they will communicate to the periphery by orthosympathetic and endocrine channels. Clinically, 'escape' of the subcortical centres from under the action of the cortex is manifested by exaggerated osteotendinous reflexes, agitation, insomnia, reduced perception, etc.

In the neocortex synchronisation reflects inhibition whereas in the limbic formations it represents a state of activity. The limbic system thus ensures an accentuated increase in the functional tonus of the *posterolateral ergotropic hypothalamus*, of the *midbrain reticular substance* (that will control the sympathoadrenergic effector pathways) and the *tuberohypophyseal segment* (that primes the endocrine reaction).

Against the background of active cortical inhibition there remains, notwithstanding, a dominant focus, upon which the harmful stimuli coming from the persistent shock lesion by circumscribed route, are projected. Gradually, in this overstrained cortical focus the known paradoxical phases of parabiosis will develop.

Finally, after the shock-inducing action has been removed and the organism is able to equilibrate itself haemodynamically and metabolically, the *internal active inhibition* will manage to bring the cortex back to physiological conditions.

If this does not take place and the cortical metabolic reserves are exhausted a phase of passive *external inhibition* develops, followed by *anarchy* and the appearance of neuronal structural lesions (death of the brain). The limit of reversibility of corticosubcortical disorders seems to appear at the moment in which active internal inhibition is replaced by passive external inhibition [65, 101]. When this moment is not exceeded there follows a long period of recovery of the functional lesions of the CNS. It is known that no close relationship exists between the state of consciousness and the severity of the neuronal lesions. Therefore behavioural alterations (known as reaction neurosis or disorganisation of the ego) may often accompany injuries, reflecting possible remaining intracellular lesions and disturbing the neuronal cybernetic circuits [57, 113].

4.7 NEUROHORMONAL STRUCTURES

Claude Bernard assumed many years ago that a control system must exist in the composition of the fluid bathing the cells of pluricellular organisms. The stability of this internal medium fluid offers to the organism a certain degree of freedom with regard to the cosmic environment.

The regulation system of the internal medium is based upon nervous and endocrine mechanisms, that only apparently differ from one another. Between these two mechanisms there are certain fine degrees. For instance the secretion of norepinephrine by the terminations of the postganglionic fibres and the control of smooth muscle cell contraction in the arteriolar wall are predominantly neuronal mechanisms. Splanchnic stimulation of the adrenal medulla and systemic liberation of catecholamines (the true hormones of autonomic life) is carried out by a mixed neuroendocrine mechanism, the same as hypothalamohypophyseal neurocrinia. On the other hand release into the blood of aldosterone and parathormone are purely endocrine mechanisms.

Table 4.2. gives the most important neurohormonal structures, their substances released systemically and their area of effectiveness.

Table 4.2 NEUROHORMONAL STRUCTURES

Neuronal structure	Hormonal substance	Symbol	Tropic hypophyseal hormones (intercourse hormone)	Efficiency hormone (peripherial hormone)	Receptor-effector space
Adrenal medulla	epinephrine (adrenalin) norepinephrine (noradrenalin)				most cells (α and β receptors)
Neurohypophysis	vasopressin (antidiuretic hormone)	ADH			distal nephron
	oxitocin	OTC			uterus, mammary gland
Epiphysis	melatonin, serotonin, etc.				?
Subcommissural organ	Reissner fibres (controlling hydroelectrolytemia)	SCO			?
	corticotropin realising hormone	CRH	ACTH	cortisol	all cells
	tyreotropin realising hormone	TRH	TSH	T_3-T_4	all cells
	growth (somatotropin) realising hormone	GRH	GH		all cells
Hypothalamic median eminence	folliculostimulant realising hormone	FRH	FSH	estrogens	follicular maturation
	luteostimulant realising hormone	LRH	LH ICSH	progesterone androgens	ovary, uterus, vagina Leydig cells
	prolactin inhibitor hormone	PIH	LTH		mammary gland
	melanotrop inhibitor hormone	MIH	MSH		melanocytes

4.7.1 HYPOTHALAMUS

The intricacy of the neurohormonal mechanisms occurring in the hypothalamus is highly suggestive. This nervous tissue, with a particular functional irrigation, anchored at the base of the diencephalon and implanted in the midst of the limbic rings, represents a crossroads of homeostasis. Through the reticular substance, it receives information regarding the entire neuronal afference and is the only part of the brain that can confront this information with the humoral messages culled from its vascular network, that perforates its very cells. In its connection with the thalamus it encounters a second time the nervous afferent pathways, whose messages it adopts to its own humoral messages (Dorfman, Williams) [42, 123].

The hypothalamus deciphers the semantics of chemical messages whose support it changes by recoding them in impulses whose meaning it maintains however. Its particular neuroglandular structure allows it to unify endocrinoautonomic life with nervous life, 'internal' life with the life of 'relations'.

Its efferences are extremely complex. The hypothalamus obeys the orders of the neocortex and is alone able to unify them with the older structures of the paleo- and archi cortex to which it is very closely knit; moreover, it is the common, mixed autonomic and hormonal efferent pathway. Hence the hypothalamus may be considered the dispatcher of all the parameters of homeostasis that insure the life of the organism. The control of the hypothalamus is not 'conscious', it is a dispatcher able to regulate only what is programmed in the genetic code of its neuronal structures without having the mobility of adaptation of the neocortex to unexpected environmental situations. The chemical standards of the hypothalamus are narrow and rigid, its adaptation slow and with great sacrifices. The natural history of the species stands witness to the selection imposed at times too brutally by the environment.

Although during the last few years the interest of neurophysiologists has almost exclusively been focussed on this 'narrow plateau of the third ventricle' there are still many unknown facts. Almost all the vital 'neuroendocrinostats' (glycostat, oxystat, thermostat, etc.) are billeted here. All physiological metabolic oscillations, as well as the pathological ones in states of shock, are rhythmised by the hypothalamus [17, 18]. The thermostat appears to be regulated by the intrahypothalamic change in Na^+ and Ca^{2+} concentrations [92], and the centres of hunger and satiety are equipped with specific neurochemical receptors: α for the hunger centre and β for the satiety centre [40, 81].

All the hypothalamic control devices are intensely disturbed in shock. The hypothalamic neurons undergo the same biochemical impairment specific of shock — electron depletion with the accumulation of uncompensated oxidising charges — as the entire cellularity of the organism [8]. Hence, a regeneration of the hypothalamic circuits can be obtained by the administration of reducing compounds [41].

The hypothalamus records the rhythmicity of the functioning waves of the endocrine systems: circadian rhythm for the adrenocortex, weekly rhythm for progesterone, monthly rhythm for the ovary, pluricircadian rhythm for sexoids in general, etc. Shock goes counter to all these rhythms but especially to the circadian rhythm of the rapid hormonal lines (ADH, ACTH-cortisol, GH, TSH-T_3T_4, etc.). Evidence of the transformation connection implemented by the hypothalamus between the nervous and hormonal impulses was furnished in man by injecting a cholinergic substance that was rapidly received at the central pole of the diencephalon and was immediately followed by increase in GH and cortisol concentrations in the plasma [123].

Apart from the hypothalamus, the overlying centres also play a part in neuroendocrine relations. Thus:

(i) α-catecholamine stimulation of the centrencephalic neuronal structures in the baboon brought about an increase in GH concentration and a decrease in insulin, whereas β-stimulation lowered GH levels and increased insulin and FFA levels [102, 120].

(ii) stimulation of the hippocampus inhibited the release of gonadotropins favouring the increase of ACTH [116];

(iii) stimulation of the anterior hypothalamus, preoptical area septum and complex of the medioarcuate eminence increased LH (ICSH) after 30 min and FSH after 3 hours [121];

(iv) in shock ACTH is produced by the antehypophysis in response to the stimuli that reach the hypothalamus through the reticular substance or limbic projections; epinephrine also stimulates ACTH, alerting the entire posterior hypothalamus [48];

(v) progesterone depresses the posterior hypothalamus, limbic system and neocortex, whereas the administration of testosterone, cortisol and DOCA stimulates unspecific activation of the hypothalamus and limbic system [55, 65, 69, 123].

4.7.2 THE PITUITARY BODY

The anterior hypophysis bears the name of 'dispatcher gland' in the endocrine hierarchy. It regulates the glands 'with a peripheral effectiveness' which it subordinates by its tropic hormones. The present interpretation assumes however subordination of the hypophysis to the hypothalamus within the endocrine system. The hypophysis is likewise considered a target organ, controlled by the polypeptides sent by the hypothalamus through the peri-infundibular portal network or directly through the pituitary stalk (Figure 4.9).

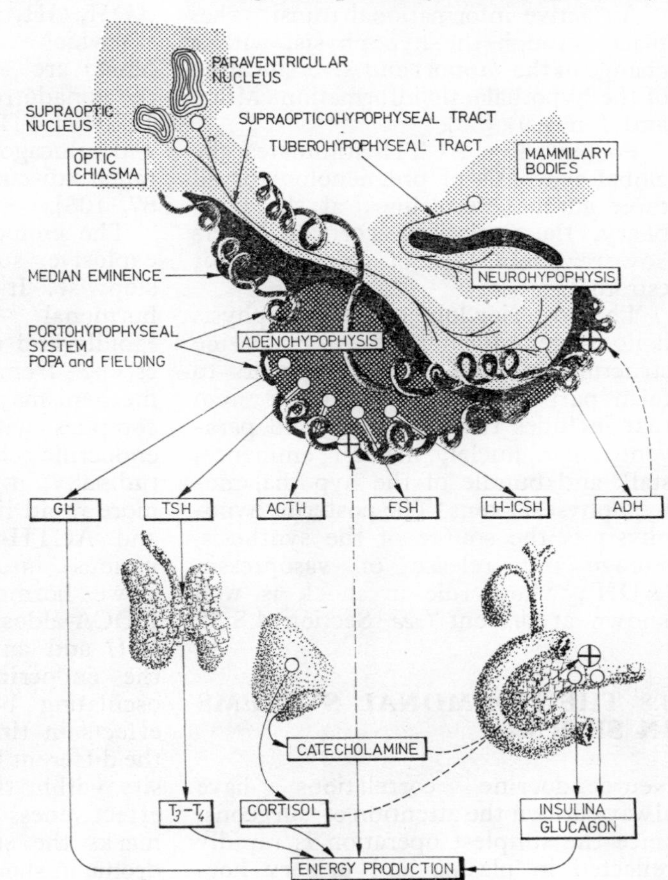

Figure 4.9. Adenoneurohypophyseal structures and tropic hormones in the energogenesis process; in this complex cellular event catecholamines and the pancreatic hormones also participate.

Isolation and the knowledge gained concerning the chemical structure of the glycoproteins representing almost all the substances by means of which the hypothalamus regulates the whole endocrine system have changed their name from releasing factors to releasing hormones. Surrounding the stalk, a vascular shunt conveys the hypothalamic neuromessages and induces in the hypophysis the synthesis and liberation of tropic hormones for the relay endocrine glands, which elaborate hormones with an increased specificity received by the effector tissular structures.

An active informational transit takes place through the hypophysis, with a change of the support and diversification of the hypothalamic information (Milcu and Ionescu) [86].

For instance ACTH stimulates the global synthesis of pregnenolon for all three adrenal sterol lines; at the periphery, the action of cortisol differs however from that of aldosterone or estradiol.

The posterior lobe of the hypophysis is no longer considered as an endocrine structure in itself. It only appears to form part of a neurosecretory system that includes the supraoptic and paraventricular nuclei, median eminence, stalk and bundle of the hypothalamo-hypophyseal axons. The posterior hypophysis is the source of the synthesis, storage and release of vasopressin (ADH), whose role in shock is well known at present (*see* Section 4.8.3).

4.8 THE HORMONAL SYSTEMS IN SHOCK

Neuroendocrine correlations have always drawn the attention of surgeons, since the simplest operation is rapidly reflected in plasma and urinary hormonal concentrations. At 4-6 hours after a major abdominal operation plasma cortisol increases more than 30 times as well as GH, aldosterone and iodine [67, 129]. The anxiety and immobility of the patient, anaesthetics, stimulation of the nerve filaments and receptors by the surgical manœuvres, the liberation of aminated tissular factors (stimulating catecholamines directly), volaemia and ion changes are only some of the factors triggering the postoperative endocrine response.

In all states of shock the global endocrine response contains increased concentrations of ACTH, cortisol, ADH, GH, prolactin and aldosterone to which catecholamines and angiotensin are added. On the other hand the gonadotropic hormones (FSH, LH-ICSH, LTH) are inhibited and insulin and glucagon present a varied and much discussed behaviour [46, 61, 87, 108].

The endocrine reaction in shock is explosive, successive, oscillating, rising stepwise. It is explosive because the hormonal concentrations increase rapidly and reach huge amounts that escape from under the autocontrol mechanisms; it is stepwise because it complies with the hierarchy of the endocrine relays that are alerted longitudinally; it is successive because the more rapid lines (catecholamines ADH and ACTH-cortisol) are followed at various intervals by the somewhat slower hormonal lines (TSH-T_3T_4 and DOCA-aldosterone) or still slower ones (GH and anabolising sterols). Finally the endocrine reaction in shock is oscillating because of its peripheral effects in time, between the tonus of the different hormonal lines, or in intensity within the same hormonal line of effectiveness. Hormonal discharge marks the start of the reaction syndrome in shock and controls the totality of the circulatory and metabolic events dealt with in Chapters 5, 6, 7, 8 and 9.

4.8.1 HORMONAL MECHANISMS; CYCLIC AMP

The 'classical' hormones are structures produced by the endocrine glandular tissue, but in the broader acceptance of free circulating messenger—the carrier of information — catecholamines can also be considered hormones. Being closely linked to the effector transmission of the adrenergic neurotissular synapses they have been discussed in Section 4.2.2.

The hormones of mammals form part of three large biochemical classes: steroids, polypeptides and amines. The endocrine cell releases the hormone directly along the common informational humoral channel; the hormone is captured either by a large peripheral cellular population (GH, T_3 and T_4, insulin, catecholamines) or a very small number of cells (oxytocin, vasopressin, gonadotrops). This suggests the existence of cellular receptors for hormones, which can be 'read' at low physiological concentrations of 10^{-7} to 10^{-12} M [7, 12, 121]. In the blood the hormones are transported on specific transport proteins; their function consists of metabolic regulation, their effect in the appraisable intensity of the induced phenomenon [35, 50, 123].

The following sites of the hormonal mechanisms of action have been suggested:

(i) the membranes (the hormones act upon the membrane receptors);

(ii) the enzymes (the hormones are assumed to be allosteric modulators);

(iii) the nucleic acid genes (the hormones stimulate the operons directly);

(iv) the oscillating ionic systems or micromolecules (histamine, acetylcholine, etc.) which are assumed to mediate the hormonal effect.

In fact, the mechanisms are always intricate: cortisol may act upon an intramembranous enzyme, determining ionic translations, or the enzymes of gluconeogenesis, but also upon RNA-polymerase, etc. (Westphal) [121].

At present most arguments lend support to a secondary intracellular messenger as hormonal mechanism of action, the messenger being cyclic AMP (3′5′-AMP) (Drummond et al., 1970; Morel, 1969) [44, 88].

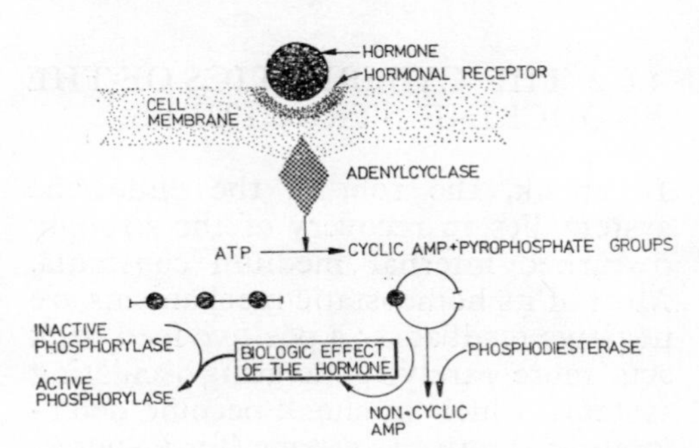

Figure 4.10. In this example, cyclic AMP activates phosphorylasis (in other instances it triggers epinephrine glycogenolysis).

Attachment of the hormone (first messenger) to its specific receptor activates adenylcyclase, which in turn induces cyclic AMP (the second messenger) from ATP. Cyclic AMP then carries into effect the biological scope of the initial hormone (it activates an enzyme, operon, etc.) (Figure 4.10). This explains why in shock when ATP deficiency becomes serious, the hormonal systems of regulation become anarchic following upon degradation of the relay monoadenylate.

There are many proofs lending support to the intervention of cyclic AMP in the hormonal mechanism of action. The action of parathormone and the prostaglandins in the skeletal tissue is mediated by cyclic AMP [26]. Hepatocyte adenylcyclase is stimulated by Ca^{2+}, epinephrine and glucagon [43]. Cyclic AMP is the mediator of vasopressin activity in the renal tubules [9]. Bovine hypothalamus extract stimulates *in vitro* the formation of cyclic AMP in the adenohypophysis. A similar stimulation was obtained with epinephrine and prostaglandin [130, 28]. In shock hypoxia always lowers catecholamine and cyclic AMP concentrations in the hypothalamus [39].

4.8.2 THE CYBERNETICS OF THE ENDOCRINE STRUCTURES

In shock, the role of the endocrine system lies in recovery of the strongly disturbed internal medium constants. Most of its homeostatic mechanisms are negative feedbacks; a positive feed back acts more rarely, generating oscillating systems which in shock become deteriorating runaway systems (for instance, the autocatalytic inrush of coagulation, glycolysis or glycogenolysis, etc.).

The vertical organisation of the endocrine system complies with a certain hierarchy and graded complexity (Figure 4.11). The simplest system in which the hormone acts upon specific cells (variably controlled, regulating release of the hormone from the gland directly) takes place by direct adjustment between insulin — glycemia, glucagon—glycemia, parathormone — calcemia, aldosterone — natremia. These simple systems also act in the absence of a direct hypothalamic or hypophyseal control, not obeying the central endocrine control of the reaction syndrome in shock.

A more complicated system, acting through the intermediacy of peripheral regulating substances, is aldosterone secretion under angiotensin control (*see* Section 4.8.5).

A typical model of rapid endocrine organisation is hypothalamic control of GH and ADH secretion. Characteristic of this type is the presence of a single extrahypothalamic gland and a converse intrahypothalamic connection.

One of the most complex endocrine models from the viewpoint of vertical organisation is that of the hypothalamus and two endocrine structures (the CRH → ACTH → cortisol; TRH → TSH → $T_3(T_4)$, and FRH → FSH → sexoid hormones). The rapid hormonal lines have thus an increased rate of peripheral effects, they penetrate deep within

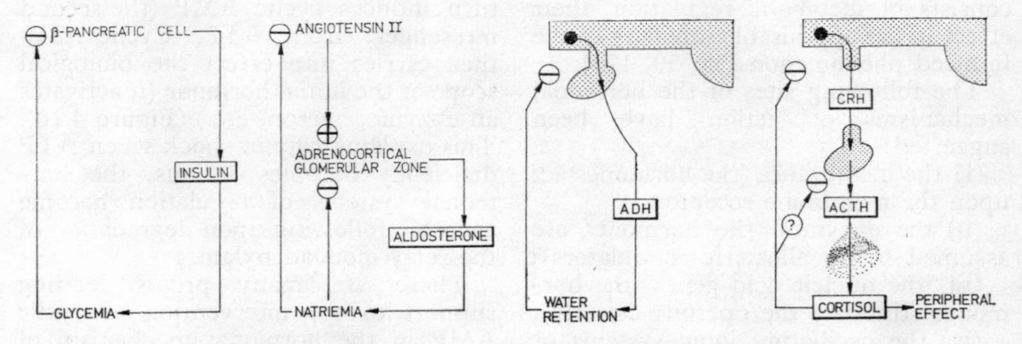

Figure 4.11. The fundamental models of the 'vertical hierarchical' organisation of the hormonal lines (after *Ionescu*, 1973).

the intimacy of the shock cell metabolism.

The endocrine system also has a horizontal organisation which may be illustrated by the control of glycemia or the electrolytes. Control implies the participation of a group of antagonist and agonist hormones that influence at least two variables (for instance insulin and glucagon interfere simultaneously with glycemia and FFA levels) (*see* Section 8.3).

In the economy of the human organism the endocrine glands and hormones are the principal means for realising the genetic programme efficiently [64] (Figure 4.12). An abstract initiative (information) is materialised by the cellularity subsystems of the organism, the effect consisting in their functional solidarity (Milcu and Ionescu) [86]. The receptors induce particularisation of the peripheral hormonal effects. They decipher only a certain information which they take up from the systemic humoral torrent. For instance, the efficiency hormone may interfere directly in a metabolic chain (the thyroid hormones are utilised as a redox system) or act upon the enzymatic chains through the intermediacy of cyclic AMP.

The endocrine network has a high functional elasticity, yet shock inducing disturbances exceeding its threshold of adaptability, will cause deterioration of the system. In shock the hypothalamus may use some of its physiological procedures: it may discharge a 'jet' of endocrine information, chanelled towards a single endocrine performance gland, by the monophasic shift mechanism (Ionescu) [64]. Actually, it is an increase in the information output along one channel, used at a given moment to the detriment of the others that are temporarily ignored.

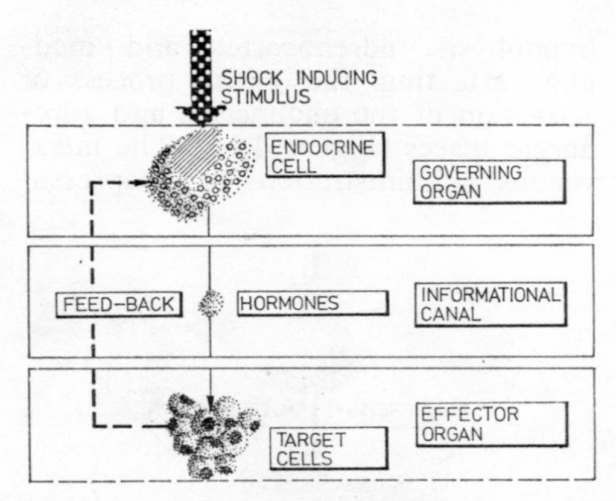

Figure 4.12. The 'endocrinon' (after *Milcu* and *Ionescu*, 1970).

This shift in the programme may act by a compensatory system, such as the overlap phenomenon described above, or by simultaneous hypertonia and automaintenance of two or more hormonal lines (escort phenomena) [64, 86]. This explains the initial ADH hypertonia and ACTH-cortisol hypertonia, accompanied by that of TSH-T_3 (T_4) and the temporary silence of the gonadotropic line (Figure 4.13).

4.8.3 HYPOTHALAMO-HYPOPHYSEAL HORMONES

ADH (vasopressin). The onset of the reaction syndrome in shock is marked in the hypothalamus by a decrease of the neurosecretion granules in the supraoptic and paraventricular nuclei [3, 75]. The granules represent the substrate — vasopressin — an octopeptide which apart from increasing the permeability of the distal renal tubules to water also proves to be a very efficient activator of the neuroendocrine structures. In shock this hormone produces functional hypertonia of the entire hypothalamic space, of the ARAS and mesencephalic decussation, adenohypophysis, adrenocortex and medulla, affecting the entire process of activation of the cholinergic and adrenergic spaces (Figure 4.14). The intravenous administration of vasopressin doubles the amount of CRH and respectively ACTH and cortisol [125]. ADH is one of the levers of the osmostatic system. A specific stimulus of ADH liberation is the increase in osmolarity, a constant phenomenon in the advanced stages of shock. The administration of hypertonic NaCl primes polydypsia by direct stimulation of the Verney osmoreceptors [6]. In shock the increase in osmolarity brought about by decompensation of the electrolytes and hyperaldosteronism takes place slowly (at least several hours), so that the initial hypersecretion of ADH must be understood as a direct result of hypertonia of the centrencephalic structures; the positive retroaction of ADH on the upper neuroendocrine levels triggers a typical runaway phenomenon (Yates) [125].

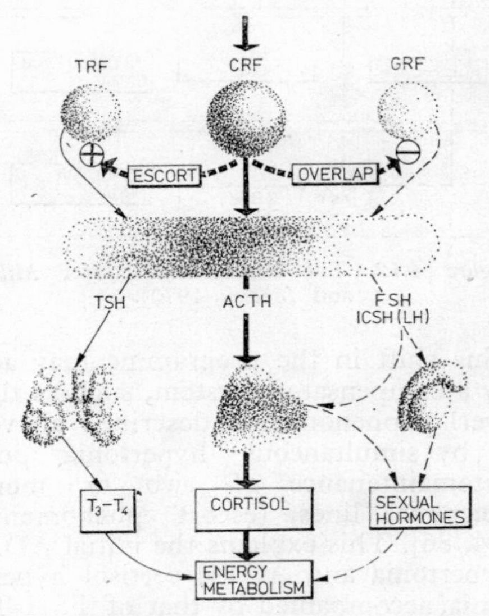

Figure 4.13. Hypertonia of the ACTH-cortisol line is accompanied by hyperfunction of the thyroid hormones and inhibition of the gonadotropical hormones (after *Ionescu*, 1969).

ACTH (corticotropin) seems to be released by the β_1 basophil cells of the hypophysis; it is a polypeptide (39 amino acids) with a molecular weight of 3 500.

Any shock stimulus releases the reticulolimbic formations from under the control of the cerebral cortex, triggering hyperfunction of all hypothalamic tuberoinfundibular centres where neurosecretion produces CRH in excess; the latter reaches through portal pathways the basophil hypophyseal cells eliciting the production of ACTH. The half-life of plasma ACTH is 5-10 min; it is taken up selectively by the adrenals and kidneys, stimulating preferentially synthesis of the glucocorticoid sterol line. The servomechanism of the glucocorticoid efficiency line is maintained under

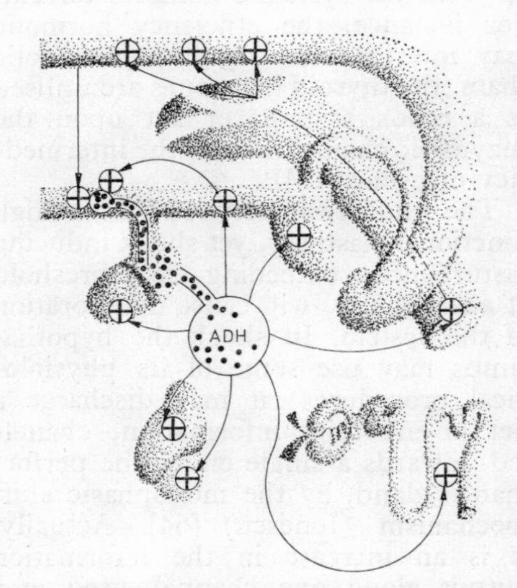

Figure 4.14. Positive feedback by excessive ADH secretion initiated by hypertonia of the anterior hypothalamus.

physiological conditions by the direct inhibitory action of the hypothalamus on cortisol, whose effectiveness disappears in shock, the hypothalamus being hyperactivated (Figure 4.15).

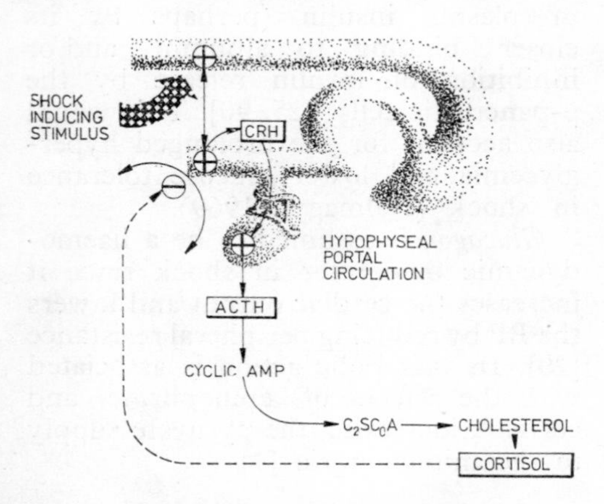

Figure 4.15. Increased cortisol synthesis stimulated by hypertonia of the entire regulation axis can no longer be stopped because the negative feedback mechanism of cortisol is no longer efficient; in shock the hypothalamic centres are governed by the overlying, stronger neuronal stimuli.

GH (growth hormone; formerly STH), in primates, initially decreases in shock probably owing to the central inhibition induced by catecholaminic hyperglycemia. Catecholamines also inhibit directly the hypothalamohypophyseal release of SRH—GH [25, 120]. This phenomenon may be interpreted as a shift effect of the releasing factor [64], the rapid energy releasing hormonal lines being favoured in the first stages of shock. In the anabolic waves of the stages of recovery from shock the same shift system will commute hypothalamic stimulation along SRH—GH lines.

Gondadotropic hormones (FSH, LH—ICSH, LTH) are released by the adenohypophysis under control of the respective releasing factors, governed especially by cholinergic mechanisms. In shock, the adrenergic system has the same shifting effect, skirting the gonadotropic lines. Almost all modern data concerning the function of the limbic and hypothalamohypophyseal neuroendocrine systems were obtained by research work on the gonadotropic hormones and sexoid sterols, which are however of secondary importance in shock; their function is converted so as to save and maintain the individual as a whole. Hence, in the initial stages of shock the sexoid sterols (adreno-gonads) first act synergically with glucocorticoids; after recovery from the critical stages of shock their role and concentrations will increase, taking on dominant anabolising effects.

4.8.4 THYROID AND PANCREATIC HORMONES

Hypothalamophypophyseal hypertonia of the CRH→ACTH line also includes, by an escort effect, hypersecretion of the TRH → TSH → $T_3(T_4)$ axis (Figure 4.16). Cold stimulates the

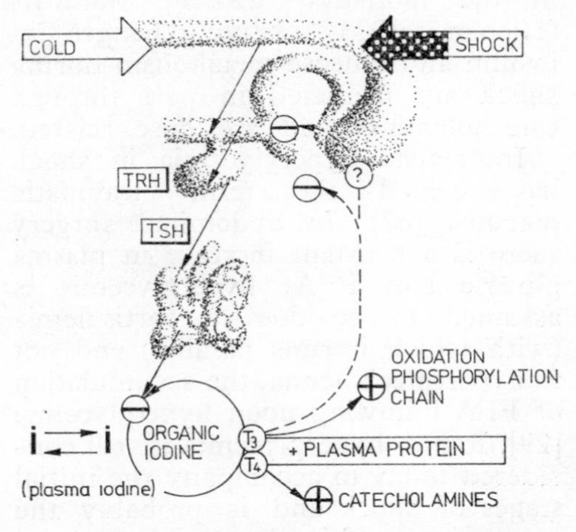

Figure 4.16. Stimulation of the thyroid hormonal system in shock.

thyrotropic axis, accentuating the initial hypercatabolism of shock. Thyroxin (T_4) and triiodothyronine (T_3) stimulate the loops of the oxidation chain directly, in the course of which they may even be consumed as redox systems.

Having common precursors, T_4 and T_3 stimulate the synthesis of catecholamines (*see* Section 8.3.6). Postoperatively T_4 rapidly increases in the plasma by its immediate release from its transport protein. Emergent labilisation of the plasma hormonal stocks is a mechanism of hyperconcentration of the free active fractions, sterols included.

Insulin is synthetized in the β-pancreatic cells; the membrane of the insulin granules fuses with the cell membrane, the granules being released into the blood flow by emiocytosis [7,123]. The presence of glucose in the blood depolarises the β-pancreatic cells triggering the release of insulin, an effect blocked by epinephrine [38]. Insulin appears to stimulate the synthesis *de novo* of hexokinase [100].

Significant hyperglycemia was always found in haemorrhagic shock but not in the monkeys without adrenals (Hiebert, 1973). Acute changes in insulin and glucose metabolism during shock are mediated in part through catecholamines and are dose related.

Immediate hyperglycemia in shock has coined the term traumatic diabetes [67]. In abdominal surgery there is a constant increase in plasma glucose and FFA; hyperglycemia is assumed to be due to cortisolemia (with which it runs parallel) and not to hypercatecholemia, the accumulation of FFA following upon hyperglycemia [29]. Rapid hyperglycemia is still considered today to accompany the initial stages of shock and is probably the consequence of hypercortisolemia and of epinephrine phosphorylase through cyclic AMP or calcium for the myofibrils [44, 47]. Inaccurate however, at least for primates is the increase in plasma insulin concentration. On the contrary, in shock there is an immediate decrease of plasma insulin (perhaps by its closer binding to albumin and/or inhibition of insulin release by the β-pancreatic cells) [25, 90]. This would also account for the prolonged hyperglycemia and lower glucose tolerance in shock (Halmagyi, 1969).

Glucagon is assumed to be a haemodynamic moderator in shock since it increases the cardiac output and lowers the BP by reducing peripheral resistance [20]. Its metabolic action is associated with the effects of epinephrine and cortisol, increasing the pyruvate supply to the mitochondria [1].

4.8.5 STEROID HORMONES

The adrenals are symmetrical glands with a reddish-brown medulla weighing about 1-2 g and a three-layered yellowish cortex (4-5 g). The adrenals have a dense vascular network and sinusoidal tissue. The rich blood supply from the aortic, diaphragmatic and juxtarenal segments, as well splanchnic hypertonia (to which the intestine is also subject) are essential 'anatomic advantages' for the function of these glands. Similar to the hypophysis the adrenals have a preferential perfusion: at a haemorrhage of 40% of the circulating blood volume, perfusion falls below 50% in the stomach, kidneys, skin and muscles and increases up to 108% in the adrenals [8]. Moreover, the adrenal cortex is very resistant to hypoxia; however, it requires an adequate perfusion volume that may be realised even with small amounts of substitutes, of the simplest kind, since adrenocortical perfusion will always be given preference.

All the steroid hormones are finally derived from cyclopentanperhydrophenanthrene and its methylated compound, steran. C-21 steroids contain gluco- and mineralocorticoids. C-19 derivatives contain androgens and C-18 (with an A aromatic ring) estrogens. Sterol synthesis in the adrenal cortex is controlled by ACTH (Figure 4.17). Under normal conditions, as well as in shock, ACTH acts rapidly (under 30-40 min) mediated by cyclic AMP that activates cholesterol conversion into pregnenolon. Actually, cyclic AMP stimulates the pentose pathways, the supplier of $NADPH_2$, necessary for hydrogenation. The catabolic nucleotide $NADH_2$ (the reduced form) favours the synthesis of 17-hydroxycorticosteroids, and NAD (oxidated form) that of 17-deoxycorticosteroids. Ascorbic acid oxidises $NADH_2$ (lowering the production of 17-hydroxycorticoids) in order to reduce NADP to $NADPH_2$ (that will act in favour of 17-deoxycorticoids) [70]. This is of particular interest in shock when $NADH_2$ accumulates and brings about a deleterious catabolic positive feedback along the 17-hydroxy line. Oxidation of $NADH_2$ is equivalent to metabolic reanimation of the cellular energy processes.

Among the sterols released following ACTH stimulation, cortisol alone has the role of bringing negative feedback to a close. Plasma cortisol is regulated by this glucocortistat up to the moment when shock situations induce the increase of ACTH liberation independently of the circulating cortisol concentration [98]. Four hours after severe injuries the plasma cortisol level increases very much, no longer in keeping with its circadian rhythm [71].

In the three stratified cortex over 40 different steroids are synthetised, few of these however are active systemically [42, 66, 99, 121].

Glucocorticoids (17-hydroxy) include cortisol (15-20 mg are produced daily) and 11-deoxycortisol (2-5 mg daily). In shock (at BP levels of 35-40 mmHg) 10 mg cortisol is produced per hour, starting 3-4 hours after the onset of shock.

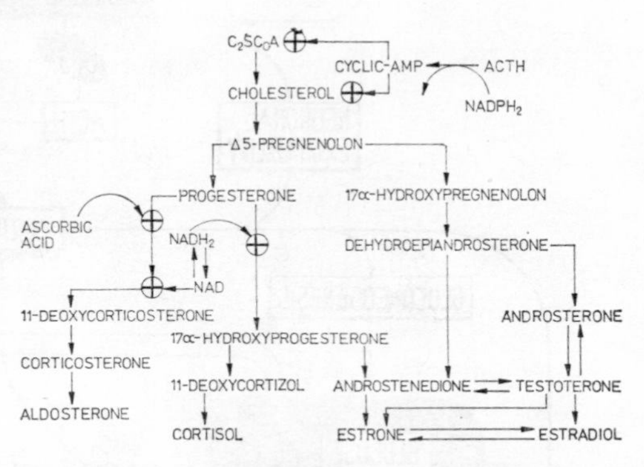

Figure 4.17. Synthesis of the steroid hormones.

Mineral corticoids (17-deoxy) include aldosterone (75-125 μg daily output), corticosterone and 11-deoxycorticosterone.

Androgens (17 keto) are represented in a proportion of 75% by dehydroepiandrosterone and androstenedione (25-30 mg daily).

Cortisol represents 80% of the total amount of 17-hydroxycorticosteroids. In the plasma it is bound especially to α_2-globulin (transcortin), a rapid source of the circulating hormone [12, 35]. The half life of plasma cortisol is 2 hours. For cortisol all the mechanisms of hormonal action discussed are possible, particularly worthy of the note being its direct insertion on RNA-polymerase or stabilisation of microsomal RNA. Tritiated cortisol is retained in all the tissues in increased concentrations in the neuronal structures [118]. The physiological effects of cortisol are illustrated in Figure 4.18. The most important pharmacological effects due to its direct metabolic insertion are the following: (i) acceleration of the transformation of inorganic phosphate into ATP;

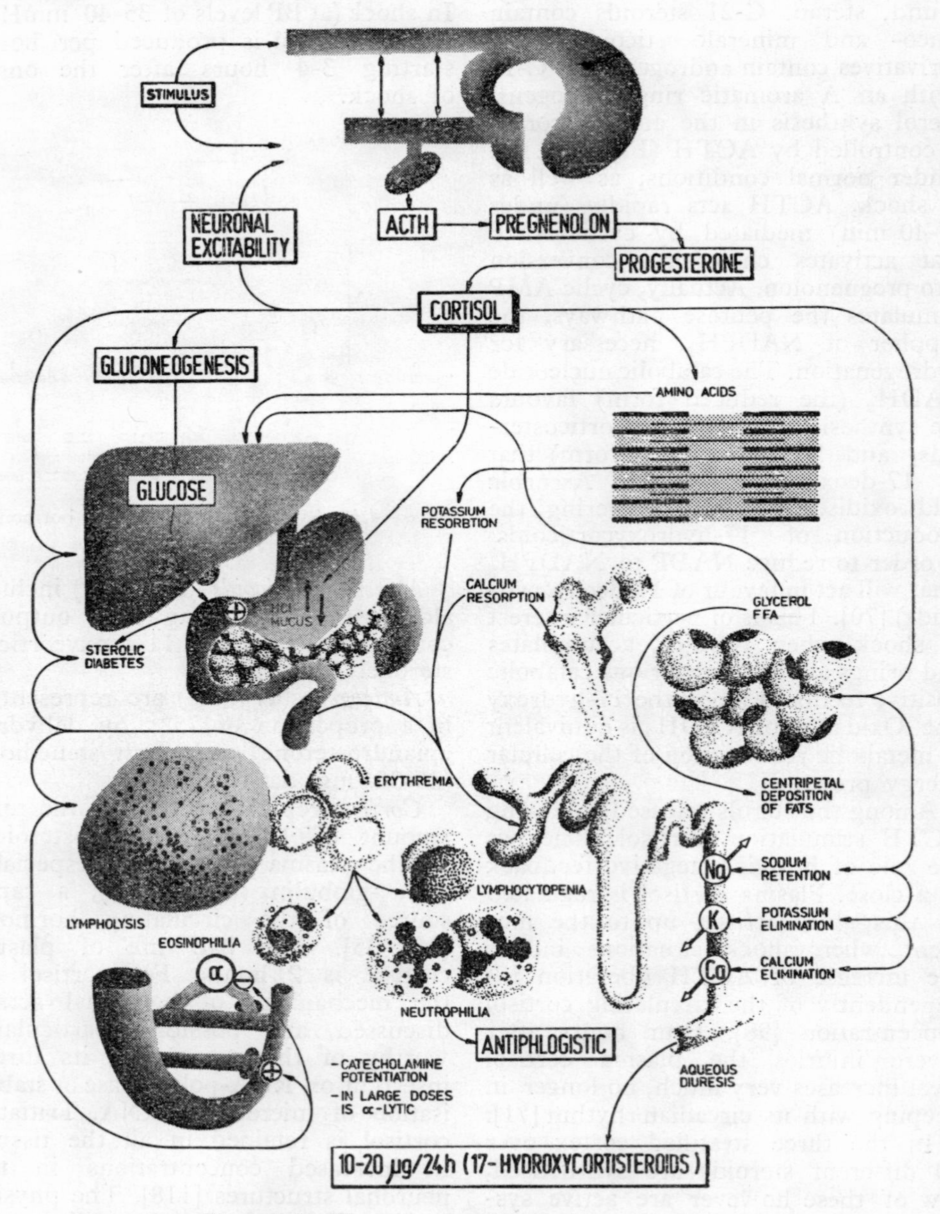

Figure 4.18. The role of cortisol in human body.

(ii) channelling of the amino acids in neoglucogenesis;

(iii) decrease in lactic acidosis opposed to the glycolytic and lipolytic effects of epinephrine;

(iv) a shift to the right of the oxyhemoglobin dissociation curve;

(v) stability of the capillary membrane and in general of the entire intracellular membrane system with its enzymatic content.

Aldosterone is one of the vital hormones that has been carefully investigated in shock [91]. Aldosterone deficit produces sodium and volume losses with decrease in the ECBV, and excess sodium retention, increase in voalemia and hypertension. A decrease in osmolarity raises hormonal synthesis from 100-200 μg to 1000 μg per day; in the plasma aldosterone circulates bound to albumin and also in very small amounts to transcortin. It was observed some time ago that aldosterone hypersecretion is accompanied by an increase in renin concentration and decrease in norepinephrine and sodium in the plasma, renal cortex and vascular walls. An increased G-6-PDH activity in the adrenal glomerulus may be considered as an enzymatic test of aldosterone hyperproduction [59]. Aldosterone secretion does not seem to depend upon the hypophysis and neither on epiphyseal adrenoglomerular corticotropin (Farrell, 1959). On the other hand it has been established that the epiphysis and subcommissural organ are structures that interfere actively in the regulation of volaemia and electrolytaemia.

Aldosterone is the most important pace-maker of plasma Na^+ and K^+ concentrations. Today, it is admitted that there are two systems for the stimulation and control of aldosterone synthesis and release [91].

The proximal system in the glomerular area, for the conversion of cholesterol into pregnenolon (Figure 4.19): stimulation is achieved by ACTH (in doses ten times greater than the physiological ones, so that this system may only be considered efficient in states of shock), ammonium ions, rubidium, cesium, cyclic AMP, serotonin and especially angiotensin II. The unspecific, proximal system is controlled by angiotensin II, whose hyperproduction may stimulate that of aldosterone,

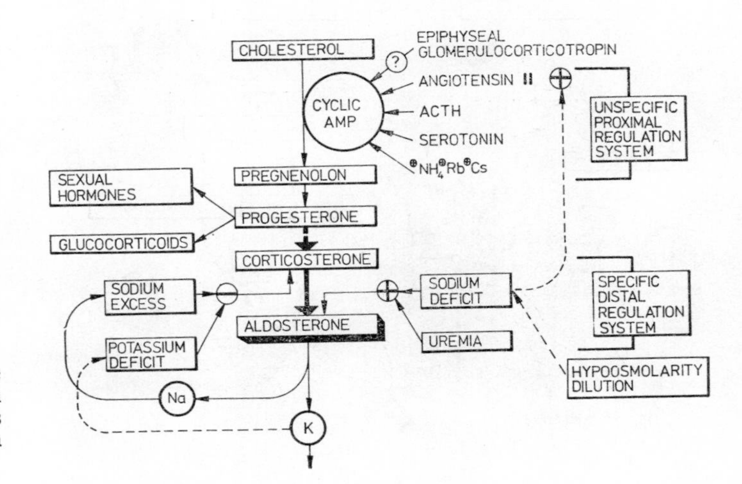

Figure 4.19. In shock the distal system of regulation of aldosterone synthesis is stimulated by the dilution caused by excess ADH.

not however in the presence of hypernatremia or hypopotassemia which are much stronger stimuli of the distal system (Müller, 1971) [91].

The distal system controls the decisive line: progesterone-deoxycorticosterone-corticosterone-aldosterone by a simple chemical feedback loop that contains the key stages in the synthesis of aldosterone.

4.8.6 RENIN AND ERYTHROPOIETIN

Renin is an enzyme produced by the cells of the renal juxtaglomerular system: from plasma α_2-globulins it splits off decapeptide fragments — angiotensin I — which is then converted into an octopeptide (angiotensin II) by other plasma or tissular enzymes. Angiotensin II is the most powerful known vasoconstrictor (for its haemodynamic effects see Figure 4.20). Renin is released especially by the kidney in shock, and is to be found in numerous humoral control circuits. A resonance organ of hypoxia, the kidney liberates several kinds of secretory granules into the juxtaglomerular system. Gallagher asserts that these granules are the precursors of *erythropoietin*, which accelerates the red series in haematopoiesis. In fact, erythropoietin is an active plasma 'hormone' and is produced by the action of a 'renal erythropoietin factor' likewise from α_2-globulin, the same as angiotensin I [51]. Increase in the erythron will enable the transport to the tissues (including the juxtaglomerular system), even under hypoxic conditions, of a sufficient amount of oxygen, thus closing the chemical feed-back circuit. Erythropoietin produced by the hypoxic kidney can build up again the erythron within a

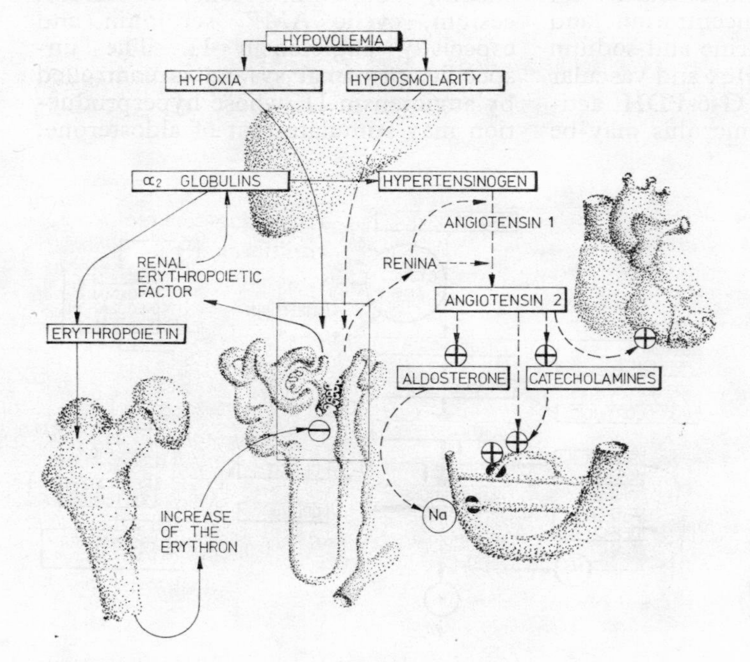

Figure 4.20. The release and roles of angiotensin and erythropoietin.

few weeks. Moreover, it is assumed that the spleen ceases to secrete the 'hormone' which inhibits cytodiabasis, thus allowing incomplete erythrocytic forms (reticulocytes, etc.) to pass into the circulation at moments of great stress in shock (Finch) [51].

Goormaghtigh suggested that the granules of the juxtaglomerular system (or eventually another type of granule) also contain renin, whose liberation is triggered especially by a decrease in natremia; renin then accentuates the synthesis of aldosterone, which has also been stimulated in the meantime by hyponatremia.

4.9 COMPLEX HORMONAL REGULATION

The hormones of several lines associate agonistically or antagonistically to regulate the principal functions of the body. A suggestive example is the hormonal sequence that controls sexualisation and perpetuation of the especies.

In states of shock, however, the pooling of hormonal action becomes of great importance for saving the life of the individual (blood rheodynamics, haemostasis, volaemia, electrolytaemia and especially the production of energy).

4.9.1 BLOOD RHEODYNAMICS

The hormonal effects on haemodynamics are summed up in table 4.3.

A group of estrogen and thyroid hormones act upon the coagulolytic system. The feminising effects of estrogens is only part of their role; just as important are their other functions: metabolism, electrolyte activity and haemostasis.

Estrogens control indirectly the coagulostat, acting upon the ground substance of the capillary walls (and their vicinity) in which the synthesis of acid mucopolysaccharides increase, thus transforming the tissues in gel and ensuring normal tightness. The antihaemorrhagic effect of estrogen products is well known.

The thyroid likewise seems to play an important role in the mechanism of fibrinolysis [14]. In shock an excess

Table 4.3 THE HAEMODYNAMIC EFFECT OF SOME HORMONES

	Heart rate	Cardiac force	Cardiac output	BP	Arteriole tonus	Osmolarity	ECBV
Norepinehrine	—	+++	±	++++	+++	—	—
Epinephrine	+++	++++	++	+	±	—	—
Cortisol	±	+	+	+	——	+	+
Aldosterone	±	±	+	++	+	++	++
Angiotensin	±	±	±	++++	+++	+	+
T_3-T_4	+++	+	+++	+	—	—	++
ADH	±	±	+	+	—	—	+++
Glucagon	±	±	+	—	——	+	+

of thyroid hormones (T_3-T_4) inhibits plasminogen activators and stimulates their inhibitors, thus increasing the danger of disseminated intravascular coagulation (*see* Section 7.4).

4.9.2 VOLAEMIA AND ELECTROLYTAEMIA; PORTAL OSMOSTAT

The hormones influence the hydroelectrolytic turnover directly through effector structures, or indirectly by metabolic effects. The action of hormones with a role in hydroelectrolytic balance is shown in table 4.4.

The hydroelectrolytic balance may be influenced by:
(i) increase in hydric reabsorption (ADH);
(ii) antagonising the transtubular ionic transport (aldosterone);
(iii) inhibition of the pathological stimulation of aldosterone (glucocorticoids);
(iv) tissular catabolisation (T_3-T_4 and most sterols).

Water is indispensable in chemical metabolic reactions. The amount of dissolved substances and the volume of the solvent are monitored by different receptors: osmoreceptors for the concentration of the solvent and volume receptors for its volume (Figure 4.21).

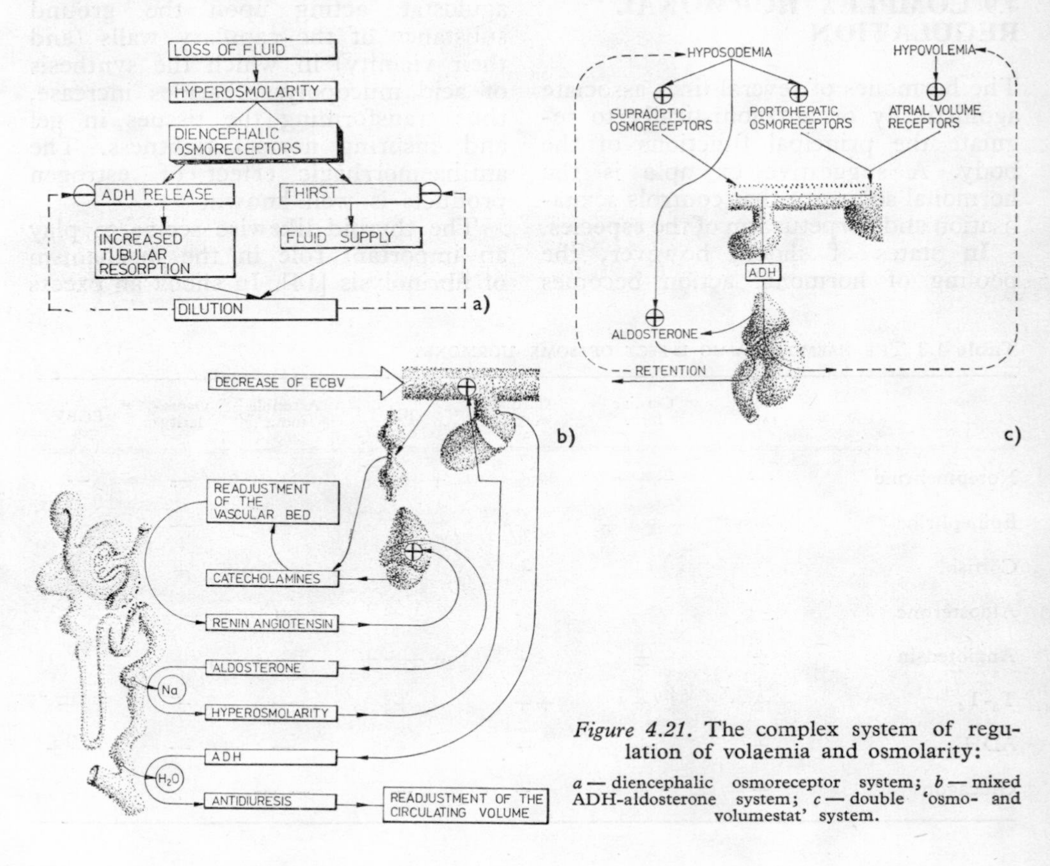

Figure 4.21. The complex system of regulation of volaemia and osmolarity:

a — diencephalic osmoreceptor system; *b* — mixed ADH-aldosterone system; *c* — double 'osmo- and volumestat' system.

In the osmostat and volumestat system not only the Verney osmoreceptors, but also the atrial volume receptors, ADH, aldosterone, the kidney, natriuretic hormone (hypothetical) and the portal osmoreceptors interfere.

The role of osmoreceptors has been postulated for some time. The portal blood has a specific homeostasis and its osmoreceptors have a priority over the hypothalamic ones, acting within less than 20 min, therefore before the affection of systemic osmolarity (after 20-30 min). The portal barrier in shock increases osmolarity of the pooled fluid, the effect being a state of antidiuresis. The enteral administration of hypotonic solutions may dialyse intraportal electrolytes and decrease of portohepatic osmolarity may release diuresis (Figure 4.22).

Table 4.4 HORMONES, WATER AND ELECTROLYTES

	Na^+	K^+	Cl^-	Plasma volume	Renal filtration	Diuresis	Cerebral œdema	Extracellular œdema
GH (STH)	↑	↑	↑	↑	↑	↑	↑	↑
ADH	—	—	—	↑	—	↓↓↓	↑	↑↑
Aldosterone	↑↑↑	↓↓	↑	↑↑	—	↓↓	↑↑	↑↑
Cortisol	↑	↓	↑	↑	↑↑	↑↑	↑	↑↑
Progesterone	↓	↑	↓	↓	↑	↑	—	—
T_3-T_4	↓	↓	↓	↓	↑↑	↑↑	↓	↓
Estrogens	↑↑	—	—	↑	—	↓	↑	↑
Androgens	↑	↓	↑	↑	↑	—	↑	—
Angiotensin	↑	↓	—	↑	↓	↓	--	—

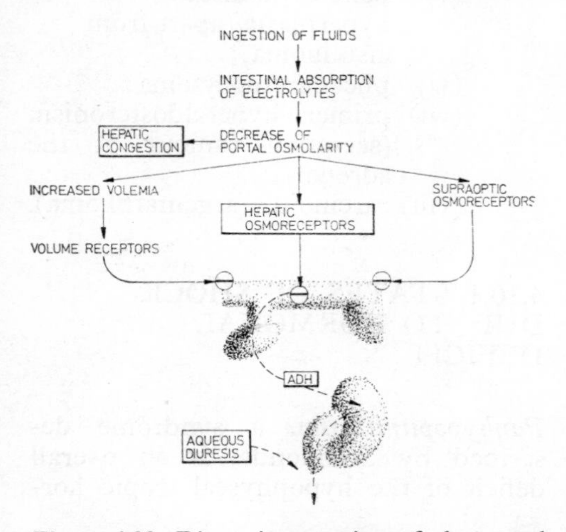

Figure 4.22. Direct intervention of the portal osmoreceptors in the inhibition of ADH production and precipitation of aqueous diuresis.

4.9.3 METABOLISM REGULATION

The hormones are the chemical signals of metabolic adaptation to chronic or acute disturbances [85, 119].

They act upon the 'strategic' enzymatic chains, modifying their rhythm and intensity, as well as the sense of metabolic pathways *(see* Chapters 8 and 9). In shock, the coupled metabolic systems that are the first to undergo complex metabolic regulation are glycogenolysis and glyco-

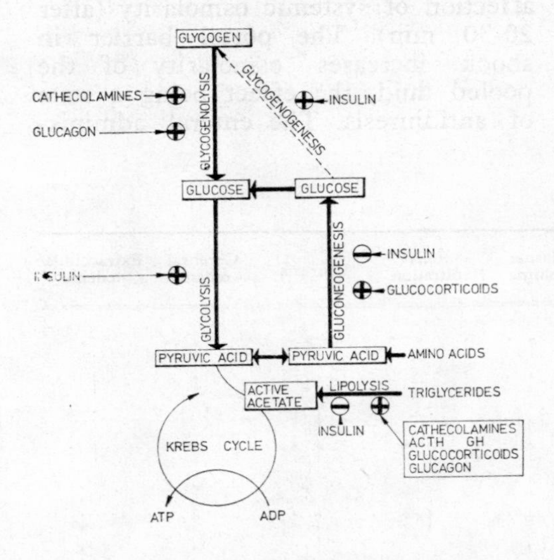

Figure 4.23. The main enzymatic chains stimulated by the hyperactive hormonal lines in shock.

lysis (stimulated), glycogenosynthesis (inhibited) and gluconeogenesis (stimulated (Figure 4.23).

4.10 ACUTE HORMONAL DYSFUNCTIONS

The role of hormones consists in the regulation of general homeostasis. Circulation of the fluids and the integtrity of the metabolic structures is heir main task. Therefore any excess or deficit of these chemical messengers may bring about homeostatic disorders up to the intensity of phenomena specific to shock. These are true endocrinological emergencies [110], also known as attacks, coma, insufficiency or acute hyperfunction (Ionescu, 1970) [86].

Acute hormonal dysfunctions [1] are classified as follows:

(a) *States of shock due to hormonal deficit*
- (i) acute panhypopituitarism (hypophyseal coma);
- (ii) acute hypothyroidism (mixedematous coma);
- (iii) hypoinsulin coma (diabetic);
- (iv) acute adrenal insufficiency (Addison attack);
- (v) hypoaldosteronism (destructive tumours of the adrenals).

(b) *States of shock due to hormonal excess*
- (i) thyrotoxic attack;
- (ii) thymotoxic attack (myasthenia) (*see* Section 4.11.2);
- (iii) hyperparathyroidism;
- (iv) hyperinsulin coma (insulinoma);
- (v) pancreatic insular hyperplasia (apart from insulinoma);
- (vi) pheochromocytoma;
- (vii) primary hyperaldosteronism (secreting tumour of the adrenals);
- (viii) carcinoid (argentaffinoma).

4.10.1 STATES OF SHOCK DUE TO HORMONAL DEFICIT

Panhypopituitarism, a syndrome described by Simmonds, is an overall deficit of the hypophyseal tropic hor-

[1] Shock is only an evolutive possibility of great intensity of these acute dyscrinias.

mones manifested peripherally by multiple secondary insufficiencies of the target glands (Figure 4.24). The dominant clinical elements in states of acute hypophyseal insufficiency are

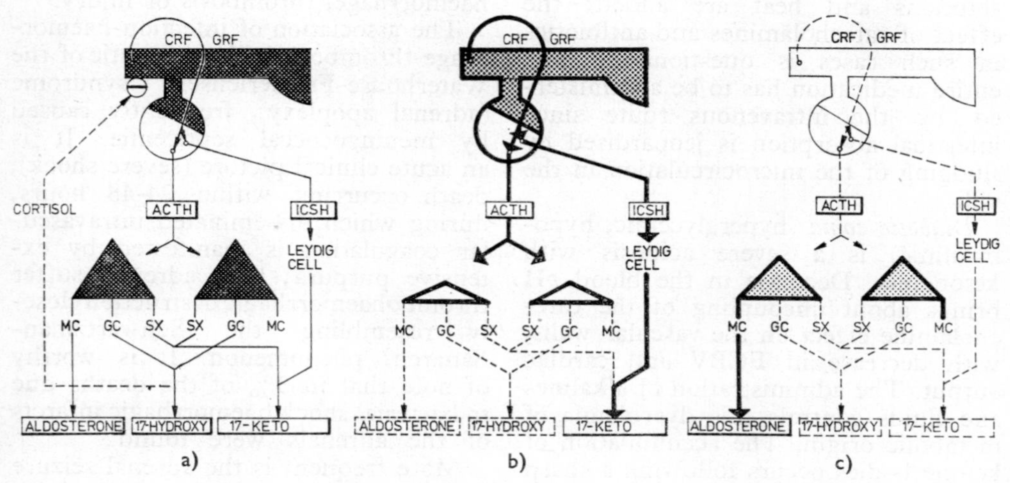

Figure 4.24:

a — normal situation in man. Physiological production of mineralocorticoid (MC), glucocorticoid (GC) and sexoid hormones (SX) in the adrenocortex; *b* — adrenocortical insufficiency; excess ACTH secretion owing to the cortisol deficit that can no longer close the negative feedback; 17-ketosteroid secretion controlled by the sex glands; *c* — hypophyseal insufficiency; the hypophyseal tropic hormones present an overall deficit and only aldosterone secretion is possible in the glomerulus.

generated by a deficit of the ACTH-cortisol axis that results in a sharp decrease of ECBV, endangering tissular perfusion. After therapeutical removal of the hypophysis, the cardiac output rapidly falls by 25%; after 3-4 days, signs of adrenocortical insufficiency develop and within 4 weeks polyuria and hypothyroidism. The main causes of the brutal onset of global hypophyseal insufficiency are:

(i) infarction of the gland in head injuries (fractures of the base);

(ii) intratumoral haemorrhage (pituitary apoplexy);

(iii) postpartum pituitary necrosis (Sheehan syndrome) due to prolonged vasospasm in the peri-infundibular network or disseminated vascular coagulation in the adenohypophysis;

(iv) subacute and chronic causes: craniopharyngiomas and chromaphobe adenomas, aneurysms, sarcoidosis, metastases, etc.

Mixedematous coma is the last stage of severe hypothyroidism. It occurs during the cold months, with loss of consciousness, hypothermia (32 to 27 °C skin temperature), bradycardia, hypotension, hyperviscosity of the blood and blocking of the microcirculation, thus representing one of the most particular and difficult to treat pictures of shock. In addition to cold and infections, hypoventilation and hyponatremia may precipitate the phenomena. The determination of iodemia (very low) helps to differentiate between mixedematous coma and cerebral vascular coma with hypotension [82].

The treatment is based on the administration of rapid-acting thyroid hormones: Liothyronin (25 μg at 12 hours), Thyroton or Levothyroxin (500 γ daily), administered by the intravenous route, to which cortisone, hypertonic saline solutions and heat are added; the effect of catecholamines and antibiotics in such cases is questionable. The entire medication has to be administered by the intravenous route since intestinal absorption is jeopardised by sludging of the microcirculation in the villi.

Diabetic coma (hyperglycemic, hypoinsulinic) is a severe acidosis with ketonemia. Decrease in the blood *p*H brings about uncoupling of the catecholamine effect on the vascular walls, with decrease in ECBV and cardiac output. The administration of alkalines and fluids controls this dysvolemia of metabolic origin. The accumulation of ketone bodies occurs following a sharp fall in the utilisation of glucose by the EMK cycle. Clinically, the initial anorexia is followed by vomiting, abdominal and/or thoracic pains, dry skin, tachycardia, hypotension, cyanosis, headache and progressive coma. The ketone bodies have values over 300 mg %, the *p*H is about 7.2 and glycemia 400-800 mg %. The differential diagnosis must be established with cerebral vascular accidents, uremia and hypoglycemic coma.

Hyperosmolar coma (non-ketotic) is a particular variant of diabetic coma characterised by an impressive polydipsia, polyuria and accentuated neurological signs; it is associated with hyperglycemia of over 500 mg % (as a rule between 1000 and 2000 mg %), sodium values that exceed 150 mEq per thousand and a normal *p*H. The treatment consists in the administration of large amounts of hypotonic fluids and antialdosterone drugs.

Acute adrenal insufficiency. The cardinal signs are: headache, vomiting and diarrhœa, abdominal and/or costovertebral pains, circulatory failure; it develops following acute destruction of the glandular cortex by infections, haemorrhage, thrombosis or injury.

The association of infection-haemorrhage-thrombosis is characteristic of the Waterhouse-Friederichsen syndrome (adrenal apoplexy), frequently caused by meningococcal septicemia. It is an acute clinical picture (severe shock), death occurring within 24-48 hours, during which disseminated intravascular coagulation is manifested by extensive purpura; both adrenals suffer thrombohaemorrhagic destruction closely resembling the Schwartzman-Sanarelli phenomenon. It is worthy of note that in 3% of the deaths due to bacterial shock haemorrhagic infarcts of the adrenals were found.

More frequent is the adrenal seizure triggered by surgery, traumatic, bacterial shock, etc., especially in individuals with subclinical chronic adrenal insufficiency after partial adrenalectomy or corticotherapy.

Within about 10 hours after these shock stimuli act acute peripheral circulatory failure develops, endangering perfusion of the microcirculation, hyperthermia and progressive coma. In these cases pigmentation of the skin and mucosa (the classical sign of clinicians) is minimal or absent. The diagnosis cannot wait for hormonal determinations; eosinophilia and hypopotassemia are sufficient. The administration of cortisone should start, even empirically, on a simple presumption, as corticotherapy of short duration does not present practically any danger.

Chronic adrenal insufficiency (Addison's disease) is caused by destructive atrophy, by toxic or autoimmune factors (55%), tuberculosis (40%) and other causes (5% metastases, amyloidosis, etc.). It has, an incidence of over 0.4% and a characteristic clinical picture, against the background of which an acute Addison seizure may develop, precipitated by relatively minor stimuli [13].

Secondary adrenal insufficiency is the foremost deficit of panhypopituitarism. It is almost always accompanied by insufficiency of the thyrotropic and gonadotropic lines (pluriglandular insufficiency of the thyrotropic and gonadotropic lines (pluriglandular insufficiency). Acute decompensation minimises the states of shock described, the treatment always consisting of broad substitution of all the deficient hormonal lines.

Primary adrenocortical insufficiency (genetic). A deficit in 21-hydroxylase turns Δ 5-pregnenolon and progesterone exclusively towards the synthesis of sexoid sterols. The consequence is severe insufficiency of glucocorticoids and virilisation of children under the age of 5, with excessive loss of water and salt, dehydration and hypovolaemia.

Adrenocortical insufficiency secondary to corticotherapy. Patients treated with cortisone or one of its numerous homologues may develop a secondary adrenocortical insufficiency with the possible appearance of 'acute' attacks precipitated by infection or trauma. This may even lead to atrophy of the gland, that no longer responds to stimulation by ACTH. In addition even the synthesis and release of endogenous ACTH may be arrested. This inhibition of the synthesis of endogenous steroids is considered to appear at daily doses larger than 50 mg cortisol, irrespective of whether the treatment lasts a week or eight years [35]. The intermittent administration of cortisone has not this drawback but when it appears a permanent substitution is necessary. Even if the endogenous sterol insufficiency is not clear-cut in practice the possibility of a subclinical adrenal insufficiency has to be taken into account at least six months after interrupting the therapy with exogenous cortisone.

4.10.2 STATES OF SHOCK DUE TO HORMONAL EXCESS

Thyrotoxic attack is a severe complication in the evolution of Graves' disease, of uni- or multinodular goitre or after subtotal thyroidectomy. It is precipitated by infection, irradiation, parturition, surgical trauma, cold. The onset is sudden, progresses rapidly towards coma and takes on the aspect of shock, dominated by hypercatabolic phenomena, with acidosis, peripheral fluid pooling and acute myocardial failure. The diagnosis is based upon certain major signs: hyperthermia, profuse perspiration, sinusal or ectopic tachyarrhythmias, acute left ventricular failure with pulmonary œdema, neuropsychical and digestive signs. The laboratory offers no immediate pertinent data and an emergency treatment must be instituted, which reduces the lethality rate today to only 20%. The following are administered: antithyroid drugs (1000 mg propylthiouracyl/day), by oral route or still better by gastric intubation, Lugol per os or by intravenous route, reserpine to counter the catecholamine effect of thyronine (2.5 mg i.m. doses, repeated every 4 hours), cortisone (more than 300 mg/day), glucose, antithermal drugs, digitalis and rapid diuretics.

Hyperparathyroidism develops a typical picture of shock when hypercalcemia exceeds the critical level of 17 mg%. Cardiac arrhythmia appears and is followed by cardiogenic shock. Cortisone in large doses and sodium sulphate perfusion are administered.

Insulinoma is produced by neoplasia of the β-pancreatic cells and is benign in 90% of the cases. Most tumours measure less than 3 cm in diameter and are found by the surgeon at the first operation in 77% of the cases and by the histologist in 9%, at the second operation in 5% and at necropsy in 4% of the cases. The diagnosis implies spontaneous glycemia of less than 50%, a positive response to the stimulation tests and insulinemia greater than 10-15 microunits. The hypoglycemic accident has a more rapid onset than hypoglycemic coma, but there is no vomiting, tachypnea or dehydration; in exchange there is perspiration, trembling, tachycardia and other signs of sympathetic excitation. The treatment is surgical.

Non-β-pancreatic cellular neoplasia (nesidioma) is often accompanied by hypoglycemia. The stimulating effect of gastrin, pancreozymin and secretin on insulin secretion is well known. Adenoma of the Δ-pancreatic cells also produces gastric hypersecretion, diarrhœa and steatorrhœa, with dehydration and acute hypovolaemia *(Zollinger-Ellison syndrome)*. Pancreatic adenomas and carcinomas with aqueous diarrhœa and hypopotassemia without peptic ulcer have also been described *(Matsumoto syndrome)*.

Pheochromocytoma is a catecholamine-secreting tumour localised in 90% of the cases in the adrenal medulla. In 15% of the cases the tumour is palpable. The incidence at autopsy is 0.1% [123]. The clinical picture is often that of hypertensive shock. Intermittent hypertensive paroxysms are noted in 20-50% of the cases, in the rest of the cases hypertension is a persistent clinical background. The hypertensive episode lasts 15-30 min and is accompanied by perspiration, headache, pallor, anxiety, tremor, tachycardia, palpitations, precordial, lumbar and epigastric pains. The patients are weak and nervous, with hypoglycemia between and hyperglycemia during the episodes.

The blood pressure paroxysms may be accompanied by coronary or cerebral vascular accidents or renal failure. The hypertensive episode is followed by a fall in BP, since prolonged vasoconstriction of the microcirculation results in extracapillary loss of fluid with a decrease in the ECBV. Shock is a much feared accident, even after ablation of the secreting tumour and necessitates the administration of huge amounts of catecholamines.

Diagnostic tests (regitin, histamine) and the determination of plasma and urinary catecholamines are useful. Urinary values of more than 250 μg catecholamines and 12 mg vanylmandelic acid in 24 hours are to be found. The treatment is surgical and during the hypertensive episode phentolamine is administered in perfusions (50-100 mg every 4 hours).

Primary hyperaldosteronism (Conn syndrome) is caused by a small adenoma in the fasciculate zone and may give a clinical picture of 'hypertensive shock', the same as pheochromocytoma, but with a major distinctive sign: hypopotassemia and alkalosis.

The carcinoid syndrome. The source of this syndrome is considered classically to be proliferation of the enterochromaffin cells. The tumours are yellowish and often millimetrical in size; they rarely grow sufficiently to obstruct the lumen. One of the clinical pictures of this syndrome is 'acute abdominal attack' with the symptoms of a hypertensive shock, with flush and perspiration as particular features.

Apart from serotonin, other substances, such as plasma kinins, have been recently included in its pathological physiology. These carcinoid tumours may be integrated in states of pluriglandular adenomatosis. They can secrete ACTH and insulin-like substances and are associated with mastocytosis, in which case a wide range of biogenic amines appear *(see* Section 6.6).

The physiological interdependence of vasoactive substances is emphasised by multiple associated pathological situations: pheochromocytoma, sympathoblastoma, ganglioneuroma, carcinoid, etc. (Figure 4.25).

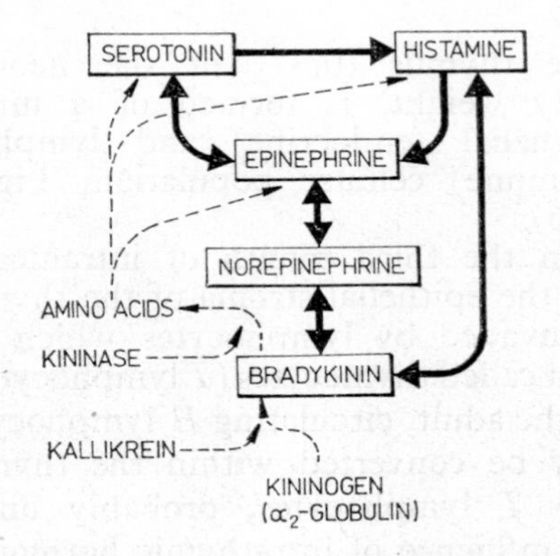

Figure 4.25. Metabolic transformation relationships of the principal biogenic amines.

4.11 **REGULATION OF THE IMMUNE SYSTEM**

4.11.1 GENERAL MECHANISMS

For its conflicts with environmental factors, the human body also has a particular system — the immune system — consolidated phylogenetically and equipped according to the principles of the genetic code files. In its huge 'archives' it keeps records of all 'personal' traits and the entire experience of the immune contacts along phylogenetic lines.

At first sight, this system whose mechanism of action demands a certain interval, does not appear to have too many points in common with the phenomenology of shock, especially when shock is understood simply, in its immediate traumatic or haemorrhagic facet. Even in bacteriemic shock, the rapid onset of the rheodynamic and metabolic events has led to their interpretation as the direct effect of bacterial antigens on vascular receptivity, coagulation, intracellular enzymes, etc. However, the conflict of any antigen with the human organism is always reflected at the level of the immune system, expressing the true and age-old 'shock' between the normal existence of the organism and all that comes counter to it.

If we admit the term of chronic shock, the immune phenomena (in which the local and systemic events generated by rejection of grafts might also be included) acquire an importance that may be readily understood. Moreover, in any state of shock, following disintegration of the cellular protein structures, waste matter appears with antigenic properties that will have to be treated during the long period of recovery after the critical stages have been solved *(see* Section 1.5.3). These modified self-structures sometimes deceive the immune system, causing the production of autoantibodies. The possible appearance of 'autoaggressive diseases' was noted by Selye as an eventual unfortunate termination of the 'adaptation syndrome'. This aspect is however insufficient to motivate the importance of the immune system in states of shock if during the last decade the role of central dispatcher of immune memory and the possibility of its rapid intervention had not been made clear.

In any kind of shock, the central mechanisms of the immune system are disturbed from the very beginning; what was known as self is now denied, the interdicted syntheses being spontaneously deinhibited by the state of shock of the genetic apparatus, which can no longer stop the operonal registers of autoantibodies. A dangerous autoimmune attack is directed upon the DNA, that is exactly against the controlling factor that is assumed to organise sooner or later the struggle of the entire cellular 'redoubt' to annul shock. In various types of shock anti-DNA antibodies have been isolated at least 6 min after the onset of shock (*see* Chapters 3 and 6).

Our present understanding of the immune system assumes that the whole range of antibodies which the body is able to synthetise pre-exist in the files of immune memory. This has been asserted by Ehrlich (1900), theorised by Jerne (1955), and discussed by Burnet (1957), and has proved exact and constantly confirmed by new findings, although immunologists still consider it today a moot point. When a new antigen appears, a certain lymphocyte able to synthesise the corresponding antibody is selected and multiplied, generating the antibody clone against the respective antigen.

Two distinct lymphocyte families are known: type *T* (according to their thymic origin) and type *B* (originating in the bone marrow). Type *B* lymphocytes are responsible for humoral immunity and produce rapid antibodies. Type *T* lymphocytes are responsible for tissular immunity and 'cooperate' with the *B* lymphocytes in the production of late antibodies; without their help the *B* lymphocytes could not synthesise ultraspecific antibodies. Both lymphocyte types have receptors for antigens on their surface. The *T* lymphocyte is assumed to react with a determinant receptor on the surface of lymphocyte *B*, thus activating the rest of the receptors. In addition, the immune competence acquired by *B* lymphocytes demands an obligatory passage through the primary immune organs: the thymus and probably the tonsils, Peyer plaques and appendix (equivalent of the Fabricius sac in birds) [6].

4.11.2 THYMUS

The thymus (0.8% of the neonate body weight) is formed of a mixed epithelial (endocrine) and lymphoid (immune) cellular population (Figure 4.26).

In the third month of intrauterine life the epithelial stroma of the thymus is invaded by lymphocytes which are then called thymocytes (*T* lymphocytes). In the adult, circulating *B* lymphocytes may be converted within the thymus into *T* lymphocytes, probably under the influence of intrathymic hormones.

The data concerning the thymus immune dispatcher are however fairly confused. Its hormones only act within the gland and thymine is also assumed to reach the neuromuscular plates.

By its endocrine function (thymine is considered to be the effector hormone) the thymus interferes in neuromuscular transmission, thus accounting for myasthenia gravis by its 'thymotoxicosis' effect (autoimmune thymitis with excess hormonal secretion).

Several substances with hormonal value have been extracted from the thymus (thymine, thymosine, thymotoxin, etc.), and a 'peripheral thymic factor' also appears to exist.

At present the double endocrine and immune role of the thymus in states of shock acting by fairly rapid mechanisms can no longer be questioned. Thymocytic hyperactivity in periods of aggression has been postulated some time ago. In contrast however, precise proof exists that inhibition of oxidative carbohydrate metabolism, specific of shock, lowers the amount of ATP in the thymocytes [127]. In shock, the immune memory, recognising what is self, is disturbed from the very beginning. The thymus is also an organ, which is affected in shock by the general metabolic impairment, the cause of all immune 'mistakes' made by the thymus; it can no longer maintain inhibition of the 'interdicted' clones, which trigger a direct attack upon the morphologic structures of the individual.

If the patient survives the critical state of shock and manages to reorganise homeostasis, including the immune one, typical immune phenomena slowly develop generated by the antigenicity of the destroyed tissues and/or initial shock factors (bacteria, foreign proteins, etc.).

Which is the actual space where immune events take place? Although lymphocytic proliferation is present in the inflamed tissue spaces, the actual antibody synthesis occurs in the secondary lymphoid organs: the lymph nodes and spleen *(see* Section 6.7).

Apart from the lymphocyte (the pure immunoglobin 'factory') there are also two types of cell-stores for antigens. In the first stages of shock the macrophages rapidly envelop the antigens(by various endocytosis means), which are digested in the phagolysosomes. Notwithstanding antigen particles resist for many months in the

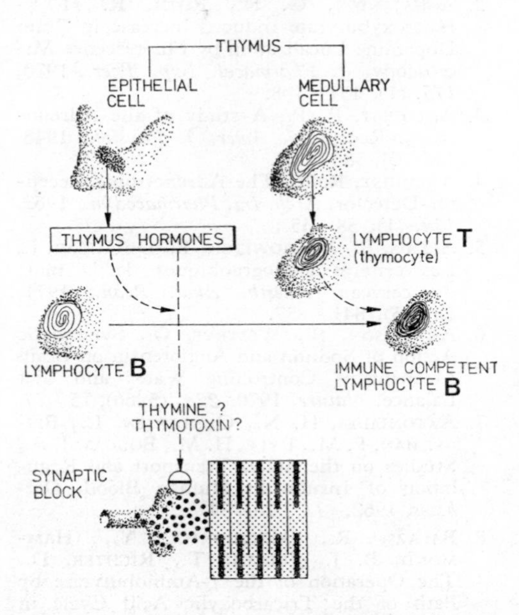

Figure 4.26. The immunological and hormonal roles of the thymus.

macrophages of the lymph nodes, continuously offering to the lymphocytes a fragmentary antigenic material. In the lymphoid germinative centres there are also dendritic cells that rapidly attach to their surface large amounts of antigenic particles, without endocytosis and digestion.

Capture of the antigens by the lymphocytes then takes place in the lymphoid follicles and medullary sinuses of the lymph nodes and spleen. Here, for six months and even several years the lymphocytes of the selected clone will synthesise hetero- and/or autoantibodies.

SELECTED BIBLIOGRAPHY

1. ADAM, P. A. J., HAYNES, Jr. R. C., Control of Hepatic Mitochondrial CO_2 Fixation by Glucagon, Epinephrine and Cortisol, *J. Biol. Chem.*, 1970, *244*, (23), 6440 – 6450.
2. AGHAJANIAN, G. K., ROTH, R. H., γ-Hydroxybutyrate-Induced Increase in Brain Dopamine Localization by Fluorescence Microscopy, *J. Pharmacol. Exp. Ther.*, 1970, *175*, (1), 131–138.
3. AHLQUIST, R. P., A study of the Adrenotropic Receptors, *Amer. J. Physiol.*, 1948, *153*, (5), 586–600.
4. AHLQUIST, R. P., The Adrenotropic Receptor-Detector, *Arch. Int. Pharmacodyn.*, 1962, *139*, (1), 38–45.
5. ALLAIS, B., VLAHOWITCH, B., DUCAILAR, J., Les critères angiographiques de la mort du cerveau, *Anesth. Anal. Réan.*, 1971, *28*, (5), 841–857.
6. ANDERSSON, B., WESTBYE, O., Synergistic Action of Sodium and Angiotensin on Brain Mechanisms Controlling Water and Salt Balance, *Nature*, 1970, *228*, (5266), 75–77.
7. ANTONIADES, H. N., GUNDERSEN, L., BEIGELMAN, P. M., PYLE, H. M., BOUGAS, J. A., Studies on the State, Transport and Regulation of Insulin in Human Blood, *Diabetes*, 1962, *(11)*, 26–33.
8. BALAZS, R., MACHIYAMA, Y., HAMMOND, B. J., JULIAN, T., RICHTER, D., The Operation of the γ-Aminobutyrate by Path of the Tricarboxylic Acid Cycle in Brain Tissue in Vitro, *Biochem. J.*, 1970, *116*, (3), 445–467.
9. BARRACLOUGH, M. A., JONES, N. F., Effects of Adenosine 3′, 5′-monophosphate on Renal Function in the Rabbit, *Brit. J. Pharmacol.*, 1970, *40*, (2), 334–341.
10. BARTHOLINI, G., GEY, K. F., PLETSCHER, A., Enhancement of Tyrosine Transamination in vivo by Catecholamines, *Experientia*, 1970, *26*, (9), 980–981.
11. BAUE, E. A., JONES, F. E., PARKINS, M. W., The Effects of Beta-Adrenergic Receptor Stimulation on Blood Flow, Oxidative Metabolism and Survival in Haemorrhagic Shock, *Ann. Surg.*, 1968, *167*, (3), 403–412.
12. BAULIEU, E. E., Les 'récepteurs hormonaux'. Mise en évidence de la liaison spécifique de l'oestradiol des protéines utérines, *Ann. Endocr.*, 1968, *29*, 1 bis, 131–140.
13. BELLOIU, D. D., ZAMFIRESCU, STEFANIA, Insuficienţa corticosuprarenală acută. Patogenie şi tratament (Acute Adrenocortical Insufficiency. Pathogeny and Treatment), *Viaţa Medicală*, 1970, *XVII*, (24), 110–115.
14. BENETT, N. B., OGSTON, C. M., MCANDREW, G. M., The Thyroid and Fibrinolysis, *Brit. Med. J.*, 1967, (5572), 147–148.
15. BENKERT, O., MATUSSEK, N., Influence of Hydrocortisone and Glucagon on Liver Tyrosine Transaminase and on Brain Tyrosine, Norepinephrine and Serotonin, *Nature*, 1970, *228*, (5266), 73–75.
16. BERK, J. L., HAGEN, J. F., BEYER, W. H., DOCHAT, G., R., The Treatment of Endotoxin Shock by Beta Adrenergic Blockade, *Ann. Surg.*, 1969, *169*, (1), 74–81.
17. BLACK, I. B., AXELROD, J., Inhibition of Tyrosine Transaminase Activity by Norepinephrine, *J. Biol. Chem.*, 1969, *244*, (22), 6124–6129.
18. BONDY, Ph. K., ROSENBERG, L. E., *Duncan's Diseases of Metabolism Endocrinology*, W. B. Saunders Company, Philadelphia, London, Toronto, 1974.
19. BOSMANN, H. B., HEMSWORTH, B. A., Intraneural Mitochondria. Incorporation of Amino Acids and Monosaccharides into Macromolecules by Isolated Synaptosomes and Synaptosomal mitochondria, *J. Biol. Chem.*, 1970, *245*, (2), 363–371.
20. BOWER, M. G., OKUDE, S., YOLLEY, W. B., SMITH, L. L., Hemodynamic Effects of Glucagon Following Haemorrhagic and Endotoxic Shock in the Dog, *Arch. Surg.*, 1970, *101*, (3), 411–415.
21. BREZENOFF, H. E. JENDEN, D. I., Modification of Arterial Blood Pressure in Rats Following Micro-Injection of Drugs into the Posterior Hypothalamus, *Int. J. Neuropharmacol.*, 1969, *8*, (6), 593–600.
22. BRISTOW, M., SHERROD, T. R., GREEN, R. D., Analysis of Beta Receptor Drug Interactions in Isolated Rabbit Atrium, Aorta, Stomach and Trachea, *J. Pharmacol. Exp. Ther.*, 1970, *171*, (1), 52–61.
23. CAILLARD, C. ROSSINGNOL, P., Une nouvelle méthode d'étude de l'adrénergie: utilisation de l'association désipraminepyrogallol pour la révélation de mécanismes adrénergiques '*in vivo*', *Path. Biol.*, 1971, *19*, (3–4), 149–156.
24. CAREY, L. C., LOWERY, B. D., CLOUTIER, C. T., Blood Sugar and Insulin Response of Humans in Shock, *Ann. Surg.*, 1970, *172*, (3), 342–350.
25. CERCHIO, G. M., Serum Insulin and Growth Hormone Response to Haemorrhagic Shock, *Endocrinology*, 1971, *88*, (1), 138–143.
26. CHASE, L. R., AURBACH, G. D., The Effect of Parathyroid Hormone on the Concentration of Adenosine 3′,5′-monophosphate in

Skeletal Tissue *'in vitro'*, *J. Biol. Chem.*, 1970, *215*, (7), 1520–1526.

27. Cheng S. C.; Nakamura R., A Study on the Tricarboxylic Acid Cycle and the Synthesis of Acetylcholine in the Lobster Nerve, *Biochem. J.*, 1970, *118*, (3), 451–455.
28. Christ, E. J., Nugteren, D. H., The Biosinthesis and Possible Function of Prostaglandins in Adipose Tissue, *Biochim. Biophys. Acta*, 1970, *218*, (2), 296–307.
29. Clarke, R. S. J., Johnston, H., Sheridan, B., The Influence of Anesthesia and Surgery on Plasma Cortisol., Insulin and Free Fatty Acids, *Brit. J. Anesth.*, 1970, *42*, (2), 295.
30. Clemens, J. A., Shaar, C. J., Kleber, J. W., Tandy, W. A., Areas of the Brain Stimulatory to LH and FSH Secretion, *Endocrinology*, 1971, *88*, (1), 180–184.
31. Dairman, W., Gordon, R., Spector, S., Sjoerdsma, A., Undenfriend, S., Increased Synthesis of Catecholamines in the Intact Rat Following Administration of α-Adrenergic Blocking Agents, *Molec. Pharmacol.*, 1968, *4*, (5), 457–464.
32. Dairman, W., Undenfriend, S., Effect of Ganglionic Blocking Agents on the Increased Synthesis of Catecholamines Resulting from α-Adrenergic Blockade or Exposure to Cold, *Biochem. Pharmacol.*, 1970, *19*, (8), 979–984.
33. Danysz, A., Buczko, W. L., Wisniewski, K., The Influence of Somatotropin on the Chlorpromazine Distribution in Rat Tissues, *Agressol.*, 1969, *10*, (3), 391–397.
34. DeMeis, L., Rubin-Altchul, B. M., Machado, R. D., Comparative Date of Ca^{++} Transport in Brain and Skeletal Microsomes, *J. Biol. Chem.*, 1970, *245*, (8), 1883–1889.
35. DeMoor, P., Steeno, O., Heyns, W., Van Baelen, H., The Steroid Binding β-Globulin in Plasma: Pathophysiological Data. *Ann. Endocr.*, 1969, *30*, (1 bis), 233–239.
36. DeRobertis, E., Studies on Neurochemical Changes in Experimental Epilepsy, *Triangle*, 1971, *10*, (3), 93–98.
37. DeWied, D., Weijne, J. A. W.N., Progress in Brain Research, volume 32: *Pituitary, Adrenal and the Brain*, Elsevier, Amsterdam-London-New York, 1970.
38. Dean, P. M., Matthews, E. K., Glucose-Induced Electrical Activity in Pancreatic Islet Cells, *J. Physiol.*, 1970, *210*, (2), 255–264.
39. Debijadji, R., Petrovik, L., Varagic, V., Stosik, N., Effect of Hypoxic Hypoxia on the Catecholamine Content and some Cytochemical Changes in the Hypothalamus of the Cat, *Aerospace Med.*, 1969, *40*, (5), 495–499.
40. Devenport, L. D., Balagura, S., Latera Hypothalamus: Reevaluation of Function in Motivated Feeding Behavlour. *Science*, 1971, *172*, (3984), 744–746.
41. Dilman, V., M., Berstein, L., M., Vassilieva, I. A., Krylova, N. K., Ostroumova, M. N., Tsyrlina, E. V., Recherche de médicaments faisant baisser le seuil hypothalamique de freinage homéostatique. I. Résultats des études du produit Agr. 307, *Agressol.*, 1971, *12*, (4), 261–268.
42. Dorfman, R. I., Ungar, F., *Metabolism of Steroid Hormones*, Academic Press. New York-London, 1965.
43. Drummond, G. I., Duncan, L., Adenyl Cyclase in Cardiac Tissue, *J. Biol. Chem.*, 1970, *245*, (5), 976–983.
44. Drummond, G. I., Harwood, J. P, Powell, C. A., Studies on the Activation of Phosphorylase in Skeletal Muscle by Contraction and by Epinephrine, *J. Biol. Chem.*, 1969, *244*, (15), 4235–4240.
45. Eccles, J. C., *The Physiology of Nerve Cells*, John Hopkins Press, Baltimore, 1957.
46. Egdahl, R. H., Drucker, W. R., Trauma Workshop Report: Endocrine Response, *J. Trauma*, 1970, *10*, (11), 1050–1051.
47. Entman, M. L., Levey, G. S., Epstein, S. E., Mechanism of Action of Adrenaline and Glucagon on Canine Heart. Evidence for Increase in Sarcotubular Calcium Stores Mediated by Cyclic 3',5'-AMP, *Circ. Res.*, 1969, *25*, 429–483.
48. Epstein, M., Hollenberg, N. K., Merrill, J. P., The pattern of the Renal Vascular Response to Epinephrine in Man, *Proc. Soc. Exp. Biol., (N.Y).*, 1970, *134*, (3), 720–724.
49. Euler, U. S. von, *Noradrenalin*, Charles C. Thomas, Springfield, Illinois 1955.
50. Ferriman, D. G.; Gilliland, I. C., *A Synopsis of Endocrinology and Metabolism*, John Wright and Sons, Bristol, 1968.
51. Finch, C. A., L'érythropoïétine, *Triangle*, 1969, *9*, (4), 127–131.
52. Foldes, F. F., Effets de l'association des myorésolutifs dépolarisants, *Anesth. Anal. Réan.*, 1971, *28*, (5), 859–869.
53. Gaches, J., Caliskan, A., Contribution à l'étude du coma dépassé et de la mort cérébrale, *Sem. Hôp. Paris*, 1970, *46*, (22), 1487–1497.
54. Gastaut, H., Lammers, H. J., *Les grandes activités du rhinencéphale*, vol. I; *Anatomie du rhinencéphale*, Masson, Paris, 1961.
55. Goldstein, M., Ohi, Y., Backstrom, T., The Effect of Ouabain on Catecholamine Biosynthesis in Rat Brain Cortex Slices, *J. Pharmacol. Exp. Ther.*, 1970, *174*, (1), 77–82.

56. Groat, W. C., The Actions of γ-Aminobutyric Acid and Related Amino Acids on Mammalian Autonomic Ganglia, *J. Pharmacol. Exp. Ther.*, 1970, *172*, (2), 384–396.
57. Harris, P., Current Concepts in the Management of Brain Trauma, *J. Roy Coll. Surg. Edinburgh.*, 1970, *15*, (5), 268–282.
58. Hasselbach, W., Taugner, G., The Effect of a Crossbridging Thiol Reagent on the Catecholamine Fluxes of Adrenal Medulla Vesicles, *Biochem. J.*, 1970, *119*, (2), 265–271.
59. Hayduk, K., Brecht, H. M., Vladutiu, A., Rojo-Ortega, J. M., Boucher, R., Genst, J., Effects of Sodium Restriction on Renin, Norepinephrine and Cation Content of Cardiovascular Tissues of Dogs, *Proc. Soc. Exp. Biol.*, 1970, *135*, (2), 271–274.
60. Heidger, Jr., P. M., Miller, F. S., Miller Jr. J. A., Cerebral and Cardiac Enzymic Activity and Tolerance to Asphyxia during Maturation in the Rabbit, *S. Physiol.*, 1970, *206*, (1), 25.40.
61. Herman, A., Mak, E., Egdahl, R. H., Adrenal Cortical Secretion Following Prolonged Haemorrhagic Shock, *Surg. Forum*, 1969, *10*, (1), 5–10
62. Hiebert, J. M., Sixt, N., Soeldner, J. S., Egdahl, R. H., Altered Insulin and Glucose Metabolism Produced by Epinephrine during Hemorrhagic Shock in the Adrenalectomised Primate, *Surgery*, 1973, *74*, (2), 223–234.
63. Hostetler, K. Y., Landu, B. R., White, R. Y., Albin, M. S., Yashon, D., Contribution of the Pentose Cycle to the Metabolism of Glucose in the Isolated Perfused Brain of the Monkey, J. *Neurochem.*, 1970, *17*, (1), 33–39.
64. Ingvar, D.H., Lassen N. A., Irrigation cérébrale et métabolisme du cerveau, *Triangle*, 1970, *9*, (7), 234–243.
65. Ionescu, B., Maximilian, C., Bucur, A., Two Cases of Transsexualism with Gonadal Dysgenesia, *Brit. J. Psychiat.*, 1971, *119*, (550), 311–314.
66. Jenkins, L. C., *General Anesthesia and the Central Nervous System*, The Williams and Wilkins Co., Baltimore, 1969.
67. Johannisson, E., The Foetal Adrenal Cortex in the Human. Its Ultrastructure at Different Stages of Development and in Different Functional States, *Acta Endocr.*, 1968, *58*, suppl. 11, (130), 11–18.
68. Johnston, D. A. I., The Role of the Endocrine Glands in the Metabolic Response to Operation, *Brit. J. Surg.*, 1967, *54*, Lister Centenary Number, 438–441.
69. Karli, P., Système limbique et processus de motivation, *J. Physiol.*, 1968, *60*, suppl. 1, 3–148.
70. Kato, R., Onoda, K., Studies on the Regulation of the Activity of Drug Oxidation in Rat Liver Microsomes by Androgen and Estrogen, *Biochem. Pharmacol.*, 1970, *19*, (5), 1649–1660.
71. King, L. R., McLaurin, R. L., Lewis, H. P., Knowles, Jr. H. C., Plasma Cortisol Levels after Head Injury, *Ann. Surg.*, 1970, *172*, (6), 975–984.
72. Kirpekar, S. M., Prat, J. C., Yamamoto, H., Effects of Metabolic Inhibitors on Norepinephrine Release from the Perfused Spleen of the Cat, *J. Pharmacol. Exp. Ther.*, 1970, *172*, (2), 342–350.
73. Klensch, H., Gott, V., Liquor Adrenalin und Noradrenalin im Operationsstress, *Klin. Wochschr.*, 1970, *48*, (14), 853–855.
74. Kong, Y., Lunzer, S., Heyman, A., Thompson, H. K., Saltzman, H. A., Effects of Acetazolamide on Cerebral Blood Flow of Dogs during Hyperbaric Oxygenation, *Amer. Heart J.*, 1969, *78*, (3), 229–237.
75. Kux, M., Holmes, D. D., Hinshaw, L.B., Masion, W.H., Effects of Injection of Live E. coli Organisms on Dogs after Denervation of the Abdominal Viscera, *Surg.*, 1971, *69*, (3), 392–398.
76. Laborit, H., Baron, C., Action comparée de différents précurseurs des catécholamines et de certains cofacteurs de leur synthèse et de leur métabolisme sur la restauration des réponses aux stimulations électriques centrales après choc hémorragique, *Agressol.*, 1969, *10*, (3), 217–239.
77. Laborit, H., Baron, C., Laborit, G., Action de la L-tyrosine sur la réponse cardio-vasculaire à l'adrénaline, à la stimulation du bout péripherique du vague au cou et du splanchnnique sur l'animal normal et en état de choc hémorragique (lapins), *Agressol.*, 1969, *10*, (3), 241–248.
78. Laborit, H., Corrélations entre synthèse protéique et sérotonine dans diverses activités du système nerveux central, *Agressol.*, 1971, *12*, (1), 24.
79. Laborit, H., Baron, C., Lond, A., Olympie, J., Activité nerveuse centrale et pharmacologie générale comparée du glyoxylate, du glycolate et du glycolaldéhyde, *Agressol.*, 1971, *12*, (3), 187–212.
80. Lamothe, C., Thuret, F., Laborit, H., Action de l'acide glyoxylique, de l'acide glycolique et du glycolaldéhyde, 'in vivo et in vitro', sur quelques étapes du métabolisme énergétique de coupes de cortex cerebral, de foie, et de myocarde de rat, *Agressol.*, 1971, *12*, (3), 233–240.
81. Leibowitz, S. F., Hypothalamic β-Adrenergic 'Satiety' System Antagonises on α-

Adrenergic 'Hunger' System in the Rat, *Nature*, 1970, *226*, (5249), 963–964.
82. LUTON, J. P., Le coma myxoedémateux, *Rev. Prat.*, 1968, *18*, (14), 2177–2184.
83. MACK, E., EGDAHL, R. H., Adrenal Blood Flow and Corticosteroid Secretion in Haemorrhagic Shock, *Surg. Gynec. Obst.*, 1970, *131*, (1), 65–71.
84. MANNI, C., BONDOLI, A., SCRASELA, E., NASO, O., RUSSO, V., L'equilibrio acido-base del liquor e del sangue nel circolo cerebrale durante l'annegamento e la rianimazione, *Min. Anest.*, 1970, *36*, (3), 214–219.
85. MARRACK, D., Regulation of Glycogen Metabolism in Man, *Amer. J. Clin. Path.*, 1968, *50*, (1), 12–19.
86. MILCU, ST. M., MAXIMILIAN, C., IONESCU, B. *Endocrinopatiile genetice* (Genetic Endocrinopathies), Editura Academiei Republicii Socialiste România, Bucureşti, 1968.
87. MOORE, F. D., Effects of Haemorrhage on Body Composition, *New Eng. J. Med.*, 1965, *273*, (11), 567–577.
88. MOREL, F., L'adénosine monophosphate cyclique, médiateur intracellulaire de l'action de nombreuses hormones, *Triangle*, 1969, *9*, (4), 119–126.
89. MORI, S., SIMEONE, F. A., Haemorrhagic Shock and Central Nervous Functions: Spinal and Spino-Bulbo-Spinal Reflexes, *Surg.*, 1970, *68*, (5), 870–877.
90. MOSS, G. S., CERCHIO, G. M., SIEGEL, D. G., POPOVICH, P. A., BUTLER, E., Serum Insulin Response in Haemorrhagic Shock in Baboons, *Surg.*, 1970, *68*, (1), 34–39.
91. MULLER, J., *Regulation of Aldosterone Biosynthesis*, Springer Verlag, Berlin-Heidelberg-New York, 1971.
92. MYERS, R. D., VEALE, W. L., Body Temperature: Possible Ionic Mechanism in the Hypothalamus Controlling the Set Point, *Science*, 1970, *170*, (395), 95–97.
93. NAKAMURA, R., CHENG, S. C., NARUSE, H., A Study on the Precursors of the Acetyl Moiety of Acetylcholine in Brain Slices. Observations on the Compartimentalization of the Acetyl-coenzyme as a Pool, *Biochem. J.*, 1970, *118*, (3), 443–450.
94. NARA, Y., GEHA, A. S., BAUE, A. E., Blood and Cerebral Lactate Metabolism during Sustained Hyperventilation in Anesthetized Intact Dogs, *Surg.*, 1971, *69*, (6), 940–946.
95. NAUTA, W. J. H., Neural Associations of the Amygdaloid Complex in the Monkey, *Brain*, 1962, *85*, 505–520.
96. NEFF, N. H., NGAI, S. H., WANG, C. T., COSTA, E., Calculation of the Rate of Catecholamine Synthesis from the Rate of Conversion of Tyrosine-^{14}C to Catecholamines. Effect of Adrenal Demedullation on Synthesis Rates, *Molec. Pharmacol.*, 1970, *5*, (1), 90–99.
97. OSWALD, I., Psychoactive Drugs and Sleep; Withdrawal Rebound Phenomena, *Triangle*, 1971, *10*, (3), 99–104.
98. OYAMA, T., TAKIGUCHI, M., KUDO, T., Effects of Hydroxybutyrate on Plasm Levels of ACTH and Cortisol in Man, *Agressol.*, 1969, *10*, (5), 411–414.
99. PASQUALINI, J. R., JAYLE, M. F., *Structure and Metabolism of Corticosteroids*, Academic Press, New York-London, 1964.
100. PILKIS, S. J., Hormonal Control of Hexokinase Activity in Animal Tissues, *Biochim. Biophys. Acta*, 1970, *215*, (3), 461–476.
101. PINCUS, J. H., CO_2 fixation in Rat Brain: Relationship to Cerebral Excitability. *Exp. Neurol.*, 1969, *24*, (2), 339–347.
102. REES, L., BUTLER, P. W. P., GOSLING, C., BESSER, G. M., Adrenergic Blockade and the Corticosteroid and Growth Hormone Response to Methylamphetamine, *Nature*, 1970, *228*, (5271), 565–566.
103. ROTH, R. H., ALLKMETS, L., DELGADO, I. M. R., Synthesis and Release of Noradrenaline and Dopamine from Discrete Regions of Monkey Brain, *Arch. Int. Pharmacodyn. Ther.*, 1969, *181*, (2) 273–282.
104. ROTH, R. H., SUHR, Y., Mechanism of the γ-Hydroxybutyrate-Induced Increase in Brain Dopamine and its Relationship to 'Sleep', *Biochem. Pharmacol.*, 1970, *19*, (12), 3001–3012.
105. ROTH, R. H., Formation and Regional Distribution of γ-Hydroxybutyric Acid in Mammalian Brain, *Biochem. Pharmacol.*, 1970, *19*, (12), 3013–3019.
106. RUBIN, R. P., The Role of Energy Metabolism in Calcium-Evoked Secretion from the Adrenal Medulla, *J. Physiol.*, 1970, *206*, (1), p. 181–192.
107. SHASKAN, E. G., SNYDER, S. H., Kinetics of Serotonin Accumulation into Slices from Rat Brain: Relationship to Catecholamine Uptake, *J. Pharmacol. Exp. Ther.*, 1970, *175*, (2), 404–418.
108. SIMIONESCU, N., Les particularité d'organe des proliférations pathologique hormonales actives des glandes endocrines, *Rev. Roum. d'Endocrinol.*, 1966, *17*, (3), 209.
109. SOMJEN, G. G., Evoked Sustained Focal Potentials and Membrane Potential of Neurons and of Unresponsive Cells of the Spinal Cord *J. Neurophysiol.*, 1970, *33*, (3), 562–582.
110. STOICA, T., *Urgenţele în endocrinologie* (Emergencies in Endocrinology), Editura Medicală, Bucureşti, 1968.

111. TEODORESCU-EXARCU, I., *Agresologie chirurgicală generală* (Surgical General Agressology), Editura Medicală, Bucureşti, 1968.
112. TCHERDAKOFF, P., SAMARCO, P., ALEXANDRE, J. M., IDATTE, I. M., MILLIEZ, P., Hypertension artérielle par phéochromocytome: à propos de 20 cas, *Sem. Hôp. Paris*, 1967, *43*, (31), 2011–2041.
113. TITCHENER, J. L., Management and Study of Psychological Response to Trauma, *J. Trauma*, 1970, *10*, (11), 974–980.
114. UNGAR, G., *Excitation*, Charles C. Thomas, Springfield, Illinois, 1963.
115. VANDERWENDE, C., JOHNSON, J., C., Interaction of Serotonin with the Catecholamines. I. Inhibiton of Dopamine and Norepinephrine Oxidation. II. Activation and Inhibition of Adrenochrome Formation, *Biochem. Pharmacol.*, 1970, *19*, (6), 1991–2007.
116. VELASCO, M. E., TALEISNIK, S., Effect of Hippocampal Stimulation on the Release of Gonadotropin, *Endocrinology*, 1969, *85*, (6), 1154–1160.
117. VILLIERS, J. C., Shock and the Central Nervous System. *Med. Proc.*, 1970, *16*, (25), 412–423.
118. WALKER, M. D., HENKIN, R. I., HARLAN, A. B., CASPER, A. G. T., Distribution of Tritiated Cortisol in Blood, Brain, CSF and Other Tissues of the Cat, *Endocrinology*, 1971, *88*, (1), 224–232.
119. WEBER, G., Hormonal Regulation and Liver Enzymes, *Gastroenterology*, 1967, *53*, (6), 984–988.
120. WERRBACH, J. H., GALLE, C. C., GOODNER, C. J., CONWAY, M. J., Effects of Autonomic Blocking Agents on Growth Hormone, Insulin, Free Fatty Acids and Glucose in Baboons, *Endocrinology*, 1970, *86*, (1), 77–82.
121. WESTPHAL, U., *Steroid-Protein Interactions*, Springer-Verlag, Berlin-Heidelberg-New York, 1971.
122. WHITSETT, T. L., HALUSHKA, P. V. GOLDBERG, L. I., Attenuation of Postganglionic Sympathetic Nerve Activity by L-dopa, *Circ. Res.*, 1970, *27*, (4), 561–570.
123. WILLIAMS, R. H., *Textbook of Endocrinology*, W. B. Saunders Comp., Philadelphia-London-Toronto, 1968.
124. YAMORI, Y., LOVENBERG, W., SJOERDSMA, A., Norepinephrine Metabolism in Brainstem of Spontaneously Hypertensive Rats, *Science*, 1970, *170*, (3957), 544–546.
125. YATES, F. E., RUSSEL, S. M., DALLMAN, M. F., HEDGE, G. A., MCCANN, S. M., DHARIWAL, A. P. S., Potentiation by Vasopressin of Corticotropin Release Induced by Corticotropin-Releasing Factor, *Endocrinology*, 1971, *88*, (1), 3–15.
126. YEH, B. K., MCNAY, J. L., GOLDBERG, L. I., Attenuation of Dopamine Renal and Mesenteric Vasodilatation by Haloperidol: Evidence for a Specific Dopamine Receptor, *J. Pharmacol. Exp. Ther.*, 1970, *168*, (2), 303–309.
127. YOUNG, D. A., Glucocorticoid Action on Rat Thymus Cells. II. Interrelationships between Cortisol and Substrate Effects on These Metabolic Parameters '*in vitro*', *J. Biol. Chem.*, 1970, *245*, (10), 2747–2752.
128. ZALIS, D. E., ROSALES, C. D., PRINCE, M. A., Injuria termica experimental. Estudio de la acción de las gonadotrofinas serica y corionica, *Bol. Soc. Argent. Ciruj.*, 1970, *31*, (7), 161–171.
129. ZIMMERMAN, B., Pituitary and Adrenal Function in Relation to Surgery, *Surg. Clin. N. Amer.*, 1965, *45*, (4), 299–315.
130. ZOR, U., KANEKO, T., SCHNEIDER, H. P. G., MCCANN, S. M., FIELD, J. B., Further Studies of Stimulation of Anterior Pituitary Cyclic Adenosine 3′,5′-Monophosphate Formation by Hypothalamic Extract and Prostaglandins, *J. Biol. Chem.*, 1970, *245*, (11), 2883–2888.

5 VASCULAR DYNAMICS

Why in the evolution of man did this vasoconstrictive response to stress survive if it is harmful?

HENRY LABORIT

5.1 GENERAL DATA
- 5.1.1 Haemodynamostat
- 5.1.2 Immediate vascular reactions
- 5.1.3 Oscillations of the haemodynamostat in shock

5.2 THE HEART IN SHOCK
- 5.2.1 Intrinsic factors
- 5.2.2 Metabolism of the heart in shock
- 5.2.3 Regulation of cardiac performance
- 5.2.4 Impairment of cardiac performance

5.3 REGIONAL FLOW
- 5.3.1 Flow-tissular perfusion relationship
- 5.3.2 Features of regional flow
- 5.3.3 The significance of arterial pressure

5.4 VENOUS RETURN
- 5.4.1 The significance of the central venous pressure
- 5.4.2 Characteristics of portal circulation

5.5 THE MICROCIRCULATION
- 5.5.1 The sphincteral apparatus and the receptors
- 5.5.2 Extrinsic and intrinsic regulation
- 5.5.3 Regional morphofunctional features
- 5.5.4 The microcirculation and the stages of shock

5.1 GENERAL DATA

Circulatory disturbances in shock are only partly expressed at clinical level. Pallor, hypotension and, when detected, oligoanuria are important clinical features, but incomplete and general. It is only what occurs within the depth of the tissues that is of importance in the actual metabolic distress of the shocked organism. Catheterization and flowmetry have always detected a decrease in cellular perfusion (*see* Chapters 8 and 9).

Any pluricellular system has an available cannular network and a fluid that flows through it and bathes every cell. The cardiac pump, vascular tubular system and blood content form the links of a regulatory system that may be designated as the *haemodynamostat*. The essential haemodynamic parameters of the haemodynamostat (Figure 5.1) may be expressed by the equation:

$$Q = \frac{BP}{R}$$

where BP is the mean arterial pressure and R the total peripheral resistance; the ratio of these factors determines the values of the blood volume output, Q, i.e. ECBV, which expresses the functional efficiency of the entire system. Maintenance of this vital relationship depends upon the cardiac pump, the dynamics of the vascular walls and the properties of the circulating fluid.

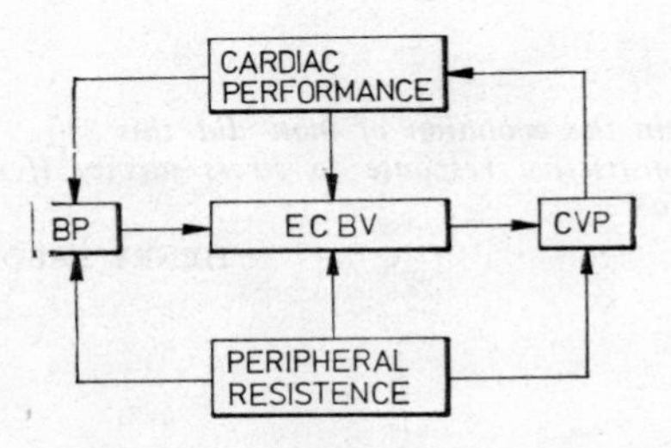

Figure 5.1. The essential parameters of the haemodynamostat.

But the above formula must integrate the continuously changing values. The term, *haemometakinesis*, [91], suggests the need for successive integration of both the blood flow and the vascular dynamics. The laws of blood flow *(haemorheology)* form the subject of a separate chapter (*see* Chapter 7). This chapter is concerned only with cardiac performance and the dynamics of the macro- and microcirculation.

Almost any change in the flow of the intravascular fluid is subordinated to the dynamics of the walls of the vascular tree. This tubular system embodies a valvular apparatus that channels the flow of blood in a given direction, and an unequally developed muscular tunica that organizes the myocardial pump, arteriolar barrier and network of sphincters in the microcirculation.

The first response of the organism to shock is a change in the calibre of the entire vascular network.

The whole vascular system (the container) is readjusted. All rheologic blood flow and cellular metabolic changes (although they do not comply uniformly since they often have a start of their own) immediately 'shadow' the *readjustment haemodynamic reaction*, always present in, and characteristic of shock (Figure 5.2).

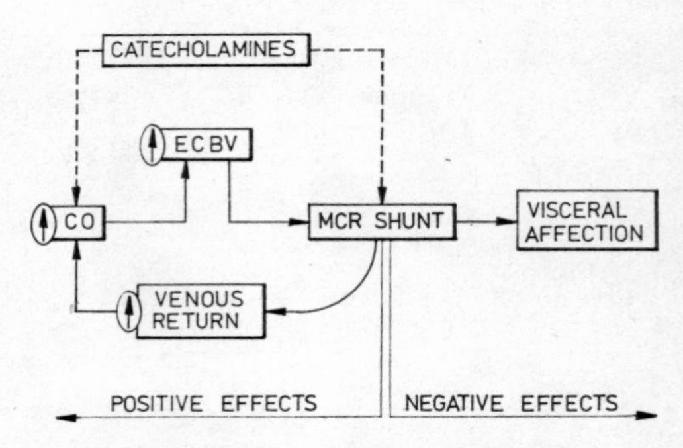

Figure 5.2. The immediate readjustment reaction of the vascular network.

Although the 'asymmetrical regulation' of perfusion of the visceral vascular beds is, on general lines, well understood there are still several points of uncertainty. Apart from systemic, neuroautonomic and hormonal regulation, a cardiac autoregulator also intervenes, and there is a local humoral regulation of the microcirculation (Figure 5.3).

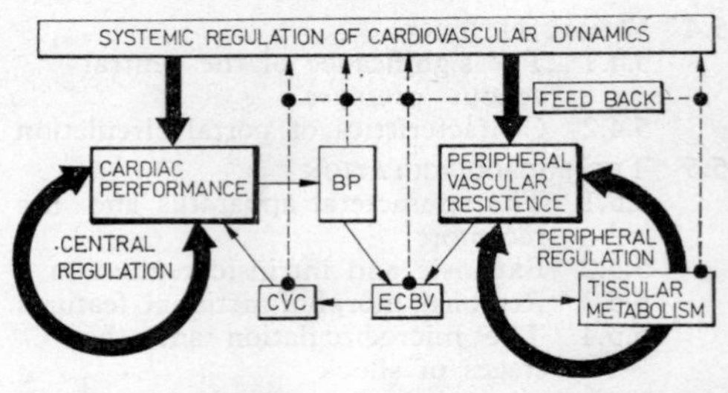

Figure 5.3. The three general types of regulation of the haemodynamostat: systemic neuroautonomic regulation, cardiac intrinsic regulation and tissular regulation of the microcirculation.

An attempt at systematizing the factors on which haemodynamic performance in shock depends, might reveal the following:

(i) the contractile cardiac tissue, upon which an optimal cardiac output depends;
(ii) arteriolar resistance, which dictates regional flow and pressure;
(iii) the microcirculation, upon which cellular immersion depends, named, by De Takats (1966), as the third haemodynamic force;
(iv) the venous sector controlling the blood return;
(v) the low pressure sector of the functional pulmonary circulation;
(vi) the lymphatic return.

Fundamental among these factors that work together, appears to be *cellular perfusion at the level of the microcirculation*, which also expresses the efficiency of the other factors. It is necessary in shock, therefore, to assess repeatedly at least the following values: cardiac output, ECBV, central venous pressure, arterial blood pressure peripheral vascular resistance and, especially, the dynamics of the regional flow and microcirculation. Part of them are monitored clinically and the others are calculated.

5.1.1 HAEMODYNAMOSTAT

As a rule, the organism receives information on the degree of perfusion of the tissues by physiologic monitoring of several parameters:

(i) mean BP values, continuously recorded by the sinusal *baroreceptors;* blood volume, perceived by the
(ii) *volume receptors* which record the lateral pressures produced by venous and atrial distension at optimal ECBV;
(iii) osmolarity, controlled by the hypothalamic and portal *osmoreceptors* summing up the flow properties of the circulating fluid.

Thus, there are several physiological control systems (barostat, volumestat, osmostat) which have at their disposal two important mechanisms that may correct the ECBV promptly: performance of the cardiac pump and diameter of the blood vessels.

In its entirety this regulatory system is sufficient to supply an optimal circulation in *the interest of the organism as a whole* but often insufficient for maintaining *the regional interests of the viscera.*

All these receptors mentioned may record optimal systemic data but in point of fact the tissues may be insufficiently perfused.

This 'betrayal' is possible since the parameters studied are exclusively haemorheodynamic; they do not also express the metabolic effectiveness of the haemodynamostat. The first perceptible signs of actual tissular affection are *hypoxia* and *hypercarbia.* It is in this instance that mixed adaptive mechanisms will be set off: circulatory, respiratory, renal and metabolic intracellular. Finally, another metabolic echo of maximum gravity and sensitivity is *acidosis.* It is worthy of note that a decrease of the systemic *p*H does not mark the start of acidosis, but the moment when the latter has exceeded the capacity of the buffering systems and has become decompensated. Therefore exhaustion of the buffer systems gives an early and valuable warning. which, as a matter of fact, the hypothalamus 'hears' before the onset of decompensated acidosis.

In shock, the two great initial stimuli for the readjustment of the vascular bed are *hypotension* and *hypovolaemia* (decrease in ECBV). The changes in *osmolarity*, *hypoxia*, *hypercarbia* and *acidosis* represent a second sequence of stimuli that further stress the mechanisms of the haemodynamostat and local and regional regulation (Figure 5.4).

The general effects of readjustment are asymmetrical and partial; the organism cannot save the whole and, therefore, protects what is the more precious, which is obviously also more sensible. Moreover, sacrifices in the perfusion of certain areas not only compensate the blood volume in other areas, but also seem to be highly logical: cutaneous vasoconstriction stops caloric losses, and splanchnic vasoconstriction arrests hydroionic depletion by the exocrine glands. On the other hand, the hypoxic kidney takes up its role of 'peripheral endocrine gland' reflected in its reponse to hypovolaemia (renin secretion) and hypoxia (erythropoietin secretion).

5.1.2 IMMEDIATE VASCULAR REACTIONS

In shock, the immediate haemodynamic response is the increase of cardiac performance (*see* Section 5.2) and reduction of the vascular tree.

The dynamics of the vascular network tonus is in keeping with the specific evolutive sinusoid of shock.

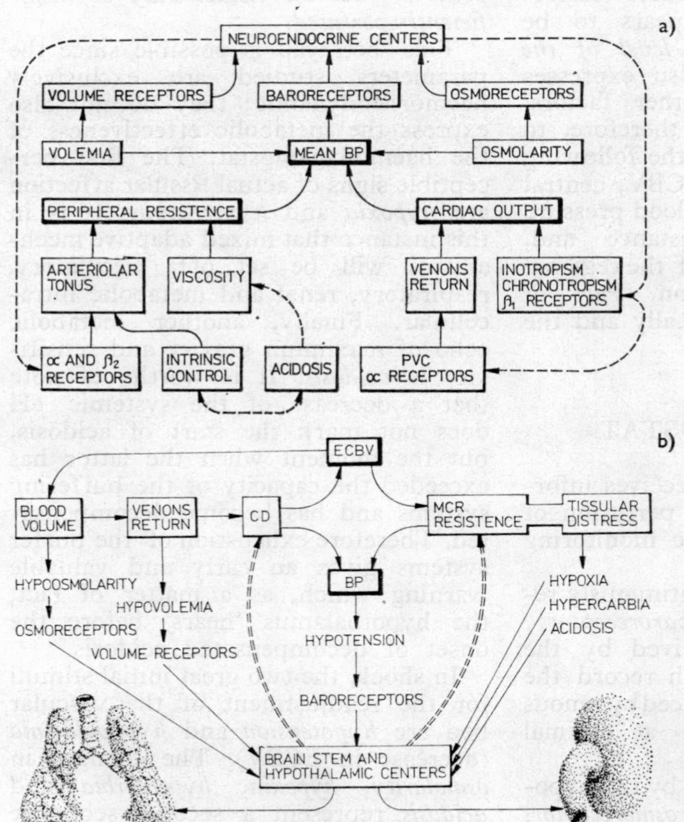

Figure 5.4:
a — regulation mechanisms of the complex haemorheodynamic constants; b — the haemorheodynamic parameters do not directly reflect cellular impairment which demands a general adaptation with the participation of another two organs: the lung and the kidney.

The kinetic mechanism of readjustment of the vascular bed is governed by an initial discharge of catecholamines (*see* Chapters 3 and 4) and performed in accordance with the histopharmacodynamic distribution of the receptors.

Vasocontriction, generalized in extent but unequal in regional intensity and duration, develops.

This initial haemodynamic reaction has been described in many ways: centralization or axialization of the circulation, preferential circulation, redistribution of the visceral flow, collateral vasoconstriction, etc. Perhaps the term *unequal systemic vasoconstriction* is better than the others.

Initial spasm occurs where there are smooth muscular cells (especially in the arteriolo-microcirculation-venule compartment), but there are regional features. Spasms skirt the cerebral, coronary, hypophyso-thyroid-adrenal and diaphragmatic circulation but are severe in the kidneys, skin, portal visceral space and skeletal musculature (Figure 5.5). A first consequence is mobilization of the stagnant blood (which generally forms pools in the subdermal, muscular and splenohepatoenteral microcirculation 'sponge') and the second, shunting of the blood from areas that 'withdraw' a good part of the aortic flow (kidneys). This rapid haemodynamic step represents a veritable 'autotransfusion'.

Another consequence of persistent arteriolar constriction is a fall in the hydrostatic pressure on entering the microcirculation space. Colloido-osmotic pressure remaining, in general, at the same level, the *isosphygmic* point is displaced towards the arteriolar end of the capillaries, and 'suction' of the interstitial fluid takes place towards the venular end of the capillaries, thus producing a veritable 'autoperfusion' (Figure 5.6).

In this way, the areas first sacrificed by the decreased perfusion are then squeezed out of their intercellular (interstitial) fluid.

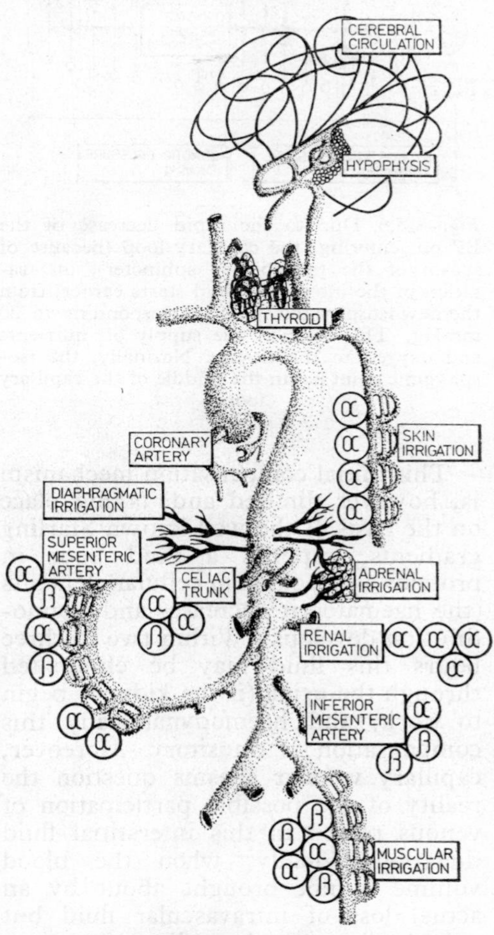

Figure 5.5. The coronary, cerebral, hypophyseal, thyroid, diaphragmatic and adrenocortical circulation has a microcirculation network that is not equipped with alpha-adrenergic receptors and is skirted by the initial vasoconstriction. The sacrificed areas are the skin and kidney that have only alpha-receptors. The next to be affected by vasoconstriction are the splanchnic circulation (where alpha-receptors are predominant on the small intestine and beta-receptors on the colon) and the muscular mass (where beta-receptors are predominant).

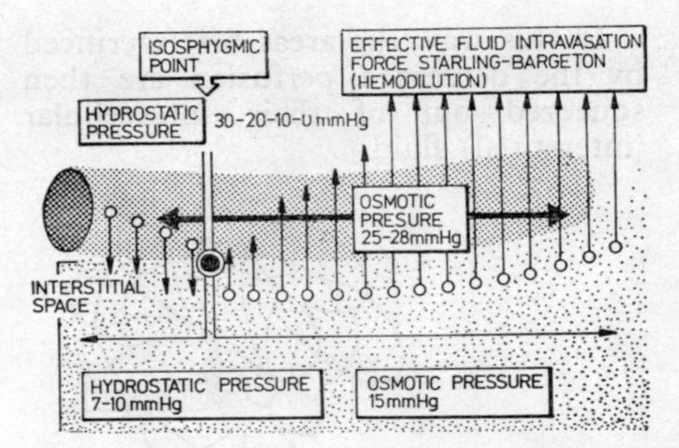

Figure 5.6. Due to the rapid decrease of the BP on entering the capillary loop (because of spasm of the precapillary sphincter), intravasation of the interstitial fluid starts earlier, from the new isosphygmic point corresponding to 20 mmHg. This restricts the supply of nutrients and oxygen to the tissues. Normally, the isosphygmic point lies in the middle of the capillary loop.

This initial compensation mechanism is, however, limited and, taking place on the basis of the well known Starling gradients, supplies a fluid poor in proteins and lacking in cellular elements (the haematocrit decreases and haemodilution develops). Within two to three hours this fluid may be eliminated through the urine (if the kidneys begin to work) and haemodynamically this compensation is transitory. Moreover, capillary venular spasms question the reality of the possible participation of venous return in this interstitial fluid donation. Finally, when the blood volume is not brought about by an actual loss of intravascular fluid but only by interstitial pooling (in some types of shock), these mechanisms of action are blocked from the start. In all situations with a real loss of fluid, penetration of the interstitial fluid into the microcirculation is in fact a biphasic phenomenon. The first fluid phase has been described; it is transitory. The second is a sustained phase which travels by the lymphatic route, bringing back to the systemic circulation an important volume of fluid rich in proteins (*see* Section 6.8).

5.1.3 OSCILLATIONS OF THE HAEMODYNAMOSTAT IN SHOCK

Haemodynamic alterations in the first phase are exclusively controlled by catecholamines. Their action is then completed and enhanced by a central and peripheral discharge of hormones in an attempt by all the mechanisms to readjust the ECBV (Figure 5.7).

In the meantime, the lymphatic flow has time to transport the proteins and water through the interstitial space (*see* chapter 6). In the first phase, triggering of the tissular mechanisms, which increases oxygen extraction, takes place, also the intracellular effort systems that activate metabolic shunts in order to adapt themselves to the severe restrictive hypoxic regime imposed upon some of the organs (*see* Chapters 8 and 9).

The entire chain of adaptive vascular events in the first phase is not obviously specific of shock. The human body will respond in a similar way to a stimulus whose lesional characteristics may give rise, after an injury, to an harmonious oscillating reaction; however, it differs from this when the lesional characteristics of the stimulus are shock-inducing (that is, the seriousness of the injury exceeds the response threshold of the body), homeostasis then being brought into balance haphazardly at the cost of irreversible peripheral scars.

In shock, the immediate vascular reactions in the first phase are more violent, unsynchronized, and incapable of smooth adaptation. This first phase can clearly be classed as the haemodynamic stage initiated by shock since it differs sharply, by its obvious *chaotic* and *unsystemic* character, from the type of mild injury that only produces an harmonious oscillating reaction.

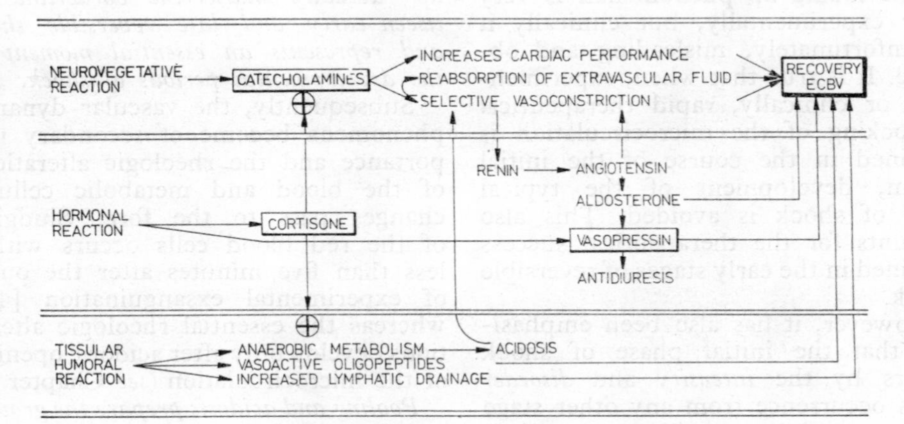

Figure 5.7. The mixed mechanisms of the immediate compensation reaction of ECBV decrease.

Aggravation of shock is due to the fact that the human body cannot repair the initial lesion without certain sacrifices. Catecholamines, as well as the other active messengers (cortisol, ADH, T_3-T_4, etc.) will continue to stimulate the shocked body. When the ECBV cannot be compensated for by means of endogenous rapid readjustment the vascular spasm continues in the sacrificed spaces. This persistent vascular spasm, which takes place in the microcirculatory areas, can be lysed only by a steady and dangerous increase in acidosis. However, for this acidotic deblocking of the microcirculation too high a price may be paid — cellular metabolic damage — which is often irreducible.

Opening of the microcirculation by a relatively rapid remission of the ECBV, owing to the mechanisms of immediate haemodynamic readjustment, should result in a return to the normal rhythmic after-effects of an injury — the well known post-injury oscillating reaction (post-injury systemic syndrome). In this case, deep acidosis does not appear and the tissues receive a thorough cleansing.

In case of shock, delayed acidotic opening of the precapillary sphincters has the effects of a dangerous trap: *it opens an immense sluice which remains obstructed at the venular end and will hold back an amount of fluid three to four times greater than under physiologic conditions; moreover, this occurs in full blood volume crisis and postcapillary venular arrest*. This is the 'pooling' phenomenon, the 'lake stage' (Knisely) or 'stagnant anoxia' phase (Lillehei). Under experimental conditions (the shock with reservoir model) opening of the large sluice of the microcirculation is marked by spontaneous withdrawal of the blood from the reservoir, the 'taking up' phenomenon (*see* Chapter 2).

Many authors consider that this event marks the actual onset of shock. In fact, it is only a crucial moment when the vascular and rheologic disorders of shock become more and more unpredictable and the depth of metabolic lesions dangerous *quad ad vitam*.

The taking-up phenomenon is very clear experimentally, but clinically it is, unfortunately, misleading and obscure. It is true that when, experimentally or clinically, rapid therapeutical deblocking of the microcirculation is obtained in the course of the initial spasm, development of the typical state of shock is avoided. This also accounts for the therapeutical success obtained in the early stages of reversible shock.

However, it has also been emphasized that the initial phase of shock differs by the *intensity* and *disorder* of its occurrence from any other stage of the post-injury oscillating reaction, which is always *spontaneously* harmonious, without therapeutical assistance (PIOR or PISS). It is time to remember that PIOR, PISS, PAOR and PASR are very similar terms that all significate a particular post-injury state not so dangerous, severe and complex as shock *(see* Section 1.4.3).

In the clinic, the gravity of a lesion must be appraised from the very beginning, assessing its potentiality for generating shock as opposed to only a post-injury oscillating reaction.

To believe that asymmetrical opening of the microcirculation is the onset of shock is to regard shock as only a *complication* of the post-injury oscillating reaction and to deny its aetiological pathogenic roots, which go much deeper. In other words, shock resembles, in its first steps, a post-injury oscillating reaction, but very quickly its characteristics shift and it takes on a severe pathophysiological and clinical aspect entirely its own. In fact, opening of the microcirculation sluice is performed by acidosis which already reflects intense metabolic impairment characteristic of shock.

Therefore, *pooling* (and its corresponding experimental phenomenon *taking-up) actually marks the borderline between early and late reversible shock and represents an essential moment in the developmental periods of shock.*

Subsequently, the vascular dynamic phenomena become of secondary importance and the rheologic alterations of the blood and metabolic cellular changes pass to the fore. Sludging of the red blood cells occurs within less than five minutes after the onset of experimental exsanguination [47], whereas the essential rheologic alterations develop only after acidotic opening of the microcirculation (*see* Chapter 7).

Pooling and acidosis prepare the ground for disseminated intravascular coagulation, which will draw the second borderline between the evolutive stages of shock: between the late reversible stage and the refractory reversible stage.

This second crucial physiopathological borderline is clinically as blurred as was that mentioned above.

Disseminated intravascular coagulation will result in thrombi blocking the microcirculation and a worsening of the cellular affliction, with intravisceral clots and progressive cellular necrosis.

There immediately follows a stage of *consumption coagulopathy*, almost always mixed with hyperfunction of the *fibrinolytic system* that may render the microcirculation patent again even in the postcapillar-venular section. This phenomenon occurs too late, however, since the phase of lysis of the clots takes place beyond the final threshold of the stage of shock (with an irreversible trend), because in the meantime the cellular lesion has produced irreparable local and systemic scars (*see* chapter 9).

5.2 THE HEART IN SHOCK

In shock the heart either cannot pump the amount of blood received (cardiogenic shock) or does not receive sufficient fluid to introduce it into the circulation (non-cardiogenic shock) (*see* Section 12.6). The heart is always the centre of all important macrohaemodynamic loops in any kind of shock (*see* again Figure 3.1).

In an organism in shock, the heart is of interest both as a systemic pump and as a viscus shocked in its turn.

The cardiac output depends upon the following factors, which separately or grouped together may be affected in shock:

- venous return;
- myocardial force (inotropism);
- efficiency of the valvular system; frequency (bathmotropism);
- cardiac rate (chronotropism);
- conductibility of the nodal system (dromotropism);
- peripheral resistance.

To this list may be added the drugs administered to the patient previously. Intrinsic or extrinsic, all these factors may be reflected pathologically by a decrease in cardiac performance and its immediate corollary, a decrease in the ECBV.

5.2.1 INTRINSIC FACTORS

The intrinsic factors must be understood in their deep infracellular functional meaning.

The contractile substrate is, as in the skeletal striated muscle, a protein tissue of *myosin* and *actin*, to which is added *troponin* (as allosteric regulator of the action of Ca^{++} on the contractile system) and *α-actinin*. In the myocardium, contractile proteins are organized in myofilaments, homogenized within a structure crossed by mitochondrial rows and sarcoplasma septa; a particular complex spatial geometry is construed whose energy rationale is only guessed at present. Macroscopically, by sliding, spiralling, torsion, lateral motion, etc. the myofilaments (depending upon the accepted theory) produce a reversible shortening of the muscular segments, i.e. *contraction.*

Of greater interest is the mechanism of the excitation-contraction coupling, which may be summed up as follows:

The action potential generated by the pace-maker of the sinoatrial node is generalized by the nodal tissue and propagated by ionic currents along the sarcolemma penetrating through their tubular invaginations; at this level, calcium ions are released from the sarcoplasmatic cisternae and, especially for the myocardium, from the intercalar disks and even the mitochondria (Figure 5.8). This 'dust' of calcium ions is bound to troponin which initiates the actomyosin changes that elicit contraction. The Ca^{++} pump now brings the calcium ions back again into the reticulum; therefore, the calcium ions couple the electric and mechanical cardiac systems, and myosin ATP-ase activity supplies the source of energy for the complex function of contraction.

The myocardium benefits by certain ionic, metabolic, circulatory and regulatory characteristics. Contractility requires certain Ca^{++} ionic concentrations, rhythmicity Ca^{++} and K^{+}, excitability Na^{+}, conductibility K^{+} and certain *p*H values. Hence, in shock, the myocardium will suffer from numerous asymmetrical systemic ionic losses.

It is known that cardiac automatism depends upon an increased permeability for Na^+ of the nodal cell membranes, allowing for spontaneous, rhythmic decrease of the resting potential up to the critical level of endogenous triggering of the action potential. This automatism might also be balanced by the permanent oscillation of the calcium ions between the mitochondria and myofilaments. The positive inotropic effect of cardiac glycosides and epinephrine is due not only to anti-ATP-ase action but also to the increase in calcium affinity of the intra-sarcoplasmatic storage system (specific β_1-effect). The effect of β_1-receptor inhibitors is assumed to be induced by a decrease in membrane permeability to Ca^{++} [75]. The acetylcholine quanta liberated by the vagus nerve, as well as norepinephrine and epinephrine released by the postganglionic sympathetic filaments, act upon the sinoatrial node influencing these ionic concentrations.

The ionic particularities of the myocardium also confer upon it its sensitivity to acidosis which lowers the performance of the heart (by Ca^{++} blocking), to sodium retention, which accelerates the heart rate, and to potassium retention, which slows it down.

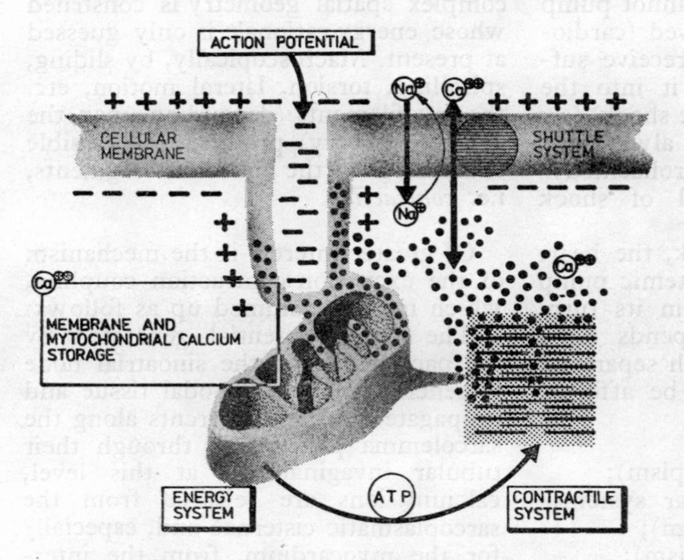

Figure 5.8. The action potential determines the entrance of Na^+ in the cell and inverts polarization of the membrane. The Ca^{++} ions are rejected by Na^+ ions from their storage points (reticular cisternae and mitochondria) flooding the contractile system supplied with energy by ATP. After contraction there is a rapid interference of the calcium ion transport system towards the storage points and of sodium towards the exterior of the cell (instead of sodium, extracellular calcium is introduced, replacing the losses of the entire calcium cycle).

5.2.2 METABOLISM OF THE HEART IN SHOCK

To continue contraction in shock — as in the physiological state — the heart needs a coronary flow of at least 250 ml/min blood with at least 10-12 g Hb% (one twentieth of the cardiac output). This will allow for a metabolism with numerous characteristics:

- as a rule glucose is a minor source of the cardiac metabolism (11 g glucose/day is burned) whereas lipids furnish over 60 per cent of the heart's daily energy;
- for the heart the first energy source is the oxidation of FFA;
- pyruvatekinase is inhibited by FFA, which thus impose their metabolic pathway towards active acetate;

- glycerol cannot be used since there is no glycerolkinase;
- the hexokinase system is stronger than phosphorylase with a prevalent use of exogenous glucose (in shock, hypoxia stimulates the myocardial hexokinase still more);

cyclic AMP, which, in turn, is activated by epinephrine and glucagon, Ca^{++} and an as yet unknown protein factor.

These features govern the metabolism of the heart muscle allowing it to

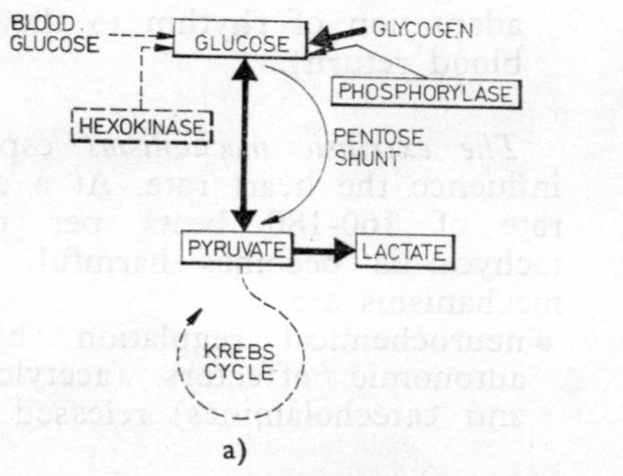

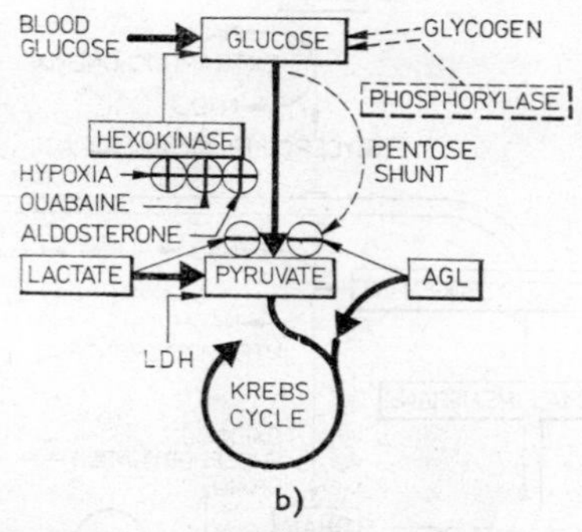

Figure 5.9. The metabolic features in shock of: *a* – the skeletal muscle; *b* – the cardiac muscle.

- *LDH is represented by a tetrameric isoenzyme more apt to transform lactate in pyruvate than the contrary* (in shock, the initial acidotic period does not affect the heart energy metabolism) (*see* Section 9.3);
- lactate and pyruvate freely cross the cell membrane and do not depend upon insulin; *the heart utilizes 10 g lactate per day and ten times less pyruvate.*
- the pentose cycle, utilized for synthesis, is but poorly represented in the heart; nevertheless the heart is gradually enlarged by protein synthesis, a phenomenon that can be inhibited by immunosuppressive drugs when it is undesirable, as in transplants [4, 53, 106];
- glycogen (9 mg/g fresh tissue) is closely bound to proteins and hydrolyses more rapidly in shock, giving ester glucose-1-phosphate, the reaction being controlled by a not too abundant phosphorylase activated by

perform for a long period under the hypoxic conditions of shock (Figure 5.9). Moreover, benefiting by its preferential circulation, the heart disposes of an EMK cycle that is active for a long time and can use as source of energy lactate, which is a dangerous residue for the other cells of the body.

In addition, glycerol-1-P-dehydrogenase (stimulated in shock by the thyroid hormones T_3 and T_4) is able to furnish intramitochondrial $NADH_2$ by shorter pathways. This will be used in the oxidation-phosphorylation chain as long as hypoxia does not also affect the myocardium (Figure 5.10). It is a very rapid source of energy, specific of the myocardium.

5.2.3 REGULATION OF CARDIAC PERFORMANCE

As a rule the heart increases its output by two economical systems (lengthening of the myocardial fibres and improvement of the metabolic effectiveness) and a non-economic mechanism (increased heart rate). These modalities are under intrinsic and extrinsic control. In shock, the heart adapts its performance by exaggerating the same ex-

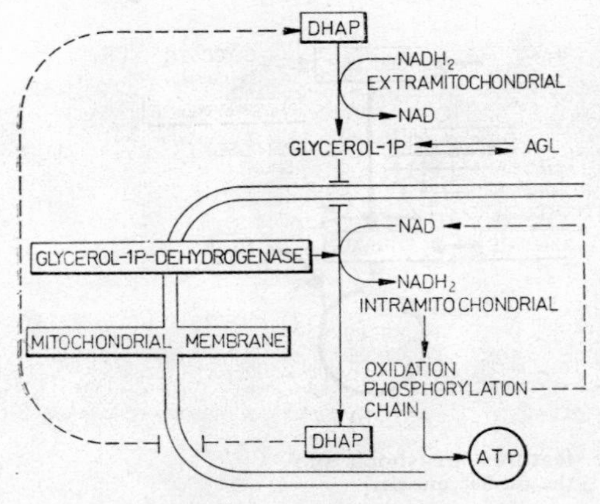

Figure 5.10. Dehydroxyacetonphosphate (DHAP) is reduced to glycerol-1-P, which is oxidised in the mitochondria by glycerol-1-P-dehydrogenase, the hydrogens obtained being carried by $NADH_2$ in the oxidative chain.

trinsic and intrinsic, nervous and humoral mechanisms it generally uses (Figure 5.11).

The intrinsic mechanisms that exercise a particular influence on the force of the myocardium are:

- Franck-Starling heterometric autoregulation ('the law of the heart'), which increases blood ejection (stroke output) by lengthening the myocardial fibres (tonogenic dilatation) and enhancing the contraction force of the muscle. Excessive cardiac stress will then lead to myogenic stretching with increase in the diastolic volume and decrease in stroke output;
- homeometric autoregulation involves changes in the maximum rate of shortening of the fibres, accounted for by an increase in the metabolic rate due to K^+ losses from the cells and a richer Ca^{++} supply;
- intrinsic autoregulation of rhythmicity: venous return compresses the sinoatrial node and probably increases the permeability of the node cell membranes (the well known Bainbridge reflex; this is an economic adaptation of rhythm to the actual blood return).

The extrinsic mechanisms especially influence the heart rate. At a cardiac rate of 160-180 beats per minute tachycardia becomes harmful. These mechanisms are:

- neurochemical regulation by the autonomic effectors (acetylcholine and catecholamines) released parti-

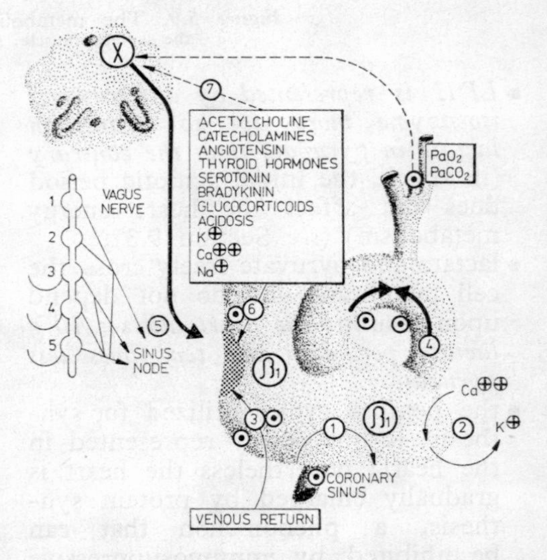

Figure 5.11. The regulation mechanisms of cardiac performance. The site of the right atrium, vena cava, carotid glomus, coronary sinus, arch of the aorta and pulmonary vein receptors are shown:

1 — heterometric Franck-Starling autoregulation mechanism; *2* — homeometric autoregulation; *3* — intrinsic autoregulation of the rhythm (Bainbridge reflex); *4* — dynamic mechanism of regulation by variation of the resistance to ejection; *5* — direct neuronal autonomic sympathetic regulation; *6* — direct neuronal autonomic parasympathetic regulation; *7* — indirect mixed neuronal regulation.

cularly in the region of the sino-atrial node by the vagal fibres and
- Wrisberg sympathetic cardiac plexus;
- direct humoral regulation of the cardiac fibre by plasma concentrations in catecholamines, T_3-T_4, H^+, lactic acid, O_2, CO_2, K^+, Na^+, Ca^{++}, etc.;
- indirect humoral regulation following the influence exercised upon the nervous centres on the floor of the 4th ventricle (the interobex-vestibular area) and hypothalamic space by catecholamines, angiotensin, histamine, serotonin, kinins and plasma H^+, CO_2 and O_2 concentrations.

5.2.4 IMPAIRMENT OF CARDIAC PERFORMANCE

In shock, disorders in the coronary flow are interpreted differently by the two great schools, according to whether heart failure is considered as the initial factor of irreversibility in shock, or as a subordinate factor depending upon the decrease of ECBV which is the *primum movens*. Although running in different pathophysiological directions it is a question of the same fundamental haemodynamic looping circuit of which only an isolated segment can be studied artificially and, therefore, uselessly (*see* again Figure 3.1.) [5, 15, 21, 72].

In practice it is advisable to consider as transitory and incomplete the initial haemodynamic reaction (favouring the heart-brain axis) and to protect therapeutically, from the beginning, the coronary flow and cardiac electromechanical system. Persantin (elective coronary dilator) administered from the onset in dogs with haemorrhagic shock significantly prolonged their survival [45]. Many experimental models lend support to the same idea. The metabolic defect in the myocardium in haemorrhagic shock may be attributed to the effects of hypoxia. The presence in the blood of a myocardial depressant factor may also be contributary (Heimbach, 1973) [41].

In the rhythm and the conduction disturbances of shock the way in which *cardiac arrest* occurs is of great importance; it implies a sudden, unexpected cardiac arrest, irrespective of the pathophysiologic process, thus excluding the terminal arrest in chronic progressive diseases.

The mechanisms of cardiac arrest are the following:
- asystolia (no ECG activity);
- electromechanical dissociation, that corresponds to certain severe morphofunctional lesions, is difficult to control (ECG activity is present);
- ventricular fibrillation, manifested by uncontrolled ECG activity; apart from ventricular fibrillation, ventricular flutter and ventricular paroxysmal tachycardia must be considered of equal importance since they lower the cardiac output sharply and rapidly and result in cardiac arrest.

Electronic monitoring is the only seismograph of the premonitory signs of severe rhythm disturbances [10, 14, 26].

In shock, the causes of cardiac arrest are very many: hypovolaemia (decrease in venous return), hypoxia, insufficient or excess ventilation (hyper- and hypocarbia), pulmonary embolism, coronary thrombosis, acidosis (that lowers calcaemia, e.g. inotropic force, resulting in ventricular fibrillation and electromechanical dissociation), increased corticoid production that raises the cardiotoxicity of catecholamines, electrolytic imbalance, drugs, etc. In brief, this emphasizes especially the syntropic gravity of the three great chemical pathogenic coefficients: hypoxia, hypercarbia and acidosis.

The tonus and the activity of the cardiac fibre depend upon intra- and extracellular ionic concentrations, especially Na^+, K^+ and H^+. Increase in H^+ concentration (metabolism acidosis in shock) causes a decrease of K^+ in the blood and cells that leads to depolarization with atonia and exhaustion of the cardiac fibre. Hypovolaemia, hypoxia and hypercarbia have the same effect. In such cases the following repolarization drugs are used: glucose, insulin, neurogangliopleglics, β-stimulants, cytochrome C, etc.

Hypocapnoeic alkalosis only produces hyperpolarization and lowers tonicity by relaxation of the cardiac fibre. In such instances norepinephrine alone may be useful (by depolarization).

The existence of direct cardiac lesions must be looked for in all categories of shock; wounds are readily recognized and can be readily dealt with in emergency. It is more difficult to recognize myocardial contusion, which must be suspected in all cases of thoracic trauma. Inexplicable pains in the chest, tachycardia, ECG alterations and, possibly, increased transaminases may point to the diagnosis, the treatment being that of a typical myocardial infarct, apart from the anticoagulants [9, 38].

In the course of shock with a non-cardiogenic onset, coronary embolism and thrombosis may develop. At present, when faced with a myocardial infarct with a state of shock considered as refractory to any therapy, surgery of the necrotic lesion or revascularization is performed, with good results in about 50% of the cases [7] (*see* Section 12.6).

Some time ago, it was thought that certain toxic factors in the systemic circulation, among which are the myocardial depressors, are the most dangerous, as their appearance is constantly confirmed by their isolation in almost all forms of shock. Their origin may be the reticuloendothelial system, the blood plasma or the pancreas. Specific of the heart appears to be an octopeptide with a molecular weight ranging between 800 and 1000, called *the myocardiac depressor factor* (MDF), probably produced by the ischaemic pancreas [58]. Starting from this well known polypeptide 'reactor' the MDF follows lymphatic pathways, as the external drainage of the lymphatic duct protects the heart of shocked animals. Similarly, the MDF may be cleared by dialysis, being found in high concentrations in dialyzed fluid [70]. MDF forms part of the lysosomal hydrolase family and is not similar to any other known vasoactive polypeptide (*see* Section 7.4.5).

A simple experiment recently suggested a parodoxical idea: when severe gastrointestinal ischaemia was produced in dogs, myocardial lesions were likewise depicted; a parallelism was noted between the mortality rate and myocardial haemorrhagio-necrotic lesions that were found. It was assumed that a factor useful to normal functioning of the myocardial system and splanchnic microcirculation disappears in the late phases of shock, resulting in irreversibility. The idea that, in addition to the appearance of toxic products in shock the disappearance of certain normal tropic factors also occurs must be emphasized (Dogru, 1969) [18].

It has been shown experimentally that there are two types of myocardial lesion: haemorrhagic and subendocardial necrosis, and intramyocardial zonal lesions. Moreover, it has been observed that the administration of oxygen under high pressure prevents the increase of lactate and pyruvate in the coronary sinus but does not improve the myocardial lesions. It has been assumed that forcing of the myocardial inotropism, together with tachycardia and decrease in the ventricular volume, are responsible for the cardiac tissular lesions rather than hypoxia [78], against which the heart may fight metabolically for some time.

5.3 REGIONAL FLOW

The vascular network has a parallel arborization that is in keeping with phylo-ontogenetic angiomeria, so that any damaged compartment may even be abandoned for a time without endangering the existence of the whole. Several suggestions have been made concerning the redistribution of visceral flow in shock:

- decrease in the cardiac output is the primary cause and the effect upon the vascular pedicles depends only secondarily upon specific regional resistance [15];
- changes in the regional resistance to flow by selective vasoconstriction is the principal factor [83, 85];
- perturbation of the visceral circulation begins before vasoconstriction, due to rheologic alterations [66, 86].

Actually, what happens is due to the combined action of all these factors each one of them playing perhaps a more important role in a particular kind of shock.

Distribution of the intravascular fluid depends upon a multitude of biologic factors among which are the anatomic ones, i.e. histologic distribution of the muscle-cell receptors of the vascular bed and their pharmacologic behaviour (Figure 5.5).

The blood flow is controlled by central pressure that forces the blood downstream, where peripheral resistance prevents it from continuing its course. Millions of capillaries that represent the points of exchange with the cells, are like fine lace with only a few alternately open loops. Normally, the arterial blood contains 15% of the circulating blood and the venous system 80%; the microcirculation (whose cross-sectional area is approximately a thousand times greater than that of the aorta) contains 5% of the circulating blood volume.

The regional distribution of the blood depends upon the vital importance of the organs. The liver and splanchnic areas contain about 20% of the circulating blood volume; the heart, lungs, superior vena cava and thoracic aorta contain together about 25%. The blood in these two spaces represents *the central volume* which, under shock conditions, maintains an adequate venous return and cardiac output, compensating, for a limited time, eventual acute ECBV losses. The two blood compartments, arterial and venous (also called the high and low pressure systems) respond quite differently to the mass of blood supply: the arterial system increases its *pressure* following a moderate supply of fluids, whereas the venous system increases its *capacitance*, its pressure increasing very slowly.

In the arterial system regional pressure and distribution of the flow are regulated by changes in peripheral resistance. As the systemic pressure is almost the same, the flow in one visceral pedicle will reflect the total resistance of each organ.

The capacitance of the arterial elastic circulatory system is defined by the relationship between the blood volume in the system and its mean pressure. It may be just as important as the absolute values of the blood volume and the relationship between the latter and capacitance is essential for maintaing the ECBV. This is illustrated by the so-called isovolaemic shock in which the blood is dammed up in the visceral vascular beds. In this case the ECBV decreases but the capacitance is, in fact, the same.

The capacitance of the vascular system varies. In special situations (physical training, altitude, pregnancy) there is an increase in vascular capacitance, whereas in prolonged clinostatism (chronic shock) the capacitance decreases.

It is considered that as long as central and peripheral vasomotricity is maintained no significant variations in flow or pressure occur as long as the blood volume is not reduced by more than 10 to 15%. This only is the critical haemodynamic shock level.

5.3.1 FLOW-TISSULAR PERFUSION RELATIONSHIP

Ejection of the content of the two ventricles takes place in circuits with different pressures, areas, lengths and resistance, according to their histofunctional characteristics. Therefore, in order to obtain a constant flow the rate of flow must vary very much along their length.

The relation:

$$\frac{\text{volume}}{\text{output}} = \frac{\text{mean pressure}}{\text{resistance}} \text{ (rate of flow)}$$

in the systemic circulation presents the changes shown in Figure 5.12. The mean pressure constantly falls, the most abrupt gradient being that caused by the arteriolar barrier; 80% of the total peripheral resistance is produced by the arterioles and microcirculation, and 20% by the venules and veins. In the arteriolar region, where pressure decreases and resistance increases, the velocity increases almost tenfold to maintain the flow leaving the left

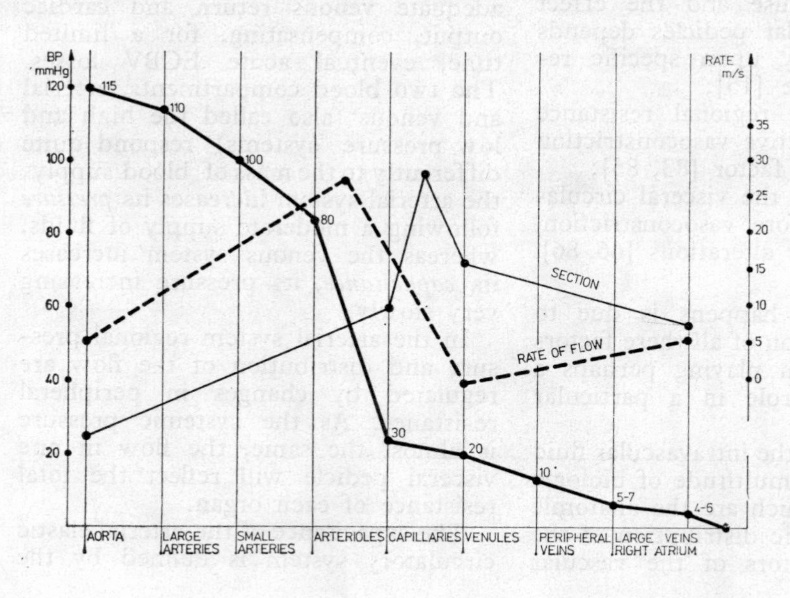

Figure 5.12. Variation of some general haemodynamic parameters.

ventricle. In the capillaries, whose walls have no muscle cells, resistance is minimal and the velocity of the blood flow is slowest, thus facilitating exchanges. Resistance also decreases owing to the much broader cross-sectional area in the region of the capillaries. In the venulo-venous sector the cross-sectional area is smaller and resistance gradually increases (without attaining the values of the arteriolar system), and consequently also the velocity.

For the systemic circulation, total peripheral resistance is 1512 ± 418 dyne/s/cm^5 and for the pulmonary circulation 250 dyne/s/cm^5 [101].

For the sake of simplicity, resistance may be expressed as follows:

$$\text{resistance} = \frac{\text{mean blood pressure}}{\text{cardiac output}}$$

which for the systemic circulation means $\frac{100 \text{ mmHg}}{100 \text{ ml/s}} = 1$ *unit peripheral resistance*, and for the pulmonary circulation 0.12 *units peripheral resistance* $\left(\frac{12 \text{ mmHg}}{100 \text{ ml/s}}\right)$ [74, 101].

The value of peripheral resistance cannot express regional readjustment, with a plus or minus balanced in parallel circuits, that is not reflected universally. Therefore, violent redistributions in the first phases of shock may readily occur in the organism. Only in a few regions is there a serial resistance (*rete admirabilis* in the kidneys and *portal circulation* in the liver) that produces marked local particular features (*see* Section 5.5.3).

Experimental and clinical studies have gathered data that at first seemed disparate: the regional flow was visualized by direct intra-operative observation and with micro-cinematography; visceral blood pressure values were recorded by transducers adapted directly to catheters; the flow was measured by flowmeters; the wash-out capacity of tissular hydrogens was determined in order to establish the effectiveness of perfusion.

There are, however, certain differences between the animal species and between the methods used and types of experimental shock. Of major importance also is the exclusiveness of the microcirculation and resistance to ischaemia of each tissue (Table 5.1).

Table 5.1 HAEMODYNAMIC AND VISCERAL METABOLIC CHARACTERISTICS

	Blood flow		Oxygen consumption ml/100 g/min	Total proportion in terms of		Onset of necrosis in anoxia. Interval in minutes
	ml/min	ml/100 g/min		stroke volume	total O_2 consumption	
Heart	250	84	9.7	7.7	11.6	20 (90 minutes in hypothermia)
Brain	750	55	3.3	13.9	18.4	3 (at 10″ loss of consciousness)
Kidneys	1200	420	6	23.3	7.1	60
Liver	1500	57.7	2	27.8	20.4	30
Skeletal muscle	850	2.7	0.2	15.6	20	60
Skin	450	12.8	0.3	8.6	4.8	—
Pancreas	—	—	—	—	—	30–60
Thyroid	—	500	—	—	—	—

Nevertheless, several facts may be mentioned in the light of what we know today:

The systemic pressure criterion is indirect and imperfect but is still maintained for its orientative clinical value. Below 75 mmHg the kidney and intestine are affected, below 35 mmHg the lung and heart, below 20 mmHg perfusion cannot be maintained in any of the tissues [32]. In shock, the perfusion of three organs is obligatory — the brain, heart and lung — and it is therefore possible even under 75 mmHg BP (Figure 5.13). At the moment when the fall in cardiac output is greater than 50%, systemic blood pressure values

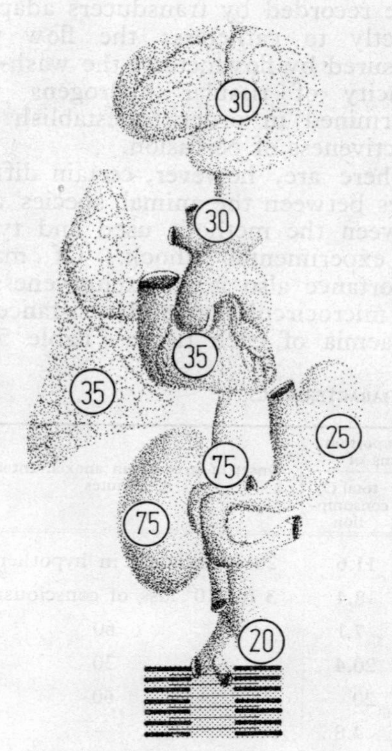

Figure 5.13. 'Critical closing pressures' (in mmHg) of the different viscera.

are so randomly distributed in the visceral pedicles that no correlation can be established [82]. This implies a need to monitor tissular flow.

The tissular flow criterion is necessary because in addition to the blood pressure the *total tissular resistance* is also introduced, a relation that depends upon the 'critical closure pressure' of the regional circulation, as demonstrated by the values mentioned. Within a tissular area, the flow falls in its arterial pedicle far more rapidly than the blood pressure, emphasizing the importance of tissular resistance (which includes decrease in the vessel diameter due to vasoconstriction, increase in transmural pressure and the viscosity of the fluid). A total loss of blood of 20 to 40% (corresponding to an up to 50% decrease in the cardiac output and a fall in the BP to 75-50 mmHg) results in:

- 33% decrease of the celiac flow (the gastric mucosa appears to be the most sensitive);
- a diminution greater than 35% of the flow in the superior mesenteric artery (from 11% of the cardiac output to 7%);
- decrease of the renal flow by 50%;
- increase of the coronary flow by 30% (Golden, 1970).

The criterion of tissular perfusion efficiency. Numerous experiments have shown that when the flow decreases by 20-30% in the mesenteric, renal, femoral and axillary arteries, tissular oxygenation falls by 40-60% [80].

Therefore, the flow is not reflected in blood pressure values nor is tissular oxygenation reflected with fidelity in the flow values.

Determination of the hydrogen ion concentration with a platinum electrode alone permits evaluation of the tissular clearance, i.e. the effectiveness of the acid wash-out of the functions eliminated by cellular metabolism [7, 105].

5.3.2 FEATURES OF REGIONAL FLOW

Current techniques used for quantitative appraisal of the regional blood flow are difficult to apply in shock. A series of data: ECG, EEG, urinary elimination, peripheral plethysmography, central temperatures, central venous pressure, etc., corroborated with the cardiac output and ECBV, may give sufficient information concerning regional flow. Several specific systems of autoregulation of the regional flows have been elucidated (Figure 5.14).

Splanchnic circulation. Increase in P_aCO_2 in the superior mesenteric artery and celiac trunk increases the blood flow in these areas (determined by ^{85}Kr clearance of the microcirculation in the gastric mucosa [2]. This is a direct effect not mediated by a decrease in the *p*H. Excessive and sudden therapeutical diminution of the P_aCO_2 may aggravate splanchnic hypoxia.

The administration of glycol in perfusions or of blood haemolysate has a dissociating effect: the flow increases in the superior mesenteric artery and decreases in the renal artery, worsening the state of the kidney. In the hepatic artery the flow decreases more slowly than in the superior and inferior mesenteric arteries. In the celiac trunk (in the neighbourhood of the diaphragmatic arteries that are always privileged) the flow is less influenced and only begins to be affected at the begining of the renal and mesenteric arteries.

The question now arises of whether vasoconstriction may be attributed only to the classical 'excessive discharge of catecholamines' or whether the cortical adrenals, the kidneys (the source of renin) and hypophysis (where vasopressin is stocked) do not also play an important and rapid part. For the

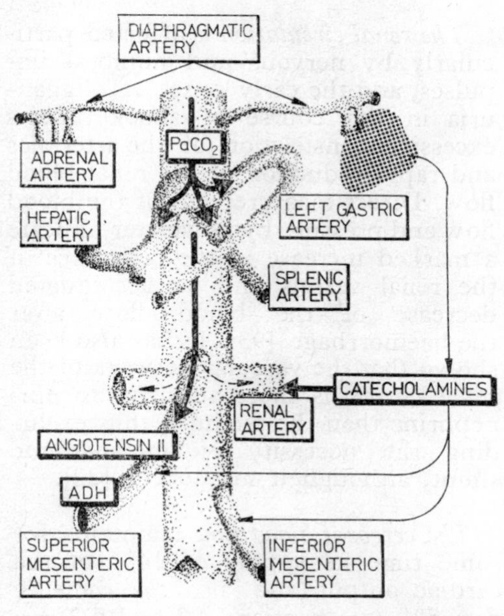

Figure 5.14. Elevation of the carbon concentration in the aorta increases the flow in the celiac trunk and the superior mesenteric artery, withdrawing part of the flow from the renal arteries. Vasoconstriction of the superior mesenteric artery is also dictated by angiotensin II and vasopressin.
The first to be affected in hypercatecholaemia is the renal circulation, then the infra-mesocolic splanchnic circulation (superior and inferior mesenteric arteries). The celiac trunk suffers to a lesser extent and in the following order: gastric microcirculation (the most sensitive), splenic artery, hepatic artery.

superior mesenteric artery, it was demonstrated in the cat (by successive ablation) that angiotensin and vasopressin control the increase of resistance in the intestinal microcirculation, with a lesser participation of the autonomic fibres and adrenals. In bacteriemic shock, mesenteric vasoconstriction is dominated by autonomic neuro-effectors, since blocking with xylocaine of the celiac area almost always protects the splanchnic microcirculation [6, 88].

The renal circulation is affected particularly by nervous and humoral impulses, and the early onset of oligoanuria in the course of shock reflects excessive constriction of the arterioles and rapid reduction of the renal blood flow. Direct measurement of the blood flow and pressure by flowmeter indicate a marked increase in the resistance of the renal vessels and an accentuated decrease of the blood flow after the haemorrhage [93]. It has also been shown that the vascular network of the renal cortex is more sensitive to epinephrine than the medulla, thus excluding the necessity of an anatomic shunt, although it may occur [22].

The cerebral blood flow maintains for some time its supply of 14% of the cardiac output and also *the coronary flow* 5% (*see* Sections 4.3 and 5.2).

The functional pulmonary circulation is governed by the systemic circulation. It works under a low pressure system, the pulmonary circulation being perfused at a pressure of 25 mmHg achieved by the right ventricle. In the pulmonary artery and arterioles the pressure reaches 15 to 25 mmHg and in the pulmonary venules and veins 9 to 12 mmHg, reflecting the pressure in the right atrium (4 to 12 mmHg). The low pressure gradient (15 to 9 = = 6 mmHg) along the pulmonary capillary network allows the fluid column to advance without passing through the capillary walls which at this level are only permeated by gases.

Anastomoses exist between the pulmonary arterioles and venules, as well as at the level of the preterminal and subsegmentary branches, thus shunting the blood over wide pulmonary areas.

As a rule, the pulmonary vascular bed is not contractile but extensible. Its participation in haemodynamic alterations is passive and subordinated to intracardiac pressures [98]. Hypoxia and hypercarbia in shock regulate the output of the pulmonary artery, increasing vascular resistance in the pulmonary bed [3] (*see* Sections 5.5.3 and 10.4).

5.3.3 THE SIGNIFICANCE OF ARTERIAL PRESSURE

Shock is accompanied by a decrease in blood pressure but a fall in BP is not always indicative of shock, nor are the terms hypotension and shock synonyms; for instance, hypotension due to fever is accompanied by an *increase* in the peripheral flow, not a decrease. Conversely, a deficient flow in the microcirculation may be accompanied by almost normal BP. Study of the relationship between blood pressure and capillary flow has shown in fact *that there is no linear correlation between BP and capillary flow* (Golden and Jane, 1970; Kirimli et al., 1970; Matsumoto et al., 1967; Simon and Olsen, 1969).

In haemorrhagic shock, microcinematography of the splanchnic microcirculation established the general limits, between which BP values may, however, be considered as an indirect indication of tissular flow [64]. Up to a BP of 80 mmHg the capillary flow exhibits no alterations; below 70 mmHg sludging appears at first in the venules, then at systemic values below 40 mmHg it becomes general throughout the capillary venular system, and under 30 to 20 mmHg the capillary flow ceases.

5.4 VENOUS RETURN

All causes of shock lower the venous return, because all haemodynamic disturbances in the vascular tree are reflected directly or indirectly in the venous flow. The real or relative intravascular fluid losses, as well as acute obstruction of the vessels reflect diminution in the venous flow. Decrease in myocardial work *(vis a tergo)* and loss of the intrathoracic subatmospheric pressure gradient *(vis a fronte)* due to direct thoraco-pleuro-pulmonary lesions (e.g. pneumo-, hydro-, pyo-, haemo-chylothorax) or abdominal distension will result in a diminution of the venous return. Ligature of the inferior vena cava (for recurrent pulmonary embolism) is followed in more than two-thirds of the cases by shock; pooling of the plasma in the lower limbs and of erythrocytes in the thrombus that forms below the ligature diminish venous return.

In the dynamics of shock, venous return depends upon:

- the fluid flow through the microcirculation (passive factor);
- the venous tonus (active factor).

The flow in the microcirculation is lowered in the spasm and pooling stages and by disseminated intravascular clotting.

The venous tonus depends upon contraction of the smooth muscle fibre in the vein walls, provided with α-receptors upon which dopamine and norepinephrine have a verified constrictor effect [61]. Bradikinin has a net vasoconstrictor effect in the rabbit, an effect that could not be obtained with histamine, serotonin, acetylcholine or catecholamines [8].

Among the first intrinsic factors that control venous dynamics next to the valvules, is the structure of the walls: the fibrous veins (dural and cerebral) and fibro-elastic veins (coronary, pulmonary) are passively in keeping with the pressure of the neighbouring tissues, whereas the fibromuscular veins (portal system) and connective muscular veins (lower cava system) have a motility and receptivity of their own. In the early phase of shock, spasm in the microcirculation is reflected by a decrease in the venous return and a marked increase in central venous pressure. This increase in venous resistance produced by constriction of the walls, which cannot be lysed by α-blocking drugs, suggests the existence of other substances with a vasoconstrictive action (especially bradykinin).

The absolute value of the venous return has many reasons to decrease in shock. Only in the first phase of shock, due to the attempt at rapid compensation (squeezing of the parenchymatous storage sponges and venular refilling), venous return is, for a short time, greater than the fluid losses will permit, which is a favourable effect but only in the presence of an intact heart. Increase in the flow follows upon an increase in pressure, which may maintain an acceptable flow even with a low return but increased velocity. In fact, venoconstriction and arteriole constriction are not always synchronous; for instance the administration of Levarterenol often elevates the central venous pressure more than the arteriolar BP [8].

5.4.1 THE SIGNIFICANCE OF THE CENTRAL VENOUS PRESSURE

In the dynamics of the onset of shock the blood volume in the large veins may decrease three to four times more than in the venules, reflecting a decrease in the cross-sectional area of the central veins, i.e. a selective central venous constriction with increase in the central venous pressure, the venular sector maintaining its role of blood 'reservoir' [46]. Apart from the increase in central venous pressure, the pressure falls in the right atrium so that the filling gradient of the right heart is doubled in an attempt to maintain the work of the heart under accentuated hypovolaemic conditions.

Phlebo-constriction (the great venulo-venous sphincter) appears as a fundamental element in the first phases of shock; it is considered by some investigators as the essential decompensatory factor and even the cause of irreversibility in shock.

As the central venous pressure is influenced by the blood volume, tonus of the venous walls, function of the myocardial pump, thoraco-pulmonary ventilation and even therapeutical drugs [19, 20], it must be closely monitored since it offers two essential elements: continuous assessment of the state of shock and evaluation of therapeutical efficiency.

The central venous pressure is closely connected with the real haemodynamic alterations since it reflects the microcirculation in its totality, whereas the BP only offers data on a single dynamic parameter of the microcirculation — overall resistance (Kerr and Kirklin, 1970). In other words CVP is a posterior, i.e. a more complete parameter than BP, which is an anterior, i.e. an incomplete haemodynamic parameter. BP is a proximal haemodynamic parameter; CVP is a distal haemodynamic parameter (Şuteu et al., 1972).

The central venous pressure reflects filling of the venous tree to a lesser extent but is a reliable indication of the venous tonus and of the ability of the right heart to pump blood into the pulmonary circulation; but it offers no information concerning the systolic volume or cardiac output, which belong to the performance of the left heart. Therefore, hypovolaemia may be masked by a normal central venous pressure in the presence of right or general myocardial failure.

In all cases, refilling with fluids must be done before the central venous pressure falls to 25 mmHg since acute pulmonary œdema develops at central venous pressure values over 30 mmHg. When high values are reached, acidosis has to be corrected for deblocking the heart, and myocardial inotropic substances administered.

In practice, measurement of the central venous pressure permits:

- evaluation of the ECBV;
- evaluation of cardiac performance (especially for the right heart);
- introduction of fluids and drugs directly into the central return circulation (hypertonic solutions are more readily tolerated);
- rapid fluid extraction in threatened myocardial failure;
- collection of samples for laboratory tests.

A low central venous pressure implies failure of the right ventricle and, only after that, a low blood volume. Conversely, at high central venous pressures the administration of fluids, leading to an increase in coronary perfusion, may improve myocardial performance and then lower the central venous pressure.

If, during perfusion of 500 ml fluid, the central venous pressure does not increase by more than 5 cm H_2O above the initial level (and the increase becomes stable at approximately 25 cm H_2O) it means that the heart is tolerating well the increase in intravascular volume. If the initial increase is of more than 20 cm H_2O the heart will probably not bear the surplus of fluids.

The information furnished by measuring the central venous pressure is more accurate than that supplied by the peripheral venous pressure because:

- the peripheral vein is not advantageous for measuring venous pressure in practice;
- during shock segmentary constriction of the peripheral veins accounts for local alterations in the venous pressure;
- the pressure in a vein segment does not represent in all instances the pressure in the large veins, right atrium or ventricle (DuCailar et al., 1970; Riordan, 1969).

5.4.2 CHARACTERISTICS OF PORTAL CIRCULATION

Mobilization of the portal blood (Figure 5.15) takes place in the following order:

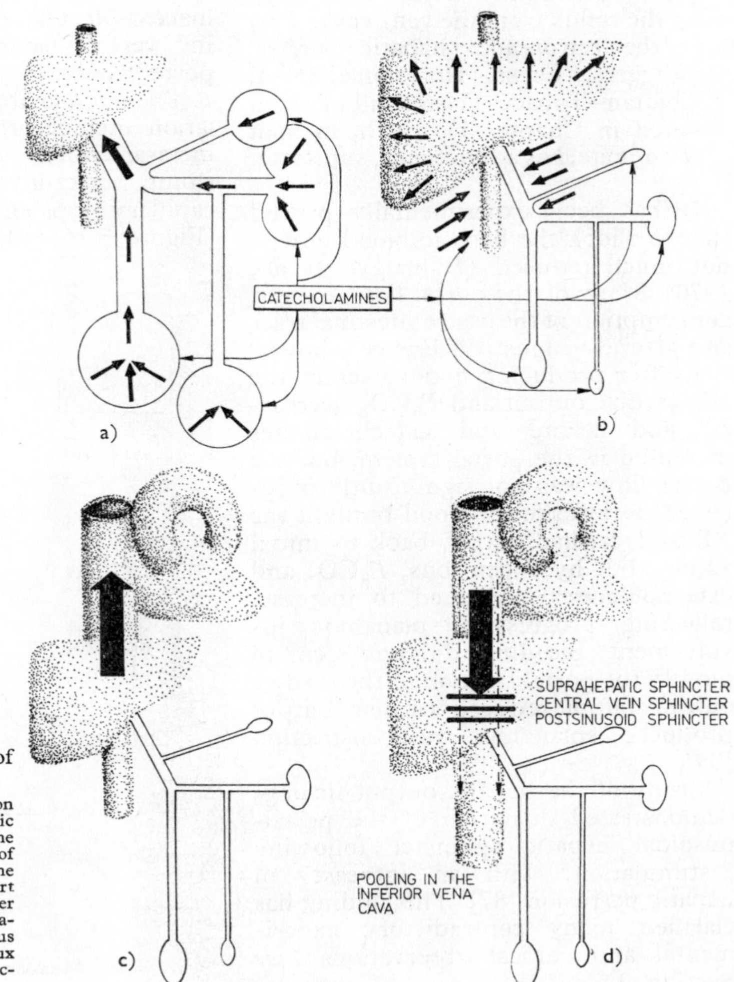

Figure 5.15. The stages of portal evacuation:

a — spleno and vasoconstriction 'squeezes out' the splanchnic microcirculation and loads the portal axis; *b* — contraction of the portal axis then loads the liver; catecholamines support these two stages; *c* — the liver increases the flow in the suprahepatic veins improving venous return; *d* — the tendency to reflux is blocked by a complex sphincteral system.

phase 1: splenocontraction and spasm of the gastrointestinal microcirculation;
phase 2: hepatic filling (portal stage) ensured by the muscular walls of the portal confluent;
phase 3: evacuation into the lower vena cava (hepatic stage of evacuation);
phase 4: contraction of the postsinusoidal sphincters of the centrolobular and suprahepatic veins that dam up the reflux from the vena cava. This is the retrograde 'hepatic barrier' (Wiggers, Leger, Hortolomei-Bușu, Chiotan -Tabacu-Cristea, all of them noted in Chapter 1) which, in man at any rate, has a functional substrate.

It has been experimentally proved that in shock the hepatic blood flow is not much reduced (Halmagyi et al., 1970). Study of the portal flow, oxygen consumption of the gastrointestinal tract and arterio-venous difference showed that after producing hypovolaemia the BP, cardiac output and P_aCO_2 decreased, and lactate and catecholamines increased in the portal system, but the portal flow was not significantly influenced. Reinfusion of blood brought the BP and cardiac output back to initial values, but hydrogen ions, P_aCO_2 and catecholamines continued to increase, reflecting progressive splanchnic involvement. Bleeding did not seem to modify the portal fraction of the cardiac output but a reduced cardiac output produced splanchnic vasoconstriction [39].

Study of the hepatic output in dogs demonstrated dilatation of the postsinusoidal hepatic sphincter following β-stimulation, with an increase in hepatic perfusion [87]. This finding has clarified many contradictory experimental and clinical observations *(see* Section 2.5).

5.5 THE MICROCIRCULATION

The microcirculation represents the network of the vessels with a diameter of less than 100 μ (with variations, according to certain authors, of 20 to 250 μ) [11, 25, 29, 31, 33, 60]. Owing to its rheodynamic function the microcirculation has also been called 'the third force' of the cardiovascular system (De Takats, 1966). This sector represents the space of gas, hydroionic, micro- and macromolecular exchanges, the remaining vessels having the role of transporters only.

The microanatomy of the microcirculation includes the arterioles, venules, meta-arterioles, arterio-venular (AV) shunt, meta-arteriolar and terminal capillary loops and a sphincteral system (Figure 5.16) [40, 42, 107]. The histo-

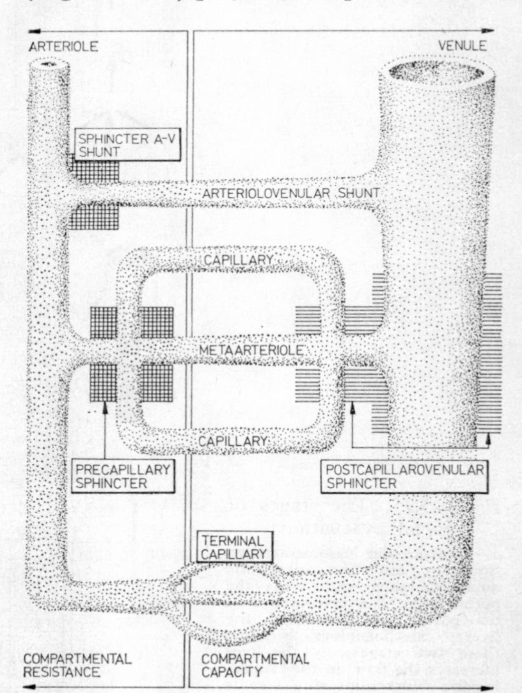

Figure 5.16. Microanatomy of the microcirculation.

logy of the capillary loop is illustrated in Figure 5.17 [49, 52, 55].

The term 'AV shunt' is complex and must be understood both anatomically and functionally, the peripheral effect being the same. In shock, the functional shunts act with certainty in the lungs and probably also in the somatic muscles [94].

Anatomically, AV shunts have been described before entrance into the microcirculation area in the lungs, kidneys and heart, and are likewise known as anatomic pathologic shunts of the

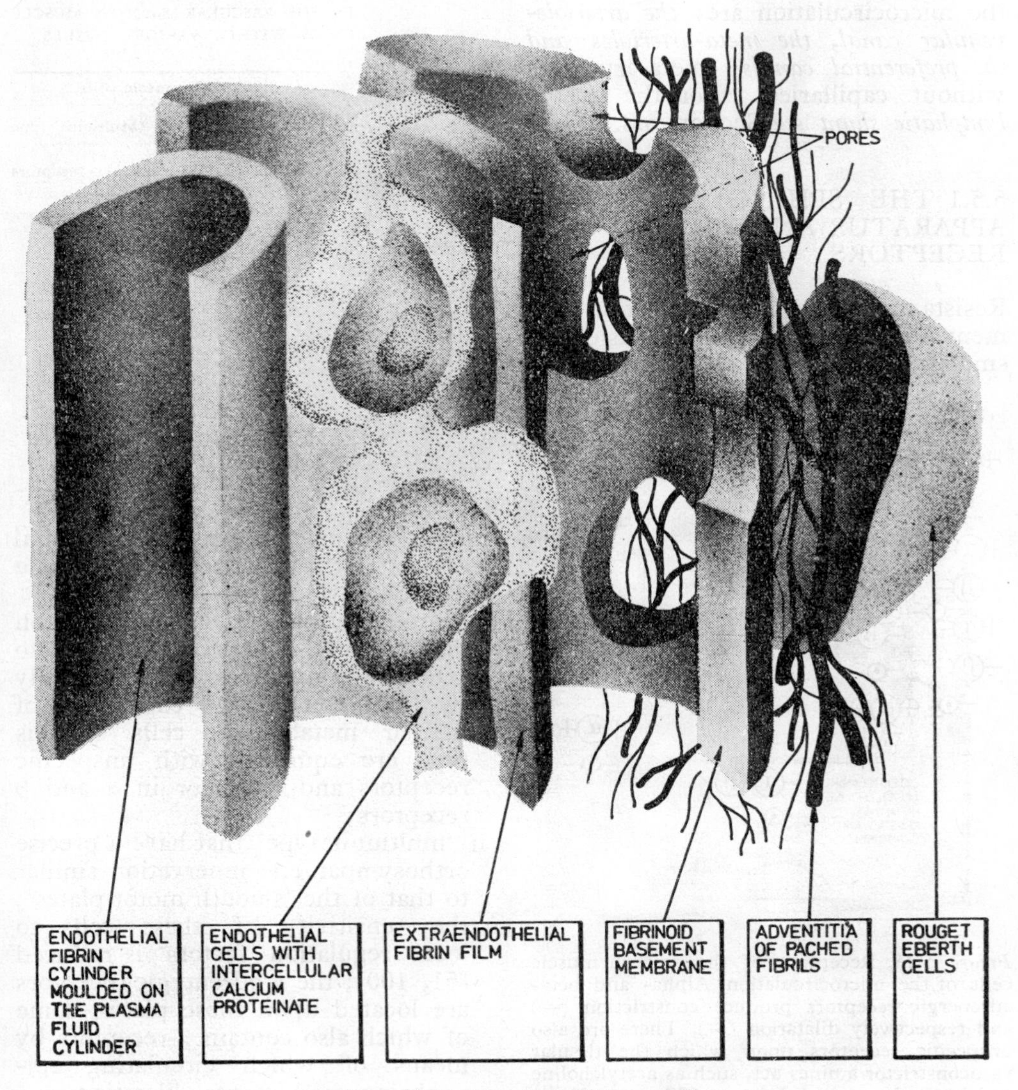

Figure 5.17. Histology of the capillary.

large and medium vessels (mixed aneurysms, Paget's disease, cirrhosis, multiple myeloma, etc.).

There are also AV shunts with particular cybernetic regulatory features: the carotid glomus and the renal juxtaglomerular apparatus.

Always present in the AV shunts of the microcirculation are: *the arteriole-venular canal, the meta-arterioles and the preferential canals* (meta-arterioles without capillaries). For the *venulo-lymphatic shunt see* Section 6.8.

5.5.1 THE SPHINCTERAL APPARATUS AND THE RECEPTORS

Resistance of the microcirculation segment is conditioned by the tonus of the smooth muscle cells dispersed in the vascular walls or concentrated in the sphincters. The muscle cells in the microcirculation area embody a number of unspecific receptors and at least two types of receptors for catecholamines: α and β (Figure 5.18 and Table 5.2).

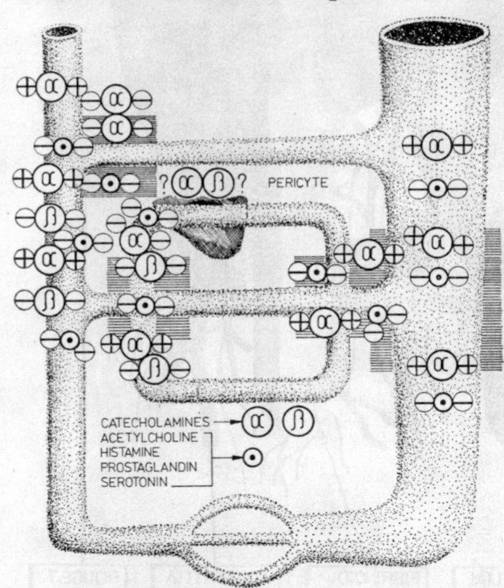

Figure 5.18. Receptors of the smooth muscle cells of the microcirculation. Alpha- and beta-adrenergic receptors produce constriction (+) and respectively dilatation (−). There are also unspecific receptors upon which the tissular vasoconstrictor amines act, such as acetylcholine and histamine, which have vasodilatory effects.

Table 5.2 DISTRIBUTION OF THE RECEPTORS IN THE VASCULAR SMOOTH MUSCLE CELLS WITHIN VARIOUS TISSUES

	Smooth muscle cells		
	'Visceral' type	'Multiunit' type	
Microcirculation	Unspecific receptors	Specific receptors α	Specific receptors β
Cardiac muscle	+	+?	+
Skeletal muscle	+	++	++++
Nervous tissue	+	−	−
Splanchnic area	+	+++	++
Kidneys	−	++++	−
Skin	−	++++	−
Lungs	+	+?	++
	Intrinsic regulation	Extrinsic regulation	

From the point of view of functional organisation the vascular smooth muscle cells are of two types:

(i) 'visceral type', in which excitation may be diffusely spread from cell to cell. The tonus is regulated especially by local factors, the expression of tissular metabolism; cells of this type are equipped with unspecific receptors and are poor in α and β receptors;

(ii) 'multiunit type' that have a precise orthosympathetic innervation similar to that of the 'smooth motor plates'; the sensitivity of these cells to local regulation factors is reduced [51, 100]; the α-adrenergic receptors are located upon these plates, some of which also contain β-receptors by means of which circulating epinephrine produce vasodilatation.

The splanchnic and muscular microcirculation combine in complex regulations since they have an entire range of muscle cells all of them equipped with specific and unspecific receptors (*see* Figure 5.5).

In the microcirculation the dynamics of the blood flow presents the following features:
(i) the capillary flow is determined by the arteriole-venular gradient (10 mmHg);
(ii) the velocity of the blood flow through the capillary segment is very small (1 mm/min) favouring the exchanges;
(iii) flow through the microcirculation lasts about 1 to 2 sec;
(iv) the total area of the capillary network is 6200 m^2 and its length approximately 100 000 km;
(v) the capillary walls are half permeable; exchanges take place according to the gradients shown in Figure 5.6;
(vi) the red blood cells progress in axial groups diminishing the viscosity (*see* Section 7.3);
(vii) under physiological conditions the blood flows through the meta-arterioles and some of the capillary loops by *rotation* (the sphincters of the microcirculation contract rhythmically 6 to 12 times per minute, so that only one-twentieth to one-fiftieth of the total capillary surface is alternately supplied with blood) (*see* Figure 5.20);
(viii) the distance between the capillary and the cells is at most, 25 to 50 μ — this minute space favouring the exchanges — and in the hypothalamus the microcirculation network is so rich that the neurons are perforated by the capillaries [13, 30, 34, 74, 84, 90].

Owing to these features the closest relationship between the dynamics of the vascular wall and the rheologic behaviour of the plasma and red blood cells is to be found in the microcirculation. This is the 'rheologic miracle' due to which the red blood cells measuring 6 to 7 μ can pass through the narrowest capillaries (1 to 2 μ).

5.5.2 EXTRINSIC AND INTRINSIC REGULATION

The tonus of the smooth muscle cells throughout the microcirculation is controlled by extrinsic and intrinsic mechanisms. The disseminated smooth muscle cells form cuffs at the bifurcations, representing, at least functionally, true sphincters. This is particularly evident in the postcapillary venular sphincter; the precapillary sphincter has been demonstrated histologically.

Induced by stimulation of the α-receptors, the initial and prolonged spasticity in the sector of resistance of the microcirculation is a fundamental, initial phenomenon in shock.

But, apart from the preferential spasm of the precapillary sphincter, the lateral derivation of the group of capillary loops may explain the shunt of the circulation in the first phases of shock. Arteriolar vasoconstriction increases the rate of flow through the microcirculation with a decrease in the lateral pressure that can no longer press the fluid into the capillaries; consequently the AV shunts will be given preference (Figure 5.19) [62, 104].

The capillary loops are successively perfused by local feedback (Figure 5.20). Under physiological conditions, 5 to 7% of the ECBV passes through the microcirculation at each moment, yet if the hepatic microcirculation 'sponge' alone were filled completely it

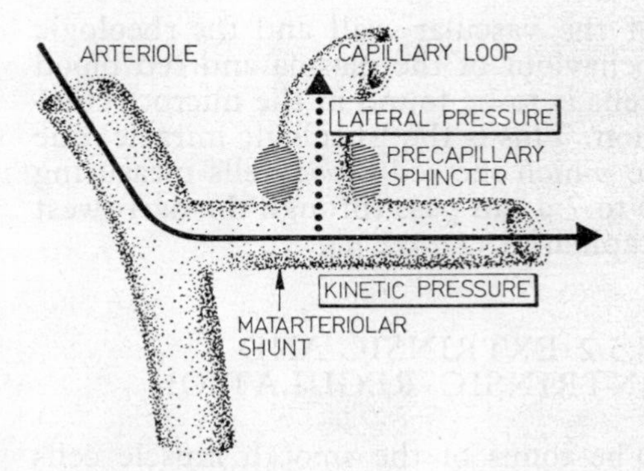

Figure 5.19. Arteriolar vasoconstriction increases the velocity with which the blood passes through the microcirculation, the lateral pressure decreasing.

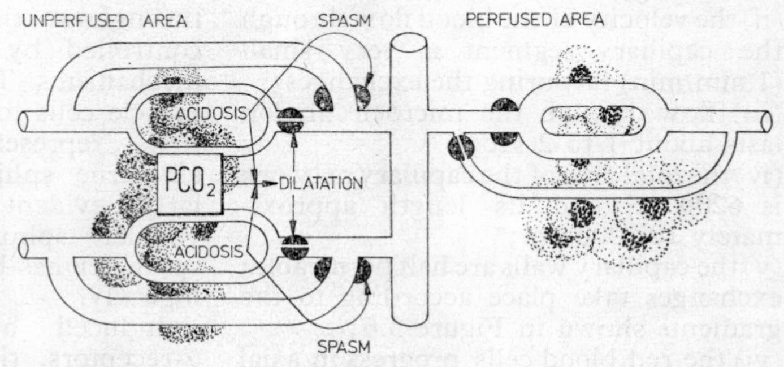

Figure 5.20. The accumulation of carbon gas and protons in the areas that are not perfused will open the precapillary sphincters and close the sphincters of the AV shunts. The anterior perfused area will then be avoided for a short interval.

could still absorb the entire volume of the circulating blood.

Arteriole-venular shunts are rarely used, but become important in the initial phases of shock. If, in general, one-third of the capillary loops open alternately, in some regions (muscles at rest, hypoderm) the proportion falls to between one-twentieth to one-fiftieth of the whole capillary network (but the intracerebral capillaries are almost all permanently open). Sudden pooling of the entire visceral microcirculation space increases threefold (12 to 18%) the proportion of fluid taken from the ECBV, resulting in non-functional stagnation of the blood.

Regulation of the microcirculation is therefore of the utmost importance; it is controlled by an extrinsic neurogen tonus and an intrinsic myogen tonus and modelled by renal (angiotensin) hormonal (steroids) and electrolytic factors.

Along general lines vasoconstriction may be said to be controlled systemically, and vasodilation locally.

The modes of regulation may be summed up as follows:

- Regulation of the microcirculation tonus
  - systemic neurochemical regulation
    - catecholamines
    - acetylcholine
  - local regulation
    - chemical
      - hormonal
      - humoral (oligopeptides, gases, ions)
    - physical (*see* Section 7.3)

Extrinsic neurogenic control is carried into effect by vasoconstriction and vasodilatation fibres (the latter, although not yet fully established are assumed to be both parasympathetic and sympathetic but cholinergic in effect, especially as regards the skeletal muscles) (Lacroix, 1970).

The tonus of the multi-unit smooth muscle is regulated by effector autonomic impulses expressed, in the last instance, by catecholamine quanta. The arterioles react by constriction to stimulation of the α-andrenergic receptors within their musculature. Stimulation of the β-receptors produces dilatation of the arterioles. The meta-arterioles, precapillary sphincters and capillary pericytes react in the same way to α and β stimulation.

The postcapillary sphincters and venules are believed to have only α-receptors, stimulation of the β type not producing vasodilatation of these structures. Arteriovenous anastomoses are both α and β sensitive but stimulation of both types appear to induce opening of the shunts. The pre- and postcapillary sphincters also differ in their sensitivity to hypoxia and changes in *p*H of the internal medium; thus, acidosis produces rapid relaxation of the precapillary sphincter and, only much later, of the postcapillary one.

In addition, the segments described also have cholinergic receptors whose stimulation produces dilatation of the respective vessels but also opening of the arterio-venous shunt sphincters.

General extrinsic regulation of the vascular diameter depends upon certain complex neurochemical factors, shown in Figure 5.21 [73, 79, 101].

The tonus of the visceral smooth muscles is equilibrated by two tendencies: under the influence of the perfusion pressure (mean BP) the tonus tends to rise (autoregulation phenomenon) whereas the locally accumulated metabolites tend to lower the tonus in order to increase the fluid column that supplies their own 'washing' (Lareng, 1969; Simon and Olsen, 1969).

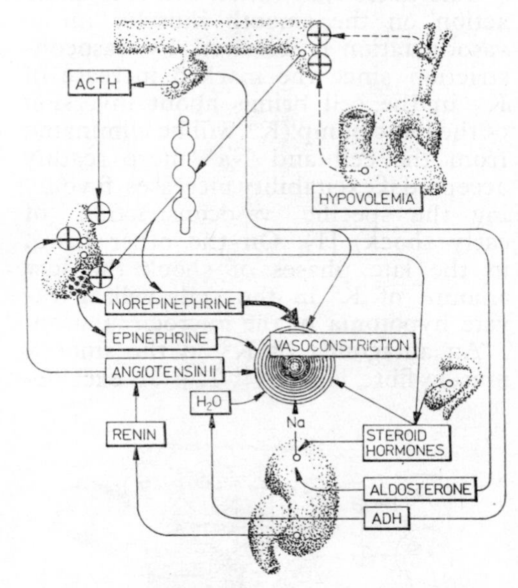

Figure 5.21. Extrinsic regulation of the arteriolar tonus.

Intrinsic myogenic control of the tonus involves the totality of the tissular vasotropic factors with an action on the specific and unspecific receptors illustrated in Figure 5.22. The capillary tonus is especially under the influence of P_aO_2 [48]. When P_aO_2 falls from 70 to 60 mmHg the entire capillary bed begins to respond, increasing tissular extraction of oxygen. When the values fall below 35 to 30 mmHg the means of tissular adaptation are insufficient. Moreover, in local regulation of the tonus there is interference by electrolytic factors of great sensitivity, of which the most important is the sodium ion gradient in the vascular walls. Sensitivity to catecholamines increases when Na^+ concentration increases in the vascular wall. Sodium retention is controlled hormonally (aldosterone, but also cortisol, œstrogens and progesterons) [28, 65, 69, 92].

The potassium cation has a biphasic action on the smooth muscle: initial vasodilatation is followed by vasoconstriction since the marked increase of K^+ in the cell brings about inversion of the ionic pump (K^+ will be eliminated from the cell and Na^+ more readily accepted if excitability increases, favouring the specific vasoconstriction of early shock) [1]. On the other hand, in the late phases of shock the low amount of K^+ in the cells will aggravate hypotonia of the microcirculation.

An antagonist of K^+ in the smooth muscle fibre is Mg^{++}. It is an excellen

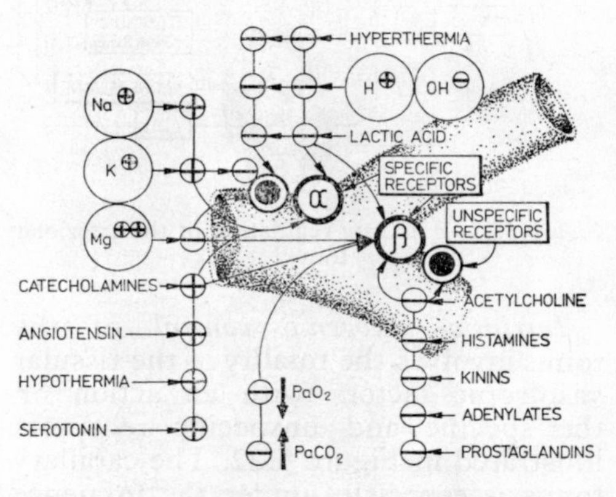

Figure 5.22. Intrinsic regulation of the tonus in the microcirculation.

inhibitor of regional vasoconstriction if perfused in the artery of the respective visceral pedicle (0.8 to 4.0 mEq/min). The magnesium ion opposes the action of catecholamines, angiotensin, vasopressin or K^+ by at least two mechanisms: one similar to the action of curare and the other with a direct effect on the membrane potential of the muscle cell [54].

5.5.3 REGIONAL MORPHOFUNCTIONAL FEATURES

The microcirculation is an 'homogenized' system because its segments act, in turn, by rotation. Alternation is achieved by regional autoregulation, i.e. by the intrinsic tendency of each organ to maintain a constant blood flow in spite of changes in the systemic BP.

To the same stimuli the vascular areas will react differently from one organ to another and even within the same organ.

The angioarchitectonics of the microcirculation is specific of each organ, forming morphofunctional units (the 'modules' of Hershey) that leave their imprint upon the anatomic organization of the organ (hepatic, renal, pulmonary or splenic lobe, etc.) [42].

In the 'vital' organs the microcirculation has a smooth muscle reinforcement of the visceral type only, hence the regulation is local, always to the benefit of autochthonous metabolism; the systemic severe constriction effect in shock skirts these networks (brain, heart, hypophysis) [23, 89, 99, 104].

The extensive network of the microcirculation in the skeletal and splanchnic muscles (which at rest receive half the cardiac output) is equipped with both visceral smooth muscles and the multiunit type with α and β receptors. The gastric wall has a microcirculation sensitive to shock, confirming once again that the stomach is the resonance organ of all neurohormonal stress.

The kidney and skin have a microcirculation with a smooth musculature of the multi-unit type, in which α-receptors are predominant, and receive, as a rule, a blood supply that far exceeds their trophic requirements because they concomitantly carry out general homeostatic functions (glomerular filtration and thermoregulation); their vascular network can be, and is, rapidly sacrificed under stress conditions (Figure 5.23).

The intrahepatic microcirculation has a complex sphincteral system whose role is to homogenize the blood flow that arrives under different pressure from the arterial and portal sectors (Figure 5.24) [76].

In the liver, in shock, the sphincteral system is blocked, rheologic disturbances appear and a *sinusoidal barrier* accentuated by two mechanisms of the liver develop:

- hypertrophy of the Kupffer network;
- lymphatic œdema of Disse's space (*see* Sections 6.8 and 9.5).

The dynamic relationship of Disse's space with the biliary system explains the possibility of cholostasis and jaundice brought about by high lymphatic pressure, especially in bacterial, thermal, heterotransfusional, traumatic shock, etc.

Study of the intrahepatic microcirculation regulation in the isolated dog liver perfused intra-arterially and intraportally with different drugs gave the results shown in Table 5.3 [59].

The features of the pulmonary microcirculation are manifest in the anatomic shunts, in the possible appearance of physiologic shunts, and the merging of the functional with the nutrient circulation [36]. When ventilation is interrupted in a pulmonary area, the blood remains venous and 'contaminates' the arterialized blood in the other areas, having the effect of a right-left shunt. On the other hand, the bronchial circulation brings arterial blood, producing a left-right shunt, but the blood that nourishes the pulmonary tissue becomes venous and will empty into the arterialized blood of the pulmonary veins, which is equivalent

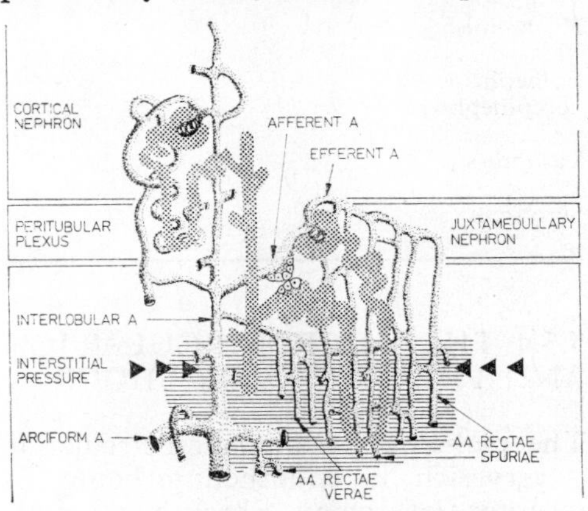

Figure 5.23. Intrarenal circulation shunt. Increase in the interstitial pressure will block the filtering system of Henle's loop, accentuating the antidiuresis state.

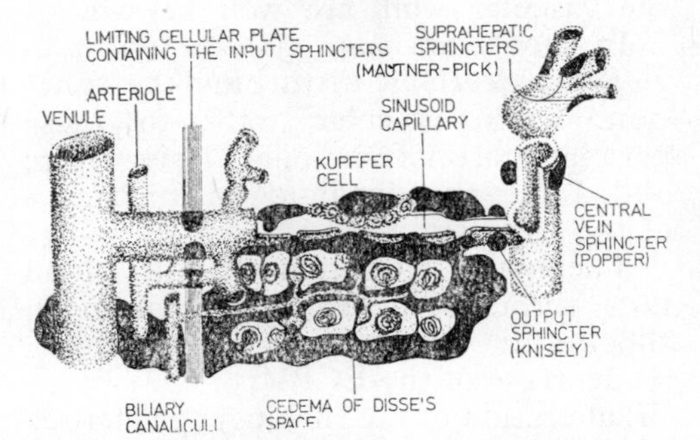

Figure 5.24. Hypertrophy of Kupffer's cells, oedema of Disses's space and hypertonia of the sphincters produce an intrahepatic sinusoidal barrier and cholostasis.

to a right-left shunt. Although the lung has at its permanent disposal an oxygenated medium (necrosis is not characteristic of the pulmonary tissue in shock) it may be considered in man the 'critical organ of shock', undergoing a veritable fluid interstitial flooding (*see* Section 10.4).

Table 5.3. CHANGES IN HEPATIC HAEMODYNAMICS UNDER THE INLFUENCE OF CERTAIN DRUGS (*after Mahfouz*)

Drug used	Portal flow	Arterial flow	Arterial hepatic resistance	Venous hepatic resistance	Hepatic outflow
Histamine	—	—	↑	↑	↑
Epinephrine & norepinephrine	—	—	↑	↑	—
Vasopressin	↓	↓	—	—	↑
Serotonin	↓	↓	—	—	↓

5.5.4 THE MICROCIRCULATION AND THE STAGES OF SHOCK

The pathology of the microcirculation is very rich, its endothelium being the receptor of many chronic systemic diseases.

Acute, localized microangiopathies (chilblains, limited burns, etc.) affecting the vascular wall, are well known.

In shock, an *acute systemic microangiopathy develops,* disturbing the function of the vascular wall; to these parietal morphofunctional lesions are added the rheodynamic disturbances of the circulating fluid.

The major alterations in the blood flow through the microcirculation in shock are:
(i) decrease of the ECBV;
(ii) alteration of the rheologic properties of the blood *(see* Chapter 7);
(iii) changes in the extrinsic and intrinsic regulation of the muscle cell tonus.

In the first phase of shock *(early reversible shock)* catecholamines prime the spasticity of the arterioles, metaarterioles and precapillary sphincters, acting upon the α-receptors (Figure 5.25 *b*). Venulospasm follows and, in the venous trunks, increase in the CVP and venous return is brought about by central phleboconstriction. Although spasm plays a marked compensatory role *(see* Section 5.1), when it is prolonged it becomes an essentially harmful factor for the tissues, which are bypassed through the AV shunts. The first rheologic event that diminishes the flow in the microcirculation is sludging *(see* Section 7.4.1).

In the next stage of shock *(late reversible shock)*, spasm lessens in the microcirculation resistance area due to the α-lytic action of acidosis (produced in the meantime by hypoxic metabolism) [13, 37, 41, 77, 95]. The postcapillary venular sphincter is, however, less sensitive to the accumulated protons; even under normal conditions it works at a low *p*H. Thus, the microcirculation will fill with a pool of acidotic fluid, forming a closed dam (Figure 5.25 *c*) [96]. The result is an overflow of the fluid towards the interstitium (the selective permeability of the capillary wall likewise diminishes), and the collagen network is filled with interstitial fluid; there are rheologic disturbances, and a decrease in the venous return becomes more accentuated [6, 27, 35, 43].

This is the stage of transition between late and refractory reversible shock.

The deteriorating course of shock is then marked by the following rheody-

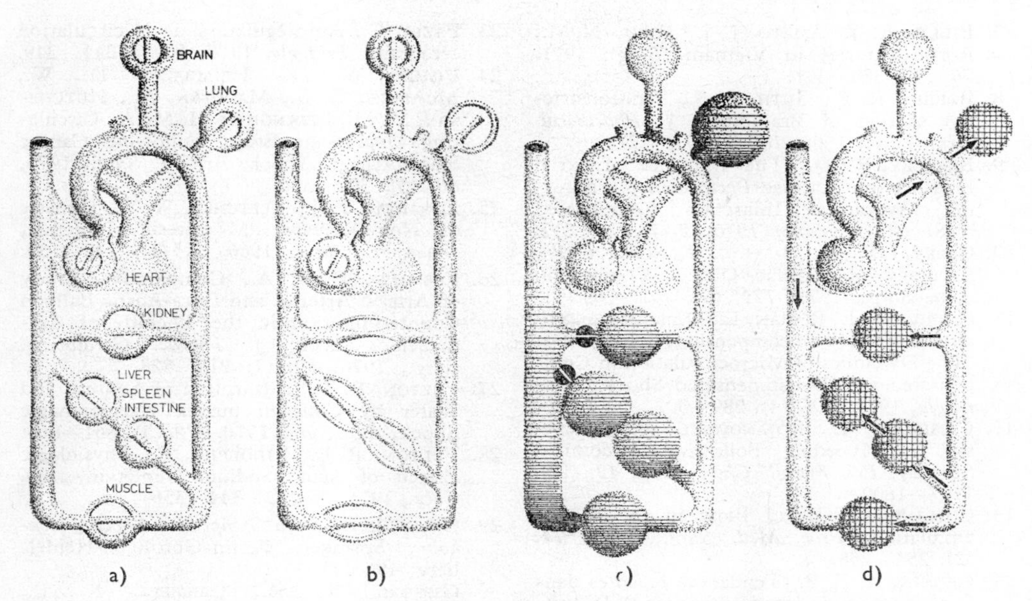

Figure 5.25. The dynamics of the microcirculation in shock:
a — normally, most microcirculation networks are only partly open; *b* — at the onset of shock all the microcirculation networks, except the coronary and cerebral circulation, are spastic and shunted; *c* — acidotic opening of the precapillary sphincters transforms the microcirculation spaces into pools of dammed up fluid (the postcapillary sphincters remain closed); *d* — the onset of disseminated intravascular coagulation totally dams the microcirculation.

namic events: blocking of the microcirculation by *disseminated intravascular coagulation,* enormous increase in the interstitial acidotic hydric pool, and steady decrease in the venous return (Figure 5.25 *d) (see* Chapters 6 and 7). This is the stage of *refractory reversible shock* which, without treatment, very soon turns into the stage of *irreversible shock.*

SELECTED BIBLIOGRAPHY

1 BALEY, S. J., KAMIL, N., LEVOWITZ, B. S., TREIBER, W. F., AGRAWAL G. P., WEITH, F. J., Potassium Vasoactivity. A Biphasic Effect, *Surg.*, 1970, *67*, (2), 350—354.

2. BALEY, S. J., COHEN, M. I., WINSLOW, P. R., BECKER N. H., TREIBER, W. F., McNAMARA HELEN, WEITH, F.J., GLIEDMAN, M. L., Mezenteric Ischemia: a Cause of Increased Gastric Blood Flow Hyperacidity and Acute Gastric Ulceration, *Surg.*, 1970, *68*, (1), 222 — 230.

3. BARER, G. R., HOWARD, P., SHAW, J. W., Stimulus Response Curves for the Pulmonary Vascular Bed to Hypoxia and Hypercapnia, *J. Physiol.*, 1970, *211*, (1), 139—155.

4. BARNER, H. B., JELLINEK, H., KAISER, G. C., Effects of Isoproterenol Infusion on Myocardial Structure and Composition, *Amer. Heart J.*, 1970, *79*, (2), 237—243.

5. BASSIN, R., VLADECK, B. C., KARK, A. E., SHOEMAKER, W. C., Rapid and Slow Haemorrhage in Man: I. Sequential Haemodynamic Responses, *Ann. Surg.*, 1971, *173*, (3), 325—330.

6. BELL, P. R. F., BATTERSBY, A. C., The Effect of Arterial pCO_2 on Gastric Mucosal Blood Flow Measured by Clearance of ^{85}Kr., *Surg.*, 1967, *62*, (3), 468—474.

7. Billy, L. J., Amato, J. J., Rich, N. M., Aortic Injuries in Vietnam, *Surg.*, 1971, *70*, (3), 385–391.
8. Bobbin, R. P., Guth, P. S., Venoconstrictive Action of Bradikinin, *J. Pharmacol. Exp. Ther.*, 1968, *160*, (1), 11–21.
9. Bradley, R. D., The Influence of Atrial Pressure on Cardiac Performance Following Myocardial Infarction Complicated by Shock, *Circulation*, 1970, *42*, (7), 827–837.
10. Carey, J. S., Williamson, H., Scott, C. R., Accuracy of Cardiac Output Computers, *Ann. Surg.*, 1971, *174*, (5), 762–768.
11. Chiricuţă, I., Roman, L., Dinamica microcirculaţiei în şocul compensat şi decompensat (The Dynamics of Microcirculation in Compensated and Decompensated Shock), *Chirurgia*, 1971, *XX*, (4), 289–302.
12. Ciesielski, L., Rojanowski, W., Zielinski, A., Bleeding Following Abdominal Surgery, *Prl. Przegl. Chir.*, 1970, *42*, (12), 1797–1802.
13. Collins, G., Blood Flow in the Microcirculation, *Proc. Med. Surg.*, 1966, *74*, (2), 254–258.
14. Cronin, R. F. P., Tendances récentes dans le traitement du choc survenant à la suite d'un infarctus aigu du myocarde, *L'Union méd. du Canada*, 1967, *96*, (2), 134–139.
15. Crowell, J. W., Guyton, A. C., Further Evidence Favouring a Cardiac Mechanism in Irreversible Haemorrhagic Shock, *Amer. J. Physiol.*, 1962, *203*, (2), 248–252.
16. Dedichen, H., Worthington, G. S., Hemodynamics of Endotoxin Shock in the Dog, *Arch. Surg.*, 1967, *95*, (6), 1013–1016.
17. DeTakats, G. Microcirculation, *J.A.M.A.*, 1966, *195*, (4), 302–304.
18. Dogru, M. A., Akut mide barsak iskemisi soku ve irreversibil sokta anatomopatolojik bulgular, *Tip Fakult. Mecmuasi*, 1969, *22*, (3), 429–439.
19. DuCailar, J., Triadou, Ch., Roquefeuil, B., Malzac, P., Effets de la ventilation artificielle sur la circulation, *Agressol.*, 1969, *10*, (1), 55–64.
20. DuCailar, J., Roquefeuil, B., Lefebvre, F., Kienlen, J., Les difficultés d'interprétation de la pression veineuse centrale en réanimation et plus particulièrement chez le choqué, *Ann. Anesth Franç.*, 1970, *11*, (2), 253–259.
21. Edelman, N. H., Experimental Cardiogenic Shock: Pulmonary Performance after Acute Myocardial Infarction, *Am. J. Physiol.*, 1970, *219*, (12), 1721–1730.
22. Epstein, M., Holeenberg, N. K., Merrill, J. P., The Pattern of the Renal Vascular Response to Epinephrine in Man, *Proc. Soc. Exp. Biol.*, 1970, *134*, (3), 720–724.
23. Fazio, C., Autorégulation de la circulation cérébrale, *Triangle*, 1970, *9*, (7), 244–249.
24. Fisher, W. D., Heimbach, D. W., McArdle, C. S., Maddern, M., Hutcheson, M. M., Ledingham, I. Mc A., Circulating Depressant Effect Following Canine Haemorrhagic Shock, *Brit. J. Surg.*, 1973, *60*, (5), 392–394.
25. Fogarty, T. J., Fletcher, W. S., Genesis of Non-Occlusive Mesenteric Ischemia, *Amer. J. Surg.*, 1966, *3*, (1), 130–135.
26. Fleming, W. H. A., Comparative Study of Arterio-Arterial and Intra-Aortic Balloon Counterpulsation in the Therapy of Cardiogenic Shock, *J. Thorac. Cardiovasc. Surg.*, 1970, *60*, (7) 818–828.
27. Fulton, R. L., Adsorption of Sodium and Water by Collagen during Haemorrhagic Shock, *Ann. Surg.*, 1970, *172*, (5), 861–869.
28. Fulton, R. L., Ridolpho, P., Physiologic Effects of Acute Sodium Depletion, *Ann. Surg.*, 1971, *173*, (3), 344–356.
29. Gelin, L. E., *Ciba Symposium Shock Stockholm*, Springer, Berlin-Göttingen-Heidelberg, 1962.
30. Gifford, R. M., Carmell, K. L., McNay, J. L., Haas, A., Change in Regional Blood Flows Induced by Dopamine and by Isoproterenol during Experimental Haemorrhagic Shock, *Canad. J. Physiol.*, 1968, *46*, (6), 847–851.
31. Giurgiu, T., Ionescu, P., Cafriţă, A., Hemodiluţie cu soluţii coloidale în şocul hemoragic; cercetări experimentale (Haemodilution with Colloid Solutions in Haemorrhagic Shock, Experimental Investigations), *Revista Sanitară Militară*, 1972, *LXXV*, (3), 347–356.
32. Golden, P. F., Jane, J. A., Survival in Profound Hypovolaemia: Differentiation of the Role of Heart, Lung and Brain, *Ann. Roy. Coll. Surg.*, 1970, *3*, (1), 63–64.
33. Greenfield, L., Blalock, A., Effect of Low Molecular Weight Dextran on Survival Following Haemorrhagic Shock, *Surg.*, 1964, *55*, (5), 684–691.
34. Grover, F. L., Newman, M. M., Paton, B.C., Beneficial Effect of Pluronic F. 68 on the Microcirculation in Experimental Haemorrhagic Shock, *Surg. Forum*, 1970, *21*, (1), 30–32.
35. Gruber, U. F., Volume Expansion and Flow Promotion in Shock, *Postgrad. Med. J.*, 1969, *45*, (526), 534–538.
36. Gump, F. E., Moshima, Y., Jørgensen, S., Kiuney, Y. M., Simultaneous Use of Three Indicators to Evaluate Pulmonary Capillary Damage in Man, *Surg.*, 1971, *70*, (2), 262–270.

37. GUTELIUS, J. R., SHIZGAL, H. M., LOPEZ, G., The Effect of Trauma on Extracellular Water Volume, *Arch. Surg.*, 1968, *97*, (2), 206–214.
38. HADDY, F. J., Pathophysiology and Therapy of the Shock of Myocardial Infarction, *Ann. Intern. Med.*, 1970, *73*, (7), 809–827.
39. HALMAGYI, D. F. J., GOODMAN, A. H., LITTLE, M. J., Foetal Blood Flow and Oxygen Usage in Dogs after Haemorrhage, *Ann. Surg.*, 1970, *172*, (2), 284–290.
40. HARDAWAY, R. M., Microcoagulation in Shock, *Amer. J. Surg.*, 1965, *110*, (3), 298–301.
41. HEIMBACH, D. W., FISHER, W. D., HUTTON, I., MCARDLE, C. S., LEDINGHAM, I. McA., Myocardial Blood Flow and Metabolism during and after Haemorrhagic Shock in the Dog., *Surg. Gynecol. Obstet.*, 1973, *137*, (2), 243–252.
42. HERSHEY, S. G., *Shock* (Ed. Hershey), Little Brown Comp., Boston, 1964.
43. HINSHOW, L. B., Mechanism of Decreased Venous Return. Subhuman Primate Administered Endotoxin, *Arch. Surg.*, 1970, *100*, (4), 600–606.
44. HOPKINS, R. W., Septic Shock: Hemodynamic Cost of Inflammation, *Arch. Surg.*, 1970, *101*, (2), 298–307.
45. JONES, C. E., BETHEA, H. L., SMITH, E. E., CROWELL, J. W., Effects of a Coronary Vasodilator on the Development of Irreversible Haemorrhagic Shock, *Surg.*, 1970, *68*, (2), 356–362.
46. KERR, A. R., KIRKLIN, J. W., Changes in Canine Venous Volume and Pressure during Haemorrhage, *Surg.*, 1970, *68*, (3), 520–527.
47. KIRIMLI, B., KAMPSCHULTE, S., SAFAR, P., Pattern of Dying from Exsanguinating Hemorrhage in Dogs, *J. Trauma*, 1970, *10*, (5), 393–404.
48. KITTLE, C. F., AOKI, H., BROWN, E. B., The Role of pH and CO_2 in the Distribution of Blood Flow, *Surg.*, 1965, *57*, (1), 138–146.
49. LABORIT, H., Corrélations entre les structures morphologiques, métaboliques et enzymatiques du capillaire et sa physiopathologie, *Agressol.*, 1969, *10*, (4), 291–302.
50. LABORIT, H., BARON, C., WEBER, B., PAVLOVITCHOVA, H., Antagonisme de la L-tyrosine sur la chute progressive de la pression artérielle (tachyphylaxie) au cours de la perfusion d'adrénaline et sur l'hypotension secondaire à l'arrêt de la perfusion. Recherche du mécanisme d'action, *Agressol.*, 1970, *11*, (1), 25–53.
51. LACROIX, E., Fysiopathologische beschouwingen over Shock, *Acta Chir. Belgica*, 1970, Suppl. 1, 30–41.
52. LARENG, L., Rappel de physiologie du capillaire, *Anesth. Anal. Réan.*, 1969, *26*, (5), 701–706.
53. LEVITSKY, S., PARKS, L. C., WILLIAMS, W. H., MARROW, A. G., Inhibition of Experimental Cardiac Hypertrophy by Azathioprine, *Surg.*, 1970, *68*, (3), 536–540.
54. LEVOWITZ, B. S., GOLDSON, H., RASHKIN, A., KAY, N., VALCIN, A., MATHUR, A., LA GUERRE, J. M., Magnesium Ion Blockade of Regional Vasoconstriction, *Ann Surg.*, 1970, *172*, (1), 33–40.
55. LITARCZEK, G., Medicaţia vasoactivă şi cardiotonică în tratamentul tulburărilor circulatorii în şoc (Vasoactive and Cardiotonic Medication in the Treatment of Circulatory Disturbances in Shock), *Revista Sanitară Militară*, 1973, *LXXVI*, (4), 371–377.
56. LITTON, A., Microcirculatory Effects of Endotoxin, *Bibl. Anat.*, 1969, *10*, (2), 334–339.
57. LOEB, H. S., PIETRAS, R. J., NINOS, N., Haemodynamic Responses to Chlorpromazine in Patients in Shock, *Arch. Intern. Med.*, 1969, *124*, (4), 354–358.
58. LOVETT, W. L., WANGESTEEN, S. L., GLENN, T. M., LEFER, A. M., Presence of a Myocardial Depressant Factor in Patients in Circulatory Shock, *Surg.*, 1971, *70*, (2), 223–231.
59. MAHFOUZ, M., AIDA, G., Pharmacodynamic of Intrahepatic Circulation in Shock, *Surg.*, 1967, *61*, (5), 775–762.
60. MAREŞ, E., ŞUTEU, I., BĂNDILĂ, T., CÎNDEA, V., ILIE, A., Date noi în fiziopatologia şi tratamentul şocului (New Data in the Pathophysiology and Treatment of Shock), *Revista Sanitară Militară*, 1963, *LIX*, (6), 887–895.
61. MARK, A. L., IIZUKA, T., WENDLING, M. G., ECKSTEIN, J. W., Responses of Saphenous and Meseteric Veins to Administration of Dopamine, *J. Clin. Invest.*, 1970, *49*, (2), 259–266.
62. MARKS, L. S., KOLMEN, S. N., Tween 20 Shock in Dogs and Related Fibrinogen Changes, *Amer. J. Physiol.*, 1971, *220*, (1), 218–221.
63. MATHESON, N. A., The Microcirculation in Shock, *Postgrad. Med. J.*, 1969, *45*, (526) 530–533.
64. MATSUMOTO, T., HARDAWAY, R. M., MCCLAIN, J. E., Microcirculation in Haemorrhagic Shock with Relative Shift to Blood Pressure, *Arch. Surg.*, 1967, *95*, (6), 911–917.
65. MCLEAN, F. C., HASTINGS, A. B., The State of Calcium in the Fluids of the Body. The Conditions Affecting the Ionization of Calcium, *Clin. Orthop.*, 1970, *69*, (1) 4–27.

66. Melrose, D., G. Circulation, *Brit. J. Surg.*, 1967, *54*, Lister Centenary Number, 447–449.
67. Milstein, B. B., Acute Circulatory Arrest, *Brit. J. Surg.*, 1967, *54*, Lister Centenary Number, 471–473.
68. Motsay, G. J., Duettnab, R. H., Ersek, R. A., Lillehei, R. C., Hemodynamic Alterations and Results of Treatment in Patients with Gramnegative Septic Shock, *Surg.*, 1970, *67*, (4), 577–583.
69. Motsay, G. J., Laho, A., Jaeger, T., Schultz, L. S., Dietzman, R. H., Lillehey, R. C., Effects of Methylprednisolone, Phenoxybenzamine and Epinephrine Tolerance in Canine Endotoxin Shock: Study of Isogravimetric Capillary Pressures in Forelimb and Intestine, *Surg.*, 1971, *70*, (2), 271–279.
70. Mundth, E. D., Sokol, D. M., Levine, F. H., Austen, W. G., Evaluation of Methods for Myocardial Protection during Extended Periods of Aortic Cros Clamping and Hypoxic Cardiac Arrest, *Bul. Soc. Int. Chir.*, 1970, *29*, (4), 227–235.
71. Neely, W. A., Cardiac and Renal Blood Flow. Comparison of the Effects of Isoproterenol, Phenozybenzamine and Levarterenol, *Arch. Surg.*, 1970, *100*, (2), 249–252.
72. Nickerson, M., Vascular Adjustments during the Development of Shock, *Canad. Med. Ass. J.*, 1970, *103*, (8), 852–859.
73. Olsen, W. R., Capillary Flow in Haemorrhagic Shock. 3. Metaraminol and Capillary Flow in the Nonanesthetized and Anesthetized Pig, *Arch. Surg.*, 1969, *99*, (5), 637–640.
74. Papadopol, S., *Tulburările de hemodinamică* (Haemodynamic Disturbances), Editura medicală, București, 1971.
75. Pîrvu, V., Mecanismul celular al contracției miocardului (The Cellular Mechanism of Myocardial Contraction), *Viața Medicală*, 1970, *XVII*, (22), 1017–1024.
76. Potvin, P., Les effets de modifications de la pression sushépatique sur la circulation du foie chez le lapin, *Agressol.*, 1969, *10*, (1), 45–49.
77. Race, D., Cooper, E., Hemorrhagic Shock: the Effect of Prolonged Low Flow on the Regional Distribution and Its Modification by Hypothermia, *Ann. of Surg.*, 1968, *167*, (4), 454–460.
78. Ratliff, N. B., Hackel, D. B., Mikat, E., Myocardial Carbohydrate Metabolism and Lesions in Hemorrhagic Shock. Effects of Hyperbaric Oxygen, *Arch. Path.*, 1969, *88*, (5), 470–475.
79. Reed, R. R., Owens, G., Simultaneous and Instantaneous Blood Fow and Tissue Gas Measurements in Experimental 'Stroke' Therapy, *Surg.*, 1971, *70*, (2), 254–261.
80. Reich, M. P., Eiseman, B., Tissue Oxygenation Following Resuscitation with Crystalloid Solution Following Experimental Acute Blood Loss, *Surg.*, 1971, *69*, (6), 928–931.
81. Riordan, F., The Significance of Central Venous Pressure in Cardiac Output Measurements in Shock, *Postgrad. Med. J.*, 1969, *45*, (526), 506–511.
83. Roding, B., Worthington, G. S. Jr., Mesenteric Blood Flow after Hemorrhage in Anesthetized and Unanesthetized Dogs, *Surg.*, 1970, *68*, (5), 857–861.
82 Rothe, C. F., Heart Failure and Fluid Loss in Haemorrhagic Shock, *Fed. Proc.*, 1970, *29*, (11), 1854–1860.
84. Salsbury, A. J., Clarke, J. A., The Surface Appearance of Blood Cells, *Triangle*, 1968, *8*, (7), 260–266.
85. Sandol, E., Kardiogent Shock, *Nord. Med.*, 1970, *83*, (22), 681–689.
86. Scheidt, S., Ascheim, R., Killip, T., Shock after Acute Myocardial Infarction; A Clinical and Hemodynamic Profile, *Amer. J. Cardiol.*, 1970, *26*, (6), 556–564.
87. Schon, G. R., Labat, R., The Hepatic Outflow Tract in Dogs, *Surg.*, 1971, *69*, (5), 748–754.
88. Shires, T., Coln, D., Carrico, J., Lightfoot, S., Fluid Therapy in Haemorrhagic Shock, *Arch. Surg.*, 1964, *88*, (6), 688–696.
89. Simeone, F. A, The Central Nervous System in Experimental Haemorrhagic Shock; The Cerebrospinal Fluid Pressure, *Amer. J. Surg.*, 1970, *119*, (3), p. 427–432.
90. Simon, M. A., Olsen, W. R., Capillary Flow in Hemorrhagic Shock, *Arch. Surg.*, 1969, *99*, (5), 631–636.
91. Skinner, D. B., Camp, T. F., Austen, W. G., Use of Vasopressor Agents to Increase Somatic Blood Flow, *Arch. Surg.*, 1967, *94*, (5), 610–618.
92. Smith, R. H., *Pathological Physiology for the Anesthesiologist*, Charles C. Thomas, Springfield, Illinois, 1966.
93. Stone, M. A., Sthal, M. W., Renal Effects of Hemorrhage in Normal Man, *Ann. Surg.*, 1970, *172*, (5), 825 – 836.
94. Strock, P. E., Majno, G., Microvascular Changes in Acutely Ischemic Rat Muscle, *Surg. Gynec. Obst.*, 1969, *129*, (6), 1213–1224.
95. String, T., Robinson, A. J., Blaisdell, F. W., Massive Trauma. Effect of Intravascular Coagulation on Prognosis, *Arch. Surg.*, 1971, *102*, (4), 406–411.

96. ŞUTEU, I., CAFRIŢĂ, A., BUCUR, I. A., Progrese Recente in Fiziopatologia Şocului (Recent Progress in the Pathologic Physiology of Shock), *Revista Sanitară Militară*, 1972, *LXXV*, (3), 267–276.
97. TAKAORI, M., SOFAR, P., Acute Severe Hemodilution with Lactaed Ringer's Solution, *Ann. Surg.*, 1967, *94*, (1), 67–73.
98. TIEFENBRUN, J., SHOEMAKER, W. C., Sequential Changes in Pulmonary Blood Flow Distribution in Haemorrhagic Shock, *Ann. Surg.*, 1971, *174*, (5), 727–733.
99. TRAGUS, E., PARKINS, W., BAUE, A. E., The Effects of Sequential Buffering, Extracellular Fluid Replacement and Hemodilution in Haemorrhagic Shock, *Surg.*, 1967, *61*, (5), 795–801.
100. TRINKLE, J. K., Mechanical Support of the Circulation: A New Approach, *Arch. Surg.*, 1970, *101*, (6), 740–743.
101. VLADECK, B. C., BASSIN, R., KARK, A. E., SHOEMAKER, W., C., Rapid and Slow Haemorrhage in Man: II. Sequential Acid-Base and Oxygen Transport Responses, *Ann. Surg.*, 1971, *173*, (3), 331–336.
102. WANGENSTEEN, S. L., KIECHEL, S. F., LUDEWIG, R. H., MADDEN, J. J., The Role of Vasoconstriction in the Suppression of Haemorrhage from Arteries. I. The Completely Severed Artery, *Surg.*, 1970, *67*, (2), 338–341.
103. WEIL, M. H., Circulatory Shock. A Symposium on Advances in the Understanding of Mechanism and Treatment, *Calif. Med.*, 1967, *106*, (1), 4–7.
104. WEIL, M. H., SCHUBIN, H., *Diagnosis and Treatment of Shock*, Williams and Wilkins Comp., Baltimore, 1967.
105. WEIL, M. H., AFIFI A. A., Experimental and Clinical Studies on Lactate and Pyruvate as Indicators of the Severity of Acute Circulatory Failure (Shock), *Circulation*, 1970, *41*, (6), 989–1001.
106. ZÜHLKE, V., DuMESNIL, W., GUDBJORNASON, S., BING, R. J., Inhibition of Protein Synthesis in Cardiac Hypertrophy and its Relation to Myocardial Failure, *Circ. Res.*, 1966, *18*, (5), 558–563.
107. ZWEIFACH, B. W., Microcirculation, *Sc. Amer.*, 1959, *200*, (1), 56–64.

6
INTERSTITIAL COMPARTMENT

The cells congregate in societies, which are called tissues and organs,... and cell sociology is more advanced than the science of the structure and functions of the cell as an individual.

ALEXIS CARREL

6.1 Collagen
6.2 Interstitial hydroelectrolytic oscillations
6.3 Interstitial cells
6.4 Glycolysis and transcapillary transport
6.5 The reticuloendothelial system
6.5.1 Phagocytosis
6.5.2 Antibody genesis
6.6 Interstitial vasoactive substances
6.6.1 Histamine
6.6.2 Endogenous heparin
6.6.3 Serotonin
6.6.4 Prostaglandins
6.7 Metabolic oscillations of the interstitium
6.8 The lymphatic circulation
6.8.1 Lymphatic rheodynamics
6.8.2 Lymphatic transport in shock
6.8.3 Lymphatic flow oscillations in shock

The interstitium is the transit space between the organized epithelium and the tubular circulation system. As it contains a large amount of fluid, in shock it is the 'recoil space' or 'primary buffer' of blood volume losses, after which an oscillating system of exchanges with the intravascular and intracellular system is set-off. In addition, the interstitial space is provided with a 'safety valve' permitting readjustments: the lymphatic system (Figure 6.1). Its texture is a dense collagen grid populated with polyglot cells, possessing an essential functional competence for the organism, mobilized in states of shock for a struggle the complexity of which warrants a special study.

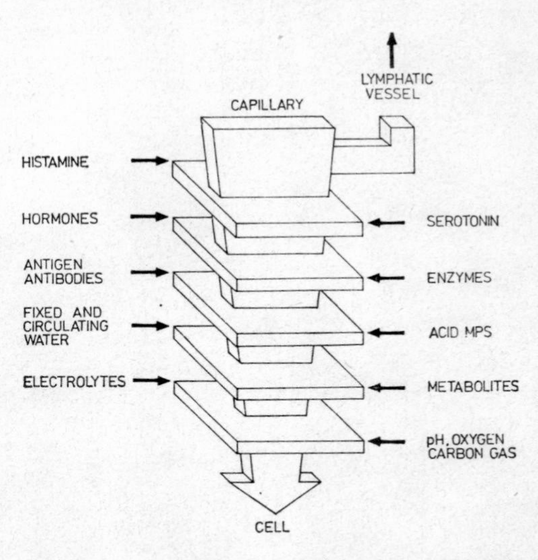

Figure 6.1. Some of the elements active in the interstitial space.

6.1 COLLAGEN

Collagen represents a woof pattern of protein fibres in a colloid matrix in which cations are included (Na^+, Ca^{++}, K^+, Mg^{++}). The collagen, together with the elastin fibres, form the 'soft skeleton' of the human organism [20, 68]. Produced by the fibrocyte, the collagen fibre consists of a bundle of polypeptides (tropocollagen) formed, in turn, by the twisting together of three amino-acid chains. The collagen fibrils, representing about 6% of the human body weight, weigh approximately 4200 g (for a body weight of 70 kg) and have a total surface of over one million square metres [12, 26].

These morphologic characteristics of the collagen network point to its sorption properties for water and electrolytes, which may interfere promptly in shock.

Lactacidaemia has proved a constant stimulent of the collagen capacity for pooling water and electrolytes [12]. The interstitial collagen network may rapidly take up 0.50 ml water/1 g collagen tissue (i.e. a total of 2000 ml water) and 0.08 mEq sodium (i.e. a total of over 330 mEq Na). The collagen 'sponge' is therefore able quickly to dam up a hypertonic fluid (with more than 160 mEq Na) (Fulton, 1970) [26].

It must be understood that this outstanding hydroelectrolytic retention in states of shock takes place at the expense of the plasma fluid; *an intervascular tissular hypertonic pool* then forms, representing a harmful medium for the cells whose imbalance rapidly follows.

How do these important hydroelectrolytic oscillations take place in the course of shock?

In experimental shock, interstitial œdema of the ground substance reaches a peak at four hours (measured by increase in the radioactivity induced by specific tracers) [12]. Oedema, exclusively extracellular, appears to be closely linked to decrease of the *p*H in the ground substance. By interrupting the vascular connexions it was demonstrated that interstitial œdema cannot be attributed only to the classical phenomena of increased vascular permeability; it is only due to the mechanism of *active sorption* exercised by collagen. Similarly, the rapid damming up of electrolytes is not brought about by hyperaldosteronism, since the latter only becomes manifest four hours after the onset of shock (Broido et al., 1966).

6.2 INTERSTITIAL HYDROELECTROLYTIC OSCILLATIONS

The dynamics of interstitial hydroelectrolytic oscillations is represented in Figure 6.2.

In the first stage of shock, the microcirculation spasms lower the blood output and, by osmosis in the venular sector, 'absorbs' interstitial fluid, giving rise to the well known phenomenon of haemodilution *(see* Section 5.1); following this process of intravasation the lymphatic flow tends to decrease.

However, acidosis of the hypoxic interstitium rapidly opens the dam of the microcirculation pool asymmetrically and, due to the sorption properties of collagen, which increases with the decrease in *p*H, there begins an inverse hydroelectrolytic transfer from the microcirculation towards the interstitium, aggravating intravascular hypovolaemia. The lymphatic flow continues to decrease, but the venulo-lymphatic shunts begin to open wide, similar to the 'safety valves' of a

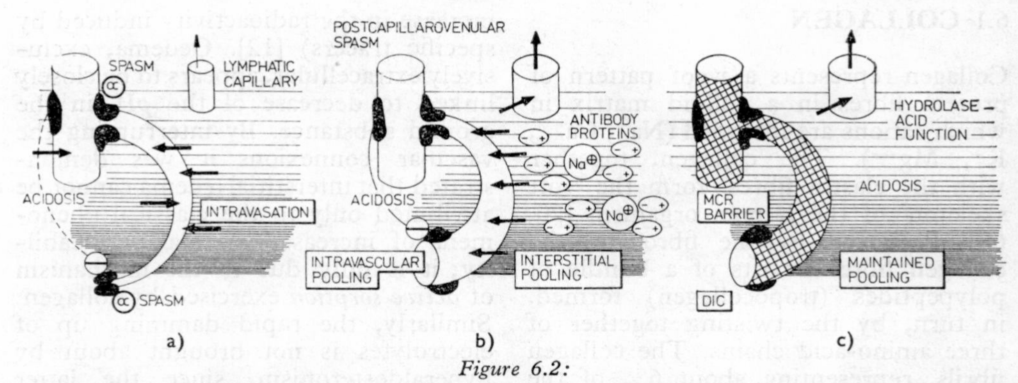

Figure 6.2:
a — temporary intravasation of fluids; *b* — mixed intra- and extravascular (collagenic) pooling; c — blocking by disseminated interstitial coagulation results in tissular impairment expressed in the lymphatics by a flow of noxious substances.

microcirculation overloaded with fluids (*see also* Figure 6.11).

It is a critical moment in the evolution of shock that adds collagen hydro-electrolytic retention to acidotic blood pooling in the microcirculation.

There appears, therefore, a mixed intra- and extravascular fluid sequestration with a severe decrease of the ECBV. This aggravation corresponds to the taking-up phenomenon in experimental shock and is expressed in the human clinic by the onset of accentuated hypotension. Due to the loss of active circulating fluid, only pharmacodynamic active mobilization of the microcirculation is insufficient and fluids in amounts exceeding the actual calculated losses have to be administered.

During these moments (which may be several hours in the clinic) the essential struggle takes place; decisive therapeutical steps, which have not yet been precisely coded in practice, must be taken.

Which is the best category of fluids? A first and classical answer would be colloids, which would bring back the fluid from the interstitium by an increase in intravascular oncotic pressure. Opponents of this therapy bring cogent arguments:

- the fluid pooled in the interstitium being hypertonic, the intravascular electrolytic deficit must first be rapidly corrected and this cannot be realized with colloids;
- the addition of colloids would continue to fix intravascular electrolytes, accelerating their plasma depletion;
- acidosis blocks the osmotic effects of colloids five times more than the osmotic effects of crystalloids;
- the decrease in vicosity of the blood is more efficiently obtained with crystalloids, which means that crystalloids may avoid DIC (*see* Section 7.1.5).

In fact, both colloids and crystalloids are useful but in optimal therapeutical dynamics; the question is when should the one or the other be used?

In practice, the *optimal therapeutic moment* in the course of shock must be clearly understood. Unfortunately, this 'moment' is difficult to establish as the current paraclinical data reflect only statically and indirectly isolated disturbances, that may often lead to untoward interpretations.

The interstitial compartment is the most frequent and difficult trap for our clinical and therapeutical understanding of the dynamics of shock.

In general, an initial treatment with crystalloids and/or colloids is applied, but when the microcirculation has become a fluid pool with persistent postcapillary venular spasm the colloids can no longer be of any use.

Lysis of the spasm at the venous end of the microcirculation, blood hyperviscosity and acidosis have to be controlled and the intravascular electrolyte depletion compensated. This may be solved by a constant flow of crystalloid solutions. Only after opening the microcirculation and 'washing' the acid functions can colloids again be brought into consideration for their remaining osmotic effects.

When the microcirculation remains blocked and disseminated intravascular coagulation develops the exchanges between vessel and interstitium are abolished. Under these conditions, the only spillway of the interstitial space is the lymphatic system; the lymphatic flow increases, draining the collagen hypertonic pool, but carrying with it the dangerous dregs of the cells in full acidosis. Thus, apart from its useful volume of water, electrolytic, protein and antibody supply, the lymphatic flow also brings a lot of substances harmful to the organism (*see* Section 6.6.2).

The therapy specific of this stage (when it has not been prevented by anticoagulants) consists in thrombolytics; if, however, endogenous proteolysis becomes excessive and superposed upon the consumption coagulopathic effects, thrombolytics have no longer any purpose, a haemorrhagic syndrome of extreme gravity having to be solved (*see* Section 11.4.3).

When nothing can arrest the effects of shock, then the interstitial space together with all the other neighbouring structures, will pass into the irreversible stage of shock whose dark prognosis often renders inefficient but not useless any attempt at treatment. The interstitial compartment is transformed into a deposit of hydroelectrolytic enzymes released by the disintegrating cells; an immense 'acid pool' appears in which the cellular residues become confluent in necrotic masses.

6.3 INTERSTITIAL CELLS

The interstitium contains cells of its own, and transit cells. The cells that populate the interstitial space derive from the mesodermal embryonic layer. All these cells serve the general system functions or local functions of the associated cells derived from the ecto-endodermal layers and forming the parenchyma and mucosa.

Among these cells that float in the ground substance, armed with collagen and elastin fibres, are some with an active role in shock (Figure 6.3). The fact that they have a common forerunner, the primary mesenchymal cell, through which they are related to the blood cells and the cells of the reticulo-endothelial system, facilitates frequent histofunctional exchanges.

The intima cell (endothelium cell) is poor in mitochondria and rich in microvesicles; it is separated from the vascular basement membrane by an extensible subendothelial space and plays an important role in intravascular rheologic processes.

The pericyte has similar features to those of the smooth muscle cell in the walls of the arterioles and venules; it is enclosed in folds of the basement membrane and has abundant mitochondria and fibrillary elements. Serotonin and histamine trigger off phagocytotic activity in the pericyte. The basement membrane is discontinuous and at the site of these gaps the proces-

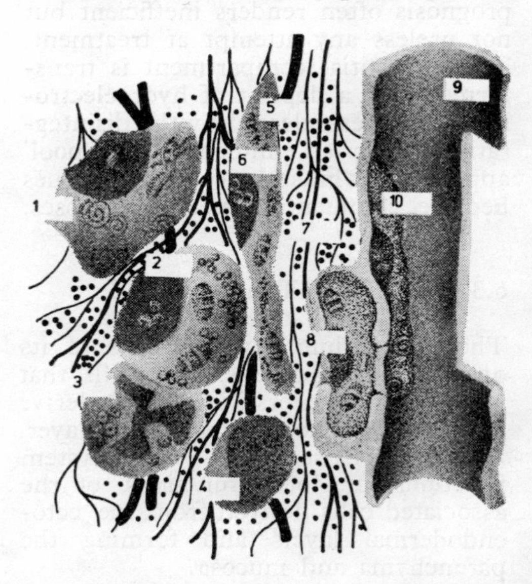

Figure 6.3. The cells populating the interstitium:
1 — macrophage; *2* — mast cell; *3* — leucocyte; *4* — lymphocyte; *5* — fibrils; *6* — fibrocyte; *7* — mucopolysaccharide; *8* — pericyte; *9* — intima cell; *10* — capillar.

ses of the intima cell come in contact with the pouches containing pericytes.

The macrophage has an abundant Golgi apparatus, vacuoles and lysosomes, being prepared for phagocytosis.

The mast cell is the factory and reservoir of serotonin, histamine and heparin, contained in lysosome-like cytoplasmatic granules.

The fibrocyte synthesizes the collagen fibres and mucopolysaccharides of the ground substance.

The lymphocyte is not only the final cell of a linear cycle of development; it is capable of cytofunctional 'transfiguration', giving rise to the 'immunoblast' whenever unspecific or specific stimuli (hetero- or autoantigens) impose an increase in immune competence.

According to their metabolic characters, Laborit has called the cells lacking intramitochondrial enzymatic chains, *type A cells*, which appear earlier on the phylogenetic scale; these have a role in hydrolytic intracellular digestion (phagocytosis), taking the energy from glucolysis.

Type B cells produce far greater energy by intramitochondrial oxidative processes *(see* Section 8.2.1).

The thrombocytes, neuroglia, intima cells and most of the RES and interstitial cells belong to cellular type A.

6.4 GLYCOLYSIS AND TRANSCAPILLARY TRANSPORT

The capillary wall is readily crossed by liposoluble substances as it is in all lipoprotein membranes. But for glucose, amino acids or urea to pass through the basement membrane of the capillary wall the pores ought to be at least 60 Å in diameter, which has not yet been proved.

The possible active crossing of the transintima cell has been proposed but this ought to take place within about 0.012 seconds, the period during which the blood flow passes in the microcirculation.

Based upon many analogies, Laborit proposed a particularly active mechanism sustained by glycolysis of the intima cell. It has been ascertained that glycolysis controls phagocytosis and that the agents blocking glycolysis always inhibit the phagocytic function of the RES. The pentose cycle is likewise involved in phagocytosis: normally, in aerobiosis, $NADPH_2$ is a dehydrogenizer and forms H_2O_2, which oxidises the lyosomal lipoprotein membranes, where the included particles are digested. Moreover, activation of the pentose pathway also governs the synthesis of lipoprotein membranes necessary for the vesicular engulfment of the particles. Complete phagocytosis involves both engulfment and intralysosomal digestion.

Comparing phagocytosis (the ingestion of solids) to pinocytosis (the ingestion of fluids), Laborit postulates the same glycolytic energy support for pinocytosis; he likewise considers the transit of various vesicles through the intima cells similar to pinocytosis, but as the phenomenon is merely a passage (without engulfment and digestion) he finds the term of *cytopempsis* more suitable. For this transport the intima cell does not need the pentose pathway, only glycolysis (the particles in transit are not digested).

Hence, any blocking of glycolysis will also arrest crossing of the intima cell through the capillary wall [33].

The fact that carbohydrate metabolism intervenes in the active transcapillary exchanges is also emphasized by interference with the ion transfer; blocking of glycolysis results in a total ATP deficit which is accompanied by sodium retention and cellular œdema of the intima cells.

6.5 THE RETICULOENDOTHELIAL SYSTEM

6.5.1 PHAGOCYTOSIS

Phagocytosis (endocellular engulfment and digestion of different particles) is sustained by the energy produced by glycolysis and the pentose pathway. It is an old phylogenetic phenomenon and does not necessitate the presence of a mitochondrial apparatus, which developed later. Phagocytosis is characteristic of the metabolic A type cells *(see* Section 8.2.1) *Engulfment* also occurs for a time in anaerobiosis, sustained by the low energy output of glycolysis, but *digestion* demands aerobiosis for a good performance of the pentose cycle.

The action of histamine on the microcirculation favours the passage of leucocytes into the interstitium, where it starts phagocytosis next to the tissular histiocyte and the rich RES cellular population [50, 52].

The endocellular engulfed particle represents the *phagosome*, which fuses with the 'virgin' lysosomes, forming the *phagolysosome* and, after digestion of the particle, the secondary lysosome.

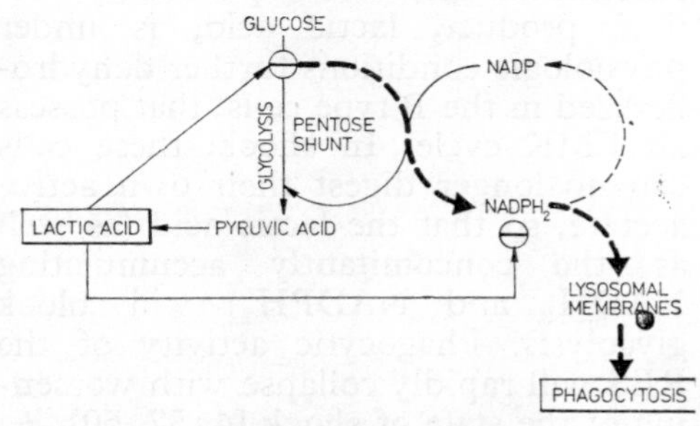

Figure 6.4. Blocking of glycolysis and of pentose cycle inhibits phagocytosis.

Fusion demands oxidation of the lysosomal membranes, achieved by peroxidase functions. All antoxidating or reducing substances (vitamin E, cysteine, glutathione, levomepromasine, phenergan, nupercaine, etc.) protect the lysosomal membrane. The stimulants of the pentose pathway (among which is sodium gamma-hydroxybutyrate) supplying $NADPH_2$, likewise protect the membranes.

In shock, the accumulation of lactate in excess will reduce NAD to $NADH_2$ and NADP to $NADPH_2$, which inhibits glycolysis (Figure 6.4). In this way the engulfment of particles is blocked, i.e. the phagocytic function of the RES. Dehydroxyacetone — a speccific accelerator of glycolysis — may re-establish the phagocytic activity of the RES.

In the first stage of shock, glycolysis is stimulated, that being the only way in which the cell can produce energy in the absence of oxygen (*see* Section 8). One of the most important consequences of the initial hyperglycolysis is the stimulation of the RES, also revealed by an increase in the peroxidase activity of the serum [32]. Experimental blocking of the RES constantly aggravates shock, especially bacteriemic shock [3, 11, 13, 35].

Hyperglycolysis is, however, an autolimited metabolic possibility. Its final product, lactic acid, is under physiologic conditions further dehydrogenized in the B type cells, that possess an EMK cycle. In shock, these cells can no longer digest their own active acetate, so that the lactic acid (as well as the concomitantly accumulating $NADH_2$ and $NADPH_2$) will block glycolysis. Phagocytic activity of the RES will rapidly collapse with worsening of the state of shock [4, 57, 60].

The phagocytic activity of the RES is sustained or even stimulated by dehydrooxyacetone, glycolytic acid, triolein, etc. [3, 34]. Experimentally, after the administration of triolein, hyperphagocytosis was demonstrated by hypertrophy of the Kupffer cells and splenic macrophages but the lung became fatally charged with microemboli. Cellular hypertrophy might be due only to the engulfment of triolein which would have an inhibiting and not a stimulating effect on phagocytosis (Altura and Hershey, 1970; Lazar et al., 1969).

Collapse of the RES functions also affects the coagulolytic system, for which the RES is the clearance organ (*see* Section 7.4). The RES avidly takes up colloids with a molecular weight of over 50 000 if the latter are administered during its period of initial phagocytic hyperactivity.

The activity of the RES as a clearance organ is evident in the Schwartzman-Sanarelli phenomenon when, after a second administration of the endotoxin, disseminated intravascular coagulation develops (the first administration only blocks the RES). A similar result is obtained by the perfusion of lipids. The administration of endotoxin may also produce an increase in FFA and hyperlipaemia, influencing the reticuloendothelial clearance system.

6.5.2 ANTIBODY GENESIS

Both antibody synthesis and synthesis of the numerous components of the C' system require a sequence of genetic operations that in shock no longer have a correct regulation.

A denatured immune response appears and the genetic code gradually deteriorates, fairly promptly followed by the appearance of anti-DNA autoantibodies, very often detected in various antibodies, very often detected in various forms of shock (*see* Section 4.11).

The immune activity of the RES also appears to persist after collapse of its phagocytic function [17, 29, 39, 44, 57]. It is known that some of the shocked patients with blood volume and metabolic re-equilibration are then lost by infections. Decrease in the resistance to infections is due both to collapse of the plasma opsonin index and to prolonged contact of the bacteria with the hypoxic phagocytic cell. As soon as bacteria invade the blood or tissues they are encompassed by opsonins (natural or immunization or hyperimmunization antibodies, together with complement) and directed towards the phagocytes, which engulf them and destroy them by digestion. Triggering of both the 'opsonin flow' and of phagocytosis is controlled by the RES. In shock, the RES is hypoxic and washed by a slow blood flow in the immune competent organs (liver, spleen, lung, bone marrow and lymph node stroma) [42, 43].

Under normal conditions, circulation of the 'travelling' lymphocytes, towards and among the cuboid cells of the postcapillary venules in the lymph nodes, is stimulated by histamine; this is produced locally following histidin decarboxylase activation by the antigen-antibody complex [46]. The functional local hyperaemia produced induces an increase in the permeability of the membrane, permitting recognition of the antigen and the production of antibodies. Activation of the discharge of the complement components (especially C'3 and C'5) in the presence of the antigen-antibody complex results in stimulation of histidin decarboxylase with increase in the blood and lymphatic flow in the lymph nodes [61].

Complement complex (C') seems to be involved in the general pathologic physiology of shock. It is known that C' is reduced, especially in bacteriemic shock probably because of its rapid consumption.

After the experimental blocking of C' (with a purified cobra poison), the experimental injection of bacterial liposaccharide no longer caused the abrupt hypotension that always develops after blocking the C' system [25]. This suggested that initial hypotension in shock is mediated by C' system.

The immune response in several kinds of states of shock was investigated by determining the immunoglobulins. In general, IgG decreases in shock and returns slowly to initial values after one to two months; IgA and IgM appear to be less influenced [47].

6.6 INTERSTITIAL VASOACTIVE SUBSTANCES

Among the vasoactive substances there are many amines, prostaglandins, kinins and heparin (catecholamines have been discussed in Section 4.2.2 and kinins in Section 7.4).

The peripheral levels of these substances do not necessarily reflect their true effect. A substance may exert its effect locally and be quickly degraded. Furthermore, many of these agents act on the same receptor site.

Irrespective of the site where they are produced, vasogenic amines (catecholamines, serotonin and histamine) require an enzymatic chain that also contains a decarboxylase, whose indispensable cofactor is pyrodoxal-5-phosphate, produced by intestinal bacteria.

Histamine and serotonin are produced chiefly by the mast cells lying along the vessels of the microcirculation. The blood platelets, however, are also prolific producers, particularly of serotonin. The massive production of these vasogen amines in the platelet clusters that flood the lung may be arrested by methysergid (a specific serotonin antagonist), heparin, salicylic acid (which inhibits the release of serotonin by the platelets) and especially by destroying the intestinal bacteria (for instance by polymixin) (Figure 6.5).

6.6.1 HISTAMINE

Histamine is synthesised and stocked in the mast cells, in a granular complex with heparin and a storage protein

[21, 51]. Degranulation of the mast cells follows upon the interstitial accumulation of cations (mostly K^+) or of CO_2 and lactic acid. By causing vasodilatation of the microcirculation, hista-

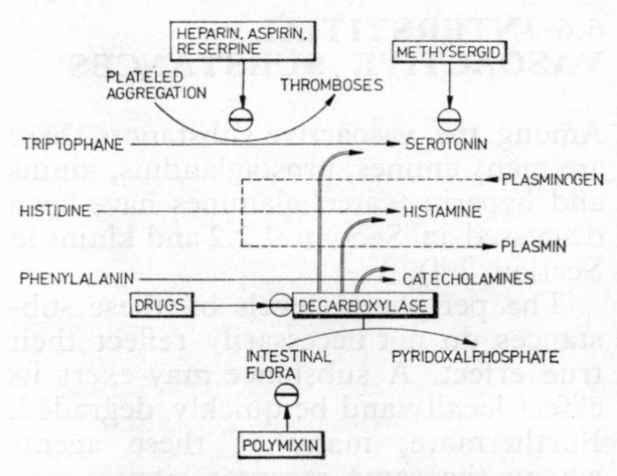

Figure 6.5. Blocking of the decarboxylating reaction stops the synthesis of vasogenic amines whose liberation also brings about the activation of fibrinolysis.

mine increases the flow and washes the catabolites that start the degranulation process of the mast cells; tissular autocontrol of cellular perfusion, nutrition, clearance is a known fact.

In 1912 Dale and Richards discovered that histamine was a vasodilator that increased tissue permeabiliy and, when given to cats in large doses, produced severe hypotension.

In general, histamine is an antagonist of catecholamines at the level of the smooth cells in the vascular walls (capillary vasodilatation with hyperpermeability) but in other instances may act synergically (arteriolar vasoconstriction). Its own characteristic effect, however, is vasodilatation of the small vessels in the microcirculation, whereas serotonin remains the efficient vasoconstrictor of the arteriolar and, particularly, venous walls (Gecse et al., 1969).

In the human body histamine may be found in a labile or extrinsic form (released by the mast cells), in a combined or an intrinsic form and in a tissular or free form [55].

In the pathologic physiology of shock, histamine has long ago been considered as the major cause of hypotension; the presence of histamine was investigated in the blood of subjects with all kinds of shock, and an experimental model of histaminic shock was developed. Today it has been proven, however, that histamine becomes instantly inactive (and, in fact, is inactivated) in the systemic circulation in the tissues.

Antihistaminics do not prevent and do not control hypotension, and histaminic shock produced by blocking the pyridoxal source with polymixin, affects, to the same extent, all three generations of vasogen amines, not only histamine (Figure 6.5).

In fact, the role of histamine in shock as well as in general pathology, which has been studied in detail, is still subject to controversy today as it was two decades ago.

Mention, however, should be made of the latest hypothesis on the major role of histamine in the adaptation of organisms in shock; a particular substance, *resistin* (adaptation factor), which can be transferred, together with the plasma of animals adapted to repeated shock, has proved capable of inhibiting the synthesis and release of histamine (inhibiting the histidin decarboxylase) [27, 35].

6.6.2 ENDOGENOUS HEPARIN

Endogenous heparin is a mucopolysaccharide sulphate with an appreciable negative charge, synthesized by the mast cells. The positive charges (with a plasma or cellular source) cause, in the interstitial ground substance, the release of mast cell granules which contain, apart from heparin, histamine and serotonin. The two last named act upon the microcirculation, whereas heparin buffers the positive charges in the interstitium, being taken up by the fibroblasts that metabolize it.

6.6.3 SEROTONIN

Serotonin (5-hydroxytriptamine) is produced mostly by mast cells associated with the microcirculation but also by argentaffin cells in the Lieberkühn glands (Kultschitzky-Masson cells) and the platelets. At present, its role in shock seems clearer even than that of histamine. It was found in higher concentrations in the portal blood, particularly in cases of shock with severe intestinal involvement (obstruction, peritonitis, pancreatitis, intestinal infarct, etc.). Its pro-sludge effect, as well as the increased viscosity and platelet clumping, renders it undesirable in the pathogeny of shock *(see* Section 7.4). It has the same effects on the microcirculation as histamine whose production it appears to stimulate. Its constrictive effect on the venous wall accentuates the postcapillary venular spasm.

But serotonin may, nevertheless, be considered useful in some of the phases of shock since it ameliorates venous return. Serotonin acts metabolically as a reducing agent and may be used in most of the phases of shock, when a great 'hunger for electrons' exists. It activates phosphorylation (energogenesis) and inhibits the pentose cycle (synthesis).

In shock, histamine and serotonin are switched from mast cell to fibroblast in a cycle of events superposed upon the known stages of shock. Owing to their particular sensitivity to hypoxia (and perhaps also directly to catecholamines), the mast cells are the first to break down, spreading histamine and serotonin granules in the fibroblast mass where they are absorbed (Figure 6.6 *a)*. If the catecholamine concentration and acidosis increase, then the fibroblasts likewise break down and histamine and serotonin are again released into the interstitial tissue and flood the postcapillary venular segment. The postcapillary venular vasoconstriction, in turn, accentuates blood pooling in the microcirculation (Fig. 6.6 *b)*. Starting from this phase, intracellular methanolic acidosis will produce an explosion of the lysosomal sacs with

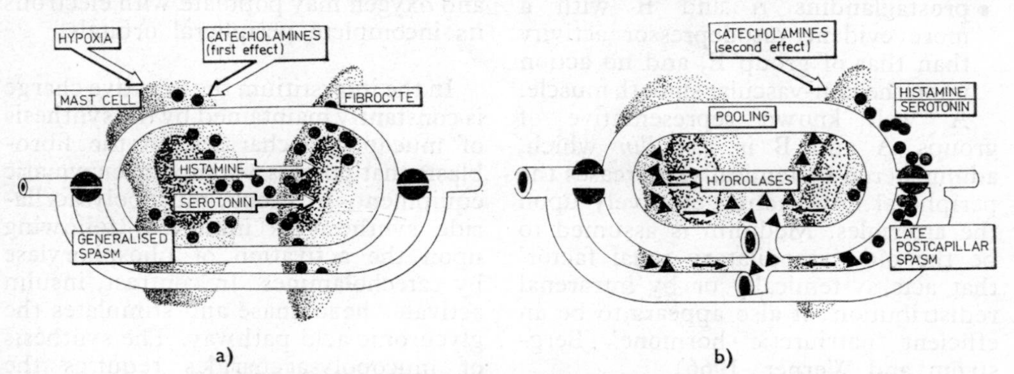

Figure 6.6. Participation of the mast cell and fibrocyte in shock.

the liberation of hydrolases, which along cellular lines will result in autodigestion, and systemically in activation of the fibrinolytic and kinin mechanisms (*see* Section 7.4).

6.6.4 PROSTAGLANDINS

Since 1933 when prostaglandins were first isolated from the seminal fluid, they have been found in almost all the tissues in laboratory animals and man. The chemical structure of 15 components of this class has been determined, all being unsaturated ketons, synthesized biologically by essential fatty acids (with a prostanoic acid as basic ring). They have been classed into four main groups according to their dominant action:

- prostaglandins E with a vasodepressor activity but stimulating the non-vascular smooth muscle;
- prostaglandins F which are vasopressors and also stimulate the non-vascular smooth muscle;
- prostaglandins A and B with a more evident vasodepressor activity than that of group E, and no action upon the non-vascular smooth muscle.

A well known representative of groups A and B is *medullin* which, administered systemically, increases the peripheral flow, acting electively upon the arterioles. Medullin is assumed to be the old 'vasodilatory renal factor' that acts systemically or by intrarenal redistribution; it also appears to be an efficient 'natriuretic hormone' (Bergström and Werner, 1966).

Prostaglandins inhibit the lipolytic effect of epinephrine, are the antagonists of glucagon and ACTH, and may be considered as intracellular metabolic modulators [7]. Because of their effect on non-vascular smooth muscle they have been included in the pathologic physiology of the carcinoid syndrome, duodenal and gastric ulcer, etc.; their presence in the autonomic neurons suggests that they play a role in the transmission of synaptic information, always increasing in case of stimulation of the neuroendocrine axis (and therefore also in shock).

6.7 METABOLIC OSCILLATIONS OF THE INTERSTITIUM

The oscillating hydroelectrolytic phenomena in the interstitium are superposed over the metabolic oscillations of the connective tissue [33].

The metabolic interstitial oscillations have four phases: of the mast cell and fibrocyte, the ground substance, fibrils, water and electrolytes.

Normally, the parenchymatous and mucosa cells deploy their activity in a medium rich in electrons, supplied by the interstitial paravascular space. Under these conditions the hydrogen ions readily pass from capillary to cell, and oxygen may populate with electrons its incomplete peripheral orbitals.

In the interstitium the negative charge is constantly maintained by the synthesis of mucopolysaccharides, in the fibroblasts that possess a complete enzymatic equipment. In shock, mucopolysaccharide synthesis is inhibited, following upon the activation of phosphorylase by catecholamines. In contrast, insulin activates hexokinase and stimulates the glycuronic acid pathway. The synthesis of mucopolysaccharides requires the presence of NAD (in shock there is, however, an accumulation of $NADH_2$) and $NADPH_2$ (which is reduced in shock). Moreover, in shock due to the excess release of thyroid hormones, hypercatabolism of the mucopolysaccharides occurs.

In the first phase of shock, catecholamines degranulate the mast cells, liberating serotonin, heparin, histamine and kinins (*see* Section 7.4.5). Endogenous heparin is rapidly consumed but the synthesis of mucopolysaccharides is abolished. Initially, an invasion of the interstitial connective tissue occurs with positive charges (equivalent to a lack of electrons or electron drainage). This is the oxidating phase, encountered in any inflammatory reaction, that stimulates phagocytosis, induces labilization of the lysosomes and capillary hyperpermeability. In this phase large amounts of oxygen are still found, required by the pentose pathway for realizing the peroxide function and lability of the lysosomes. Cellular hyperglycolysis ensures phagocytosis and antibody genesis (in the cells belonging to the RES) and supplies the energy necessary for the transport membrane systems, i.e. the essential mechanisms for maintaining intercompartmental functional selectivity and preventing cellular anarchy (Figure 6.7).

The therapeutical administration of electron donors protects the membranes, including the lysosomal ones, reforms the SH groups and favours the synthesis of mucopolysaccharides, as reducing elements.

But in the absence of an adequate therapy (if a PAOR and not actually a state of shock) there also follows a spontaneous reducing phase, dominated by the presence of endogenous glucocorticoids that act chiefly on the fibroblasts, stimulating the synthesis of mucopolysaccharides (Figure 6.8).

The third phase is, again, oxidizing, slow reconstructive cicatrisation. It is also controlled by glucocorticoids which are, in the meantime, oxidized and begin their activity as mineralocorticoids, ensuring the reconstructive phlogistic events. Therefore, in order to exercise its reducing role in the second phase, glucocorticoids have to be efficiently hydrogenated in the liver, otherwise their peripheral oxidation will become predominant.

The oscillating evolution mentioned is reconstructive, harmonious and economic.

In shock, however, the metabolic phenomena in the interstitial space

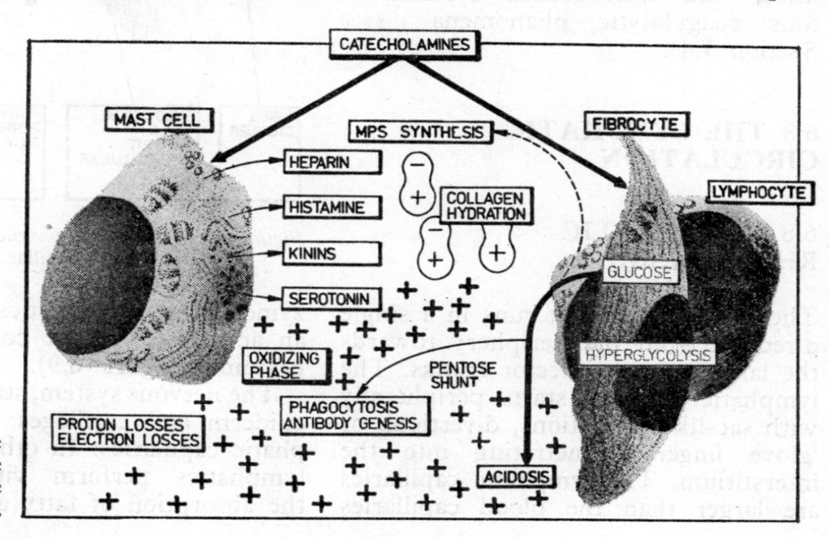

Figure 6.7. The oxidizing phase of shock in the interstitial space.

exhibit chaotic development, because the initial invasion of the oxidizing substances cannot be equilibrated. In shock, the evolution of the microcirculation-interstitial space system is aggravated by progressive impairment of the two components. The interstitial acidotic cells release lysosomal hydrolases, with intravascular dysharmonious coagulolytic phenomena *(see* Section 7.4).

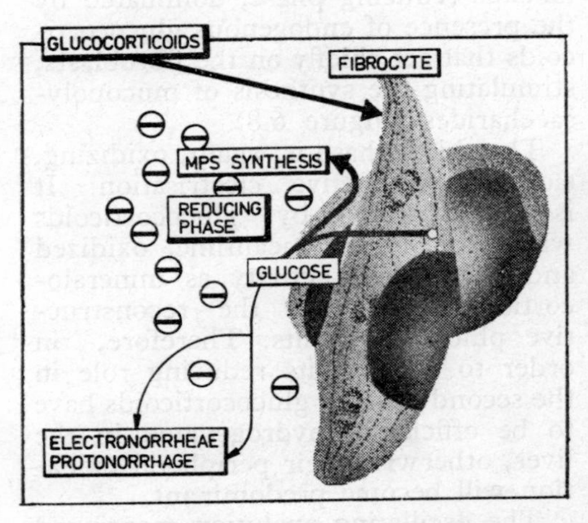

Figure 6.8. The reducing phase of shock in the interstitial space.

6.8 THE LYMPHATIC CIRCULATION

6.8.1 LYMPHATIC RHEODYNAMICS

The lymphatic circuit runs in a single direction from the periphery towards the large venous collector trunks. The lymphatic network starts peripherally with sac-like formations, diverticuli or 'glove fingers' penetrating into the interstitium. The lymphatic capillaries are larger than the blood capillaries (having a diameter of 20-100 μ); their wall is formed of discontinuous epithelial cells with 'oak-leaf' edges. The walls of the lymphatics have large pores, like open windows, as the endothelial cells do not adhere to one another. The intima of the lymphatics has histioenzymologic characteristics that suggest an active role in the concentration of proteins (Figure 6.9).

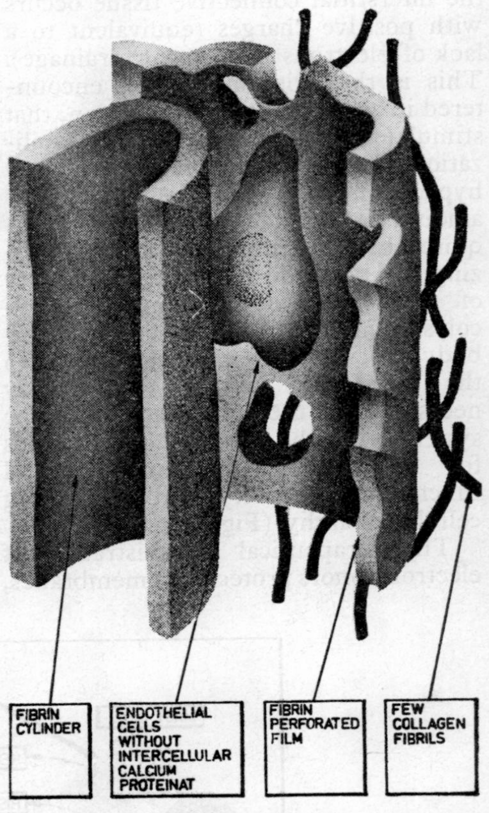

Figure 6.9. The intimate structure of the lymphatic capillar.

The nervous system, striated muscles, epiderm and cartilages have no lymphatic capillaries. In other regions the lymphatics perform vital functions: the absorption of fatty acids from the intestine, and countercurrent multiplication in Henle's loop.

The lymphatic capillaries coalesce into bead-like lymphatic venules (with supravalvular swellings) and then into larger and collector lymphatic trunks, whose walls are lined with smooth muscle fibres. The fluid flows through the lymphatics at the rate of 3 to 4 litres/day (a volume equivalent to the whole plasma fluid of the body) and empties into the superior cava system by 'compliance waves' at the upper thoracic aperture [23]. The rate of the lymph flow is approximately 100 ml/h or 1.3 ml/h/kg body weight at rest and 4 ml/h/kg body weight during digestion [15, 16].

The liver and kidney are the great lymph producers. The kidney produces in one minute an amount of lymph equal to that of urine. Obstruction of the renal lymphatics blocks the countercurrent mechanism of the loop of Henle and within two to three days the tubulocyte is impaired. Affection of the intrarenal lymphatic network plays a major role in the onset of shock kidney.

The lymphatic output of the liver is about 1 ml/min, covering about half the amount of lymph that passes through the thoracic duct. The chief haemolymphatic crossroads is to be found at the level of Disse's space [38, 56, 65].

The lymphatic system passes through the organs with a primary and secondary immune competence, thus rendering them functionally solid (*see* Section 4.11).

Lymph. The main function of the lymphatics is to return proteins to the circulation [8, 69]. Several hypotheses have been put forward concerning the permeability of the lymphatic walls, all subordinated to the general concept of crossing of the membranes. As long ago as 1961 it was postulated that permeability was possible due to the pores of the lymphatic capillaries; but it was also assumed, at about the same time, that a substance passes through the wall by becoming soluble in the membrane, or by its chemical interaction with the latter (a slow, unwieldy method).

Electron microscopy has confirmed the existence of pores at endothelial level in the small lymphatics and has likewise shown that there are no pores in the larger lymphatics so that the exit of proteins can occur only in the areas of blood and lymph microcirculation [45].

The lymph flow starts in the interstitium by imbalance of the same Starling forces which also act in the blood microcirculation (Figure 6.10). Since the lymphatics take up a large amount of interstitial fluid they act as a mechanism compensating circulating blood volume (this accounts for the different reaction of the lymphatic

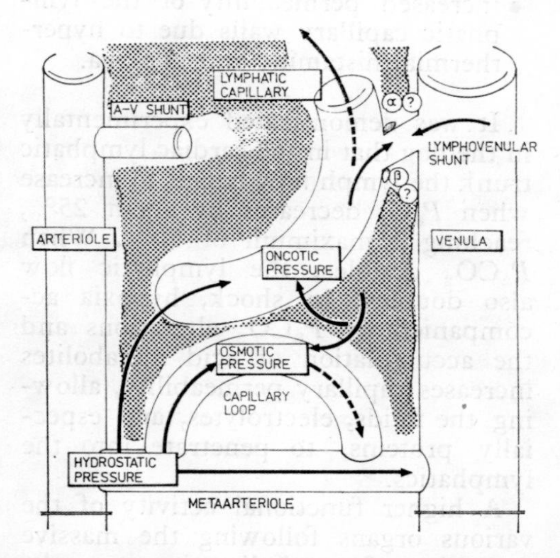

Figure 6.10. The relationship between the lymphatic microsystem and the blood microcirculation.

system in non-cardiogenic and cardiogenic shock (*see* Figure 6.13).

Films taken of the lymph flow have demonstrated the existence in humans of a contractile activity of the pelvic and inguinal lymphatics and of the thoracic duct. The smooth muscles of the lymphatics are richly innervated and histologic findings have revealed the presence of a sympathetic texture in the upper layer of the adventitia that forms a plexus closely connected with the smooth musculature of the thoracic duct.

The lymph flow is indirectly influenced by:

- the factors that increase the rate of filtration of the fluid from the blood capillaries;
- elevation of the venous pressure (due to compression of the lower cava above the site at which the hepatic veins empty) increases the hepatic lymphatic flow; ligature of the vena porta increases the lymph flow in the intestinal area;
- increased permeability of the lymphatic capillary walls due to hyperthermia, histamine and hypoxia.

It was demonstrated experimentally in the dog that in the cardiac lymphatic trunk the lymph flow begins to increase when P_aO_2 decreases by about 25%, reaching a maximum of 50%. When P_aCO_2 doubles, the lymphatic flow also doubles. In shock, hypoxia accompanied by P_aCO_2 alterations and the accumulation of acid metabolites increases capillary permeability, allowing the fluids, electrolytes, and especially proteins, to penetrate into the lymphatics.

A higher functional activity of the various organs following the massive formation of metabolites increases the osmotic pressure of the interstitium and draws the fluid from the blood vessels; under these conditions the lymphatic circulation likewise increases.

Increase in the rate of flow of the lymph is also influenced by active and passive thoracic movements (pulmonary compliance waves).

A problem that has not yet been elucidated is that of the action of catecholamines on the lymphatic vessels [36, 37]. It has been assumed, however, that vasoconstriction of the lymphatics, with diminution of the lymph production, is the result of catecholamine stimulation, in keeping with the response of the entire vascular system to this stimulation.

Lymphovenous communications. Today the existence of lymphovenous shunts is no longer questioned [9, 41, 54]. Studies in embryology have shown that, initially, the lymphatics and venules form a single network and some of the connections appear to persist for a lifetime. Clinical and experimental investigations with radioactive tracers proved that these shunts have a reduced activity in the normal human or animal, *but may become permeable again in certain pathologic states, when the intercompartmental flow of fluids is abnormal* (Teodorescu-Exarcu et al., 1972).

There are numerous lymphovenous shunt pathways, especially in the peritoneum, renal and suprarenal veins, azygos, portal area and mediastinum. Worthy of particular note are the lymphovenous shunts in the lymph nodes, whose existence was revealed by microscopic examination of the nodes blocked with ultrafluid lipiodol after lymphography.

The lymphovenous shunts respond to a complex neurohormonal regulation and may be set off, even in the absence of any ligature or obstruction of the lymphatics, by the injection of procaine or alpha-blocking agents. A series of investigations have shown that epinephrine increases the output and permeability of the lymphovenous communications but only within the superior cava area.

The lymphovenular shunts represent a possible route of evacuation of the lymph into the venules. The circulation of fluid may, however, also take place in a contrary direction, which explains the existence of red blood cells in the lymph of the thoracic duct (portal hypertension syndrome). In cirrhotic patients, both the thoracic duct and the visceral lymphatics are distended by the pressure of the excess fluid, dilating the lymphatic collaterals and rendering the lymphovenular shunts functional. When the collateral circulation develops by functioning of the porto-azygo-caval shunts, the pressure rises in the azygos system, opening the venolymphatic shunts and allowing the red blood to penetrate into the thoracic duct. Lending support to the existence of these two-way shunts is the clinical observation of the œsophageal varices which cease to bleed following lymphatic drainage by cannulation of the thoracic duct [30]. The existence of venulolymphatic shunts in the lungs has likewise been demonstrated by experimental stenosis of the pulmonary artery.

6.8.2 LYMPHATIC TRANSPORT IN SHOCK

The correlation of the lymphatics with the interstitial cellular population and the various products of cellular metabolism is very close, particularly in shock. It has been established that in the onset of shock the action of catecholamines and adrenohypophyseal hormones brings about a lytic and lymphopenic effect in the lymphoid formations. Apart from the release of specific and unspecific antibodies necessary for an immediate anti-infectious defence, this action also favours the appearance of nucleoproteins and amino acids, which are the raw material required for tissular repair, when eventually the shock-inducing factors are removed.

The lymph is the elective carrier of the lymphocytes. When a chronic fistula of the thoracic duct occurs and the lymph is reintroduced after hypotonic lysis of the lymphocytes has been contained, lymphocytic depletion develops, especially in the thymodependent areas (spleen, intestine and lymph node medulla) [24]. 'Immunocyte' depletion of all the lymphatic structures takes place in shock extremely rapidly by lymphatic pathways [5, 14].

Study of the disorders occurring in the synthesis of nucleic acids of lymphocytes after thermal shock have shown that they may affect both humoral mechanisms and cellular immunity [19]. In order to analyse the lymphocytic defect in 15 cases of burns covering from 60 to 95% of the body surface, the rate of DNA and RNA synthesis was determined. On the 2nd and 19th day after the burn, lymphocytes were isolated from the peripheral blood, cultivated at 37°C, and labelled with tritiated uridin at 23 hours and with tritiated thymidine at 31 hours. The increased nucleic acid synthesis may be considered as a response to the action of the products released by the burns which by immediate, intense, postaggressive lymphocytolysis will subsequently jeopardize the cellular defence of the organism due to the anarchic synthesis of autoantibodies [40, 53].

In the late phases of shock, drainage of the interstitial fluid through the lymphatics will supply the necessary volume, proteins and electrolytes but will also bring into the circulation the metabolic residues of the acidotic cells. Owing to the renal shunt these residues cannot be evacuated and the hypoxic liver is not able to neutralize them, thus completing a vicious circle with a lymphatic component that worsens the state of shock. Serotonin transported by the lymphatics accentuates the capillary venular spasm and most of the enteral toxins likewise take the lymphatic route.

At 8 hours after intraperitoneal injection (in the dog) of ^{51}Cr tagged endotoxin, 15% of the amount injected is to be found in the thoracic duct, 6% in the left duct and only 14% in the blood, liver, spleen, lungs and kidneys [18]. This confirms the 'preferential infection' of the lymphatic spillway in bacteriemic shock.

Lymphatic return transports a dangerous concentration of bradykinin [62] and lysosomal hydrolases collected from the large retroperitoneal spaces [22]. Other authors, determining acid phosphatase and betaglucuronidase by cannulation of the thoracic duct found increased concentrations in the plasma and lymph in haemorrhagic shock [6, 31]. The lymph also appears to be the preferred transporter of the myocardial depressor factor (MDF) whose synthesis occurs in the pancreatic space; exteriorization of the lymphatic flow always protected the heart of the experimental animals (*see* Section 7.4).

The thoracic duct is considered in man, too, as a derivative channel of the hepatosplanchnic interstitial fluid [2]. Since, during the period of intervascular cellular pooling flow is only possible through the thoracic duct, the composition of the lymph reflects more reliably the metabolism of the splanchnic tissues than the arterial and venous fluids do (Barankay et al., 1969; Glenn and Lefer, 1970).

Oxygen pressures are variable, particularly in the veins, consequently, in order to determine splanchnic oxygenation O_2 pressure has been measured in the thoracic duct [48, 70]. The lymphatic fluid appears to play a role in some of the stages of shock, acting for a short time as a neutralizing alkaline buffer of systemic acidosis. Hence, the study of gas exchanges, acid-base balance and lysosomal enzymes in the lymphatic sector have furnished useful data for our understanding of the pathology of shock. It should not be forgotten, however, that investigation of the lymphatic system in general, and in shock in particular, is still in its early stages (*see* Section 2.11).

6.8.3 LYMPHATIC FLOW OSCILLATIONS IN SHOCK

Similar to the other phenomena of shock, rheodynamics of the lymphatic fluid presents oscillations [10, 59, 66].

In *the stage of early reversible shock* hypercatecholamine stimulation has a constrictive effect upon the lymphatic vessels; probably equipped with a contractile cellular mechanism, with alpha and beta receptors, the lymphatics likewise become spastic. Initially, in the thoracic duct, the lymph flow rises slightly until the lymph contained at the onset of shock is eliminated, after which the flow decreases. During this period the sympathoadrenergic reaction triggered off tries to achieve a rapid haemodynamic equilibration. Intravasation of the interstitial fluid is at first accounted for on the basis of the Starling pressure forces (Figure 6.11 *a*). The intravasation fluid, however, lacks proteins and will be maintained in the vascular space only as long as the filtration pressure remains low; the intravasation fluids are isotonic, without proteins and, due to haemodilution, are eliminated through the

the tubular return route that is still permeable.

The fluid from the microcirculation cannot pass beyond the postcapillary venular spasm and forms a pool in

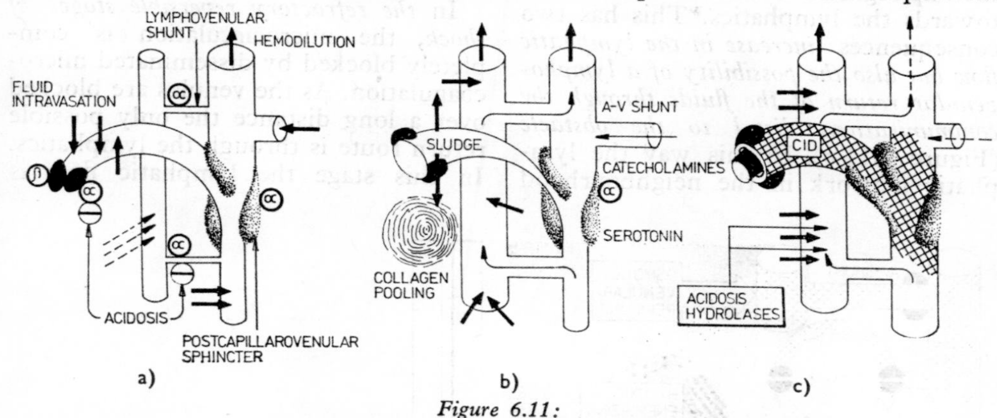

Figure 6.11:
a — the relationship between the lymphatic system and the microcirculation in the early phase of reversible shock; *b* — the late phase; *c* — the refractory phase.

urine within approximately two hours. In haemorrhagic shock, intravasation starts at the same time as the bleeding and continues after it has ceased, reaching a peak after 30 to 45 minutes [66].

In *late reversible shock*, owing to metabolic disturbances, acidosis of the interstitium develops and the precapillary barrier breaks down. Fluid pooling appears in the microcirculation and hydroelectrolytic retention in the interstitial space.

The dominant event in this stage is the prolonged spasm of the postcapillary venular segment, since it now directs all the possible fluid circuits. This spasm, due to activation of the alpha-receptors, perhaps also of the beta receptors [49], or also to serotonin and bradykinin [58], is a pathophysiologic reality although it is not yet known how it is maintained.

In this stage of shock, lymphatic drainage plays its major role, being

the vessels and the interstitial space where the collagenic 'sponge' becomes hypertrophic (*see* Section 6.1). The interstitial hydrostatic pressure increases, bringing about penetration into the lymphatic network of a fluid rich in proteins and electrolytes, either through the pores of the lymphatic capillaries (broadened by the presence of tissular catabolites), or by chemical interactions with the constituents of the lymphatic capillary basement membrane. The process is slow but efficient as the fluids that have penetrated into the lymphatic vessels have a high protein concentration and play the role of *plasma autoperfusion.*

Moreover, the lymphatic flow brings large amounts of water electrolytes and proteins to the return circulation, due especially to a simple mechanical event of particular dynamic importance, *the opening of the venulolymphatic shunts* [66, 67]. Their presence offers the possibility of by-passing the postcapillary venular spasm by regurgitation. Owing to their multiple arrangement, the shunts proximal to the obstacle drain the fluid from the dammed up segment of the microcirculation towards the lymphatics. This has two consequences: *increase in the lymphatic flow but also the possibility of a lymphovenular return of the fluids through the communications distal to the obstacle* (Figure 6.11 *b*). In this way the lymphatic network in the neighbourhood of the microcirculation may improve the rheodynamic blood failure in the late reversible stage of shock, acting as a multitubular sponge, with many compensatory possibilities.

In *the refractory reversible stages of shock*, the microcirculation is completely blocked by disseminated microcoagulation. As the venules are blocked over a long distance the only possible return route is through the lymphatics. In this stage the lymphatic flow is

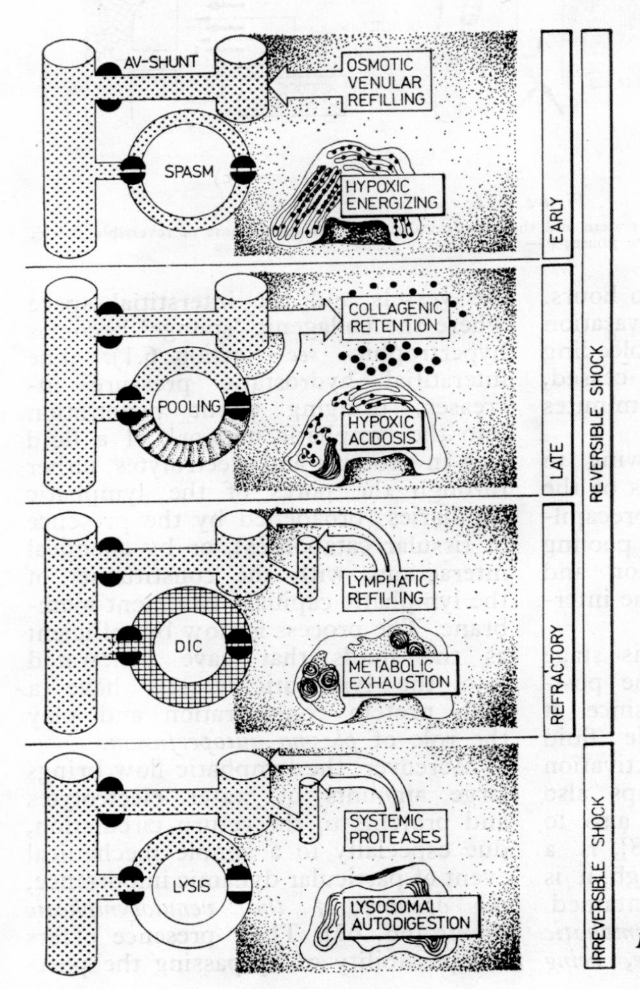

Figure 6.12. **The stages of shock with lymphatic participation.**

maintained but brings within the entire range of deleterious products generated by prolonged affection of the cells.

In *the final, irreversible stage of shock*, although fibrinolysis has washed the microcirculation, functional and cellular morphologic anarchy have developed in the meantime in the interstitium. The lymphatic flow gradually falls to zero, with disappearance of all the factors that supported it. These events are also illustrated in figure 6.12, which resembles figure 1.13 but also shows the lymphatic participation.

Particular mention should be made of the adaptive reaction of the lymphatic system in cardiogenic and non-cardiogenic shock [1, 66]. In non-cardiogenic shock the lymphatic flow increases by indirect mechanisms:

- increase in capillary permeability;
- a rise in the total interstitial pressure;
- decrease in the oncotic pressure of the plasma;
- elevation of the interstitial protein concentration;
- dilatation of the pre-microcirculation area and venular spasm caused by histamine, facilitating filtration towards the interstitium and pushing off the lymph column.

In cardiogenic shock these peripheral mechanisms are diminished and the lymphatic flow decreases.

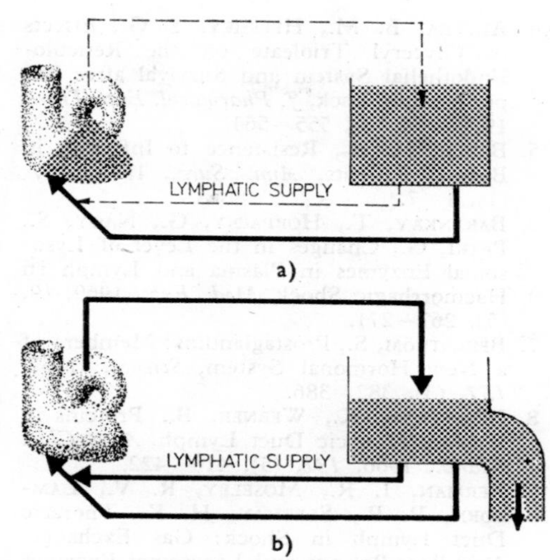

Figure 6.13. **Limphatic output in cardiogenic shock and in non-cardiogenic shock:**

a — in cardiogenic shock the diminished lymphatic supply does not overload the venous return; *b* — in non-cardiogenic shock the lymphatic flow increases, improving the venous return.

It has been demonstrated experimentally and clinically that in non-cardiogenic shock the CVP decreases and the lymphatic output increases whereas in cardiogenic shock the CVP increases and the lymphatic output decreases (Figure 6.13). Decrease of the lymphatic output in cardiogenic shock emphasizes the effort of the organism to maintain homeostasis (the 'wisdom of the human body'), decrease of the lymphatic flow appearing to protect an inefficient heart that can no longer pump all it receives.

SELECTED BIBLIOGRAPHY

1. Adams, J. T., Germon, R. H., Schwartz, S. I., Dynamics of Lymph Flow in Cardiac Shock, *Ann. Surg.*, 1969, *170*, (2), 193–198.
2. Alicon, F., Hardy, J. D., Mechanism of Shock as Reflected in Studies of Lymph of Abdominal Organs, *Surg. Gynec. Obst.*, 1961, *113*, (5), 734–739.
3. Altura, B. M., Hershey, S. G., Res Phagocytic Function in Trauma and Adaptation to Experimental Shock, *Amer. J. Physiol.*, 1968, *215*, (6), 1414–1419.

4. ALTURA, B. M., HERSHEY, S. G., Effects of Glyceryl Trioleate on the Reticulo-Endothelial System and Survival after Experimental Shock, *J. Pharmacol. Exp. Ther.*, 1970, *175*, (3), 555–564.
5. BALCH, H. H., Resistance to Infection in Burned Patients, *Ann. Surg.*, 1963, *157*, (1), 1–7.
6. BARANKAY, T., HORPACSY, G., NAGY, S., PETRI, G., Changes in the Level of Lysosomal Enzymes in Plasma and Lymph in Haemorrhagic Shock, *Med. Exp.*, 1969, *19*, (5), 267–271.
7. BERGSTRÖM, S., Prostaglandins: Members of a New Hormonal System, *Science*, 1967, *157*, (3), 382–386.
8. BERGSTRÖM, K., WERNER, B., Proteins in Human Thoracic Duct Lymph, *Acta Chir. Scand.*, 1966, *131*, (5), 413–422.
9. BERMAN, I. R., MOSELEY, R. V., LAMBORN, P. B., SLEEMAN, H. K., Thoracic Duct Lymph in Shock: Gas Exchange, Acid Base Balance and Lysosomal Enzymes in Haemorrhagic and Endotoxin Shock, *Ann. Surg.*, 1969, *169*, (2,) 202–209.
10. BERNARD, J. P., BIRON, A., Variations des liquides extracellulaires rapidement échangeables estimés par l'EDTA ^{51}Cr au cours du choc hémorragique, *Agressol.*, 1971, *12*, (4), 275–281.
11. BIOZZI, G., BENACERRAF, B., HALPERN, B. N., Quantitative Study of the Granulopectic Activity of the Reticulo-endothelial System, *J. Exp. Path.*, 1953, *34*, (3), 441–457.
12. BROIDO, P. W., BUTCHER, H.R., MOYER, C.A., The Expansion of the Volume Distribution of Extracellular Ions during Haemorrhagic Hypotension and Its Possible Relationship to Change in the Physical-Chemical Properties of Extravascular-Extracellular Tissue, *Arch. Surg.*, 1966, *93*, (4), 556–661.
13. CHIRICUŢĂ, I., TEODORUŢIU, C., SIMU, G., MULEA, R., Modificările sistemului reticulohistiocitar în şocul prin arsură (Alterations of the Reticulohistiocytic System in Burn Shock), *Revista Sanitară Militară*, 1965, *LXI*, (special no.), 183–187.
14. CLEUDINNEN, B. G., BARNES, A. D., WILLIAMSON, N., A Study of Antibodies to DNA in the Sera of Shocked Patients, *Ann. Surg.*, 1969, *170*, (5), 1021–1024.
15. COPE, O., LITWIN, S.B., Contribution of the Lymphatic System to the Replenishment of the Plasma Volume Following a Haemorrhage, *Ann. Surg.*, 1962, *155*, (4,) 655–660.
16. COURTICE, F. C., Lymphatic Function, *Med. J. Austr.*, 1968, *1*, (10), 379–385.
17. DAMESHEK, W., Autoimmunity, Theoretical Aspects, *Ann. N.Y. Acad. Sc.*, 1965, *106*, (6), 124.
18. DANIELE, R., SINGH, H., APPERT, H. E., PAIRENT, W. F., HOWARD, J. M., Lymphatic Absorption of Intraperitoneal Endotoxin in the Dog, *Surg.*, 1970, *67*, (3), 484–487.
19. DANIELS, J. C., COBB, E. K., LYNCH, J. B., Altered Nucleic Acid Synthesis in Lymphocytes from Patients with Thermal Burns, *Surg. Gynec. Obstet.*, 1970, *130*, (5), 733–788.
20. DIMITRIU, C. GH., *Bolile colagenului (Collagen Diseases)*, Editura Medicală, Bucureşti, 1968.
21. DOMBRO, C.R., RAGINS, H., Histamine Metabolism in the Dog Stomach as Determined by Intra-Arterial Infusion of Histamine ^{14}C, *Surg.*, 1971, *69*, (4), 504–509.
22. DUMONT, A. E., WEISSMANN, G., Lymphatic Transport of beta Glucuronidase during Haemorrhagic Shock, *Nature*, 1964, *201*, 1231–1232.
23. EL-GENDI, M. A., ZAKY, H. A., Thoracic Duct Lymph Flow: A Theoretical Concept, *Surg.*, 1971, *68*, (5), 786–790.
24. FISH, J. C., BEATHARD, G., SARLES, H. E., RENMERS, A. R., RITZMANN, S. E., Circulating Lymphocyts Depletion, *Surg.*, 1970, *67*, (4), 658–666.
25. FROM, A. H., GEWURZ, H., GRUNINGER, R.P., Complement in Endotoxin Shock: Effect of Complement Depletion on the Early Hypotensive Phase, *Infection Immunity*, 1970, *2*, (1), 38–41.
26. FULTON, L. R., Adsorption of Sodium and Water by Collagen during Haemorrhagic Shock, *Ann. Surg.*, 1970, *172*, (5), 861–869.
27. GECSE, A., HORPACSY, G., KARADY, S., Histamine Liberation and Histamine Formation in Traumatic Shock and in Shock-Resistance. I. Histamine Release, *Med. Exp.*, 1969, *19*, (2), 272–278.
28. GLENN, T. M., LEFER, A. M., Protective Effect of Thoracic Lymph. Diversion in Haemorrhagic Shock, *Amer. J. Physiol.*, 1970, *219*, (1), 10–16.
29. GLYNN, L. E., HALBORW, E. Y., *Autoimmunity and Disease*, Blackwell Scientific Publications, Oxford, 1965.
30. GODART, S., COLLETTE, J., DOLEM, J., Pathologie chirurgicale des vaisseaux lymphatiques, *Acta Chir. Belg.*, 1964, *6*, (suppl. 1).
31. JACOBS, F. A., LARGIS, E. E., Effect of Protein Inhibitors on Protein and Amino acids in Mesenteric Lymph, *Proc. Soc. Exp. Biol.*, 1969, *130*, (3), 697–702.
32. JANSA, P., Herabsetzung der Peroxydaseaktivität bei Ratten mit gehemmter Funktion des Res, *Folia Haemat.*, 1969, *91*, (4), 401–412.

33. LABORIT, H., Régulations métaboliques dans le tissu conjonctif, *Agressol.*, 1969, *10*, (1), 11–31.
34. LAMOTHE, C., THURET, F., LABORIT, H., Influence de l'acide glycolique sur la fonction phagocytaire du système réticuloendothélial. Etude cinétique, *Agressol.*, 1971, *12*, (3), 213–216.
35. LAZAR, G., SÁNDOR, TH., KARÁDY S., WEST, G. B., Reticulo-Endothelial System (RES) – Function and Shock-Sensitivity, *Med. Exp.*, 1969, *19*, (5), 287–292.
36. LEANDOER, L., Coagulation and Fibrinolysis in Blood and Lymph after Massive Haemorrhage in Dogs, *Acta Chir. Scand.*, 1968, *134*, (8), 597–605.
37. LEVIT, D. G., Theoretical Model of Capillary Exchange Incorporating Interactions between Capillaries, *Amer. J. Physiol.*, 1971, *220*, (1), 250–255.
38. MALLET-GUY, M., MANDOIRON, J., Recherches expérimentales sur la circulation lymphatique du foie, *Lyon Chir.*, 1967, *63*, (5), 652–660.
39. MANDACHE, F., *Fiziopatologia circulației și imunitatea în șoc (The Pathophysiology of Circulation and Immunity in Shock)*, Editura Academiei Republicii Socialiste România, București, 1966.
40. MARKLEY, K., SMALLMAN, E., EVANS, G., Antibody Production in Mice after Thermal and Tourniquet Trauma, *Surg.*, 1967, *61*, (6), 896–903.
41. MCKINLAY, A. T., SANGUINETTI, F. A., ROSAS, G., Drainage and Collection of Lymph in the Dog, *Argent. Chirurg.*, 1968, *15*, (5–6), 144–146.
42. MCKNEALLY, M. F., SUTHERLAND, E. R., GOOD, R. A, The Central Lymphoid Tissues of Rabbits. I. Functional Studies in Newborn Animals, *Surg.*, 1971, *69*, (2), 166–174.
43. MCKNEALLY, M. F., SUTHERLAND, E. R., GOOD, R. A., The Central Lymphoid Tissues of Rabbits. II. Functional and Morphologic Studies in Adult Animals, *Surg.*, 1971, *69*, (3), 345–353.
44. MESROBEANU, I., BERCEANU, ST., *Imunologie și imunopatologie (Immmunology and Immunopathology)*, Editura Medicală, București, 1968.
45. MIHAI, C., FILIPESCU, Z., *Metabolismul normal și patologic al apei, sării și potasiului (Normal and Pathologic Metabolism of Water, Salt and Potassium)*, Editura Medicală, București, 1955.
46. MOORE, T. C., A Theory of the Role of Histamine Metabolism in Transplant Rejection, *Surg. Gynec. Obst.*, 1971, *132*, (3), 489–491.
47. MUNSTER, A. J., HOAGLAND, H. C., PRUITT, Jr. B. A., The Effect of Thermal Injury on Serum Immunoglobulins, *Ann. Surg.*, 1970, *172*, (6), 965–969.
48. NAGY, S., BARANKAY, T., TARNOKY, K., Effect of Haemorrhagic Shock on Oxygen Tension of Thoracic Duct Lymph, *Physiol. Acad. Sc. Hungary*, 1969, *35*, (1), 87–92.
49. NELSON, M. C., NELSON, J. L., WELDON, W. E., VELA, R. A., Mechanisms of Change in Thoracic Duct Lymph Flow during Hypotensive States, *Ann. Surg.*, 1970, *171*, (6), 883–891.
50. NIKULIN, A., PIKULA, B., GMAZNIKULIN, E., PLAMENAC, P., Changes in the Endothelial Cells of Blood Vessels over the Recovery Phase Following Histamine Shock, *Folia Med. Fac. Univ.*, 1969, *4*, (1), 49–79.
51. NOSAL, R., SLORACH, S. A., UNVAS, B., Quantitative Correlation between Degranulation on Histamine Release Following Exposure of Rat Mast Cells to Compound 48/80 in vitro, *Acta. Physiol. Scand.*, 1970, *80*, (2), 215–221.
52. OLLODART, R., MANSBERER, A. R., The Effect of Hypovolemic Shock on Bacterial Defense, *Amer. J. Surg.*, 1965, *110*, (3), 302–307.
53. POPESCU, E. A., *Patologie și terapeutică imunitară (Immune Pathology and Therapy)*, Editura Medicală, București, 1971.
54. PRESSMAN, J. J., DUN, R. F., Lymph Node Ultrastructure Related to Direct Lymphatico-Venous Communication, *Surg. Gynec. Obst.*, 1967, *124*, (5), 963–973.
55. RAAB, W., The Effect of Tritoqualine on Anaphylactic and Anaphylactoid Shock in Rats, Evaluated by Renal Enzyme Excretion Studies, *Med. Exp.*, 1969, *19*, (5), 301–311.
56. REDGRAVE, T. G., Inhibition of Protein Synthesis and Absorption into Thoracic Duct Lymph of Rats, *Proc. Soc. Exp. Biol.*, 1969, *180*, (3), 776–780.
57. RITTENBURY, M. S., The Response of the Reticulo-Endothelial System to Thermal Injury, *Surg. Clin. N. Amer.*, 1970, *50*, (6), 1227–1234.
58. ROCHA E. SILVA, M., The Physiological Significance of Bradykinin, *Ann. N. Y. Acad. Sc.*, 1963, *104*, (2), 190–197.
59. ROTH, E., LAX, L. C., MALONEY, J. V. S., Ringer's Lactate Solution and Extracellular Fluid Volume in the Surgical Patient: A Critical Analysis, *Ann. Surg.*, 1969, *169*, (2), 149–164.
60. RUTENBURG, S. H., SCHWEINBURG, F. B., FINE J., In vitro Detoxification of Bacterial Endotoxin by Macrophages, *J. Exp. Med.*, 1960, *112*, (8), 801–807.

61. SCHAYER, R. W., Relationship of Stress-Induced Histidine Decarboxylase to Circulatory Homeostasis and Shock, *Science*, 1960, *131*, (2), 226–227.
62. SHAH, J. P., Studies on the Release of Bradykinin by the Splanchnic Circulation during Endotoxin Shock, *J. Trauma*, 1970, *10*, (2), 255–259.
63. SHANON, A. D., LASCELLES, A. K, Lymph Flow and Protein Composition of Thoracic Duct Lymph in the Newborn Calf, *J. Exp. Physiol.*, 1968, *53*, (4), 415–421.
64. SUTHERLAND, N. G., BOUNOUS, G., GURD, F. N., Role of Intestinal Mucosal Lysosomal Enzymes in the Pathogenesis of Shock, *J. Trauma*, 1968, *8*, (3), 350–380.
65. SYLVEN, C., BORGSTROM, R., Absorption and Lymphatic Transport of Cholesterol in the Rat, *J. Lipid Res.*, 1968, *9*, (5), 596–601.
67. TEODORESCU-EXARCU, I., CÂNDEA, V., IORDAN, T., BUCUR, I. A., Şocul hipovolemic şi circulaţia limfatică (Hypovolemic Shock and the Lymphatic Circulation), *Revista Sanitară Militară*, 1972, *LXXV*, (3), 337–342.
66. TEODORESCU-EXARCU, I., *Agresologie chirurgicală generală (Surgical Aggressology)*, Editura Medicală, Bucureşti, 1968.
68. VIIDIK, A., *Function and Structure of Collagenous Tissue*, Göteborg, Dissertation, 1968.
69. WERNER, B., The Biochemical Composition of the Human Thoracic Duct Lymph, *Acta Chir. Scand.*, 1966, *130*, (1–2), 63–76.
70. WITTE, C. L., COLE, W. R., CLAUS, R. H., DUMONT, A. E., Splanchnic Tissue Oxygenation Estimation by Thoracic Duct Lymph pO_2, *Lympho.*, 1968, *1*, (4), 109–116.
71. ZWEIFACH, B. W., Microcirculation, *Sc. Amer.*, 1959, *200*, (1), 56–64.

7 ELEMENTS OF HAEMORHEOLOGY

Although many questions have been answered, at least an equal number have been raised.

BLAISE PASCAL

7.1 PLASMA
7.1.1 The functions of the intravascular fluid
7.1.2 Intercompartmental biological barriers
7.1.3 The biology of water
7.1.4 Macromolecular systems
7.1.5 Osmosis and polyelectrolytes
7.2 BLOOD CELLS
7.2.1 The red blood cell
7.2.2 The thrombocyte
7.3 THE BLOOD
7.3.1 Viscosity
7.3.2 The laws of intravascular flow
7.3.3 Closing pressure
7.4 THE COAGULOLYTIC SYSTEM
7.4.1 The sludge phenomenon
7.4.2 Disseminated intravascular coagulation
7.4.3 Platelet agglutination and lipid embolism
7.4.4 The plasmin system
7.4.5 The kinin system

The discipline of rheology is not new; it was initiated in the past century by Poiseuille. The flow of a fluid through a tubular system with a variable resistance, ΔR, presupposes a pressure gradient, ΔP, with an *output of volume*, Q_v, according to the relation:

$$Q_v = \frac{\Delta P}{\Delta R} \qquad (1)$$

and a *fluid mass output*, Q_m, corresponding to the relation:

$$Q_m = \rho Q_v \,, \qquad (2)$$

where ρ is the density of the fluid.

But the flow of a fluid cannot be explained only by the difference in pressure. Actually, the movement is ensured by the *total energy* of the fluid which depends upon the dynamics of all its state functions (internal energy, enthalpy and entropy), expressed by the appearance of pressure, kinetic, gravitational, osmotic, electric and thermal gradients.

The blood is a very special fluid; it is a complex biological tissue whose rheological behaviour is different from a newtonian fluid.

The blood, considered as a whole, is a colloid solution: the red blood corpuscles are the particles, and plasma the medium of dispersion. On the other hand, plasma is a colloidal solution, too, the medium of dispersion being water, and plasmatic proteins the particles. Therefore, blood is a very particular fluid whose rheology is not only complex but still very confusing.

Among the features of the blood flow are, also: the periodical *pulsatile character* of the flow; *vasomotricity*, with continuous changes in the tubular diameters under systemic and local regulation; *viscosity* that only relatively complies with the laws of blood flow, especially in the microcirculation; *elasticity* of the vascular walls, and figured circulating elements which express the intrinsic metabolic characteristics.

But, in shock, it is not only the blood that must be studied as a fluid; the interstitial, intracellular and lymphatic fluids have also to be considered from a rheologic point of view.

In order to understand the behaviour of fluids in shock the phenomena that take place within the intercompartmental frontiers must be studied. Then, considering the blood as a fluid tissue made up of plasma and cells, these two components must be analysed separately, as the one cannot in fact exist without the other.

The plasma, an aqueous solution, contains dissociated crystalloids (the ions of different salts), undissociated micromolecules (glucose, FFA, etc.) and macromolecules.

It will therefore be necessary to discuss several of the biophysical properties of pure water and then of the macromolecules that confer upon water its plasma quality. We shall likewise sum up some of the rheologic properties of the blood cells and, finally, the general features of *haemorheology*.

7.1 **PLASMA**

7.1.1 THE FUNCTIONS OF THE INTRAVASCULAR FLUID

The human organism may be considered as a non-homogeneous system of macro- and micromolecules in an aqueous solution. Maintenance of the fluids in the various compartments is achieved by osmotic and electric asymmetry as well as by a biologic activity specific of the elements of each compartment (Table 7.1 and Figure 7.1).

The intravascular fluid compartment is distinguished by its high mobility and the presence of circulating cellular elements; the flow phenomena in this compartment forms the subject of study of haemorheology.

The circulation of fluids in the interstitial, lymphatic and intracellular space (in the tubular system of the endoplasmatic reticulum), although very slow cannot be ignored in shock.

If the first wave of shock affects the intravascular fluid, the overall intercompartmental circulation is very soon taken up in the pathologic circulation of blood volume disturbances.

It is these very intercompartmental blood volume changes that produce the pathogenic axis of shock in which

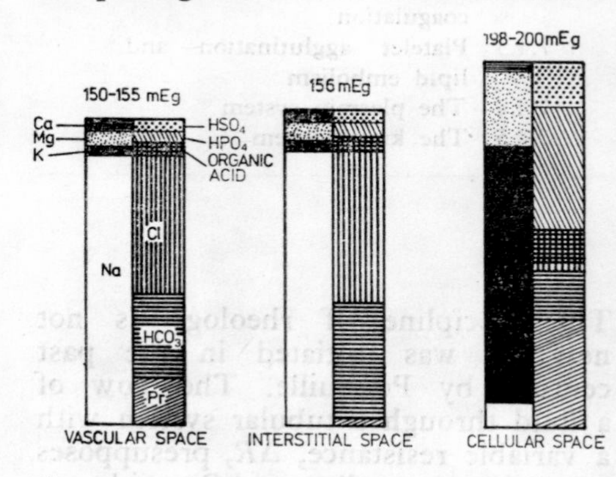

Figure 7.1. The distribution of electrolytes.

Table 7.1 YDRIC AND ELECTROLYTIC COMPARTMENTAL DISTRIBUTIONS *
(Modified by combining the data of Bland, Gruber, Allgöwer, Fleischer and Fattorusso)

	Volume/70 kg body weight (ml)	Cations (mEq/1000)					Anions (mE/1000)				
		Na^+	K^+	Mg^{++}	Ca^{++}	Proteins	HCO_3^-	Cl^-	Organic acids	$H_2PO_4^-$	HSO_4^-
Plasma	3 500	142	4–5	2	5	16	27	101	6	2	1
Interstitial fluid	15 000	135–145	4	2	5	1	30	115	7	2	1
Intracellular fluid	30 000	10	150–160	26–35	2	65	10	3	5–20	100	20
CSF	200–300	144	2– 3				24	125			
Sweat fluid	0–700	2–120	3–12					2–125			
Saliva	1 000–1 500	20–35	3–50				10–45	30–50			
Gastric juice	2 000–3 000	20–70	5–15					80–160			
Gastric juice (without acid)	2 000–3 000	70–150	5–15				25–40	80–120			
Bile	500–1 000	130–165	3–10				30–40	80–100			
Pancreatic juice	500–1 000	100–150	2–15		4	1,2	60–120	50–100			
Intestinal juice	2 000–3 000	80–150	6–15		4		30	40–130			
Faeces	100	50–150	7–15		300						
Ileostomy	variable		3–10					20–130			
Cecostomy	variable	40– 50	8–15					40			
Gastrointestinal aspiration	variable	150	10					60–120			
Intestinal aspiration	variable	110	5					100			
Urine (g/1000)	1500	5–6	2–6	0.1	0.2	0.02	0.1	7–9	0.8–1	1.2	2.7

* Note the ionic features of the digestive fluids

both the aspects of hydric imbalance — œdema and dehydration — may be encountered. In states of shock the character of these fluctuations in blood volum lead to almost all the known clinical forms of dehydration (*see* Figure 7.2).

The intravascular fluid transports throughout the entire macrosystem energy, substance and information, achieving functional synchronization of all the cellular systems.

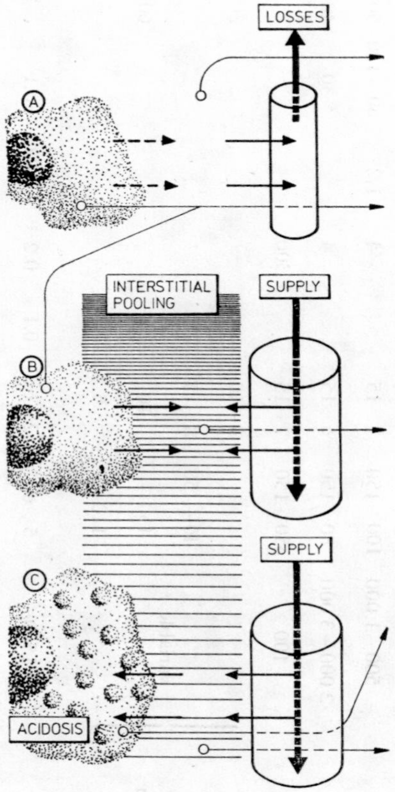

In any type of shock the first decrease takes place in the amount of intravascular fluid, with disturbance of the vital functions, which can only be carried on by a blood possessing optimal quantitative and qualitative functions.

The blood homogenizes the following homeostatic constants:

- *isotonia*, or maintenance at constant values of plasma osmotic pressure; both the plasma crystalloids and colloids bring their contribution;
- *isovolaemia* or maintenance in each compartment of the optimal fluid volume;
- *isoionia* or maintenance of the plasma ion concentrations at constant levels;
- *isohydrogenaemia* (isoprotonaemia) or maintenance of the *p*H at physiologic values;
- *isothermia* or constant maintenance of the absolute temperature at which human metabolism is possible.

In addition, the blood:

- transports energy by flow of the particles of nutritive substrate (glucose, FFA, amino acids), supplying the cells with protons;
- transports gases representing the flow of oxygen towards the cells where it is charged with electrons, and the flow of CO_2 and organic substrate residues towards the elimination systems;
- represents the information circuit by means of the glycoprotein, sterol or aromatic messengers that bathe the cellular spaces, modulating their activity to the advantage of the whole organism; i.e. the blood ensures the informational unity of the human body as a whole.

Figure 7.2. Types of dehydration: A. mixed (cellular and extracellular) dehydration in the first stages of shock; B. interstitial pooling results in extracellular hyperhydration and cellular dehydration ('thirsty oedematous patient'); C. metabolic acidosis brings about cellular hyperhydration accompanied by extracellular hyperhydration (excessive supply) or extracellular dehydration (insufficient supply: 'dehydration with disgust for water').

Almost the totality of these functions is performed by the circulating blood plasma (the red blood cells only intervene in the transport of gases and, to a lesser extent, in maintaining the *p*H). The particular functional value of *plasma* is illustrated by the fact that 70% of the physiologic value of plasma is necessary in order to maintain life; but one can live with only 15% of the hepatic tissue, 25% of the renal tissue, 35% of the total volume of the red blood cells, 45% of the pulmonary tissue and even in the absence of the spleen, pancreas, stomach and over half of the enteron (Figure 7.3). *Therefore, up to 65% of the total volume of the red blood cells can be lost but not more than 30% of the plasma volume.*

The human body contains 60 to 80 ml/kg body weight integral blood

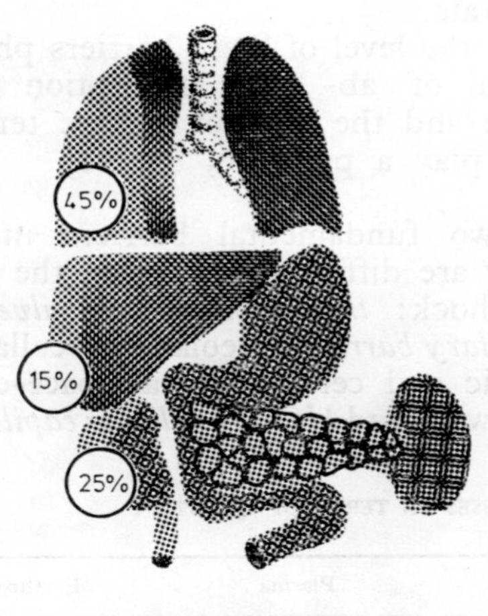

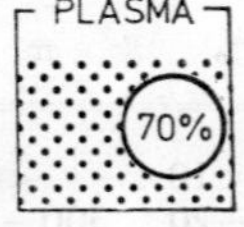

Figure 7.3. The proportion of normal tissue necessary for maintaining the life of some viscera, compared with the erythrocytic and plasmatic volume (modified after Weil and Shubin).

of which 30 ml consists of red blood cells and 40 ml of plasma. These values vary with the state of hydration of the organism, adiposity, body surface, etc.; corrections may be made in terms of the parameters given in standard tables and nomograms [9, 19, 24, 28, 80, 84]. The distribution of the blood volume is illustrated in Figure 7.4.

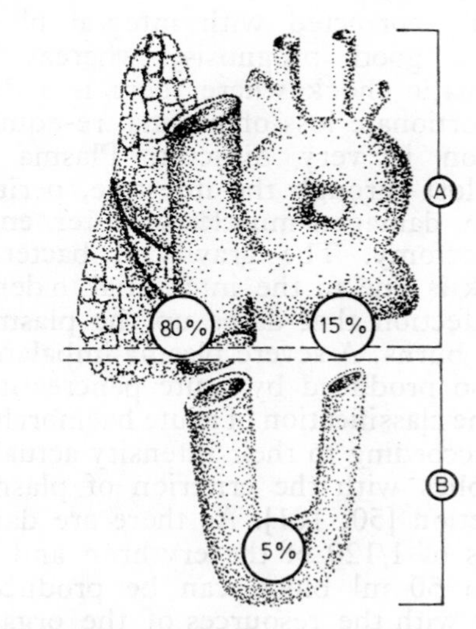

Figure 7.4. The arterial (15%) and capillary (5%) effective perfusion spaces are exceeded by the capacity of the venous space (80%). On the other hand the central viscera, *A*, contain 75% of the blood volume and the periphery, *B*, only 25%.

For a man weighing 70 kg, the dangerous 30% loss of the plasma volume represents 900 ml plasma. A loss of more than 900 ml plasma (over 1600 ml blood for a haematocrit of 45%) may be considered as *critical* in a man weighing 70 kg. As already mentioned, up to 65% of the erythron might be lost (corresponding to 1.5 million erythrocytes per ml of remaining blood), i.e. over 1000 ml of the erythrocyte mass, representing a loss of over 2200 ml blood which is a *critical* situation because of the consecutive plasma depletion (1200 ml plasma).

Hence, the repeated erythrolytic attacks in haemolytic anaemia are comparatively well supported whereas burns covering more then 15% of the body surface cause an impressive imbalance because of the important loss of plasma. Similarly, a purely haemorrhagic shock, rapidly corrected with integral blood has a good prognosis, whereas in traumatic shock where there is a disproportionate loss of plasma, re-equilibration is very difficult. Plasma is also lost through the intestine, peritoneum, damaged muscles or after endarterectomy. The gravity of bacterial shock is due to the interstitial œdema of infection that dams up the plasma, as in burns. A severe plasma imbalance is also produced by acute pancreatitis.

The classification of acute haemorrhages according to their intensity actually complies with the criterion of plasma depletion [50, 101]. As there are daily losses of 1/120 of the erythron and as 40 to 50 ml blood can be produced (only with the resources of the organism), a haemorrhage of up to 500 ml blood can be *autocompensated* within 10 days, and a severe (up to 1000 ml) blood haemorrhage within 20 to 30 days. *A critical haemorrhage* means a blood loss of up to 1500 ml, i.e. a plasma loss of up to 900 ml. This haemorrhage must be controlled therapeutically. *A very severe haemorrhage* is that where there is a loss of over 900 ml plasma; therapeutical control must be very prompt and may sometimes remain without effect.

Table 7.2. gives orientative values concerning the gravity of acute blood volume losses.

7.1.2 INTERCOMPARTMENTAL BIOLOGICAL BARRIERS

The biologic systems present a great physico-chemical heterogenicity: crystalline structures next to amorphous, fluid or semicrystalline ones. Here and there frontiers appear, whose features are expressed in the composition asymmetry of the compartments they separate.

At the level of these barriers phenomena of ab- and adsorption take place and the laws of surface tension also play a part [26, 84, 91].

Two fundamental barriers in the body are difficult to cross in the state of shock: *the blood-air or alveolar-capillary barrier* (alveolar wall-collagen-elastic and cellular septal space-capillary wall) and *blood-cellular or capillary-*

Table 7.2 THE CLASSIFICATION OF BLOOD VOLUME LOSSES IN TERMS OF QUANTITY

Fluid volume losses (oedema, peritonitis, vomiting, haemorrhage, etc.)	Total fluid volume/ 70–75 kg body weight		Blood		Plasma		Erythron	
	%	ml	%	ml	%	ml	millions/ml	Ht %
Moderate	<3	<2000	<10	< 500	<10	<300	>3	>30
Severe	3–6	2000–4000	10–20	500–1000	10–20	300–600	3–2	30–20
Critical	6–9	4000–6500	20–30	1000–1500	20–30	600–900	2–1	20–10
Very severe	>9	>6500	>30	>1500	>30	>900	<1	<10

tissular barrier (capillary walls-interstitial space-cellular boundaries).

There are as many variants of these biologic barriers as there are functional systems in the organism. As the impairment of some of these variants in shock is of particular importance they are worthy of additional mention:

The glomerulocapsular barrier, reduced to a double filter system, by uncentred superimposed unequal meshes of two membranes (membrana densa and membrana fenestrata), together with a subpodocyte labyrinth in which phenomena of dialysis and ultrafiltration take place;

The blood-brain barrier where the interstitium likewise disappears, but between the capillary and the neuron the neuroglia layer appears to play the role of metabolic censor (obviously to protect the functional sensitivity of the neuron).

The endocrine-capillary barrier that exists within the intimacy of the capillary-cellular endoplasmatic reticular system (as in the hypothalamus).

The obligatory element in any barrier is the presence of at least one *biologic membrane*. This consists of a mixed triple suture (the same as in a semiconductor) formed of a phospholipid bi-molecular layer in the interior and two protein layers in the exterior. The total thickness is 75 to 100 Å, dotted here and there with pores measuring 3 to 30 Å in diameter. Some of the pores may measure up to 100 Å (in the kidneys) or more than 300 Å (in the capillary sinusoids) according to their functional state at the moment. A recent report mentions a hexagonal arrangement of the pores on the nuclear cellular membrane [100].

The biologic membrane forms the basement of all the cellular epithelia, part of the structure of the barriers mentioned and, by tridimensional folding, the entire intracellular architecture (reticulum, mitochondria, lysosomes); i.e. the whole human body is a mixture of fluids and membranes. The membrane is not an inert pellicle but a biologic functional complex with a decisive role in maintaining life that has drawn the attention of numerous research workers in various domains. Today, the biologic membrane with its autonomic and general activity becomes an exciting interdisciplinary subject (Pullman and Weissbluth, 1965). This led to the discovery of its semipermeability and selectivity which permits the maintenance of an asymmetry that is not in keeping with the usual and trivial gradients of the physico-chemical laws; the phenomena that determine polarization, the resting potential, and conduction of the action potential were investigated. Finally, thermodynamic studies have shown the biologic membranes to be fundamental sources generating entropy in the organism, sources that are the last to be overcome in states of shock [21, 91, 94] (*see* Section 9.6.2).

The biologic membrane can only be crossed with the participation of its energy (active mechanism) or by virtue of the common physico-chemical laws (passive mechanism).

Active crossing may take place by:

- *a 'pumping system'* or *'carriers'*, both implying an energy consumption of ATP and enzymatic intervention; the mechanism consists in a reversible binding of the particle to one surface of the membrane and its release on the other. In this way glucose, some hormones and most metallic ions pass through the membrane;
- *phagocytosis* consists in engulfment of solid particles by the membrane vesicles, as a rule endocellular;

- *pinocytosis,* the drinking of fluid particles of the order of 10 000 to 20 000 Å, taken up by the reflected exocellular membranes;
- *rhopheocytosis* consists in engulfment of very small particles by invaginations with a diameter of only 200 Å.

Passive crossing is governed by the laws of diffusion, osmosis, electrolysis, etc. Particularly worthy of note are:

- *osmotic pressure* (*see* Section 7.1.5);
- *diffusion* that governs especially the penetration of O_2 and CO_2 in the blood. It is known that these gases are taken up and transported to a great extent by the erythrocytes, as their partial pressure only permits dispersion of very small amounts by simple diffusion into the plasma (Figure 7.5).

The laws of diffusion tend to govern all passive exchanges between the compartments separated by a barrier. Nevertheless, some phenomena may go counter to these laws and gain priority:

- *one-way permeability* that admits the passage of substance in a single direction (for instance the intestinal epithelium and nephronic epithelium);
- *permeability through the polarized membrane* (does not depend upon the laws of osmosis), when exchanges take place according to Donnan's equilibrium principle; in various forms of shock, imbalance of the Donnan phenomenon has been observed with increase in the permeability of the polarized membranes for Na, reflecting a loss in physiologic ionic selectivity [15, 17, 68, 73].

In shock, the permeability of all the barriers loses its selectivity; numerous factors (which are in fact the well-known shock-inducing factors)

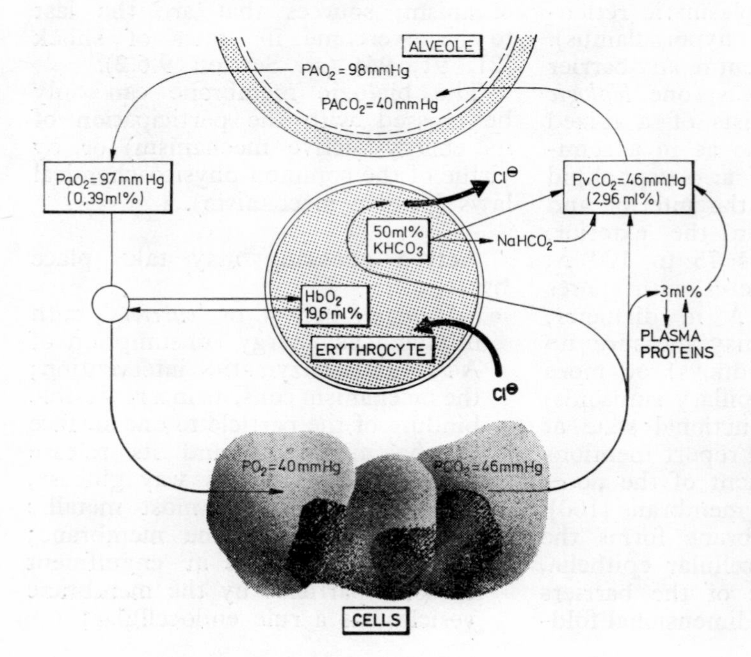

Figure 7.5. The transport of blood gases. Only 0.39 ml% oxygen and 2.96 ml% carbon dioxyde are dispersed in plasma by simple diffusion. The principal part of both gases is carried by intra-erythrocytic mechanisms.

determine a rapid, simplified but always harmful crossing of the biologic membranes (Figure 7.6).

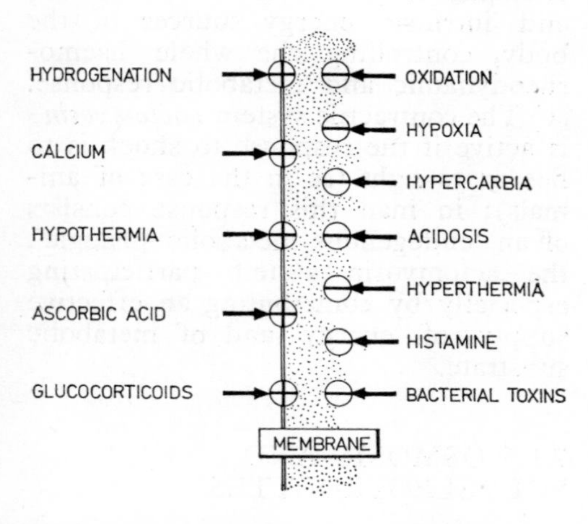

Figure 7.6. Factors that degrade the biologic membrane (−) and factors stabilizing it (+). The stabilizing factors are real therapeutical possibilities to help a shocked organism.

7.1.3 THE BIOLOGY OF WATER

In shock, the prejudice caused by the overall loss of blood (hydrocolloid, crystalloid and cellular elements) can be practically appraised only if the totality of the gross disturbances that appear clinically are taken into consideration. But the fine processes of the vital phenomenon, which normally take place in an aqueous solution must be analysed, beginning with the biology of water.

In the state of liquid aggregation water has most of the qualities that confer upon it the role of a vital matrix. Bound by hydrogen bridges to the polar groups of the glucidolipoprotein macromolecules, or free, water forms the continuous micro-network in which metabolic activity takes place [19]. The proportion of 90% water in the plasma attributes to the latter its metabolic role and known sensibility.

Water has its diffusion coefficients and surface tension, which are, however, modified by crystalloids, colloids and blood cellular elements (*see* Section 5.3.1).

The thermal characteristics of water confer upon it the role of fluid thermostat of the organism. In shock, due to dysvolaemia — especially with actual loss of fluid — there is no longer a strict compliance with thermal homogeneity. In the viscera with intense metabolic requirements there is a rise in local temperature up to deleterious values, bringing about enzymatic exhaustion accompanied by denaturation of the proteins and death of the cell.

Water is the *dispersion*, *solving* and *ionizing* medium of the organism. It is the compulsory medium of all hydrolytic and redox reactions, all the enzymes acting by intermediate kinetic stages with hydronium ions. Moreover, water stabilizes in its microcrystalline network collagen and nucleic acid macromolecules; in shock, water real volume depletion is rapidly reflected directly upon the genetic code, modifying its energy and its informational constants.

7.1.4 MACROMOLECULAR SYSTEMS

The macromolecules are particles of a substance with a mass greater than 5 000 μ; in water they form a microheterogeneous solution with a degree of subdivision of less than 10^{-5} (colloid solution).

Proteins (bound or not in glucidolipid synapses) with nucleic acids and mucoitin acids represent biologic macromolecules. Apart from the colloid solutions formed in the blood plasma or cytoplasm the macromolecules also organize a structural tissular network.

Between their circulation and fixed state there is a permanent exchange of macromolecules. In addition, macromolecules have a dynamic steric structure and may undergo reversible changes (the most common examples are the allosteric enzymatic transitions), justifying their functionality (Osteen and Klebanoff, 1971; Fleischer and Frohlich, 1965).

Macromolecules have optical, magnetic and thermodynamic properties fully studied in various fields. However, what must be recalled and fully understood in the human clinic is the fact that in a shocked organism the clinical data are the gross global expression of intimate atomic-molecular alterations of the live matter. From macromolecular changes to clinical signs there is still an obscure pathway that must always be kept in mind.

Macromolecules form polymacromolecular structures joined together in *biologic systems*, with compartmental biophysical properties, specific of certain functions. These systems play, in shock, many temporary roles that are dealt with separately in sections 3.4 and 8.9. These *biologic systems* organized by macromolecular properties are as follows:

(i) The genetic information storage and carrier system *(nucleic acids)* regulate the entire human response to shock.
(ii) The receptor analyser and transmitter of biologic information systems *(neuronal proteins)* generate the corrector response with regard to the shock-inducing lesion.
(iii) The system of passive and active passage of the substances into the human body homeostasis *(membrane lipoproteins)* tries to change the usual selectivity temporarily, but this effort is almost always insufficient in the abrupt variations of the various factors that interfere in states of shock.
(iv) The biocatalytic systems *(hormones, vitamins, enzymes)* are the semantic and intrinsic energy sources of the body, controlling the whole haemorheodynamic and metabolic response.
(v) The contractile system *(actomyosin)* is active if the response to shock is to flee or to fight (as in the case of animals); in man this response consists of an 'endogenous metabolic struggle', the actomyosin system participating especially by contributing an effective supply of energy and of metabolic substrate.

7.1.5 OSMOSIS AND POLYELECTROLYTES

In the phenomenology of shock the most important biophysical properties of the macromolecules is the development of osmotic pressure and the formation of polyelectrolytes.

Osmotic pressure. The osmotic pressure of colloids (i.e. oncotic pressure) is far smaller than that of crystalloids, osmotic pressure being the greater the larger the number of particles that develop it, according to Van't Hoff's relation.

The total osmotic pressure of plasma is 7.5 to 8.1 atmospheres (5500-6100 mmHg) corresponding to a Δ-cryoscopic point of —0.55 up to —0.60 [12, 28].

Proteins and lipoproteins represent 9% of the plasma volume and develop an osmotic pressure of 20-40 mmHg (i.e. 0.5% of the total plasma osmotic pressure). The crystalloids make up only 1% of the plasma volume but develop an osmotic pressure of 5530 mmHg. When the osmotic gradient of the electrolytes is readily and quickly equilibrated on both sides of the capillary wall, the osmotic pressure of intravascular proteins supplies an asymmetry of forces that brings about capillary tissular exchanges. A decrease of plasma proteins of less than 5 g% (less than 3 g% albumins) causes a leakage of intravascular fluids towards the interstitium.

Albumins furnish about four-fifths of the osmotic pressure (at the common *p*H of the blood they are more dissociated than the globulins).

If, however, the *p*H of the blood falls, the dissociation of albumins decreases and osmotic pressure diminishes.

Acid blood loses more water towards the interstitium and, consequently, its viscosity increases.

A decrease of the *p*H breaks down the Donnan balance, causing a loss of electrolytes in the interstitium. Acidosis thus controls the extravascular exit of water and electrolytes, bringing about an increase in the viscosity of the blood with all its rheologic consequences in the microcirculation (*see* Section 7.4).

Polyelectrolytes. The macromolecules have many ionized groups (COO^-, HO^-, NH_3^+, HPO_4^{--}, etc.) and behave like polyelectrolytes. A similar behaviour in water is that of the ground substance acids (mucoitin and condroitin sulphuric or hyaluronic acids), chromosomes and heparin.

An increase in the viscosity and coagulability of the blood was also obtained by electric stimulation with a bi-metallic microgenerator [88]. The electric stimulation probably modifies the macromolecular polyelectrolytes of the thrombocyte membrane, decompensating them and thus increasing the adhesiveness of the platelets, or producing the elimination of procoagulating substances from the thrombocytes and erythrocytes by concomitant oxidation of the SH groups in the membrane macromolecules. The reducing effect of heparin constantly arrests the tendency to coagulation by recompensation of the plasma polyelectrolytes and platelet-erythrocytic membranes.

When the electric charges are not compensated, as happens in the asymmetrical intercompartmental hydroelectrolytic losses in shock (especially in acidosis), the viscosity of the plasma increases because the uncompensated charges repel one another, releasing the macromolecules. In contrast, neutralization of the charges will result in the packing of macromolecules and, owing to the lesser friction between them, in diminution of the viscosity (Figure 7.7). In this way the reducing effect of heparin (compensatory action) brings about rapid neutralization followed by decrease in plasma viscosity.

Of particular importance is the fact that the uncompensated polyelectrolytes (due to acidosis) also give a high osmotic pressure, often very dangerous because manifested extravascularly as well. Any electric compensation of the polyelectrolytes gives a loss in osmotic pressure. Therefore, if, in an acidotic state, compensation of the polyelectrolytes is obtained by administration of colloidal solutions, viscosity will only temporarily decrease but colloids will lower the osmotic pressure of the plasma thrity-fold, *because acidosis continues to block the osmotic properties of the colloids.*

Reduction of the viscosity and compensation of the polyelectrolytes may, however, also be obtained with crystalloids, after which a plasma osmotic pressure decreases only five-fold. From this point of view, too, the crystalloid solutions are more advantageous, both in the first phase of shock and, especially, in the acidotic phase. In acidosis, the paradox of the more efficient osmotic effect of crystalloids than of colloids stands true. The dextrans alone manage to be both efficient antiviscosity agents and strong osmotic substances [22, 33, 103].

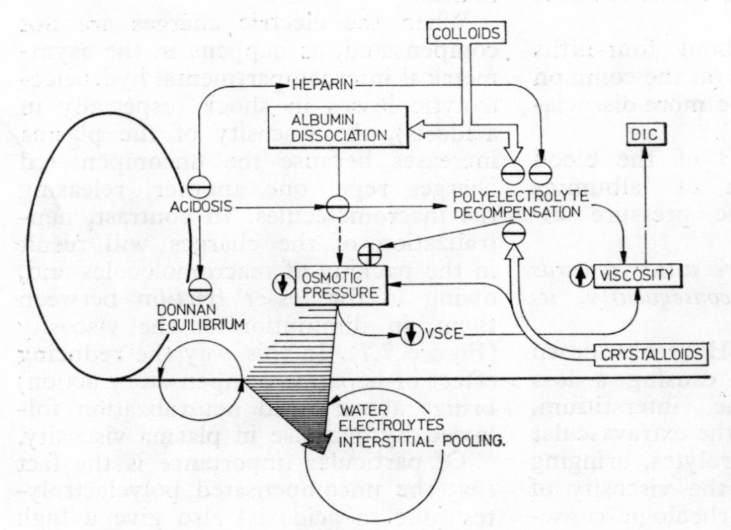

Figure 7.7. The increase of osmotic pressure by colloids is blocked by acidosis; nevertheless the colloids may reduce the viscosity of the blood by compensation of the polyelectrolytes. A better effect is induced by crystalloids both in the decrease of viscosity and in the increase of osmotic pressure (acidosis does not block the osmotic effect of crystalloids). The same final effect as that of heparin and all reducing substances.

7.2 BLOOD CELLS

7.2.1 THE RED BLOOD CELL

The red blood cell (RBC) participates in almost all the important moments of shock:

- the viscoelastic and electric properties of the cell membrane interfere in haemolysis, sludge and coagulation (*see* Section 7.4);
- as an assumed producer of lactate, a residue of intra-erythrocytic hyperglycolysis, it interferes in the acid-base balance and Cori cycle;
- by the intra-erythrocytic reducing systems it participates in the mechanisms of release of oxygen to the tissues.

Some twenty years ago the RBC was still considered a simple, inert 'haemoglobin sac'. Today, the life of the RBC no longer seems elementary. Although it has no nucleus and mitochondria (and therefore no Krebs cycle) the RBC starts its 120 days journey with a complete glycolytic enzymatic equipment that supplies it with the necessary energy for permanently pumping out Na^+ and taking in K^+, and especially for maintaining haemoglobin in a reduced state that is suitable for the transport of oxygen. The three dimensional images of the RBC in the stereoscanner confirm its bi-concave discoid shape and its well-known dimensions [8, 100].

The smooth surface of the EBC rapidly shrivels in the presence of agglutinins, the divisions and shadows that appear revealing the presence of different electrical properties. These membrane irregularities (digitations) have also been observed in aged platelets and leucocytes.

The RBC spends a large amount of energy in maintaining its osmosis, which appears to be an essential condition for its function as a gas transporter. An erythrophilic gamma globulin, probably produced by the spleen, adheres to the cell membrane, giving it its normal shape. It has likewise been assumed that beta-receptor blocking substances are taken up selectively by the RBC which rapidly diminishes in size and loses large amounts of potassium [68].

The fundamental rheologic property of the erythrocyte is its *plasticity*, which sums up several biophysical properties. Alteration of its rheologic properties increases the fragility of the RBC, as revealed by mechanical resistance and osmotic or filtrability tests. In states of shock, osmotic haemolysis is always preceded by oxidation of the membrane GSH groups (reduced glutathione) with altered permeability, particularly for cations [77].

Rheologic investigation of the RBC may be carried out with the Voigt-Kelvin osmolysis model or the Mitchinson micropipette suction model [56]. On 'swelling' the erythrocyte takes on the shape of a sphere, and from this Young's modulus of elasticity (G) can be calculated, as well as the viscosity module of Newton (η), from the formula:

$$\tau = G\gamma + \eta \frac{d\gamma}{dt} \qquad (3)$$

where τ is the contraction of the membrane, γ its deformation, and ratio $\frac{d\gamma}{dt}$ is the rate of deformation. Calculation of these modules by various authors has always shown the rheologic behaviour of the RBC to be that of a visco-elastic body (Larcan et al., 1971; Plouvier et al., 1969).

Investigations carried out on the plasticity of RBC by the filtrability test have shown that the blocking of glycolysis, *p*H variations, the accumulation of reduced haemoglobin, and the prolonged storage of blood at +4°C alter the plasticity of erythrocytes, modifying their normal rheologic behaviour with regard to filtration, favouring aggregation and then destruction [111].

Intra-erythrocytic metabolism. The features of intra-erythrocytic hydrocarbonate metabolism are shown in Figure 7.6. The Embden-Meyerhof pathway and the pentose cycle inevitably lead to lactic acid. The intra-erythrocytic enzymatic chains function in order to produce energy (ATP) and maintain haemoglobin iron in the ferrous state ($HbFe^{++}$); that is, in a state permitting the transport of oxygen. Small amounts of methaemoglobin ($HbFe^{+++}OH$) are permanently reduced by the hydrogen supplier ($NADH_2$) or reduced glutathione (GSH).

Figure 7.8 suggests that in order to reduce ferric iron to ferrous iron two complete dehydrogenations must take place to obtain $NADH_2$. Stages *a* and *b* in the figure result in:

- the possible permanent hydrogenation of methaemoglobin (reduction);
- the production of GSH that can likewise be used for the reduction of methaemoglobin;
- the loading with inorganic phosphorus of two high energy acids: 1,3 and 2,3-diphosphoglyceric acid (DPG) that will supply the phosphorus necessary for the production of ATP (stage *c*).

The accumulation of 2,3-DPG points to a deficit in free inorganic phosphorus without which dehydrogenation (stage *a)* cannot take place; this leads to a decrease in $NADH_2$ and the impossibility of maintaining the reduction of haemoglobin iron, which loses its property of fixing oxygen, rapidly releasing it into the tissues (the accumulation of 2,3-DPG normally occurs in the erythrocyte in its peripheral excursion).

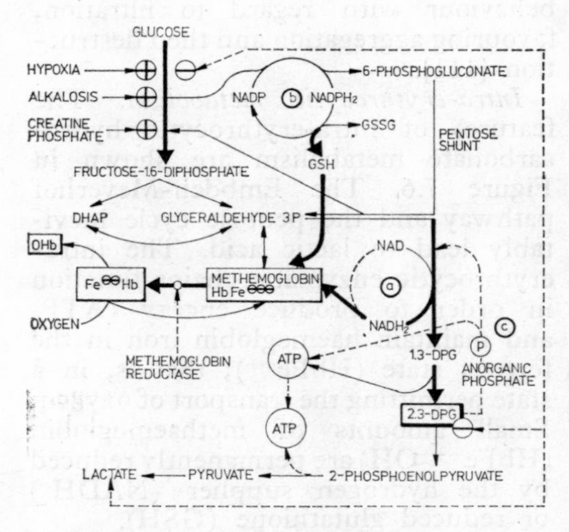

Figure 7.8. When 2,3-DPG values fall, inorganic phosphorus is offered to the reactions on point *a*, methaemoglobin is reduced and oxygen is fixed in the red blood cells (the alveolar phase). The accumulation of 2,3-DPG blocks the *a* stage and restores methaemoglobin, releasing oxygen (the tissular phase). In shock, the accumulation of lactate will block both the distal and the proximal mechanism of methaemoglobin hydrogenation. An accumulation of $HbFe^{+++}$, which cannot transport oxygen, takes place.

The phenomenon might be briefly described as an increase in the affinity of 2,3-DPG for $HbFe^{++}$ from which it detaches its oxygen, yielding it to the tissues, i.e. *an increase in intra-erythrocytic 2,3-DPG lowers the affinity of $HbFe^{++}$ for oxygen*. The accumulation of GSH has the same effect. It may be considered as a shift to the right of the haemoglobin dissociation curve by intra-erythrocytic metabolic mechanisms (Proctor et al., 1971).

After the decrease of the erythron at the onset of shock the adaptive mechanisms that come into action seek to satisfy the immediate oxygen requirements of the tissues; apart from an improved flow in the microcirculation there is also a shift to the right of the oxyhaemoglobin dissociation curve.

The RBC is able, by its own intrinsic mechanisms, to monitor tissular oxygen requirements, investigation of its metabolic competence in the different phases of shock permitting evaluation of its actual tissular oxygenation possibilities [38].

The mechanisms of peripheral monitoring of oxygen release are summed up in Table 7.3.

The Bohr effect (acceleration of O_2 release to the tissues, following an increase in the local accumulation of CO_2) and the Haldane effect (accelerated fixation of CO_2 on plasma bicarbonates with increased liberation of plasma O_2) have been known for some time, but the intra-erythrocytic mechanisms are not yet fully understood.

It was noted that in any depletion of the erythrocytic mass, as in shock, there is an intra-erythrocytic accumulation of 2,3-DPG, GSH, ATP and creatine. This reveals an RBC in the *stimulation phase* specific to the first stages of shock when hyperglyco-

Table 7.3 THE MECHANISMS OF PERIPHERAL CONTROL OF OXYGEN RELEASE

Tissular control of oxyhaemoglobin dissociation	intra-erythrocytic mechanisms	2,3-DPG and $HbFe^{++}$ relationship GSH and $HbFe^{++}$ relationship
	extra-erythrocytic mechanisms	Bohr effect Haldane effect

lysis takes place. What stimulates this hyperglycosis with its accumulation of large amounts of 2,3-DPG, GSH and ATP? Several mechanisms have been proposed: hypoxia, alkalosis (due to hyperventilation) or an accumulation of creatine-phosphate released by the hypoxic tissues [38, 78].

The progress of shock brings to the fore, however, acidosis which at certain *p*H values appears to inhibit this adaptive intra-erytrocytic oxygen-donor mechanism. In bacteriemic shock, for instance, there exists from the very beginning a shift to the left of the oxyhaemoglobin dissociation curve due to the initial decrease of intra-erythrocytic 2,3-DPG. This leads to blocking of the release of O_2 to the tissues with an increased oxygenation of the venous blood, which may be considered as equivalent to a real left-right shunt [78].

The intra-erythrocytic mechanisms of O_2 release offers a new explanation to the physiologic shunts so much discussed in shock (particularly in the lungs).

It has been proved experimentally that animals adapted to high altitudes have an increased tolerance to various types of shock (bacteriemic, haemorrhagic, thermal, tourniquet, etc.) due to increase in the intra-erythrocytic amount of 2,3-DPG and facilitation of the O_2 release mechanisms to the tissues. The importance of physical exercise in the mountains thus has an additional motivation.

A decrease in erythrocytic 2,3-DPG after massive blood transfusions has been noted, with a return to normal within one to two days, followed by a hypercompensation increase on the fifth day. In stored blood containing citrate, which lowers 2,3-DPG, the Hb dissociation curve shows a shift to the left [90].

7.2.2 THE THROMBOCYTE

Although it lacks a nucleus, the thrombocyte is well equipped with enzymatic chains (except for the intra-mitochondrial ones). The thrombocyte is a reactor of haemostatic factors: the factor that catalyses the splitting of prothrombin into thrombin; the factor favouring the transformation of fibrinogen into fibrin; an adjuvant factor of intrinsic thromboplastin; a heparin neutralizing factor; a factor controlling the permeability of the membranes; a clot retraction factor; serotonin and platelet antifibrinolysin. Moreover, the thrombocyte is charged with calcium ions [87, 125].

Hypoxia in itself stimulates platelet aggregation. Even after transitory post-operative hypoxia the number of platelets and their tendency to aggregation increases up to day 20, thus maintaining the thromboembolic danger [2].

Aggregation depends upon the electric surface charges of the platelets which are normally negative. The alpha-carboxylic groups of neuraminic acid contribute to a large extent to these electric charges [76].

When the negative charges are blocked or neutralized the platelets aggregate.

Catecholamines increase the number of platelets, as well as alkaline phosphatase activity in the capillary endotheliocytes; this enzyme diminishes the negative charge of the platelets, which come more easily in contact with the capillary wall. Moreover, increase in the plasma concentration of FFA accentuates platelet aggregation, FFA probably acting as a dielectric layer that annuls the negative charge of the thrombocytes [39]. ADP, epinephrine and collagen strongly enhance platelet aggregation; in an intermediate stage the platelets form a complex with the bi-valent ions of the blood fluid [55].

Actually, Ca^{++}, stocked in the platelets, is released due to polarizing restructuring of the membrane; this calcium cation release is accentuated by thrombin and depends strictly upon intra-thrombocytic synthesis of ATP, since blocking of the Embden-Meyerhof and Krebs cycles with N-ethyl maleimide or aspirin arrests the evolution of this process [81].

Platelet aggregation is a specific function the thrombocyte exercises when exterior factors modify its visco-electric configuration. The substances that cause platelet aggregation are collagen, catecholamine, calcium, ADP and thrombin.

Aggregation is proceeded by *platelet adhesion; dispersion* is controlled by the negative charges of the platelets. Their adhesion therefore, is a direct consequence of a decrease in their electrostatic charge by depolarization. The latter occurs when the denuded basement membrane (collagen) comes in contact with the œdematous and depolarized intimacytes, or by intra-platelet metabolic disturbances.

The aggregation phase proper is under extrinsic ADP control and of the ADP released by the platelets following ATP-ase hydrolysis in the platelet reticulum. As platelet aggregation forms part of the general after-shock reaction it takes place under acidotic conditions, accompanied by glycolytic blocking, a decrease in ATP and excess increase of ADP.

The last phase is that of *viscous metamorphosis* which is assumed to have a lysosomal mechanism. The platelet lysosomal granules broken down by acidosis liberate hydrolases and disintegrate the platelets which release a cluster of nucleotides, amino acids, peptides, serotonin, etc.

All the reducing agents can protect the thrombocytic membrane, arresting the entire process of platelet aggregation.

7.3 THE BLOOD

Whereas *biorheology* deals with the general study of the flow and deformation of live matter, *haemorheology* is concerned with the kinetics of the blood fluid and stability of the blood as a system [54]. In other words, rheology of the blood is a study of the phenomena of viscosity and sedimentation, with their pathologic effects so common in states of shock: clumping and agglutination of the RBC (sludge phenomenon), disseminated intravascular coagulation, defibrination, fibrinolysis, etc.

7.3.1 VISCOSITY

A decisive factor in tissular perfusion is the viscosity of the blood, which assumes greater importance as the rate of flow becomes slower and the tubes through which it flows become narrower.

The blood may roughly be considered as a colloidal suspension, the suspended particles being the RBC and the suspension medium the blood plasma.

At a given temperature, for a certain value of molecular movement, the most important factor upon which viscosity depends is, therefore, the concentration of the particles in the suspension, i.e. of the RBC in the blood and of the macromolecules in the plasma. An increase of the haematocrit from 33 to 53% increases viscosity two- and even threefold [6, 63].

Viscosity of the blood, in general, also depends upon changes in plasma viscosity, considered apart from the blood plastic cellular elements. Plasma in itself is a colloidal suspension, its disperse phase being represented by macromolecules.

During venous punctures in patients with severe shock following multiple injuries, every practitioner has had the opportunity to observe increases in the viscosity of the blood.

Stasis and acidosis in the microcirculation confer upon the stagnant blood the viscoelastic properties of a gel. Being a non-newtonian fluid the blood has a viscosity with numerous anomalies, depending upon the haematocrit, plasma proteins and lipids, erythrocytic elasticity, aggregation of all the cellular and molecular elements suspended in the plasma, *p*H, temperature, etc.

Viscosity is characterized by a dynamic viscosity coefficient η, expressed in *poises* in the CGS system and a cinematic viscosity coefficient ν, measured in *stokes*. The relation between the two coefficients and density ρ, is:

$$\nu = \frac{\eta}{\rho} \quad (4)$$

Viscosity exists in all fluids in motion and is caused by internal friction between the molecules.

The rheologic behaviour of plasma macromolecules is a constant equilibrium between the diffusion force which depends upon molecular movement and the tendency to sedimentation, which, in turn, depends upon viscosity. When the viscosity of a colloidal system increases, the stability of the suspensions diminishes. Stokes' formula establishes the following relation: the greater the sedimentation force F_s the greater the values of the viscosity coefficient η, of the radius of the colloidal particles, r, and of the sedimentation rate, v_s:

$$F_s = 6\,\pi\,r\,\eta\,v_s \quad (5)$$

In turn, the sedimentation rate increases as the number of particles becomes greater.

In shock, mobilisation of the immunoglobulins, representing immediate antibody genesis, and of the lipoproteins from the fracture focus or the attritional tissues brings about a numerical increase of the larger particles, dangerously accentuating plasma viscosity [6]. The immediate response of the body to a loss of intravascular fluid is, wisely, a rapid haemodilution that lowers, but only for a short time, blood viscosity. Haemodilution is obtained with a fluid poor in proteins or depleted of proteins from the interstitial space [34, 112]. The gravity of bacteriemic and traumatic shock lies also in the fact that haemoconcentration appears with the increase in viscosity, in spite of the activation of fibrinolysis and decrease in fibrinogen levels. It has been suggested that any source of neuramidase (for instance intestinal bacteria) may trigger off splitting of neuraminic acid from serum proteins, inducing a marked increase in plasma viscosity [97].

Viscosity of the blood thus appears as a complex phenomenon resulting from the double equilibrium: plasma-figured elements and water-macromolecules.

The viscosity of plasma is only 1.8 times greater than that of water and, consequently, its properties are those of a newtonian fluid. The addition of erythrocytes, however, produces numerous anomalies, the overall viscosity of the blood being five times that of water, and total blood behaving like a non-newtonian fluid (Gelin and

Zederfeldt, 1971; Larcan et al., 1971; Litwin et al., 1970; Rosato et al., 1968).

In man, the normal viscosity values are 6.7 ± 1.1 centipoise for total blood, 82.2 ± 7.5 for the blood cells, and 1.4 ± 0.1 for plasma [64].

At very low rates of flow the blood reaches the behavioural limits of visco-plastic fluids (Bingham bodies). Any lowering of the rate of flow no longer complies with axialization of the erythrocytes and therefore increases viscosity in the microcirculation, exactly where it should normally be diminished in order to facilitate exchanges. The harmfulness of a slow peripheral flow in shock is once again evident.

In practice, the following factors are clinically investigated in order to assess the viscosity of the blood:

- the BP values;
- haematocrit (up to minimal values of 20%, from which viscosity no longer depends upon haematocrit);
- the platelet count;
- the absolute amounts of albumin, fibrinogen and macroglobulins;
- lipoprotein concentrations.

In addition, for the vessels with a diameter smaller than 0.5 mm, the *sigma* effect has to be taken into consideration, and for those with a diameter of less than 20 μ the rate of flow *per se* also (*see* Section 7.3.2).

In newtonian fluids viscosity does not depend upon flow pressure values. With the blood, which is a non-newtonian fluid, an increase in the flow pressure is followed by a disproportionate decrease in viscosity and increase in the volume output; *the lower the blood pressure and the slower the volume output, the greater is the viscosity of the blood* (and excessively high the BP values necessary for remobilization of the flow).

Increase in plasma viscosity brings about an increase in the affinity of the erythrocytes that clump together or aggregate, especially in the microcirculation where the flow is slowed down by the sphincteral system of the capillaries (*see* Section 7.4.1).

7.3.2 THE LAWS OF INTRAVASCULAR FLOW

Being a polyphasic system, from the point of view of the molecular theory, the rheologic behaviour of the blood also depends upon the plasma blood cell relationship and upon their individual viscoelastic characteristics (*see* Sections 7.1.4 and 7.2).

As compared to flow in technology, the flow of intravascular fluids shows marked differences since the blood behaves like a non-newtonian fluid and the vessels are not rigid tubes, hence there is a permanent variation in viscosity and in the radii of the tubular system [85, 112].

Fluid flowing through a vessel is slowed down, mainly along the walls, by the friction between the blood molecules and the vascular wall. For this reason the velocity of the blood flow in the middle of the vessel is considerable while along the surface it is very slight. The flow of the blood is *laminar*, the shearing forces between the layers acting upon the macromolecules and red blood cells and deforming them elastically. In the arterioles the red blood cells are centred along the axis of the vessel according to Bernoulli's laws, and the plasma with a low haematocrit (25%) will be pushed towards the walls and pass, for the most part, into the capillaries arising on both sides of the meta-arterioles, a phenomenon known as 'plasma skimming'.

The equilibrium of a laminar flow depends upon the radius r of the vessel, the velocity of the flow v, viscosity η and density of the blood ρ:

$$\boxed{\text{Laminar flow} = f(r, v, \eta, \rho)} \quad (6)$$

In shock, these factors undergo marked changes and the flow becomes turbulent. The radius of the vessels is strongly modified by catecholamines in the microcirculation. The velocity of the flow increases more and more, as well as the blood viscosity owing to the decompensation of plasma poly-electrolytes.

Turbulence appears at a critical velocity v_c, whose values depends on viscosity η, density ρ, and the tubular radius r, according to the equation:

$$\boxed{v_c = \frac{K\eta}{\rho r},} \quad (7)$$

where K is Reynolds constant.

From relation (4) it results that:

$$\boxed{v_c = \frac{K\nu}{r}} \quad (8)$$

i.e. turbulent flow develops with increase in the diameter and decrease in cinematic viscosity; the flow in the capillaries is as a rule smooth, in a laminar regimen. Only in state of shock does the turbulent flow invade the microcirculation space.

Normally, in the microcirculation, the total cross sectional area increases considerably and resistance decreases; the velocity of the flow and friction likewise decrease and lateral pressure increases (Bernoulli effect), with a sharp fall in pressure of 15 to 20 mmHg per mm along the capillary wall. These physiological phenomena facilitate the exchange of gases and nutrient substances, to which the decrease in viscosity with the reduction of haematocrit (25% in the capillaries) also contributes.

The laminar volume output Q is given by the Hagen-Poiseuille relation:

$$\boxed{Q = \frac{\pi (p_1 - p_2)\, r^4}{2\,\eta l}} \quad (9)$$

In shock, the decrease in the fluid output of the microcirculation is caused by the very terms of the relation:

- increase in the viscosity of the blood (η), a common event in shock and decisive for the decrease in the perfusion output of the microcirculation;
- opening of the arteriole-venular shunts offering a shorter by-pass, l, with increase of the output in these direct shunts and its decrease in the capillary loops;
- the BP (the pressure gradient p_1—p_2) decreases in shock due to well-known causes and will obviously bring about a diminution of the output;
- the 'geometrical factor' (r) interferes exponentially, illustrating the major importance of the sphincters, whose constriction drastically lowers the output. The vascular radius reduced by half lowers the output by one sixteenth of the initial values.

A decrease in the vascular diameter increases the velocity of the flow, but in shock, near total constriction of the sphincters in the damaged area actually dams up the blood with its consequent pooling in the microcirculation.

The reduced output and increased viscosity are the essential physical causes of the two rheologic events at the level of the microcirculation: sludge and disseminated intravascular coagulation.

Today it is known that the flow is actually pulsatile, not only in the large vessels but also in the arteriolar system; use of the Hagen-Poiseuille formula is, therefore, merely conventional.

In the microcirculation marked anomalies intervene owing to the particular behaviour of viscosity. The output no longer has precise mathematical relations with the BP, vascular diameter or vessel length along which the blood travels. In vessels with a diameter of less than 1 to 0.5 mm the viscosity is always less than that calculated for the large vessels; for instance in the arterioles viscosity is normally one-third of that in the aorta.

It is the Fahreus-Lindquist phenomenon explained, by Scott-Blair and Magnus, by the *sigma effect: axialisation of the erythrocytes with decrease of haematocrit in the microcirculation.*

Therefore, the notion of 'apparent viscosity' (η') has been introduced for the blood flow in the microcirculation; this 'apparent viscosity' depends upon the temperature T, velocity v, radius r and haematocrit Ht, according to relation:

$$\boxed{\eta' = f(T, v, r, Ht)} \qquad (10)$$

But, at a diameter of less than 20 μ there is no longer any proportionality between η' and the velocity, and at less than 20% the haematocrit ceases to influence η' [37].

In tubes with a diameter of 10 μ (therefore closely resembling those of the microcirculation) the η/η' is very near to unity, therefore the viscosity of the blood takes on again the calculated values and no longer depends upon blood viscosity and haematocrit and only *depends upon the viscosity of the plasma.*

Hence, the colloidal qualities of the macromolecules come again to the fore (*see* Section 7.1.5).

7.3.3 CLOSING PRESSURE

Systemic blood pressure, measured classically, supplies information on the gradient between the ejection force of the left heart and arteriolar resistance, i.e. the perfusion pressure that makes the blood flow through the microcirculation.

The difference between the BP and central venous pressure represents the 'real active pressure', the total pressure required for maintaining circulation of the ECBV [85].

In addition, it has been established that in the microcirculation the relationship between the BP and volume output depends upon a further factor, not yet mentioned, i.e. the *transmural pressure* (the difference between the intracapillary and tissular pressures) exercised upon the thin walls of the capillaries and even of the arterioles. This pressure actually expresses capillary tissular exchanges; acting from outside it may modify the geometric factor, lowering it below the critical values (the 'closing pressure' specific of each tissue), arresting the flow in the microcirculation even if the systemic BP still has fairly high values.

In shock, the accumulation of fluid and electrolytes in the interstitial spaces results in extrinsic compression of the microcirculation network that collapses even after the constrictor effect of catecholamines has diminished.

Therefore, to the therapeutical effort to open up the microcirculation must be added an attempt to release the dammed up hypertonic fluid from the extracellular space.

The transmural pressure values are governed by Laplace's law, according to which for a pressure P within a tube to overcome the tension T outside the tube it must be directly proportional to it and inversely proportional to the radius of the tube:

$$P = \frac{T}{r} \quad (11) \text{ and } \quad T = Pr \quad (12)$$

Parietal pressure is therefore directly proportional to intravascular pressure and the radius. If, in the capillary ($r = 3\mu$) the parietal pressure is 16 dyne/cm, in the aorta ($r = 1.3$ cm $= 13\,000\ \mu$) it is 170 000 dyne [54, 85].

The muscular wall of the arteries is less distensible, explaining why 70 to 80% of the blood is to be found in the venous system, and only 8 to 10% in the arteries, of which 1% is in the arterioles; after transfusion only 1% of the perfused blood passes into the arterial system, the rest penetrating into the systemic veins, in the pulmonary circulation, and the stagnant areas of the microcirculation.

Sympathetic excitation lowering the vessel diameter in the microcirculation may increase the closing pressure from 20 mmHg up to 100 mmHg (which means that the microcirculation network collapses) whereas alpha-blocking drugs may reduce it to 5 to 10 mmHg, permitting refilling of the microcirculation even at lower BP values.

Increase in the critical closing pressure by alpha-stimulation is a fundamental noxious phenomenon in arresting the visceral peripheral flow in shock.

Transmural pressure plays a particular role in 'spontaneous haemostasis' (Wangesteen et al., 1970).

The bleeding output Q may be calculated according to the relation:

$$Q = A\sqrt{\frac{2\,\Delta P}{\rho} + v^2} \quad (13)$$

where the output is directly proportional to the surface of the lacerated or sectioned area A, of an artery or vein of any size. The transmural pressure gradient ΔP, reduces the output at lower values. Hence, if the BP falls and the interstitial fluid tension increases, bleeding is slight if the flow velocity v is likewise reduced; finally, the output is less when the density of the blood increases [116]. In haemorrhagic shock, therefore, systemic hypotension, vasoconstriction (which modifies especially the lacerated area and flow velocity), and the increased viscosity (and, proportionally, the density) are the immediate measures the organism adopts for protection and haemostasis.

7.4 THE COAGULOLYTIC SYSTEM

The fluidity of the blood is the result of the equilibrium between two contrary phenomena: *coagulation* and *coagulolysis*. Both processes have a regulation system formed of activator and inhibitor couples, with a double source: plasmatic (intrinsic) and tissular (extrinsic). A veritable *coagulostat* exists within which the same effect may be obtained by a contrary sense, imbalances belonging to the opposed subsystems (Figure 7.9).

The coagulostat is affected in three great biologic processes that sometimes overwhelm it by their size: *haemostasis*, *thrombosis* and *proteolysis*.

Haemostasis (spontaneous, physiologic arrest of bleeding) is a coagulolytic imbalance, directed towards hypercoagulability, localized, harmonious, transitory and reversible [92].

the clinic must try to break down the vicious circles of autoexcitation by prompt and sometimes adverse measures that often seem paradoxical to the uninitiated.

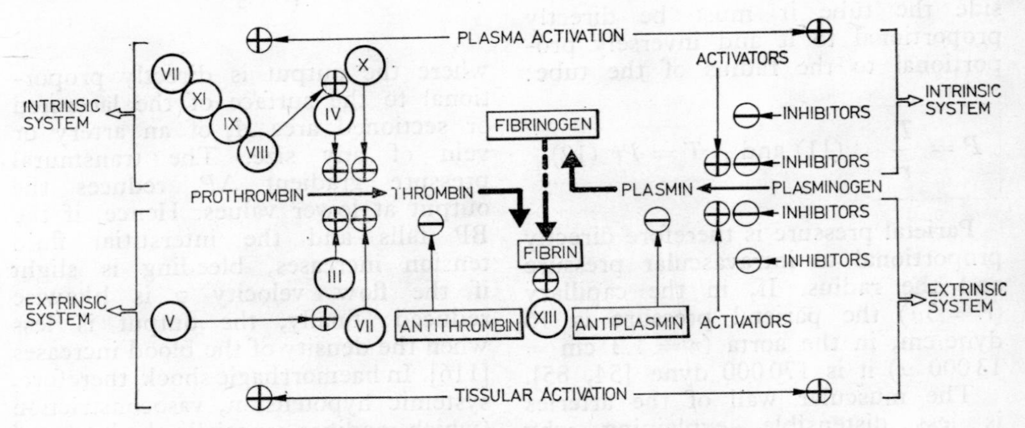

Figure 7.9. The coagulostat symmetrical mechanism.

Thrombosis (pathologic intravascular coagulation) is the result of an anarchic, extensive imbalance tending towards hypercoagulability and irreversibility.

Proteolysis has a physiologic aspect maintaining the permanent fluidity of the blood (the same as intracellular lysosomal circulation and digestion), but it may sometimes exceed its scope and take on a pathologic aspect, producing dangerous, acute fibrinolysis.

The same harmful overreaching of the initial scope appears following *disseminated intravascular coagulation*, a state that is common in shock when, due to the excessive exhaustion of the factors of the procoagulating system, haemorrhage is brought about by *consumption coagulopathy*, as a rule accentuated and with *secondary reaction fibrinolysis*. Taking account of the exaggerated consumption and correction reactions, therapeutical intervention in

Physiologic haemostasis is a complex of favourable events; in shock it is frequently encountered as a primary adaptive phenomenon in direct vascular lesions. Physiologic haemostasis complies with the vascular parietal time (vasoconstriction), tissular and plasmatic time (the formation of clots) and the thrombodynamic time (retraction and lysis of the clot). But shock is far from a physiological state.

In shock, there is an anarchic avalanche of pro- and anticoagulating phenomena that are almost always present [49, 58, 62, 86, 105]. The sequence of the haemorheologic pathologic phenomena condition one another and are self-maintained in the course of shock: *disseminated intravascular coagulation (DIC)*, *consumption coagulopathy and fibrinolysis*.

DIC is, in fact, a massive thrombosis which can affect not only the microcirculation, but also the arteries and veins. Venous thrombosis is the same as arterial thrombosis, the structure of the thrombi being similar in the electron and the optical microscope [89]. Shock produces the essential five factors contributing to thrombosis: infection, anaemia, dehydration, stasis and hyperviscosity of the whole blood.

In the veins, the haemodynamic factors are more important (in clinostatism a three-layered flow may be observed in the large veins, the end layer containing a stagnant, cellular 'mud'); in the arteries the parietal factor is essential (the presence of atheroma plaques forms the starting point of the thrombus). In the mesenteral or coronary arteries the periosteal or longitudinal plaques offer a further explanation of the distress of the intestine and heart in shock, as the flow which is not yet dangerous for an intact vessel may become inefficient when faced with arteries impaired by atheromatous plaques.

In the microcirculation, coagulolytic balance disturbances accompany parietal dynamic alterations (*see* Section 7.5) and tissular metabolic disorders (*see* Sections 7.6 and 7.8), taking on an oscillating character (Figure 7.10).

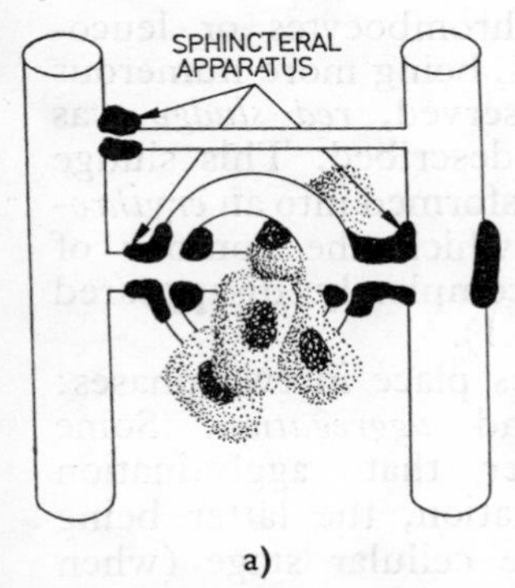

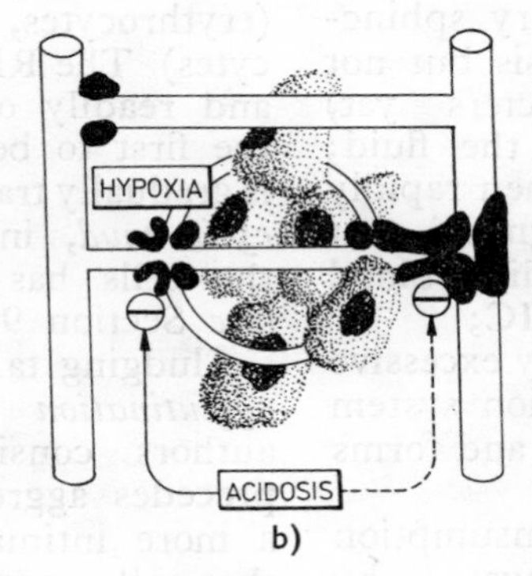

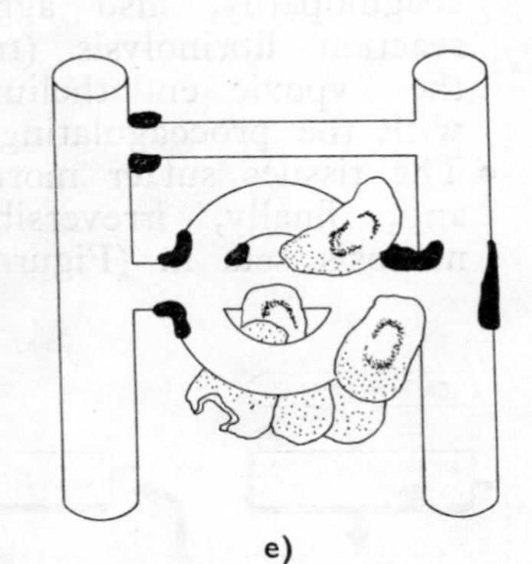

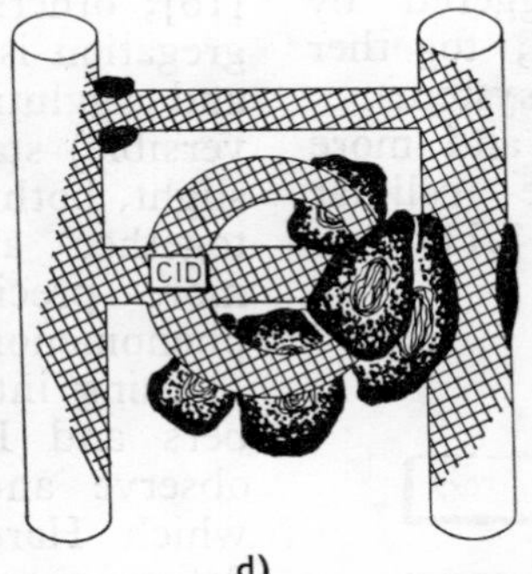

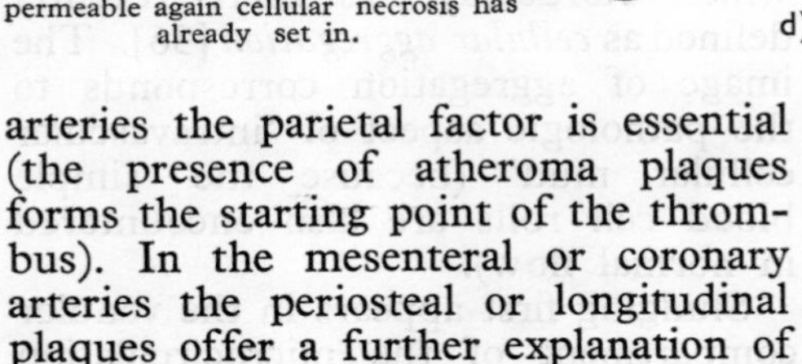

Figure 7.10. Dynamic rheologic disturbances in the microcirculation space: *a* — normal state with an alternating circulation through the functional capillaries; the arteriole-venular shunt is closed; *b* — generalised spasm; the arteriole-venular shunt is opened; the cell begins to be affected and acidosis appears; *c* — acidosis has opened the proximal dams of the microcirculation and intracapillary sludging of an acidotic blood sets in, which will be followed by coagulation; the postcapillary sphincter is still closed; *d* — blocking by DIC of the microcirculation accentuates cellular distress; *e* — when fibrinolysis renders the microcirculation permeable again cellular necrosis has already set in.

Part of the events that modify the coagulolytic balance in shock may be listed as follows:

- vasoconstriction of the microcirculation and decrease in the flow below the critical closing level for each tissue is the onset element of rheologic disorders;

- the pooled blood cells adhere to one another in rolls, spheres and/or clumps, i.e. sludging;
- hypoxia of the endothelial intimacyte and neighbouring tissues, as well as decrease of the clearance activity of the coagulolytic factors in the RES inclines the balance towards hypercoagulability;
- progressive metabolic acidosis increases the viscosity of the blood;
- asymmetrical opening of the microcirculation (the precapillary sphincters are relaxed by acidosis but not the postcapillary sphincters yet) brings about pooling of the fluid; the initial blood pool is then rapidly transformed into a 'quagmire' that is the harbinger of immediate blood jellification caused by DIC;
- the onset of DIC proper by excessive activation of the coagulation system plugs the microcirculation and forms a peripheral barrier;
- there follows a manifest consumption coagulopathy, also aggravated by reaction fibrinolysis (triggered by the hypoxic endothelium, together with the procoagulating system).
- The tissues suffer more and more and, finally, irreversible cellular necrosis sets in (Figure 7.11).

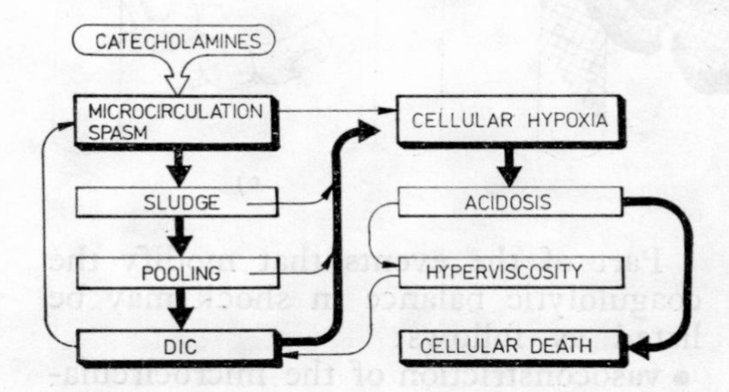

Figure 7.11. The rheologic and coagulation vents in the space of microcirculation in shock.

7.4.1 THE SLUDGE PHENOMENON

Sludge is a fundamental rheologic phenomenon in states of shock and, in short, cellular aggregation in the blood.

Decrease of the flow in the microcirculation network below the critical level of laminar flow clusters the blood cells into *rolls* and *spherules* in the capillary loops, and in *clumps* in the venules; when stasis lasts longer the blood cells adhere to one another (erythrocytes, thrombocytes or leucocytes). The RBC being more numerous and readily observed, *red sludge* was the first to be described. This sludge is gradually transformed into an *erythrocytic mud*, in which the contour of the cells has completely disappeared (*see* Section 9.5.1).

Sludging takes place in two phases: *agglutination* and *aggregation*. Some authors consider that agglutination precedes aggregation, the latter being a more intimate cellular stage (when the cell contours have disappeared) [16]; others [31, 53] sustain that aggregation is the first (reversible) phase and agglutination the second (irreversible) stage. This difference is but slight, both words meaning a 'sticking together' and cannot help us to a more precise understanding of the phenomenon.

Using intravital microscopy, Chambers and Knisely were the first to observe and describe sludging [29], which Hardaway has more recently defined as *cellular aggregation* [36]. The image of aggregation corresponds to the pathologic aspect of 'intravascular cellular mud' (because the simple blood cell rolls are also encountered in normal flow).

Sludging first appears in the venular compartment of the microcirculation and sinusoid capillaries where the slower rate of flow prevails over cellular adhesion [16]. The tendency to adhesiveness of the RBC may be detected in practice by measuring the sedimentation rate. The known antiphlogistic agents (salicylates, aspirin, corticoids) are also useful antisludge agents.

Hardaway compared sludge to DIC and found a close resemblance between these two phenomena [35]. The tendency of the blood to hypercoagulate seems to act in the first phase of shock, fine fibrin films are deposited on the surface of the RBC, which they smother (like a dielectric isolator) and conglomerate into globular masses. These augment, by apposition, and plug the small vessels. Proximal or distal to these erythrocytic microemboli DIC begins, the smothered RBC in the globular masses disintegrating and releasing Hb and erythrocytic thromboplastin. The same peri-erythrocytic fibrin film would also account for the roll-like arrangement of the cells, i.e. DIC fibrin films trigger sludge phenomenon and sludge cellular masses augment DIC.

Sludge is the prophase of DIC, both phenomena being due to a tendency of the blood to hypercoagulate.

It is worthy of note, however, that, together with and behind the rolls of RBC, platelet aggregation occurs, which is important for the dynamics of the coagulolytic system. In point of fact, the appearance of *white bodies* [30] marks the onset of intrinsic coagulation mechanisms, following the release of platelet thromboplastin (*see* Section 7.4.3).

Laborit gave a metabolic explanation for the phenomenon of sludge, which he produced experimentally by simple perfusion of sodium lactate until lactacidaemia of 80 mg% was obtained [52]. Acidosis produces œdema of the intima cells and depolarization of the membrane, triggering both erythrocytic and platelet aggregation. In shock, metabolic acidosis is the result of hypoxic disturbances of the intracellular enzymatic chains. Hyperlactacidaemia is the main reason for the accumulation of reduced $NADH_2$ with all its untoward consequences (*see* Section 8). Lactacidaemia is therefore an important pro-sludge factor [11, 48]. An efficient antisludge agent is dihydroxyacetone which oxidizes $NADH_2$ and unblocks glycolysis.

Opening of the precapillary sphincter and persistence of the postcapillary venular spasm dams up the blood in the microcirculation.

This pooling phenomenon seems to be the forerunner of DIC, because the pooled, acidotic blood is hypercoagulable [35].

The RBC disappear from the circulation either due to the formation of rolls and initial globular masses, or their being trapped in the microcirculation which is closed at the venular end, or, finally, to their taking-up in DIC. Double isotopic tagging (the host RBC with ^{32}P and the perfused cells with ^{51}Cr) experimentally demonstrated rapid sludging of the perfused RBC, especially in the hepatic microcirculation [43, 52, 102]. Other authors using only ^{51}Cr labelled RBC found evidence in experimental traumatic shock of RBC sludging in the liver and intestine, 30 minutes after the onset of shock (Şuteu and Cafriţă, 1968).

In shock, the plasma-RBC balance is abruptly disturbed and, as the vital importance of plasma is greater than that of the cellular elements, the first spontaneous reaction of the organism is to correct the plasma deficit. Experiments carried out on baboons subjected to exsanguination followed by perfusion with human RBC showed rapid recovery of the plasma (plasmaphoresis), up to 90% of the initial volume, and recovery of the proteins in a proportion of 50% within the first hour, owing to venular return and not through the thoracic duct [83].

The relationship between intra- and extravascular proteins in shock has not yet been fully elucidated. Displacement of the interstitial proteins towards the vascular space in response to haemorrhage or other forms of protein losses appears to be the initial but not the only possible element; another way may be through the thoracic duct, lymphatic return being very important, especially for proteins (*see* Chapter 6).

7.4.2 DISSEMINATED INTRAVASCULAR COAGULATION

Disseminated intravascular coagulation (DIC) has been defined as an acute temporary coagulation of the blood in the entire vascular space, obstructing, in particular however, the microcirculation [35, 74].

DIC marks a critical moment in the evolution of shock, which it maintains and aggravates, generating disease and sometimes directing it towards irreversibility.

DIC is a phenomenon that may appear in the course of shock but may in itself be a cause of shock. Although not all types of shock necessarily include a stage of DIC, the onset of DIC almost always gives rise to a state of shock. Similarly, although not all coagulation defects in the course of shock are brought about by a DIC episode, the latter will always generate an entire and very dangerous series of complex coagulostatic disturbances.

DIC consists of a coagulation process produced in its usual dynamic sequence but of disproportionate intensity and extent, causing the coagulostat mechanisms to deteriorate.

Part of the 13 factors of the coagulating system (Figure 7.12) have been studied separately in shock.

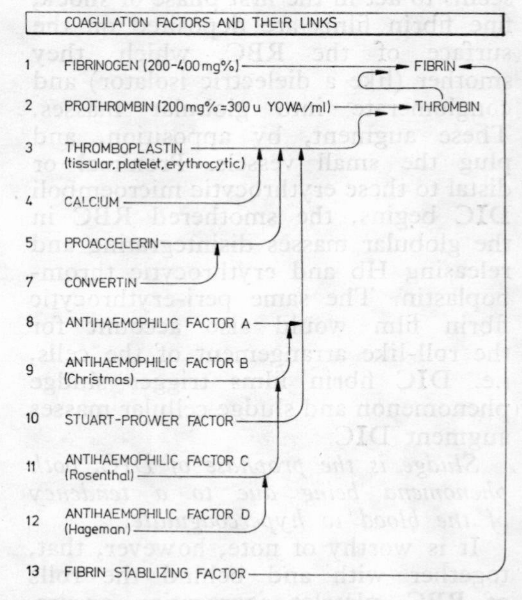

Figure 7.12. Coagulation factors and their links.

Fibrinogen is a fibrous euglobulin of the Cohn fraction 1; it migrates electrophoretically between the beta and gamma globulins and has a molecular weight of 400 000. It is synthesized in the liver cell ribosomes and appears to be identical with fibrin, from which, however, it differs by its degree of polymerization, brought about by oxidation of the sulphhydryl groups under the influence of thrombin and of the platelet factor II. The total amount of fibrinogen in the organism is 15 to 16 g. The normal half-life of fibrinogen is 6 days, during which it is constantly being jellified into a film of endothelial fibrin, and to the same extent removed from the walls of the vessels by the RES cells or dissolved by the plasmin system.

In the varied forms of shock, the catabolism of labelled fibrinogen increases two to three -fold/24 h, irrespective of age [24]. Initially, in shock there is a *hyperinosis* due to hepatic stimulation of fibrinogen synthesis under the action of adrenal steroids. Increasing to 50 mg/h, the production of fibrinogen rapidly induces plasma concentrations above the upper critical level (800 mg%) at which blood viscosity increases dangerously. More recently it has been shown that the increase in viscosity is also due to the accumulation of factor VII.

Following the onset of the DIC episode, fibrinogen is rapidly consumed and falls to lower critical haemorrhagic levels (below 100 mg%). This shift in plasma fibrinogen concentration (calculating the correction with regard to proteinaemia imposed by eventual haemodilution) may clinically bear witness to a DIC episode that has passed.

Another proof is the appearance of *cryofibrinogen* (a monomer with a shorter chain) which forms a fibrin precipitable at low temperatures. Under normal conditions cryofibrinogen is not to be found in the blood; it only appears following ultrarapid conversion of fibrinogen into fibrin in the course of DIC.

The intense action of thrombin also favours the appearance of a large amount of peptide A and B *(cofibrin)*, the intermediary products of fibrinogen conversion into fibrin.

Apart from DIC, exaggerated fibrinolysis also interferes in the incomplete digestion of fibrinogen (from 100 mg% to zero). In excessive amounts plasmin loses its specificity for fibrin and also attacks fibrinogen.

Afibrinogenaemia becomes an indirect proof of a DIC episode followed by reactive hyperfibrinolysis.

Plasmin digests fibrinogen in two stages: the first gives two fragments with different molecular weights *X* (240 000-300 000) and *Y* (155 000); the second gives fragments *D* (83 000) and *E* (50 000). Fragments *X* and *Y* produce complexes with fibrinogen monomers that have a strong anticoagulant action (greater than that of fragments *D* and *E)* and account for the tendency to bleeding in the course of fibrinolysis by an aggravating selfmaintenance mechanism [69, 70].

In shock plasma, fibrinogen appears to be accompanied by an increase in its concentration in the lymph [71]. Thus, an extravascular fibrinogen deposit is assumed to exist that might improve the plasma deficit by an increase in the return lymph flow (*see* Section 6.8).

Calcium. Of vital importance not only for coagulation, calcium has been exhaustively investigated in states of shock [4, 35]. The determination of total calcium, ionized calcium and the *p*H revealed a triphasic oscillating variation, especially of ionized calcium: decrease, increase and then slow decrease during a 48-hour interval. However, these calcium variations do not reach critical values in shock and, as a matter of fact, follow the variations of acidosis, plasma proteins and inorganic phosphate.

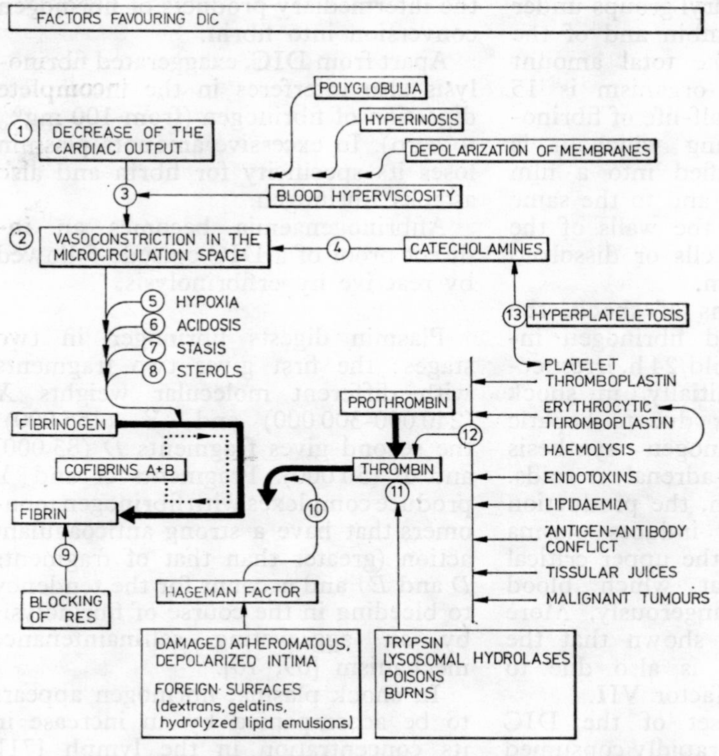

Figure 7.13. Factors favouring DIC.

The mechanism triggering DIC depends upon various factors, thirteen of which are shown in figure 7.13. Because, in shock, the treatment of DIC is of primary importance we have given in figure 7.14. a list of eleven factors that may prevent or control DIC.

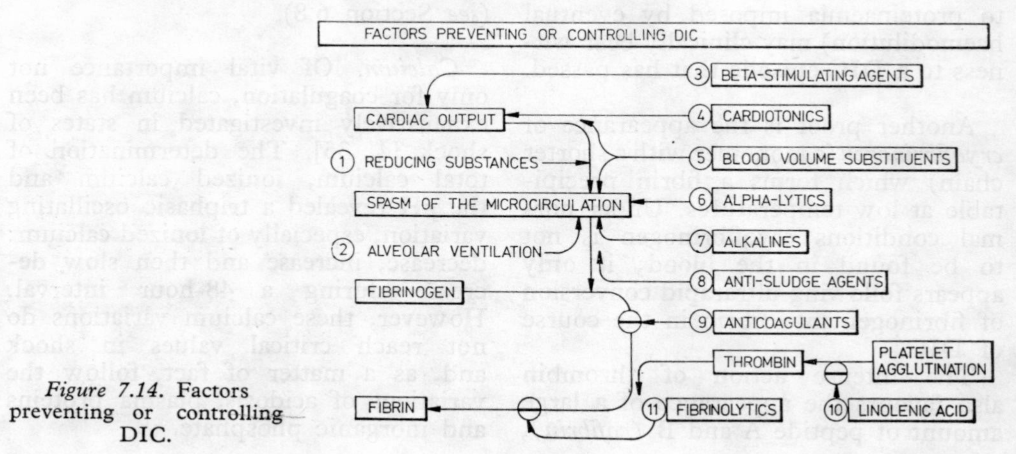

Figure 7.14. Factors preventing or controlling DIC.

DIC is a phenomenon that may be listed in the general tendency of the organism to arrest bleeding by hypercoagulation; but the *moment* and the *way* in which this occurs render DIC the most severe event in the worsening of shock.

DIC develops consequent to the preparatory rheodynamic phenomena in the microcirculation (Figure 7.15). lation that gives rise to hypoxia of the capillary intima cells. In this, intima cell depolarization and œdema immediately follow, with parietal hydrophil zones, due to discontinuity of the fibrin film and endothelium which denude the collagen fibrils. This results in adherence of the marginal platelets and activation of the Hageman contact factor (XII). Coagulation is triggered

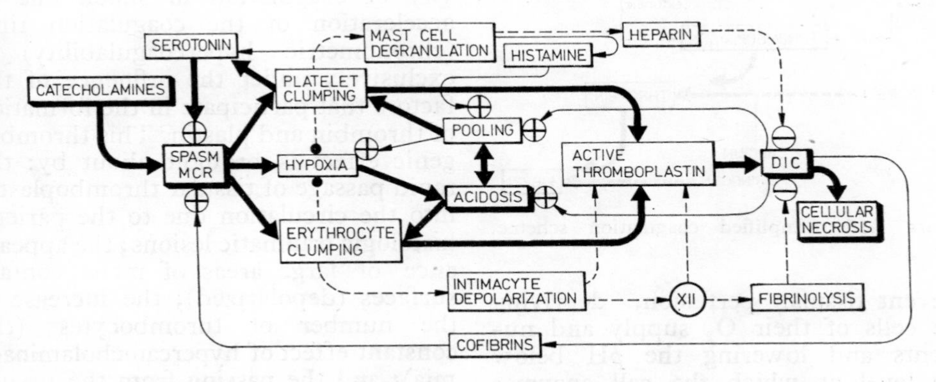

Figure 7.15. The chain of events that generate DIC.

The slow flow allows histamine, released by the paracapillary mast cells, to persist in the microcirculation, which it opens wide, doubling the containing space; this is followed by pooling of an acidic (below *p*H 7.20 is the critical level) hypercoagulable fluid in the dammed up microcirculation.

Even in pure haemorrhagic shock the acid, pooled blood in the microcirculation may begin coagulating after any incentive: a few haemolysed RBC (haemolytic activity increases between days 3 and 5, and 7 and 10, after any haemorrhage) are sufficient to release erythrocytic thromboplastin just as actively as the platelet one (Hardaway, 1966).

The DIC phenomena are precipitated by the initial spasm of the microcircu- off but also fibrinolysis (the coagulating system controls the final clot from the very beginning).

This initial synchronism may be considered as 'intentional only' in cases of shock.

The spasm in the microcirculation → sludging → pooling → DIC sequence is followed by an auto-aggravating return: DIC → spasm in the microcirculation, because the disintegrated platelets continue to liberate serotonin (Hardaway, 1966).

Peptides A and B (cofibrins), that appear as intermediate factors under the action of thrombin on fibrinogen, exercise the same stimulating effect on the smooth capillary muscle as catecholamines and serotonin. As a matter of fact, the arterial spasm that accompanies both peripheral and central thrombophlebitis is well known. The vasoconstriction-coagulation phenomena seems to be a very close correlation, the two processes potentiating one another. The consequence for the tissues is disastrous. Obstruction of the microcirculation by DIC will prevent tissular perfusion, depriving the cells of their O_2 supply and nutrients and lowering the *p*H below the level at which the cell enzymes can activate.

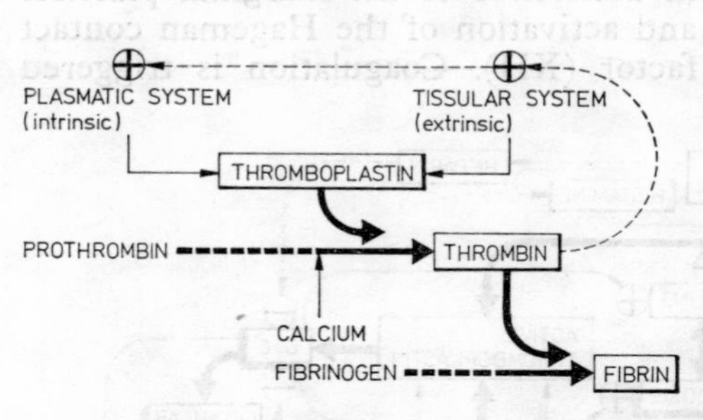

Figure 7.16. Simplified coagulation scheme.

DIC develops according to the usual pattern of the stages of coagulation (Figure 7.16). Fibrinogen, soluble in the plasma, is converted into an insoluble reticulum (fibrin) that reinforces the clot. The reaction is triggered off by thrombin, generated by prothrombin under the action of thromboplastin and calcium. Substances with a thromboplastin activity are numerous in the plasma, platelets and tissues.

Once initiated, the system is accelerated by autocatalysis, by the 'enzymatic cascade', the principal autocatalyzer being thrombin (Figure 7.17). This mechanism is a typical example of positive feedback, with deteriorating runaway.

At 37°C the blood clot develops, as a rule, after 7 minutes. The first stage (the formation of thromboplastin) lasts 6 minutes and 45 seconds, 5 minutes of which are necessary for contact activation of factors XI and XII, the formation of thromboplastin actually demanding only 1 minute and 45 seconds.

Then, in only 15 seconds, the other two stages of coagulation take place: the formation of thrombin and fibrin.

Therefore, the pathologic hyperactivity of coagulation in shock due to acceleration of the coagulation time (chronometric hypercoagulability) is exclusively under the influence of the factors that participate in the formation of thrombin and plastin. This thrombogenic effect is brought about by: the rapid passage of tissular thromboplastin into the circulation due to the parietal histologic traumatic lesions; the appearance of large areas of moist contact surfaces (depolarized); the increase in the number of thrombocytes (the constant effect of hypercatecholaminaemia); and the passing from the tissues

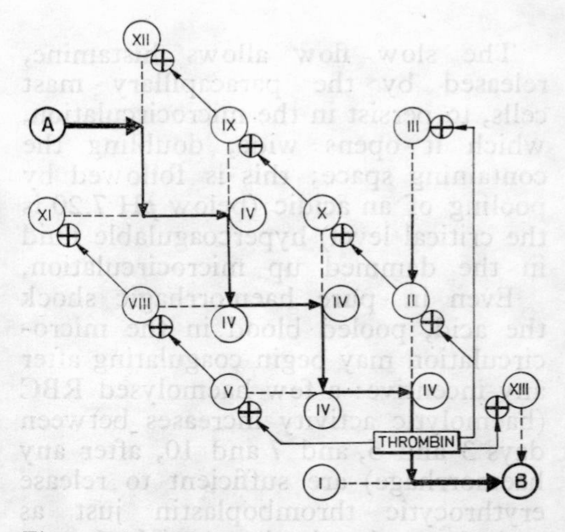

Figure 7.17. Triggering of the coagulating system; step *A*, attains its result, step *B* (fibrin), through the intermediacy of thrombin, which will then autoamplify the rhythm of the entire system (positive feedback).

into the blood of proteases that rapidly activate the contact factors and, at the same time, the fibrinolytic system.

In shock, two of the laws of coagulation are important: the *potentiation law* which acts by association with the thrombogenic factor complex, and the *substitution law* according to which hypocoagulability of the factors in the second and third stage (their consumption in the course of DIC) may be compensated for by hyperactivity of the dominant thromboplastin-forming complex (hyperplateletosis).

Once the clot is consolidated under the control of the platelets factor XIII and calcium, it exhibits a certain resistance to the proteolytic enzymes. In shock, hypofribrinogenaemia, deficit in factor XIII and thrombocytes and/or anaemia result in the development of an inefficient clot (structural hypocoagulability). The chronometric and structural aspects of coagulation may by studied by thrombodynamography [92].

The thromboplastin-forming stage may take place in two distinct ways which almost always coexist (Figure 7.18):

- the tissular juices together with calcium and factors V, VII and X form the extrinsic thromboplastin activator;
- the cascade of factors XII, XI, IX, VIII and X, together with calcium and platelet thromboplastin (III) form the intrinsic thromboplastin activator. This first stage can only be induced by existing factors in the systemic blood.

The platelets have a complex behaviour: they liberate ADP and Ca^{++}, performing a 'viscous metamorphosis' whose result is the platelet white thrombus. Within its core, serotonin, epinephrine, histamine, the platelet lipid

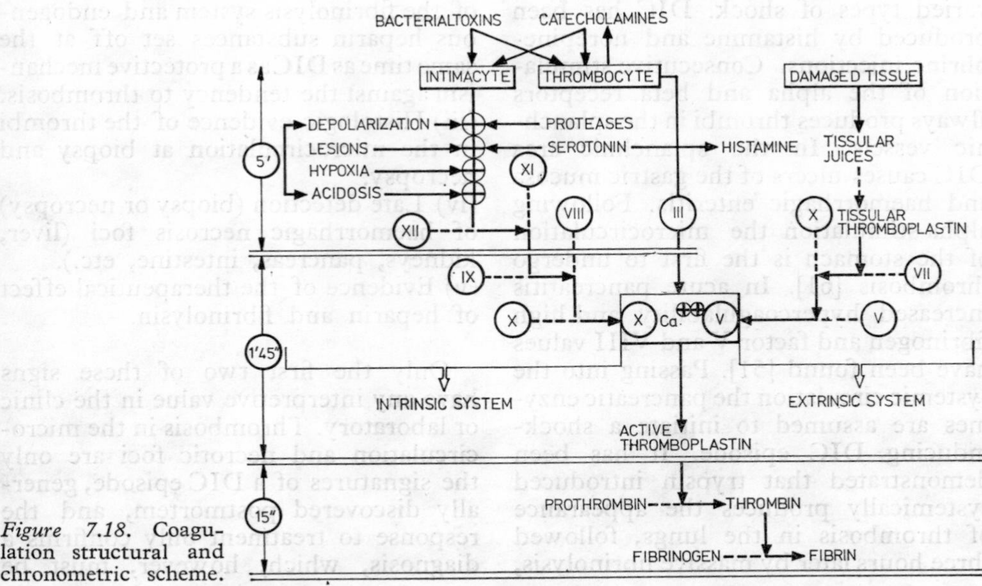

Figure 7.18. Coagulation structural and chronometric scheme.

factor II, etc. are released, rendering the white thrombus impermeable. Together with the extrinsic factor (that has formed in the meantime) and the cascade of endogenous factors (intrinsic activator), the platelet lipid factor III will trigger the formation of thrombin and then of fibrin, in the network in which the RBC will be caught.

Proof of the DIC episode is not easily found in the clinic.

Histologically it may always be very clear but hardly of practical use; moreover, although the thrombi have often been seen in the light microscope they may have been caused merely by postmortem necrosis.

Other thrombi cannot be found because of their transitory nature due to the concomitant fibrinolysis, since coagulation from the very beginning always prepares its own dissolution.

The experimental data that confirm the actual existence of the pathogenic DIC sequence have been reported in varied types of shock. DIC has been produced by histamine and norepinephrine injections. Consecutive stimulation of the alpha and beta receptors always produces thrombi in the splanchnic vessels. In the splanchnic area DIC causes ulcers of the gastric mucosa and haemorrhagic enteritis. Following alpha stimulation the microcirculation of the stomach is the first to undergo thrombosis [61]. In acute pancreatitis increased hypercoagulability and high fibrinogen and factor V and VIII values have been found [51]. Passing into the systemic circulation the pancreatic enzymes are assumed to initiate a shock-inducing DIC episode. It has been demonstrated that trypsin introduced systemically produces the appearance of thrombosis in the lungs, followed three hours later by massive fibrinolysis, with decrease of the thrombocytes, fibrinogen, factors V and VIII, and increase in plasminogen.

In dogs, in endotoxin shock, thrombi appear after 15 minutes in the renal capillaries; after 20 to 60 minutes fibrinolysis interferes and the thrombi disappear. Actually, fibrinolysis is always exacerbated when the end approaches. In humans this might explain the paucity of capillary thrombi in routine autopsies.

Diagnosis of a DIC episode is clinically based on a group of symptoms and signs:

(i) Abrupt onset of an inexplicable hypotension, accompanied by cyanosis and accentuated dyspnoea, abdominal pains and vomiting.

(ii) A tendency to haemorrhage, ranging from dramatic clinical bleeding up to subclinical laboratory signs of hypocoagulability. Therefore, DIC can indirectly be made evident clinically by two of its consequences: consumption coagulopathy and exaggerated activation of the fibrinolysis system and endogenous heparin substances set off at the same time as DIC as a protective mechanism against the tendency to thrombosis.

(iii) Histologic evidence of the thrombi in the microcirculation at biopsy and necropsy.

(iv) Late detection (biopsy or necropsy) of haemorrhagic necrosis foci (liver, kidneys, pancreas, intestine, etc.).

(v) Evidence of the therapeutical effect of heparin and fibrinolysin.

Only the first two of these signs have any interpretive value in the clinic or laboratory. Thrombosis in the microcirculation and necrotic foci are only the signatures of a DIC episode, generally discovered postmortem, and the response to treatment only confirms a diagnosis, which, however, must be established from the beginning since no therapeutical risks should be taken.

According to their evolution in time the manifestations of DIC may be grouped as immediate and late (Hardaway, 1970).

Immediate manifestations include the clinical symptomatology mentioned in the above paragraphs. Within a few minutes after the onset of DIC coagulation defects develop due to acute depletion of the pro-coagulating factors. Consumption hypocoagulability in many instances does not reveal itself clinically and only a longer coagulation time suggests it. In other cases refractory bleeding of a traumatic, operative wound, etc., or ecchymoses, haemarthroses, gastro-intestinal bleeding, haematuria, nose bleeding or haemorrhagic exudate of the serosa may appear.

Late manifestations of a DIC episode are represented by their tissular morphologic effect. The sequence of events is as follows: obstruction of the microcirculation by DIC → interrupted perfusion → anaerobic metabolism → acidosis → lysosomal autodigestion → cellular necrosis. If perfusion is not taken up within a given interval, which depends upon the sensitivity of each parenchyma, necrotic cicatrices appear that, by summation, may result in major organ insufficiences. Renal or acute hepato-renal failure, pancreatic necrosis, haemorrhagic gastro-enteritis or even acute myocardial failure may be the manifestation of a severe DIC episode with micronecrosis summation. As a rule, necrosis is not to be found in the lung but the alveoles are filled with fluid and suprainfected due to blocking of the pulmonary microcirculation by DIC.

Except for the brain, no other viscus appears to be protected; in almost all types of shock evidence has been found of necrosis foci in the liver, kidneys, pancreas, intestine, heart, skin, spleen, etc. The flow in the microcirculation may be taken up spontaneously by activation of endogenous fibrinolysis. It is evidently necessary that this lysis should occur before the onset of tissular necrosis and should not be exaggerated.

A real DIC episode can be demonstrated by histologic, haematologic and direct proof.

Histologic proof. If carefully looked for in the microcirculation, thrombi are found in 95% of cases [3, 35]. The association with œdema, focal areas of necrosis and polymorphonuclear infiltrate confirm the pre-mortem formation of thrombi, which may sometimes also be found circulating in the blood or fixed in the biopsy material obtained intraoperatively.

The thrombi may be formed by:

- platelet fusion and disintegration;
- fibrin, as a matter of fact the pericellular fibrin coat often seen in the electron microscope;
- red blood cells, likewise common (erythrocyte masses mixed with fibrin);
- globular masses formed by red blood cells covered with fibrin and conglomerated, frequent in bacteriemic shock;
- white blood cells in compact, bead-like rows on the endothelium and which appear very early (leukopenia is also a proof of DIC);
- the classical thrombi in the veins and arteries are not typical of DIC as they cover wide spaces; the classical thrombus is, however, also present in shock and may develop into a massive cavo-renal thrombosis.

The haematologic proof is summed up in Table 7.4. In shock, each of the factors that intervenes in the dynamics of the coagulostat may diminish due

Table 7.4 THE LABORATORY DIFFERENTIAL DIAGNOSIS OF COAGULATION DEFICIENCIES

	Consumption coagulation	Acute secondary proteolysis (accompanying fibrinolysis)	Acute primary fibrinolysis
Fibrinogen	Hypofibrinogenaemia (<100 mg%)	Afibrinogenaemia	↓
Platelet count	<60,000 risk of haemorrhage	decreased	normal
Quick time	prolonged	normal	normal
Factors II, V, VII, VIII, IX	↓ <20%	↓	↓
Leucocytes	↓	↓	normal
Haematocrit	↓	↓	normal
Euglobulin lysis (Von Kaulla test)	prolonged	curtailed	very short
Degraded fibrinogen	no	present	present
Fibrinopeptides	present	no	no
Plasminogen	normal	curtailed	curtailed
Antiplasmins (Nilsson test)	normal	curtailed	curtailed
Resistance of the clot to trypsin (Raynaud test)	negative	positive	positive

to increased catabolism, dilution, proteolytic destruction, insufficient synthesis and excessive consumption in DIC.

Direct proof is, as a rule, experimental, by cinephotography of the microcirculation in the serosa or by marking certain coagulation factors (fibrinogen, platelets, etc.) and finally by histographic evidence (Knisely et al., 1945; Larcan and Stoltz, 1970).

In vivo the most useful proof is obtained by the coagulation tests combined with the clinical signs. The presence of thrombi is variable and even contested [3, 36], as the fibrinolytic mechanism is at least as active as that of DIC and almost always manages to dissolve the clots pre- or immediately post-mortem (Attar et al., 1970; Markwardt, 1971).

7.4.3 PLATELET AGGLUTINATION AND LIPID EMBOLISM

Intravital microangiography has revealed in shock the formation of platelet thrombi in the mesenteral venules [2, 7]. Platelet agglutination may be reversible (breaking down, detaching or being caught up in the systemic torrent) or irreversible (strongly fixed to the vascular wall obstructing the vessel). The circulating platelet thrombi were detected by determining the plasma filtration pressure through a diaphragm.

Clumped platelets were also observed after administration of endotoxin or histamine both in the light and electron microscopes [23, 29, 60]. In endotoxin and histaminic shock, increase in the resistance within the lesser circulation appears to be due to blocking of the pulmonary circulation, especially by platelet thrombi. Lending support to this hypothesis is the finding that after remission of the pulmonary hypertension the agglomerated platelets disappear from the pulmonary circulation and reappear in the systemic one. Endotoxin electivity for the platelets is emphasized by the fact that 97% of the amount of endotoxin injected experimentally is to be found in the platelet clumps (Rosoff et al., 1971).

At present, the importance of the white platelet thrombi has been fully recognized: they are assumed to be responsible for *hepatic barrage* with portal hypertension and especially *pulmonary barrage* with hypertension in the lesser circulation (Berman et al., 1971; Shoemaker and Iida, 1962).

Platelet agglutination is also precipitated and favoured by: diminution of the blood flow, accumulation of the platelets and leucocytes in the lateral branches of the capillary network, the presence of catecholamines and ADP in the haemolysed cells, as well as of biogenic amines liberated into the systemic circulation (histamine and serotonin) and the procoagulating substances in the traumatized, ischaemic tissues and vascular walls. Under the influence of all these factors the agglutinated platelets pass into *viscous metamorphosis*. Contact of the circulating platelets with the collagen exposed by the traumatic lesion triggers off, by electrostatic exchanges through the membranes, an extraplatelet efflux of ADP, serotonin and Ca^{++}, which results in agglomeration of the platelets *(see* Section 7.2.2). Once these white masses have formed they may be mobilized and produce platelet embolism. The platelets also expel simultaneously the lipid factor III (platelet thromboplastin), which is added to the intrinsic plasma activator, whose formation was likewise precipitated by traumatic parietal discontinuity.

In point of fact, the white thrombus is initiated by the red one, but may also exist independently of it. White platelet masses have constantly been found in the tissues post-mortem. It is the pulmonary filter, particularly, that allows particles with a diameter of between 10 and 75 μ to pass and retains those greater than 75 μ, which only pass into the systemic circulation in the course of local pulmonary congestion, hypercatecholaemia, etc. The lung of the primate plays the role of a sponge and sieve for platelets and leucocytes. The aspect of DIC in the lungs is dominated by the presence of obstructive platelet white emboli.

The platelet emboligenic material may release serotonin from the moment of pooling in the pulmonary filter, increasing capillary permeability. With ^{14}C labelled serotonin it was demonstrated that the lung is the *target organ* of platelet emboli [7]. Breakdown of the platelets at pulmonary level releases ADP, serotonin and lysosomal hydrolases whose effects were foreseen by Gelin (1962) (Figure 7.19).

Pulmonary illness that develops in the absence of a direct thoraco-pulmonary lesion has been observed in almost all forms of shock *(see* Section 10.4).

Lipid embolism. Apart from the *platelet microemboli* the lung is also subjected to a further bombardment, that of the *lipid microemboli*. Conglomerated chylomicrons have been found in the platelet masses and it is known that FFA initiate platelet aggregation [119].

Essential phospholipids are a permanent constituent of biologic membranes; they follow the portal route and their polynonsaturated fatty acids may form

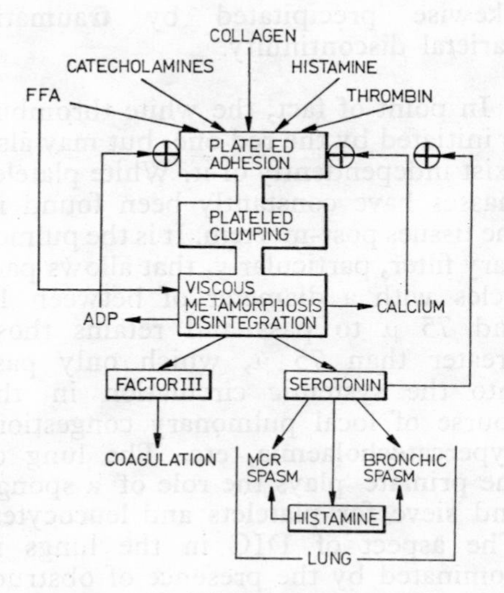

Figure 7.19. The platelet 'reactor'.

active acetate before the saturated fatty acids of triglycerides. Essential phospholipids emulsify the triglyceride droplets, aiding lipoproteinlipase to transport them in the form of α- and β-lipoprotein chylomicrons and introduce them into the metabolic cycle (King et al., 1971).

Today, lipid embolism is perceived in its microscopic dimensions, in its contribution to the organization of blood clots [18, 40, 75]. Evidence of triglyceride droplets has always been found in the clots of shocked patients with fractures, burns, crushed tissues, etc. Lipid emboli have two target organs: the lung and the kidney. It has been assumed that by the changes produced in plasma viscosity, hypertonia of the thrombin system and initial hypotonia of the fibrinolytic system create the conditions for the appearance of *lipid droplets* and, to a lesser extent, for increase in FFA or decrease in lipoproteinlipase [60]. Attention has recently been drawn to patients with severe war injuries not affecting the chest and lungs who stand up to surgical intervention fairly well but die from pulmonary complications that take on the clinical picture of the classical fatty embolism: disorientation, coma, dyspnoea with agitation, tachycardia, haemorrhage and high fever [23, 47].

Lipuria, increase in serum lipase and decrease in P_aO_2 have been considered useful in the prognosis and diagnosis of lipid embolism, but several systematic studies have refuted these correlations [23].

However, the correlation between lipid embolism and the coagulolytic mechanism has been clearly proven. Injection of oleic acid into the circulatory system of dogs produced DIC and consumption coagulopathy with fibrinolysis, and analysis of the thrombi always showed encompassed lipid droplets [46, 47].

7.4.4 THE PLASMIN SYSTEM

In shock, the intima cell, platelets, erythrocytes and leucocytes are the endovascular (intrinsic) source of hydrolases liberated from their lysosomal sacs [95, 96]. Depolarization of the membranes and acidotic metabolic impairment brings about the enzymatic efflux of hydrolases that may simultaneously trigger activation of the essential plasmatic systems — the *thrombin*, *plasmin* and *kinin* systems. The lysosomal hydrolases contained in the tissular juices have a similar effect, resulting in disintegration of the mesodermal or epithelial cells, which represent the exogenous (extrinsic) source of activation of the same three plasmatic systems.

Mention has already been made of the balance-imbalance of the thrombin and plasmin systems in shock; the kinin system, more heterogeneous, containing a variety of polypeptides, and with vasoactive properties, only influences the coagulostat indirectly *(see* Section 7.4.5).

Although in purpose a protective mechanism, plasmin digestion of fibrin exceeds its usual role in shock, passing on to the digestion of fibrinogen and factors V and VIII. The defibrination syndrome develops due to two major causes:

- *primary fibrinolysis* or activation of the fibrinolytic system at a rate that exceeds its inactivation system;
- excessive activation of the thrombin system that exhausts large amounts of fibrinogen in the DIC phenomenon. Severe hypofibrinogenaemia may develop (less than 100 mg%), representing *consumption coagulopathy*, the inevitable consequence of DIC. Consumption up to afibrinogenaemia progresses only when consumption coagulopathy is made worse by the onset of *reactive fibrinolysis* [44].

It is worthy of note that *subacute defibrination syndromes* also exist (cancer, the administration of dextran, leukaemia, etc.), in which only laboratory tests offer data for diagnosis.

Primary fibrinolysis is relatively rare (in cirrhosis and portal surgery), whereas *reactive fibrinolysis* (almost always associated with *consumption coagulopathy)* is relatively frequent; a common cause is shock, besides cardiopulmonary, urologic, obstetrical surgery, organ transplants, etc. [21, 25, 79].

The blood, lymph and cerebrospinal fluid contain *plasminogen*, a *plasmin* precursor that continuously dissolves the intra- and extravascular fibrin stocks. The activators of plasminogen conversion into plasmin are either endogenous (tissular fibrokinase activator and urinary urokinase activator) or exogenous (streptokinase, endotoxins, poisons, etc.). It has recently been proved that the intestine is the main 'reactor' of these activating materials that pass through the liver under certain conditions (cirrhosis, hepatectomy, etc.). In shock, the hypoxic liver acts in the same way. There is no general consensus concerning the production of activators in the intestinal wall or in the endoluminal bacterial flora, but it has been demonstrated experimentally that enterectomy almost totally lowers plasma concentration in activators of the plasmin system; neomycin brings about a partial reduction. The fibrinolytic system is schematically represented in Figure 7.20.

The activator system may be expressed by:

- tissular route, when the transformation of plasminogen into plasmin is activated by tissular *fibrokinases,* especially urokinase (these activators exist in the lung, uterus, prostate, ovaries, kidneys, thyroid);
- plasmatic route, mediated by a proactivator present in the plasma which is transformed into an activator under the influence of *lysokinases* of the figured or bacterial elements.

The inhibitory system acts either against plasminogen activation (by *antiactivators)* or directly on plasmin (by *antiplasmins).* Plasmin inhibitors are proteins belonging to the group of α-globulins (α_1-slow globulin and α_2-rapid globulin). Normally, they are to be found in the circulating blood and slow down the continuous physiologic fibrinolysis. Apart from the endogenous inhibitors there are also exogenous ones: ε-amino caproic acid (opposed to the generation of plasmin, inhibiting urokinase), Trasylol (Frey-Werle) and Iniprol (Kunitz-Northrop), opposed to the protease action of plasmin [10, 110].

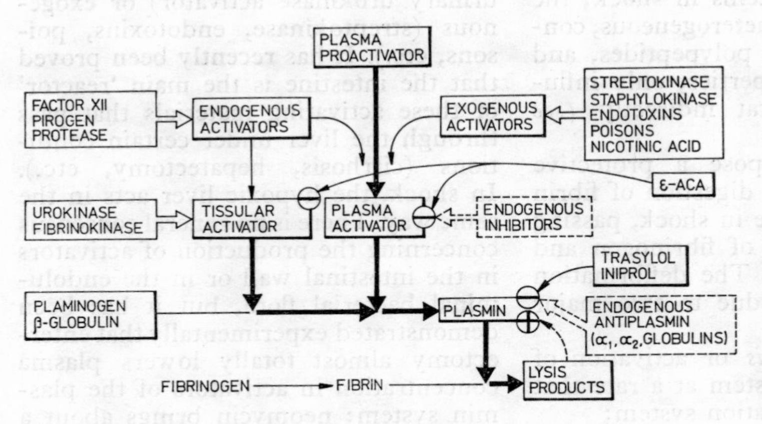

Figure 7.20. Fibrinolytic system. Endogenous and exogenous activators may be intercepted in the initial phase of fibrinolysis by ε-aminocaproic acid. (Trasylol and Iniprol act after the formation of plasmin).

In the case of defibrination due to excess consumption in DIC the antithrombins are exceeded (in primary fibrinolysis the antiplasmins are exceeded) and a paradoxical association develops: *multiple visceral thromboses with haemorrhage of variable intensity.*

In fact, it seems that with the onset of DIC the protective fibrinolytic system is also triggered. But the intensity of both processes escapes from under the control of the coagulostat and their exaggerated progress finally leads to an auto-aggravating superposal of the haemorrhagic effects: *consumption coagulopathy appears against a background of hyperfibrinolysis.* Consequently the gravity of the haemorrhagic syndrome in the course of shock becomes impressive.

Two problems arise: which of the two phenomena is predominant at a given moment in the pathologic physiology of bleeding, and in what direction must the treatment be applied? The cases with defibrination syndromes, clearly discernable in the course of various states of shock are statistically impressive (Broido et al., 1966; Josso, 1968; Thal and Sardesai, 1965; Zucker, 1971).

The following clinical aspects of primary fibrinolysis may be encountered:

- *Marked fibrinolysis* which develops following major surgery, complicated with shock. The volume and rate of the pulse fall sharply and haemorrhage is diffuse. Intra-operatively, bleeding is massive and surgical haemostasis cannot be performed. Fibrinolysis may also develop post-operatively, when it becomes manifest by bleeding at the level of the sutures, operative wound, ecchymoses, nose bleeding, etc.
- *Attenuated fibrinolysis* which appears as a spontaneous haemorrhagic syndrome and is represented by large ecchymoses with an irregular contour (the geographical map type).
- *Subclinical fibrinolysis* has no clinical signs.

Table 7.4. gives the tests that permit by comparison the differentiation between:

- *Consumption coagulopathy without reactive fibrinolysis* (a transitory phase in the course of shock).
- *Endogenous hyperheparinaemia*. Activation by DIC of certain endogenous anticoagulants. It is a situation that is difficult to control, but the fact remains that in shock acidosis inactivates heparin and the heparinoid substances should not be overlooked.
- *Afibrinogenaemia prior to shock*. A rare event with a genetic determinism.
- *Consumption coagulopathy combined with accompanying reactive fibrinolysis; the most frequent situation in shock.*
- *The haemorrhagipar syndrome of the liver in shock*, a phenomenon that always participates in shock and may at times even dominate the clinical picture.
- *Primary fibrinolysis* which develops in particular types of shock or after hepatoportal surgery. The experimental injection of histamine likewise produces anaphylactic shock with primary fibrinolysis, the target organ proving to be the liver [42].

The fibrinolytic system also activates directly the conversion of kininogen into bradykinin and opposes DIC; some authors therefore avoid administration of ε-aminocaproic acid which, by blocking fibrinolysis, favours the progress of DIC [3, 9]. These controversies outline once again the question of the *optimal therapeutical moment* in the state of shock (*see* Sections 9.6.3 and 11.4.3).

7.4.5 THE KININ SYSTEM

Vasoactive substances belong to two large groups:

(i) *Biogenic amines* (catecholamines, histamine and serotonin) whose synthesis is predominantly extravascular (*see* Sections 4.2.2 and 6.6). They have as biochemical root a single amino acid and are released by degranulation of the mast cells, adrenomedullar cells or postganglionic terminal neuronal knobs.

(ii) *Plasma kinins*, a polymorphous group of polypeptides, generated intravascularly by degradation of the plasma proteins. This is caused by certain *proteases* of intrinsic origin (thrombin, plasmin, circulating pancreatic enzymes, blood cell lysosomal hydrolases, etc.) or of extrinsic origin (lysosomal hydrolases liberated by the tissular juices). In addition, the exogenous factors with protease value (poisons, endotoxins, etc.) may also precipitate the formation of plasma kinins.

In shock, catecholamines, by the sequence of: *hypoxia → acidosis → lysosomal degradation → protease activation*, simultaneously stimulate all the three essential plasma systems (Figure 7.21).

In the microcirculation large space plasmakinins produce vasodilatation, œdema and pain, and in the systemic circulation transitory hypotension, due to the decrease in peripheral resistance, and increase in the cardiac output by acceleration of the pulse rate. Their role in the secretion of the exo- and endocrine glands is well known [106]. Evidence of an increase in plasma kinins was first found in bacteriemic shock but at present their role in all types of shock has been recognized.

Activation of the plasma kinin system. The Hageman factor and/or plasmin activate *prekallikreinase* (the permeability factor) which transforms *prekallikrein* into active *kallikrein;* the latter, in turn, separates *kinin* from *kininogen.* The fibrinopeptides A and B, resulting from the breakdown of fibrinogen, also activate this transformation, in the same way as plasmin, the process being equilibrated by a plasma inhibitor *(carboxypeptidase B)*.

Plasma contains 10 to 20 mg per thousand kininogen which appears to regulate the regional circulation by autonomic mechanisms. Plasma kinins interfere in shock, infections, pancreatitis and haemorrhage. Shock is accompanied by the diminution of kininogen (whose values remain low throughout the duration of shock) and increase of kinins. Haemodynamically, the fall in kininogens runs parallel to severe hypotension.

The following is a partial list of the plasma kinins whose role in shock is, however, but partly understood [1, 20, 27, 41, 45, 107, 113, 120, 122, 123]:

(i) *Kallidin II* is a segment of 10 amino acids produced by the action of kallikrein on α_2-globulins (Figure 7.22). An aminopeptidase splits of a lysine molecule and forms bradykinin.

(ii) *Bradykinin (Kallidin I)* is a short-living nonapeptide, being inactivated by splitting of an arginin molecule by carboxypeptidase.

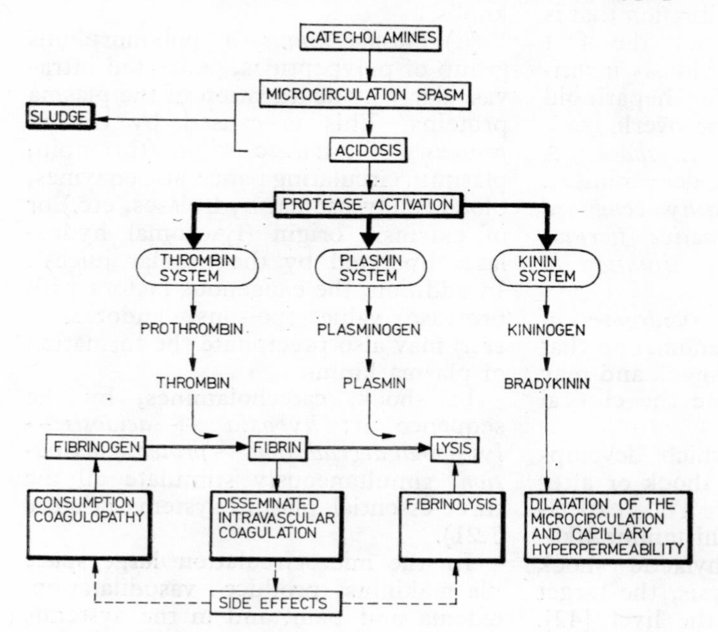

Figure 7.21. In shock protease activation triggers the three plasmatic rheodynamic systems.

Bradykinin and Kallidin II produce arteriole capillary vasodilatation and venoconstriction, increase capillary permeability, trigger allergic manifestations, and cause bronchi constriction.

These are the two main plasma kinins produced in states of shock, especially in the intestine. It has been demonstrated, however, that simple systemic injection of bradykinin does not induce shock [115].

Bradykinin seems to be a more potent vasodilator than histamine; but it is rapidly degraded by a carboxypeptidase (bradykininase) which means it does not produce many systemic effects.

(iii) *Vasodilator material.* A vasodilator material (VDM) from the blood of animals in an advanced stage of shock was isolated by Chambers in 1943. Although it gave rise to many arguments in the pathologic physiology of shock, nowadays VDM has been forgotten, as also has the 'vasoexcitatory material' (VEM). Both terms are unspecific and in point of fact mask tow substances that have now been isolated in the pure state and classed in one of the known plasma systems. These substances are worthy of note since they have been permanently considered to be of polypeptide origin and to play fundamental roles in the pathology of shock. Even if it is admitted that VDM is represented by *ferritin* (Mazur and Shorr, 1948), this complex with its vasodilator properties (apoferritin + ferric iron) released by the hypoxic RES, is at present considered to represent the intrinsic redox mechanisms of the iron ion.

In the first phase of shock, ferric iron acquires an electron from plasma proteins and forms an inactive ferrous iron; in the course of shock, iron then loses the acquired electron and in the form of ferric iron (together with a kelating agent) will oxidize epinephrine into adrenochrome, arresting its vasoconstrictive action. Not only the iron of ferritin but also that of the terapyrol rings (cytochrome, haem, etc.) is involved in this mechanism.

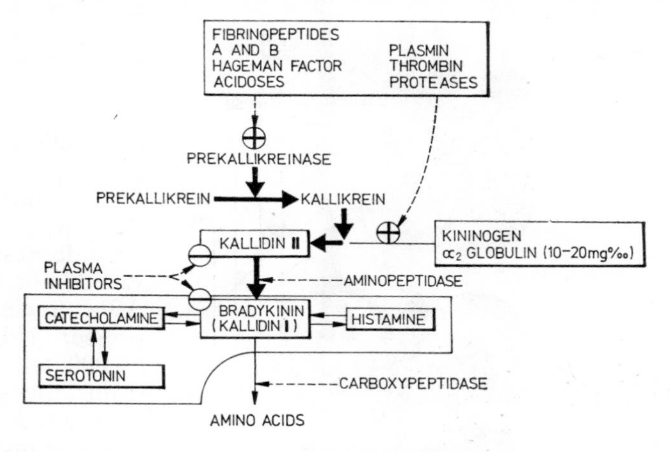

Figure 7.22. The plasma kinin system and its relations with proteases and vasogenic amines.

(iv) *Myocardial depressor factor.* Another plasmakinin is the myocardial depressor factor (MDF), an octo- or nonapeptide with a molecular weight of 800 to 1000, and different from bradykinin and angiotensin. It belongs to the class of lysosomal cathepsins and always appears in the plasma following protease activation (Brandt et al., 1970; Lefer et al., 1971; Lovett et al., 1971; Wangesteen et al., 1971).

The MDF is inotropically negative on the papillary muscle of the cat [59, 67]. Its presence has been noted in all types of shock especially in the shock of acute pancreatitis. The pancreas is considered to be its main producer and its route of penetration into the blood is predominantly lymphatic since overflow of the thoracic duct lymph significantly increases the survival of experimental animals [14]. The blood MDF levels run parallel to the increase in pancreatic trypsin and amylase. The

MDF appears to be antagonized by bradykinin, which attenuates its negative inotropic effects (perhaps by competition for the cardiac β_1-receptors) [13]. Isolated in a great number of cases of haemorrhagic shock, intestinal obstruction, septicaemia and acute necrotic pancreatitis, MDF was closely correlated with increase in lysosomal hydrolases (β-glucuronidase, acid phosphatase, cathepsin D, etc.) and aggravation of shock (fall in the urinary output, acute myocardial failure and death) [117, 118]. Today the importance of the MDF in the pathology of shock is generally admitted (Figure 7.23).

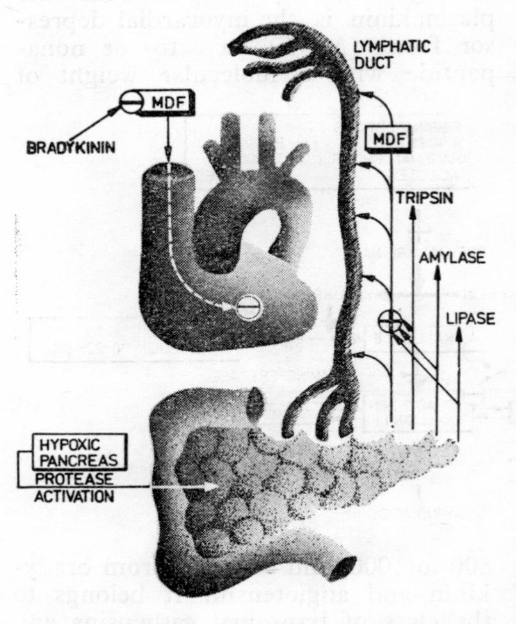

Figure 7.23. Synthesis, release and circulation of the myocardial depressor factor in shock.

(v) *Angiotensin II* is an octopeptide and has been discussed in chapter 4 together with its intimate relationship with aldosterone *(see* Section 4.8.6).

In the group of plasma polypeptides there are also a number of factors that cannot be classified and are still imperfectly understood, among which are the 'antisurfactant factor', the 'pulmonary depressor factor', 'the declamping vasodilator factor', etc. [15, 93, 114]. After unclamping the aorta, a vasodepressor substance (molecular weight 60 000 to 100 000) has been isolated from the peripheral blood and is considered to be responsible for 'declamping shock'. This substance may be a precursor of the kallidin-bradykinin system or a self-standing polypeptide, which can be removed by dialysis [15].

SELECTED BIBLIOGRAPHY

1. ARTURSON, G., The Plasma Kinins in Thermal Injury, *Scand. J. Clin. Lab. Invest.*, 1969, *24*, (107), 153–162.
2. ATKINS, P., LEMPKE, R. E., The Effect of Hypoxia on the Platelet Count, *Brit. J. Surg.*, 1970, *57*, (8), 583–586.
3. ATTAR, S., HANASHIRO, P., MANSBERGER, A., MCLAUGHLIN, J., FIRMINGER, H., COROLEY, R. A., Intravascular Coagulation-Reality or Myth?, *Surg.*, 1970, *68*, (1), 27–33.
4. BARRY, G. D., Plasma Calcium Concentration Changes in Haemorrhagic Shock, *Amer. J. Physiol.*, 1971, *220*, (4), 874–879.
5. BATEY, N. R., RAJASINGHAM, M. S., SINGH, C. M., FLEAR, C. T. G., The Effect of Escherichia Coli Endotoxin Membrane Permeability in Skeletal Muscle, *Brit. J. Surg.*, 1970, *57*, (11), 851.
6. BERGENTZ, S. E., GELIN, L. E., RUDENSTAM, C. M., ZEDERFELD, B., The Viscosity

of Whole Blood in Trauma, *Acta Chir. Scand.*, 1963, *126*, (3), 289–295.
7. BERMAN, I. R., SMULSON, M. E., PATTENGALE, P., SCHOENBACH, S. F., Pulmonary Microembolism after Soft Tissue Injury in Primates, *Surg.*, 1971, *70*, (2), 246–253.
8. BEUTLER, E., Clinical Aspects of Enzymatic Deficiencies of Erythrocytes and their Detection, *Triangle*, 1968, *8*, (7), 267–272.
9. BLAND, J. H., *Disturbances of Body Fluids*, W. B. Saunders Comp., London, 1956.
10. BOGE, J. C., FAVAREL GARRIGUES, J. C., De JOIGNY, C., Syndromes de défibrination observés dans un centre de réanimation. La place de la coagulation intravasculaire dans l'évolution des états de choc, *Bordeaux Méd.*, 1970, *3*, (11), 2593–2640.
11. BOHR, D. F., MCVAUGH, R. B., Influence of pH on Vascular Response, *Physiologist*, 1959, (2), 12–17.
12. BOYD, D. R., ADDIS, H. M., CHILIMINDRIS, C., Utilization of Osmometry in Critically Ill Surgical Patients, *Arch. Surg.*, 1971, *102*, (4), 363–372.
13. BRANDT, E. D., LEFER, A. M., Myocardial Depressant Factor in Plasma from Cats in Irreversible Postoligemic Shock, *Proc. Soc. Exp. Biol.*, 1966, *122*, (2), 200–205.
14. BRANDT, E. D., COWGILL, R., LEFER, A. M., Further Characterization of a Myocardial Depressant Factor Present in Haemorrhagic Shock, *J. Trauma*, 1969, *9*, (2), 216–223.
15. BRANT, B., ARMSTRONG, R. P., VETTO, R.M., Vasodepressor Factor in Declamp Shock Production, *Surg.*, 1970, *67*, (4), 650–653.
16. BRÎNZEU, P., *Tulburările microcirculației în șoc* (Microcirculation Disturbances in Shock), Editura Medicală, București, 1971.
17. BROIDO, P. W., BUTCHER, H. R., MAYER, C. A., The Expansion of the Volume Distribution of Extracellular Ions during Haemorrhagic Hypotension and its Possible Relationship to Change in the Physical-Chemical Properties of Extravascular-Extracellular Tissue, *Arch. Surg.*, 1966, *93*, (4), 556–561.
18. BUCHERU, D., SĂNDULESCU, D., BĂLAN, A., SÎRBULESCU, I., Embolia grăsoasă posttraumatică (Posttraumatic Fatty Embolism), *Chirurgia*, 1970, *XIX*, (2), 129–134.
19. BURTON, A. C., *Physiology and Biophysics of the Circulation*, Year Book Medical Publ. Inc., Chicago, 1965.
20. CAILAR, J., L'enzyme et le choc, *Jrn. Méd. Mont.*, 1969, *4*, (5), 206–211.
21. CAMPION, D. S., LYNCH, L. J., RECTOR, Jr., F. C., Effect of Transmembrane Potential, *Surg.*, 1969, *66*, (6), 1051–1059.
22. CAREY, J. S., SCHORSCHMIDT, B. F., CULLIFORD, A. T., Hemodynamic Effectiveness of Colloid and Electrolyte Solutions for Replacement of Simulated Operative Blood Loss, *Surg. Gynec. Obst.*, 1970, *131*, (4), 679–686.
23. CLOUTIER, C. T., LOWERY, B. D., STRICKLAND, T. G., CAREY, L. C., Fat Embolism in Vietnam Battle Casualties in Haemorrhagic Shock, *Milit. Med.*, 1970, *135*, (5), 369–373.
24. DAVIES, J. W. L., LILJEDHAL, S. O., REIZENSTEIN, P., Fibrinogen Metabolism-Following Injury and its Surgical Treatment, *Injury*, 1970, *1*, (3), 178–185.
25. DINBAR, A., DECIO, M. R., FONKALSRUD, E. W., Effects of Hepatic Ischemia on Coagulation in Primates. Application to Liver Transplantion, *Surg.*, 1970, *68*, (1), 269–276.
26. ECCLES, J. C., *The Physiology of Nerve Cells*, The John Hopkins Press, Baltimore, 1957.
27. EIGLER, F. W., KRISTEN, H., STOCK, W., HÖFER, J., Der Proteinaseinhibitor Trasylol als therapeutisches Prinzip beim Tourniquet-Syndrom, *Bul. Soc. Int. Chir.*, 1970, *XXIX*, (4), 197–203.
28. FLEISCHER, V., FRÖHLICH, E., *L'eau et les électrolytes dans l'organisme*, Masson, Paris, 1965.
29. GELIN, L. E., Intravascular Aggregation and Capillary Flow, *Acta Chir. Scand.*, 1957, *113*, (4), 463–470.
30. GELIN, L. E., ZEDERFELDT, B., Experimental Evidence of the Significance of Disturbances in the Flow Properties of Blood, *Acta Chir. Scand.*, 1971, *122*, (4), 336–347.
31. GELIN, L. E., *Ciba Symposium Shock*, Stockholm, Springer, Berlin-Göttingen-Heidelberg, 1962.
32. GRÖNWALL, A., *Dextran and its Use in Colloidal Infusion Solution*, Almquist and Wiksell, Uppsala, 1957.
33. GRUBER, U. F., ALLGÖWER, M., Infusion Therapy in Surgery, *Triangle*, 1968, *8*, (7), 273–283.
34. GUTELIUS, J. R., SHIZGAL, H. M., LOPEZ, G., The Effect of Trauma on Extracellular Water Volume, *Arch. Surg.*, 1968, *97*, (2), 206–214.
35. HARDAWAY, R. M., *Syndromes of Disseminated Intravascular Coagulation with Special Reference to Shock and Haemorrhage*, Charles C. Thomas, Springfield-Illinois, 1966.
36. HARDAWAY, R. M., The Significance of Coagulative and Thrombotic Changes after Injury, *J. Trauma*, 1970, *10*, (4), 354–357.
37. HARDERS, H., *Advances in Microcirculation*, vol. 3, Karger Edit., Bâle-New York, 1970.
38. HERMAN, C. M., RODKEY, F. L., VALERI, C.R., FORTIER, N. L., Changes in the Oxyhemoglobin Dissociation Curve and Peripheral Blood after Acute Red Cell Mass Depletion

and Subsequent Red Cell Mass Restoration in Baboons, *Ann. Surg.*, 1971, *174,* (5), 734–743.

39. HOAK, J. C., SPECTOR, A. A., FRY, G. L., WARNER, E. D., Effect of Free Fatty Acids on ADP-induced Platelet Aggregation, *Nature,* 1970, *228,* (5278), 1330–1332.
40. HUAMAN, A., NICE, W., Fat Embolism Syndrome: Premortem Diagnosis by Cryostat Frozen Sections, *Bull. Amer. Coll. Chest Phy.*, 1970, *9,* (2), 20.
41. IONESCU TÎRGOVIŞTE, C., Chininele plasmatice (Plasma Quinines), *Viaţa Medicală,* 1970, *XXII,* (3), 103–108.
42. ISHIZU, H., Relationship between Anaphylactic Shock and Fibrinolysis, *Jap. J. Leg. Med.*, 1969, *23,* (3), 218–229.
43. JAMES, N. J., MATSUDA, L., SHOEMAKER, W. C., Red Cell Equilibration in Liver Tissue during Experimental Shock, *Ann. Surg.*, 1970, *170,* (1), 109–115.
44. JOSSO, F., Traitement des syndromes de défibrination, *Rev. Prat.*, 1968, *18,* (31), 181–184.
45. KATZ, V., SILVERSTEIN, M., KOBOLD, E. E., THAL, A. P., Trypsin Release, Kinin Production and Shock, *Arch. Surg.*, 1964, *89,* (3), 322–330.
46. KING, E. G., WEILY, H. S., GENTON, E., ASHBAUCH, D. G., Consumption Coagulopathy in the Canine Oleic Acid Model of Fat Embolism, *Surg.*, 1971, *69,* (4), 533–541.
47. KING, E. G., NAKANE, P. K., ASHBAUCH, D. G., The Canine Oleic Acid Model of Fibrin Localization in Fat Embolism, *Surg.*, 1971, *69,* (5), 782–787.
48. KITTLE, C. F., AOKI, H., BROWN, E. B., The Role of pH and CO_2 in the Distribution of Blood Flow, *Surg.*, 1965, *57,* (1), 138–146.
49. KNISELY, M. H., ELIOT, T. S., BLOCH, E. H., Sludged Blood in Traumatic Shock I. Microscopic Observations of the Precipitation and Agglutination of Blood Flowing through Vessels in Crushed Tissues, *Arch. Surg.*, 1945, *51,* (3), 220–228.
50. KRAGELUND, E., Loss of Fluid and Blood to the Peritoneal Cavity during Abdominal Surgery, *Surg.*, 1971, *69,* (2), 284–287.
51. KWAAN, H. C., ANDERSON, M. C., GRAMATICA, L. A., Study of Pancreatic Enzymes as a Factor in Pathogenesis of Disseminated Intravascular Coagulation during Acute Pancreatitis, *Surg.*, 1971, *69,* (5), 663–672.
52. LABORIT, H., Corrélations entre structures morphologiques, métaboliques et enzymatiques de capillaire et sa physiopathologie, *Agressol.*, 1969, *10,* (4), p. 291–302.
53. LARCAN, A., Définition du sludge, *Ann. Anesth. Franç.*, 1965, (4), 148–152.
54. LARCAN, A., STOLTZ, J. F., *Microcirculation et Hémorhéologie,* Masson, Paris, 1970.
55. LARCAN, A., STREIFF, F., STOLTZ, J. F., ALEXANDRE, P., Influence du liquide sur la mobilité électrophorétique et certaines propriétés plaquettaires en corrélation avec l'action sur la circulation mésentérique du rat, *Agressol.*, 1971, *12,* (1), 49–56.
56. LARCAN, A., STREIFF, F., STOLTZ, J. F., VIGNERON, C., A propos des techniques d'étude des propriétés mécaniques et rhéologiques de la membrane de l'hématie, *Path Biol.*, 1971, *19,* (5–6), 303–309.
57. LARENG., L., Rappel de physiologie du capillaire, *Anesth. Anal. Réan.*, 1969, *26,* (5), 701–706.
58. LASCH, H. G., Coagulation Disturbances in Shock, *Postgrad. Med.*, 1969, *45,* (526), 539–542.
59. LEFER, A. M., GLENN, T. M., O'NEILL, T. J., LOVETT, W. L., GEISSINGER, W. T., WANGENSTEEN, S. L., Inotropic Influence of Endogenous Peptides in Experimental Haemorrhagic Pancreatitis, *Surg.*, 1971, *69,* (2), 220–228.
60. LILIENBERG, G., RAMMER, L., SALDEEN, T., Intravascular Coagulation and Inhibition of Fibrinolysis in Fat Embolism, *Acta Chir. Scand.*, 1970, *136,* (2), 2, 87–90.
61. LINDER, M. M., MCKAY, D. G., An Experimental Study of Thrombotic Ulceration of the Gastrointestinal Mucosa, *Surg. Gynec. Obst.*, 1971, *133,* (1) 21–29.
62. LITTON, A., Haemovascular Changes in Septic Shock, *Postgrad. Med.*, 1969, *45,* (526), 551–554.
63. LITWIN, M. S., Blood Viscosity in Shock, *Amer. J. Surg.*, 1965, *110,* (3), 313–316.
64. LITWIN, M. S., CHAMPMAN, K., STOLIAR, J. B., Blood Viscosity in the Normal Man, *Surg.*, 1970, *67,* (2), 342–345.
65. LITWIN, M. S., CHAMPMAN, K., Physical Factors Affecting Human Blood Viscosity, *J. Surg. Res.*, 1970, *10,* (9), 433–436.
66. LITWIN, M. S., CHAMPMAN, K., Comparison of Effects of Dextran 70 and Dextran 40 Infused into Postoperative Animals, *Europ. Surg. Res.*, 1970, (2), 125–126.
67. LOVETT, W. L., WANGENSTEEN, S., GLENN, T. M., LEFER, A. M., Presence of a Myocardial Depressant Factor in Patients in Circulatory Shock, *Surg.*, 1971, *70,* (2), 223–231.
68. MANNINEN, V., Movements of Sodium and Potassium Ions and their Tracers in Propranololtreated Red Cells and Diaphragm Muscle, *Acta Physiol. Scand.*, 1970, Suppl. 353, 76–86.
69. MARDER, V. J., SCHULMAN, N. R., CARROLL, W. R., High Molecular Weight Derivatives of Human Fibrinogen Produced by

Plasmin. I. Physicochemical and Immunological Characterization, *J. Biol. Chem.*, 1969, *244*, (8), 2111–2119.
70. MARDER, V. J., SCHULMAN, N. R., CARROLL, W. R., High Molecular Weight Derivatives of Human Fibrinogen Produced by Plasmin. II. Mechanism of their Anticoagulant Activity, *J. Biol. Chem.*, 1969, *244*, (8), 2120–2124.
71. MARKS, L. S., KOLMEN, S. N., Tween 20 Shock in Dogs and Related Fibrinogen Changes, *Amer. J. Physiol.*, 1971, *220*, (1), 218–221.
72. MARKWARDT, F., *Anticoagulatient* (In edit. EICHLER, O., FARAH, A., JERKEN, H., WELCH, A. A.: *Handbuch der experimentellen Pharmakologie*, Band XXVII), Springer-Verlag, Berlin-Heidelberg-New York, 1971.
73. MCEWAN, A. J., MCLEDINGHAM, A. I., Total Blood-flow and Tissue Anoxia in Apparently Ischaemic Feet, *Brit. J. Surg.*, 1970, *57*, (11), 850.
74. MCKAY, D. G., *Disseminated Intravascular Coagulation*, Harper and Row, New York, 1965.
75. MCNAMARA, J. J., MOLOT, M. D., STREMPLE, J. F., Screen Filtration Pressure in Combat Casualties, *Ann. Surg.*, 1970, *172*, (3), 334–341.
76. MEHRISHI, J. N., Phosphate Groups (Receptors?) on the Surface of Human Blood Platelets, *Nature*, 1970, *226*, (5244), 452–453.
77. MEZICK, J. A., SETTLEMIRE, C. T., BRIERLEY, G. P., BAREFIELD, K. P., JENSEN, W. N., CORNWELL, D. G., Erythrocyte Membrane Interaction with Menadione and the Mechanism of Menadione-induced Hemolysis, *Biochim. Biophys. Acta*, 1970, *219*, (2), 361–371.
78. MILLER, L. D., OSKI, F. A., DIACO, J. F., DAVIDSON, D., MARIA DELIVORIA PAPADOPOULOS, The Affinity of Hemoglobin for Oxygen: Its Control and in vivo Significance, *Surg.*, 1970, *68*, (2) 187–195.
79. MILLER, R. D., ROBBINS, T. O., TONG, M. J., BARTON, S. L., Coagulation Defects. Associated with Massive Blood Transfusions, *Ann. Surg.*, 1971, *174*, (5), 794–801.
80. MOORE, D. F., Body Composition and its Measurement 'in vivo', *Brit. Surg.*, 1967, *54*, Lister Centenary Number, 431–435.
81. MURER, R. H., HOLME, R. A., Study of the Release of Calcium from Human Blood Platelets and its Inhibition by Metabolic Inhibitors, N-ethyl-Maleimide and Aspirin, *Biochim. Biophys. Acta*, 1970, *222*, (1), 197–205.
82. NAHAS, G. G., CAVERT, H. M., Cardiac Depressant Effect of CO_2 and its Reversal, *Amer. Physiol.*, 1957, *190*, (5), 483–490.
83. OSTEEN, R. T., KLEBANOFF, G., Early Plasma Protein Recovery after Total Volume Plasmapheresis in Baboons, *Surg.*, 1971, *69*, (2), 276–283.
84. PANAITESCU, GH., OLTEANU, D., *Apa şi electroliţii* (Water and Electrolytes), Editura Medicală, Bucureşti, 1969.
85. PAPADOPOL, S., *Tulburările de hemodinamică* (Haemodynamic Disturbances), Editura Medicală, Bucureşti, 1971.
86. PĂUN, L., Sindromul de coagulare diseminată intravasculară. Coagulopatia de consum. (Disseminated Intravascular Coagulation Syndrome. Consumption Coagulopathy), *Viaţa medicală*, 1969, *XVI*, (8), 505–512.
87. PLOUVIER, S. R., MOINE, M., LATOUR, J., Libération par choc hypotonique des facteurs plaquettaires procoagulants. Intérêt en thrombélastrographie, *Agressol.*, 1969, *10*, (1), 33–43.
88. PLOUVIER, S. R., MOINE, M., LATOUR, R., Effect activateur de charge électrique sur les tests de coagulabilité globale du sang, *Agressol.*, 1969, *10*, (4), 309–316.
89. POOLE, J. C. F., The Pathogenesis of Venous Thrombosis, *Brit. J. Surg.*, 1967, *54*, Lister Centenary Number, 463–466.
90. PROCTOR, H. J., OARJER, J. G., JOHNSON, G. Jr., Alterations in Erythrocyte 2,3-Diphosphoglycerate in Postoperativ Patients, *Ann. Surg.*, 1971, *173*, (3), 357–362.
91. PULLMAN, B., WEISSBLUTH, M., *Molecular Biophysics*, Acad. Press New York, London, 1965.
92. RABY, C., Les bases physiologique et physiopathologique d'une thérapeutique anticoagulante rationnelle, *Maroc. Méd.*, 1970, *533*, (1), 129–145.
93. RANGEL, D. M., DINBAR, A., STEVENS, G.H., BYFIELD, J. E., Cross Transfusion of Effluent Blood from Ischemic Liver and Intestines, *Surg. Gynec Obst.*, 1970, *130*, (6), 1015–1024.
94. RASHEVSKY, N., *Mathematical Biophysics. Physico-Mathematical Foundation of Biology*, Dover Publications Inc., New York, 1960.
95. REED, P. W., Glutathione and the Hexose Monophosphate Shunt in Phagocytizing and Hydrogen Peroxide-treated Rat Leukoctyes, *Biochim. Biophys. Acta*, 1970, *222*, (1), 53–64.
96. REICH, T., DIEROLF, B. M., REYNALDS, B. M., Plasma Cathepsin-like Acid Proteinase Activity during Haemorrhagic Shock, *J. Surg., Res.*, 1965, *5*, (1), 116–121.
97. ROSATO, F. E., MILLER, L. D., HEBEL, M., A Biochemical Definition of Blood Viscosity; its Possible Significance in the Pathophysiology of Shock, *Ann. Surg.*, 1968, *167*, (1), 60–66.

98. ROSOFF, C. B., SALZMAN, E. W., GUREWICH, V., Reduction of Platelet Serotonin and the Response to Pulmonary Emboli, *Surg.*, 1971, *70*, (1), 12–19.
99. RUSSE, GH., Aspecte biofizice ale microcirculației (Biophysical Aspects of the Microcirculation), *Viața Medicală*, 1970, *XVII*, (7), 291–296.
100. SALSBURY, A. J., CLARKE, J. A., The Surface Appearance of Blood Cells, *Triangle*, 1968, 8, (7), 260–266.
101. SHIRES, G. T., Initial Management of the Severely Injured Patient, *J. A. M. A.*, 1970, *213*, (11), 1873.
102. SHOEMAKER, W. C., IIDA, F., Studies on the Equilibration of Labeled Red Cells and T-1824 in Haemorrhagic Shock, *Surg. Gynec. Obst.*, 1962, *114*, (5), 539–545.
103. SIEGEL, D. C., MOSS, G. S., COCHIN, A. T., DAS GUPTA, T. K., Pulmonary Changes Following Treatment for Haemorrhagic Shock; Saline versus Colloid Infusion, *Surg. Forum*, 1970, *21*, (1), 17–19.
104. SIMON, M. A., OLSEN, W. R., Capillary Flow in Haemorrhagic Schock, *Arch. Surg.*, 1969, *99*, (5), 631–636.
105. STRING, T., ROBINSON, A. J., BLAISDELL, F. W., Massive Trauma. Effect of Intravascular Coagulation on Prognosis, *Arch. Surg.*, 1971, *102*, (4), 406–411.
106. STUART, F. P., MOORE, F. D., Effects of Single, Repeated and Massive Manitol Infusion in the Dog, *Ann. Surg.*, 1970, *172*, (2), 190–196.
107. ȘUTEU, I., CÂNDEA, V., TOMA, T., Sindromul proteolitic în chirurgie (The Proteolytic Syndrome in Surgery), *Revista Sanitară Militară*, 1965, *LXI*, (3), 325–329.
108. ȘUTEU, I., CAFRIȚĂ, A., Studiul experimental al evoluției și tratamentului șocului traumatic cu ajutorul radioizotopului ^{51}Cr (Experimental Study of the Evolution and Treatment of Traumatic Shock by ^{51}Cr), *Revista Sanitară Militară*, 1967, *LXII*, (4), 575.
109. ȘUTEU, I., CAFRIȚĂ, A., Eficiența comparată a unor agenți farmacodinamici în combaterea tulburărilor hemodinamice din șocul traumatic experimental, studiate cu radioizotopul ^{51}Cr (Comparative Effectiveness of Some Pharmacodynamic Agents in the Control of Haemodynamic Disturbances in Experimental Traumatic Shock Studied by ^{51}Cr), *Revista Sanitară Militară*, 1968, *LXIV*, (6), 821–830.
110. TAN, B. H., MORI, K., RICHTER, D., Study of the Defibrination Syndrome Associated with Acute Hepatic Failure, *Surg., Gynec. Obst.*, 1971, *132*, (2), 263–274.
111. TEITEL, P., Probleme de patologie moleculară în hematologie. In Probleme de biologie și patologie moleculară (Problems of molecular pathology in haematology. In *Problems of biology and molecular pathology*), Editura Medicală, București, 1964.
112. TEODORESCU-EXARCU, L., *Agresologie chirurgicală generală* (General Surgical Aggressology), Editura Medicală, București, 1968.
113. THAL, A. P., SARDESAI, V. M., Shock and the Circulating Polipeptides, *Amer. J. Surg.*, 1965, *110*, (3), 308–312.
114. TICE, D. A., WORTH, M. H., Cellular Damage Associated with Extracorporeal Perfusion: Plasminogen Activation and Lysosomal Enzymes Activity, *Surg.*, 1968, *63*, (6), 669–676.
115. URBANITZ, D., SAILER, R., HABERMANN, E., In vivo Investigations on the Role of the Kinin System in Tissue Injury and Shock Syndromes, *Adv. Exp. Med. Biol.*, 1970, *8*, (3), 343–353.
116. WANGENSTEEN, S. L., KIECHEL, S. F., LUDEWIG, R. H., MADDEN, J. J., The Role of Vasoconstriction in the Suppression of Haemorrhage from Arteries. I. The Completely Severed Artery, *Surg.*, 1970, *67*, (2), 338–341.
117. WANGENSTEEN, S. L., DE HOLL, J. D., DIECHEL, S. F., MARTIN, J., LEFER, A. M., Influence of Haemodialysis in a Myocardial Depressant Factor in Haemorrhagic Shock, *Surg.*, 1970, *67*, (6), 935–943.
118. WANGENSTEEN, S. L., GEISSINGER, W. T., LOVETT, W. L., GLENN, T. M., LEFER, A.M., Relationship between Splanchnic Blood Flow and a Myocardial Depressant Factor in Endotoxin Shock, *Surg.*, 1971, *69*, (3), 410–418.
119. WARNER, W. A., Release of Free Fatty Acids Following Trauma, *J. Trauma*, 1969, *9*, (6), 692–702.
120. WEBSTER, M. E., WILLIAM, R. C., Significance of the Kallicrein-Kallidinogen-Kallidin System in Shock, *Amer. J. Physiol.*, 1959, *197*, (3), 406–412.
121. WEIL, M. H., SCHUBIN, H., *Diagnosis and Treatment of Shock*, Williams and Wilkins Comp., Baltimore, 1967.
122. WESTERFIELD, W. W., WEISINGER, J. R., FERRIS, B. G., HASTINGS, A. B., The Production of Shock by Kallicrein, *Amer. J. Physiol.*, 1944, *142*, (5), 519–523.
123. WIEDMEIR, V. T., STEKIEL, W. J., SMITH, J. J., The Mechanism of Colloid Shock, *Research New York*, 1970, *7*, (6), 716–730.
124. WINTROBE, M. M., *Clinical Hematology*, Lea, Philadelphia, 1956.

8
METABOLISM

There is a group of organs whose function is the conversion of potential to kinetic energy.
This kinetic system includes the brain, thyroid, suprarenals, muscle and liver.

GEORGE W. CRILE, 1914

8.1 METABOLIC PRIORITIES IN STATES OF SHOCK
8.2 ENERGY YIELDING FUNCTION
8.2.1 Cellular metabolic typology
8.2.2 Oxygen and carbon dioxide cycle
8.2.3 Proton cycle; acidosis and alkalosis
8.2.4 The redox systems and adenylates
8.2.5 Electrolytes
8.2.6 Vitamins and enzymes
8.3 METABOLIC LINES
8.3.1 Organic fuels
8.3.2 The muscles
8.3.3 The liver cell and the glycostat
8.3.4 Adipocyte and the lipostat
8.3.5 The nitrate functions
8.3.6 *L*-tyrosine metabolism
8.3.7 Regulation of the enzymatic chains

8.1 METABOLIC PRIORITIES IN STATES OF SHOCK

In the human organism, shock may cause the most intense metabolic disturbances, which do not, however, occur with the same intensity, at the same time, in the different viscera because hypoxia is not equal throughout the human body. In the make up of the body, however, metabolic alterations continue for a long time, and are progressive and total (summative).

Hypoxic distress is of variable intensity even within various areas of the same organ. Significant differences may be found on determining the parameters of the acid-base balance in the venous flow of the various organs, that is, the particular output of catabolites characterizing each organ [6]. The events take place as if the active space of the macrosystem is traversed by actual *metabolic waves* (Şuteu et al., 1975).

Metabolic reorganization is governed by the peripheral oxygen deficit and controlled by hormonal messengers; hence, many investigators consider them as exclusively and secondary to rheodynamic failure in the microcirculation. (The classical expression is: metabolic disturbances are the 'shadow' of rheologic events).

Nevertheless numerous arguments show that the metabolic events in shock are not altogether subservient to perfusion. In point of fact, cellular metabolism is reorganized according to its own laws in the various stages of shock, taking on specific characteristics that in fact determine the evolution of the stages.

Even if not yet defined, the borderline between reversibility and irreversibility may be better understood at metabolic level (Schloerb, 1969; Shires, 1967) [64, 69].

Although insufficient oxygen supply to the cells is actually the *primum movens*, chemical messengers and bacterial toxins insert themselves directly into the cells and the enzymatic chains, *independent of the gas supply*, perform a readaptation according to the principles of autonomic metabolic regulation [7]. Following different experiments in which hypotension was produced by haemorrhage and by ganglioplegics (up to the same BP values) there was an increased adenylate consumption and an accumulation of lactate only in the case of haemorrhagic hypotension [61], proving the direct insertion into the intracellular enzymatic network of the catecholamines released in haemorrhagic shock (and not in hypotension caused by ganglioplegics, when catecholamines are not released into the blood stream). Similarly, in bacteriemic shock, lactacidaemia is more acute than when caused only by hypoxia [62]. Actually, direct metabolic blocking of aerobic glycolysis takes place due to the bacterial toxins and/or their degradation products.

The pH_a, P_aO_2 and P_aCO_2 triad values form the *gas index* and the triad of lactate, inorganic phosphate and plasma α-amino acid concentrations represent the *biochemical index* of the state of shock. At one medical centre it was observed that in bacteriemic and haemorrhagic shock changes in the biochemical index preceded, on average by 50 minutes, the haemodynamic alterations, expressed by the pulse rate, BP and central venous pressure. In the autoanalyzer the three prognostic markers of the biochemical index can be determined in only 15 minutes [67]. The gas index values precede the clinical haemodynamic signs but come after changes in the biochemical index (Kinney, 1967; Polonovski, 1969; Schumer, 1968) [25, 56, 66].

Energy metabolism is also reflected in the general parameter of temperature. A fall in temperature is indicative of a reduced total energy metabolism. The temperatures in the various viscera may behave differently. In tourniquet shock, for instance, the temperature in the liver and brain of mice rapidly falls before the blood flow is reduced [20]. Therefore, the inhibition of carbohydrate oxidation starts before the diminution of perfusion. Hence, temperature may be considered to reflect reliably the functional energy-yielding complex of the macrosystem, being closely correlated with the intensity and severity of shock.

The metabolic characteristic of shock states is a general hypercatabolism, whose intensity and duration depend upon the gravity of the initial lesion. This hypercatabolism persists for a variable period after the critical moments: several months in burns, several weeks after a perforated ulcer, etc.

Starting from a daily fluctuation in the body weight of the normal adult of ± 1%, major surgery may bring about a loss of 2 to 5%, and injuries and infections of over 10%. Resting metabolism is increased by 10 to 25% in a single diaphyseal fracture, by 30 to 60% in a medium burn, and by 40 to 100% in peritonitis [56].

8.2 ENERGY YIELDING FUNCTION

Being the result of an energy conflict between the human organism and the ambient factors, shock stresses *the general yielding function* whose substrate and mechanism are metabolic. The energy apparatus of the macrosystem includes the mechanisms of oxygen and organic fuel supply to the cell, the oxide reducing enzymatic chains and the mechanisms utilizing the energy produced.

In the *chloroplast* of the plant cell a porphyrin with a magnesium redox system 'breathes' carbonic gas which it then 'packs' into carbohydrates by the following reaction:

$$6CO_2 + 6H_2O + 688.5 \text{ Kcal (free solar energy)} = C_6H_{12}O_6 + 6O_2 \quad (1)$$

In the *mitochondria* of the animal cell a porphyrin chain whose redox system consists of iron atoms, does not breath CO_2, but oxygen. Reaction (1) is reversed and offers more than 40% of its energy reserve, hidden in the organic substrate of its carbohydrates.

Oxygen, indispensable for the function of the human cell, is mainly responsible for taking up the electrons circulating on the intramitochondrial cytochrome belt, which, finally, joining a pair of protons, produces about 300 ml water daily for a human body of average weight. The protons of metabolized organic substances are thus discharged, biologic oxidation (cellular respiration) actually being an uninterrupted *dehydrogenation*.

In the absence of oxygen the cells of the human organism offer very little resistance and try to procure their energy by fermentation. The phylogenetically older enzymatic fermentation system is located in the *cytosol* which offers less energy and for a shorter interval, because it loads the cell with protons that accumulate, while waiting for their final obligatory binding to oxygen. All the human cells have an enzymatic fermentative apparatus in the cytosol but not all are also equipped with the mitochondrial systems of biologic oxidation.

8.2.1 CELLULAR METABOLIC TYPOLOGY

According to their enzymatic equipment the cells have been classed by Laborit [29, 30] into three metabolic types:

Cellular metabolic type A, the oldest phylogenetically, involves the presence of glycolysis (EM cycle) and direct oxidative pathways (pentose shunt) in the absence of intramitochondrial cycles. The cells with this enzymatic type in the cytosol are resistant to hypoxia and carry out fundamental functions in the economy of the organism like haemostasis, phagocytosis, hormonogenesis, etc., showing a predilection for reduction and synthesis.

Cellular metabolic type B, rich in mitochondria, carries out oxidation and phosphorylation by a complete Embden-Meyerhof-Krebs cycle and an oxidation-phosphorylation chain. These cells are very sensitive to hypoxia and have an essential role in the production of large amounts of energy, with output superior to that of the enzymatic processes.

Cellular metabolic type C is represented by well-balanced cells, possessing almost all the metabolic pathways.

The metabolic wastes of the A type cells are further utilized in the mitochondria of type B cells, attaining a bioenergetic best functional association: neuroglia-neurone, nodal myocardial tissue-contractile myocardial tissue, Kupffer cell-liver cell, etc. (*see* Section 9.3.1 and Figure 8.1).

In the absence of oxygen, type B cells that can no longer utilize their

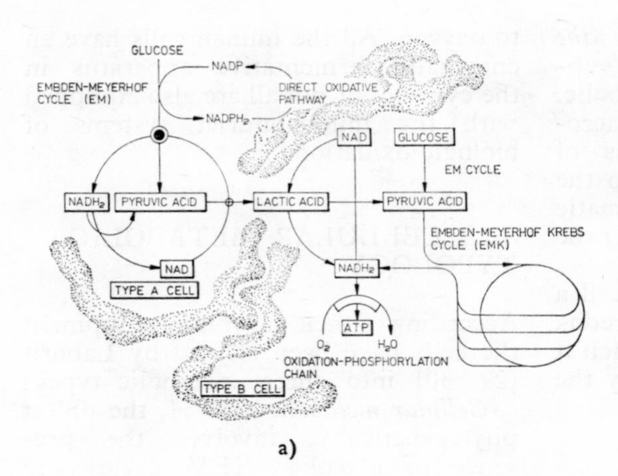

Figure 8.1:
a — normal metabolic coupling between cells of the A and B type; lactic acid produced by A cells is burned in B cells that possess a Krebs cycle and an oxidation-phosphorylation chain; *b* — initial catecholamines stimulate the total energy producing catabolism of carbohydrates in both A and B cells; *c* — hypoxia blocks oxidation-phosphorylation and then the accumulation of $NADH_2$ stops the Krebs cycle: the accumulated lactic acid will inhibit the energogenetic enzymes in both cell types.

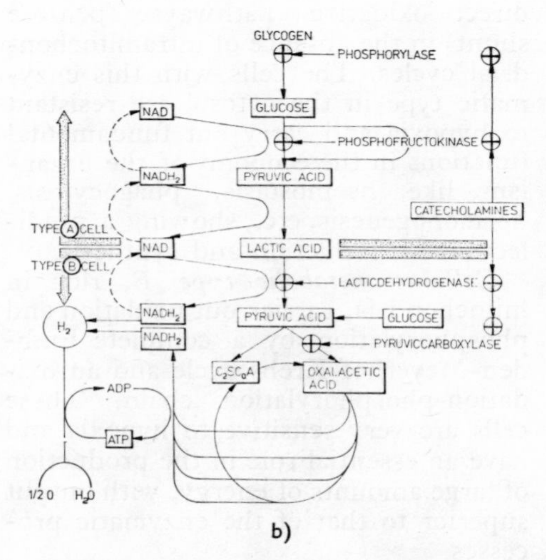

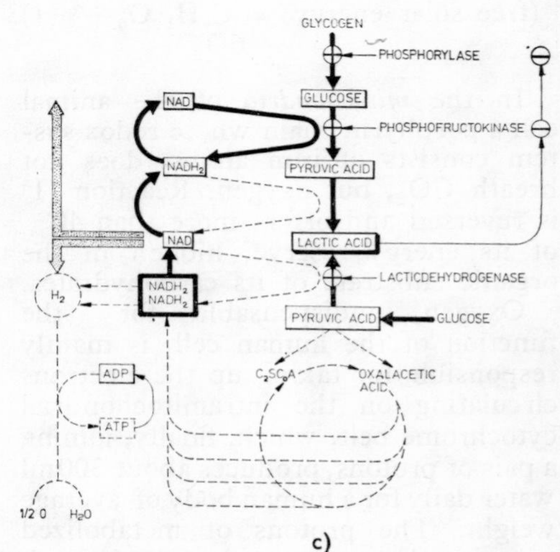

own active acetate (C_2SCoA), i.e. can no longer *breathe*, are equally loaded with the waste material of the A cells. Mitochondrial oxidation is blocked. In this way, large amounts of lactate accumulate intra- and extracellularly, inhibiting fermentation in all three cell types. When glycolysis ceases in all three cell types, with excessive accumulation of protons, cellular metabolism breaks down completely and the rheodynamic and tissular consequence will be: sludging, œdema of the neuroglia, depolarization of the intima cell (followed by coagulation), blocking of the RES, corrosion of the lysosomal sacs with liberation of hydrolases, etc.

Between fermentation and respiration there is a fundamental metabolic balance brought about by a negative retroaction: exacerbation of respiration inhibits fermentation *(Pasteur effect)* and hyperfermentation (hyperglycolysis) inhibits respiration *(Crabtree effect)* [52].

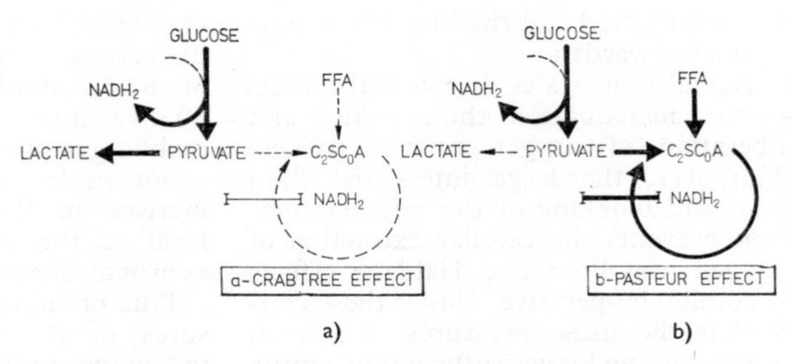

Figure 8.2:
a — fermentation activity is predominant; *b* — respiration is predominant.

In states of shock, hypoxia forces the *Crabtree effect* since, in the absence of oxygen, enzymatic activity in the fermentation process is for a time, and in many tissular spaces, the only possible way of producing energy (Figure 8.2).

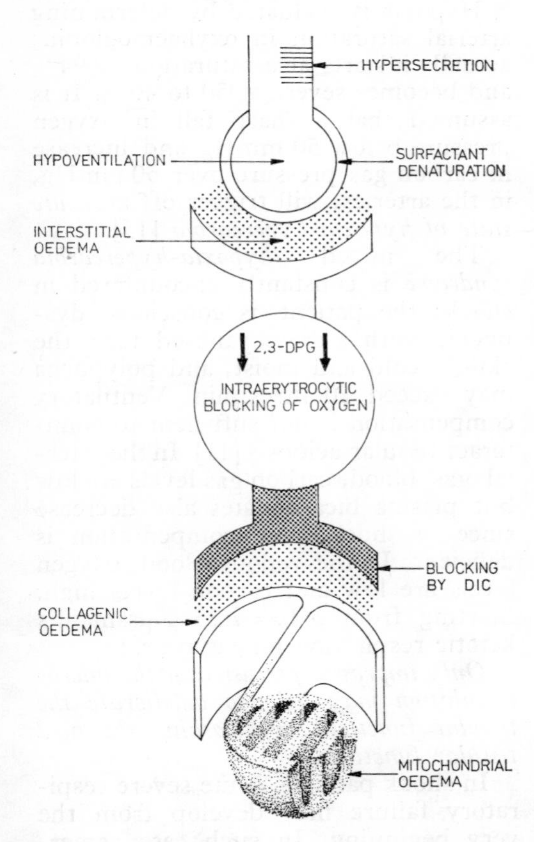

Figure 8.3. The morphological and functional obstacles in the way of oxygen from alveoles to mitochondrial enzymes.

8.2.2 OXYGEN AND CARBON DIOXIDE CYCLE

At rest the total oxygen consumption per minute is 300 ml, in a normal human body. Actually, through the 90 m² of the alveolar-capillary surface about 1 200 to 3 600 ml of oxygen diffuse per minute under conditions where there is a physiologic dead space of 25% and a ventilation/perfusion ratio of 0.8, offering a large safety margin [9]. Notwithstanding the excessive supply of oxygen to the tissues, in shock, intracellular oxygen deficit is the essential loop in the pathogenic chain [9, 15, 76, 78]. Numerous functional and morphologic obstacles are raised in the pathway of oxygen from the alveoli to the mitochondria (Figure 8.3). The disturbances arising on passing the alveolar-capillary barrier are discussed in section 10.4.

It has been demonstrated, however, that the marked fall in P_aO_2 does not in itself generate peripheral cellular damage. Following a prolonged fall of P_aO_2 by 75%, the cardiac output only decreases by 30% and peripheral resistance increases by 35% [26, 38, 39]. In shock, it seems that the peripheral release and reception of oxygen are at first obstructed and the blood transport only afterwards.

Actually, in states of shock the RBC proves inadequate in the retention and liberation of oxygen to the tissues. Moreover, the large interstitial fluid pool and blocking of the microcirculation prevents the cellular extraction of oxygen, the Bohr and Haldane effects becoming inoperative. Under these conditions the usual pressures of arterial oxygen are no longer sufficient to supply the tissues. Hence, only high pressure oxygen therapy and the intravascular administration of peroxides are able to raise the pressure of oxygen in the tissues (measured by polarographic electrodes) [1].

Normally, the tissues extract 25% of the oxygen brought by the RBC to the periphery; in shock, however, there is a temporary extraction of 60% or even more. But the tissular mechanisms of oxygen extraction are rapidly exhausted, especially by intraerythrocytic blocking *(see* Section 7.2.1).

Intracellular hypoxia soon becomes the major stimulus of the reorganization of the enzymatic chains.

The accumulation of carbon gas runs parallel to O_2 deficit. Intracellular CO_2 pressure is an important regulator of hydrocarbonate metabolism; when the intracellular concentration of carbon gas increases, neoglucogenesis from glycerol is stimulated in the liver cell [36]. Carbon dioxide is also used in the synthesis of oxalacetate from pyruvate (by malate pathway), thus stimulating the Embden-Meyerhof-Krebs cycle. But the fleeting positive metabolic effects of carbon dioxide are soon overcome by the drawbacks of its excessive accumulation, which modifies the intracellular *p*H.

If the normal P_aCO_2 of 40 mmHg is exceeded by only 2 mmHg, the rhythm and depth of breathing increases twofold in order to eliminate the dangerous carbon excess. At P_aCO_2 of 60-80 mmHg, ventilation reaches 60 litres/min; starting from these values carbon gas pressure begins its narcotic action on the cortical neurons (the initial increase in P_aCO_2 is only felt at the level of the bulbopontine respiratory neuronal centres).

Plus or minus variations in the pressures of the respiratory gases result in pathophysiologic situations with a separate or associated clinical manifestation as shown in Table 8.1. These disturbances are usually mixed.

Hypoxia is evaluated by determining arterial saturation in oxyhaemoglobin; actually it starts at a saturation of 94% and becomes severe at 50 to 40%. It is assumed that a sharp fall in oxygen pressure below 50 mmHg and increase in carbon gas pressure over 50 mmHg, in the arteries, will trigger off an *acute state of hypoxia-hypercarbia* [15].

The mixed *hypoxia-hypercarbia syndrome* is constantly encountered in shock; the patient is conscious, dyspnoeic with a grey cyanosed face, the skin is cold and moist, and polypnoea may exceed 20 litres/min. Ventilatory compensation is not sufficient to counteract tissular acidosis [11]. In the arterial gas blood, carbon gas levels are low but plasma bicarbonates also decrease since in shock renal compensation is deficient. In the venous blood, oxygen levels are low and carbon levels high. Starting from $pH = 7.25$ Kussmaul's ketotic respiration may appear.

Only improved perfusion of the microcirculation may help to ameliorate the tissular functions, by washing the acid tissular functions.

In shock patients, acute severe respiratory failure may develop from the very beginning. In such cases emergency treatment must be instituted as, according to Mundeleer, 'such a patient

Table 8.1 Some of the clinical features of the imbalance in the blood concentrations of respiratory gases

Hypercarbia	Hypocarbia	Hypoxia	Hyperoxia
Hyperpnoea	Hypoventilation	Tachypnoea	Variable
BP rises	BP falls	BP rises, then falls	BP falls very low
Drowsiness → coma → cardiac arrest	Tetany → coma → ventricular fibrillation	Convulsions → coma → ventricular fibrillation	Convulsions → coma → cardiac — arrest
Hot, congested skin	Pinkish, pale cold skin	Cyanosis	Pallor
Perspiration, hypersecretion	Dry skin	Perspiration	Variable
Acid urine	Alkaline polyuria		
Ventilation acidosis	Ventilation alkalosis	Induced metabolic acidosis	
Cellular hyperhydration			Cytotoxicity

has no time to wait; the game of life and death is played within a few minutes', and the role of the lung in this game is of prime importance *(see* Section 10.4).

8.2.3 PROTON CYCLE; ACIDOSIS AND ALKALOSIS

The first elements connected with the acid-base balance were furnished by determination of the decrease in $NaHCO_3$ in the faeces of cholera patients (O'Shaughnessy, 1830). Then there followed the works of Henderson (1909) and of Van Slyke (1936). The *p*H of each intracellular organelle has now been determined [18, 44, 79].

The acid-base balance depends on the proton concentration. Shock is characterized by an accumulation of protons, a dangerous condition rich in protons and very poor in electrons *(pronorhoea* and *electronorrhage).* This accumulation of protons, expressed by a decrease in the symbol of the acid-base balance *(p*H*)* also becomes manifest in the redox balance *(r*H*)* by a deficit of the electron donors (reducing substances) whose therapeutic role is well known.

All types of acid-base imbalance may be encountered in shock *(acidosis* at *p*H < 7.35 and *baseosis* at *p*H > 7.55). Characteristic, however, is the deep metabolic acidosis that is difficult to reduce because the compensatory systems are also affected (Figure 8.4).

The sources of proton accumulation in shock are varied. Intense cellular metabolism produces the non-volatile organic acids or mineral acids (the catabolism of sulphurated amino acids produces sulphuric acid; phosphoproteins and nucleoproteins release phosphoric acid and uric acid; lipids produce FFA; carbohydrates produce lactic and pyruvic acid). Although organic, lactic acid is a fairly 'strong' acid, at 26°C it has a *p*H of 3.73 [18].

To the endogenous protons are added the exogenous ones brought about by massive blood transfusions (conserved in a citrate solution at *p*H up to 6.20) [14], or by ethylene or propyleneglycol intoxications. The loss of bases by

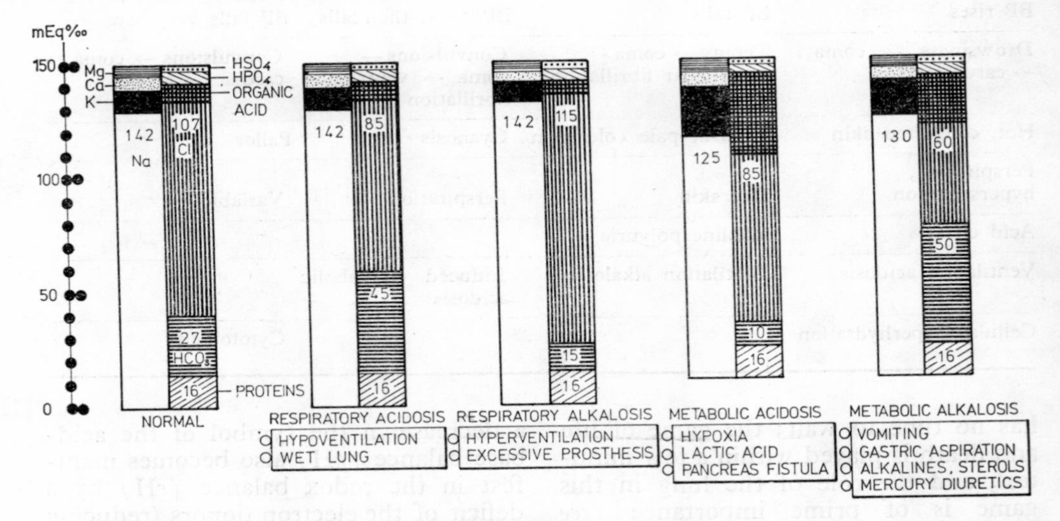

Figure 8.4. Blood ionograms in several types of acid-base imbalance. Note the overall decrease o electrolytaemia in metabolic imbalance.

diarrhoea and postpyloric digestive fistulae further enhance the acidotic state of the organism in shock.

Owing to its sensitivity and mobility in the plasma, the carbon acid/bicarbonate ratio is the main indirect marker of proton circulation in the human body fluids. Its value,

$$\frac{1.35 \text{ mEq per thousand}}{27 \text{ mEq per thousand}} = \frac{1}{20} \quad (2),$$

must be almost constant for the *p*H to vary within physiologic limits (7.40 ± ± 0.04) [18, 19, 48, 56].

Metabolism is the great generator of protons, and the lung and kidney are the main correctors of the *p*H; the lung modifies carbon gas concentrations and the kidney those of bicarbonate.

The presence in all states of shock of *metabolic acidosis* imposes a correcting alveolar hyperventilation. In rare instances, however, excessive hyperventilation, but only in the initial stages of shock, manages to neutralize acidosis, or even the transitory onset of respiratory alkalosis. In most cases the 'wet lung' in shock soon proves inadequate to correct metabolic acidosis, to which a grave *respiratory acidosis* is added (*see* Section 10.4).

The kidney, initially, tries to economize on bicarbonate, eliminating the protons in acid urine; this well-known mechanism, of particular value, ceases, however, when the BP in the renal artery falls below 70-60 mmHg, as in shock, when, as a rule, the kidney is affected.

The accumulation of protons in shock is classed in the large group of *metabolic acidoses*, that includes four categories:

- diabetic ketoacidosis;
- acidosis due to the retention of non-volatile acids;
- acidosis due to alkaline losses;
- acidosis due to an exogenous supply of acid substances.

In shock, all four types of metabolic acidosis occur in different proportions. Specific of acidosis in shock is, however, the accumulation of lactic acid, known to be a great proton donor.

Almost all metabolic intracellular lines are directed in shock towards this 'blind alley' (Fig. 8.5). Pyruvic acid can no longer take the decisive step as in the usual aerobic metabolism: the conversion into active acetate and entrance into the Krebs cycle. In shock, pyruvic acid has a single choice: conversion into lactic acid, whereas, by contrast, all the intramitochondrial metabolites (eventually also deriving from FFA) return to lactic acid by pathways that are but seldom used under normal conditions, but made permeable under shock conditions.

Schumer considered the fundamental 'metabolic anoxic block' in shock to be situated between pyruvic acid and active acetate [66, 67]. It has been assumed that lactic dehydrogenase no longer affects the dehydrogenation of lactic acid produced in excess by *A* cells but also by *B* and *C* cells *(see* Section 9.3.1).

The huge accumulation of lactate has a rheodynamic cause as well: the slow flow in the microcirculation no longer washes the tissular atmosphere rich in protons. Hence, shock acidosis has been designated by several different terms: *hypoperfusion acidosis*, *hypoxic acidosis*, *stagnant lactic acidaemia* [42], *hypooxidosis* [28], *enzymatic energogenetic acidosis* [6].

In the slightly increased initial plasma concentrations, lactic acid is a mild vasoconstrictor of the smooth muscle cell, thus accentuating the deficit in

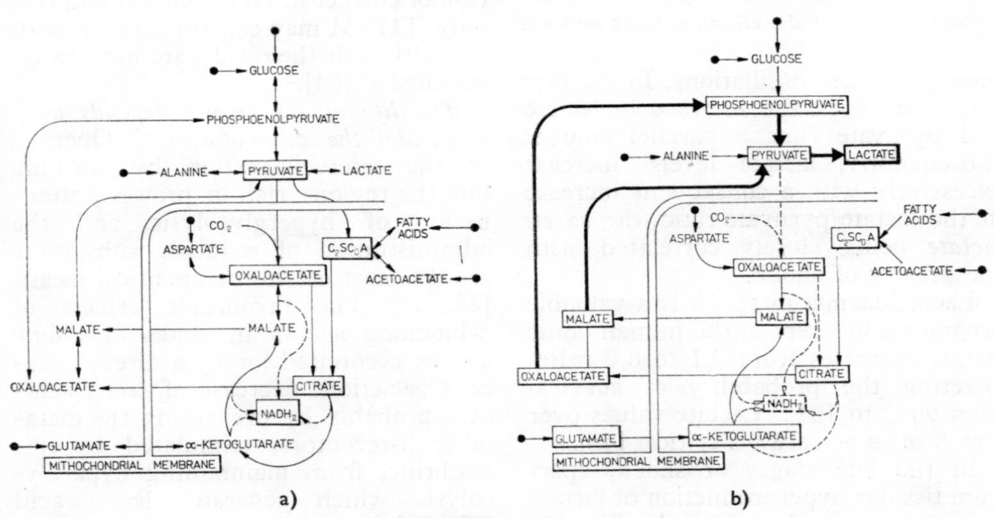

Figure 8.5:

a — normal cell: glucose, amino acids and fatty acids enter the Krebs cycle accumulating $NADH_2$ which will be dehydrogenated by oxidation-phosphorylation; b — in shock the Krebs cycle is blocked and all the metabolic pathways are reversed towards the accumulation of lactic acid.

oxygen supply to the cell, i.e. the very cause of lactic acid accumulation. In higher concentrations, which are rapidly reached, the condition rich in protons will now have an uncoupling effect on the action of catecholamines upon the α- and β-receptors of the microcirculation. The initial vasoconstriction and terminal refractory vasodilatation are progressive situations characteristic of shock in which variations in lactic acid concentrations impose their effects. Monitoring of plasma *p*H permits correction of these oscillations. In the first stages of shock the increase in lactate and pyruvate runs a parallel course; subsequently, lactate levels increase excessively with a consequent increase in the lactate/pyruvate ratio, the *excess lactate* being closely correlated with the gravity of shock.

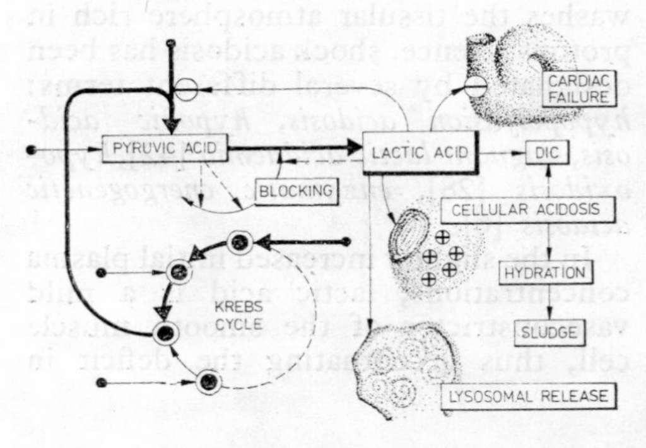

Figure 8.6. The cellular effects of lactic acidosis.

Lactacidaemia in shock is a valuable prognostic marker; in the human clinic lactate increases from 2.1 to 8.0 mEq, lowering the probability of survival from 90% to 10%. Lactate values over 7 to 8 mEq are always critical [48].

In the late stages of shock, apart from tissular hyperproduction of lactate there appears a hepatic subutilization of lactate, by hypoperfusion, the Cori cycle no longer being operative.

High lactic acid levels may, however, coexist with metabolic alkalosis (the administration of alkalines) or respiratory alkalosis (hyperventilation).

Among the noxious consequences of acidosis the following should be noted asymmetrical opening of the microcirculation dam, myocardial toxicity, cellular hyperhydration, precipitation of the sludge phenomenon and DIC, inhibition of the synthesis and dehydrogenation enzymes, as well as the liberation of lysosomal hydrolases (Figure 8.6).

Nevertheless, *acidosis in itself is not dangerous.* A low *p*H is only a symbol, and correction of the acid blood *p*H by simple alkalinization has never influenced the patient's recovery [47]; *p*H values must obviously be brought within physiological limits, bearing in mind, however, that the *p*H only indirectly reflects the effects of hypoperfusion, hypoxia and hyperglycolysis, which have to be corrected. Moreover, it is known that $NaHCO_2$ cannot correct intracellular *p*H and that only THAM may control cellular acidosis although the results are not always conclusive [64].

The life of the tissues depends upon removal of the causes of acidosis. Opening of the microcirculation by washing out the regions rich in protons, attenuation of hyperglycolysis, and the administration of reducing substances are the most useful therapeutical means [46, 59]. The favourable effects of β-blocking agents in states of shock can be accounted for by a direct intracellular action: decrease of lactic acidosis, probably by preventing the metabolic β-receptors, stimulated by epinephrine, from maintaining hyperglycolysis which generates lactic acid [40, 73].

Since it has been concluded that lactic acidosis is not in itself dangerous,

Ringer's lactate solution has again been administered in states of shock [16, 69, 79, 81]. Actually, when the circulation in the liver is within normal limits lactate is an excellent precursor of neoglucogenesis owing to repermeabilization of the Cori cycle [65].

Therefore, systemic *p*H may be considered only as a rough indicator of the regional cellular acid-base balance, and it is of importance in shock to determine separately the parameters of the acid-base balance in the venous pedicles of the key viscera [6]. This multivisceral venous monitoring will supply better evidence of the correcting role of the kidney in the first stage of shock and especially of the liver in buffering portal acidity, a role also maintained in the late stages of shock *(see* Section 2.9).

8.2.4 THE REDOX SYSTEMS AND ADENYLATES

Proteins and lipids are subordinated to the catabolism of carbohydrates since they are inserted at key points in their metabolic pathways (Figure 8.7).

In its overall mechanism the production of energy has a single final pathway strictly depending upon oxygen cellular supply.

As a rule, the hydrogen atoms charged with nicotinamidadenindinucleotide (NAD) and derived from the glycolysis and oxidative cycles penetrate into the mitochondria in the course of oxidation-phosphorylation. After a series of oxidations along the flavinadenindinucleotide-cytochrome chain, small but permanent amounts of energy are released; the hydrogen protons encounter oxygen which in the meantime has 'stolen' the electrons of the last cytochrome (Figure 8.8). In the oxidation-phosphorylation chain, a group of redox systems, among which serotonin [10] and the quinones may be listed, are also active [80]. The energy, liberated discontinuously, is taken up by ATP, the richest in energy of the adenylates. For each pair of hydrogens that pass through the oxidation-phosphorylation chain 3 ATP are loaded with energy.

The energy efficiency of biologic oxidation is decided in the course of oxidation-phosphorylation. Under normal respiratory conditions, the total oxidation of hexoses is:

$$C_6H_{12}O_6 + 6O_2 = 6CO_2 + 6H_2O + 38(39)\ ATP \quad (3)$$

The energy producing reaction, which actually takes place in the cytochromes, will be:

$$2H^+ + 1/2O_2^{--} \rightarrow H_2O + \text{energy} \quad (4)$$

The great dehydrogenator is the Krebs cycle, which offers four pairs of protons for each active mole acetate, according to the following reactions:

$$C_2SCoA + 3H_2O = HSCoA + 2CO_2 + 4H_2 \quad (5)$$

$$\begin{aligned} 8H^+ + 4O^{--} &= 4H_2O \\ 12\ ADP + 12\ HPO_4^{--} &= 12ATP \end{aligned} \quad (6)$$

In the absence of oxygen, energy is obtained by fermentative activity, as follows:

$$C_6H_{12}O_6 = 2\ C_3H_6O_3 + 2\ ATP \quad (7)$$

In order to obtain by enzymatic activity (fermentation) the same energy as that produced by total utilization of hexoses, 15 times more hexose molecules are necessary, and the rate of the reaction must likewise be 15 times greater.

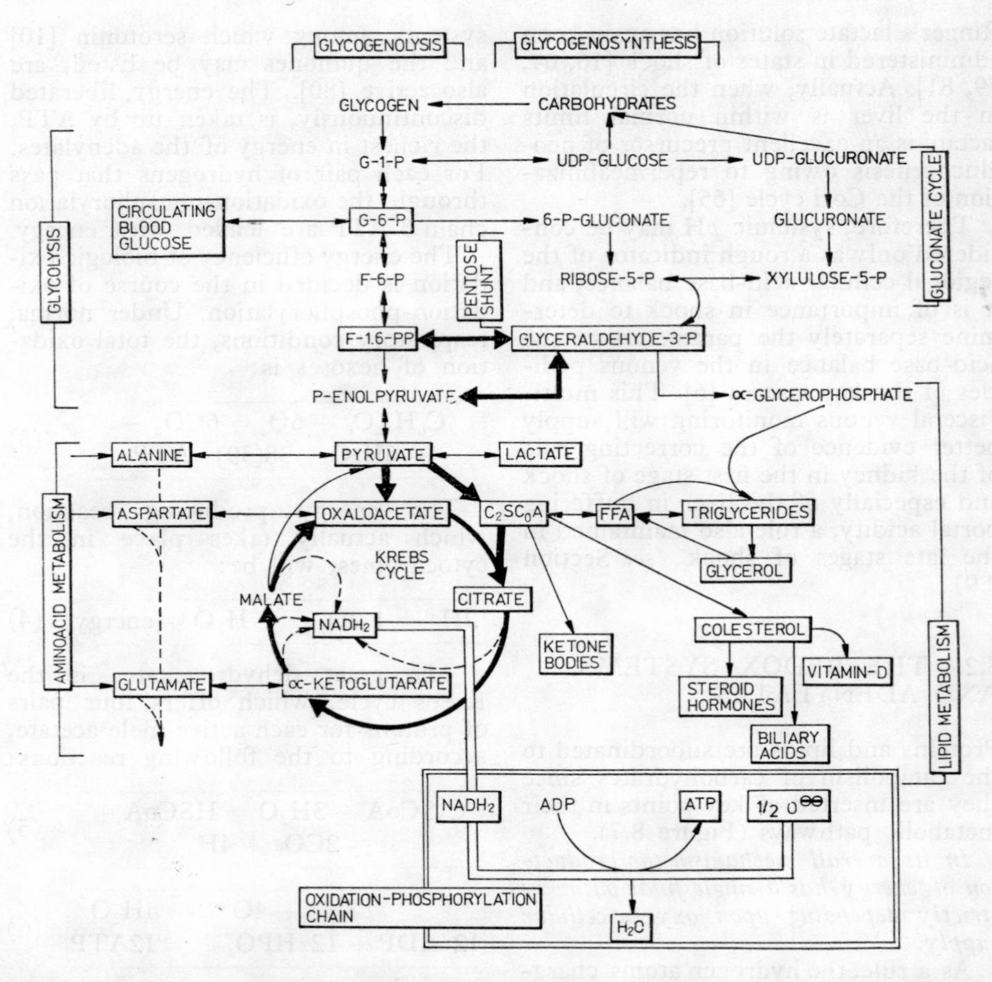

Figure 8.7. The insertion of amino acids and fatty acids on the energogenetic chain of carbohydrates.

This is what accelerated hyperglycolysis and excessive neoglucogenesis attempt in shock.

But anaerobic glycolysis brings about the accumulation of huge and irreducible amounts of lactate because this reaction is shifted to the right:

$$\text{Pyruvic acid} + NADH_2 \rightleftharpoons \text{lactic acid} + NAD \qquad (8)$$

As $NADH_2$ cannot be discharged in oxidation-phosphorylation chain (stopped by the lack of oxygen), it is accumulated in exaggerated amounts, inhibiting citratesynthase, closing the fundamental chemical feedback between fermentation and respiration *(see* Section 3.8.2 and Figure 8.9).

Thus, the excess $NADH_2$ arrests the course of the Krebs cycle, which is

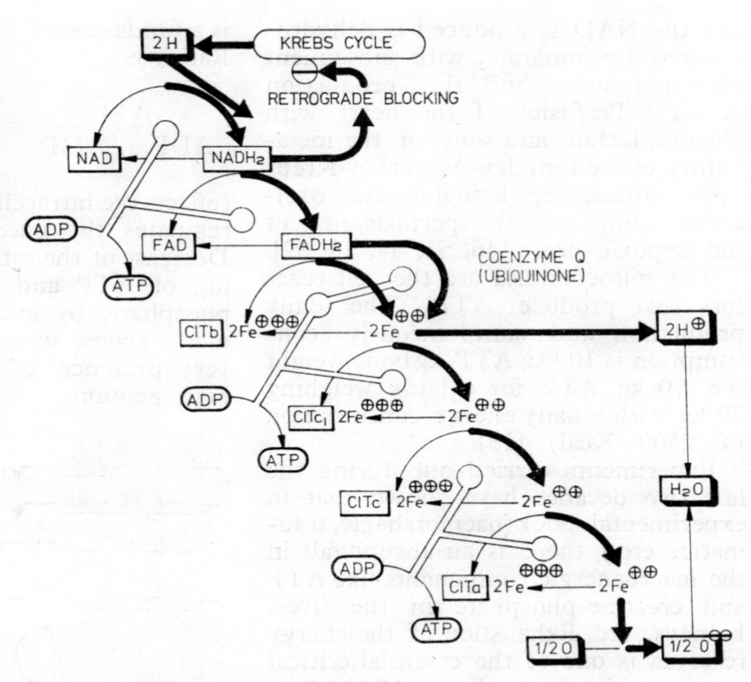

Figure 8.8. The six classically admitted steps of the oxidation-phosphorylation chain. When the last electron acceptor (oxygen) is missing all the loops of the system are stopped retrogradely and $NADH_2$, with an inhibitory role on citratesynthase, accumulates, blocking the Krebs cycle.

in fact its greatest producer. The Crabtree effect, the inhibition of respiration by fermentation characteristic of the shock cell, as well as of the cancerous cell, takes place. Finally, $NADH_2$ also inhibits glycolysis, stopping the production of lactic acid but also the last energy source of the cell [22].

Evidence was obtained by means of radioactive labelled metabolites that in shock the activity of the oxidizing process decreases and lactate accumulates. Similarly, it was found that phenoxybenzamine and cortisone reactivate the oxidizing process, with possible intrinsic metabolic effects [16].

In the hypoxic myocardium there are several bypaths by means of which ATP can still be produced under shock conditions. Glycolysis reaches a peak

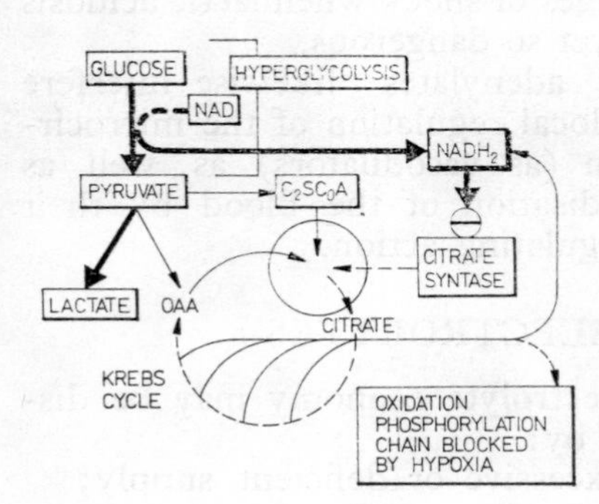

Figure 8.9. $NADH_2$ accumulated by hyperglycolysis and from the Krebs cycle can no longer be oxidized by oxidation-phosphorylation and inhibits citratesynthase; the lactic waste increases as the Krebs cycle has been stopped.

and the $NADH_2$ produced is dehydrogenated, by fumarate, with subsequent phosphorylation and the production of ATP. Perfusion of the heart with glucose, lactate and some of the metabolites of the Embden-Meyerhof-Krebs cycle (fumarate, ketoglutarate, oxalacetate) improves the performance of the hypoxic heart [46, 53, 54, 75, 82].

The mitochondria are the cell reactors that produce ATP. The daily production and simultaneously consumption is 1000 g ATP/kg body weight (i.e. 70 kg ATP for a body weighing 70 kg with a daily energy consumption of 2500 Kcal) [18].

Experiments carried out during the last two decades have shown that in experimental shock (haemorrhagic, traumatic, etc.) there is an abrupt fall in the macroenergic components like ATP and creatine-phosphate in the liver, kidneys, etc. Exhaustion of the energy reserves is one of the essential critical events in shock (Le Page, 1946; Staples, 1969) [34, 72].

The energy produced by adenylates is actually a thermodynamic function of the balance of the hydrolysis reaction splitting the pyrophosphate bond, doubly ensured by quantic participation of the *3d* next to the *3p* molecular orbitals.

ATP is the source of immediate energy for most of the endergonic systems of the cells. Quantitative determination of ATP in the various tissues and in different types of shock have shown net variations in terms of the methods used and the organs investigated [20, 34, 72]. The organs most affected are the liver and the kidney, then the lung. The results obtained in the heart and brain are contradictory; however, the energy reserves of these regions appear to be protected.

The molar intracellular concentration of adenylates ATP + ADP + AMP) is a fundamental biochemical constant. Ratio:

$$\frac{ATP}{ADP + AMP} = K \text{ (constant)} \qquad (9)$$

reflects the intracellular energy level and regulates the energy metabolic lines. Decrease of the ratio demands a reloading of ATP and AMP with inorganic phosphate, by means of certain enzymatic stages which depend upon the very presence of ADP and AMP in high amounts.

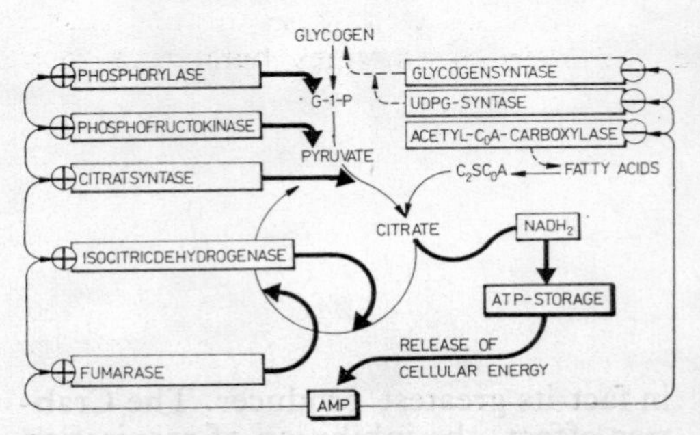

Figure 8.10. The stimulation of energy production and arrest of synthesis tends to build up again the ATP stock. Subsequently, neoglucogenesis will also be resorted to.

In shock, the rapid consumption of ATP reserves and creatine phosphate stimulates the key-enzymes of energy genesis and inhibits synthesis (Figure 8.10). These events take place in the first stages of shock when lactic acidosis is not yet so dangerous.

The adenylates likewise interfere in the local regulation of the microcirculation (as vasodilators) as well as in fluidisation of the blood by their anticoagulating action.

8.2.5 ELECTROLYTES

The electrolyte economy may be disturbed by:

- an excessive or deficient supply;

- excessive losses (diarrhoea, vomiting, perspiration, fistulae);
- intercompartmental displacement (as a rule under hormonal influence).

These possibilities of influencing the electrolyte balance are always encountered in the surgical patient [35]. It should be recalled that the daily volume of digestive secretions is double that of the plasma volume and may produce an important electrolytic imbalance, dangerous because of its clinical obscurity (80 mEq potassium are lost within the first 24 to 48 hours after a medium operation). At present it is recognized that the *postinjury antidiuretic state* (Moore and Shires) starts with aldosterone hypersecretion and hydrosaline osmotic retention approximately 2 to 4 hours after injury [42, 69].

The 'discrimination' exercised by the cellular membrane with regard to the two main alkali has a deep functional significance (Moore) [42]. Sodium gives 95% of the cation osmolarity of the extracellular space, exercising its pressor role in the circulation of fluids, and potassium promotes intracellular osmosis and especially the condition necessary for the main energy-producing enzymatic chains.

Normally, the pentose cycle controls repolarization of the membranes, supplying the cell, the electrons and the protons required for a good functioning of the electrolyte transmembrane pumping systems. The stimuli of the pentose pathways increase intracellular potassium (K_i) (lower kaliaemia) and lower intracellular natrium (Na_i) (increase natriaemia). Chlorpromazine, insulin and glucose induce hypokaliaemia and diminish spasm of the smooth musculature by stimulation of the pentose cycle [43].

The ratios K_i/K_e and Na_i/Na_e thus depend upon the intracellular redox potential. In shock, the electron deficit and the accumulation of protons in the cell induce an exodus of potassium towards the interstitium and facilitates the penetration of sodium into the cell (transmineralization). The sodium and potassium pumps (if they exist) are actually redox pumps [30]. The pentose pathways contribute to the potassium pump and the Embden-Meyerhof-Krebs pathways to the sodium pump. Because the latter pathway is arrested in shock, however, the Na pump no longer ensures its elimination from the cell. In bacteriemic shock the membrane free permeability for sodium increases [4]. Sodium is also taken up massively in the collagen 'sponge', aggravating hyponatraemia (*see* Section 6.2). Slowing down of the pentose reduction pathway also rapidly alters the potassium pump, and potassium is lost in the extracellular space (Figure 8.11).

Although *hyponatraemia* is constant in shock it does not reach dangerous levels *per se*. But *hyperkaliaemia* (more than 5.5 mEq in the plasma) is, on the other hand, one of the parameters upon which life depends. Signs of hyperkaliaemia are the first to indicate affection of the myocardium, the most sensitive neurocontractile system: on the ECG tracing the P wave becomes sharper, atrioventricular and/or intraventricular blocks develop followed by bioelectric anarchy with ventricular fibrillation and arrest in diastole at potassium concentrations higher than 7 mEq (up to 13 mEq!). The skeletal musculature and central nervous system are less often affected (paraesthesia, respiratory disturbances, obnubilation, convulsions). Many of the causes of hyperkaliaemia (Table 8.2) are encountered in shock, worsening it.

Among the bivalent cations the role of calcium has been very much discussed (*see* Section 7.4.2). Sometimes the metabolic functions of the cations condition life; for instance, iron in the tetrapyrolic rings of the cytochromes.

Calcium and potassium are necessary for phosphorylase activation by cyclic-AMP [45]; manganese favours fatty synthesis more than magnesium [68]; zinc is useful, especially in local repair processes after extensive burns by stimulating the synthesis of mucopolysaccharides; magnesium is an active protector of the cellular reducing functions [13, 37]. Magnesium is the second intracellular cation, the K/Mg ratio having constant values of between 5 and 6 [23, 24].

Magnesium is an excellent regional vasodilator that exercises its function by a double mechanism: a curare-like mechanism, on the one hand, and blocking of the postsynaptic membrane, on the other (the antagonist of potassium). The protective role of magnesium with regard to the hypoxic cell has been demonstrated on the myocardium: in primates, the association of hypothermia (28°C) and perfusion of the coronaries with a 0.75% SO_4Mg solution permit-

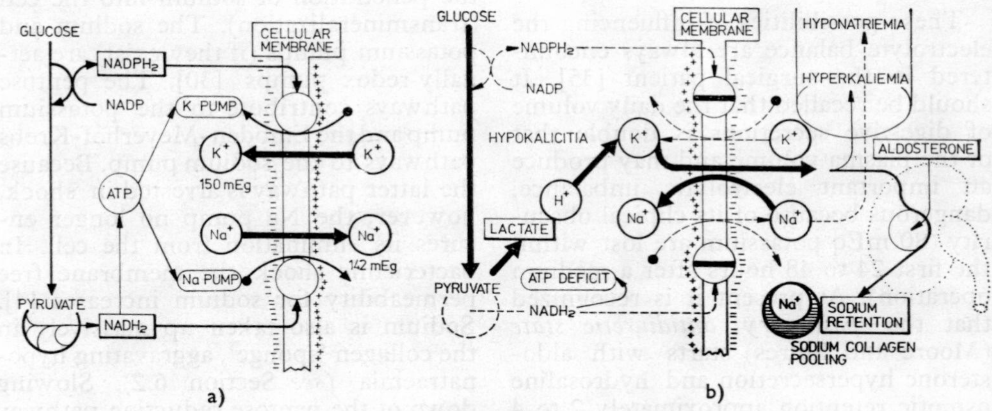

Figure 8.11:

a — normal situation: potassium is pumped and stored in the cell by the pentose cycle and sodium is pumped extracellularly by the energy offered by the Krebs cycle; b — the intracellular accumulation of protons will bring about release from the cell of the dominant cation — potassium. In addition, the normal transmembrane ionic flow slows down (together with the metabolic lines sustaining them) and sodium may enter the cell more easily. Sodium is, moreover, secluded in the collagen space and is stored to the detriment of potassium in the renal tubulocytes, with consequent blood hyponatraemia and hyperkaliaemia.

Table 8.2 THE AETIOLOGIC FACTORS OF HYPERKALIAEMIA *(The conditions more frequently encountered in shock are in italics)*

Causes	Clinical condition
1. Renal diseases	— *acute renal insufficiency* — chronic renal insufficiency — interstitial nephritis — genetic tubulocytic defects
2. Endocrine diseases	— hypoaldosteronism — *adrenocortical failure*
3. Blocking of the pentose pathway	— *metabolic acidosis* — *ventilation acidosis*
4. The administration of preserved blood	— *massive transfusions*
5. Tissular hypercatabolism	— *tissular necrosis* — *haemolysis, proteolysis*

ted clamping of the latter for 90 minutes (*see* Section 5.2). Magnesium ions interfere in a great number of processes that are affected in states of shock: the transfer of phosphates, oxidative-phosphorylation chain, nucleic acid synthesis, muscular contraction, neuronal transmission, active transmembrane transports, etc.

Although Ca^{++} and Mg^{++} have common neuronal stabilizing effects, at the level of the synapses they are antagonists: Ca^{++} helps transmission through the synaptic slit and Mg^{++} blocks it, modifying the presynaptic translation protein allosterically. In addition, Ca^{++} helps the release of norepinephrine, whereas Mg^{++} opposes it [24].

8.2.6 VITAMINS AND ENZYMES

In shock, especially after burns, the blood, tissular and urinary ascorbic acid concentrations are very low. At the same time there is a decrease in the concentration of the B complex vitamins, which are almost all strategical coenzymes with a special role in the energy producing system, some of them being indispensable redox systems [56]. In shock, determination of the plasma vitamin titres does not appear to be of any clinical use; in practice, the large amounts currently administered rapidly cover the acute depletion of the vitamin redox systems generated by shock. The question of vitamin equilibration, if we bear in mind their complex functions in the organism, appears only after recovery from the state of shock proper.

The enzymatic chains give the cellular functional stereostructure that dictates the entire metabolic reorganization of the shock cell (*see* Chapter 9). They are directly controlled by the operons of the genetic code which organizes the antishock plan for them. Since phylogenetically this *antishock plan* in man is not yet consolidated and sufficiently exercised it contains numerous errors, hence antishock action does not always give good results.

In states of shock the determination of intracellular enzymes is only of experimental interest (*see* Sections 9.3.2 and 2.10). In the clinic, enzymes are found in the plasma or urine and today evaluating the enzymatic diagnosis of the severity of shock has become obligatory.

In shock, increase in the plasma of numerous enzymes belonging to various biochemical functions has been reported: lipase, phosphatase, lactic dehydrogenase, amylase, histidin-decarboxylase, transaminase.

Determination of these enzymes may not only point to destruction of the myocardial cells (in cardiogenic shock), of the liver cells (in hepatic injury) or muscles (in crushed tissue shock) but it may also indirectly 'evaluate' cellular 'infection' in bacteriemic shock and cellular distress in shock in general. In man, GOT values were significantly elevated 2 to 5 hours after the onset of bacteriemic shock, being considered by some as a more reliable prognostic marker than are lactacidaemia, acid phosphatase or coagulation tests [71]. Transaminases play an important metabolic role in the shock cell. The transaminase reaction between alanine and pyruvate depends upon oxygen, the same as the pyruvate $\rightarrow C_2SCoA$ reaction or the Embden-Meyerhof-Krebs cycles $\rightarrow$ oxidation-phosphorylation pathway. Similarly, the transamination of oxaloacetic acid from the mitochondria to the cytosol (*via* aspartate), is necessary for neoglucogenesis from lactate [60].

Being protected in their membrane sacs mitochondrial and lysosomal enzymes are released systemically somewhat later in shock.

Of late, the determination of plasma and urinary hydrolases (acid phosphatase, β-glucuronidase, amylase, leucinaminopeptidase, etc.) has become a current laboratory practice, the hydrolase titre being closely correlated with the gravity of shock, since they are the enzymes of anarchic cellular digestion and indicate the critical moment of transition to an irreversible condition (*see* section 9.3).

8.3 METABOLIC LINES

A pattern of the characteristic metabolic cellular reorganisation exists in shock although it is not pathognomonic of *the shock cell.* All hypoxic cells behave in the same way as the shock cells, but *the intensity* of distress is specific to the shock cell. The cellular biochemical disorders are obviously produced by hypoxia due to a perfusion deficit; unsatisfactory irrigation is in fact accompanied by a series of restrictions of which the insufficient supply of oxygen is the most severe but not the only one.

The 'shock plan' of the enzymatic chains is governed directly by the genetic code operons that are the first to receive the signal of alarm of the state of shock. The informational circuit acts quickly throughout the human body. Intracellular events are still very difficult to analyse by conventional laboratory methods applied in case of shock. The biochemical constants in the blood and urine are but an indirect, little reliable and out of phase reflection of the actual intracellular disturbances [19, 21, 27, 74].

Apart from the cells anatomically destroyed by direct contact of the shock agent with the body, the whole cellular population of the body triggers off *a selective energy producing metabolic effort,* followed fairly soon, however, by exhaustion and disorganization. Integration of these biochemical cellular disturbances is subsequently expressed by tissular and visceral insufficiency: myocardial, renal, hepatic, of the RES, neuronal chains, etc., which are only general aspects of the pathological cellular metabolic processes.

8.3.1 ORGANIC FUELS

The usual organic fuels are glucose and FFA. Their utilisation differs according to the 'enzymatic anatomy' of each tissue and the alternating periods of inanition and feeding [7, 8]. Normally, during the period of food supply glucose is preferentially used (even in the adipose tissue it inhibits FFA consumption during this period). In the course of inanition FFA mobilized from the adipose tissue are used first in the vast muscular compartment where the utilization of glucose now is inhibited. This is the peripheral cycle of Randle with metabolic fatty acid-glucose autoregulation and the participation of two metabolic structures, muscles and adipose tissue [58].

But a complete energy production homeostasis also includes other metabolic structures: the red blood cell, neurone, myocardial fibre and liver cell. The metabolic characteristics of the neurone, myocardium cell and RBC were described in chapters 4, 5 and 7.

In the centre of regulation of energy-yielding homeostasis stands the *liver cell* which contains almost all the possible metabolic lines (*see* Section 8.3.3). The red blood cell, although full of oxygen, does not breath, and it procures its energy by fermentative activity, being a permanent supplier of lactate (*see* Section 7.2.1). The neurones require only glucose and cannot consume FFA. The myocardium and the diaphragm can consume only lactate to provide energy, and the muscle may burn even acetoacetate (*see* Sections 4.4 and 5.2.2).

8.3.2 THE MUSCLES

In shock, the metabolic pathways of the 'muscle carcass' are the following:

(i) The muscle cells, bathed in cortisol, break down the proteins, offering amino acids that reach the liver where neoglucogenesis from proteins arises; this metabolic pathway is stimulated in shock by hypercortisolaemia.

will be taken up by glycogen synthesis. In shock, this metabolic pathway is not available.

(iii) The muscles are the cradle of the Cori cycle which in shock supplies lactate, used in the initial stages by the liver for neoglucogenesis from lactate. Catecholamines through cyclic-AMP activate phosphorylase, and muscular glycogenolysis continues with an accelerated glycolysis whose lactic acid readily leaves the muscle and reaches the liver, where it is reconverted into glucose. This newly formed glucose is not taken up by the muscles, as at this stage the muscles are using FFA, but will be offered indirectly to the neurones (Figure 8.12).

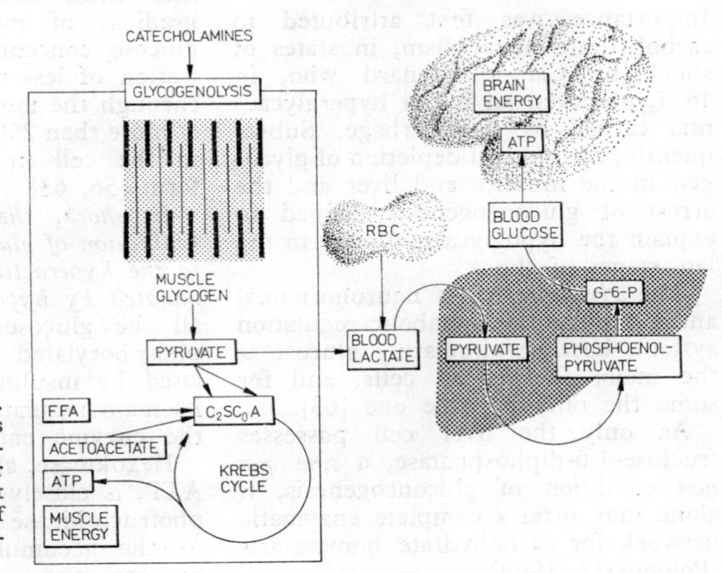

Figure 8.12. The Cori cycle. The lactate fuel represents the carbohydrate contribution of muscle and RBC for neuronal metabolism. But transhepatic conversion of lactate into glucose is obligatory.

(ii) The muscles usually store glucose by glycogen synthesis (accumulation in the muscles is greater than in the liver); extra-mitochondrial citrate and ATP cannot perform lipogenesis and inhibit phosphofructokinase which is the key of glycolysis, and glucose

(iv) In the muscles, glycolysis may release-glycerophosphate with which excess FFA are esterified, building up again the intramuscular lipid reserve; thus, the FFA, the main fuel of the muscles, will be derived not only from the adipose tissue but also from the muscles' own reserves, mobilized by catecholamines.

(v) The muscles prefer two fuels: FFA and acetoacetate. Acetoacetate does not give residues, being a 'veritable light gasoline' [8]. The preference of the muscles for FFA inhibits their use of glucose, a fuel that is thus protected for use in the structures sensitive to hypoxia. The FFA taken up in the Krebs cycle may inhibit glycolysis by the Pasteur effect. In shock this possibility is rapidly restricted by oxygen deficiency, limiting the use of the lipid fuel.

8.3.3 THE LIVER CELL AND THE GLYCOSTAT

Importance was first attributed to carbohydrate metabolism, in states of shock by Claude Bernard who, in 1877, observed an initial hyperglycaemia caused by haemorrhage. Subsequently, the gradual depletion of glycogen in the muscles and liver and the arrest of gluconeogenesis helped to explain the hypoglycaemic state in the late stages of shock.

The *glycostat* has a neurohormonal and a preferential metabolic regulation system, because circulating glucose is the major fuel of all cells, and for some the only possible one [63].

As only the liver cell possesses fructose-1,6-diphosphatase, a *sine qua non* condition of gluconeogenesis, it alone may offer a complete enzymatic network for carbohydrate homeostasis (Polonovski, 1969).

The liver is the vital crossroads of the energy-producing mechanisms. Input into the liver cell of glucose, FFA, glycerol, lactate and amino acids is followed by an output of glucose, acetoacetate and lipoproteins. For the whole energy system the liver cell performs *neoglucogenesis*, *ketogenesis*, *lipoproteinogenesis* and *urogenesis*.

Systemically, glucose is offered by the liver cell in three ways:

- the glucose assimilated in the intestine (transitive glucose);
- the glucose of intrahepatocytic glycogenolysis stimulated by catecholamines and glucagon;
- glucose synthesized by neoglucogenesis from amino acids, lactate or glycerol under the impulse of glucagon and cortisol.

Glucose enters freely the liver cell, neurone, RBC and β-islet cells. Crossing of the membrane of a muscle cell, adipocyte and enterocyte by glucose demands, however, a more complex regulation. The penetration of glucose into these cells depends upon the gradient of intra- and extracellular glucose concentrations. At a concentration of less than 60 mg% transfer through the membrane is stopped and at more than 250 mg% glucose remains in the cell in its unphosphorylated form [56, 68].

In shock, the rate of intracellular utilization of glucose is very high owing to the hyperactivity of hexokinase, stimulated by hypoxia. Hence, in shock all the glucose entering the cell is phosphorylated. Hexokinase is stimulated by insulin and aldosterone and its main regulator G-6-P inhibits it by the enzyme end-product mechanism.

Hexokinase, also inhibited by excess ATP, is closely correlated with phosphofructokinase whose blocking leads to the accumulation of G-6-P and prevents the entrance of glucose in the cell (Figure 8.13).

In the liver cell, glucose is phosphorylated and gives rise to G-6-P ester, which has at least five metabolic possibilities [56]; of these, G-6-P chooses, in shock, the following two: intrahepatocytic glycolysis and free release into the blood for the systemic utilization of glucose. The pentose pathway is likewise possible but only

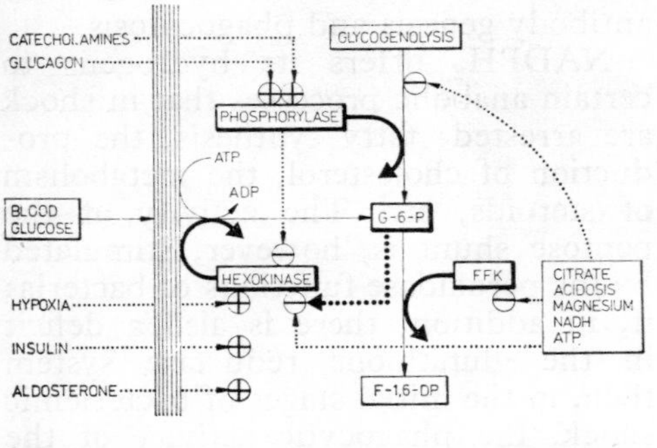

Figure 8.13. The triangle of the three enzymes that regulate the glucose supply to the cell and the onset of glycolysis. Blocking of phosphofructokinase brings about the accumulation of G-6-P, which inhibits hexokinase.

in the first stages of shock (Figure 8.14).

Glycolysis (the Embden-Meyerhof pathway) has eleven steps, with two important accelerating enzymes — phosphofructokinase and pyruvickinase. The most important step is the splitting of fructodiphosphate by aldolase into two fragments of 3 carbons each. The end product of glycolysis, pyruvic acid, tries to pass the mitochondrial barrier and supply C_2SCoA and, partially, oxaloacetic acid. In shock, this cannot take place because the Krebs cycle and oxidation chain are inhibited (Figure 8.15).

Under normal conditions, when the ATP reserve is built up again, the mitochondrial citrate passes into the cytosol and enters the malic enzymatic cycle (Young, Shrago and Lardy cycle) [32, 84]. In shock, this cycle is inoperative and, hence, also stops the following metabolic lines: the production of extra-mitochondrial C_2SCoA, which would have followed the pathway of FFA, cholesterol or bile acid synthesis, and of extra-mitochondrial oxaloacetate,

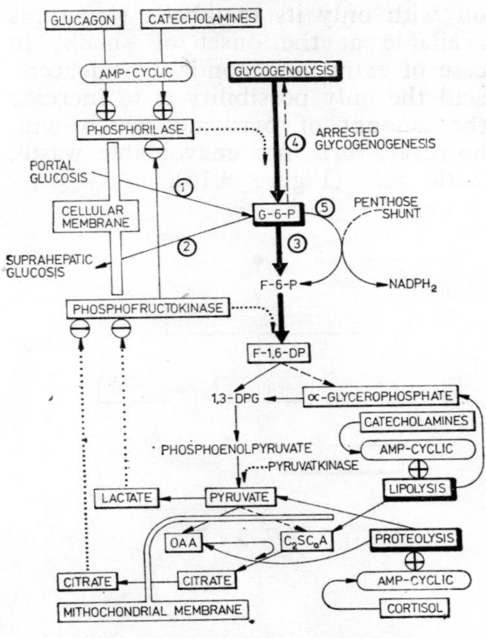

Figure 8.14. The five possible metabolic pathways of Robison ester in the liver cell in shock.

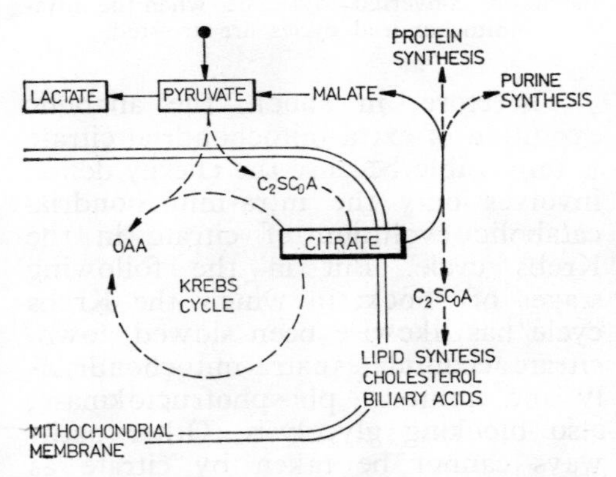

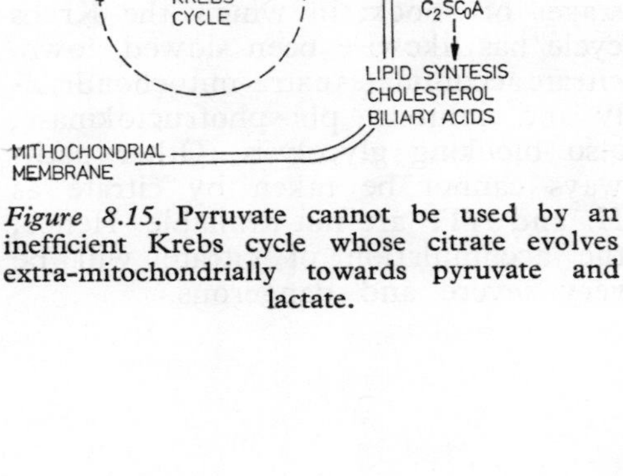

Figure 8.15. Pyruvate cannot be used by an inefficient Krebs cycle whose citrate evolves extra-mitochondrially towards pyruvate and lactate.

which would have achieved several vital syntheses: of lipids, protein, purin and pyrimidin bases.

The cell is thus forced to struggle on with only its metabolic structures available at the onset of shock; in case of extra-mitochondrial oxaloacetic acid the only possibility is to increase the amount of pyruvate which will, however, turn into unavoidable waste, lactic acid (Figure 8.16).

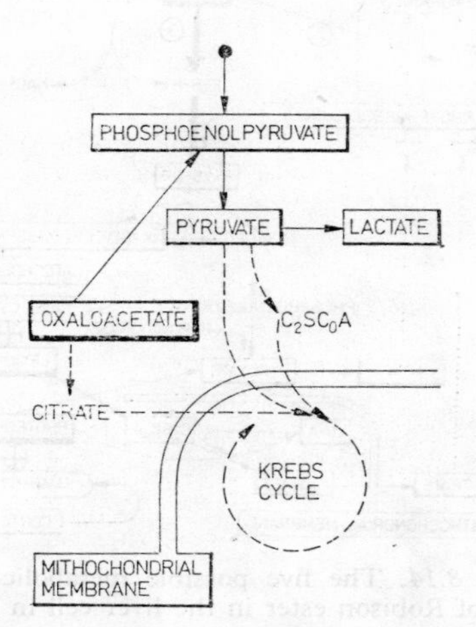

Figure 8.16. Extra-mitochondrial oxaloacetate is inevitably converted to lactate when the intra-mitochondrial cycles are arrested.

Therefore, in shock, the anabolic evolution of extra-mitochondrial citrate is impossible because the energy deficit involves only the intra-mitochondrial catabolic evolution of citrate in the Krebs cycle. But in the following stages of shock, in which the Krebs cycle has likewise been slowed down, citrate accumulates extra-mitochondrially and inhibits phosphofructokinase, also blocking glycolysis. Other pathways cannot be taken by citrate as H_2 and ATP are not available. Hence, the accumulation of citrate will be very severe and dangerous.

The pentose shunt starts from G-6-P and returns to fructose-6-phosphate ester, being in fact a bypass by which hydrogen ions are bound to NADP. The pentose shunt sustains antibody genesis and phagocytosis.

$NADPH_2$ offers its hydrogens to certain anabolic processes that in shock are arrested: fatty synthesis, the production of cholesterol, the metabolism of steroids, etc. The activity of the pentose shunt is, however, stimulated by the peroxidase functions of bacteria; if, in addition, there is also a deficit in the glutathione reductase system then, in the initial stages of bacteriemic shock, the phagocytic activity of the RES is strongly stimulated [15] (*see* Section 6.5).

Glycogenolysis takes place in the cytosol under the impulse of glucagon and epinephrine which activate phosphorylasis through cyclic AMP. Cyclic AMP inhibits the system which perform genesis of glycogen; it also stimulates lipolysis by activation of triglycerid-lipase. Glycogen is an energy stock of carbohydrates which, if continuously and exclusively utilized (125 g hepatic glycogen + 600 g muscular glycogen) would last for 48 hours; it is the great buffer of carbohydrate homeostasis (Claude Bernard). Splitting of the 1-4 and 1-6 glycoside bonds and hydrolysis of the G-6-P fragments take place in the terminal glycogen chains.

Gluconeogenesis. After depletion of the supplied glucose and stored glycogen, by the adrenaline-glucagon hormone couple, the organism will stimulate another pathway in order to avail itself of the vital glucidic material, synthesizing it from other micromolecular classes. This situation is specific of shock, as it is of starvation.

The three substances from which glucose can be obtained are:

- lactic acid, which is dehydrogenated in the liver cell by LDH in the course of the Cori cycle (which functions for quite a long time in shock);
- alanine, for the formation of pyruvic acid;
- glycerol.

As soon as it is available, pyruvate begins its intra-mitochondrial journey where it will find a large amount of C_2SCoA derived from β-oxidation of FFA. Therefore, pyruvate will no longer supply active acetate but will form oxaloacetic acid under the control of pyruvic-carboxylase, which is stimulated by cortisol. Oxaloacetic acid is converted into citric acid in slight amounts because citratesynthase is inhibited by the abundance of $NADH_2$.

This abundance of protons converts oxaloacetic acid into malate, which will pass into the cytosol, leaving the mitochondria (oxaloacetic acid cannot cross the membrane). In the cytosol oxaloacetic acid forms again and produces phosphoenolpyruvate (PEP) under the influence of a key enzyme likewise stimulated by cortisol: oxaloacetic acid-PEP-carboxykinase. PEP starts a counter current in the Embden-Meyerhof cycle, led by two 'strategic' enzymes: fructosodiphosphatase and G-6-P-ase. Many hormones stimulate or inhibit the key enzymes of the entire process (Figure 8.17).

Gluconeogenesis from amino acids is also conditioned by the ureogenesis function of the liver cell, and excessive oxidation of FFA in the mitochondria results in ketone bodies, restricting the pyruvate → oxaloacetic acid → malate transit.

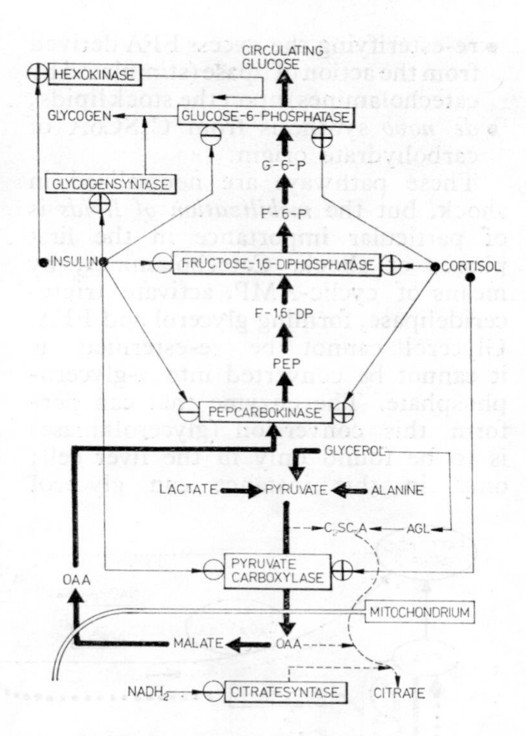

Figure 8.17. Hormonal regulation of neoglucogenesis. Insulin introduces glucose in the cell which stores it in the form of glycogen. Cortisol impulses the countercurrent of oxaloacetate and releases glucose from the cell saturated with $NADH_2$.

8.3.4 ADIPOCYTE AND THE LIPOSTAT

The adipose tissue distributes rapidly the lipid fuel, FFA, that can be used by the organism but which cannot be utilized in shock since it requires the presence of oxygen.

During the periods of food supply the adipose tissue obtains fats by:

- splitting of the intestinal chylomicrons (under the influence of lipoproteinlipase, an enzyme produced by the fat cell), the FFA being esterified with α-glycerophosphate;

- re-esterifying the excess FFA derived from the action of lipase (stimulated by catecholamines) upon the stock lipids;
- *de novo* synthesis from C_2SCoA of carbohydrate origin.

These pathways are neutralized in shock, but the *mobilization of lipids* is of particular importance in the first phases of shock. Catecholamines, by means of cyclic-AMP, activate triglyceridelipase, forming glycerol and FFA. Glycerol cannot be re-esterified as it cannot be converted into α-glycerophosphate. The enzyme that can perform this conversion (glycerolkinase) is to be found only in the liver cell; only in this instance can glycerol pass counterwise through the Embden-Meyerhof cycle in order to produce glucose (neoglucogenesis from glycerol). In shock, this pathway is not operative all the time.

The renewed esterification of FFA may take place if there is sufficient ATP available. Hence, in shock, FFA choose the second pathway: they will be mobilized in the blood and slowly burned in the muscles. The excess will, however, reach the liver, where part of it is oxidized into ketone bodies and part is esterified (hepatic lipoproteinogenesis). FFA disappear very rapidly from the blood (their half-life is 2 to 3 minutes) (Figure 8.18).

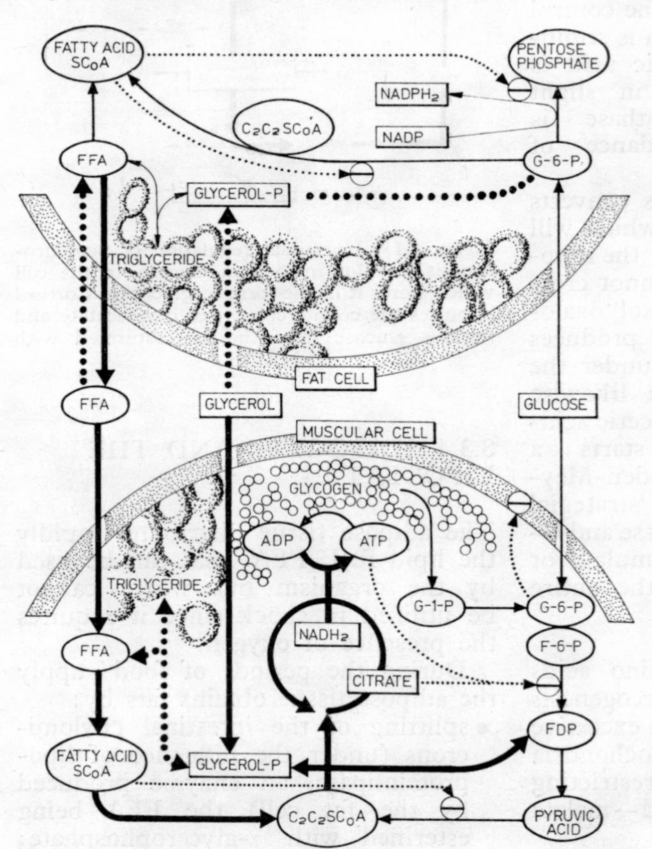

Figure 8.18. The fuels changed between muscle and fatty cells in the state of shock: FFA and glycerole are burned in muscle cell; in fatty cell lipid synthesis is arrested.

Ketogenesis initially becomes acetoacetate in the intramitochondrial atmosphere of FFA by oxidation. Acetoacetate is not used in the liver cell but is an excellent material for the muscles. In physiologic concentrations, acetoacetate is not toxic and can be utilized even by the neurones (Figure 8.19).

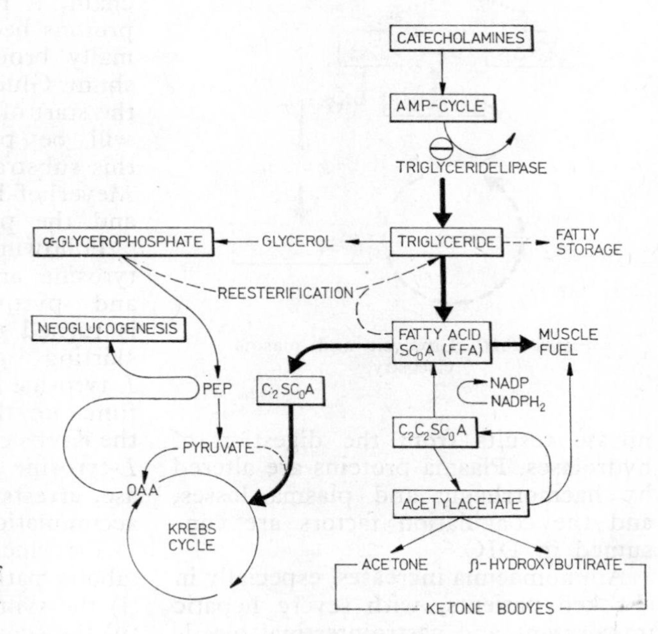

Figure 8.19. The characteristics of lipid metabolism in the liver cell in shock.

Only when it accumulates in large amounts, as occurs in diabetic coma, acetoacetate together with its other two well-known immediate derivatives becomes toxic. Similar to FFA, acetoacetate is a chemical stimulant of the β-pancreatic cells which thus release insulin in greater amounts.

The liver also exports a protein-bound lipid complex (triglycerides, phospholipids and cholesterol) which may be used in the muscles and is stored in the vascular intima and fat cells. In shock, this lipoprotein outflow augments and increases the viscosity of the blood (*see* Section 7.4.3).

Initially, hyperglycolysis inhibits lipolysis by the lactic acid that accumulates following insufficient metabolism of active acetate by the Embden-Meyerhof-Krebs cycle [5, 55]. Therefore, although catecholamines concomitantly stimulate glycolysis and lipolysis under hypoxia conditions, the inadequate metabolism by the Embden-Meyerhof-Krebs pathway also arrests the production of acetate by lypolysis (Figure 8.20).

8.3.5 THE NITRATE FUNCTIONS

Urinary nitrate sharply increases immediately after the onset of shock, reaches a peak within a few days and remains at a high level up to 6 to 8 weeks (Moore) [42]. The disintegrated proteins derive especially from the muscles. The initial increase in urinary nitrate is caused by autolysis of the cells destroyed directly by the shock agent; in the advanced stages of shock,

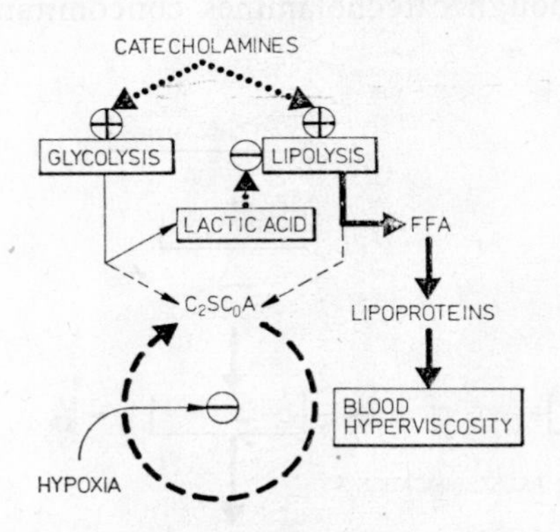

Figure 8.20. Lipolysis and plasma viscosity.

nitrate results from the digestion of hydrolases. Plasma proteins are altered by haemorrhagic and plasma losses, and the coagulation factors are consumed in DIC.

Ammoniaemia increases, especially in shocked patients, with severe hepatic impairment and gastrointestinal bleeding. The high concentration in plasma amino acids is due both to the initial tissular disintegration and to a decrease in the metabolism of amino acids in the liver cells. The liver cell in shock no longer manages urogenesis and transdeamination efficiently. At the same time, decarboxylation takes place, resulting in the appearance of biogenic amines, well known for their activity.

The metabolic characteristics of each amino acid explain the variations in their plasma levels in shock: there is an increase particularly in asparagine, leucine, isoleucine, tyrosine, phenylalanine and alanine; proline and valine diminish [56].

8.3.6 *L*-TYROSINE METABOLISM

As the Embden-Meyerhof-Krebs cycle furnishes the protons required for the redox system of the energy producing chain, a reloading of the cells with protons becomes necessary and is normally brought about by the pentose shunt. Glucose-6-phosphate is present at the start of both main pathways which will be permanently competing for this substrate [31, 50]. The Embden-Meyerhof-Krebs cycle stores $NADH_2$ and the pentose pathway $NADPH_2$.

Following the administration of *L*-tyrosine and folic acid, lactacidaemia and pyruvicaemia fall and G-6-P is turned towards the pentose shunt, skirting glycolysis [49, 80, 81]. *L*-tyrosine may supply acetoacetate and fumarate, thus supplying and sustaining the Krebs cycle. By synthesizing citrate, *L*-tyrosine inhibits phosphofructokinase, arrests glycolysis, and prevents the accumulation of lactate.

Tyrosine may follow two main metabolic pathways (Figure 8.21):
(i) the synthesis of catecholamines;
(ii) the synthesis of fumaric acid and oxaloacetic acid (Ornellas et al., 1970; Weber and Laborit, 1970) [50, 81].

The trend of both pathways is the same: to uphold an efficient Embden-Meyerhof-Krebs cycle.

If monoamino-oxidase is active the synthesis of catecholamines strictly depends upon the amount of tyrosine, and the catecholamines obtained will activate glycolysis and the Krebs cycle strongly. In shock, the maximum production of energy by catecholamines is restricted by the oxygen deficit which inevitably stops the Krebs cycle.

However, by fumaric and oxaloacetic acid synthesis the Krebs cycle may be provided directly by tyrosine. The Krebs cycle also synthesizes acetylcholine which will free catecholamines by a nicotinic effect. In this circle, ascorbic acid produces, against the background of excessive adrenergy, a favourable vagal state [49, 80], which may re-establish acetylocholine and the Embden-Meyerhof-Krebs cycle, re-equilibrating the pentose shunt and

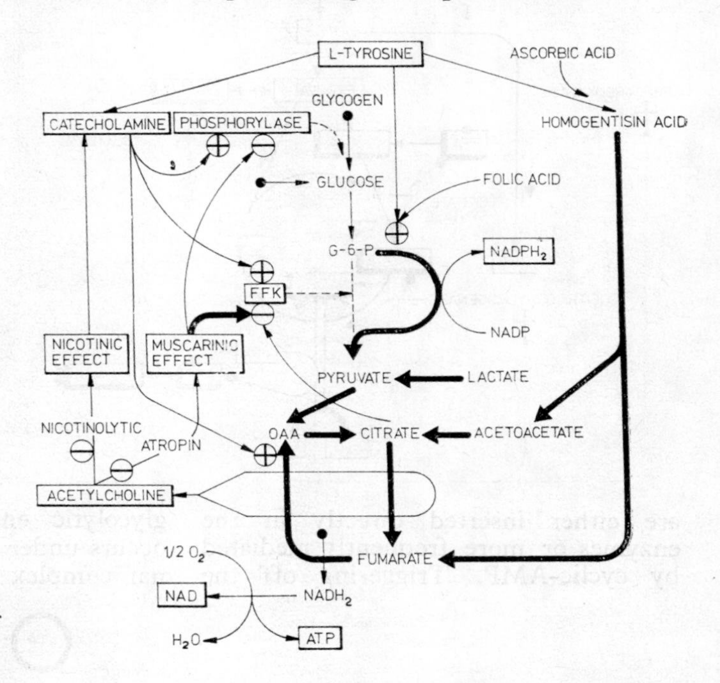

Figure 8.21. The direct and indirect metabolic insertions of *L*-tyrosine.

from which increase in catecholamine concentrations cannot be eliminated, the muscarin effect alone of acetylcholine (inhibiting phosphorylation and the Embden-Meyerhof-Krebs cycle) may contribute a fine autolimitative regulation. Nicotinolytics block the release of catecholamines by acetylcholine, and atropin its inhibitory effect on phosphorylase. Folic acid helps the hydroxylation of tyrosine and the synthesis of catecholamines, and ascorbic acid directs tyrosine towards the synthesis of acetoacetate and fumaric acid.

It results, therefore, that in shock the administration of tyrosine and excitability of the neuronal structures [76].

8.3.7 REGULATION OF THE ENZYMATIC CHAINS

In shock, simple metabolic regulation of the enzymatic chains allows for a rapid reorganization of the energy producing metabolic lines (Figure 8.22) (*also see* Section 3.8.2). However, there is also a superposal of modulated hormonal regulation with elective insertion on the enzymes of *strategic importance* [2, 3, 77]. The hormones

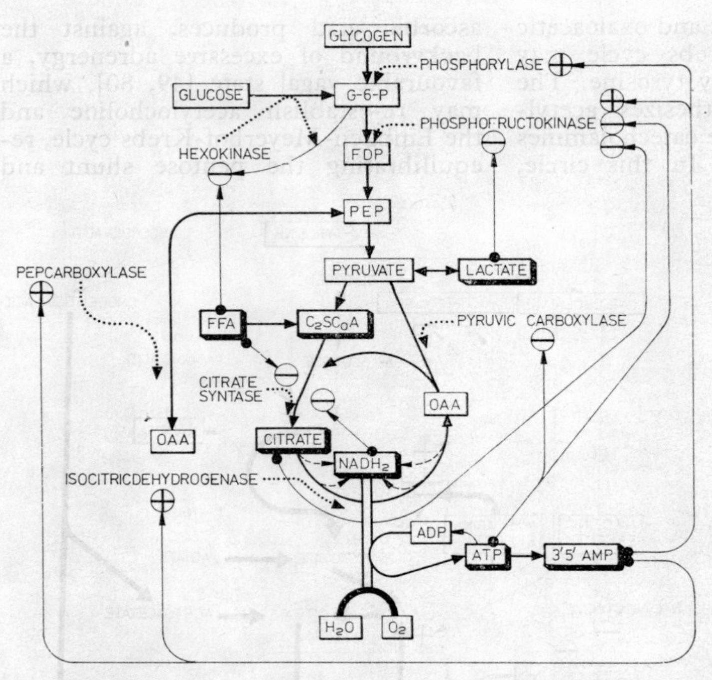

Figure 8.22. The principal chemical regulation of energy production.

are either inserted directly in the enzymes or more frequently mediated by cyclic-AMP. Triggering off the glycolytic energy producing systems occurs under the influence of a hormonal complex (Figure 8.23).

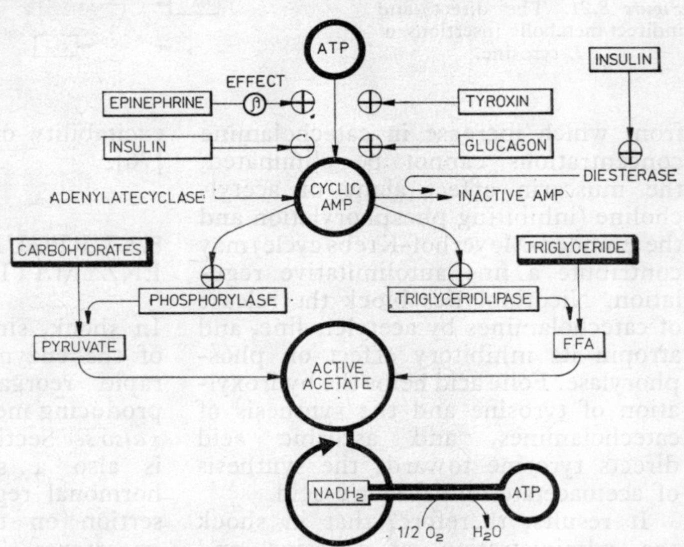

Figure 8.23. Hormonal activation of energogenesis.

Just as complex is the intervisceral hormonal regulation of the essential fuels (Figure 8.24). Owing to the proton concentration brought about, the metabolic effects of hormonal regulation are also reflected in the modelling of the hormono-enzymatic insertion. At *p*H=6.6 the action of catecholamines, ACTH and glucagon ceases [57].

In point of fact to bring about energogenesis in the *shock cell* almost all the means of intracellular regulation are used:

- the Crabtree effect overcomes the Pasteur effect;
- the dominant reaction in the output of glycolysis (phosphorylation of fructose-6-phosphate into fructose 1,6-diphosphate) is accelerated by the action of catecholamines on phosphofructokinase;
- increase in the substrate concentrations throughout the Embden-Meyerhof cycle forces, by Michaelis kinetic regulation, acceleration of the stages of this cycle;
- regulation by 'limiting' metabolites and by intracellular compartmentation is attributed especially to the adenylates (*see* Section 9.3.1).

To these must be added regulation by allosteric feedback, by changes in the concentration of enzymes and by hormonal control of the enzymatic chains (see Sections 3.8, 4.8.1, 8.2.4 and 9.3).

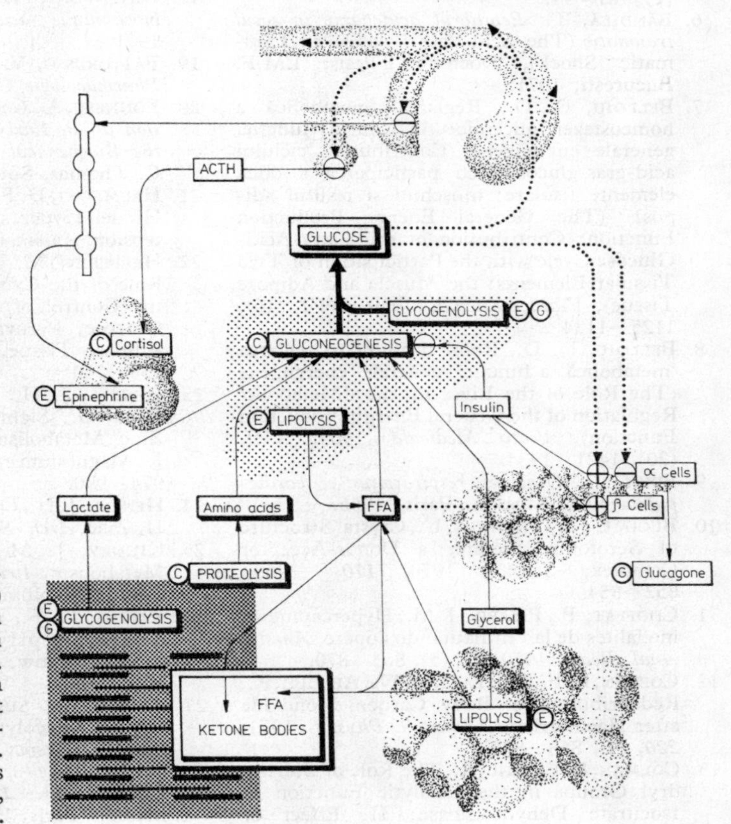

Figure 8.24. The intervisceral regulation of the metabolic substances. Glucagon (G), epinephrine (E) and cortisol (C) maintain high blood glucose levels in the first stages of shock. Hyperglycaemia will then inhibit the mechanisms that have triggered it and stimulate the release of insulin. The neurones and muscles show preference to glucose, ketone bodies and FFA.

SELECTED BIBLIOGRAPHY

1. ACKERMAN, N. B., BRINKLEY, F. B., Comparison of Effects on Tissue Oxygenation of Hydrogen Oxygen and Intravascular Hydrogen Feroxide, *Surg.*, 1968, *63*, (2), 285–290.
2. ALTMANN, F. P., ROBERTSON, A., The Effects of Various Steroids on Pentose-Shunt Dehydrogenase Activity, *Biochem. J.*, 1970, *118*, (2), 5–6.
3. AMIEL, C., In *Actualités de Physiologie Pathologique*, Masson, Paris, 1970.
4. BATEY, N. R., RAJASINGHAM, M. S., SINGH, C. M., FLEAR, C. T. G., The Effect of Escherchia Coli Endotoxin on Membrane Permeability in Skeletal Muscle, *Brit. J. Surg.*, 1907, *57*, (11), 851–858.
5. BAUM, D., The Inhibition of Norepinephrine-Stimulated Lipolysis by Acute Hypoxia, *J. Pharmacol. Exp. Ther.*, 1969, *169*, (1), 87–94.
6. BĂNDILĂ, T., *Echilibrul acidobazic în șocul traumatic* (The Acid-Base Balance in Traumatic Shock), Doctor's Thesis, I.M.F. București, 1971.
7. BELLOIU, D. D., Reglarea 'metabolică' a homeostazei glicemice în cadrul funcției generale energogene (Contribuția 'ciclului acid-gras-glucoză' cu participarea a două elemente tisulare: mușchiul și țesutul adipos) (The General Energy Production Function: Contribution of the Fatty Acid-Glucose Cycle with the Participation of Two Tissular Elements: the Muscle and Adipose Tissue), *Viața medicală*, 1969, *XVI*, (16), 1125–1134.
8. BELLOIU, D. D., Rolul ficatului în reglarea 'metabolică' a funcției generale energogene (The Role of the Liver in the 'Metabolic' Regulation of the General Energy Production Function), *Viața Medicală*, 1969, *XVI*, (20), 1421–1434.
9. BRUN, J., *Urgences respiratoires et cardiopulmonaires*, Masson, Paris, 1966.
10. BUGG, C. E., THEWALT, U., Crystal Structure of Serotonin Picrate, a Donor-Acceptor Complex, *Science*, 1970, *170*, (3960), 852–854.
11. CHOFFAT, P., PICARD, J. M., Hypercapnie et modalités de la ventilation de l'opéré, *Anesth. Anal. Réan.*, 1970, *27*, (5), 863–870.
12. COBURN, R. F., WALLACE, H. W., ABBOUD, R., Redistribution of Body Carbon Monoxide after Haemorrhage, *Amer. J. Physiol.*, 1971, *220*, (4), 868–874.
13. COLMAN, R. F., CHU, R., The Role of Sulfhydryl Groups in the Catalytic Function of Isocitrate Dehydrogenase: II. Effect of n-ethylmaleimide on Kinetic Properties, *J. Biol. Chem.*, 1970, *245*, (3), 601–607.
14. COLLINS, J. A., SIMMONS, R. L., JAMES, P. M., Acid Base Status of Seriously Wounded Combat Casualties: II. Resuscitation with Stored Blood, *Ann. Surg.*, 1971, *173*, (1), 6–18.
15. COMROE, J. H., *Physiology of Respiration*, Yearbook Medical-Publishers, Chicago-Illinois, 1965.
16. CORAN, A. G., BALLANTINE, T. V., HORWITZ, D. L., HERMAN, C. M., The Effect of Crystalloid Resuscitation in Haemorrhagic Shock on Acid-Base Balance: A Comparison between Normal Saline and Ringer's Lactate Solutions, *Surg.*, 1971, *69*, (6) 874–880.
17. CĂPÂLNĂ, S., *Biochimie dinamică* (Dynamic Biochemistry), Editura Medicală, București, 1973.
18. DAVENPORT, H. W., *ABC de l'équilibre biochimique acido-basique*, Masson, Paris, 1971.
19. FATTORUSSO, V., RITTER, O., *Dictionnaire de Pharmacologie Clinique*, Masson, Paris, 1967.
20. FONNESU, A., *Changes in Energy Transformation as an Early Response to Cell Injury in the Biochemical Response to Injury*, Charles C. Thomas, Springfield, 1960.
21. HALMAGYI, D. F. J., KENNEDY, M., VARGA, D., Hidden Hypercapnia in Haemorrhagic Hypotension, *Anesthesiol.*, 1970, *33*, (6), 594–601.
22. HALPERIN, M. L., ROBINSON, B. H., The Role of the Cytoplasmic Redox Potential in the Control of Fatty Acid Synthesis from Glucose, Pyruvate and Lactate in White Adipose Tissue, *Biochem. J.*, 1976, *116*, (2), 235–240.
23. HENZEL, J. H., DEWEESE, M. S., RIDENHOUR, G., Significance of Magnesium and Zinc Metabolism in the Surgical Patient. I. Magnesium, *Arch. Surg.*, 1967, *95*, (6), 974–990.
24. HENZEL, J. H., DEWEESE, M. S., PORIES, W. J., II. Zinc, *Arch. Surg.*, 1967, *95*, (6), 991–999.
25. KINNEY, J. M., The Effect of Injury on Metabolism, *Brit. J. Surg.*, 1967, *57*, Lister Centenary Number, 435–447.
26. KITTLE, C. F., AOKI, H., BROWN, E. B., The Role of pH and CO_2 in the Distribution of Blood Flow, *Surg.*, 1965, *57*, (1), 138–146.
27. KUELHL, L., SUMSION, E. N., Turnover of Several Glycolytic Enzymes in Rat Liver, *J. Biol. Chem.*, 1970, *245*, (24), 6616–6623.
28. LABORIT, H., *Les régulations métaboliques*, Masson, Paris, 1965.

29. Laborit, H., Olympie, J., Variations de la lactacidémie, de la pyruvicémie et de la glycémie sous l'action de la L-tyrosine combinée ou non à l'acide folique, *Agressol.*, 1969, *10*, (5), 387–390.
30. Laborit, H., Lamothe, C., Thuret, F., Action inhibitrice sur la respiration des coupes de tissus du rat (cerveau, foie) d'atmosphères oxygénées sans CO_2. Implications physiopathologiques, *Agressol.*, 1971, *12*, (2), 105–112.
31. Lamothe, C., Thuret, F., Laborit, H., Action in vitro de la L-tyrosine sur l'activité hexokinasique, *Agressol.*, 1970, *11*, (1), 21–23.
32. Lardy, H. A., Ferguson, S. H., Oxidative Phosphorylation in Mitochondria, *Ann. Rev. Biochem.*, 1969, *38*, (10), 991.
33. Larson, D. L., Maxwell, R., Abston, S. Dobrkowsky, M., Zinc Deficiency in Burned Children, *Plast. Reconstr. Surg.*, 1970, *46*, (1), 13–20.
34. Le Page, G. A., Biological Energy Transformation during Shock as Shown by Tissue Analyses, *Amer. J. Physiol.*, 1946, *146*, (2), 267–271.
35. Le Quesne, L. P., Fluid and Electrolyte Balance, *Brit. J. Surg.*, 1947, *54*, Lister Centenary Number, 449–452.
36. Longmore, W. J., Niethe, C. M., McDaniel, M. L., Effect of CO_2 Concentration on Intracelular pH and on Glycogen Synthesis from Glycerol and Glucose in Isolated Perfused Rat Liver, *J. Biol. Chem.*, 1969, *244*, (23), 6451–6457.
37. Lorber, A., Chang, C. C., Masoucka, D., Meacham, I., Effect of Thiols in Biological Systems on Protein Sulfhydryl Content, *Biochem. Pharmacol.*, 1970, *19*, (5), 1551–1560.
38. Lowery, B. D., Cloutier, C. T., Corey, L. C., Blood Gas Determinations in the Severely Wounded in Haemorrhagic shock, *Arch. Surg.*, 1969, *99*, (3), 330–338.
39. McLaughlin, J. S., Suddhimondala, C., Mech, K., Llacer, R. L., Houston, I., Blide, R., Attar, S., Cowley, R. A., Pulmonary Gas Exchange in Shock in Humans, *Ann. Surg.*, 1969, *169*, (1), 42–56.
40. Makin, G. S., Walder, D. N., The effect of Retransfusion and Beta-adrenergic Blockade on Muscle Blood-Flow and Lactic Acidosis in Haemorrhagic Shock in Dogs, *Brit. J. Surg.*, 1970, *57*, (11), 851.
41. Mongar, J. L., DeReuck, A. V. S., *Enzymes and Drug Action*, J. & A. Churchill, London, 1962.
42. Moore, F. D., *Metabolic Care of the Surgical Patient*, W. B. Saunders Comp., Philadelphia, 1968.
43. Nagy, A., Wollemann, M., Regulatory Action of Chlorpromazine on the Activity of some Dehydrogenase, *Agressol.*, 1970, *11*, (4), 327–333.
44. Nahas, G. G., In *Shock and Hypotension* Grune and Stratton, New York, 1965.
45. Namm, D. H., Mayer, S. E., Maltrie, M., The Role of Potassium and Calcium Ions in the Effect of Epinephrine on Cardiac Cyclic Adenosine 3',5'-Monophosphate, Phosphorylase-Kinase and Phosphorylase, *Molec. Pharmacol.*, 1968, *4*, (5), 522–530.
46. Neely, J. R., Whitfield, C. F., Morgan, H., Regulation of Glycogenolysis in Hearts: Effects of Pressure Development, Glucose and FFA, *Amer. J. Physiol.*, 1970, *219*, (4), 1083–1088.
47. Nelson, R. M., Poulson, A. M., Lyman, J. H., Henry, J. W., Evolution of tris (hydroxy-methyl) aminomethane (THAM) in Experimental Haemorrhagic Shock, *Surg.*, 1963, *54*, (1), 86–94.
48. Oliva, P. B., Lactic Acidosis, *Amer J. Med.*, 1976, *48*, (2), 209–225.
49. Ornellas, M. R., Thuret, F., Laborit, H., Variations dans l'orientation des voies métaboliques dues à des altérations biochimiques et ultrastructurale provoquées par des agents pharmacologiques agissant sur l'état d'oxydoréduction du foie de rat (Ag. 310, Ag. 307 et vitamine K), *Agressol.*, 1969, *10*, (3), 265–275.
50. Ornellas, M. R., Thuret, F., Lamothe, C., Laborit, H., Action de la L-tyrosine sur quelques étapes du métabolisme énergétique, *Agressol.*, 1970, *11*, (1), 15–17.
51. Paniker, N. V., Srivastava, S. K., Beutler, E., Glutathione Metabolism of the Red Cells Effect of Glutathione Reductase Deficiency on the Stimulation of Hexose Monophosphate Shunt under Oxidative Stress, *Biochim, Biophys. Acta*, 1970, *215*, (3), 456–469.
52. Pasteur, L., In *Textbook of Chemistry*, Ed. West, E. S., Tood, W. R., Macmillan Comp., New York, 1963.
53. Peach, M. J., Ford, G. D., Azzard, A. J., Fleming, W. W., The Effects of Acidosis on Chronotropic Responses, Nor-epinephrine Storage and Release in Isolated Guinea-pig Atria, *J. Pharmacol. Exp. Ther.*, 1970, *172*, (2), 289–296.
54. Penney, D. G., Cascarano, J., Anaerobic Rat Heart. Effect of Glucose and Tricarboxylic Acid-Cycle Metabolites on Metabolism and Physiological Performance, *Biochem. J.*, 1970, *118*, (2), 221–227.
55. Pinter, E. J., Peterfy, G., Cleghorn, J. M., Pattee, C. J., Influence of Emotional Stress on Fat Mobilization: Role of Endogenous Catecholemines and the Beta-Adrenergic

Receptors, *Amer. J. Med. Sc.*, 1967, *254*, (5), 634–651.
56. POLONOVSKI, M., *Biochimie Médicale*, Masson, Paris, 1969.
57. POYART, C. F., NAHAS, G. G., Inhibition of Activated Lipolysis by Acidosis, *Molec. Pharmacol.*, 1968, *4*, (4), 389–401.
58. RANDLE, P. J., GARLAND, P. B., NEWSHOLME, E. A., HALES, C. N., The Glucose Fatty Acid Cycle. Its Role, Insulin Sensivity and Metabolic Disturbances of Diabetes Mellitus, *Lancet*, 1963, *1*, 785–789.
59. ROBINSON, J. S., Acidosis in Low Flow States, *Int. Anesthesiol. Clin.*, 1969, *32*, (7), 895–903.
60. ROGNSTAD, R., KATZ, J., Gluconeogenesis in the Kidney Cortex. Effects of d-Malate and Amino-Oxyacetate, *Biochem. J.*, 1970, *116*, (3), 483–491.
61. ROTH, E., GOOS, G., LEIMBACH, H., Metabolism of Brain, Heart, Liver, and Kidney in Haemorrhagic Shock, *Europ. Surg. Res.*, 1969, *1*, (3), 181–182.
62. RUSH, B. F., HSIEH, J., In vivo and in vitro Effects of Endotoxin on Tissue Metabolism, *Surg.*, 1968, *63*, (2), 298–300.
63. SCHEUER, J., STEZOSKI, S. W., Protective Role of Increased Myocatdial Glycogen Stores in Cardiac Anoxia in the Rat, *Circ. Res.*, 1970, *27*, (5), 835–849.
64. SCHLOERB, P. R., Shock and Metabolism, *Surg. Gynec. Obst.*, 1969, *128*, (2), 315–319.
65. SCHRODER, R., GUMPERT, J. R. W., ELTRINGHAM, W. K., The Role of the Liver in the Development of Lactic Acidosis in Low Flow States, *Postgrad. Med. J.*, 1969, *45*, (526), 566–570.
66. SCHUMER, W., Lactic Acid as a Factor in the Production of Irreversibility in Oligohaemic Shock, *Nature*, 1966, *212*, (5 067), 1210–1212.
67. SCHUMER, W., Localisation of the Energy Pathway Block in Shock, *Surg.*, 1968, *64*, (1), 55–59.
68. SCORPIO, R. M., MASORO, E. J., Differences between Manganese and Magnesium Ions with Regard to Fatty Acid Biosynthesis, Acetyl-Coenzime a Carboxylase Activity and Malonyl-Coenzime a Decarboxylation, *Biochem. J.*, 1970, *118*, (3), 391–399.
69. SHIRES, G. T., Shock and Metabolism, *Surg. Gynec. Obst.*, 1967, *124*, (2), 284–287.
70. SHOEMAKER, W. C., *Shock: Chemistry, Physiology and Therapy*, Charles C. Thomas Publ., Springfield, Illinois, 1967.
71. SLEEMAN, H. K., JENNINGS, P. B., HARDAWAY, R. M., Evaluation of Biochemical Changes Associated with Experimental Endotoxemia. I. Transaminase Activity, *Surg.*, 1967, *61*, (6), 945–950.
72. STAPLES, D. A., Comparison of Adenosine Triphosphate Levels in Haemorrhagic and Endotoxic Shock in the Rat, *Surg.*, 1969, *66*, (5), 883–885.
73. STEWART, J. S. S., Tissue Hypoxia and Metabolic Acidosis; Significant Factors, *Postgrad. Med. J.*, 1969, *45*, (526), 618–622.
74. ŞUTEU, I., CAFRIŢĂ, A., BUCUR, I. A., Progrese recente în fiziopatologia metabolismului in şoc (Recent Progress in the Pathophysiology of Metabolism in Shock), *Revista Sanitară Militară*, 1975, *LXXVIII*, (3), 618–625.
75. SWYNGHEDAW, B., HATT, P. Y., Métabolisme des glucides dans le myocarde, *Press. Méd.*, 1970, *78*, (6), 211–216.
76. TRIADOU, C., Bilan du transport des gaz du sang. Intérêt de l'étude des temps de rinçage des secteurs extracellulaires, *Agressol.*, 1971, *12*, (3), 157–169.
77. TULKENS, P., TROUET, A., VAN HOOF, F., Immunological Inhibition of Lysosome Function, *Nature* (Lond.), 1970, *228*, (5278), 1282–1285.
78. UNGAR, G., *Excitation*, Charles C. Thomas Publ., Springfield, Illinois, 1963.
79. WALKER, W. F., Acid-Base Balance, *Brit. J. Surg.*, 1967, *54*, Lister Centenary Number, 452–455.
80. WEBER, B., Action de la L-tyrosine isolément ou en association sur la consommation d'oxygène du rat, *Agressol.*, 1969, *10*, (3), 261–263.
81. WEBER, B., LABORIT, H., Action de fortes doses de tyrosine sur la consommation d'oxygène du rat, *Agressol.*, 1970, *11*, (1), 67–70.
82. WILSON, M. A., CARCARANO, J., The Energy Yielding Oxydation of NADH by Fumarate in Submitochondrial Particles of Rat Tissues, *Biochim. Biophys. Acta*, 1971, *216*, (1), 54–62.
83. WELL, M. H., Experimental and Clinical Studies on Lactate and Pyruvate as Indicators of the Severity of Acute Circulatory Failure (Shock), *Circulation*, 1970, *41*, (6) 989–1001.
84. YOUNG, J. W., SHRAGO, E., LARDY, H. A., Metabolic Control of Enzymes Involved in Lipogenesis and Gluconeogenesis, *Biochem.*, 1964, (3), 1687–1692.

9
THE SHOCK CELL

Despite its minuteness, each cell is a very complex organism.

ALEXIS CAREL

9.1 POINTS OF VIEW
9.2 CELLULAR INTEGRATION OF THE MACROSYSTEM
9.3 INTRACELLULAR REGULATIONS
9.3.1 Enzymatic crossroads
9.3.2 Enzymatic dynamics of the shock cell
9.4 CELLULAR ORGANELLES
9.4.1 The nucleus and the genetic code
9.4.2 The membrane system
9.4.3 The energy-producing apparatus – the mitochondria
9.4.4 The digestive apparatus – the lysosomes
9.5 STRUCTURAL ASPECT OF THE SHOCK CELL
9.5.1 Histology of the shock cell
9.5.2 Infrastructure of the shock cell
9.6. BIOPHYSICS OF THE SHOCK CELL
9.6.1 Elements of irreversible thermodynamics
9.6.2 Thermodynamic oscillations in the shock cell
9.6.3 Therapeutical inflexion points
9.6.4 The reversible-irreversible borderline at cellular level

9.1 POINTS OF VIEW

If for the immediate diagnosis of the state of shock the haemodynamic criteria are sufficient for the clinician to be able to form a judgement, efficient treatment must from the very begining take into account the cell.

All the genetic, cybernetic and metabolic alterations discussed in chapters 3, 4 and 8 occur within the *shocked cell.* Today, it is no exaggeration to speak of a *shock erythrocyte*, a *shock neuron*, or a *shock liver cell*, etc. as these are among the cells that play a key role in shock. Directed in this way researches will gain in thoroughness and integration of the results in the total organism will be of the greatest interest.

The metabolic characteristics of some of the vital cells have already been discussed: of the neuron (*see* Section 4.4), the myocardiocyte (*see* Section 5.2.2), the erythrocyte and thrombocyte (*see* Section 7.2).

Here the cell will be studied, in a general way, as the area subjected to lesions in states of shock.

The actual struggle for annihilation of the negative shock wave takes place inside the cellular space; the genetic nuclear code expresses the supreme effort of the cell to combat shock by means of operonal units alerted by the shock-inducing stimulus. As a result, the enzymatic chains are restored, energy genesis is precipitated and every effort is made to save the whole organism.

But the cell is fighting an unequal battle because it is handicapped from the beginning by shortage of oxygen; accustomed by evolution to receive oxygen from outside and to use it in an increased respiratory output, the shocked human cell is forced *to make its effort under hypoxic conditions*. This means a phylogenetic return to fermentation, the insufficient output of which is well known. Fermentation implies inversion of the metabolic routes and dangerous accumulations of irreducible wastes.

An excess of fermentation is limited in duration and puts the life of the cell in danger and this is what the shock in fact means: *a stress beyond the limits of the adaptable resiliency of the post-injury oscillations*.

The metabolic obstructions occurring in the enzymatic pathways will reject glucose, accumulate potassium, enhance cellular and mitochondrial swelling, and ruin the thermodynamic system of the membrane, thus opening the gates for the lysosomal hydrolases which will digest the 'mother cell'. The way the cell, taken separately, performs its *endogenic metabolic struggle against shock* is the equivalent of suicide; for the person in shock, the sacrificed regional cells may not be unduly harmful provided the number of sacrificed cells does not throw the whole out of balance. But the state of shock always generates cellular lesions whose summation will develop up to the level of organ failures.

At first, the clinician's principal effort must be to sustain the shocked cell therapeutically; that is, to 'resuscitate it'. Treatment of the shock cell does not necessarily mean direct reanimation of the enzymatic intramitochondrial chains or restoring the cybernetics of the genetic code operons, since many of these tasks are, at present, still impossible. But it must be well understood that, even by mere normalisation of cellular perfusion, a target that may be achieved, the cell is enabled to discharge its protons and to 'fight shock' under less strenuous conditions.

9.2 CELLULAR INTEGRATION OF THE MACROSYSTEM

Laborit suggests that the destruction of a few cells in a tissue culture leaves the others unaffected because there is no tight connection between culture cells; nevertheless, the human organism represents a colony of cells with its own organization and hierarchy in which each specialised element is functionally connected with all the others by a cybernetic regulating system, the task of which is to provide ontogenetic and philogenetic efficiency for the whole. The thermodynamic data of the outer medium usually vary within limits the human organism is informed about phylogenetically. The permanent exchange of energy and substance, as well as the continuous readjustment of homeostasis, is the source of the most favourable steady-state that maintains life [11, 63, 64].

The totality of the material exchange is dispersed in the fluid of the *intercellular medium* that bathes the cell unit which, in turn, contains the *actual intimate intracellular medium*. It is this very intracellular medium that has to be sustained in its struggle against the state of shock, not only the classical 'internal medium' which is but a reflection — often unsynchronized and distorted — of the damage to the cells. This accounts for the fact that in no health institution up to now has it been possible to overcome a real state of shock merely by restoring the volume of the classical internal (only intercellular) medium fluid.

In functional associations the cells form, prevailingly, energetic systems (the digestive, respiratory, osteomuscular systems) or predominantly informational systems (neuro-endocrine, reticular-endothelial, collagenic, circulatory, reproductive systems) the co-ordination of which aims, with cybernetic finality, at the efficiency of the entire human organism.

The shocked cell, as a functional unit is the key to our understanding of the whole phenomenon at the level of the macrosystem. As a rule, a continuous flow of electrons and protons crosses the cellular space. By dehydrogenating its food substance the cell obtains hydrogen atoms which it oxidises by removing the electrons and leaving them in the form of protons; the electrons are loaded by the cytochromes on the oxygen that has reached the inner cell medium by a sinuous route. This flow of substances gives off the energy that will support the life of the cell.

Any violent or critically intense disturbance of this flow of substances and energy leads to a state of shock. The onset may be super-acute (for instance in the case of the cytochromes being blocked by cyanides) when shock has no time to develop, or it may be very slow (denutrition with an insufficient supply of hydrogen atoms) when the question of 'chronic shock' arises. Figure 9.1. shows that lack of hydrogen never sets in abruptly because a rich endogenic source is to be found in the cellular stores. On the other hand, as the cell is unable to store oxygen, any cause that suppresses the oxygen

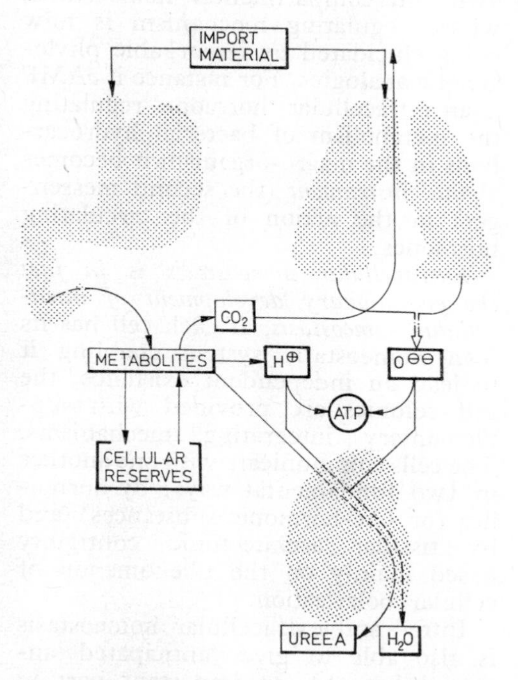

Figure 9.1. The circulation of hydrogen and oxygen ions in the human body.

supply to the mitochondria results in a deviation of the metabolic lines under anaerobic conditions, an imbalance that is limited in time.

The composition, rich in magnesium and potassium, of the pre-Cambrian oceans in which life first appeared is a milieu differing considerably from that of Bernard's 'milieu intérieur'; but it is strikingly similar in composition to the *intracellular fluid*, the actual internal medium in which the cell metabolism takes place. The cellular space is, in turn, partitioned and divided by membrane barriers into various intracellular compartments having their distinct enzymes and hydroionic equipment. The cell is provided with a complex system for the control of its own intercompartmental homeostasis, whose regulating mechanism is now being elucidated by remarkable phylogenetic analogies. For instance if *c*AMP is an intracellular 'hormone' regulating the metabolism of bacterial hydrocarbons in the macro-organism it becomes a *final instrument* (the second messenger) in the action of the circulating hormones.

Extracellular homeostasis is in fact the evolutionary development of intracellular homeostasis. If each cell has its own homeostatic system enabling it to lead an independent existence, the cell colonies are provided with supplementary integrating mechanisms. The cells communicate with one another in two fundamental ways: by hormones (or like-hormone substances) and by tissular architectonic contiguity based mainly on the phenomenon of cellular polarisation.

Intra- and extracellular homeostasis is also able to give 'anticipated answers' that play an important part in the states of shock. The best example is the release of catecholamines that precede physical effort, and the activation during shock of phosphorylase by *c*AMP, this activation starting before the appearance of catecholamines, the extra energy requirements being thus anticipated by the onset of glycogenolysis and glycolysis.

9.3 INTRACELLULAR REGULATIONS

In the enzymatic reorganisation some intracellular control systems act by priority [49, 53, 64, 77]. Some are by:

(a) control of the metabolite flow through the membrane;
(b) restoring of the hierarchy in the multi-enzymatic sequences;
(c) controlled action of the enzymatic batteries for each compartment;
(d) adjustment of the enzymatic concentration by regulation of the activity of the gene corresponding to a certain enzyme;
(e) metabolic control (presupposing transformation of the chemical regulator in the subsequent reactions);
(f) allosteric control (the regulator is not consumed);
(g) control by isoenzymes.

Hormonal control (on mitochondrial, reticular or intranuclear receptors) belongs to category (f); the production of energy by *c*AMP belongs to categories (e) and (f); the $NADH_2 \rightleftarrows NADPH_2$ relation to category (c); transmitochondrial flow of ATP, ADP and HPO_4^{--} to category (a); and so on (*see* Sections 3.8 and 8.3.7).

9.3.1 ENZYMATIC CROSSROADS

Here are some examples of regulation characteristic of the cells during shock.

Enzymatic sequences regulation. Enzymatic sequences are regulated by the combined action upon some 'key enzymes' of the end-product precursors and those of some integration substances belonging to other enzymatic lines. This complex control is possible due to the multiple allosteric sites of the enzymes; for instance, phosphofructokinase (PFK) is stimulated by the final product of its action (fructosediphosphate) by a positive feedback action. It is thus a 'runaway' enzyme, and the accumulation of fructose diphosphate, during shock, signals that glycolysis is the cellular energy system that is overstressed. The function of PFK can be arrested only by extra-mitochondrial citrate and ATP produced by a viable EMK-oxidation phosphorylation cycle, but this is not possible in the shocked cell, as the EMK-oxidative phosphorylation pathway is sacrificed (*see* Figure 8.20).

On the other hand, the enzymatic chains are controlled by hormones. Hormonal effects on the steps of energy metabolism are guided, at the key sites of major enzymatic control, by a succession of negative and positive feed-back loops, giving rise to oscillations in the metabolic systems, particularly of the pulsing or synchronized type.

It seems that hormones 'push' rather than 'pull' metabolism by regulating and directing an early key step, such as phosphorylase or phospho-fructokinase.

Correlation between the enzymatic chains and the membrane compartment is illustrated by oxidative-phosphorylation. The sequence and kinetics of electron transport reactions are governed by the arrangement of the redox systems on the mitochondrial cristae. $NADH_2$, formed in Krebs cycle, is inaccessible to the enzymatic systems of the cytosol, as it is directed only towards the oxidation chains in the mitochondria, but may be hydrogenated into intra-mitochondrial $NADPH_2$; the latter is a key cofactor in the synthesis of the lipids and the steroids. In shock, $NADH_2$ is neither admitted in the oxydative-phosphorylation chain (as this is blocked by the lack of oxygen), nor is it directed towards synthesis (as there is no ATP). The accumulated $NADH_2$ will block the Krebs-cycle and finally also glycolysis (*see* Figure 8.20).

Conversion of adenosinediphosphoric acid into adenosine triphosphoric acid

$$(ADP + HPO_4{}^{--} \rightleftharpoons ATP + OH^-)$$

also takes place in the mitochondria. ADP and inorganic phosphate are transported into the mitochondria by membrane carriers and ATP leaves by the same way. Transport through the membrane is nevertheless modified or/and arrested during shock, when the membrane transference function ceases [19].

The constant $(ATP)/(ADP+AMP)^2$ ratio is under the control of adenylatekinase. As a rule the ratio of AMP to ATP is 1:10 up to 1:100. A slight reduction in ATP will bring about a massive increase in AMP concentrations which will inhibit fructosediphosphatase and stimulate the PFK with a view to replenishing quantities of ATP. AMP also blocks phosphoenolpyruvate-carbokinase, a key enzyme in neoglucogenesis; but, on the other hand, it stimulates phosphorylase. AMP is considered to be the foremost factor in increasing the permeability of the membrane for glucose.

During shock, rapid depletion of ATP results in the accumulation of enormous quantities of AMP with all its energy stimulating properties. But the stimulation of glycolysis by AMP will aggravate the accumulation of lactic acid as a waste, and this, in turn, inhibits the glycolytic energy system (Figure 9.2).

The isozymes; they control the same reaction but in diverse situations. An example is the reversible reaction of glycolysis and of gluconeogenesis, in which either two different enzymes (PFK and fructosediphosphatase) or two aldolase isozymes take part. In shock, this central 'shuttle' of the Embden-Meyerhof pathway is overstressed in both directions *(see* Figure 8.16).

The central crossroads of the state of shock is to be found at the farthest end of glycolysis and is controlled by lactic dehydrogenase isozymes. The 5 isozymes of LDH are distributed and activated polymorphously, according to the tissue and species they belong to, but also to the functional or pathological condition of the cell. LDH_1 and LDH_2 are isozymic models of the anion type; they migrate rapidly by electrophoresis and are typical of breathing cells, which are characterized by a high redox potential (rH = 7—23), or the so-called B structures (neurons, tubulocytes, the myocardial myocyte, etc.). LDH_5 and LDH_4 are isozymic models of the cation type which migrate slowly and are typical of the cells that also cause fermentation (rH = 1 — 13); that is, the A structures (neuroglia, striated or smooth myocytes, leukocytes, etc.) *(see* Section 8.2.1). LDH is a regulating enzyme at the confines between cytosol and mitochondria, the basic crossroads of the cellular energy yielded. In aerobiosis the anionic pattern, LDH_1, provides for the dehydrogenation of lactate into pyruvate, and for the accumulation of $NADH_2$ which may be burnt in the oxidizing chain (the accumulation of lactic acid blocks LDH_1). In anaerobiosis it is mainly the LDH_5 cation pattern that acts, allowing for lactate accumulation (by hydrogenation of pyruvic acid); this cationic pattern is not inhibited by lactic acid, but by pyruvic acid and by decrease in the concentration of the Krebs cycle components (Figure 9.3). This seems to be a basic explanation for the anoxybiotic block of LDH, foreseen by Schumer [73].

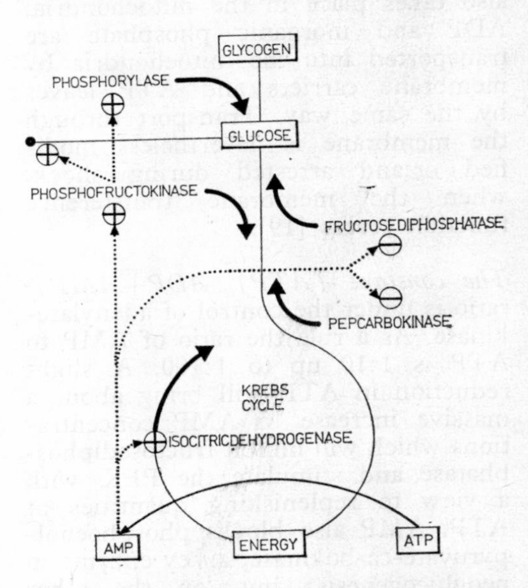

Figure 9.2. Enzymatic chains and adenylates. When ATP releases its energy, large amounts of AMP accelerate the Embden-Meyerhof-Krebs cycle to obtain another quantity of ATP.

9.3.2 ENZYMATIC DYNAMICS OF THE SHOCK CELL

Pulsatile positive and negative feedbacks act upon the enzymatic chains, the metabolic oscillations being the most obvious expression of the *biological clockwork;* shock hastens, forces or even brutalizes these oscillations, finally causing the subtle mechanisms of the vital 'clockwork' to deteriorate. In energy metabolism in the shocked cell there are several points at which the oscillations of enzymatic activity are more evident; these are also the critical biochemical steps in the evolution of shock towards irreversibility (Figure 9.4).

In dogs, mixed experimental shock (haemorrhagic and traumatic) produced by the modified Wiggers method, made

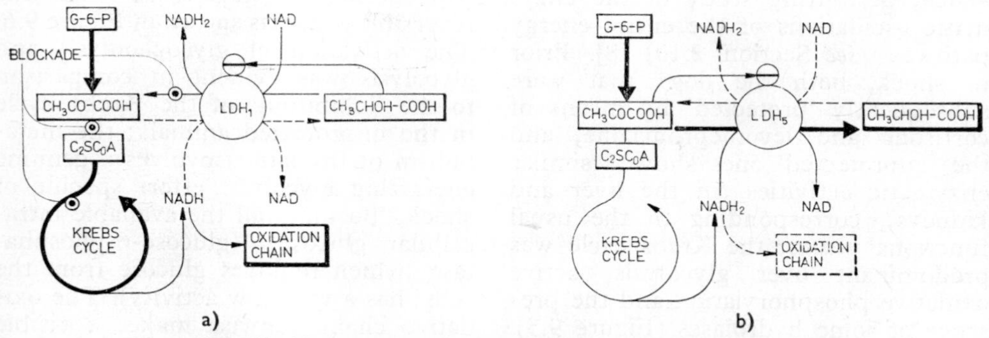

Figure 9.3. Blocking at point *a* leads to the situation illustrated at point *b*.

it possible to determine histologically and enzymologically the evolution of the enzymes representative of glycogenolysis, glycolysis, Krebs cycle and oxidative-phosphorylation in both the reversible and irreversible stages of

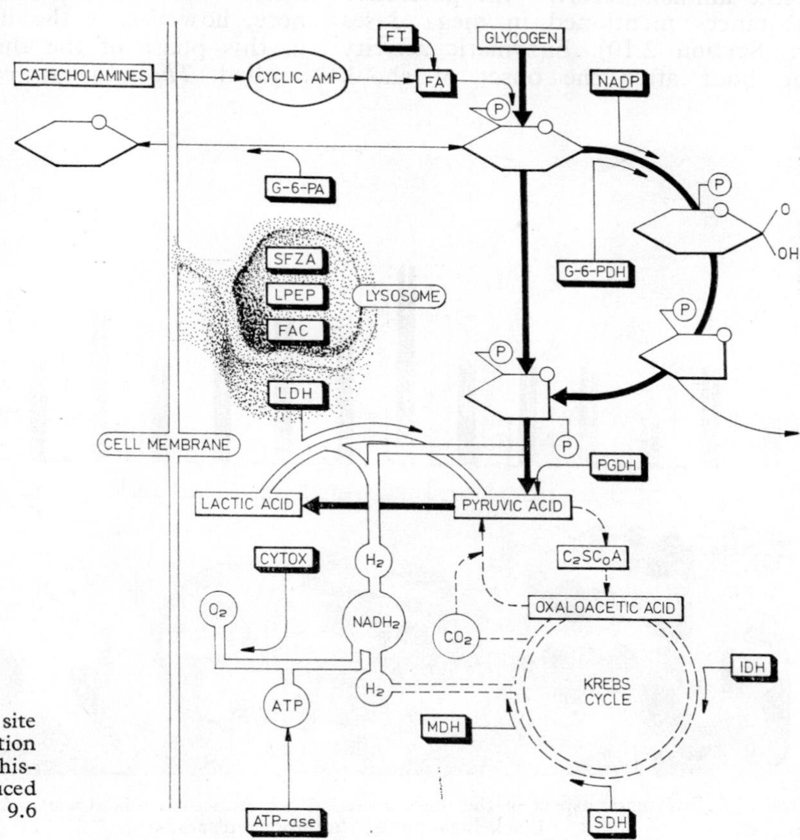

Figure 9.4. The site of enzymatic action illustrated in the histograms reproduced in figures. 9.5, 9.6 and 9.7.

shock, permitting study of the enzymatic oscillations of the entire energy pathway *(see* Section 2.10) [8]. Prior to shock, both the dogs that were subsequently 'protected' by means of cortisone and levomepromazine, and the 'unprotected' ones showed similar enzymatic activities (in the liver and kidneys), corresponding to the usual functional *status:* the Krebs cycle was predominant over glycolysis, active oxidative phosphorylation and the presence of some hydrolases (Figure 9.5). A state of shock was then developed and the blood pressure maintained between 70 and 60 mmHg for 60 minutes; after the first half to this interval some of the animals received the protective substances mentioned in mega doses *(see* Section 2.10). Enzymatic activity one hour after the onset of shook (considering this stage to be within the reversible phase) is shown in Figure 9.6. The activation of glycogenolysis and glycolysis was evident in comparison to the 'inhibition' of the Krebs cycle in the unprotected animals; the metabolism of the latter involves a genuine energizing glycolytic effort specific of shock, 'burning' all the available intracellular glucose (glucose-6-phosphatase, which removes glucose from the cells, has a very low activity). The oxidative chain likewise makes a visible effort, avidly consuming the little oxygen that can be extracted peripherally. ATP-ase is active, liberating the energy of the richest among the adenylates: ATP. Particularly worthy of note, however, is the intense activity in this phase of the three hydrolases studied. *The whole enzymatic spectrum*

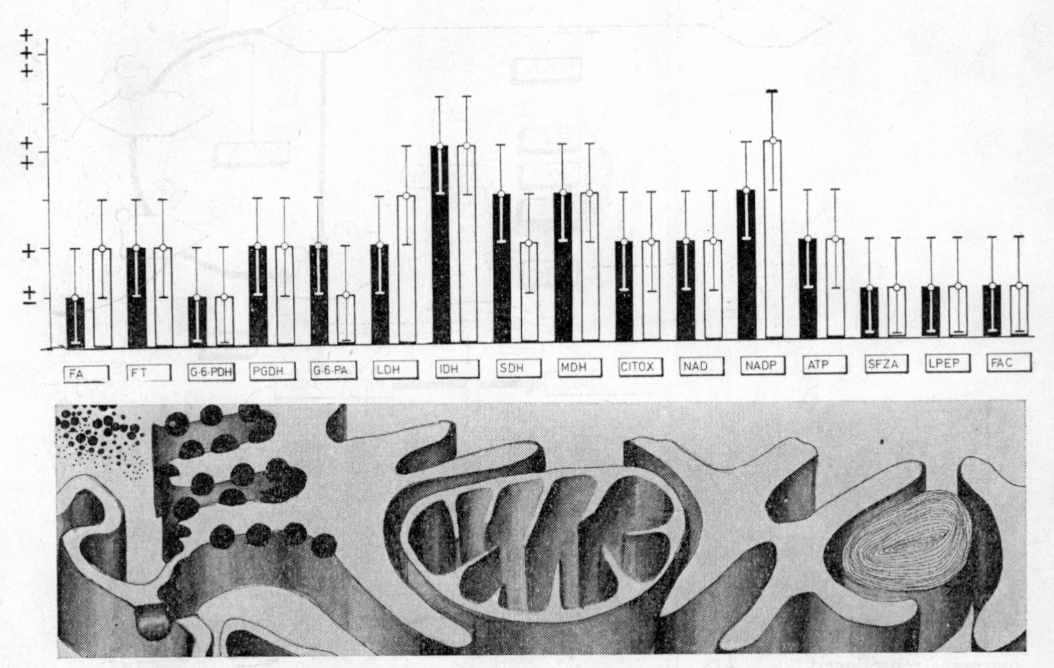

Figure 9.5. Enzymatic aspect of the onset situation (experimental, mixed acute shock in the dog). Black histogram – 'unprotected' animals.

described is characteristic of the metabolic effort of the shock cell to produce energy during the reversible phase.

In contrast to the unprotected cell is the protected shock cell which maintains its initial enzymatic balance (Figure 9.6).

After the state of shock was prolonged for another hour, at BP values of 40 to 30 mmHg, a third sampling was done (Figure 9.7). In the unprotected shock cell enzymatic activity is *exhausted:* glycogenolysis, glycolysis and oxydative-phosphorylation are insufficient. ATP-ase has nothing to split because the ATP has been entirely taken up. Glucose-6-phosphatase increases its activity, 'rejecting' glucose from the cell which is no longer able to use it. The intense, peak hydrolase activity (the phase of metabolic irreversibility of the shocked cell) when acidosis creates ideal conditions for self-digestion, is impressive.

The same significant contrast is found in the *protected cell,* in which enzymatic activity is still satisfactory. The remarkable significance of this protection does not lie so much in the drugs used, as in the fact that *it demonstrates the real possibility of therapy to sustain the membrane and enzymatic systems of the shock cell.*

Figure 9.8 shows comparative aspects in the activity of some dehydrogenases in protected and unprotected cells.

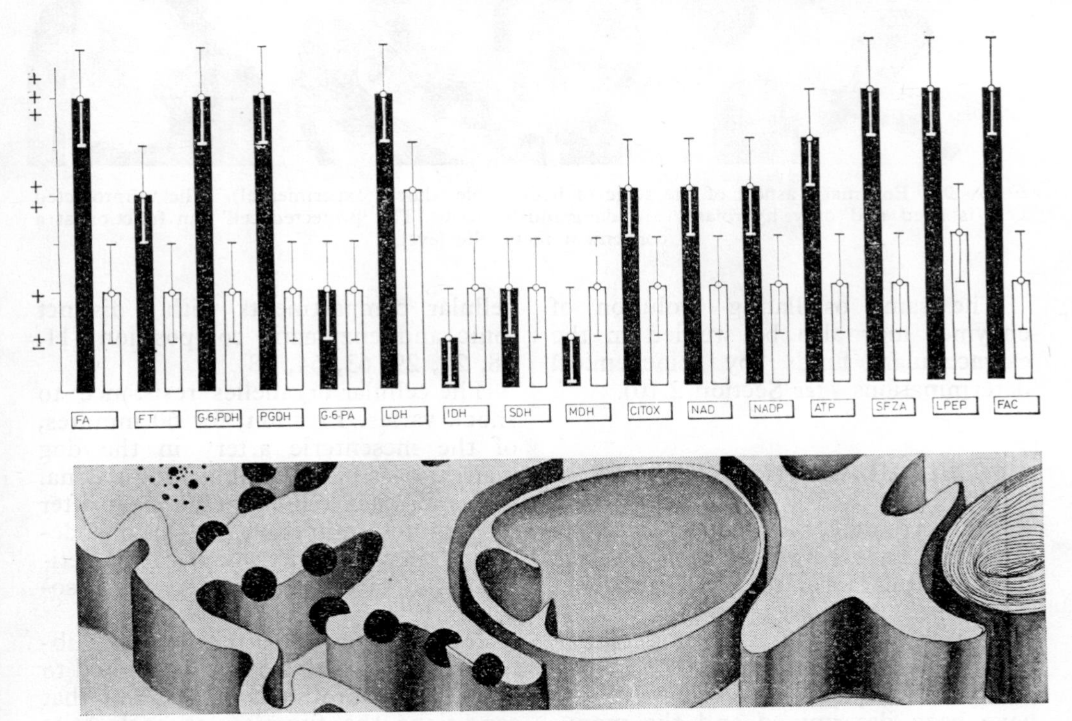

Figure 9.6. Enzymatic aspect of the stage of reversible shock (experimental). The 'unprotected' animals are in a cellular metabolic effort that is obviously exaggerated: glycogenolysis and glycolysis are exacerbated, the Krebs cycle cannot 'breathe' and hydrolases appear. The 'protected cell' does not produce this metabolic exhaustion.

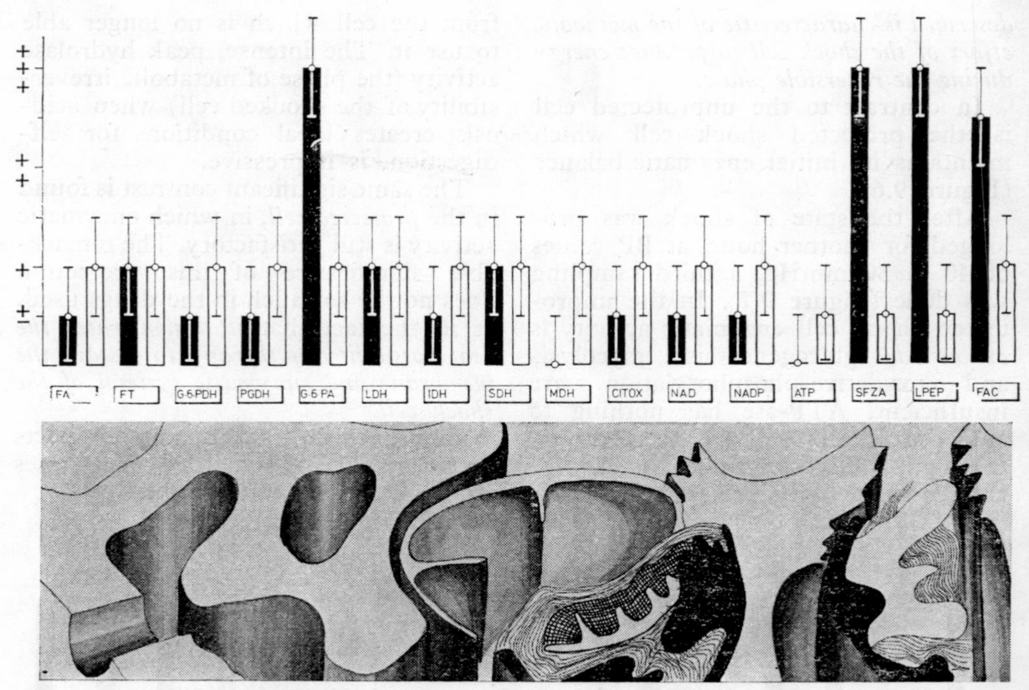

Figure 9.7. Enzymatic aspect of the stage of irreversible shock (experimental). The 'unprotected cell' is tired and only hydrolases are dangerously active. The 'protected cell' can function at a convenient metabolic level.

The same oscillating evolution of enzymes may also be studied in the extracellular fluids by biochemical determinations *(see* Section 2.10).

9.4 CELLULAR ORGANELLES

Recent cytologic, biochemical and genetic data have revolutionised the picture of cellular structures and functions. A new light was cast upon the elementary architecture of the intracellular organelles (Figures 9.9 and 9.10). Specialized functional regions of the cell have been determined and the membrane network was found to be the key to the entire intracellular organisation; the membrane folds delimit intracellular compartments with a distinct ionic and enzymatic composition [11, 18, 24, 29, 63, 64, 78].

The cellular organelles resistance to shock varies. Ligation, for 60 minutes, of the mesenteric artery in the dog made it possible to establish the lesional stages in the cellular organelles: after 30 minutes, mitochondrial lesions develop, then disorganisation of the reticulum, nucleus and, finally, the lysosomes [2].

The different sensitivity of the subcellular structures may be attributed to the variable proton concentration that conditions the function of each one; in the cytosol the *p*H is 6 to 5, the nucleic acids bring the *p*H in the karyoplasm to 5 to 4, whereas in the lyso-

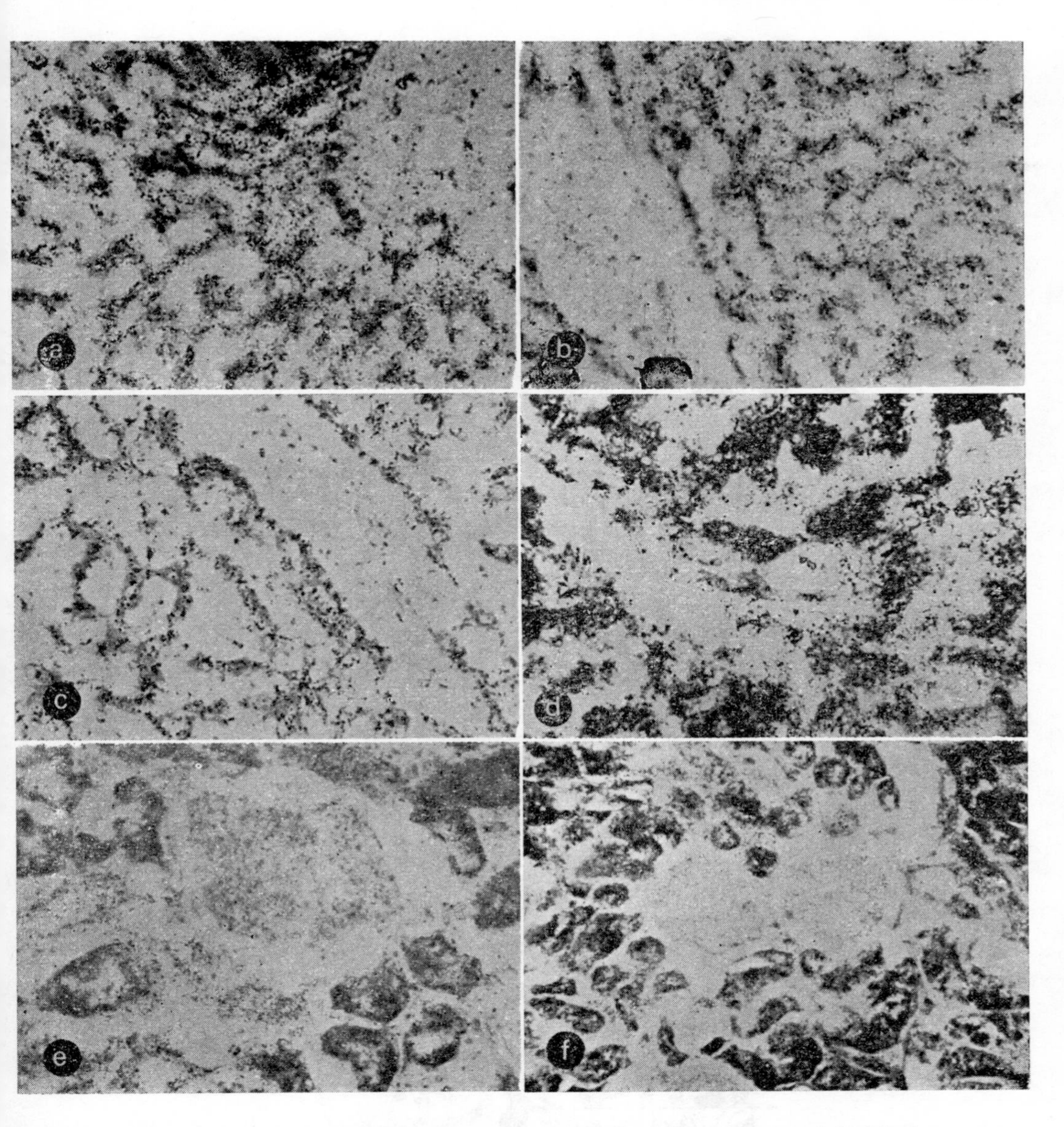

Figure 9.8: a — Liver (protected dog). Reversible shock. Intense positive succinic-dehydrogenase activity; *b*—Liver (unprotected dog). Reversible shock. Positive succinic-dehydrogenase activity; *c*—Liver (protected dog). Reversible shock. Positive and slightly positive phosphoglyceraldehyde-dehydrogenase activity; *d* — Liver (unprotected dog). Reversible shock. Intensely positive phosphoglyceraldehyde-dehydrogenase activity; *e* — Kidney (protected dog). Reversible shock. Positive (renal tubules) and slightly positive (glomerules) lactic-dehydrogenase activity; *f* — Kidney (unprotected dog). Reversible shock. Intensely positive lactic-dehydrogenase activity in the renal tubules.

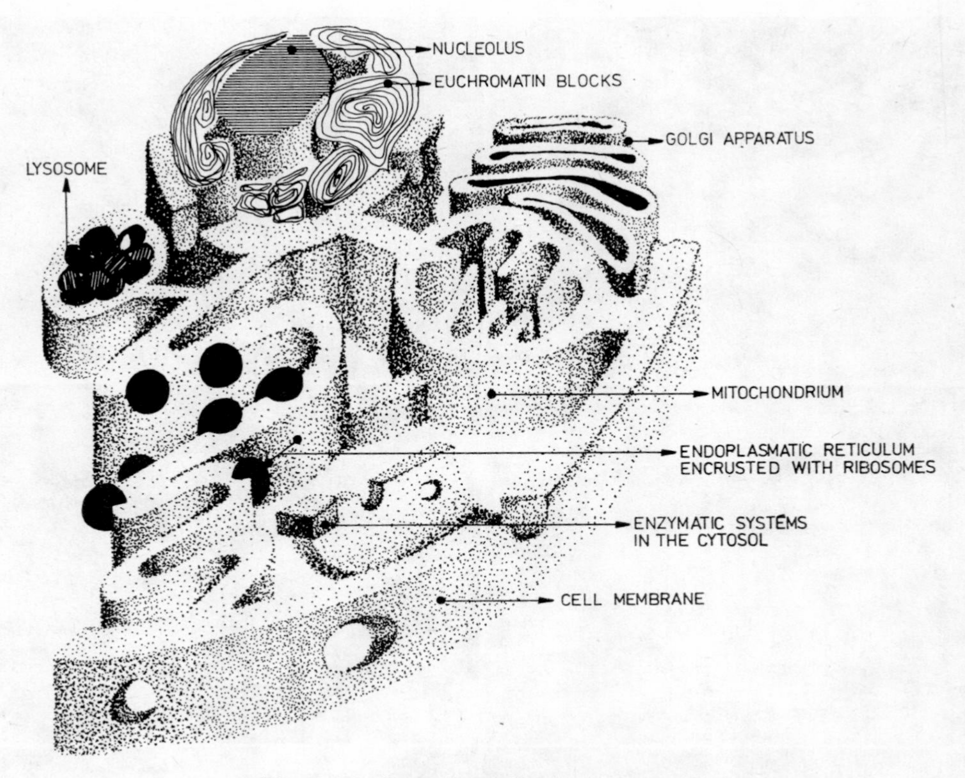

Figure 9.9. Organelles within a normal cell.

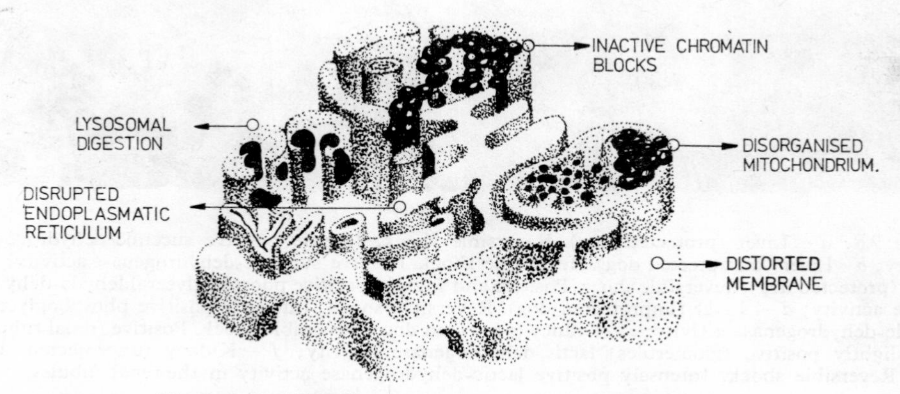

Figure 9.10. The disorganized organelles of a shock cell.

somes the *p*H has a very wide range (8.5 to 5.5). This explains why intracellular shock acidosis attacks the nucleus later, while corrosion of the lysosomal sacs, the least resistant to acidity, starts at *p*H 5.5. Paraphrasing Claude Bernard, it may be said that the cellular subsystems live in common but die separately, the death of the cell being the result of an altered cooperation of its organelles.

9.4.1 THE NUCLEUS AND THE GENETIC CODE

The genetic code is the true vital matrix. As a rule, only a small number of operonal genes are active at a given moment and in a certain cellular population. However, all the cells contain the whole genetic potential of the species they belong to and this potential may be activated at any time by modifications, etc. *(see* Sections 3.7 and 4.8). The morphologic and functional integrity of the genetic code is essential in supporting the shocked cell. In old people or in those with genetic or even chromosomal lesions, the cybernetics of the genetic code is disturbed, affecting the whole energy system of the cell [35, 82].

The genetic code is the leader of the 'federation' of intracellular organelles, sychronizing the 'enzymatic batteries' and the 'membrane lodges'; it is the genetic code that will receive the full blow of the shock agent and will make an attempt at a rapid accommodation which in most cases exceeds its abilities, forcing it into irretrievable sacrifices.

In shock, the genom increases enormously the output of its operons, forcing its energy yielding intracellular sub-systems and actually reorganizing the levels of the sources of cellular entropy.

But the alphabet of the genetic code — the nucleic acids — is promptly and severely punished; in less than 6 minutes after shock has set in, anti-DNA antibodies with a peak titre after 3 hours, have been detected by immunochemical methods. In the survivors the titres remain high for several months.

These data have been established during experimental haemorrhagic and bacterial shock as well as in the human clinic [42, 73, 78]. Among the nucleotides, cytosine and guanine proved to have the highest antigenicity. In a clinical group, in 30 out of 40 shocked patients, studied by haemagglutination-inhibition and by direct agglutination of the blood cells sensitized with individual nucleotides, the presence of specific antibodies to the tymine-containing nucleotide was frequently associated with a fatal result [42].

The appearance of anti-DNA antibodies may be due to alteration of the circulating immunoglobulins, as the promptness with which they systematically appear (before the lysosomal hydrolases) reveals the intervention of rapid plasma mechanisms.

The appearance of antibodies may also be the response to morphologic deterioration of an exhausted polynucleotide system. No matter how the events develop, the systemic circulating anti-DNA antibody titres are always connected with the severity and the duration of shock and are a more reliable prognostic *marker* than the determination of circulating hydrolases.

9.4.2 THE MEMBRANE SYSTEM

The membrane is the organelle that plays fundamental architectural and thermodynamic roles. Its folds withiu the cellular space delimit and form the intracellular functional compartments.

It is the membrane that determines and also performs intercompartmental transports, employing more than 50% of the total energy produced by the cell [60]; for instance, at least three mechanisms that insure the transport of calcium through the mitochondrial membrane have been described [15, 45].

In shock, proton accumulation in the cytosol block the release of calcium from the mitochondria causing intramitochondrial precipitation [39, 48].

The membrane liposomes act as a barrier to the cations on the water/lipid interface; at this level anaesthetics (ether, chloroform, etc.) may increase the permeability of the membrane by changing the entropy of the system [41].

During shock, the membrane potential diminishes, facilitating transmineralization; release of potassium from the cell and the admission of sodium may be due to blocking of the ion pumps or to the increased permeability for sodium, owing to a change in the membrane pores [1, 14, 18]. The loss of selectivity of the membrane is the cause of cellular and organelle œdema. In irreversible experimental shock it has been established that the transport of H^+ and Na^+ ions through the membrane definitely ceases [52].

9.4.3 THE ENERGY-PRODUCING APPARATUS — THE MITOCHONDRIA

The mitochondria are micro-ovoids with the internal membrane folded in crests within which the principal enzymatic energy chains of the cell are enclosed. The outstanding role of the mitochondrion in cell respiration and energy production is well known. The mitochondrion is very much like a bacterium; some authors indeed consider it to be an evolved derivative of some ancestral bacteria included in the cells of metazoa [58]. The number of mitochondria with which a cell is provided depends on the functional role of the latter. In man, the hepatocyte disposes of about 1 000 mitochondria that can move through the cytosol by twisting. As a rule, in the electron microscope the mitochondrion presents a shape corresponding to one of the four phases of its swelling-contracting cycle: the orthodox, the condensed, the intermediate and the swollen shape. This reversible mitochondrial cycle is based on the existence of certain contractile mechanoenzymes in the membrane and on a mechanical-chemical flexibility, which bear witness to a good function of the intra-mitochondrial enzymatic pathways.

There is, in fact, a cycle of physical aspects of the mitochondrial metabolic pulse [9]. The outer cytosol compartment appears to give up part of its volume to the mitochondrial matrix during the condensation-to-orthodox phase changes. The choking by cyanides of mitochondrial respiration ‘freezes’ the mitochondrion; the same effect is produced by endotoxins.

In the isolated mitochondria of mice with endotoxin shock, a pathological swelling is observed, which points to the inability of the mitochondrial membrane to maintain the intercompartmental ionic asymmetry (White IV, 1973).

It may be considered that hypoxia initially depresses the EMK-OP cycles, and hence the first organelle to be shocked is the mitochondrion.

The events are, however, far more complex. As the mitochondria isolated in states of shock considered to be irreversible proved able to breathe *in vitro* [4, 16, 35, 36] the question arose of whether inhibition of the Krebs cycle is but a relative state with regard to the runaway production of C_2SCoA groups by the Crabtree effect.

The intra-mitochondrial effects of *E. coli* endotoxins offers one of the most fertile experimental models. Thus, it was established that endotoxin inhibits ATP-ase as well as oxidation of the succinate glutamate and malate substrate, but that endotoxin also inhibits the accumulation of phosphorus and calcium for pyrophosphoric bonds and has an uncoupling effect on phosphorylation due to its lipid component [87].

During shock there exists an evolutive morpho-functional cycle of the mitochondria: the multi-enzymatic system of α-ketoglutaric acid is diffusely distributed in the matrix and is the first to suffer under the influence of hypoxia; on the other hand, the succinate system is more resistent as it is situated on the internal membrane of the mitochondrion. Swelling of the mitochondria would first affect the glutarate system, by dilution (or loss in the cytosol) of the cofactors of the system. Another possibility is the invasion of the mitochondria by FFA, owing to the lowered permeability of the membrane [4, 24, 37, 75].

Protection of the shocked mitochondrion implies maintainance or restoration of the selectivity of its membranes. For restoring the selectivity of the membrane it has proved useful to administer magnesium, which seems to be pathologically absorbed by a particular magnesium-avid macromolecule that appears in the shocked cell [4].

In shock, the intracellular osmotic alterations dictate a complete pathological contractile mitochondrial cycle. Acidotic hyperhydration of the cell and intra-mitochondrial blocking of calcium result in swelling of the mitochondrion, which expands 6 to 15-fold its normal volume; swelling is followed by the tearing and destruction of the cristae, by breaking of the outer mitochondrial membrane, by intra-mitochondrial precipitates and discharge of the mitochondrial matrix into the cytosol; these aspects observed in the electron microscope, confirm the irreversible destruction of the mitochondria (Strawitz and Hift, 1965; Baue, 1973; DePalma, 1970).

9.4.4 THE DIGESTIVE APPARATUS — THE LYSOSOMES

The lysosomes, phylogenetically older than the mitochondria perform the essential digestive functions of the cell. Synthesized by vesiculation from the reticular membranes, they comprise over 40 types of hydrolases (similar to those that carry out digestion in the digestive tube) and form the *endocellular digestive apparatus* [20, 21]. Since in the shock cell the mitochondria can only partly achieve the oxidative production of energy (it is not able to breathe) the digestive functions of lysosomes are likewise involved in the defence against the lesional agents.

Mobilisation of the lysosomes is exceedingly dangerous, because the hydrolases will attack not only the potential endocellular aggressors (bacteria, various antigens, etc.) but also its own cellular matrix. Although it is only during the last decade that numerous data regarding the behaviour of the lysosomal system have 'invaded' the pathologic physiology of shock, the lysosomes are old acquaintances of the haematologists, the leukocyte granules being typical lysosomes staining with Giemsa or toluidine blue, and carrying out the essential function of the phagocytes [13]. There is a physiological turnover of the endocellular wastes which are taken up by the lysosomes, where the enzymes exhausted along the metabolic cycles also find their end [45]. The lysosomal cisterns are opened by a change in the *p*H. Acidity (the same as in gastric digestion) prepares the digestive medium and staining has shown that at *p*H 5 there are no longer any granules in the leucocytes (De Duve, 1970).

During the various stages of shock, the behaviour of the lysosomes may be studied by means of biochemical hydrolase determinations — in the plasma, lymph and urine — or electron microscope, revealing both the physical aspect of lysosomes and the activity of certain hydrolases by combined histochemical and electron optical methods.

The hydrolase flow digests the cell and then penetrates into the systemic blood stream, where it triggers plasma protease activity with all its rheologic consequences (*see* Section 7.4.4).

Elevation of the serum lysosomal enzyme titres has been demonstrated in various types of shock, and is almost always related to the severity of shock and even its irreversibility [5, 6, 22, 42, 46, 69, 81]. The liver cells, enterocytes, spleen and pancreas cells seem to be the main producers of the hydrolase flow that reaches the systemic circulation by the lymphatic route [32]. Experimental enterectomy has shown that 70% of the acid phosphatase, that increases 8 to 10-fold in shock, is released by the enterocytes [17, 78]. According to the latest research it has been concluded that the cells belonging to the RES, distributed throughout various viscera, are the most sensitive to shock, releasing the first hydrolase discharges. Although the kidney is thrifty in liberating hydrolases it is 'filled' in shock with hydrolase vesicles [28].

In bacteriemic shock the lysosomes of the cells belonging to the RES rapidly phagocytize in the first phases (lysophagosomes). It seems that endotoxins have an elective insertion on the hepatic RES, attacking the lysosomal membranes inside the cells; this was confirmed experimentally by the fact that 5 minutes after inducing bacteriemic shock hydrolases massively increase in the plasma and lymph [5]. In experimental haemorrhagic shock the release of hydrolases has a latency of 1 to 2 hours, appearing after the BP values fall to 40 to 35 mmHg [5, 7, 10, 81]. In haemorrhagic shock, the part played by lysosomes in bringing forth irreversibility, has often been doubted [38, 75].

9.5 STRUCTURAL ASPECT OF THE SHOCK CELL

A comprehensive and truthful picture of shock, from the clinical and, especially, haemodynamic point of view, implies approaching the problem *via* intracellular enzymatic processes and, particularly, through the physical alterations of the subcellular organelles visualized in the optical and electron microscopes. Evaluation of the lesions of the intracellular organelles offers a reliable aetiologic basis to the clinical signs that appear subsequently following their summation.

The abundance of electron and optical pictures of the *shocked cell*, communicated in almost all the fundamental types of shock [3, 8, 11, 19, 22, 23, 25, 31, 36, 37, 52, 55, 62, 68, 69, 79, 81, 86, 87] makes it possible to discuss the architectonics of the cell in the evolutive steps of shock; however, one must not overlook the fact that an optical or electron microscopic picture shows the cell at a given moment and not in its entire evolution in shock.

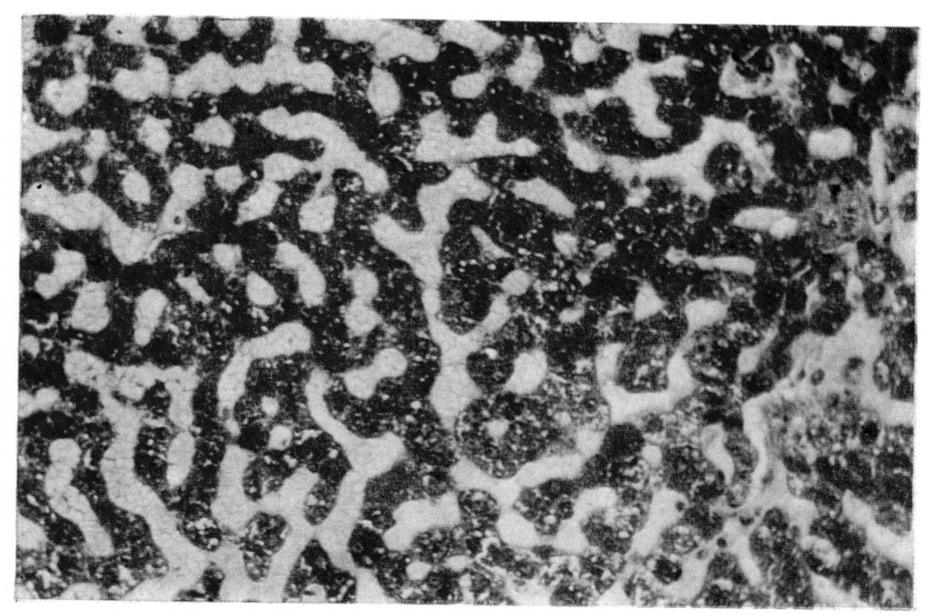

Figure 9.11a. Liver (unprotected dog) in reversible shock (H.E., ×80). Intrasinusoid erythrocytic stasis with the appearance of liver cell vacuolar dystrophies. Typical RBC clumping. Note hazy contour of some of the cells.

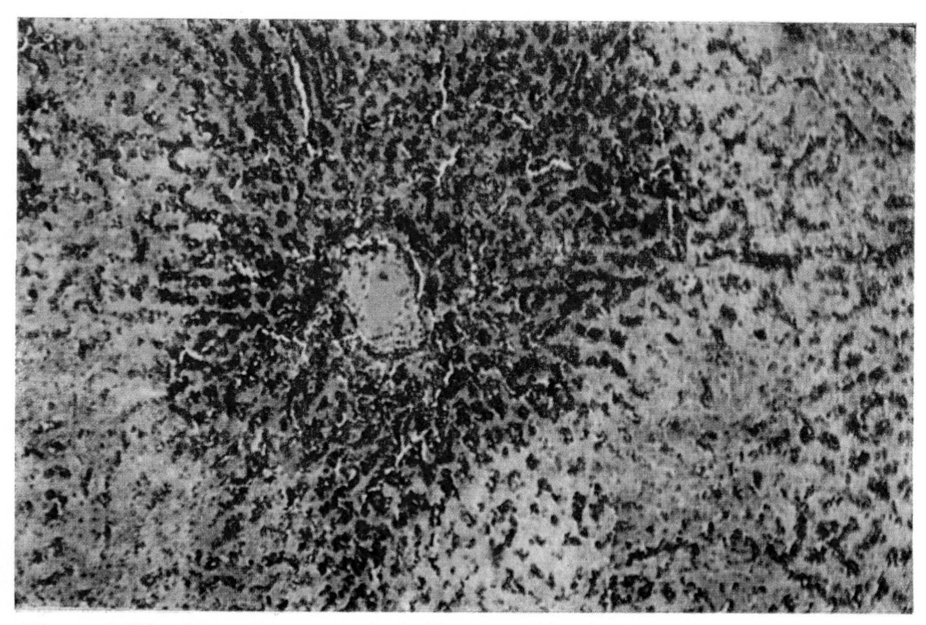

Figure 9.11b. Liver (unprotected dog)in reversible shock (H.E., ×30). Accentuated peri-centrolobular erythrocytic sludging, prior to the formation of microthrombi.

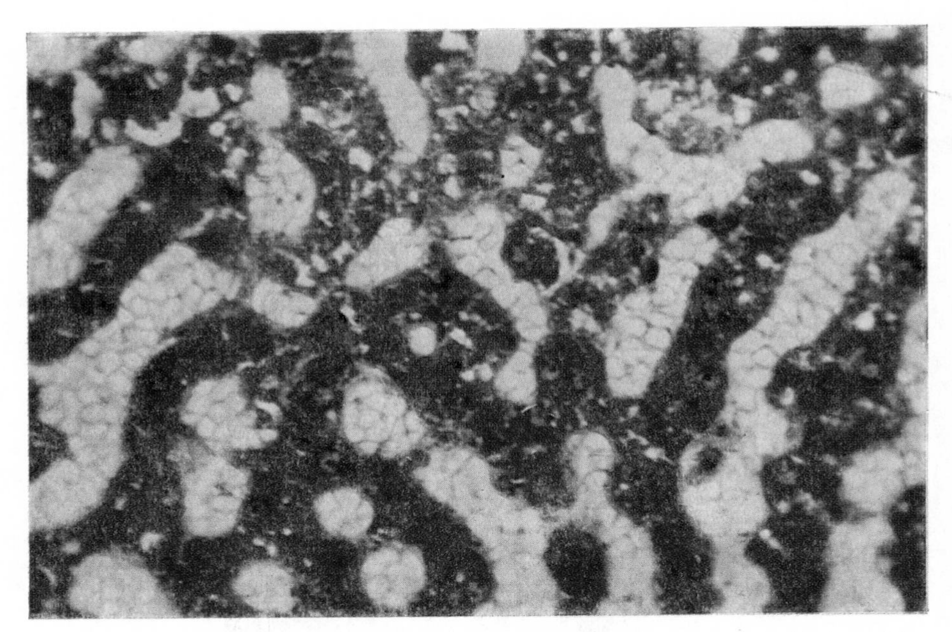

Figure 9.11c. Liver (protected dog) in shock considered to be irreversible (H.E., ×360). Intra-sinusoidal erythrocytic stasis with liver cell vacuolar dystrophy, aspect similar to that of the reversible stage in unprotected dogs. The RBC contour is more sharply outlined.

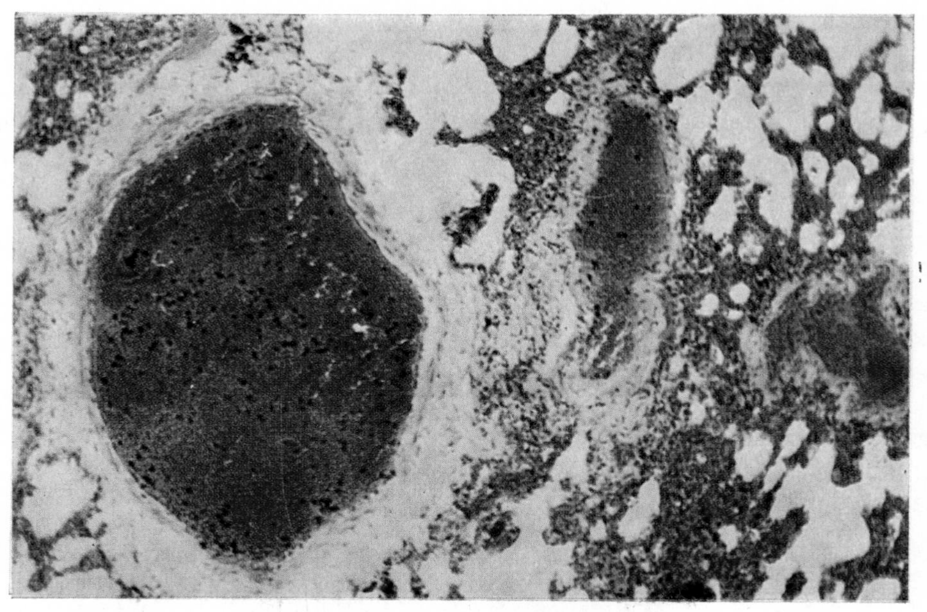

Figure 9.11d. Lung (unprotected dog) in shock considered irreversible (H.E., × 80). Massive disseminated thrombosis of mixed type.

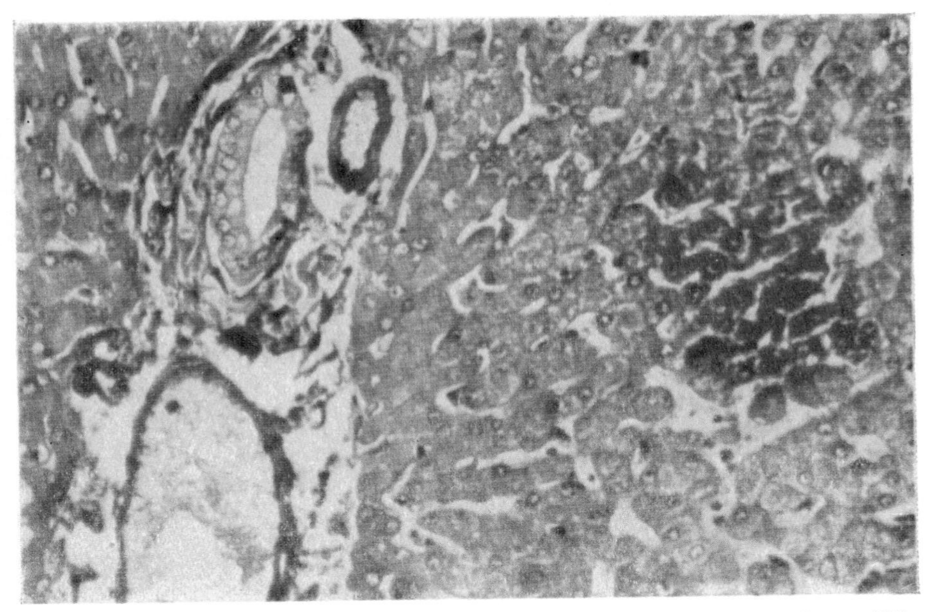

Figure 9.11e. Liver (unprotected dog) in reversible shock (Best's carmine, ×360). Note intense peri-portal 'melting' of liver cell glycogen, with persistence of a glycogen mediolobular island.

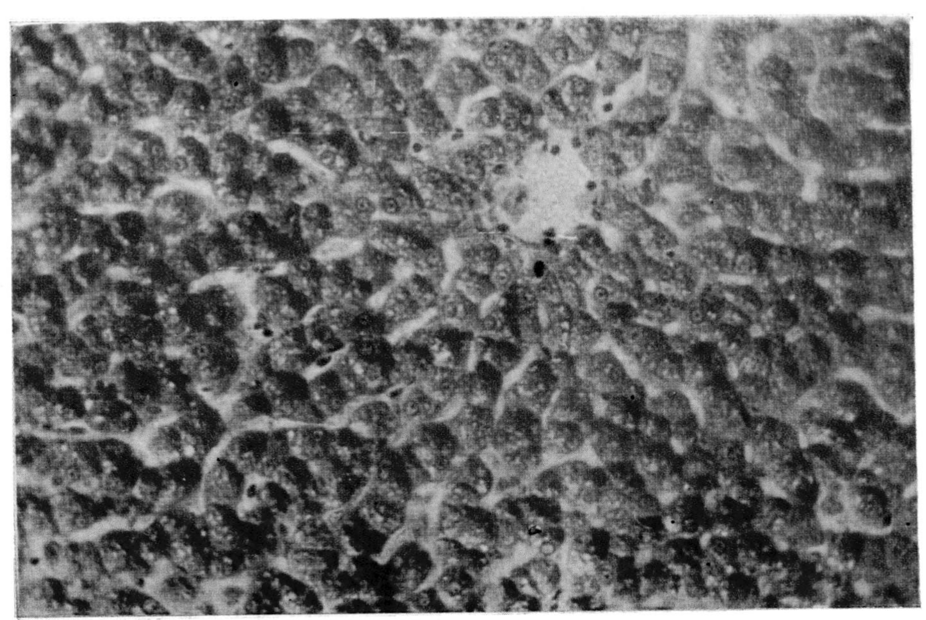

Figure 9.11f. Liver (protected dog) in reversible shock (Best's carmine, ×360). Glycogen maintains its homogeneous intra-hepatocytic distribution, consumption beginning peri-centrolobularly.

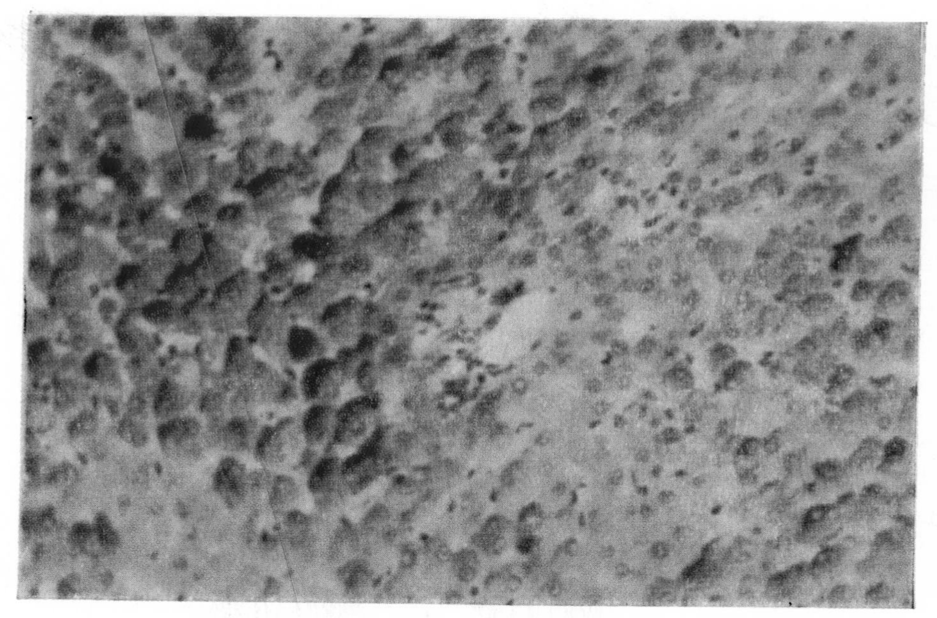

Figure 9.11g. Liver (unprotected dog) in reversible shock (Herovici stain, ×360). Asymmetrical, peri-centrolobular glycogen consumption.

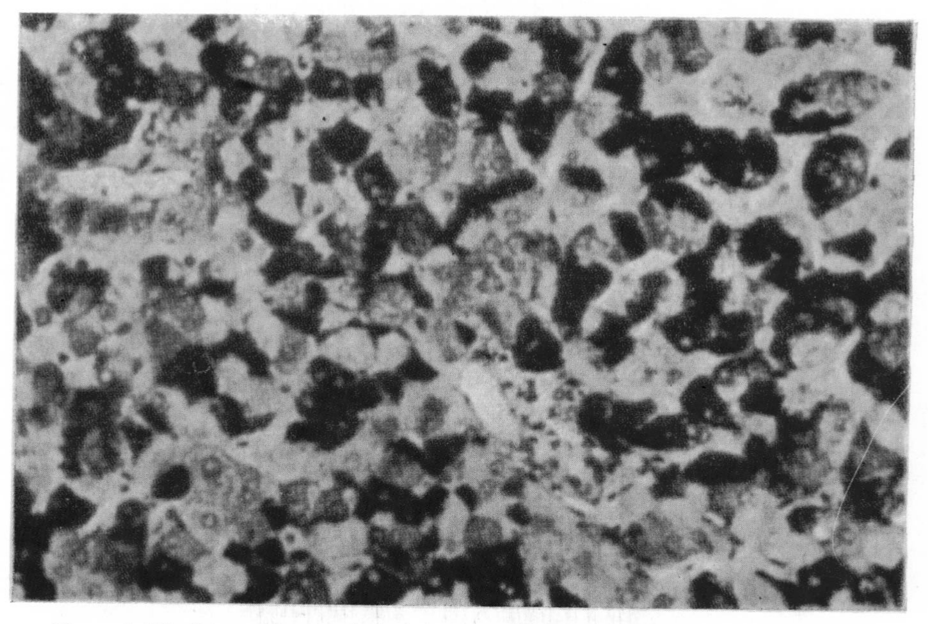

Figure 9.11h. Liver (protected dog) in reversible shock (Herovici stain, ×360). Glycogen persists in clusters, homogeneously distributed within the liver cells.

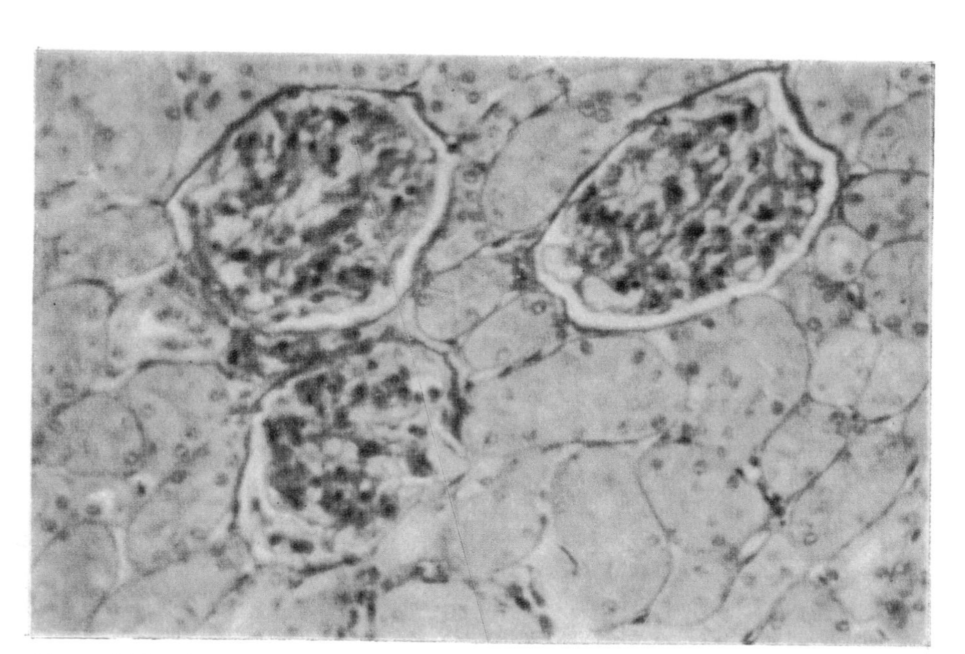

Figure 9.11i. Kidney (unprotected dog) in reversible shock (Herovici stain, × 360). Glomerular stasis and capsular oedema. Paraglomerular tubular dystrophy in the juxtaglomerular area. Diminished apical mucopolysaccharides in the tubulocytes.

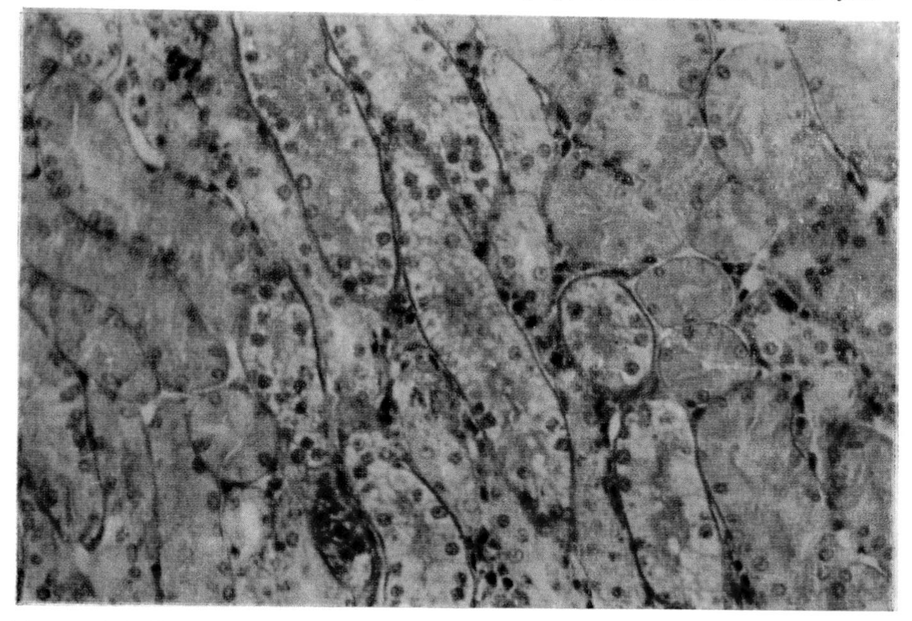

Figure 9.11j. Kidney (unprotected dog) in shock considered irreversible (Herovici stain, × 360). Cytoplasmatic vacuolar degeneration and dystrophy; nuclear ghosts and ragged basement. Detached protoplasmatic mucopolysaccharide material in the tubular lumen.

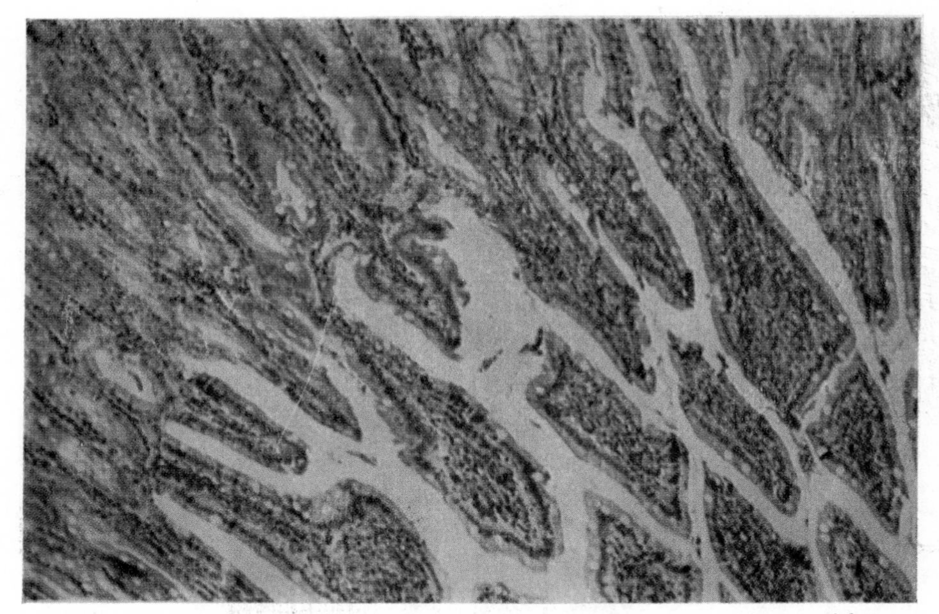

Figure 9.11k. Intestinal villi (unprotected dog) in reversible shock (Masson trichrome, × 30). Generalized haemato-lymphatic capillary stasis along the axis of the villi.

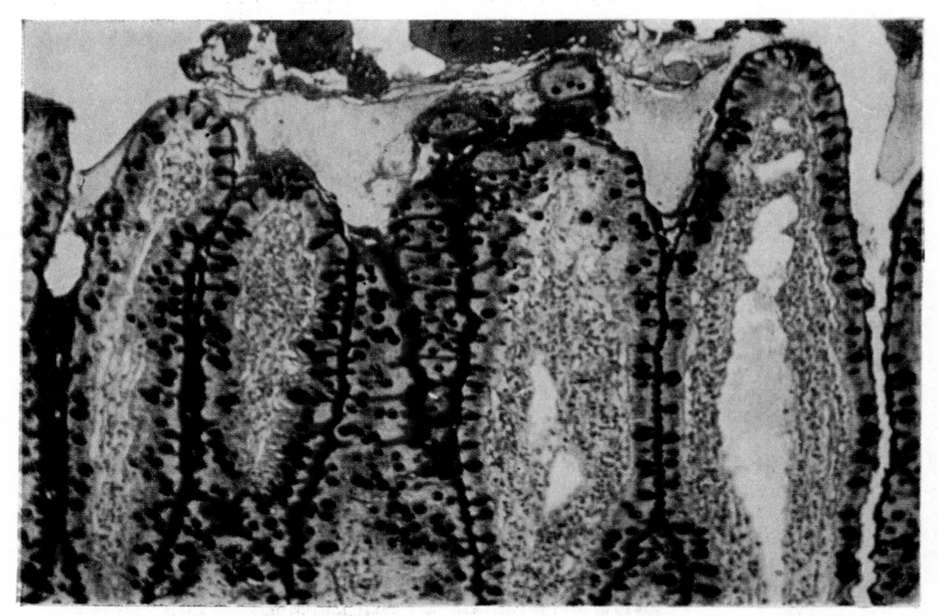

Figure 9.11l. Intestinal villi (unprotected dog) in shock considered irreversible (Masson trichrome, × 80). Pronounced catarrh of the mucosa with decapitation of the villi and expulsion *en bloc* of the mucopolysaccharide material.

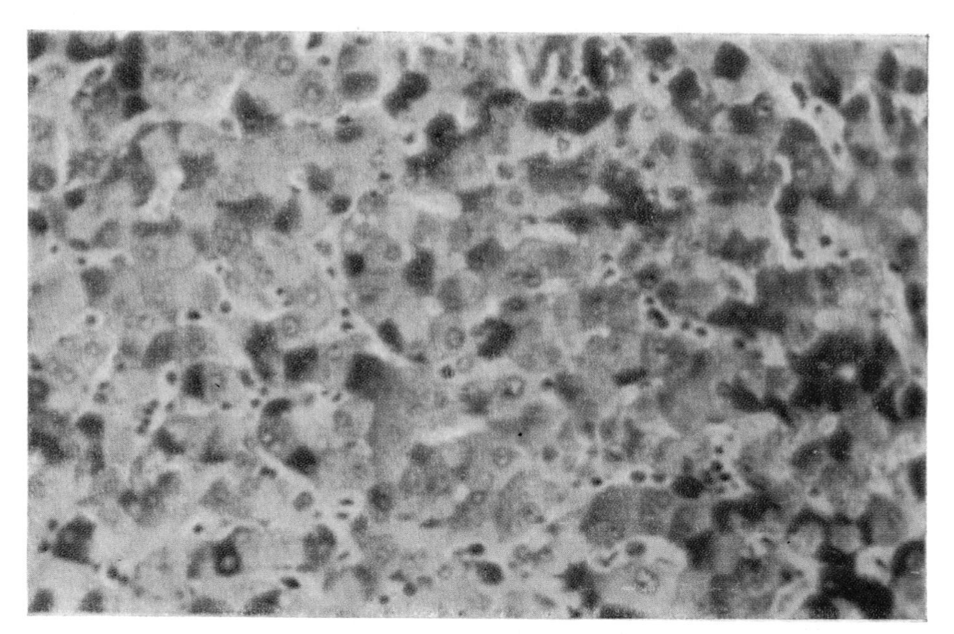

Figure 9.11m. Liver (protected dog) in shock considered irreversible (Herovici stain, ×360). Although non-homogeneous, the glycogen deposits persist in this stage.

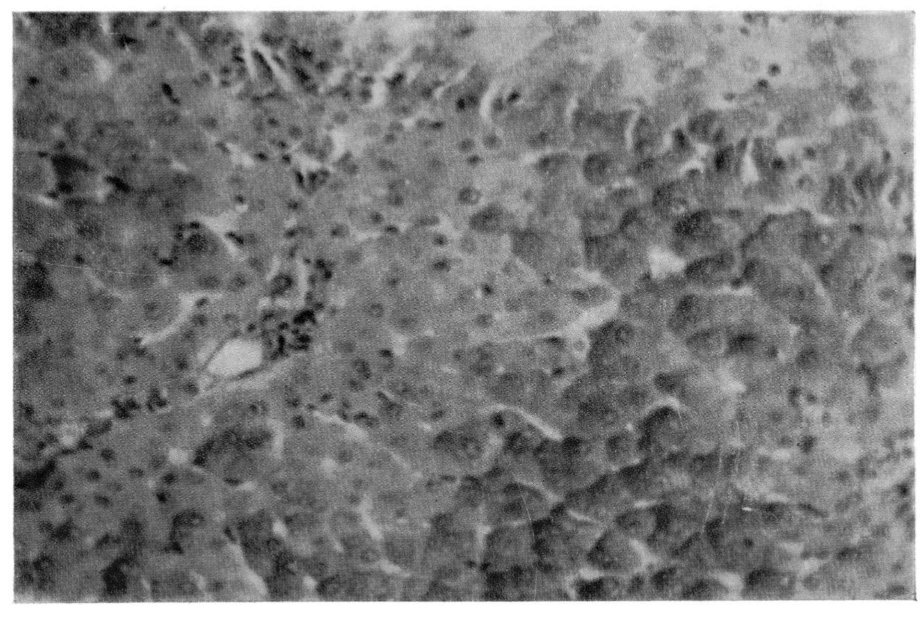

Figure 9.11n. Liver (unprotected dog) in shock considered irreversible (Herovici stain, ×360). Supra-acute evolution of shock does not always deplete the liver cell glycogen reserves.

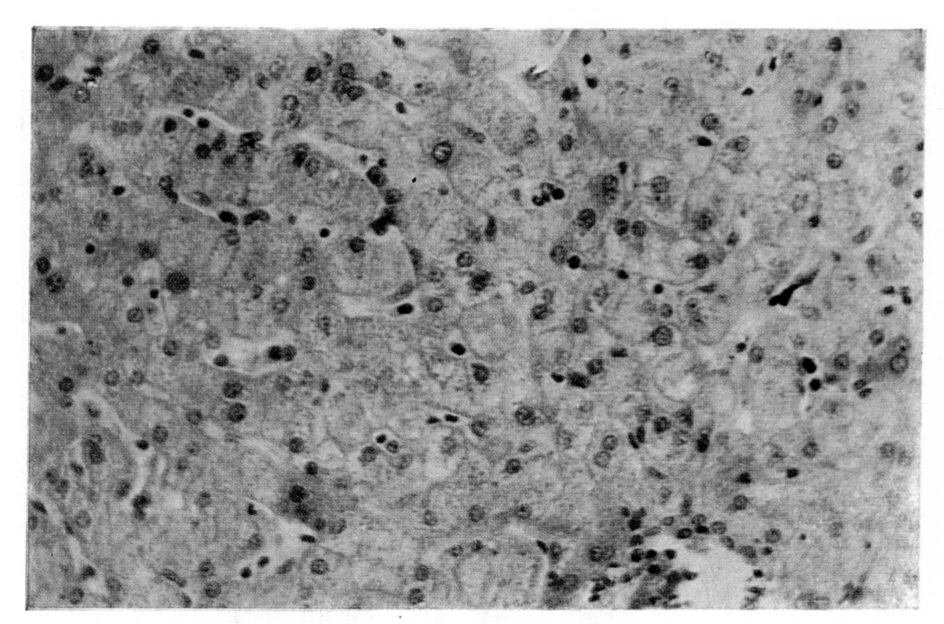

Figure 9.11o. Clinical case with upper digestive haemorrhage (duodenal ulcer) in late reversible shock. Intra-operative liver biopsy (H.E., ×360). Marked liver cell lesions of the granulovacuolar dystrophy and homogenisation type.

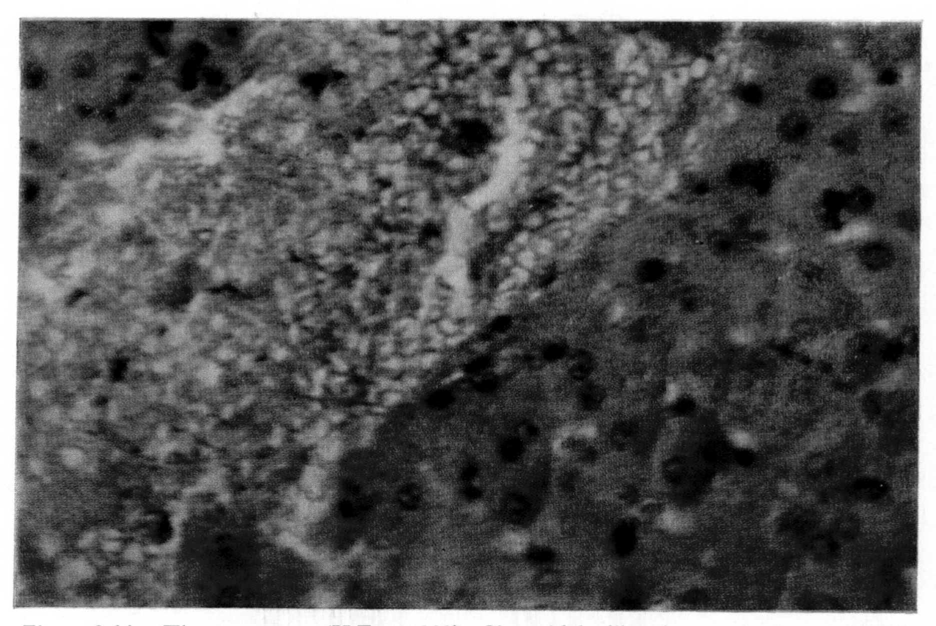

Figure 9.11p. The same case (H.E., ×320). Sinusoidal dilatation with clustered RBC with a hazy contour; intrasinusoidal thrombosis in the course of organization.

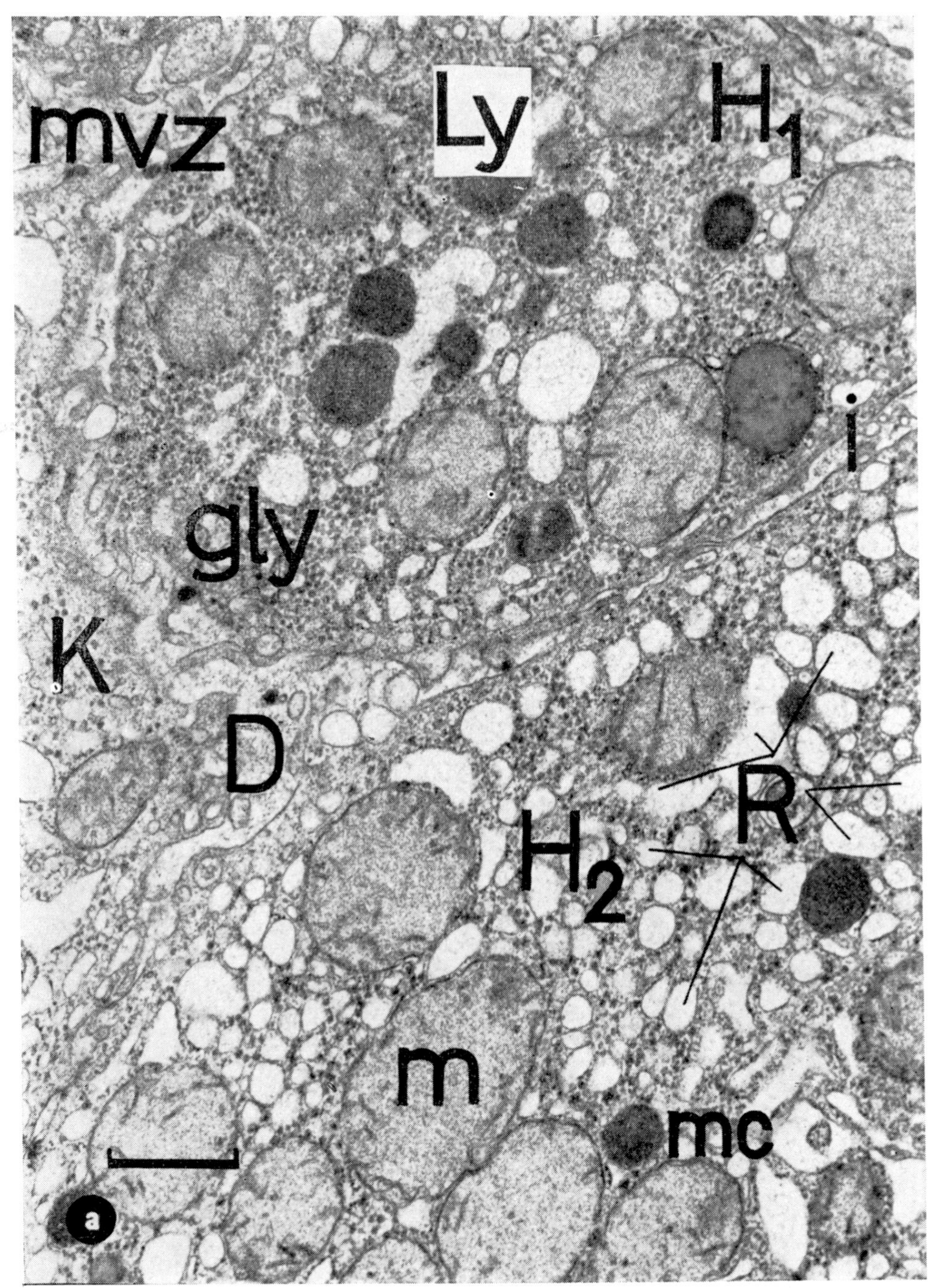

Figure 9.12a. Electron microphotograph, liver, dog, before the onset of shock (×16,000). Two liver cell areas (H_1 and H_2) separated by an intercellular space (i), exhibit microvilli (mvz) that penetrate into Disse's space (D) in the vicinity of a vesicular area in a Kupffer cell (K) – In the meshes of the smooth endoplas matic reticulum (R) several mitochondria (m) of unequal cross-section, intact lysosomes (Ly), microcorpuscles (mc) and glycogen grains (gly) are anchored.

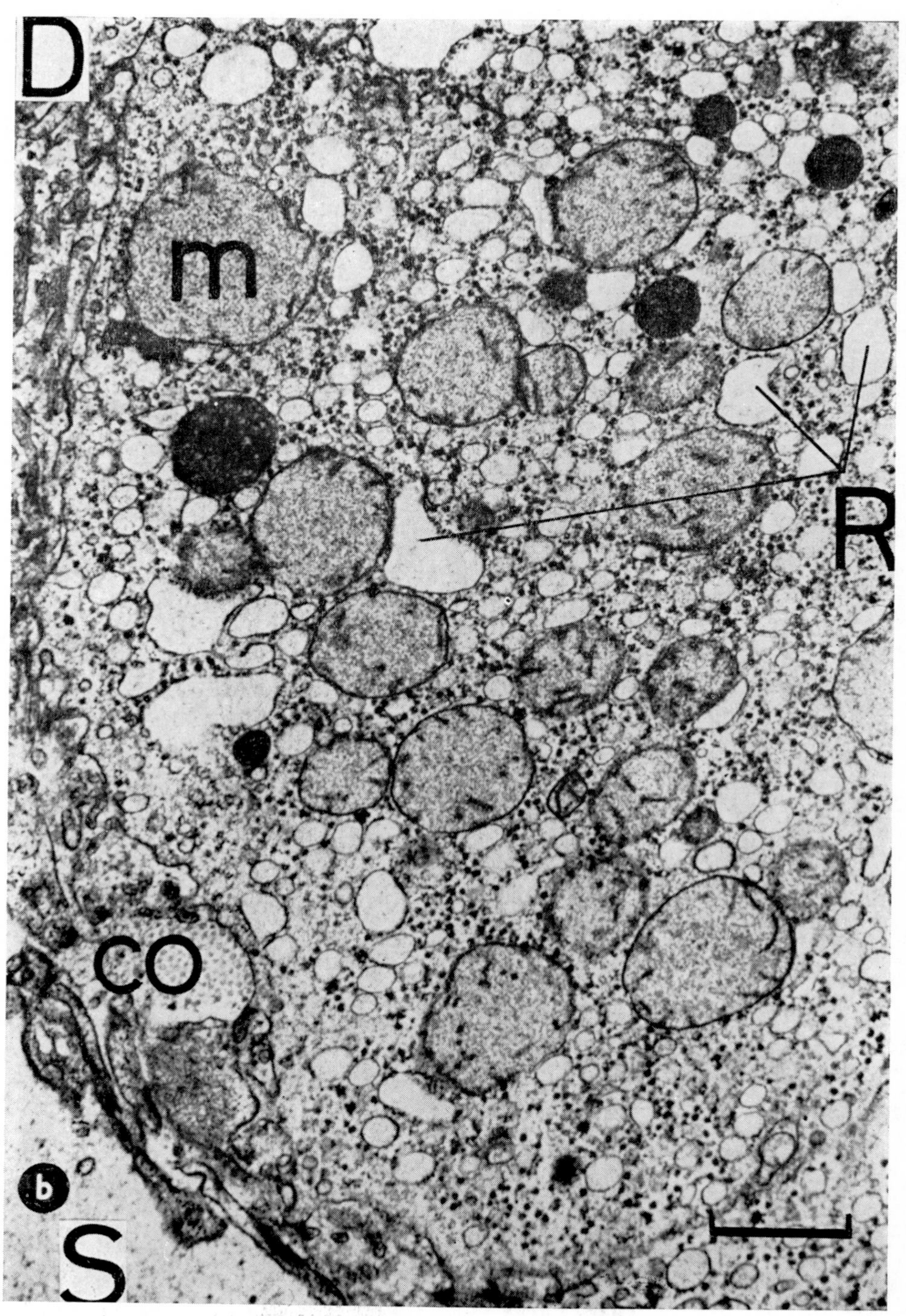

Figure 9.12b. Electron microphotograph, liver, dog in reversible shock (× 16,000). Parasinusoidal (S) liver cell. Note rarefaction of the reticulum, diminution of the glycogen, clarification of the mitochondrial matrix and amputation of the microvilli in Disse's space, where deposits of collagen material appear (co).

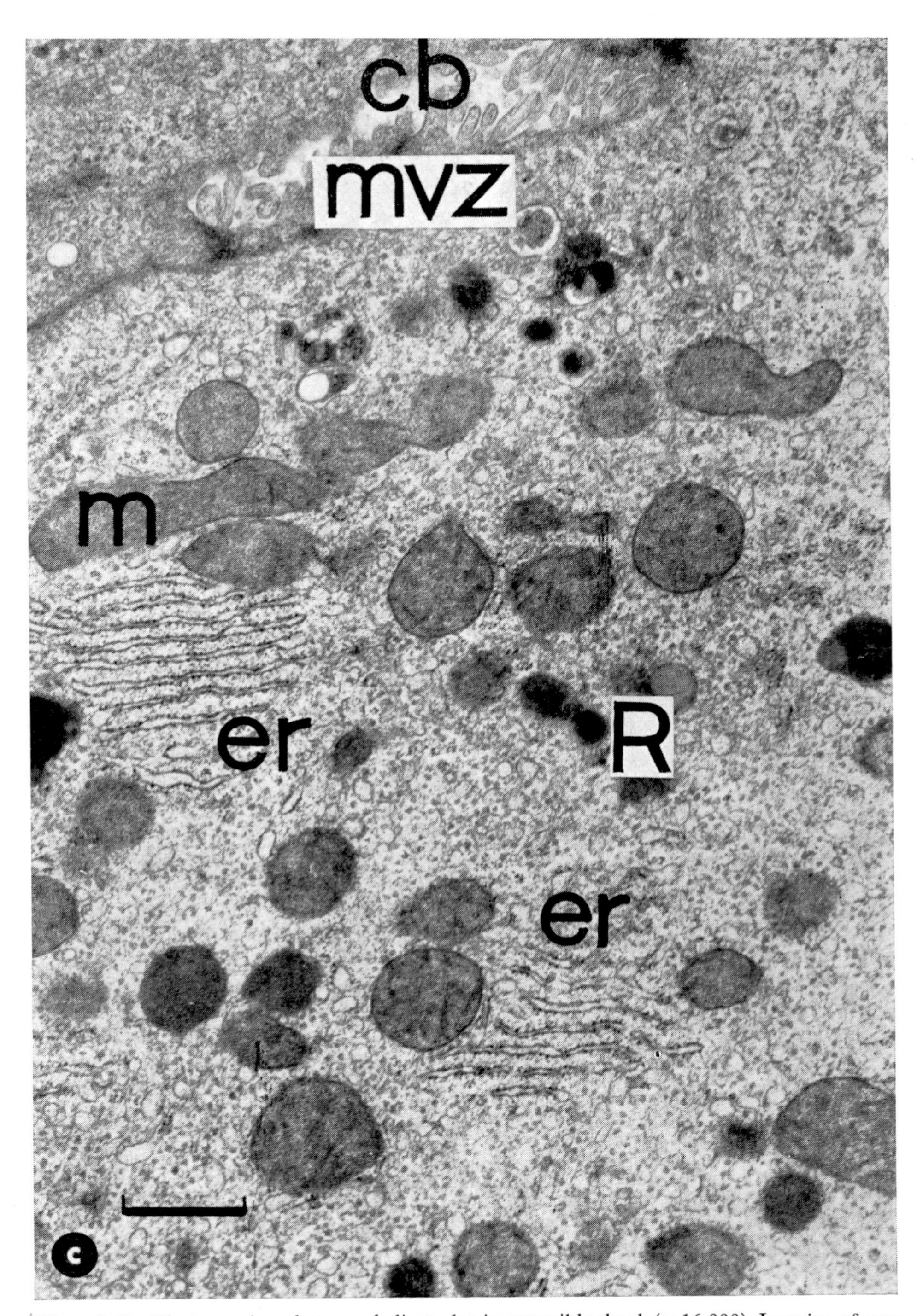

Figure 9.12c. Electron microphotograph, liver, dog in reversible shock (× 16,000). Junction of two liver cells, leaving intact a bile canaliculus (cb) full of microvilli. Within the smooth reticular field (with diluted glycogen granules), note two spots of rough ergastoplasmatic reticulum (er), suggesting hyperfunction of the cytosol. Dumb-bell-like processes may be seen in the mitochondria with a dense matrix and hazy crests.

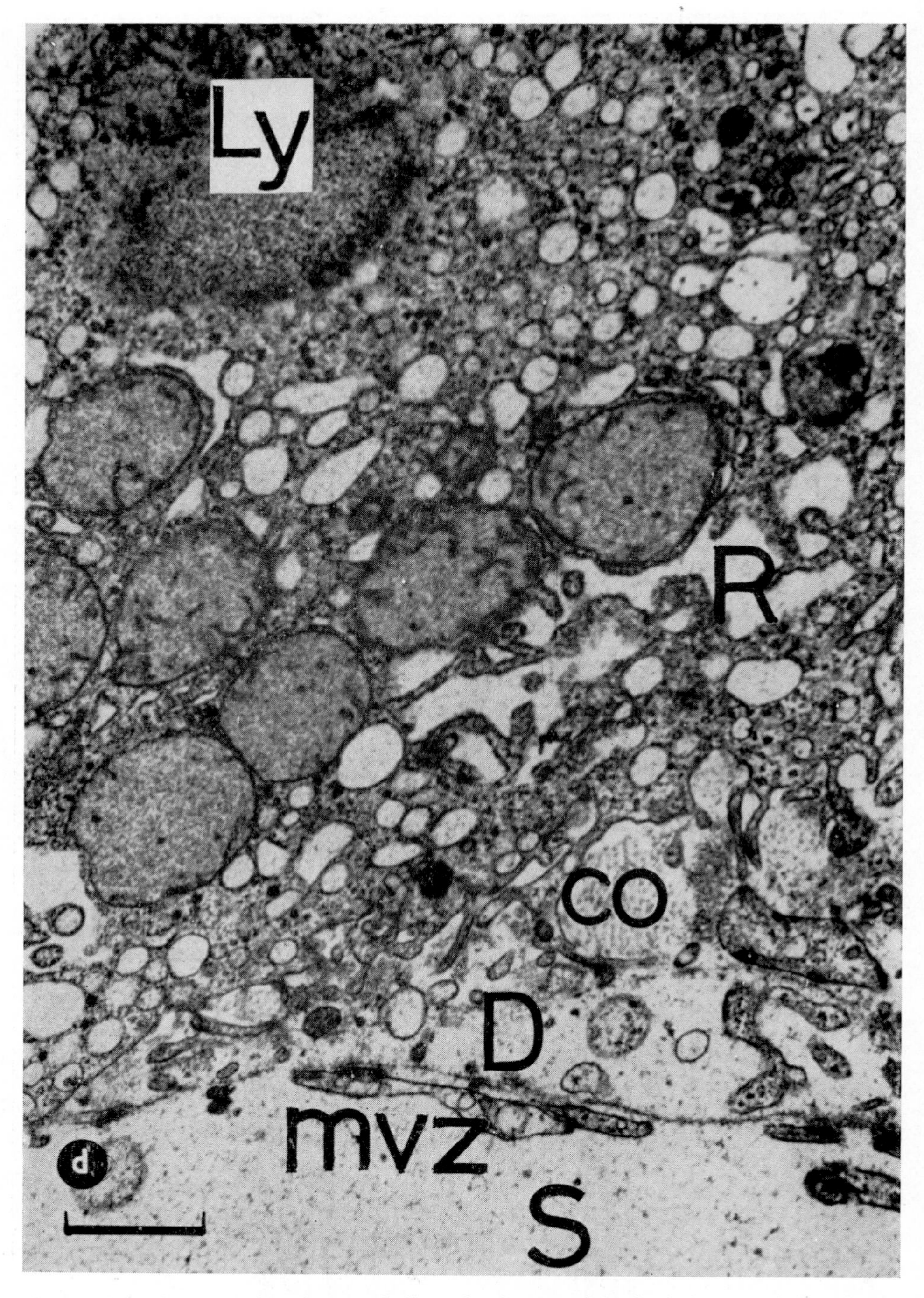

Figure 9.12d. Electron microphotograph, liver, dog in reversible shock (×16,000). Parasinusoidal liver cell. Marked rarefaction and disorganisation of the reticulum. Note a secondary giant lysosome and oedema of Disse's space with the accumulation of collagen material.

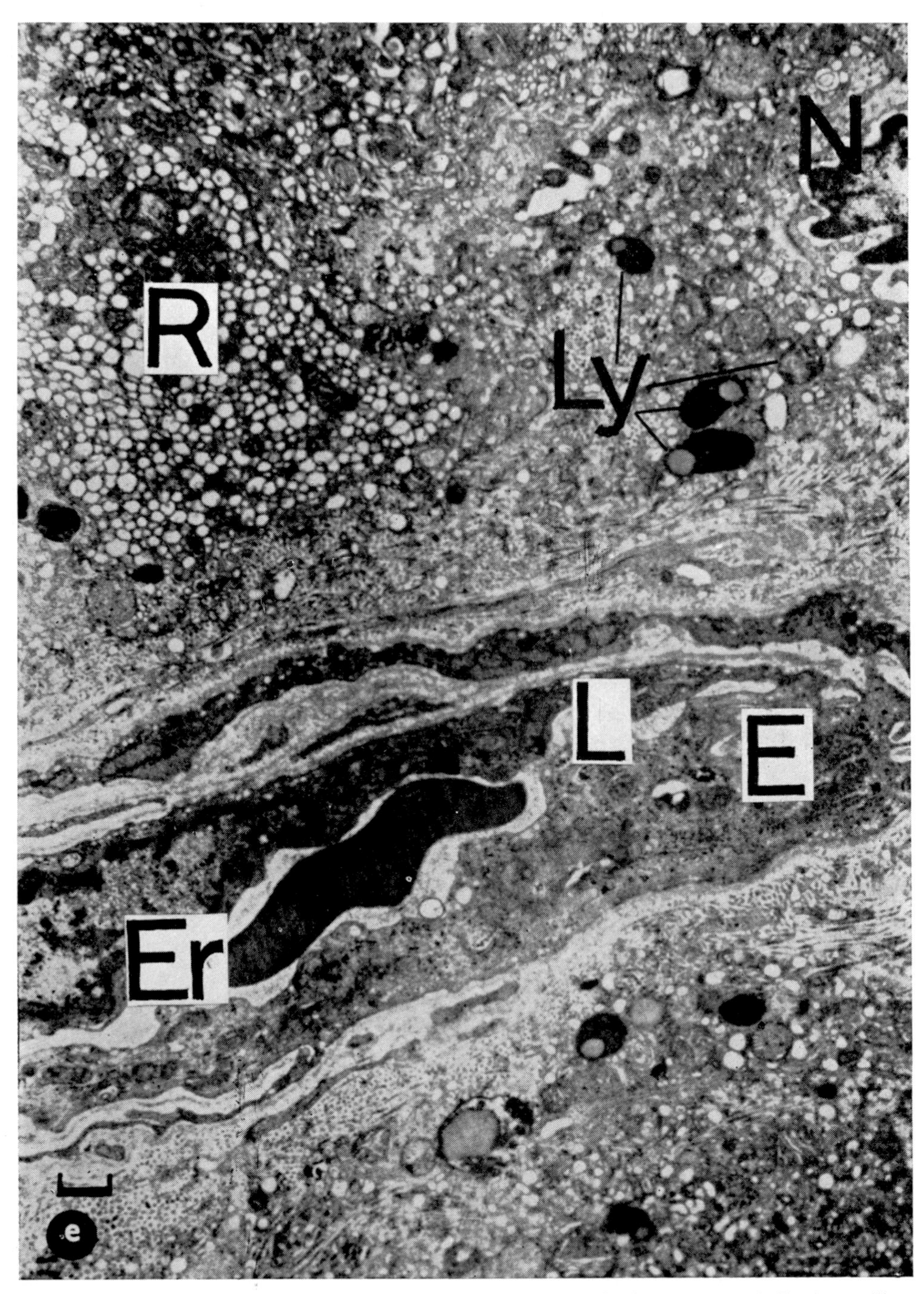

Figure 9.12e. Electron microphotograph, liver, dog in reversible shock ($\times$ 4,000). In the capillary lumen (L), plastic deformed erythrocyte (Er) due to oedema of the endotheliocyte (E). In the para-capillary field note the agglomeration of secondary lysosomes, a zone of disorganised reticulum and a deformed cell nucleus at the top right (N).

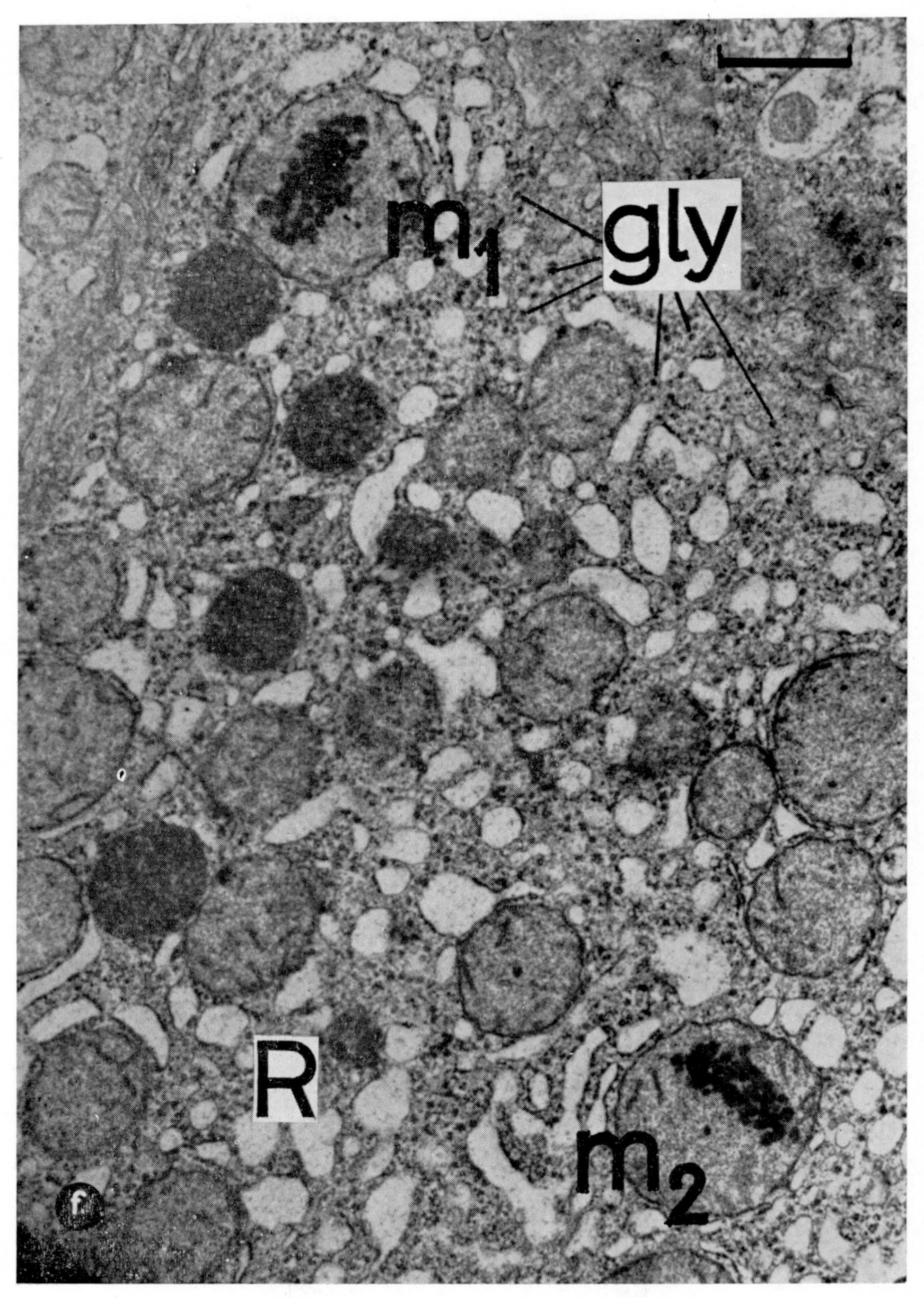

Figure 9.12f. Electron microphotograph, liver, dog in irreversible shock ($\times$16,000). In the more rarefied reticulum, which still maintains glycogen granules, mitochondria with accentuated inequalities appear, two of them exhibiting (m_1 and m_2) intramatriceal precipitates.

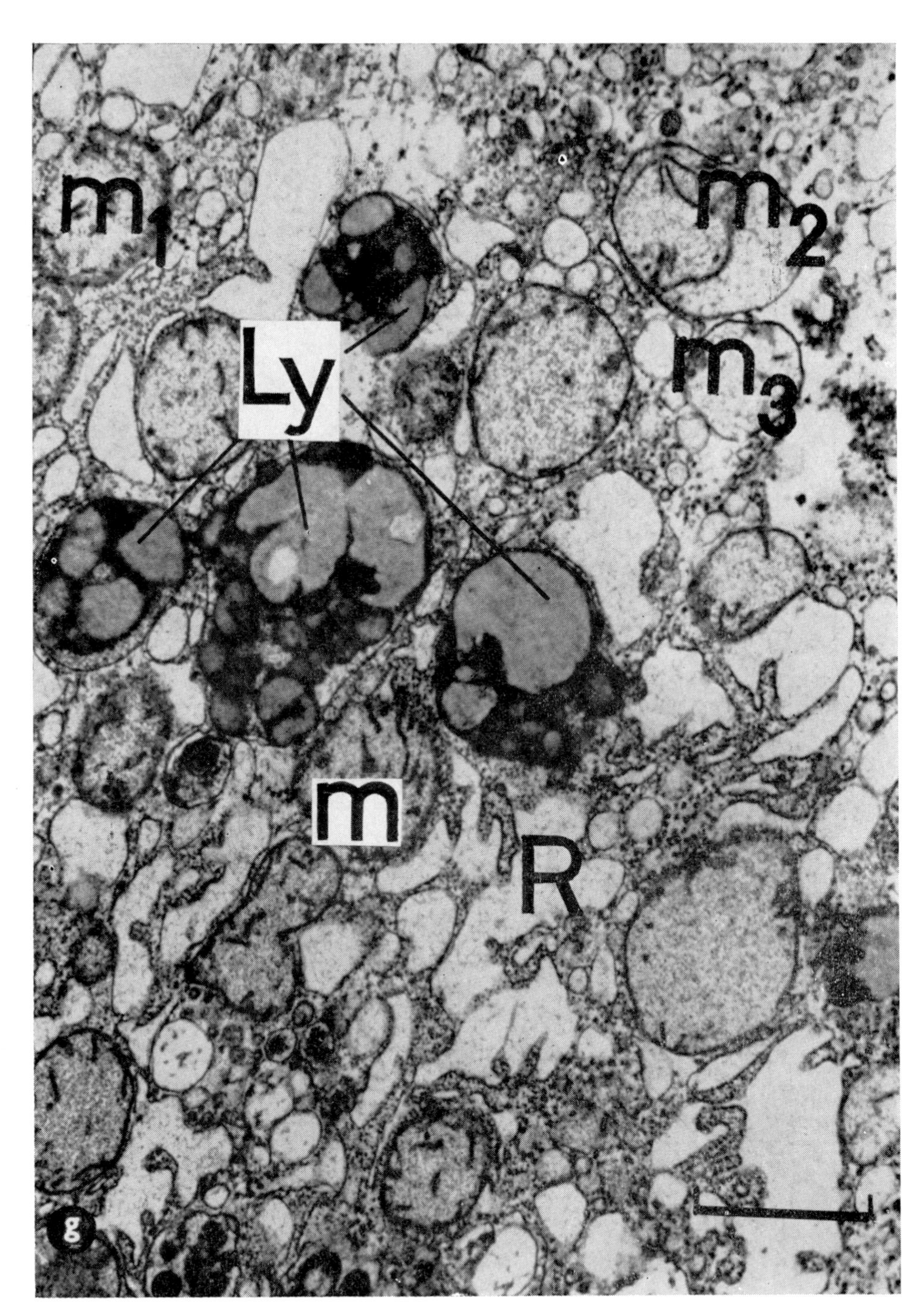

Figure 9.12g. Electron microphotograph, liver, dog in irreversible shock (×20,000). Against the background of a deeply altered endoplasmatic reticulum, a conglemerate of secondary giant lysosomes appears, encompassing in a 'common phagocytic vesicle' microparticles, reticular fragments and mitochondria. The mitochondria present fragmented crests (m_1), clear matrix (m_2) and discontinuity of the external membrane (m_3).

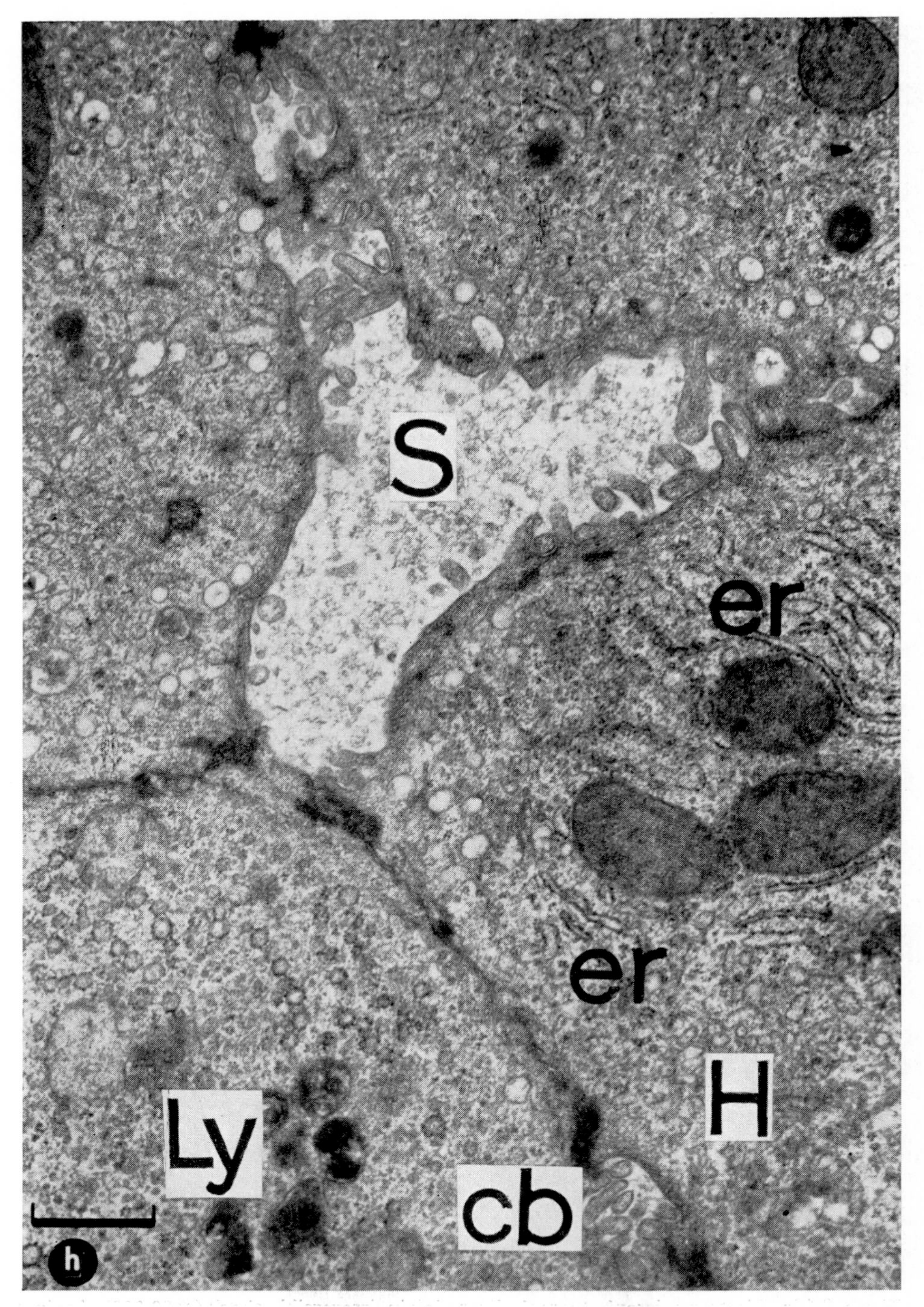

Figure 9.12h. Electron microphotograph, liver, dog in reversible shock (×16,000). Liver cell junctions leaving intact the bile ducts (cb) and cross-sectioned sinusoid (S). Note numerous zones of rough hyperfunctional reticulum (er) and almost total consumption of the glycogen granules. The mitochondria are densified and lysosomal agglomerations appear.

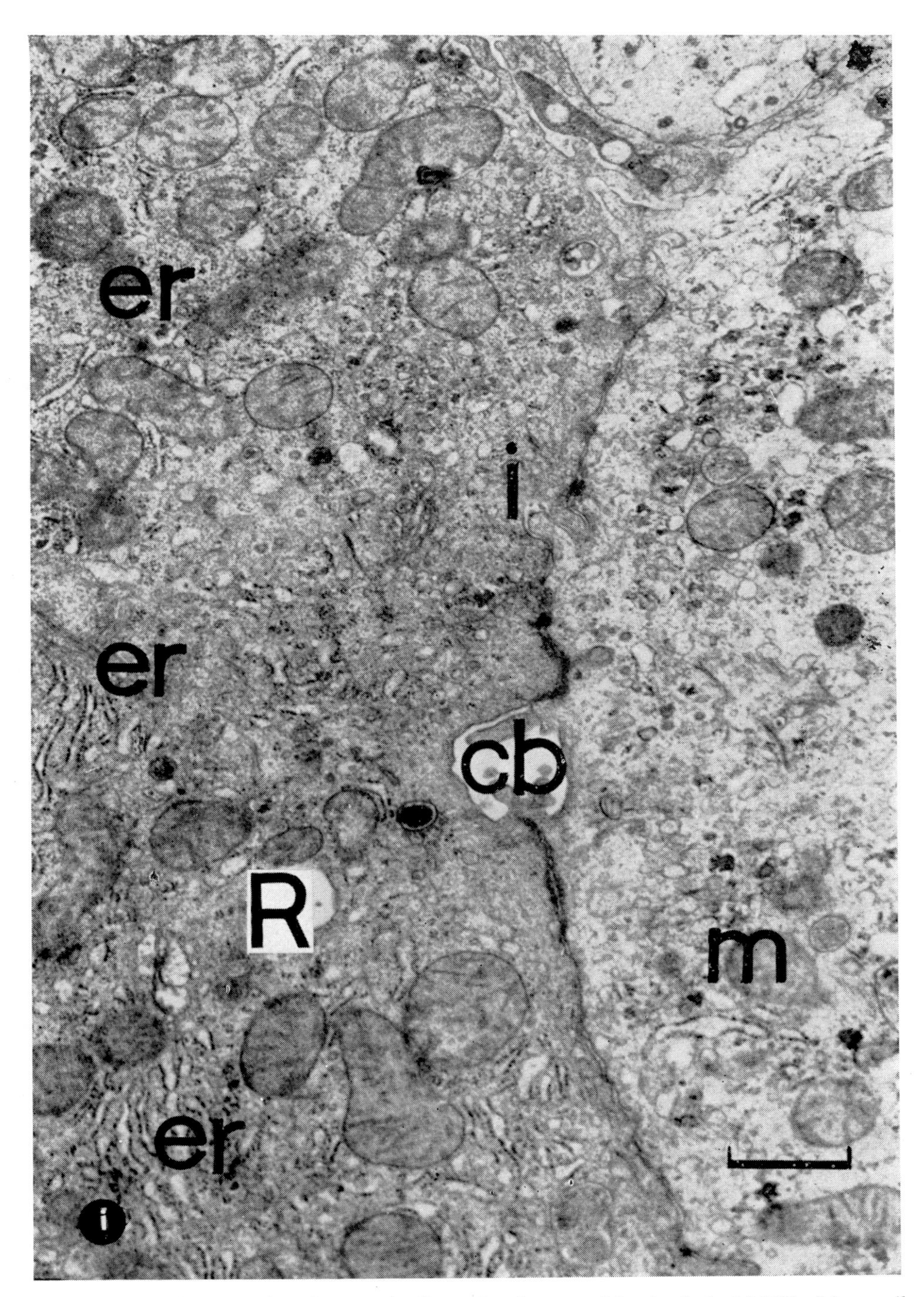

Figure 9.12i. Electron microphotograph, liver, dog in reversible shock (×16,000). Liver cell junctions containing numerous spots of rough reticulum and mitochondrial agglomerations.

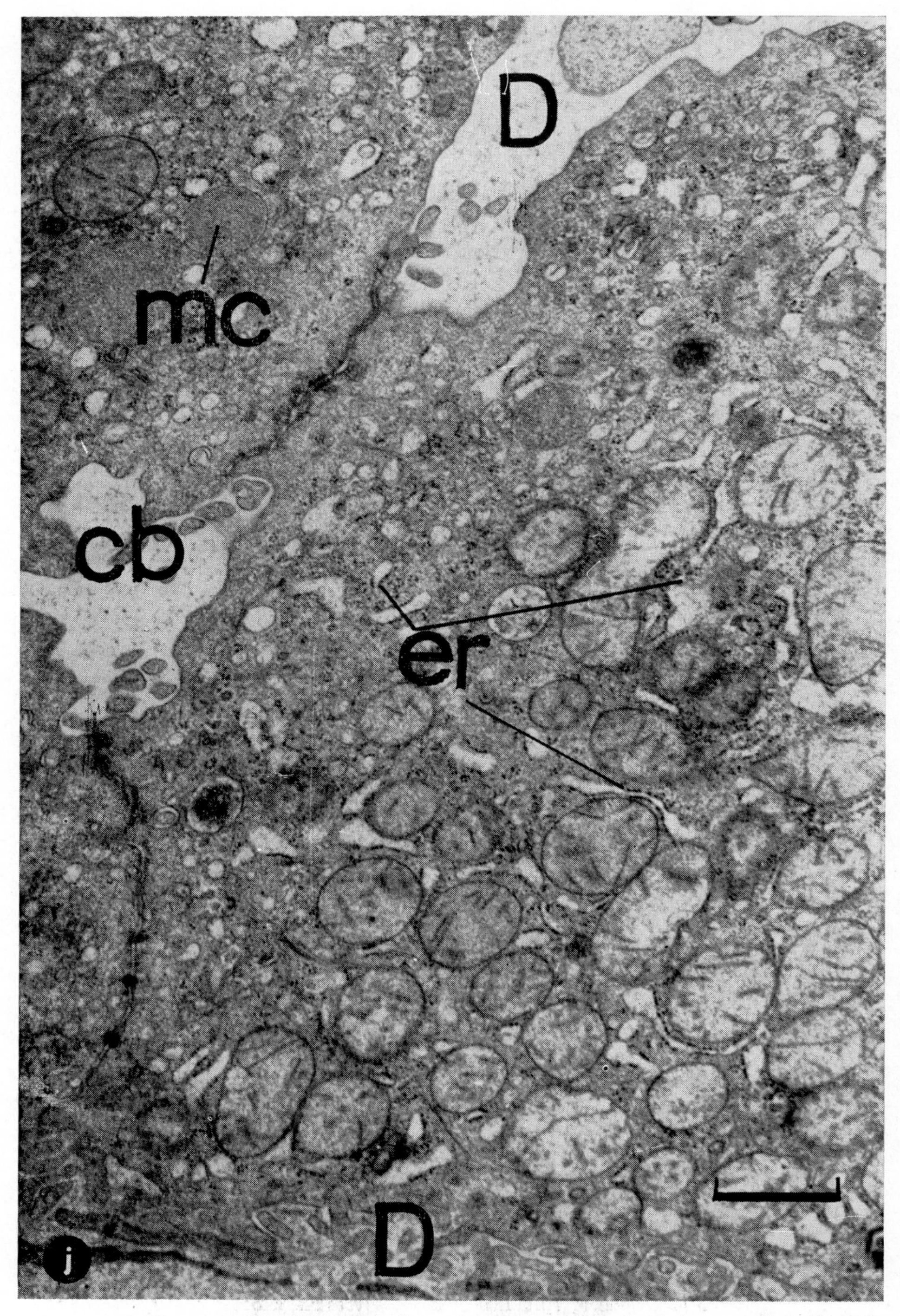

Figure 9.12j. Electron microphotograph, liver, dog in reversible shock (protected) (×16,000). Liver cell junctions marginating bile canalliculi and parasinusoidal Disse spaces. The functional reticulum is equally distributed and the mitochondria are within normal limits. Note numerous microcorpuscles (mc) and the onset of oedema in Disse's space.

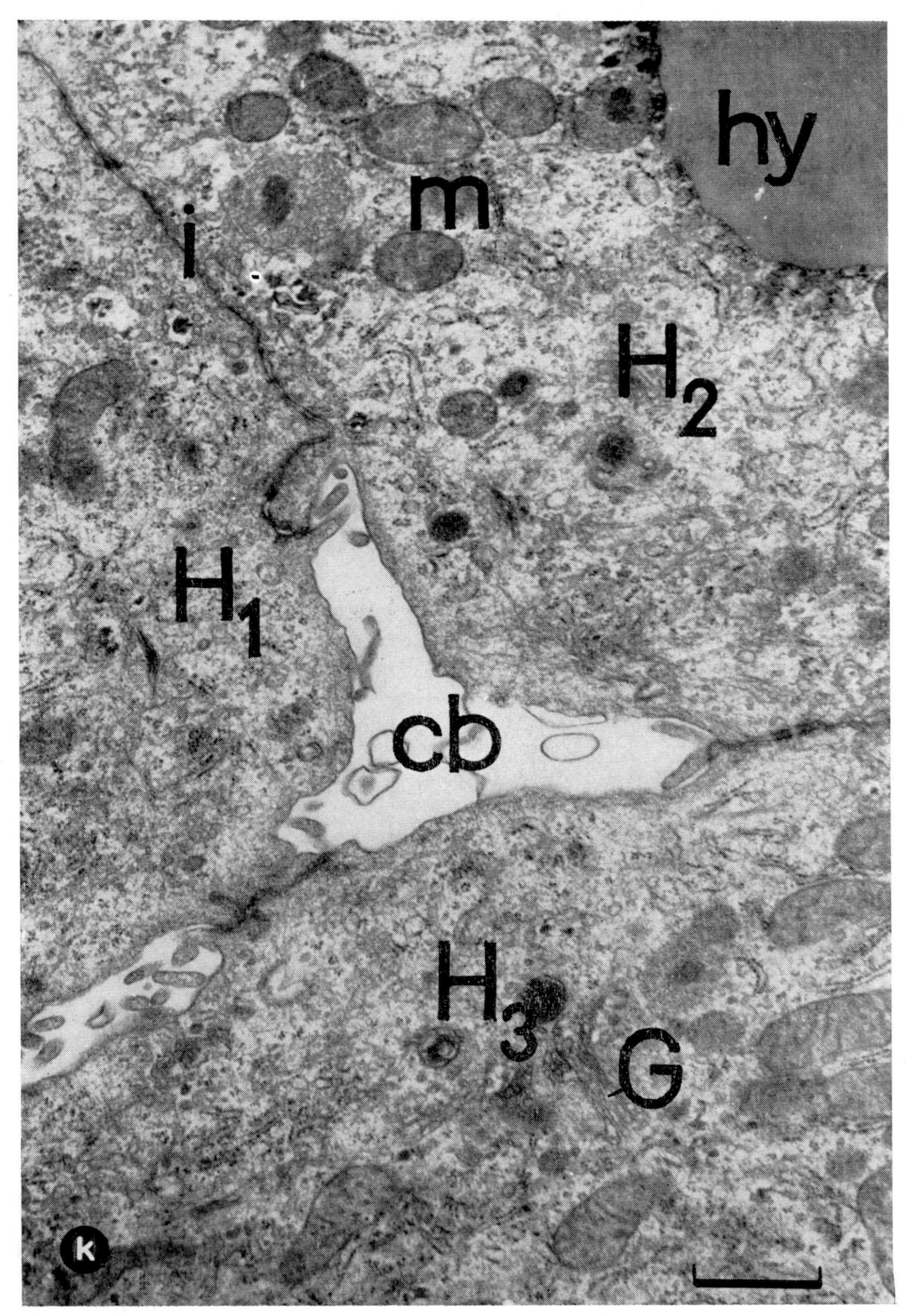

Figure 9.12k. Electron microphotograph, liver, dog in reversible shock (protected) (×16,000). Liver cell junction around a bile canaliculus. Note a hyaline block (hy) and fragmented Golgi apparatus (G).

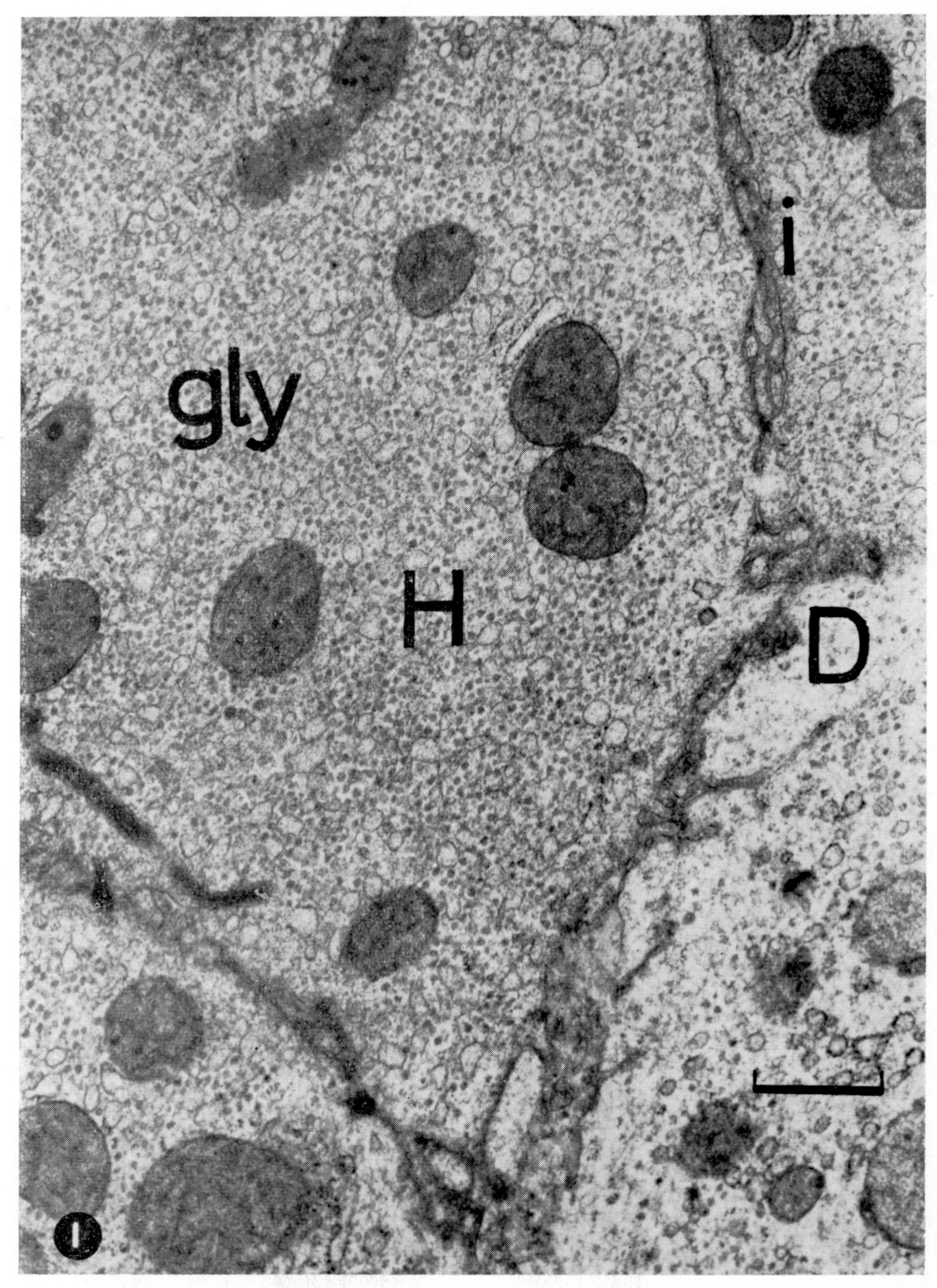

Figure 9.12l. Electron microphotograph, dog in reversible shock (protected) (×16,000). Liver cell junction with abundant glycogen granules in this reversible phase and hyperfunctional densification of mitochondria.

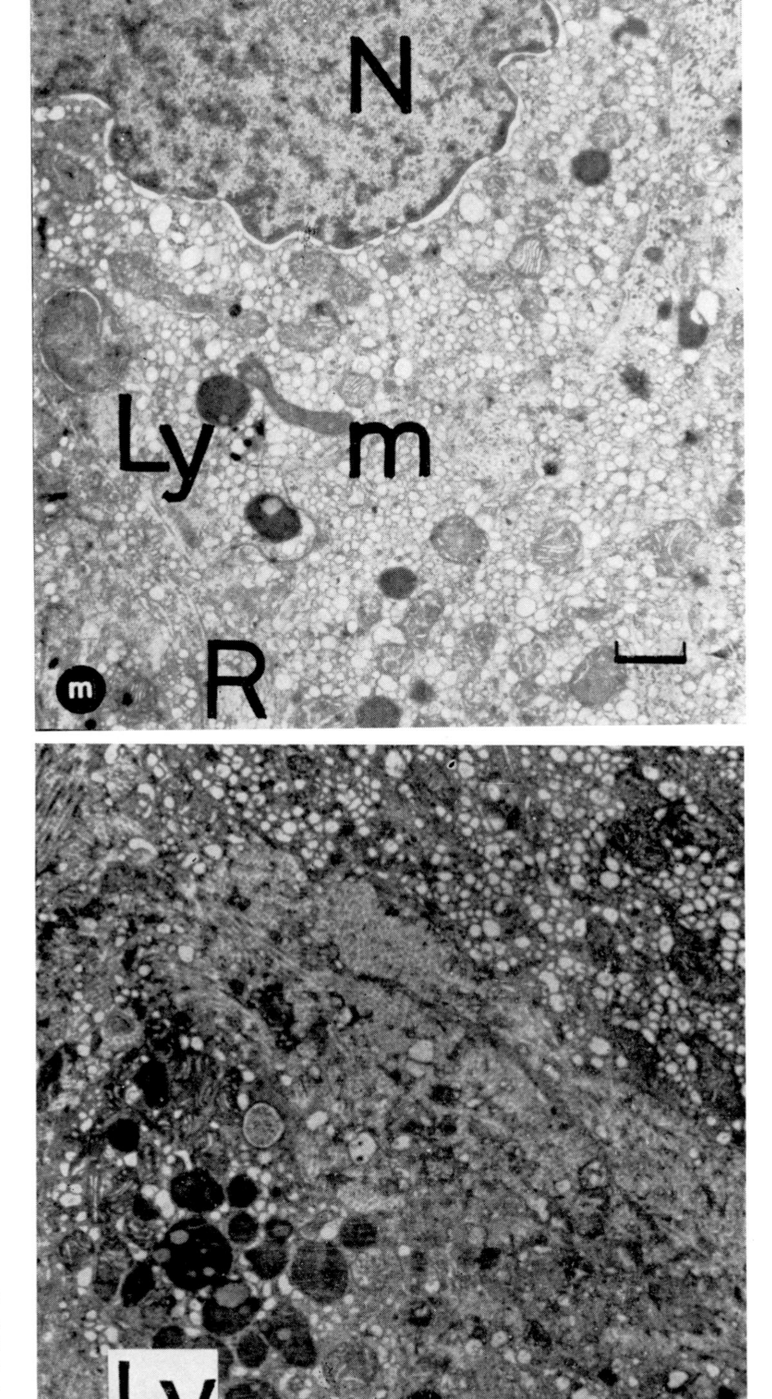

Figure 9.12m. Electron microphotograph, liver, dog in irreversible shock (×8,000). Note accentuated nuclear distress with shrivelled contour, the deposition of peripheral heterochromatin precipitate and central chromatinolysis.

Figure 9.12n. Electron microphotograph, liver, dog in irreversible shock (×8,000). Agglomeration of secondary lysosomes.

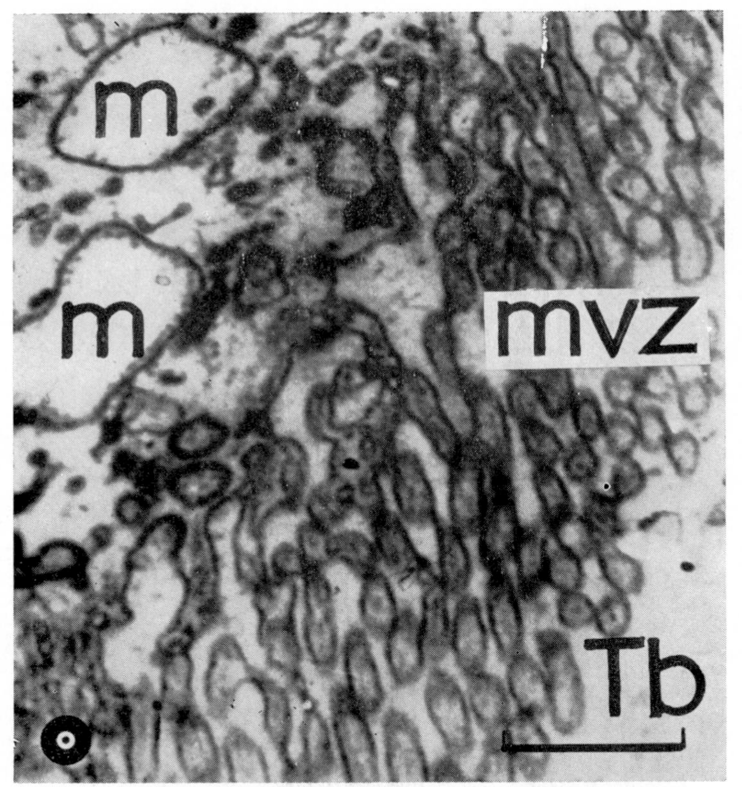

Figure 9.12o. Electron microphotograph, kidney, dog in irreversible shock (×20,000). Deep disorganisation of the tubulocyte (Tb) with densification of the 'brush border' and clarification of the mitochondria.

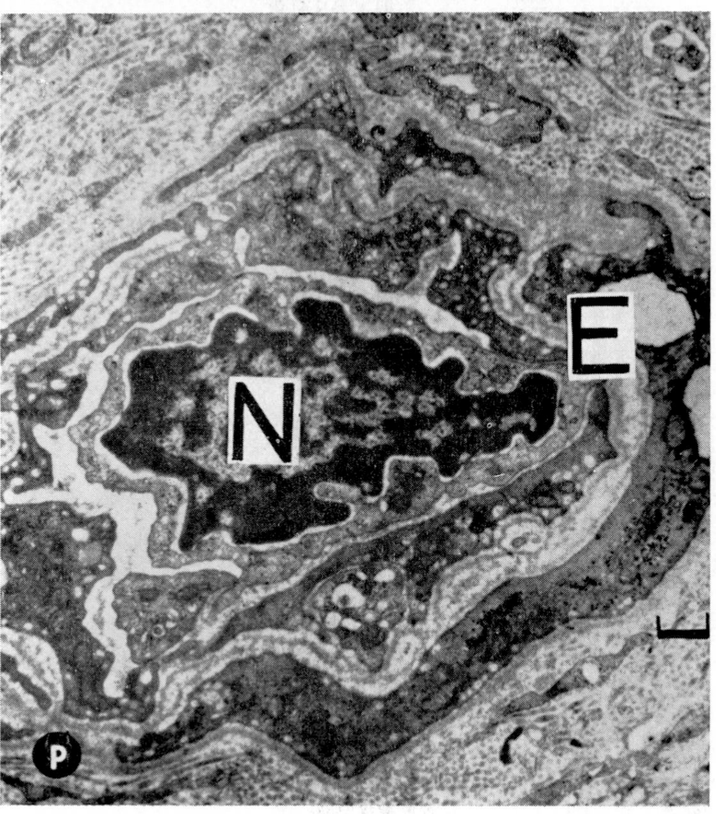

Figure 9.12p. Electron microphotograph, liver, dog in irreversible shock (×6,000). Capillary endotheliocyte (E) with shrivelled nucleus (N) and chromatin precipitate obstructing almost the entire capillary lumen.

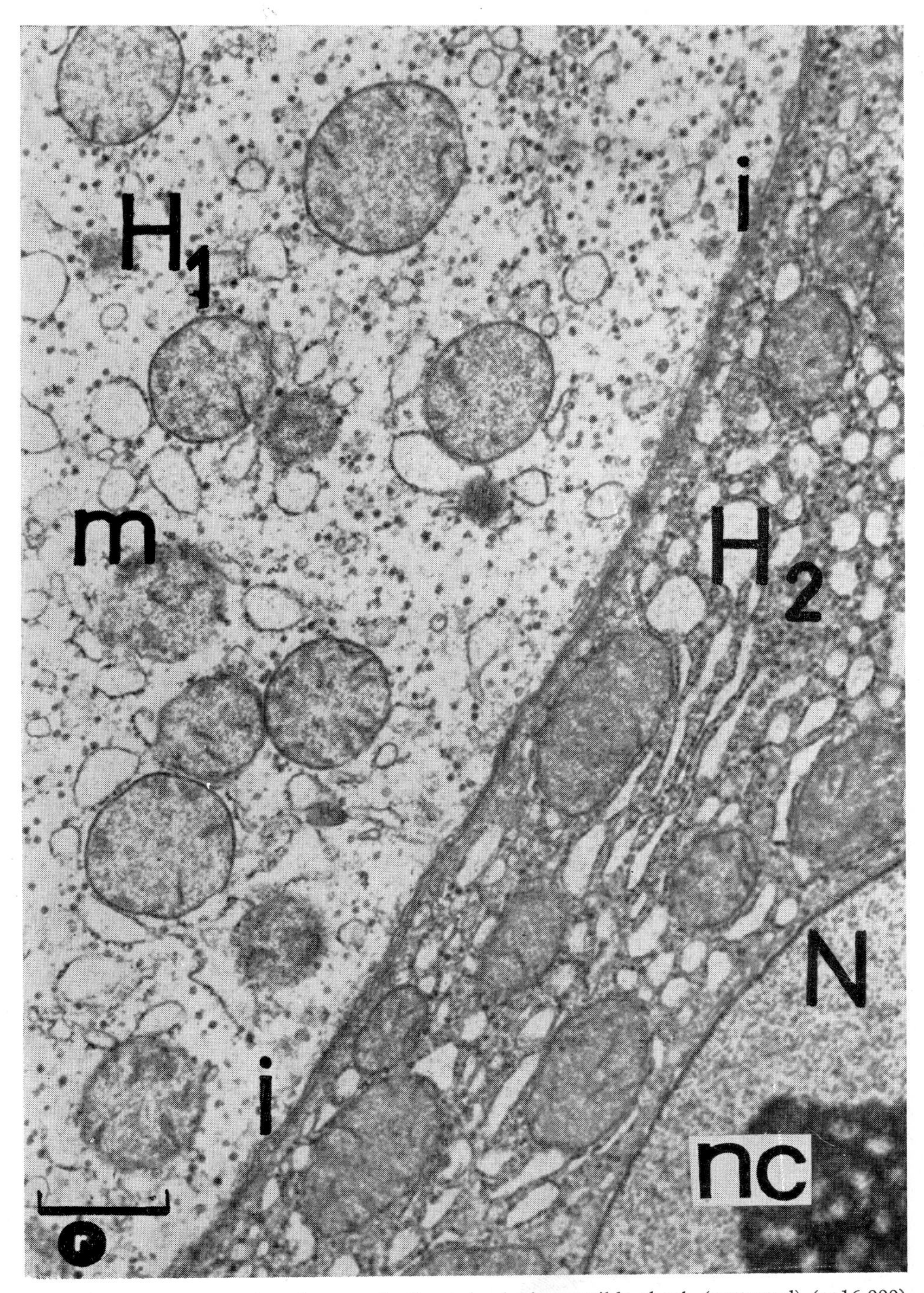

Figure 9.12r. Electron microphotograph, liver, dog in irreversible shock (protected) (×16,000). Liver cell junction; note persistence of glycogen granules in this stage, and inequality of the reticular and mitochondrial structure between H_1 and H_2 liver cell, the latter containing a nucleus (N) with a nucleolus (nc) within normal limits.

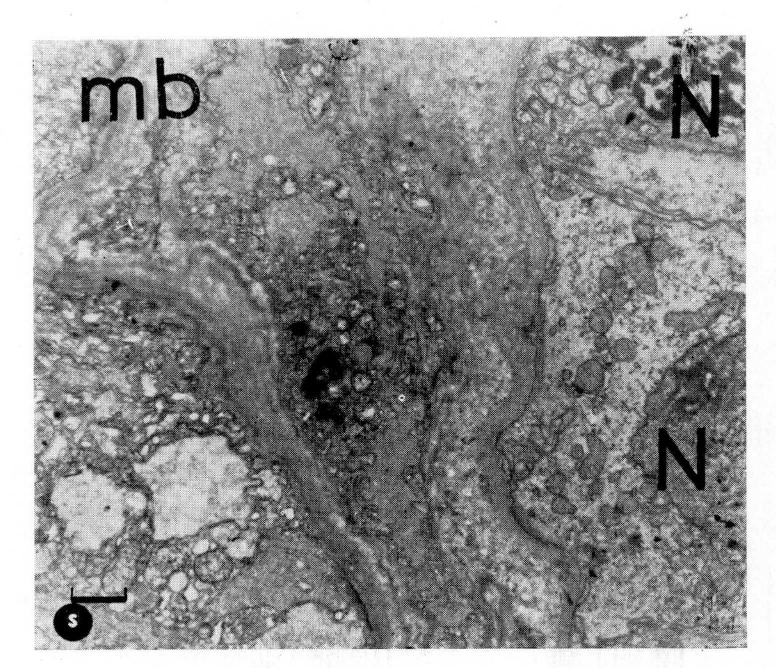

Figure 9.12s. Electron microphotograph, kidney, dog in irreversible shock (× 8,000). Cellular junctions in the kidney glomerulus area with thickness of the membrane network (mb). Against the background of a deeply altered reticulum a deformed cell nucleus (N) appears in each cell.

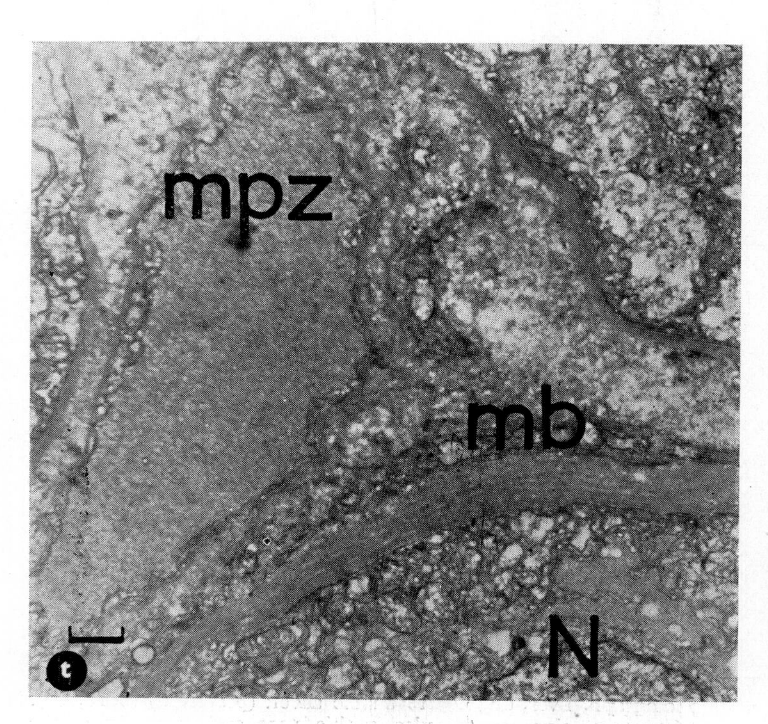

Figure 9.12t. Electron microphotograph, kidney, dog in irreversible shock (× 8,000). Monstruous membrane thickness (mb) and intra- and intercellular precipitates of mucopolysaccharides (mpz).

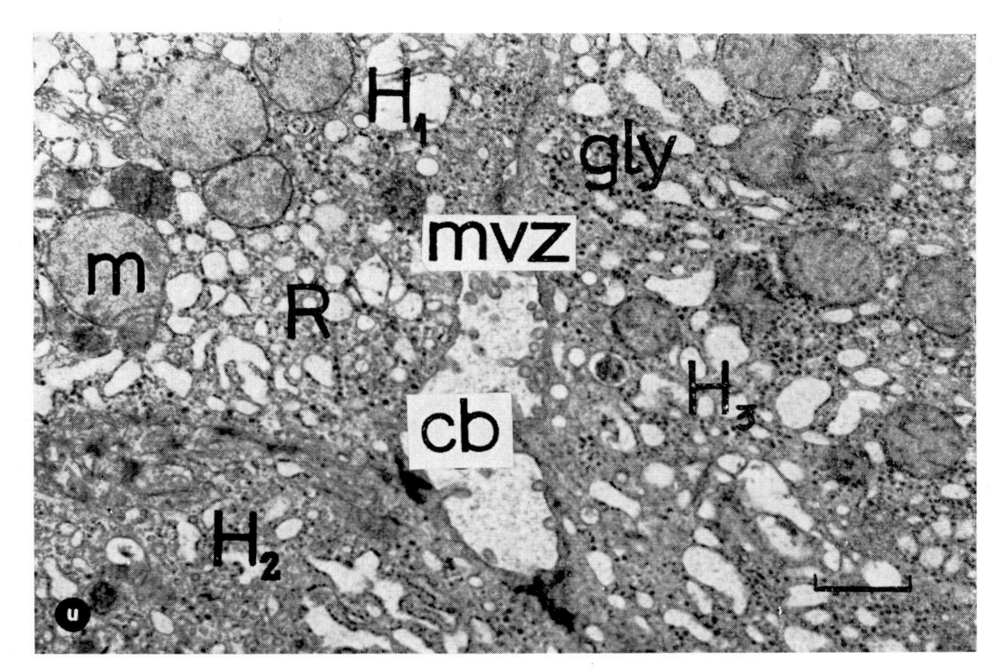

Figure 9.12u. Electron microphotograph, liver, dog in reversible shock (protected) (×16,000). Three liver cells junction (H_1, H_2 and H_3) leaving intact a bile duct (cb) with microvilli (mvz). Marked rarefaction of the reticulum in which some glycogen trains (gly) and functional mitochondria (m) still exist.

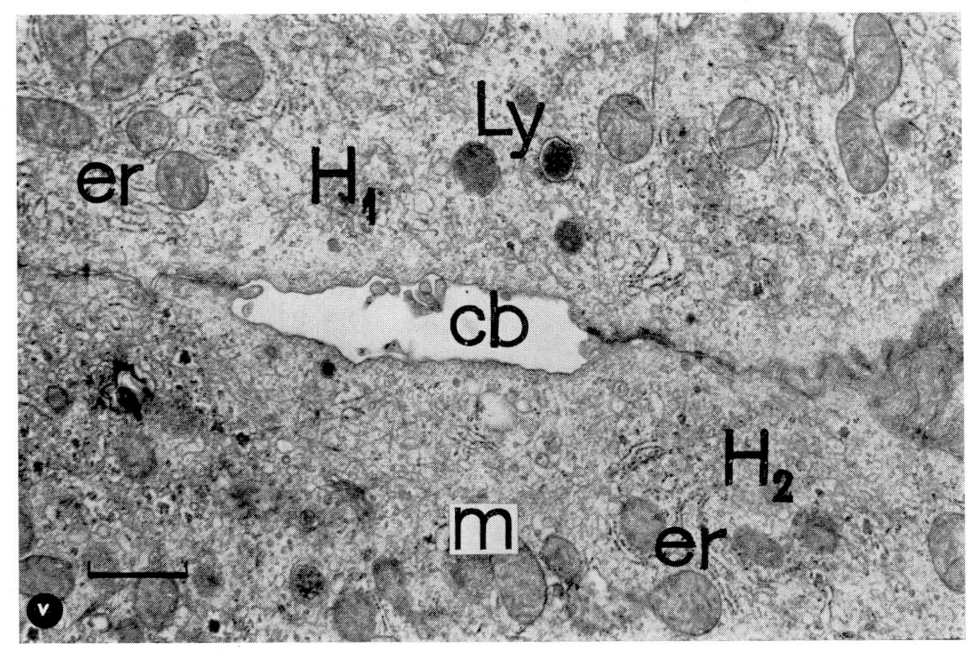

Figure 9.12v. Electron microphotograph, liver, dog in reversible shock (protected) (×16,000). Two liver cells (H_1 and H_2) with several spots of rough ergastoplasmatic reticulum (er). But some secondary lysosomes appeared (Ly) and reticulum is more disorganised.

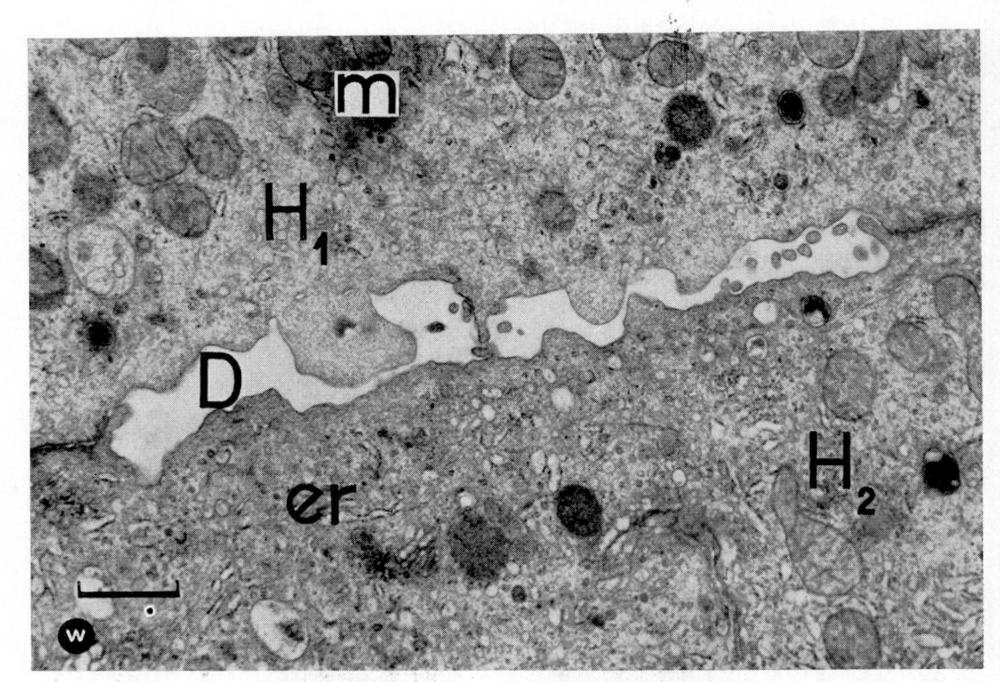

Figure 9.12w. Electron microphotograph, liver, dog in reversible shock (protected) ($\times$16,000). Two liver cells (H_1 and H_2) surrounding a Disse's space (D). Note microprecipitates and secondary lysosomes near spots of intact mitochondria and functional reticulum.

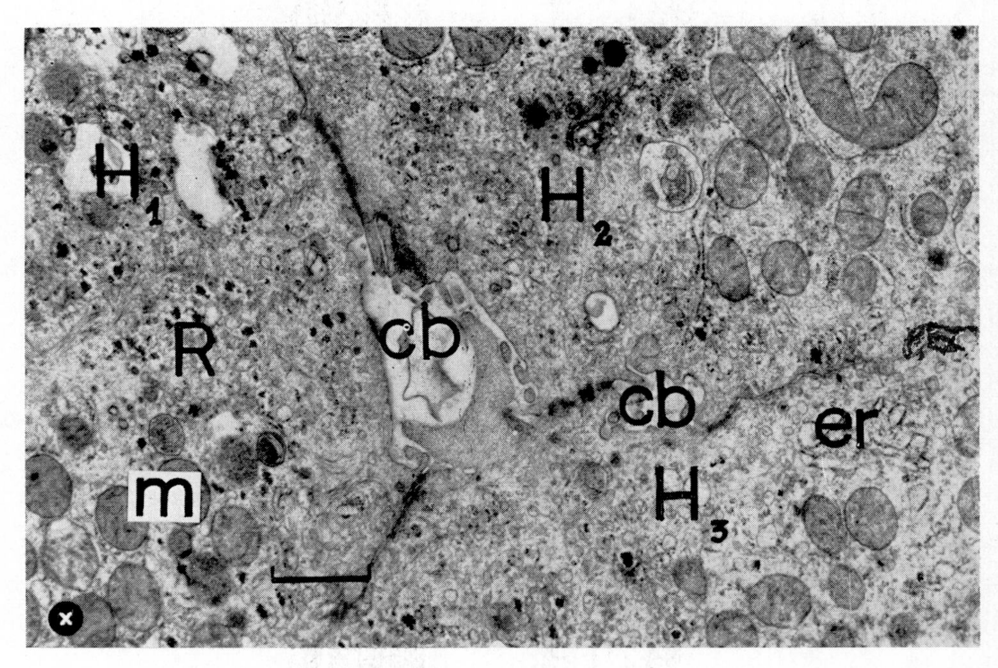

Figure 9.12x. Electron microphotograph, liver, dog in reversible shock (protected) ($\times$16,000). Three liver cells (H_1, H_2 and H_3) with two bile canaliculi (cb). Several mitochondria (m) and spots of ergastoplasmatic reticulum (er) against the background of vacuolised reticulum (R).

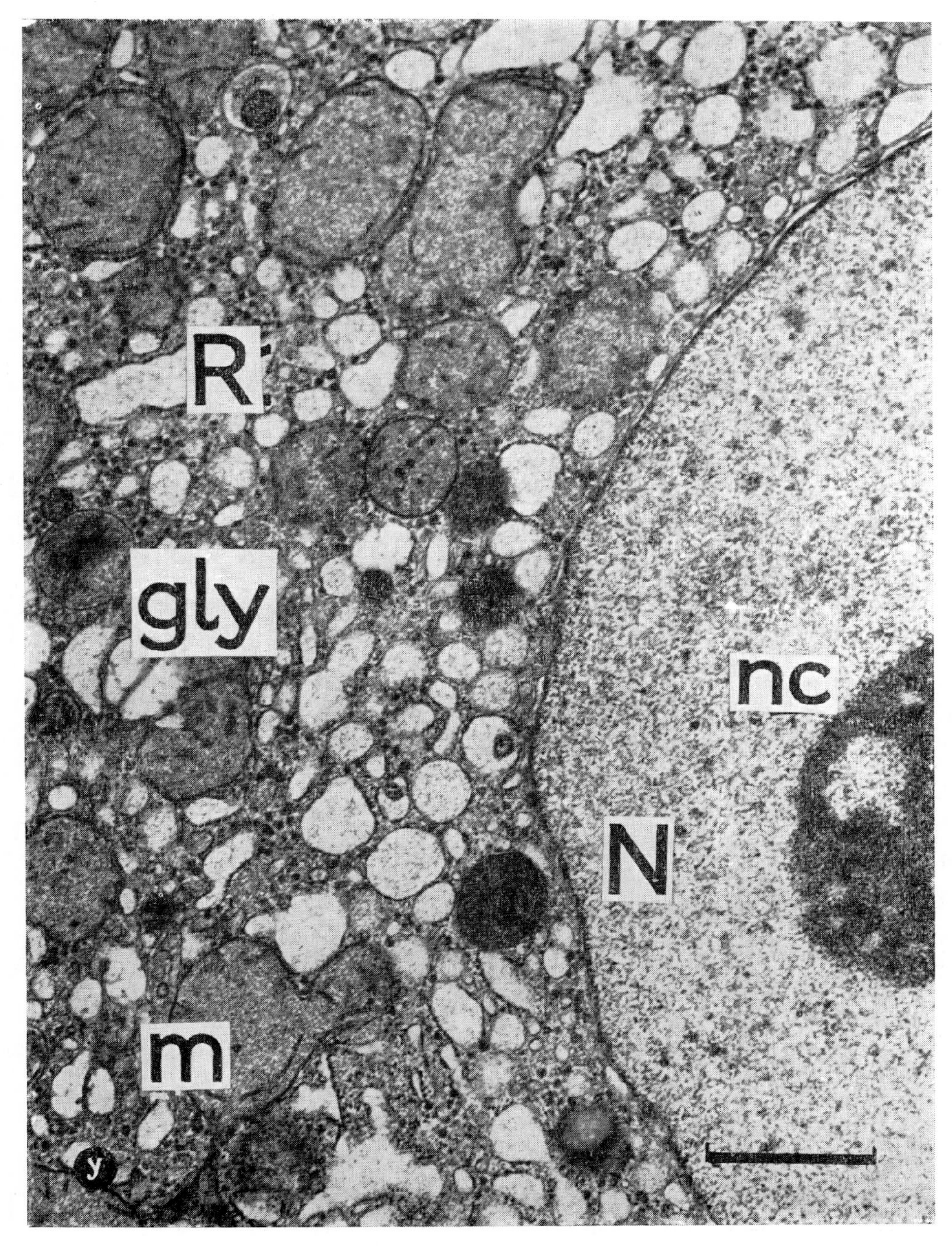

Figure 9.12y. Electron microphotograph, liver, dog in irreversible shock (protected) (×16,000). Note the integrity of the cellular nucleus (N) and nucleolus (nc) in this late stage of shock. Though reticulum is disorganised (R) there are some glycogen trains (gly) and functional mitochondria (m).

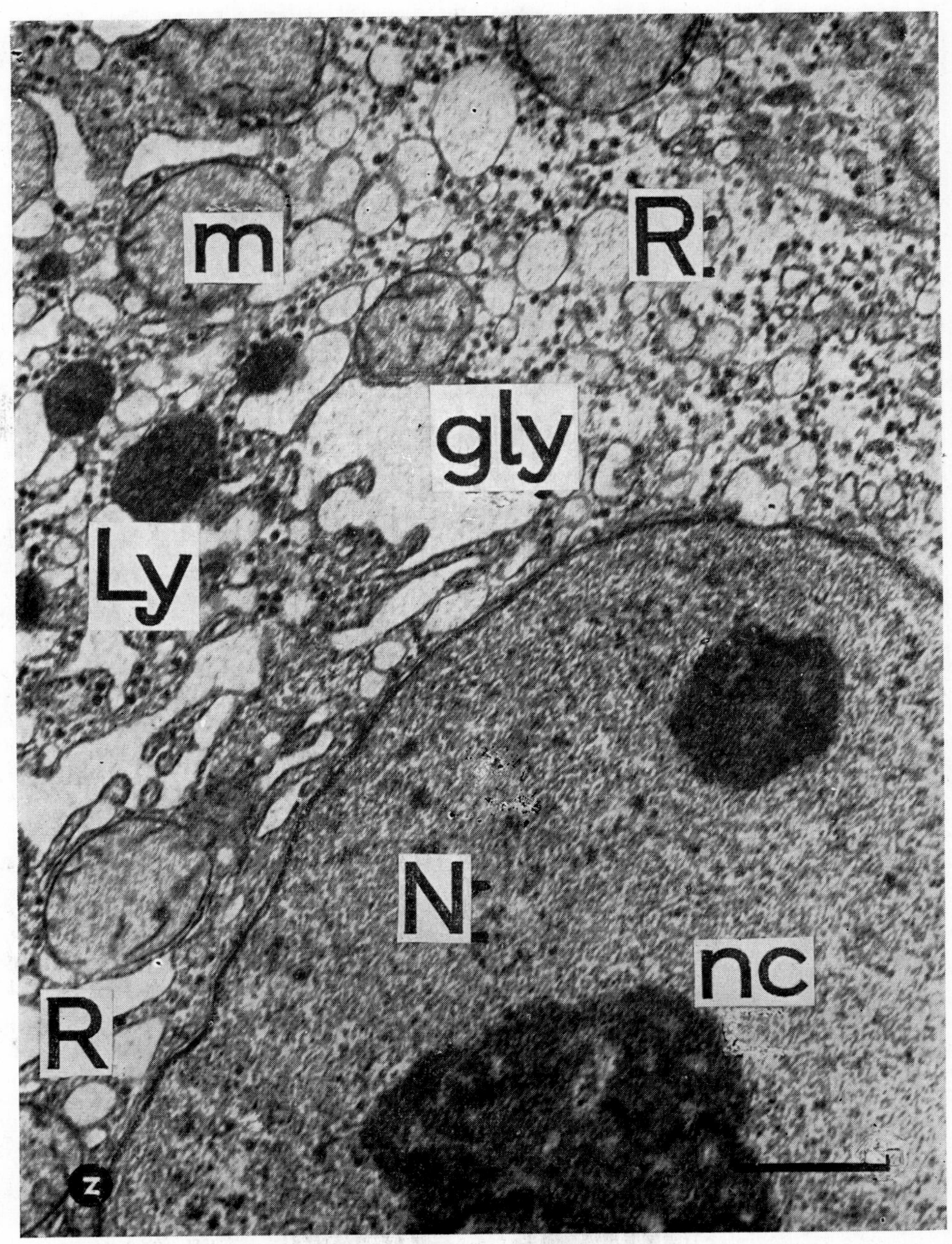

Figure 9.12z. Electron microphotograph, liver, dog in irreversible shock (protected) (×16,000). Note marked rarefaction and disorganisation of the reticulum and the presence of secondary lysosomes. In this late stage the 'protected cell' contains some glycogen (gly), intact mitochondria (m) and a like normal nucleus (N).

The lesions, therefore, have to be followed up throughout the process of shock in order to obtain proof regarding the borderline between reversible and irreversible stock at subcellular level [8].

9.5.1 HISTOLOGY OF THE SHOCK CELL

By means of the optical microscope it is possible to observe, from the erythrocytes coagulating in the capillaries, the various degrees of blood stasis leading to thrombosis an microthrombosis (Figure 11 a, b, c, d).

Cellular hypoxic and acidotic injury is revealed by the appearance of dystrophies (granular and homogenizing) as well as by cytoplasmic vacuolar degeneration of nuclear figures and phantoms, by ragged appearance of the basic membranes, etc. (Figure 9.11 a, c, e, j and o).

Rapid deletion of the hepatocyte glycogen stores takes place in the phase of reversible shock (Figure 9.11 e and g). In this phase of shock the glycogen stores are consumed to a lesser extent in the shocked animals if the cellular energy effort is protected (Figure 9.11 f and h). In other instances the brutality of the shock itself leaves part of the glycogen reserves of the hepatocytes unconsumed (Figure 9.11 m and n).

The histopathologic interpretation of the microphotographs presented in Figure 9.11 belongs to Dr V. Strîmbeanu, M. D. and Dr D. Singer, M. D. (Pathological Anatomy Laboratory, Military Hospital, Bucharest).

9.5.2 INFRASTRUCTURE OF THE SHOCK CELL

Electron microscopy reveals the most minute physical lesions, some of which are reversible and others are not, and the borderline between the two has not yet been clearly determined. Every organelle shows an initial functionaf effort during the reversible phases ol shock and exhaustion, together with damage to its leptonic structure, in the late irreversible phase (Gazzaniga, 1970; Mallet-Guy, 1971).

Figure 9.12 shows some of these infrastructural lesional oscillations. In the first stages of shock the endoplasmic reticulum displays areas of great activity with a 'rough', ergastoplasmic aspect; then the granular reticulum gradually vanishes, the ribosomes disappear and real tubovesicular islands form, with large vacuolar expansions or concentric lamellar formations that reveal disorganisation of the cytosol matrix [50, 53, 79]. In the nucleus, heterochromatin margination and subnucleolemic precipitates appear as well as blurring of the nucleoli, as observed by numerous researchers. The widest range of alterations within the shock cell is exhibited by mitochondria. They present an evolutive lesional cycle probably derived from their physiological cycle (*see* Section 9.4.3). The most diversified aspects have been reported: anarchic distribution in the cytosol with mitochondrial clustering or margination; ballooning up to voiding of the mitochondria; giant, tortuous and shrivelled mitochondria; breakdown and disorganisation of the cristae, compacting of the matrix, the formation of intra-mitochondrial inclusions and precipitation of paracrystalline precipitates.

Generalized mitochondrial hypertrophy, with increase in the number of cristae and of the matrix density, may be considered as the results of hyperactivity. The presence of inclusions and paracrystalline precipitates, the significance of wich is not yet very clear, is considered to be a degenerative lesion that may be related to the accumulation of mycelia, phospholipids, calcium and mineral phosphates [73].

From the evolutionary viewpoint, the 'swelling' and 'clarification' *(mitochondrial œdema)* as well as the paucity of the cristae are considered to be *reversible* lesions, whereas hypertrophy of the mitochondria and increase in the number of cristae are considered to be functional hyper-effective aspects. Paracrystalline precipitates, amorphous masses, compacting and shrivelling of the mitochondria and tearing of the mitochondrial edges are considered to be *irreversible* degenerative lesions.

Among the particular lesions detected in various organs during shock mention must be made of:

- fragmentation of the insertion discs and *z* bands in the myocardial cells [55];
- ruptures of the brush-like edges of the tubulocytes and basic tubular membrane (the glomerules hardly seem to be affected during shock) [20, 38, 87];
- peribiliary distribution of the hepatocyte lysosomes [38];
- œdema of Disse's space up to destruction of the sinusoid architectonics [23, 37, 87].

Electron microscopy reveals the finest fibrin particles, the distortion of erythrocytes and homogenisation of the platelets, membrane and reticular œdema, and, above all, confluence and hypertrophy of the lysosomes; the formation of giant lysosome clusters including mitochondrial elements, microbodies and fragments of reticulum, mark the irreversible onset of self-destruction of the cell [81].

9.6 BIOPHYSICS OF THE SHOCK CEL

Thermodynamic study of the shock cell was initiated by combining data belonging to various specialities [26, 33, 45, 46, 59, 61, 66, 67, 70, 71, 76, 80].

For a better understanding of cellular biophysics several ideas will have to be recalled.

9.6.1 ELEMENTS OF IRREVERSIBLE THERMODYNAMICS

Classical thermodynamics is, in fact, thermostatics, as its principles, expressed in equalities, are applied to processes considered to be in a state of equilibrium, of reversibility; actually almost all natural processes are a dynamic sequence of states of equilibrium. The whole material phenomenology is in a perennial continuity, its reversible aspect being only a temporary visualisation of a moment of encounter between processes that actually develop *irreversibly*. Modern thermodynamics is *a science of irreversible processes* whose application was soon extended to biology as well [23, 43, 29].

Evolving by elementary transformations, any system exchanges with the surroundings a certain amount of heat, Q_{el}; consequently the system modifies its entropy, S, according to the relation:

$$\boxed{\mathrm{d}S \geqslant \frac{Q_{\mathrm{el}}}{T}} \qquad (1)$$

in which T is the absolute temperature.

For a process considered reversible, relation (1) becomes:

$$\boxed{T\mathrm{d}S - Q_{\mathrm{el}} = 0} \qquad (2)$$

That is, in a reversible process the quantity of heat exchanged with the environment is equal to variation in the entropy of the system.

In contrast to the reversible processes, in the irreversible processes the quantity of heat exchanged with the environment is smaller than the variation in the entropy of the system, so that:

$$T\mathrm{d}S - Q_{el} > 0 \qquad (3)$$

or

$$T\mathrm{d}S > Q_{el} \qquad (4)$$

Hence, in an irreversible process entropy variations exceed the amount of heat exchanged with the surrounding medium.

The difference in relation (3), $\overline{Q}_{el}$ *is the quantity of uncompensated heat (Clausius)* and shows by how much the variation in the entropy of the system is greater than the quantity of heat, Q_{el}, exchanged with the medium. Hence relation (3) may be written as follows:

$$T\mathrm{d}S - Q_{el} = \overline{Q}_{el} \qquad (5)$$

Taking into account relation (5), then:

$$\mathrm{d}S = \frac{Q_{el}}{T} + \frac{\overline{Q}_{el}}{T} \qquad (6)$$

In the above relation, the Q_{el}/T ratio represents the *entropy exchanged by the system with the environment* ($\mathrm{d}S_e$), *and* $\overline{Q}_{el}/T$, *the variation in internal entropy* ($\mathrm{d}S_i$). Therefore, relation (6) becomes:

$$\mathrm{d}S = \mathrm{d}S_e + \mathrm{d}S_i \qquad (7)$$

and relation (4):

$$\mathrm{d}S > \mathrm{d}S_e \qquad (8)$$

consequently:

$$\mathrm{d}S_i > 0 \qquad (9)$$

Relation (9) expresses the reality, i.e. that *in an irreversible process* the internal entropy of the system and of its outer medium (considered as a whole) increases, or in other words *irreversible processes generate entropy in the system* [33, 66].

Classical thermodynamics admitted the possible disappearance of entropy from a certain place provided it occurred elsewhere. The thermodynamics of irreversible processes formally denies such a simple sequence of events and, instead, introduces the theory of *coupled processes*, which stipulates that in the same place several coupled processes may take place (some of which may possibly lose their entropy), provided they develop at the same time as *at least one irreversible entropy-generating process*, so that the total variation of $\mathrm{d}S$ should be positive in the considered place.

Reversible processes coupled with one or several irreversible processes (called *coupling processes)* may occur anywhere in the human body; for instance, the osmotic gradients of the transfer of water are coupled with the irreversible coupling process of the transfer of sodium ions through the membrane.

Several more theories have to be mentioned: the thermodynamic forces, X, are the gradients of certain *intensive magnitudes* that determine the irreversible processes in a given thermodynamic system (e.g. the gradients of temperature, chemical potentials, or mechanical forces of a shock-inducing agent, etc.). The variation rate of some *extensive magnitudes* generated by thermodynamic forces is known as *thermodynamic flows*, $\mathcal{J}$ (e.g. energy under its various forms of manifestation).

In the thermodynamics of irreversible processes it is postulated that *thermodynamic flows are linear functions* of the thermodynamic forces of a system:

$$\mathcal{J}_j = \sum_{k=1}^{n} L_{jk} X_k \ (j = 1,2 \ldots ,n) \quad (10)$$

where L_{jk} represents the *phenomenologic coefficients*, defined by Onsager's principle. Let us assume a given open thermodynamic system and a homogeneous linear system of phenomenological equations with a phenomenologic coefficient matrix:

$$L_{mj} = \sum_{k=1}^{n} a \ (mk) \ L_{jk} \quad (11)$$

Then, Onsager's principle states that the phenomenologic coefficient matrix is symmetrical [33, 66].

It will be noted that the irreversibility of a process is dueto those very differences between the intensive magnitudes that define the thermodynamic forces.

These forces are to be found everysiwhere in nature (thermal gradients and the gradients of material concentrations 'pulsing' in the mineral world as well as in the nucleic acids in living beings). The existence of these forces is expressed in the variation of certain extensive magnitudes (such as energy and mass) acting as thermodynamic flows.

With the aid of these ideas it is now possible to define the source of entropy, σ, as the sum of the products of thermodynamic forces and their conjugate flows, manifested by an irreversible process:

$$\sigma = \sum_{j=1}^{n} X_j \mathcal{J}_j. \quad (12)$$

Thus, the development of an irreversible process implies the existence of a thermodynamic force whose source may be the difference between mechanical, chemical, nuclear, thermal forces, etc. Once triggered, the process will manifest itself by dislocating an amount of energy or mass, that is by generating thermodynamic flows.

The process evolves towards a diminution of the forces triggered, towards a state of balance, of equilibrium, when the thermodynamic forces become equal to zero. But the events never occur in such a simple way; during the development of the initial process, several flows, acting in opposite directions, are coupled kinetically. In Figure 9.13 the following sequence is illustrated: a pri-

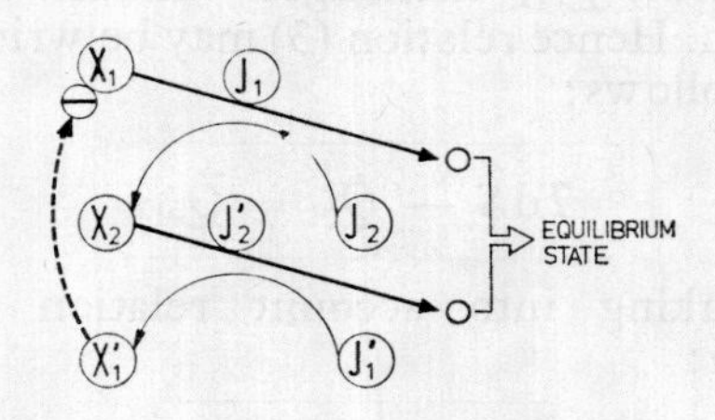

Figure 9.13. Coupled slowing down thermodynamic fluxes.

mary force X_1 produces flow $\mathcal{J}_1$ which tends to reduce force X_1 to zero, so that the system should recover its state of balance. To flow, $\mathcal{J}_1$, however, is added flow $\mathcal{J}_2$, generated by the action of flow $\mathcal{J}_1$; for instance, the transfer of cations through the membrane, owing to an electric gradient, will generate a concentration gradient and vice versa. Flow J_2 acting in the opposite direction to flow J_1 will tend to unbalance the system, at the same time developing a secondary force X_2. The latter generates its own conjugate flow J_2', which tends to reduce force X_2 to zero. Flow J_2' opposes flow J_2 and also generates a flow J_1' opposed to J_1. Flow J_1' diverts the system from its state of equilibrium and gives rise to the conjugate force X_1' which damps the initial force X_1. As a result of these synergic conjugate and coupled processes, the initial process tends irreversibly and asymptotically towards an equilibrium, being slowed down internally by reciprocal coupling of the flows.

The source of entropy thus tends slowly but steadily to decrease, and becomes zero when the system reaches its equilibrium. In this way the system may stop for a while and become stationary, maintaining its thermodynamic properties by internal adjustment of the opposing flows. As the system finds its balance in the steady phase, it has a constant value which is in fact the lowest possible value as a source of positive entropy of the system.

Analysis of irreversible processes demonstrates the complex dependance of the flow on thermodynamic forces; each flow depends on all the forces of the system, the relations being the more complex, the more 'violently' the process develops, or, in other words, the higher its 'turbulence' is.

According to equation (11) it is admitted that the flows are linear functions of all the forces of the system, also introducing in the equation the phenomenologic coefficients defined by Onsager.

Actually, the natural irreversible processes, especially those in biophysics, do not develop according to simple linear functions, but according to far more complicated ones.

Nevertheless, any irreversible biophysical process develops so that its source of positive entropy should be lowered throughout the entire system. The evolution of the process is always complicated by a runaway of flows that are coupled and conjugated with the initial forces. Onsager's phenomenologic relation can only be arbitrarily applied to these variations. If the outer disturbing forces are violent (as in the case of shock-inducing energy) the system may tend towards a stationary stage of the order zero (a state of equilibrium which in the human clinic is equivalent to death).

In a non-isolated adiabatic system (for instance the human organism and its direct environment) the exchange of entropy with the outer medium may be positive, but also negative, so that, temporarily, even in the course of an irreversible process, the total entropy of a system may increase or decrease, or remain constant if a stationary phase is reached.

A disturbing force, for instance, the force of the shock inducing agent X_S, generates a flow acting in the same direction as itself and, by generating coupled flows, will reduce that force, tending to bring the system back to its initial steady state (Le Chatelier's principle applied to irreversible processes). This means that any disturbance causes an increase in the source of entropy by priming coupled flows which tend to bring the system back to its steady state. This is how the processes take place in a shocked human organism.

According to classical thermodynamics, the live organism is an unsteady and reversible system, so that its development, accompanied by a permanent ordinance and death, appears as an explicable phenomenon.

Modern thermodynamics considers every vital phenomena as a sequence of coupled processes, obtaining entropy by irreversible coupling reactions. The vital phenomena taken individually have in fact a beginning and an end, passing through a sequence of steady states. The irreversible perennial change of the biological filum is, however, ensured by the crossroads of fecundation which generates new sources of positive entropy for each species.

The metabolic energy producing reactions are the processes that pay the cost for all the coupled vital processes in the organism.

During its normal activity the human organism acquires a physiologic steady state, characterised by a source of entropy with a minimal value (steady state). Under certain environmental conditions this steady state has the property of regenerating after disturbances caused by exterior factors, by stimulating the source of positive entropy of the organism. This extra-entropy acquired by metabolic effort, will oppose the external cause and attenuate its effects.

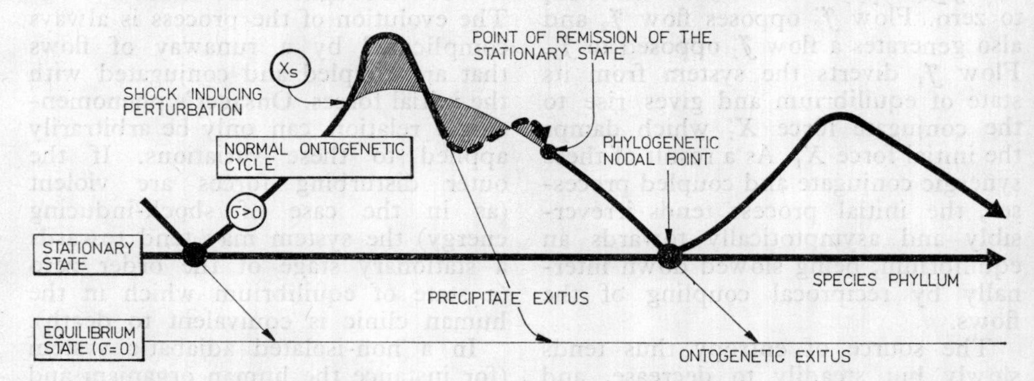

Figure 9.14. Down the evolutive filum of successive generations the shock inducing disturbance may be solved by remission of the stationary status or may cause the precipitated death of the individual (towards the equilibrium status).

A shock-inducing agent (force X_s) produces an exceptionally deep disruption of the system and the increase in positive entropy, as well as the internal braking flows, do not always have the necessary time to revert to a steady state; in these cases the shocked organism may soon reach the state of zero degree equilibrium (exitus), much before the ontogenic moment of its normal evolution. In this sense shock may be understood as an attack against the species, because it eliminates the individual from its phylogenetic chain (Figure 9.14).

9.6.2 THERMODYNAMIC OSCILLATIONS IN THE SHOCK CELL

Shock may be considered as a pathological process evolving in the space provided by the patient, and behaving as an irreversible thermodynamic non-adiabatic process open to the outer medium. The energy yielding events generated by the shock-inducing agent (lesional syndrome) and by the neuro-endocrine response of the organism (reactional syndrome) stand at the basis of the clinical symptoms and signs. Thermodynamic approach of the shocked organism as a whole, would be difficult and would not lead to useful therapeutical conclusions, but the human organism can be looked upon as a cybernetic sum of cells. In its turn, the cell is provided with subsystems that generate entropy and at the level of which irreversible processes occur throughout man's life.

There are at least two cellular subsystems generating positive entropy:

(i) *the enzymatic subsystem* which, within the network of the anabolism/catabolism metabolic pathways, has at least one irreversible nodal point on every basic pathway; for instance, the conversion of pyruvic acid into active acetate;

(ii) *the membrane subsystem* that has become, more than a decade ago, the model of a biological system, within which irreversible thermodynamic processes take place (Katchalsky, 1965).

Let σ' be a source of entropy, normally generated during the physiological steady state by the enzymatic energy yielding subsystem which, owing to force X', gives rise to flow J',

$$\sigma' = \sum_{j=1}^{n} X'_j J'_j \quad (13)$$

and σ'' a source of entropy generated under the same conditions, by the membrane subsystem, with a similar relation to (13):

$$\sigma'' = \sum_{j=1}^{n} X''_j J''_j \quad (14)$$

The source of total entropy of the macrosystem, in steady state, will be:

$$\sigma = \sigma' + \sigma'' = \sum_{j=1}^{n} X_j J_j \quad (15)$$

The total thermodynamic flow of the macrosystem in steady state will by given by relation:

$$J_j = \sum_{k=1}^{n} L_{kj} X_k \quad (16)$$

Formulas (15) and (16) are in fact only the general relations (12) and (10) applied to the human organism in its entirety (Figure 9.15).

A shock-inducing factor can be assimilated to a supplementary thermodynamic force, X_S, applied to the macrosystem in its steady state, and having a source of minimum positive entropy generated by the subsystems mentioned.

The thermodynamic force X_S, corresponding to the shock-inducing factor, will generate its conjugated flow, J_S, throughout the entire macrosystem which will pass from the steady state to a *state of shock* (Figure 9.16).

The signals of the shock disaster are first registered intracellularly by the nucleic subsystem. The genetic subsystem will reorganize the enzymatic subsystem within the field of the membrane subsystem, increasing, within the framework of the reaction syndrome, the generation of positive entropy. *Consequently, the shock-inducing flow J_S will couple with an endogenous damping flow J_S^*, that will give rise to its corresponding force X_S^*.* In fact, the coupling effects illustrated in Figure 9.13 will tend to annihilate the disturbing force X_s and bring the organism back to its steady state. During shock the source of entropy of the enzymatic subsystem is given by relation:

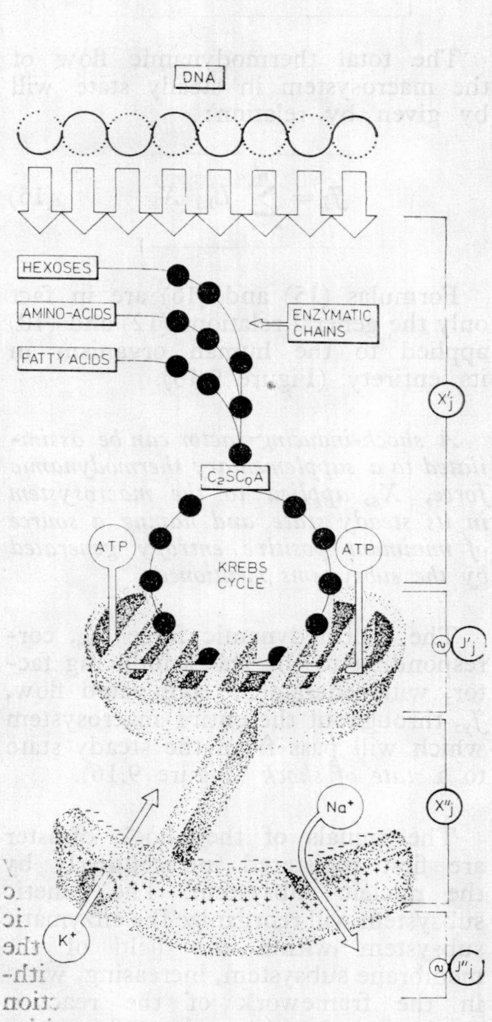

Figure 9.15. Sources of entropy in the normal cell: the enzymatic chain that governs the energy-yielding metabolic lines (especially the Embden-Meyerhof-Krebs cycle) and the membrane network.

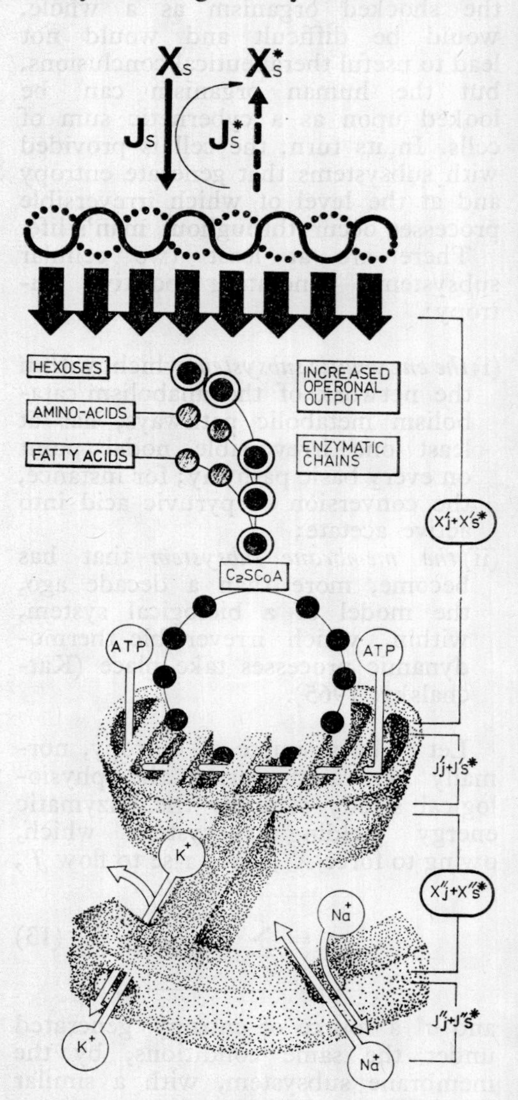

Figure 9.16. In the shock cell the antishock endogenous flux (J_s) is ensured especially by hyperglycolysis and increase in the rate of exchanges through the membrane. This is the initial maximal effort that quickly exhausts the shock cell.

$$\sigma'^* = \sum_{j=1}^{n} (X'_j + X'^*_s)(\mathcal{J}'_j + \mathcal{J}'^*_s) \quad (17)$$

and that of the membrane subsystem, by relation:

$$\sigma''^* = \sum_{j=1}^{n} (X''_j + X''^*_s)(\mathcal{J}''_j + \mathcal{J}''^*_j), \quad (18)$$

In relations (17) and (18) the subscript s merely indicates the state of shock and has no other operational significance.

For the whole shocked organism, the source of entropy will be:

$$\sigma^* = \sum_{j=1}^{n} X^*_j \mathcal{J}^*_j, \quad (19)$$

and the thermodynamic flow:

$$\mathcal{J}^*_j = \sum_{k=1}^{n} L_{kj} X_k + \mathcal{J}^*_s \quad (20)$$

From the relations (15) and (20) it is obvious that $\sigma^* > \sigma$ as a *supplementary endogenous energy flow* $\mathcal{J}^*_s$ has appeared, induced by the shock-inducing energy flow $\mathcal{J}_s$, and directed against it. This surplus of energy is obtained by the cell, especially by hyperglycolysis specific of the early stages of shock (*see* Chapter 8).

The functional reversibility of the shock cell will depend on the magnitude and duration of the supplementary thermodynamic endogenous flow, $\mathcal{J}^*_s$.

In other words, the condition for the survival of the cell, that is the minimal condition for the macrosystem to remain thermodynamically non-adiabatic (and at the new functional step imposed by force X_S), is $\sigma^* \geqslant \sigma$. This condition is ensured by $\mathcal{J}^*_s \geqslant 0$, that is by a functional level of the thermodynamic cellular subsystems at an energy rhythm at least equal to that of steady-state conditions. The eventual negativation of $\mathcal{J}^*_s$ is the consequence of functional exhaustion of the entropy-generating thermodynamic cellular subsystems, *exhaustion shortly followed by physical disorganisation of the structural elements.*

The supplementary flow $\mathcal{J}^*_s$ appears in the shock cell owing to certain time-limited phenomena that do not usually occur in the steady state.

There exists between the reversible and irreversible phases of shock an alarm interval of time during which $\sigma^* \geqslant \sigma$ is maintained. The equality $\sigma^* = \sigma$ is a signal revealing, on the one hand the functional exhaustion of the organism, and on the other the inefficiency of the treatment. From the thermodynamic point of view the limit between reversible and irreversible, in the course of shock seems to be the period during which the source of entropy, per unit of time and volume of the thermodynamic non-adiabatic macrosystem, is maintained at least equal to the source of entropy per unit of time and volume of the same system under steady-state conditions.

9.6.3 THERAPEUTICAL INFLEXION POINTS

As may be seen from Figure 9.17, after reaching a peak, flow $\mathcal{J}^*_s$ begins to fall, pointing to exhaustion of the mechanisms that attempted by a great effort to readjust the energy requirements of the shock cell ($\mathcal{J}^*_s$ is noted in figure 9.17 by $\widetilde{\mathcal{J}}$).

From this moment, t_1, the organism enters the phase considered clinically as late reversible shock and which lasts till t_2 when $\sigma^* = \sigma$. The interval between t_2 and t_3 is interpreted clinically as *refractory reversible shock* when $\sigma^* < \sigma$.

From this stage on, without adequate treatment irreversibility sets in. Bearing in mind that the loss of energy has been heralded since t_1 and begins with certainty at t_2 ($\sigma^* = \sigma$), it is worthy of note how soon the germs and harbingers of irreversibility can be detected thermodynamically (Magri, 1968; Rittenhouse, 1970).

Moreover, the thermodynamic survey of shock makes it possible to establish in time the inflexion points of a controlled, efficient therapy. During intervals t_1—t_2 and t_2—t_3 it is essential to change the treatment according to the nature of the shock-inducing agent and the patient's premorbid constitution, as well as to the treatment applied up to point t_1, although, sometimes, paradoxically but not clinically, there are no alarming symptoms or signs. Determination of the safety time limits, as well as a clear understanding of the fact that *energy disturbances*

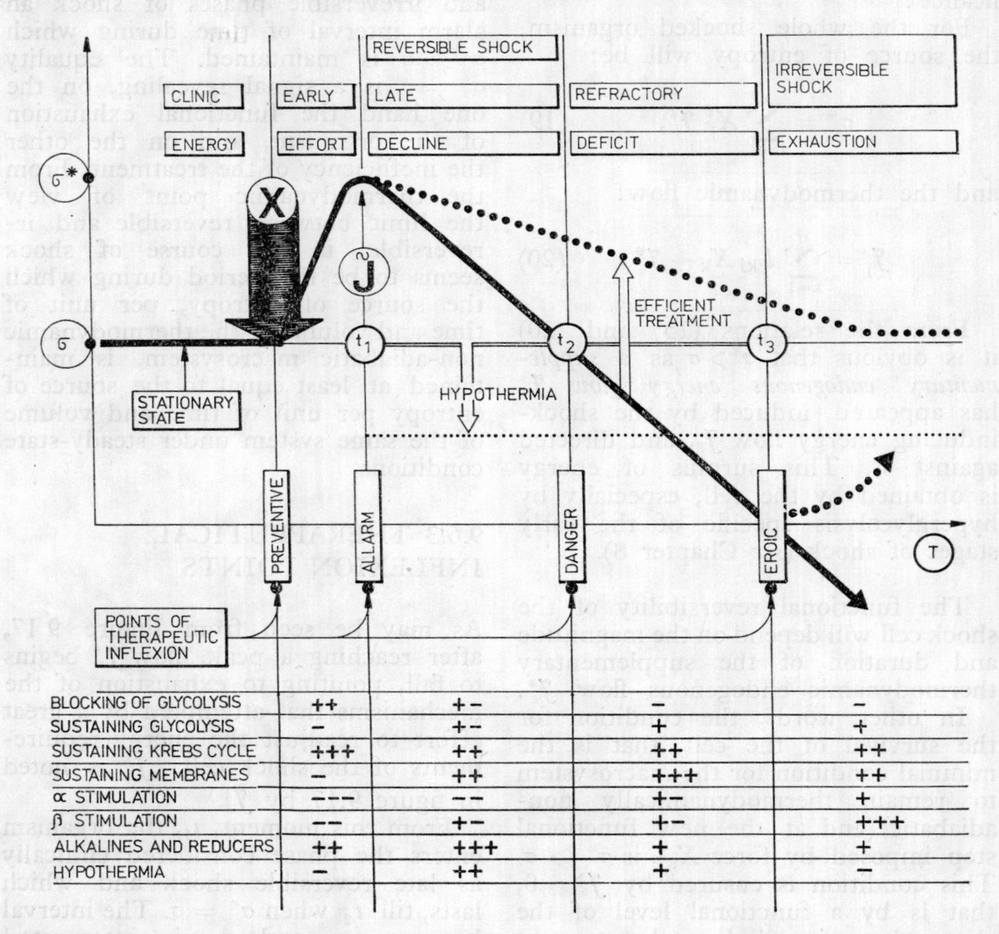

	PREVENTIVE	ALLARM	DANGER	EROIC
BLOCKING OF GLYCOLYSIS	++	+−	−	−
SUSTAINING GLYCOLYSIS	−	+−	+	+
SUSTAINING KREBS CYCLE	+++	++	++	++
SUSTAINING MEMBRANES	−	++	+++	++
α STIMULATION	−−−	−−	+−	+
β STIMULATION	−−	+−	+−	+++
ALKALINES AND REDUCERS	++	+++	+	+
HYPOTHERMIA	−	++	+	−

Figure 9.17. Nodal points in the evolution of shock are also always points of therapeutical inflexion.

precede alterations of the biological constants and clinical symptoms, might reduce the frequency of the mistakes made in the differential diagnosis of shock per stage.

Reanimation of the cell requires a therapeutical approach that is now possible: the sustaining of glycolysis (with dihydroxyacetone), buffering of lactic acidosis (with THAM), increasing the concentration of intra-mitochondrial carbonic gas for the synthesis of OAA and sustaining of the Krebs cycle (with Diamox), supplying oxygen (by hyperboric oxygenotherapy), protecting the membres, especially the lysosomal ones with antioxidants (cortisone, hydrokinone, chlorpromazine, vitamin E, or with Ag. 246-310-313 compounds, etc.), annihilating the released hydrolases (with antienzymes), hypothermia (by neuroplegia), improvement of the cellular perfusion (with dibenzylin), sustaining of the mitochondria (with magnesium), etc. [12, 46, 68, 69, 81].

In order to control hydrolase flow an antiserum made from the lysosomal fraction of the mouse liver was proposed [84].

At present, cortisone seems to be unanimously accepted as a *membrane stabilizer*, numerous data lending support to this point of view [4, 7, 12, 26, 32, 38, 50, 53, 75, 81]. Adenylates and purinic bases have become famous as 'energizing agents' in states of shock, being particularly helpful when administered in the early stages [50, 54, 56]. If the anti-enzymes are still questioned in therapy [12, 31, 52, 79, 84], chlorpromazin has been adopted for its threefold action: on the microcirculation, on the membranes, and on the dehydrogenase systems [33, 81].

Hypothermia still remains an open question [46]. The mixture of anaesthetics, cortisone and Rheomacrodex permit at 14 to 18°C surgical interventions with circulatory arrest up to 90 minutes [70, 71]. Hypothermia is able to implement one of the great desiderata in shock: to lower the usual level of entropy necessary for maintaining the steady-state, especially when the endogenous reserves of the organism begin to be depleted (Fig. 9.17). Also, hypothermia lowers the permeability of the neuronal membrane to catecholamines, impeding their release [30]. This prevents the excessive adrenergic reaction, but may also be dangerous in the phase of irreversible shock and in the early stages of shock (*see* Section 8.3.6).

Staging in shock is of the highest importance to the treatment. The oscillating evolution of the stages of shock may render a therapeutical management useful in one stage and contraindicated in another. *Therefore, the treatment has to be adapted to the stage of evolution of shock and to this end determination of its nodal inflexion points is of the utmost importance* (Figure 9.17). Chapter 11 is an attempt to present the treatment of shock dynamically.

The points at which the therapeutical management of the case has to be remodelled correspond to the inflexion points on the curve representing the endogenous thermodynamic flow, $\mathcal{J}_s^*$. Starting from this observation, it may be seen that by monitoring certain thermodynamic parameters (skin temperature, *p*H hormonal levels, cellular and plasma resistivity, creatinine and adenylate concentrations, gas pressures and membrane potentials, etc.) it is possible to determine the moments in which the treatment has to be changed [82].

Maintenance of cationic gradients between cells and their surroundings is vital for function of nerve and muscle cells and for complex activities of cells of the gastrointestinal tract and the renal tubules.

Baue et al. (1973) studied cell membrane transport in an *in vitro* system (in liver slices) to measure the capability of the transport system more directly.

The same authors found reductions in transmembrane potential in shock that may persist some time after treatment.

A progressive rise in sodium and a fall in potassium occurred in the tissues of animals in shock; the sodium content rose nearly twofold in late shock. Persistent depression of cell membrane transport may be a limiting factor in severe, prolonged shock but the changes in early shock were corrected by simple volume restoration.

The transport of electrolytes across the cell membrane in shock was studied in RBC of the monkey and man (Johnson et al., 1973). Red cell sodium increased significantly, whereas serum potassium rose markedly; there was a fall in *p*H and a rise in serum actate. These findings suggest either a failure of the RBC membrane pump, metabolic exhaustion of the RBC, an increased sodium leak into the cell or a combination of these defects in prolonged and severe haemorrhagic shock. All these shifts only occur late in shock (after at least two hours of severe hypovolaemia).

The increase in cellular water and sodium, particularly in muscle tissue as well as in kidney, and connective tissue, is in fact evidence for a failure of the sodium-potassium pump in the cell energy cycle in response to severe hypovolaemic shock (Essiet and Stahl, 1973) [25].

The study of the action potentials validate, in the living cell animals (baboons), the previously described increase in intracellular sodium, decrease in intracellular potassium and increase in intracellular water in response to haemorrhagic shock (Trunkey et al., 1973).

9.6.4 THE REVERSIBLE-IRREVERSIBLE BORDERLINE AT CELLULAR LEVEL

The problem of the limit between reversible and irreversible shock is by no means clearer at cellular level than at clinical level, since it is not yet known whether an infracellular alteration, observed in shock, is an adaptation or, on the contrary, a lesion [50, 74, 75, 78, 81]. Correlation of the biochemical alterations in the cell organelles with the structural changes observed in the electron microscope is known to be very difficult.

Nevertheless, the advanced and detailed researches of the last few years have contributed precious data concerning an assumption made some time ago: hypoxia in itself cannot account for all the damage in the shocked cell as, for a time, it may be avoided by enzymatic activity. Uncorrected blood volume, however, is the real cause of irreducible shock, because the 'suicide wastes' of fermentation cannot be washed away. The mitochondria are hardly affected by severe, pure hypoxia [37], but suffer more and are always destroyed in states of shock in which the decrease of ECBV is the essential pathogenic level of the complex neuroendocrine response (Figure 9.18).

Hypoxia itself does not appear to be responsible for disintegration of mitochondrial structure and function in shock (White IV, 1973).

Liver mitochondria showed no evidence of membrane damage after pure experimental hypoxia (Mela Leena et al., 1973). In experimental shock the findings do not conclusively prove a causal relationship between lysosomal and mitochondrial damage in the cell. They might simply be chronologically parallel cellular phenomena, caused by some unidentified membrane damage (toxin, enzyme, unbalanced electrolytes, a sharp fall in membrane energy as a thermodynamic system, etc.).

Within the framework of a biological reaction of unequal complexity, such as shock, determination of the limit between reversible and irreversible and interpretation of this limit at subcellular level is still very difficult; but it is still more difficult to establish clinically when the final stage of shock begins.

It might be assumed that the appearance of anti-DNA antibodies is an early sign of irreversibility, revealing an attack upon the genetic code, but this can only affect a limited cell population in comparison with the viability of the macrosystem. The disappearance of ATP, or rupture of the lysosomal sacs, which drive the cell to self-disintegration, may also be considered as signs of a limit between reversibility and irreversibility. But the moment glycolysis is exhausted may also announce the termination of the only energy supply the cell has overstressed in shock, so that it may be asserted that *irreversibility starts with cessation of the thermodynamic cell flows*. The loss of selectivity of the biological membrane may likewise be considered a harbinger heralding the bankruptcy of the last thermodynamic cellular subsystem generating entropy. Numerous clinical, haemodynamic or biochemical criteria could be applied for establishing this much discussed borderline, but all are subject to criticism from the very beginning because they do not coincide in time (*see* Section 3.5).

Which is therefore the actual moment at which a shocked organism advances towards irreversibility? As usually occurs in the clinic, it is very difficult to find an adequate answer. Practically, however much a patient seems to be deprived of a chance of survival, we shall never think of irreversibility in terms of therapeutical hopelessness.

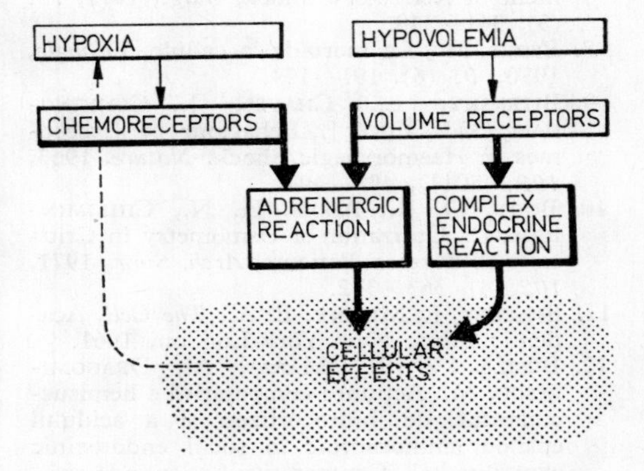

Figure 9.18. In shock, pure hypoxia cannot alone generate all the disturbances. Only the loss of circulating fluid generates the complex shock-inducing reaction.

SELECTED BIBLIOGRAPHY

1. ABOUMA, G. M., ALDERTE, J. A., STARZL, T. E., Changes in Serum Potassium and *p*H during Clinical and Experimental Liver Transplantation, *Surg.*, 1971, *69*, (3), 419–426.
2. AHONEN, J., SCHEINIR, T. M., Mesenteric Vascular Oclusion, in the Dog: Effect of Oxygen Breathing, *Europ. Surg. Res.*, 1970, *2*, (2), 88–94.
3. ASHFORD, T. P., BURDETTE, W. J., Response of Isolated Perfused Hepatic Parenchyma to Hypoxia, *Ann. Surg.*, 1965, *162*, (2), 191–207.
4. BAUE, A. E., SAYEED, M. M., Alterations in the Functional Capacity of Mithochondria

in Haemorrhagic Shock, *Surg.*, 1970, *68*, (1), 40–47.
5. BAUE, A. E., WORTH, M. A., CHAUDRY, I. H., SAYEED, M. M., Impairment of Cell Membrane Transport during Shock and after Treatment, *Ann. Surg.*, 1973, *178*, (4), 412–422.
6. BELL, D. J., SCHLOERB, P. R., Cellular Response to Endotoxin and Haemorrhagic Shock, *Surg.*, 1966, *60*, (1), 69–77.
7. BELL, M. L., HERMAN, A. H., SMITH, E. E., EGDAHL, R. H., RUTENBURG, A. M., Role of Lysosomal Instability in the Development of Refractory Shock, *Surg.*, 1971, *70*, (3), 341–348.
8. BESSIS, M., La mort de la cellule, *Triangle*, 1970, *9*, (6), 191–199.
9. BITENSKY, L., CHAYES, J., CUNNINGHAM, G. J., FINE, J., Behaviour of Lysosomes in Haemorrhagic Shock, *Nature*, 1963, *199*, (5011), 493–494.
10. BODY, D. R., ADDIS, H. N., CHILIMINDRIS, C., Utilization of Osmometry in Critically Ill Surgical Patients, *Arch. Surg.*, 1971, *102*, (4), 363–372.
11. BRACHET, J., MIRSKY, E. A., *The Cell*, Academic Press, New York-London, 1961.
12. BUŞILĂ, V. T., MÂRZA, A., DRAGOMIRESCU, M., Acţiunea comparativă a hemisuccinatului de hidrocortizon şi a acidului epsilon aminocaproic în şocul endotoxinic experimental (Comparative Action of Hydrocortisone Hemisuccinate and Epsilon Aminocaproic Acid in Experimental Endotoxinic Shock), *Medicina Internă*, 1970, *22*, (3), 361–370.
13. CALDERIO-BABUDIERI, D., Lysozyme Granules and Lysosome Structures in Cell Culture, *Nature*, 1966, *212*, (5067), 1274–1275.
14. CAMPION, D. S., LYNCH, L. J., RECTOR, Jr. F. C., Effect of Haemorrhagic Shock on Transmembrane Potential, *Surg.*, 1969, *66*, (6), 1051–1059.
15. CARAFOLI, E., Calcium Ion Transport in Mitochondria, *Biochem. J.*, 1970, *116*, (1), 2–3.
16. CHIRICUŢĂ, I., MUSTEA, I., ROGOZAN, I., Respiraţia celulară a organelor în stare de şoc (Cellular Respiration of the Organelles in Shock), *Revista sanitară militară*, 1965, *LXI*, (special number), 188–192.
17. CLERMONT, H. G., ADAMS, J. T., WILLIAMS, J. S., Source of a Lysosomal Enzymes Acid Phosphatase in Haemorrhagic Shock, *Ann. Surg.*, 1972, *175*, (1), 19–25.
18. CUNNINGHAM, J. N., SHIRES, G. T., WAGNER, Y., Cellular Transport Defects in Hemorrhagic Shock, *Surg.*, 1971, *70*, (2), 215–222.
19. DALGAARD, O. Z., An Electron Microscopic Study on Glomeruli in Renal Biopsies from Human Shock Kidneys, *Lab. Invest.*, 1960, *9*, (2), 364–370.
20. DEDUVE, C., *Lyosome Concept*. In *Lysosomes* (ed. DEREUCK, A.V.S., and CAMERON, M.R.J.), A. Churchill, London, 1963.
21. DEDUVE, C., Le rôle des lysosomes en pathologie cellulaire, *Triangle*, 1970, *9*, (6), 200–208.
22. DEPALMA, R. G., COIL, J., DAVIS, J. H., HALDEN, W. D., Cellular and Ultrastructural Changes in Endotoxemia: A Light and Electron Microscopy Study, *Surg.*, 1967, *62*, (3), 505–515.
23. DEPALMA, R. G., LEVEY, S., HALDEN, W. D., Ultrastructure and Oxidative Phosphorylation of Liver Mitochondria in Experimental Hemorrhagic Shock, *J. Trauma*, 1970, *10*, 122–134.
24. DICULESCU, I., ONICESCU, D., MISCHIU, L., *Biologie celulară* (Cellular Biology), Editura Academiei Republicii Socialiste România, Bucureşti, 1971.
25. ESSIET, G. S., STAHL, W. M., Water and Electrolyte Content of Tissues in Hemorrhagic Shock and Surgical Trauma, *Surg. Gynecol. Obstet.*, 1973, *137*, (1), 11–14.
26. FEROLDI, J., MALLET-GUY, Y., BRACONNOT, P., GUILLET, R., Etude expérimentale histologique et ultrastructurale du foie de choc, *Lyon Chir.*, 1969, *65*, (6), 720–734.
27. FODOR, O., *Biologia moleculară şi medicina modernă* (Molecular Biology and Modern Medicine), Editura medicală, Bucureşti, 1971.
28. GAZZANIGA, B. A., O'CONNOR, E. N., Effects of Intravenous Infusion of Autologous Kidney Lysosomal Enzymes in the Dog, *Ann. Surg.*, 1970, *172*, (5), 804–812.
29. GEORGESCU, L., *Morpatologia* (Pathologic Morphology), Editura didactică şi pedagogică, Bucureşti, 1971.
30. GIACCHETTI, A., SHORE, P. A., Permeability Changes Induced in the Adrenergic Neurone by Reserpine, *Biochem. Pharmacol.*, 1970, *19*, (5), 1621–1626.
31. GUILLET, R., BRACONNOT, P., BARBIER, M., STEVANOVIC, D., TRSAN, U., Etude expérimentale hémodynamique et biochimique du foie de choc, *Lyon Chir.*, 1969, *65*, (2), 182–194.
32. GLENN, T. M., Role of Lysosomes in the Pathogenesis of Splanchnic Ischemic Shock in Cats, *Circ. Res.*, 1970, *27*, (6), 783–797.
33. GUMINSKI, K., *Termodinamica proceselor ireversibile* (The Thermodynamics of Irreversible Processes), Editura Academiei Republicii Socialiste România, 1964.

34. GUMP, F. E., KINNEY, J. M., PRICE, Jr. J. B., Energy Metabolism in Surgical Patients: Oxygen Consumption and Blood Flow, *J. Surg. Res.*, 1970, *10*, (12), 613–627.
35. HIFT, H., STRAWITZ, J. G., Irreversible Haemorrhagic Shock in Dogs: Structure and Function of Liver Mitochondria, *Amer. J. Physiol.*, 1961, *200*, (2), 264–268.
36. HIFT, H., STRAWITZ, J. G., Irreversible Haemorrhagic Shock; Problem of Onset of Irreversibility, *Amer. J. Physiol.*, 1961, *200*, (2), 269–272.
37. HOLDEN, W. D., DEPALMA, R. G., DRUCKER, W. E., MCKALEN ANNE, Ultrastructural Changes in Haemorrhagic Shock. Electron Microscopic Study of Liver Kidney and Striated Muscle Cells in Rats, *Ann. Surg.*, 1965, *162*, (3), 517–536.
38. JANOFF, A., *Alterations in Lysosome (Intracellular Enzymes) during Shock: Effects of Preconditioning (Tolerance) and Protective Drugs in Shock*, (Ed. Hershey), Little Brown, Boston, 1964.
39. JENNINGS, R. B., MORRE, C. B., SHEN, A. C., HERDSON, P. B., Electrolytes of Damaged Myocardial Mitochondria, *Proc. Soc. Exp. Biol.*, 1970, *135*, (2), 515–522.
40. JOHNSON, G. Jr., BAGGETT, C., Red Cell Fluid and Electrolytes during Hemorrhagic Shock in the Monkey, *Ann. Surg.*, 1973, *178*, (5), 655–658.
41. JOHNSON, S. M., BANGHAM, A. D., The Action of Anaesthetics on Phospholipid Membranes, *Biochim. Biophys. Acta*, 1969, *193*, (1), 92–104.
42. KATAJA, J., GORDIN, R., Serum Gamma Glutamyl Transpeptidase and Alkaline Phosphatase in Severely Injured Patients, *Acta Chir. Scand.*, 1970, *136*, (4), 277–281.
43. KATCHALSKY, A., CURRAN, P. F., *Nonequilibrium Thermodynamics in Biophysics*, Harvard Univ. Press., Cambridge-Massachusetts, 1965.
44. KUELHL, L., SUMSION, E. N., Turnover of Several Glycolytic Enzymes in Rat Liver, *J. Biol. Chem.*, 1970, *245*, (24), 6616–6623.
45. KWANT, W. O., SEEMAN, P., Displacement of Membrane Calcium by a Local Anaesthetic (Chlorpromazine), *Biochim. Biophys. Acta*, 1969, *193*, (3), 338–349.
46. LABORIT, H., Perspectives d'avenir sur l'hypothermie, *Agressol.*, 1969, *10*, (4), 278–290.
47. LANGDON, R. G., BLOCH, K., *Lipid Metabolism*, John Wiley, New York-London, 1960.
48. LEHNINGER, A. L., Mitochondria and Calcium Ion Transport, *Biochem. J.*, 1970, *119*, (2), 129–138.
49. LEV, R., SIGEL, H. I., JERZY-GLASS, G. B., The Enzyme Histochemistry of Gastric Carcinoma in Man, *Cancer*, 1969, *23*, (5), 1086–1093.
50. LILLEHEI, R. C., LONGERBEAM, J. K., BLOCK, J. H., MANAX, W. G., The Nature of Irreversible Shock: Experimental and Clinical Observations, *Ann. Surg.*, 1964, *160*, (4), 682–710.
51. MAGRI, P., Ricerche sperimentali sullo shock postoperatorio, IV. Influenza della concentrazione ionica intracellulare sulla irreversibilità dello shock, *Osped. Maggiore*, 1968, *63*, (1), 1295–1304.
52. MALLET-GUY, Y., SWITALSKA, Ch., BOTEV, S., MALLET-GUY, P., Choc hémorragique et inhibiteur de protéinases, *J. Chir.*, 1971, *101*, (4), 361–376.
53. MANSOUR, A. M., NASS, S., In vivo Cortisol Action on RNA Synthesis in Rat Liver Nuclei and Mitochondria, *Nature*, 1970, *228*, (5272), 665–667.
54. MARKLEY, K., SMALLMAN, E., Protection against Burn, Tourniquet and Endotoxin Shock by Purine Compounds, *J. Trauma*, 1970, *10*, (7), 598–607.
55. MARTIN, A. M., HACKEL, D. B., KURT, S. M., The Ultrastructure of Zonal Lesions of the Myocardium in Haemorrhagic Shock, *Amer. J. Path.*, 1964, *44*, (1), 127–133.
56. MASSION, H. W., Value of High Energy Compounds in the Treatment of Shock, *Amer. J. Surg.*, 1965, *110*, (3), 342–347.
57. MELA LEENA, MILLER, L.D., BACALGO, L.V., Jr., OLOFSSON, K., WHITE, R. R., IV., Role of Intracellular Variations of Lysosomal Enzyme Activity and Oxygen Tension in Mitochondrial Impairment in Endotoxemia and Hemorrhage in the Rat, *Ann. Surg.*, 1973, *178*, (6), 727–735.
58. NASS, N.M.K., Origin of Mitochondria: Are they Descendants of Ancestral Bacteria?, *Triangle*, 1971, *10*, (1), 20–36.
59. NICOLAU, C., SIMON, Z., *Biofizica moleculară (Molecular Biophysics)*, Editura Ştiinţifică, Bucureşti, 1968.
60. NORRIS, D. M., FERKOVICH, S. M., ROZENTAL, J. M., BAKER, J. E., Energy Transduction: Inhibition of Cockroach Feeding by Naphthoquinone, *Science*, 1970, *170*, (3959), 754–755.
61. OYAMMA, T., MAEDA, A., KUDO, T., Effect of Mild Hypothermia and Surgery on Adrenocortical Function in Man, *Agressol.*, 1971, *12*, (3), 217–222.
62. PALADE, G., SIEKEVITZ, P., Liver Microsomes. An Integrated Morphological and Biological Study, *J. Biophys. Biochem. Cytol.*, 1956, *2*, (1), 171–200.
63. PILET, P. E., *La cellule. Structure et fonctions*, Masson, Paris, 1964.

64. POLICARD, A., *Eléments physiologiques cellulaires,* Masson, Paris, 1966.
65. POSTELNICU, T., TĂUTU, P., *Metode matematice în medicină și biologie (Mathematical Methods in Medicine and Biology)*, Editura tehnică, București, 1971.
66. PRIGOGINE, I., *Introduction to Thermodynamics of Irreversible Processes,* John Wiley, New York, 1962.
67. RACE, D., COOPER, E., Haemorrhagic Shock: The Effect of Prolonged Low Flow on the Regional Distribution and its Modification by Hypothermia, *Ann. Surg.*, 1968, *167*, (4), 454–460.
68. RANGEL, D. Mc, BRUCKNER, W. L., BYFIELD, J. E., ADOMIAN, G. E., DINBAR, A., FONKALSRUD, E. Q., Enzymatic and Ultrastructural Evaluation of Hepatic Preservation in Primates, *Arch. Surg.*, 1970, *100*, (3), 284–289.
69. RANGEL, D. M., BYFIELD, J. E., ADOMIAN, G. E., STEVENS, G. H., FONKALSRUD, E. W., Hepatic Ultrastructural Response to Endotoxin Shock, *Surg.*, 1970, *68*, (3), 503–511.
70. RITTENHOUSE, E. A., MOHRI, H., MERENDINO, K. A., Studies of Carbohydrate Metabolism and Serum Electrolytes during Surface-Induced Deep Hypothermia with Prolonged Circulatory Occlusion, *Surg.*, 1970, *67*, (6), 996–1005.
71. RITTENHOUSE, E. A., MOHRI, H., MORGAN, B. C., DILLARD, D. H., MERENDINO, K. A., Electrocardiographic Changes in Infants Undergoing Surface Induced Deep Hypothermia for Open-Heart-Surgery, *Amer. Heart. J.*, 1970, *79*, (2), 167–174.
72. ROUILLER, C., Les modifications ultrastructurales du cytoplasme des cellules hépatiques, *Triangle*, 1970, *9*, (6), 209–219.
73. SCHUMER, W., SPERLING, R., Shock and its Effect on the Cell (Human), *J.A.M.A.*, 1968, *205*, (4), 215–219.
74. SCHUMER, W., KAPICA, S. K., TA LEE TENG, Validity of the Lysosomal Theory in Oligohemic Shock, *Arch. Surg.*, 1969, *99*, (3), 325–329.
75. SEVITT, S., The Boundaries between Physiology, Pathology and Irreversibility after Injury, *Lancet*, 1966, *2*, 1203.
76. STEWARD, L. P., Mitochondrial Enzymes in Brains from Hypothermic Fround Squirrels and Rats, *Tex. Rep. Biol. Med.*, 1969, *27*, (3), 935–936.
77. STRAWITZ, J. G., HIFT, H., *Subcellular Changes in Haemorrhagic Shock* (ed. MILLS, L. C. and MOYER, H. J.) in *Shock and Hypotension*, Grune and Straton, New York, 1965.
78. SUTHERLAND, N. G., BOUNOUS, G., GURD, F. N., Role of Intestinal Mucosa Lysosomal Enzymes in the Pathogenesis of Shock, *J. Trauma*, 1968, *8*, (3), 350–358.
79. SZANTO, P. B., SHOEMAKER, W. C., LARSEN, K., DUBIN, A., Liver in Experimental Haemorrhagic Shock. An Electron Microscopy Study, *Amer. J. Path.*, 1966, *48*, (2), 230–236.
80. ŞUTEU, I., GIURGIU, T., IONESCO, P., SAFTA, T., L'hyperthermie induite contrôlée. Etude expérimentale, *J. Chir.*, 1968, *95*, (5–6), 669–674.
81. ŞUTEU, I., CAFRIŢĂ, A., BUCUR, I. A., Aspecte infracelulare în șocul experimental (Infracellular Aspects in Experimental Shock), *Revista sanitară militară*, 1972, *LXXV*, (3), 357–366.
82. ŞUTEU, I., CAFRIŢĂ, A., BUCUR, I. A., Progrese recente în fiziopatologia șocului (Recent Progress in the Pathophysiology of Shock), *Revista sanitară militară*, 1972, *LXXV*, (3), 267–276.
83. TRUNKEY, D. D., ILLNER, H., WAGNER, I. Y., SHIRES, T. G., Effect of Hemorrhagic Shock on Intracellular Muscle Action Potentials in the Primate, *Surg.*, 1973, *74*, (2), 241–250.
84. TULKENS, P., TROUET, A., VAN HOOF, F., Immunological Inhibition of Lysosome Function, *Nature*, 1970, *228*, (5278), 1282–1285.
85. VITALI-MAZZA, L., MISSALE, G., FERIOLI, V., La struttura ultramicroscopica del Rene Nello Shock Sperimentale, *Minerva Nefrol.*, 1964, *11*, (1), 30–37.
86. WHITE, R. R. IV, MELA, LEENA, C., MILLER, L. D., BERWICK, L., Effect of *E. Coli* Endotoxin on Mitochondrial Form and Function, *Ann. Surg.*, 1971, *174*, (6), 983–990.
87. WHITE, R. R. IV, MELA LEENA, C., BACALZO, L. V. Jr., OLAFSSON, K., MILLER, L. D., Hepatic Ultrastructure in Endotoxemia, Hemorrhage and Hypoxia: Emphasis on Mitochondrial Changes, *Surg.*, 1973, *73*, (4), 525–534.

10 THE VISCERA IN SHOCK

The various organs, the diseases of which are subdivided for treatment, are not isolated, but complex parts of a complex whole, and every day's experience brings home the truth of the saying 'when one member suffers all the members suffer with it'.

OSLER

10.1 THE GASTROINTESTINAL TRACT
10.1.1 The stomach in shock
10.1.2 The intestine in shock
10.2 THE SUBORDINATE DIGESTIVE GLANDS
10.2.1 The liver in shock
10.2.2 The pancreas in shock
10.3 THE KIDNEY IN SHOCK
19.4 THE LUNG IN SHOCK

Shock is a general disturbance of the whole organism; in the chain of pathologic events almost all the organs are affected, some correcting, others aggravating disorders which in their course become progressively worse. In the struggle against shock, almost all the viscera are overcome.

In shock, the organ deficiencies develop according to a certain order (Figure 10.1). *The neuronal system, endocrine glands* and *heart* are protected for some time, but the *kidney, intestine* and *somatic framework* are rapidly affected. The *liver* is subjected to stress for a long time and might respond adequately if it were not suffocated by persistent hypoxia. The *lung*, of excessive fragility with regard to the rheologic disturbances induced by shock, is considered at present the most critical organ of shock in man.

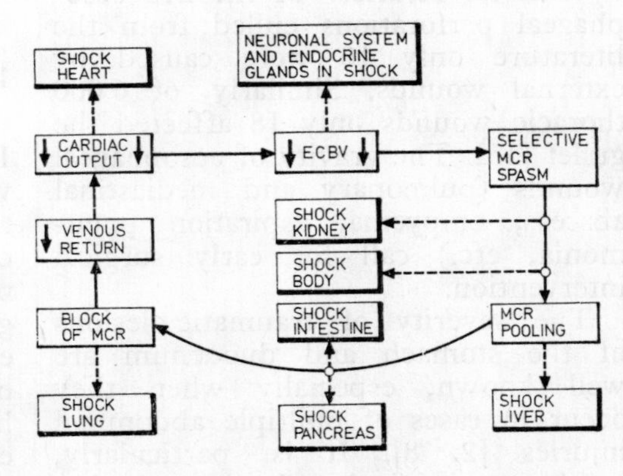

Figure 10.1. The mode and sequence of the onset of organ failure in the course of shock.

The order in which the viscera are affected may be reversed by the cause of shock (for instance, cardiogenic shock) or by pathological, visceral impairment prior to the shock episode, producing a particular sensitivity of the respective organ (cirrhosis, asthma, gastro-duodenal ulcer, etc.).

The neuro-endocrine system (*see* Chapter 4) and the heart (*see* Section 5.2) have already been discussed. The present chapter deals with the characteristics of behaviour of some of the most important viscera in the course of shock. Each section begins with a general outline of the direct visceral lesions that may produce a certain type of shock, and is followed by an analysis of the role of the respective *shocked organ* in the general pathological physiology of all states of shock.

10.1 THE GASTROINTESTINAL TRACT

The gastrointestinal tract may be directly affected in injuries giving rise to shock.

The gullet is rarely affected directly by exterior injuries. Of the 218 oesophageal perforations culled from the literature only 10 were caused by external wounds. Similarly, of 6 000 thoracic wounds only 18 affected the gullet [15]. The gravity of oesophageal wounds (pulmonary and mediastinal abscess, empyema, aspiration pneumonia, etc.) call for early surgical intervention.

The severity of traumatic lesions of the stomach and duodenum are well known, especially when they occur in cases of multiple abdominal injuries [2, 8]. It is, particularly, involvement of the duodenum that precipitates an abrupt state of shock, because the retroduodeno-pancreatic space, which is an important centre reflecting autonomic life, is extremely sensitive.

Wounds of the colon are constantly accompanied by mixed states of shock (traumatic, bacteriemic and haemorrhagic) which have a mortality of 25 to 30%. Colostomy protects for several months the mobile segments of the colon that are generally affected. Almost a quarter of 50 cases of wounds of the colon took up their normal functions after reconstruction of the continuity of the colon [1].

Rectal wounds have a longer hospitalization (on average 200 days), as a rule necessitating three or more successive operations (coccygectomy for efficient drainage and protective derivative anus, with direct remedial intervention, are recommended) [20].

In shock, however, the different segments of the gastrointestinal tract may also be impaired indirectly by the *state of shock per se* and its pathologic circuits. These cases of 'indirect lesions' may sometimes be brought to operation (ulcer due to stress) or may demand a very demanding therapy (shock intestine).

10.1.1 THE STOMACH IN SHOCK

In shock, the stomach is one of the viscera that is haemodynamically affected due to severe spasm in its microcirculatory network. Rheodynamic and metabolic disturbances of the entire gastric wall lower its resistance to the endocavitary acid medium. Erosions of the parietal structures together with haemorrhage (manifested by haematemesis and/or melena) or perforations may occur.

Our knowledge of the mechanism of destruction of the gastric wall tissues in shock has gained considerably following new findings regarding the role of gastric mucin and the DIC phenomenon. Moreover, the important role of the sterol factor is revealed by the minimal interval in which ulcer caused by stress appears (6 to 14 days), so that it must be considered as a late, and severe complication of traumatic, bacteriemic and thermal shock.

Stress ulcer. The present pathoanatomic definition of stress ulcer implies the existence of superficial and often multiple ulcerations, without induration of the neighbouring tissues and without histologic lesions of chronic inflammation.

The clinical development must have a past history of less than four weeks.

The death rate from stress ulcer is impressive: 35 to 65% or even 60 to 90%. It is difficult to assess its incidence correctly: the proportion of stress ulcers in 5 350 traumatic accidents was of 0.4% and in 2 463 burns of 11% [25, 30]. Stress ulcer was found to be the cause in 61 of 2 346 upper digestive haemorrhages [13]. But these statistics are haeterogeneous.

In general, it is assumed that in civilian hospitals there are on average about 20 stress ulcers a year (1 to 2 a month); in the evacuation hospitals of Vietnam the proportion was low (1.4% of all the laparotomies), but in the specialty hospitals the number was three times greater because stress ulcer develops for 6 to 14 days before bleeding or perforation occurs [13, 30]. Within this interval, glucocorticoids lower the volume of mucus secretion and modify its mucopolysaccharide composition (lowers sialitic acid and heoxamine) (Figure 10.2).

Erosion of the gastroduodenal mucosa brought about by extracranial injuries and infection differs from *Cushing's ulcer* which appears in head injuries; the latter ulcer is always bathed in an acidity that is encountered only in the Zollingen-Ellison syndrome (over 100 to 150 mEq). In these cases plasma gastrin also showed a marked increase [14, 37]. Central nervous lesions are the cause of these acid and gastrin hypersecretions. Typical stress ulcers (without head injuries) are not connected with increased acidity or hypergastrinaemia.

In the typical stress ulcer the essential pathogenic influences are the steroids which produce a quantitative and/or qualitative reduction of the mucin film and a relative hyperacidity.

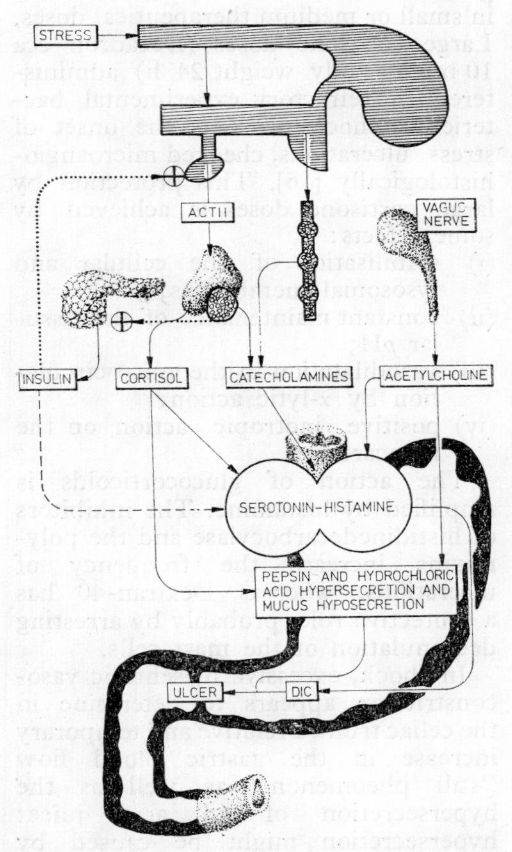

Figure 10.2. The complex pathologic physiology of 'stress ulcer'. The principal mechanism of gastric erosion is the interaction between the acid (that increases) and the mucus (that decreases). This mechanism has a systemic complex hormonal control in which cortisol plays the key role. Another important effect of the rheologic disturbances in the gastric wall is a severe spread phenomenon immediately followed by necrosis, e.g. wall ulceration.

Normal mucus may buffer an acidity of up to 50 mEq HCl, and diminution of its buffer capacity may render harmful an acidity of less than 15 to 18 mEq [26].

In addition to endogenous steroid hormones, exogenous cortisone may be deleterious, favouring the onset of gastroduodenal ulcerations, but only in small or medium therapeutical doses. Large cortisone doses (Decadron cca 10 mg/kg body weight/24 h) administered in refractory experimental bacteriemic shock prevents the onset of stress ulcerations, checked microangiohistologically [16]. This protection by large cortisone doses is achieved by some effects:

(i) stabilisation of the cellular and lysosomal membranes;
(ii) constant maintenance of the tissular *p*H;
(iii) vasodilatation in the microcirculation by α-lytic action;
(iv) positive inotropic action on the myocardium.

The action of glucocorticoids is amplified by histamine. The inhibitors of histidinedecarbocylase and the polymixins increase the frequency of ulcerations, whereas Dextran-40 has a protective role, probably by arresting degranulation of the mast cells.

In shock, excessive mesenteric vasoconstriction appears to determine in the celiac trunk a relative and temporary increase in the gastric blood flow ('still pheomenon') as well as the hypersecretion of an acid juice: hypersecretion might be caused by excess bradykinin [21]. Moreover, gastric secretion is also accentuated because in shock (as in portocaval shunts) the 'jejunal hormonal factor' is no longer inactivated by the liver [22, 28].

Linder and McKay studied the role of DIC in involvement of the stomach in shock [19]. The gastric microcirculation is known to be of exceptional fragility in shock (*see* Section 5.4). The administration of thrombin and simultaneous blocking of fibrinolysis generated, in experimental animals, numerous thromboses and haemorrhage of the duodenal mucosa; the added administration of α-stimulants changed the site of the lesions in the gastric mucosa. The onset of DIC in the gastric microcirculation was prevented by α-lytics. For DIC to appear, intervention of both the rheologic factor and the vascular factor is necessary. 'Symbiosis' of these two factors differs in the vascular micronetwork according to the nature of the viscus: in the stomach acute oval and linear ulcerations are produced but the intestinal mucosa is necrosed in strips, resulting in pseudomembranous enterocolitis.

The target space of the combined hypercoagulating action of thrombin and histamine is the duodenum and jejunum; *if to this is added a catecholamine vasoconstriction the stomach becomes the target organ (especially as the steroids have in the meantime lowered the performance of mucin)*.

In point of fact, it is known that in the pathogenic chain involving affection of the stomach in shock, the role of the vagus is not very conspicuous. Interruption of the parasympathetic effector gives free scope to the adrenergic system, the feared foe of gastric microcirculation mobility (McKay and Linder, 1971), [19].

The incidence of gastroduodenal ulcerations increases when the initial traumatic lesion is complicated with an infection. The septic factor in a severe injury or burn is always of exceptional gravity, leading to stress ulcer.

Sepsis precedes upper gastrointestinal bleeding in over half of stress ulcer patients, and ulceration occurs in about one third of surgical patients with septicaemia. Recently, in an experimental stress model (pigs) produced by endotoxin, tissue damage with ulceration appeared to be due to selective regional flow reduction in the stomach (Richardson, 1973) [32]. In addition to the gastric mucosal changes, intestinal changes also occur in response to hypovolemic shock. There is a loss of intestinal macromolecular barrier function preceding onset of deterioration of hepatic mitochondrial function (Fromm, 1973; Rhodes, 1973) [9, 31].

The preferential lesions of the stress ulcer is the gastric wall but mixed gastroduodenal lesions are frequent; many authors consider the duodenum as the initial site of ulceration. The lesions may consist in multiple punctiform erosions, diffuse erosion of the mucosa or a single ulcerous lesion without any reaction of the neighbouring tissues.

When superficial, the multiple erosions respond to medical treatment; when deeper, affecting the submucosa, they are particularly persistent. Autopsy in 281 lethal cases of burns revealed oesophageal erosions in 18%, gastric erosions in 28%, and duodenal in 11%. Actual ulcerations were found in the gullet in 7%, the stomach in 11% and duodenum in 8% of the cases. Therefore, 57% erosions correspond to 26% ulcerations, i.e. in general, erosions are twice as frequent as ulcerations [25].

The medical treatment of stress ulcers is sufficiently empirical to account for the failure. Antacids are not justified and irrigation with normal saline is inoperative in erosions through the mucosa.

Practically, surgery seems to be no more successful; the treatment begins as a rule with nasogastric aspiration (which also offers a reliable checking of the haemorrhage) and massive transfusions which may arrest the bleeding in 50% of the cases, but with many recurrences. Recent successes have been obtained in upper digestive haemorrhage caused by stress ulcer with α- and β-blocking agents [27].

Surgical treatment has not yet been standardised and the statistical arguments are inadequate. Vagotomy with pyloroplasty is only efficient in 50% of the cases but the operative risk is very small. Broader resection, although more efficient is followed by a high mortality rate. Vagotomy associated with resection is considered useful although it is not sustained by many pathophysiologic arguments [12, 24, 29]. In practice, it is preferable to begin by an explorative pylorotomy and then, if necessary, to advance towards the stomach [35]. In acute ulceration, with vascular fistula which may be ligated *in situ*, vagotomy offers a questionable protection. When bleeding due to multiple erosions occurs vagotomy is useless and, in the opinion of some surgeons, even illogical. In these cases α-lytics alone administered postoperatively may consolidate a broad gastric resection.

Preoperative selective angiography may detect the site of the haemorrhage, which is extremely useful in ulcers of the D_2—D_3 segments. Intraoperatively, when there is no blood in the stomach of the melenic patient, duodenotomy is recommended, and if a postbulbar stress ulcer is found, ligature *in situ*, raphe of the ulceration, vagotomy and poyloroplasty are performed, since the duodenal segment is the site of predilection of the effect of acetlylcholine and histaminic mechanisms [3].

In gastric stress ulcer pathophysiological arguments appear to sustain splanchnicectomy, but this has no sound experimental grounds. Until conclusive data are available, vagotomy will be the operation of choice for gastric stress ulcer. It should be recalled that there are common factors between the vagus and splanchnic and that in the choice of the therapeutical approach the modality of the haemorrhagic lesion and the digestive segment affected must be kept in view.

10.1.2 THE INTESTINE IN SHOCK

The intestine, which was initially insufficiently studied in shock, probably also neglected because its impairment was less conspicuous, has rapidly proved to play a decisive role in the irreversibility of shock, especially in experimental shock (Lillehei, 1958; Bounous, 1971).

The endotoxin theory asserted that irreversibility in shock is caused by invasion of the blood and the development of large amounts of endotoxins derived from the intestinal lumen due to hypoxic inactivation of the RES and of the physiologic bacteriolytic substance (properdine). The endotoxin theory is still valid but the role of the intestine in shock is now more broadly comprehended.

Three categories of pathologic phenomena gravitate around the intestine of the organism in shock: rheodynamic alterations, pancreatic proteolytic enzymes and the toxaemic factor.

Rheodynamic alterations. Lesions of the intestinal mucosa seem to be the first result of splanchnic catecholamine vasoconstriction. The entire splanchnic region is subjected to a severe sympathico-adrenergic reaction, but the area richest in α-receptors is that supplied by the superior mesenteric artery, this also being the area most affected by reduction of the cardiac output, increase in the resistance of the microcirculation and hypoxia. From the very onset of shock, the output of the mesenteric artery diminishes (as does that of the renal arteries), the decrease becoming more accentuated with the progress of shock [19, 32]. The usual clinical manifestation of shock-pallor and cold skin ought to be extrapolated to the level of the jejunum which suffers more even than the integuments. Ischaemia of the intestine is manifested clinically by diffuse abdominal pains, dynamic ileus, and sometimes bloody diarrhoeic stools.

In the dynamics of shock, when precapillary resistance of the intestinal microcirculation fails and constriction of the postcapillary-venular sphincter continues, extensive splanchnic pooling occurs (*see* Section 5.5).

Apart from the phenomenon of fluid pooling, extravasation of fluid into the interstitium and increase in the filtration pressure render venous return and decompression of the microcirculation still more difficult. Platelet and erythrocytic aggregation and DIC are followed by prolonged cellular hypoxia with impairment of the intracellular organelles. In this phase, even if the normal flow of the mesenteric artery is re-established the intramural lesions persist. In the dynamics of shock evidence was found of a by-pass supply to the submucosa through arteriovenous shunts, the intestinal mucosa being excluded from the circulation. According to the rheodynamic theory this explains why, in the intestine, the most extensive lesions are those of the mucosa (with centripetal progress from the submucosa towards the lumen).

Nevertheless, experiments with ^{32}P have shown that oxidative phosphorylation ceases in the enteral mucosa before the onset of rheodynamic disturbances in the submucosa microcirculation [11]. Similarly, haemorrhagic necrosis appears to advance from the tip of the villi towards the submucosa, suggesting a centrifugal hydrolasic attack. Contact with the faecal masses also plays a pathological part, since washing of the loops brings about renewed respiration of the villous enterocytes. The rheodynamic role of spasm and collapse of the microcirculation must, therefore, inevitably be associated with hydrolase digestion if we want to explain all the pathological anatomo-functional aspects of the intestine in shock.

The role of pancreatic juice proteolytic enzymes. Experimental study of the alterations produced by shock in different organs constantly reveals early necrotic lesions of the small intestine epithelium [4, 5, 18]. Trypsin is considered to play an important role in the pathogenesis of necrosis of the intestinal mucosa. Pancreatic enzymes produce haemorrhagic digestion of the villous subepithelial structures, attacking the devitalised areas and penetrating deeply into these structures. In the lumen, the pancreatic enzymes tend to transform superficial necrosis of the epithelium into parietal haemorrhagic infarction. If, before experimental shock, Trasylol is injected, the characteristic haemorrhagic necrosis of the intestinal lumen is avoided [11]. The term 'tryptic haemorrhagic necrosis' (tryptic haemorrhagic enteritis) is justified, at any rate for the rat and dog; in the intestinal chyme of man there is far less trypsin than in the dog, which explains the rare occurrence of tryptic necrosis in primates.

A close relationship exists between the plenitude of the intestines and the gravity of shock; shock developing in empty intestinal loops has a favourable prognosis at least as regards the intestinal area. A moderate degree of ischaemia (like that produced by a systemic blood pressure of 35 to 40 mmHg during two hours) may produce haemorrhagic necrosis in the intestinal areas with a high trypsin content. To cause similar lesions in the fasting dog, therefore, in the absence of intraluminal trypsin, 4-5 hours hypotension are necessary. Hence, alimentation may be considered to create conditions of relative intolerance to ischaemia, and to eliminate protein from the food intake is a good temporary measure for antishock protection.

The toxaemic factor. The hypoxic cells of the intestinal mucosa develop energy and osmotic disturbances and lose their status as a protective barrier for the content of the intestinal lumen, by permitting intra- and transepithelial penetration of intestinal toxins and proteolytic enzymes. Consequently, the biosynthesis of mucin is upset, starting from the tip of the villi, where the onset of haemorrhagic necrosis also takes place.

Decrease in the biosynthesis of the mucin-specific mucopolysaccharides may be due to an attack of intraluminal hydrolases [6], to the inhibition of oxidative phosphorylation [11], or to arrest of the ionic pump, as demonstrated in experimental hypothermia [29]. Degradation of the intraluminal mucin film denudes the *lamina propria* (so that the enterocytes are no longer solid), causing a digestive centrifugal hydrolase attack. Ballooning and detachment of the villi is assumed to be due to the inefficiency of the osmotic pump, which may be re-equilibrated by Ringer's hypertonic solution or by mannitol introduced endoluminally.

In the intestine there are countless toxic agents; in addition to the more than 60 bacterial groups and enteroviruses there are the long-chain fatty acids, the peptides and polyamines, the products of tryptic digestion of haemoglobin, etc. The colon is the most dangerous arsenal of nocuous substances. In the dog, experimental by-pass of the colon and a diet without residues controlled the irreversibility of shock but produced lesions in other viscera [5, 6]. It effect, therefore, at any rate for the dog, death is decided by haemorrhagic tryptic enteritis and pseudomembranous enterocolitis. The intestine is the most critical organ in shock for the dog as the lung is for man.

A lethal effect, however, cannot be attributed with certainty to intestinal endoluminal toxins; but they do aggravate the state of the shocked patient since they cannot be neutralized in the hepatic RES, which is also suffering from hypoxia [10].

Apart from hydrolases and the endoluminal toxins, a particular generator of toxic factors is the enterocyte *per se;* it is its cellular hydrolases, released intraparietally, that particularly leads to local structural lesions and also to systemic disturbances [34].

Some of Allgöwer's experiments demonstrated that haemorrhagic shock increases the toxic effect of the homogenates in the intestinal wall. The resistance of consanguine mice to endotoxins protected them against the toxic effects produced by intestinal lumen substances but not against those of the intestinal wall. Thus, the question arose of whether the intestinal wall toxins are produced locally or derived by resorption from the intestinal lumen? In order to check these hypotheses the two extremities of the small intestine of mice were isolated and washed with Ringer's solution and antibiotics; the mice were then subjected to haemorrhagic shock. In the mice whose resistance to endotoxins had been previously enhanced, the intestinal wall again produced the active toxin. The conclusion is obvious: the toxin does not appear by resorption from the intestinal lumen but is generated in the intestinal wall during the hypoxic phase.

In view of the rapid changes between the intestinal content and its walls in the course of shock, Lillehei managed to obtain evident protection against shock by early perfusion of the intestine [18]. It has been assumed that an 'intestinal noxious factor' exists in the mucosa; the ischaemia epithelium liberates this toxic substance which enters the vicious circle of shock. Normally, these toxins (as well as the endoluminal ones) may be neutralised by the hepatic RES, but, as the hepatic circulation is also affected in shock, the liver loses its detoxifying function.

To conclude, the intestinal mucosa is affected by the action of several factors: hypoxia, osmolysis, intestinal toxins, mucin insufficiency and, in the last instance, autolysis by the enterocytes [23, 34]. The enzymes liberated by the enterocytes start by the lymphatic route and cause, in the systemic circulation protease activation with all its evil effects (*see* Section 7.4).

This appears to be the general role of the intestine in the pathophysiology of shock, apart from its anatomic degradation proper as an organ. The intestine has also been demonstrated to be a rich source of direct activators of the fibrinolytic system [10].

Bounous used as control the isoenzyme of alkaline phosphatase specific of the enterocyte (intestinal alkaline phosphatase is electively inhibited by L-phenylalanine). Four hours after the onset of shock, alkaline phosphatase of enteral origin increases significantly, before its isoenzyme in the liver or colon [7].

In humans, the intestinal pathoanatomical alterations are uniform as regards their location and morphology. The alterations consist in disseminated petechial haemorrhage, solitary ulcerations rapidly developing into haemorrhagic necrosis of the mucosa and entire wall. Necrotic erosions may be limited to the epithelium and resemble those due to postmortem autolysis. However, before death, haemorrhage through the denuded *lamina propria* develops, an important sign for the differential diagnosis. The preferential site is the jejunum, but the stomach and duodenum also present frequent erosions; the colon alone seems to be spared, the predominance of β-receptors over α-receptors protecting it against the vascular factor.

SELECTED BIBLIOGRAPHY

1. ALDRETE, J. S., Reconstructive Surgery of the Colon in Soldiers Injured in Vietnam, *Ann. Surg.*, 1970, *172*, (6), 1007–1014.
2. ANDRIU, V., LERNER, C., HUSANU, S., Considerații asupra rupturilor de duoden (Considerations on Rupture of the Duodenum), *Chirurgia*, 1971, *XX*, (4), 369–374.
3. BLAKEMORE, W. S., BAUM, S., NUSBAUM, M., Diagnosis and Management of Massive Haemorrhage from Postoperative Stress Ulcers of the Descending Duodenum, *Surg. Clin. N. Amer.*, 1970, *50*, (11), 979–984.
4. BOUNOUS, G., CRONIN, R. F. P., GURD, F. N., Dietary Prevention of Experimental Shock Lesions, *Arch. Surg.*, 1967, *97*, (1), 46–60.
5. BOUNOUS, G., SUTHERLAND, G., MCARDLE, H., FRASER, N. G., The Prophylactic Use of an 'Elemental' Diet in Experimental Haemorrhagic Shock and Intestinal Ischemia, *Ann. Surg.*, 1967, *166*, (3), 116–122.
6. BOUNOUS, G., 'Tryptic Enteritis': Its Role in the Pathogenesis of Stress Ulcer and Shock, *Canad. J. Surg.*, 1969, *12*, (3), 397–409.
7. BOUNOUS, G., HUGON, J., Serum Levels of Intestinal Alkaline Phosphatase in Relation to Shock, *Surg.*, 1971, *69*, (2), 238–245.
8. CHRISTIAN, V., ULMET, V., LISSAI, P., TELEA, GH., GORSKI, V., ALBU, A., Posibilități terapeutice în politraumatismele de amploare deosebită (Therapeutical Possibilities in Severe Multiple Trauma), *Chirurgia*, 1971, *XX*, (3), 245–248.
9. FROMM, D., Intestinal Absorption during Hypovolemic Shock, *Ann. Surg.*, 1973, *177*, (4), 448–452.
10. GANS, H., MORI, K. QUINLAN, R., RICHTER, D., TAN, B. H., The Intestine as a Source of Plasminogen Activator Activity, *Ann. Surg.*, 1971, *174*, (5), 826–829.
11. GURD, F. N., Metabolic and Functional Changes in the Intestine in Shock, *Amer. J. Surg.*, 1965, *110*, (3), 333–336.
12. GRIFFITH, C.A., Application of Recent Research on Selective Vagotomy, *Northw. Med.*, 1969, *68*, (10), 927–931.
13. GROSZ, C. R., WU, K. T., Stress Ulcers; A Survey of the Experience in a Large General Hospital, *Surg.*, 1967, *61*, (6), 853–857.
14. HENRION, C., MAINGUET, P., Apports du tubage nocturne et des tests chimiques d'acidité gastrique dans l'orientation du traitement chirurgical de l'ulcère duodénal, *Acta Chir. Belg.*, 1969, (6), 445–458.
15. HIX, W. R., MILIS, M., The Management of Esophageal Wounds, *Ann. Surg.*, 1970, *72*, (6), 1002–1006.

16. Kawarada, J., Weiss, R., Matsumoto, T., Pathopysiology of Stress Ulcer and Its Prevention, *Amer. J. Surg.*, 1975, *129*, (3), 249–254.
17. Lawe, J., Day, S. B., McMillan, B. G., Autopsy Findings in the Upper Gastro-intestinal Tract of 81 Burn Patients. A review, *Arch. Surg.*, 1971, *102*, (4), 412–416.
18. Lillehei, R. C., McLean, L. D., The Intestinal Factor in Irreversible Endotoxin Shock, *Ann. Surg.*, 1958, *148*, (4), 515.
19. Linder, M. M., McKay, D. G., An Experimental Study of Thrombotic Ulceration of the Gastro-Intestinal Mucosa, *Surg. Gynec. Obst.*, 1971, *133*, (1), 21–29.
20. Lung, J. A., Turk, R. P., Miller, R. E., Eiseman, B., Wounds of the Rectum, *Ann. Surg.*, 1970, *172*, (6), 985–990.
21. McClelland, R. N., Shires, G. T., Prager, M., Gastric Secretory and Splanchnic Blood Flow Studies in Man after Severe Trauma and Haemorrhagic Shock, *Amer. J. Surg.*, 1971, *121*, (2), 134–142.
22. McQuarrie, D. G., Eichenholz, A., Blumentals, A. S., Vennes, J. A., Kinetics of Gastric Juice Secretion: A Correlation of Arteriovenous Differences with the Composition of the Gastric Juice, *Surg.*, 1967, *62*, (3), 475–486.
23. Mee, W. M., Endotoxin Shock, Distinct Entity or Universal Accompaniment, *Postgrad. Med. J.*, 1969, *45*, (526), 514–517.
24. Miller, R. E., Effects of Vagotomy or Splanchnicotomy on Blood Insulin and Sugar Concentrations in the Conscious Monkey, *Endocrinol.*, 1970, *86*, (3), 642–651.
25. Magel, G. B., The Nature and Treatment of Stress Ulcers. A rewiew, *Calif. Med.*, 1970, *112*, (6), 19–24.
26. O'Neill, J. A. Jr., The Influence of Thermal Burns on Gastric Acid Secretion, *Surg.*, 1970, *67*, (2), 267–271.
27. Okabe, S., Saziki, R., Takagi, K., Effects of Adrenergic Blocking Agents on Gastric Secretion and Stress Induced Gastric Ulcer in Rats, *Jap. J. Pharmacol.*, 1970, *20*, (1), 10–15.
28. Orloff, M. J., Abbott, A. G., Rosen, H., Nature of Humoral Agent Responsible for Portacaval Shunt Related Gastric Hypersecretion in Man, *Amer. J. Surg.*, 1970, *120*, (2), 237–243.
29. Parahagi, E., Popovici, Z., Ciurel, M., Rezultate imediate pe 70 de derivații portocave pentru hemoragii digestive (Immediate Results of 70 Portocaval Shunts for Digestive Haemorrhage), *Chirurgia*, 1970, *XIX*, (1), 45–52.
30. Rapin, M., Grosbius, S., Kernbaum, S., Goulon, M., Digestive Haemorrhage during Acute Circulatory Failure, *Ann. Med. Int.*, 1970, *121*, (8–9), 763–679.
31. Rhodes, R. S., DePalma, R. G., Robinson, Am. V., Intestinal Barrier Function in Hemorrhagic Shock, *J. Surg. Res.*, 1973, *14*, (4), 305–312.
32. Richardson, R. S., Norton, L. W., Sales, J. E. L., Eiseman, B., Gastric Blood Flow in Endotoxin-Induced Stress Ulcer, *Arch. Surg.*, 1973, *106*, (2), 191–195.
33. Shaffer, B. B., Gall, E. P., Shimizu, R. T., Esparza, H. S., The Pathogenesis of the Intestinal Lesion of Deep Hypothermia and a Proposed Relationship to that of Irreversible Shock, Including a Note on a Mechanism for the Normal Turnover of Intestinal Epithelium, *Surg.*, 1967, *61*, (6), 904–914.
34. Sutherland, H. G., Bounous, G., Gurd, F. N., Role of Intestinal Mucosal Lysosomal Enzymes in the Pathogenesis of Shock, *J. Trauma*, 1968, *8*, (2), 350.
35. Thomphson, B. W., Read, R. C., The Critical Angle after Vagotomy and Drainage Procedure for Haemorrhage, *Sth. Med. J.*, 1970, *63*, (8), 935–938.
36. Tibblin, S., Burns, G. P., Halmloser, P. B., Schenk, Jr. W. G., The Influence of Vagotomy on Superior Mezenteric Artery Blood Flow, *Surg. Gynec. Obst.*, 1969, *126*, (6), 1231–1234.
37. Trudeau, W. L., McHuigan, J. E., Relations between Serum Gastrin Levels and Rates of Gastric Hydrochloric Acid Secretion, *New Engl. J. Med.*, 1971, *248*, (8), 408–412.

10.2 THE SUBORDINATE DIGESTIVE GLANDS

10.2.1 THE LIVER IN SHOCK

The largest gland in the human body, the liver, is a sinusoidal, encapsulated, turgescent and friable complex exposed to injuries in 10 to 27% of the total abdominal traumatic lesions, alone or in association with other visceral traumas. Hepatic injuries are accompanied by mixed traumatic and haemorrhagic shock in half of the cases [2, 14, 21].

At present, abdominal puncture, laparoscopy and enzymatic biochemical tests, elective for the liver cell, are employed for the diagnosis of hepatic injuries. Of importance in intensive treatment is the problem of pulmonary ventilation which must be understood from the particular viewpoint of hepatic cytolysis which releases a 'specific toxic substance' for the alveolar surfactant film [2].

The surgical treatment of injuries of the liver complies with certain unanimously recognized rules: all devitalized tissues that might precipitate local fibrinolysis or sepsis are removed; no resection is performed before controlling the integrity of the hepatic pedicle; ligation of the right and left branches (or both) of the hepatic artery can be carried out and followed or not by resection; massive antibiotic therapy after any type of intervention, but especially after resection and ligation; external drainage of the bile by *T* tube in deep lesions with or without resection; Glisson's hepatic pedicle is clamped for 15 to 30 minutes, and associated clamping of the sub- and suprahepatic cava for 15 minutes. Experimentally with hypothermia, glucose and dibenzyline clamping has been possible for 45 to 60 minutes [5, 9, 11, 17, 20].

In all cases in which complex hepatic lesions appear (affecting the 'danger' areas of the parenchyma with important vasculo-biliary elements) it is recommended to perform in operative time a combined cholangio-, arterio- and portography (Mays, 1971; Nunes, 1970) [9, 10].

Even when not directly damaged, the liver may suffer in shock due to the complex disturbances. Irrespective of the type of shock, the liver is the decisive point at which an attempt is made to control all metabolic changes. Only finally, after losing all its more than five hundred functions, it will damage to the *shocked liver* ('wet liver') and will disturb the entire organism.

If it were not suffocated by the absence of oxygen supply, the huge colony of liver cells — known for its great functional capacity — would not surrender even in the end stages of shock. Some ten years ago, Hift and Strawitz demonstrated in the dog that even in the preagonic phase of shock the liver cells breath perfectly on succinate, in the Warburg apparatus. Since then, numerous findings have lent support to the assertion that 'the body dies with a liver that is still viable at cellular functional level' [1, 3, 7, 8, 16, 18, 20].

Notwithstanding in the human clinic, in many instances the functions of the liver, considered generally, are lost one by one until the onset of a typical retention jaundice, with increased bilirubin and alkaline phosphatase [10]. In shock, jaundice is a bad sign, because the biliogenetic function of the liver cell, phylogenetically the oldest function, is ontogenetically the last to suffer deprivation. The hypoxic cell liver can no longer perform glycuroconjugation and it is its bilirubin unidirectional transport enzymes that are particularly affected. Jaundice is frequent in bacteriemic shock in which the endotoxins with their fixation upon the liver cells are added to hypoxia. When the liver cell conjugation and transport enzymes of the former haem tetrapyrolic ring are no longer functioning the whole enzymatic network of the liver has already long since been exceeded *(see* Section 8.3).

Impairment of the liver develops within a few hours or days, according to the evolution of shock (Caloghera; Shoemaker) [3, 20]. It begins functionally in the cell and ends with morphologic cicatrices. Under stress, the leptonic and histologic lesions occur along enzymatic lines; biochemically, early alterations of the coagulation factors, proteins, amino acids, urea, ammonia and uric acid, etc. may be detected. Almost all the rheologic and metabolic alterations (*see* Chapters 7 and 8) depend upon the liver cell. (The dynamics of the shock cell is discussed in chapter 9 under the pattern of the liver cell).

The buffer role of the liver in shock has been made evident by simple, fundamental experiments. Study of the effluent blood in the hypoxic liver has shown a marked increase in transaminases, lactic dehydrogenase, alkaline phosphatase and potassium, and a decrease in the *p*H. Injection of blood obtained from the liver of a shocked dog into the femoral vein of a healthy dog generated a state of shock in the latter; the same blood injected into the portal vein of a healthy dog produced no disturbances [15, 16].

The role of the liver was also studied in the hepatectomised animal. Eight hours after hepatectomy, an interval during which nothing particular was noted, complex clinical disturbances set in, with an ample rheodynamic and particular biochemical substrate, identical with the shock status. At 16 hours the biligenetic function failed and the dogs died from generalized haemorrhage and neurologic signs.

An opposed experiment was carried out by Fine who, in a severe experimental haemorrhagic shock, maintained a selective flow in the hepatic artery and obtained marked improvement of the state of shock (the same procedure as that used by Lillehei for the intestine).

In the pathologic physiology of shock the liver is an 'historical viscus'. Claude Bernard detected post-haemorrhagic hyperglycaemia, and Bainbridge demonstrated increased portal pressure in prolonged shock. Today, the bromsulphalein clearance test or that with isotopes and electromagnetic flowmeter determinations have shown that arterial and portal output increase in the initial stage of shock and that in both sectors intrahepatic resistance increases with the progress of shock [3, 6, 19].

The liver is like a sponge with a particular vascular maze and a complicated system of sphincters that form homogenisation dams, equalling arterial and portal pressures.

The post-sinusoid sphincter appears to dilate under the influence of β-stimulators, which might account for the dynamic role of the portal barrier in shock (*see* Sections 5.4.2 and 5.5.3). The circulation in the portal segments also has an active role in shock owing to the osmoreceptors available (*see* Section 4.9.2). In shock, the hepatic arterial flow falls from 40 to 30% of the normal flow whereas the portal flow is relatively independent of the cardiac output [3, 7].

Intrahepatic vasoconstriction, which develops in the first phases of shock, forms a barrier that is resistant to both hepatic affluents (arterial and portal). The main spasm seems to occur at the proximal dam of the sinusoids, since it is also maintained after lysis of the post-sinusoidal spasm by β-stimulants.

In experimental animals, during intravenous perfusions the liver becomes turgescent and portal pressure increases [6]. The phenomenon is due to a rapid increase in the portal output that passes through the liver to a competent heart; the transhepatic passage presupposes lysis of the proximal spasm in the hepatic microcirculation with α-lytics and, eventually, of the distal spasm with β-stimulants. At any rate, refilling efficiency involves the absence of collapse and block of the hepatic sponge by DIC, in which case the hepatic barrier is very real (*see* Section 7.4). In addition, an inefficient heart brings about an accumulation of fluid in the liver, with dilatation of the lymphatic system in Disse's spaces, portal hypertension, and jaundice [19].

Many similar aspects may be observed between the *shock liver cell* and the hepatic alterations observed when preserving and transplanting this vital gland. The liver cell enzymes are extremely sensitive to hypoxia and endotoxins [13]. When the potential donor is in the preterminal phase shock, evaluation and correction of the liver cell function is a question of priority. The use of β-stimulants and α-blocking agents is the best means of perfusing the liver in this stage preparatory to removing the organ [12].

Determination of the arteriovenous oxygen differences, combined with that of lactacidaemia and pyruvicaemia have not yet established precisely the contribution of the liver cell to the lactic circuit in shock. The liver is a great *consumer* of lactate for a long time after the onset of shock, but from the moment when utilisation of oxygen by the liver cell decreases the liver becomes just as great a *producer* of lactate [18]. (The role of the liver in the Cori cycle and in the complex regulation of the entire glycostat was discussed in section 8.3.3 and the role of the liver in the proton balance in section 2.9).

There is a general consensus regarding the function of the liver as a great producer of potassium in states of shock. In the isolated liver, it was found that an extremely labile part of the liver potassium appears in the suprahepatic veins a few seconds after adrenergic stimulation or perfusion with blood whose oxygen saturation is less than 35%; glucose appeared only at a saturation of 30% but was promptly liberated by epinephrine stimulation [3, 8].

The liver cell has been studied both in the optical and the electron microscope (almost all the data in sections 9.3.2 and 9.5 refer to the shock liver cell) [4, 6].

Oedema of Disse's space and collapse of the sinusoids with erythrocyte sludge and the onset of DIC should again be emphasized as the factors of cholostasis. In shocked patients, jaundice has a marked retention character, impairment of the liver cell being, however, in general irreversible *per se*. In the electron microscope the hepatocyte vascular pole appears to be the elective site of the initial morphologic lesions, probably as a consequence of the sinusoidal rheologic alterations (*see* Section 9.5.2).

SELECTED BIBLIOGRAPHY

1. ALDRETE, J. A., WEBER, M., Le rôle du foie dans le métabolisme des agents anesthésiques, *Anesth. Ann. Réan.*, 1970, *27*, (2), 297–314.
2. AMERSON, J. R., STONE, H. H. Experiences in the Managements of Hepatic Trauma, *Arch. Surg.*, 1970, *100*, (1), 150–162.
3. CALOGHERA, C., BORDOS, D., DRUGARIN, D., BERGER, E., TIROCH, A., DAN, I., Cercetări experimentale asupra participării ficatului în şocul hemoragic (Experimental Investigations on the Participation of the Liver in Hemorrhagic Shock), *Chirurgia*, 1971, *XX*, (8), 755–763.
4. FEROLDI, J., MALLET-GUY, Y., BRACONNOT, P., GUILLET, R., MALLET-GUY, P., Etude expérimentale, histologique et ultra-

structurale du foie de choc, *Lyon Chir.*, 1969, *65*, (5), 720–734.
5. Fortner, J. G., Shiu, M. H., Howland, W. S., A New Concept for Hepatic Lobectomy. Experimental Studies and Clinical Application, *Arch. Surg.*, 1971, *102*, (4), 312–315.
6. Guillet, R., Braconnot, P., Feroldi, J., Mallet-Guy, Y., Barbier, M., Stefanovic, D., Trzan, U., Etude expérimentale, hémodynamique, biochimique, histologique et électromicroscopique du foie de choc, *Méd. Acad. Chir.*, 1969, *95*, (6–7), 197–204.
7. Halmagyi, D. F. J., Goodman, A. H., Little, M. J., Portal Blood Flow and Oxygen Usage in Dogs after Haemorrhage, *Ann. Surg.*, 1970, *172*, (2), 284–290.
8. Hardcastle, J. D., Ritchie, H. D., The Liver in Shock, *Brit. J. Surg.*, 1967, *54*, (8), 679–684.
9. Mays, E., T. Complex Penetrating Hepatic Wounds, *Ann. Surg.*, 1971, *173*, (3), 421–428.
10. Nunes, G., Blaisdell, F. W., Margaretten, W., Mechanism of Hepatic Dysfunction Following Shock and Trauma, *Arch. Surg.*, 1970, *100*, (5), 546–556.
11. Ochsner, J. L., Meyers, B. E., Ochsner, A., Hepatic Lobectomy, *Amer. J. Surg.*, 1971, *121*, (3), 273–282.
12. O'Connell, T., Fonkalsrud, W. E., Hepatocellular Response to Hypovolemic Shock and Vasoexcitor Drugs, *Surg.*, 1971, *69*, (3), 373–379.
13. Orloff, M. J., Chandler, J. G., Bernstein, J. E., Transplantation of the Liver, Recent Results, *Cancer Res.* 1970, *26*, (2), 213–239.
14. Pokrovski, G. A., Durnev, V. S., Shock in Incised Peace Time Wounds of the Liver (Russian), *Khirurgiya,* 1970, *9*, (1), 13–17.
15. Rangel, D. M., Byfield, J. E., Adomian, G. E., Stevens, G. H., Fonkalsrud, W. E., Hepatic Ultrastructural Response to Endotoxin Shock, *Surg.*, 1970, *68*, (3), 503–511.
16. Rangel, D. M., Dinbar, A., Stevens, G. H., Byfield, J. E., Cross Transfusion of Effluent Blood Flow in Ischemic Liver and Intestines, *Surg. Gynec. Obst.*, 1970, *130*, (6), 1015–1024.
17. Reinhardt, G. F., Hubay, C. A., Surgical Management of Traumatic Haemobilia, *Amer. J. Surg.*, 1971, *121*, (3) 328–333.
18. Schroder, R., Gumpert, J. R. W., Eltringham, W. K., The Role of the Liver in the Development of Lactic Acidosis in Low Flow States, *Postgrad. Med. J.*, 1969, *45*, (526), 566–570.
19. Setlacec, D., Stăncescu, M., Popa, G., Proinov, F., Colostaza intrahepatică (Intrahepatic Cholostasis), *Chirurgia*, 1970, *XIX*, (7), 589–602.
20. Shoemaker, W. C., Szanto, P. B., Fitch, L. B., Brill, N. R., Hepatic Physiologic and Morphologic Alterations in Haemorrhagic Shock, *Surg. Gynec. Obst.*, 1964, *118*, (6) 824–836.
21. Şuteu, I., Văideanu, C., Mancaş, O., Cândea, V., Traumatismele hepatice (Hepatic Traumas), *Revista sanitară militară*, 1973, *LXXVI*, (2), 149–155.

10.2.2 THE PANCREAS IN SHOCK

The pancreas is the principal producer of extracellular hydrolases in the human body. As a rule, hydrolases pass through the pancreatic ducts and manifest themselves in the digestive tract lumen, distal to the pylorus. The harmful influence of these endoluminal hydrolases on the intestine of shocked patients is discussed in section 10.1.2.

In shock, however, there also appears a dangerous flow of intracellular hydrolases from the broken-down lysosomes of the pancreatic cells which may take either the blood or the lymphatic route in returning slowly but certainly to the systemic circulation [14].

The excessive production of hydrolases in the abdominal viscera, and in the first place by the pancreas, is a fundamental factor in causing imbalance of the coagulolytic and kinin systems in the blood flow *(see* Section 7.4.2). It is likewise in the pancreatic space that the myocardial depressor factor (MDF) is produced, with its pernicious influence on cardiac performance (*see* Section 7.4.5).

As pancreatic hydrolases do not differ from those of other viscera in shock, due to acute pancreatitis only lipases have a certain specificity, they being released especially by the pancreatic cell [1, 16].

Jaundice that accompanies acute pancreatitis in almost one-third of the cases cannot be absolutely accounted for by biliary calculus in the ampulla. It is more frequently produced by an inflammatory ileus of Vater's ampulla caused by the transit of very small irritative calculi, which may be retained by spasm of Oddi's sphincter, or even by fragments of vesicular cholesterol crystals. It is, therefore, preferable to relieve compression of the common bile duct than to attack the ampulla crossroads directly, where the lesions are reversible in most cases [8, 9, 10].

Proteolysis and haemolysis, hypertension of the intrahepatic lymphatic system, and especialy affection of the liver cell are also some of the causes of jaundice in acute pancreatitis shock [13, 18].

The response of the pancreas to different lesions is, as a rule, the same. Impairment of the pancreas in shock resembles very much that of the pancreas that has received a direct injury or the pathophysiologic and clinical picture of acute pancreatitis. Pancreatic acini are very fragile; there are many ways of damaging them, causing pancreatic maladies that appear to have certain common elements [8]. Any factor that unbalances the usual functioning of the pancreas triggers off a chain of pathological events similar to shock and to the clinical state of acute pancreatitis (Burlui; Jones; Kwaan) [3, 7, 12].

Hypertension of the pancreatic ducts and *dyschylia* (with inversion of the functional poles of the pancreatic cells), direct injuries, but especially the subcellular mechanisms of lysosomal disruption, which may also be caused by a simple drug, will induce the first biochemical link of the so-called *kinase shock:* local at first, then generalized plasmatic protease activation [2, 3, 5, 6, 9, 15]. It is actually the phase of *acute pancreatitis enzymopathy* (Ţurai and Ciurel) which, regardless of the cause that produced it, will give rise to shock [17].

The second biochemical link in the sequence is dictated by the first: protease activation triggers both the Hageman factor and the activators of fibrinolysis. Trypsin has proved to activate factors V, VIII and X directly. In dogs, systemic injection of trypsin and phospholipase A causes the formation of thrombi, especially in the lung [12]. The appearance of microthromboses characteristic of the DIC episode is well known in acute pancreatitis. This is rapidly followed by fibrinolysis, the decrease of fibrinogen, thrombocytes and factors V and VIII. Haemorrhage, specific of severe acute pancreatitis may be controlled by antiplasminic and antiprotease drugs, but only at a certain moment and under the control of laboratory data, since the risk of thrombosis is always present [11].

In abdominal injuries, a severe development of shock must always make one aware of a possible involvement of the pancreas, as affection of the latter is often masked. Enzymatic determinations may be useful for the diagnosis of the lesion. Surgery may supply a functional-conservative solution by drainage, early excision of the necrotic zones or by caudal, cephalic or even bipolar pancreatectomy [4, 7, 8].

SELECTED BIBLIOGRAPHY

1. APPERT, H. E., YACOUB, R. S., PAIRENT, F. W., HOWARD, I. M., Distribution of Pancreatic Enzymes between Serum and Red Blood Cells, *Ann. Surg.*, 1972, *175*, (1) 10–14.
2. BOUKHAR, M., Traumatismes du pancréas, *Praxis*, 1970, *59*, (55), 813–818.
3. BURLUI, D., CONSTANTINESCU, C., TEJU, GH., STRUTENSCHI, T., Probleme de tactică terapeutică în pancreatitele acute de origine canalară coledocowirsungiană (Therapeutical Approach in Acute Pancreatitis of Choledochowirsungian Origin), *Chirurgia*, 1970, *XIX*, (6), 487–502.
4. FROGGY, J. D., HERMRECK, A. S., THAL, A. P. Metabolic and Hemodynamic Effects of Secretin and Pancreozymin on the Pancreas, *Surg.*, 1970, *68*, (3), 498–502.
5. HAIG, B. T. H., Cellular and Subcellular Mechanism in Acute Pancreatitis, *Ann. Roy. Coll. Surg.*, 1970, *3*, (3), 195–202.
6. HAIG, B. T. H., Nutritional Alteration of Pancreatic Cell Stability, *Ann. Surg.*, 1970, *172*, (5), 852–860.
7. JONES, R. G., SHIRES, G. T., Pancreatic Trauma, *Arch. Surg.*, 1971, *102*, (4), 424–430.
8. JUVARA, I., FUX, I., PRIȘCU, A., *Chirurgia pancreasului (Surgery of the Pancreas)*, Editura medicală, București, 1957.
9. JUVARA, I., RĂDULESCU, D., PRIȘCU, A., *Probleme medico-chirurgicale de patologie hepato-biliară (Medico-Surgical Problems in Hepatobiliary Pathology)*, Editura medicală, București, 1969.
10. JUVARA, I., RĂDULESCU, D., PRIȘCU, A., *Boala hepato-biliară post-operatorie (The Hepatobiliary Postoperative Disease)*, Editura medicală, București, 1972.
11. KRYUCHOC, A. G., Fibrinolysis in Acute Pancreatitis, (Russian), *Khirurgiya*, 1970, *11*, (1), 98–101.
12. KWAAN, H. C., NADERSEN, M. C., GRAMATICA, L. A., Study of Pancreatic Enzymes as a Factor in the Pathogenesis of Disseminated Intravascular Coagulation during Acute Pancreatitis, *Surg.*, 1971, *69*, (5), 663–672.
13. LABDENNOIS, S., NICAISE, H., PIRE, J. C., RIVES, J., La signification de l'ictère dans les syndromes pancréatiques aigus, *Méd. Interne*, 1970, *6*, (6–7), 579–592.
14. LEFER, A. M., GLENN, T. M., O'NEILL, T.J., LOVETT, W. L., GLISSINGER, W. T., WANGENSTEEN, S. L., Inotropic Influence of Endogenous Peptides in Experimental Haemorrhagic Pancreatitis, *Surg.*, 1971, *69*, (2), 220–228.
15. MOORMANN, S. P., Frühveränderungen bei der experimentellen partiellen permanenten Ischemie des Katzenpankreas, *Path. Anat.*, 1969, *349*, (3), 249–259.
16. OGER, A., Dosage de la lipase sérique dans les affections pancréatiques — méthodes de dosage, *Acta Gastroent. Belg.*, 1970, *33*, (4–5), 335–342.
17. ȚURAI, I., CIUREL, M., GEORGESCO, L., AMBRUS, I., Le traumatisme du pancréas. Recherches expérimentales, *Lyon Chir.*, 1965, *61*, (2), 279–289.
18. RUDLER, J. C., GIULI, R., Etude de 125 dossiers classés comme pancréatites aiguës, *Lyon Chir.*, 1969, *65*, (4), 482–498.

10.3 THE KIDNEY IN SHOCK

The kidney is one of the vital organs that does not benefit by a preferential circulation in shock. Renal involvement is rapidly expressed — the clinical symptom being oliguria — and may persist after the blood pressure and central venous pressure have returned to normal. It is not always possible to establish with certainty the evolution of the kidney in shock (Burghele, 1966).

Several studies show that at necropsy, in cases of traumatic shock, pathoanatomic lesions characteristic of the shock kidney are to be found in 39% of the cases, although at clinical diagnosis renal affection was only detected in 14% of the cases [2, 3, 6, 8, 15]. In shock, direct injury to the kidneys is a particular case; in such instances the diagnosis is in general established by X-ray examination and surgery has to be performed as early as possible. If the parenchyma is not affected, an attempt should be made to repair the renal pedicle [7, 9].

Pathologic physiology. With the exception of post-transfusional shock and of the crush syndrome both of which have a direct, primary intrarenal lesional impact, the other types of shock affect the kidney by a mechanism of prerenal ECBV insufficiency, following the systemic haemodynamic disturbances specific of shock.

In most shock-inducing situations, after a variable interval of oligoanuria, diuresis progressively reappears, but the concentration and acid-base regulation function still remains deficient for some time. Latent renal failure prior to shock favours the onset of severe shock kidney.

Decrease in the filtration pressure at the level of the glomerulus is the first pathologic link in shock kidney. From the normal values (120 to 130 ml/min) the glomerular filtrate falls from 30 to 40 ml/min or may disappear completely. Renal ischaemia reduces tubular perfusion and the concentration capacity of the kidney, which eliminates low amounts of hypo- or isostenuric urine. Para-aminohippuric acid clearance falls to one-third of its normal value.

Fick's method for the determination of the circulating volume showed that in the oliguric phase of the kidney in shock renal perfusion is only 5% of the normal values, a situation in which glomerular filtration is nil [10].

Some authors consider that a close agreement between renal blood flow values and glomerular filtration is not essential. A reduction of the renal flow to 33% still permits a certain glomerular filtration. In man, in the course of remission from shock the total renal blood flow, measured by radioactive krypton, was 190 ml/min and the endogenous creatinine clearance coefficient 66 ml/min. Similar renal blood flow values were also found in other cases in which, however, the creatinine clearance coefficient was only 0.5 ml/min.

Impairment of the renal cellular organelles, and especially of intramitochondrial metabolism, appears to justify the notion of *shock kidney* [5, 8, 13, 15].

Irrespective of whether the primary factor is acute hypovolaemia, as in haemorrhagic shock, or hyperergic vasoconstriction of the microcirculation, as in traumatic or bacteriemic shock, the intrarenal lesions are the same. Regulation of the renal circulation in shock is performed by the afferent vessels of the juxtamedullary glomeruli, arteriovenous anastomoses, and juxtamedullary by-passes. If ischaemia is prolonged, necrotic tubular alterations develop.

Acute renal failure (ARF), developing as a consequence of shock, is particularly severe due to the catabolic phase specific of the post-acute hyperergic syndrome. In this phase, the suprarenal glycoprotein corticoids increase catabolism, producing intense cellular destruction with proteolysis, manifested by the increase in blood nitrogen. Under almost normal physiologic conditions the non-protein nitrogen accumulating in the plasma is 50 mg/24 h, in violent post-acute phases with exaggerated catabolism the non-protein nitrogen may increase to over 1000 mg/24 h. In the meantime, a deleterious accumulation of potassium occurs. Mineralocorticoids in excess likewise contribute to the onset of shock kidney, due to the hyperaldosteronic syndrome with the retention of sodium ion and water. However, the retention of sodium and water may be considered as a compensatory mechanism since it has only a slight effect upon the increase of blood volume, but, on the other hand, it accentuates peritubular intrarenal œdema, causing the state of the kidney to deteriorate [10]. Extravascular intrarenal pressure shows a marked increase *(see* Section 5.3.3).

To the onset of shock kidney the antidiuretic hormone also contributes; in contrast to the mineralocorticoids whose release is slow but lasting, ADH is liberated abruptly and the organism responds immediately. The action of ADH upon the tubules is mediated by cyclic AMP *(see* Section 4.8.3).

The clinical stages of the kidney in shock are: onset, anuria, and the recovery stage of diuresis [3].

Onset. In this stage the clinical picture is dominated by the primary lesional syndrome (haemorrhage, crushing, bacteriemic shock). Systemically, the blood pressure remains almost normal but vasoconstriction of the glomerular microcirculation, manifested by oligoanuria, reduces the amount of urine eliminated by the kidneys to 500 ml/day or even less. The clinical picture is that of shock and, biologically, non-protein nitrogen oscillates around the maximum limit admitted. In this stage it is advisable to measure diuresis at least every hour; a decrease to less than 30 ml/h draws attention to the possible onset of shock kidney.

After the therapeutical measures taken to refill the vascular bed, alpha-blocking and corticoid medication lead to an increase in diuresis, and the ARF may be considered to be of a transitory functional type, its subsequent course following that of the shock.

In order to define ARF as functional, the kidney must not have lost its capacity to concentrate urea (more than 10 g/day), the urine density must be 1 012 and urinary potassium elimination must be efficient (Diomi; Powers; Stone) [5, 12, 13, 15].

Anuria is the stage in which tubular necrotic lesions develop. If the kidney excretes small amounts of urine it is isostenuric, with reduced urea values (less than 5 g/day) and low potassium levels. The anuric stage may last from 3 up to 30 days.

The most important clinical signs are: nervous signs (somnolence, obnubilation, hallucinations, delirium) due to oedema of the central nervous system; cardiovascular signs, especially myocardial due to the onset of uraemic myocarditis (manifested by cardiac arrhythmia); digestive signs (nausea, anorexia, vomiting). As, in the course of vomiting, K^+ is eliminated, together with hydrochloric acid, kaliaemia does not reach alarming values in this period. Haemorheologically, disturbance of the coagulolytic mechanism develops, with the concomitant onset of DIC.

Nitrate retention increases and urea may, within 3 to 4 days, attain values of 0.3-0.5 g%, in terms of catabolic proteolysis. The ionogram reveals a marked increase in K^+ which may attain 6 to 8 mEq, when ECG anomalies also develop. The hydric balance is perturbed by overall hyperhydration, especially when excess amounts of fluid are introduced by the parenteral route. Metabolic acidosis increases up to standard bicarbonate values of 10 to 12 mEq (haemodialysis is continued throughout this period).

Recovery of diuresis. When the tubular necrotic lesions lack an irreversible character diuresis reappears after a time (3 to 30 days).

Within the first few days, the urine is hypostenuric and urinary urea less than 1.0 g%. After several days the amount of urine increases from 2 to 4 litres in 24 hours, with a low density and reduced urinary urea values; it is the phase of polyuric renal insufficiency, during which particular attention should be paid to the maintenance of an adequate hydric balance.

Complications. Infectious complications are fairly frequent due to the use of venous and vesicle catheterisation in the presence of a medium rich in strains resistant to conventional antibiotic therapy, with the selection, especially, of pyocyaneus. Complications of the haemorrhagic type are the consequence of DIC and consumption coagulopathy. Hydroionic disturbances, especially general hyperhydration due to faulty management of the case and hyperkaliaemia dominate the clinical picture. Ion exchange resins and calcium in large doses (10 to 15 g) appear to lower the high potassium figures.

Conclusions. The clinical and biological picture differs from that observed in pure renal failure since *shock kidney* is accompanied by all the noisy symptomatology of the post-acute reaction syndrome. The rapid elevation of blood urea attests to the existence of certain extrarenal factors that produce intense tissular proteolysis; in the first phase of shock, metabolic acidosis may be accompanied by ventilation alkalosis, and hyperkaliaemia may be absent if there are extrarenal losses due to vomiting.

Two anatomical forms are characteristic of the shock kidney:

- *acute tubulopathy*, with the maintenance of diuresis; maintenance of the BP below the level required by glomerular filtration initially produces renal tubular 'stupor' with necrosis of variable extent, potentially reversible within 24 to 48 hours;
- *cortical necrosis* is a rare form but more severe due to anuria, frequently irreversible, or to morphologic sequelae.

Irreversible lesions develop if the shock-inducing factors persist. Cortical necrosis is the consequence of a spasm of the afferent glomerular arteries with upstream stasis and intracapillary fibrinous depositions [8]. The pathological histology of the shock kidney has been extensively studied [3, 12, 16].

There are certain particular aspects of the kidney in shock.

Postoperative shock kidney may be caused by a number of contributing factors with a synergic action: operative shock, anaesthesia, mild hypoxia, hypovolaemia inadequately compensated, etc.

Postoperatively, incorrect management of hydration and ionic re-equilibration may aggravate a preoperative renal affection. Shock kidney may be produced by hypohydration, the lack of administration of glucosaline solutions necessary for the control of ADH and mineralocorticoid hypersecretion, the untimely use of osmotic diuresis or rapid saline diuretics. A number of factors prior to the operation, such as chronic neoplastic shock, chronic pulmonary failure and hypovolaemia due to any cause have a high anurigenic potential.

After surgery, if the pathways of pain are not blocked, an excessive catecholamine reaction sets in followed by vasoconstriction of the glomerular microcirculation and diminution of the filtrate. Postperative ARF may develop rapidly or slowly in the course of two or three days either precipitated by preexisting lesions or by a surgical complication not solved in due time. Complete renal block by massive transfusion of an incompatible blood group is manifested immediately after the operation by the complete absence of diuresis or by small amounts of intensely haemoglobinuric urine.

The onset of postoperative ARF is marked by the triad: decrease of diuresis, increase in blood nitrogen, and decrease of urinary urea. The general phenomena are drowsiness and obnubilation and signs of intoxication of the nerve cell with the products of endogenous catabolism.

There are no biologic tests that furnish pre-operative data pointing to the risk of postoperative shock kidney. A reduced concentration capacity of the kidney in the Volhard test and delay in phenomolsulphonphthalein clearance might be considered as signs of an unfavourable prognosis. At any rate, the existence of a slow renal insufficiency that must be detected before surgery, favours the onset of postoperative shock kidney. For assessing the renal function it is necessary to add to the screening tests (total urine examination and blood urea) a number of analyses (urea clearance coefficient and urography), especially for operations on or near the urinary tract.

In principle, hydroionic and acid-base disturbances, and systemic and microcirculation haemodynamic disturbances stand at the basis of postoperative shock kidney.

The hydroionic disturbances that affect the renal function are, in particular, cellular hyperhydration, as a result of plasma osmotic hypotonia and acid-base imbalance. Haemodynamic disturbances are caused by prolonged hypovolaemia with BP values of less than 70 mmHg.

In the hyperergic phase with an almost normal BP but with vasoconstriction of the microcirculation, shock brings about a marked diminution of glomerular filtration. The administration of alpha-stimulating drugs has the same deleterious result. The intravenous administration of excessive amounts of hypertonic saline solutions may release tissular nephrotoxins by proteolysis.

To prevent postoperative shock kidney it is essential to sustain systemic haemodynamics and perfusion of the microcirculation. A correct, slightly positive water balance, maintained with glucosaline solutions to neutralize the postoperative aldosteronic syndrome, may prevent ARF in many cases. It is recommended that sufficient antialgic drugs to prevent a catecholaminic reaction should be administered. It is considered useful to determine repeatedly the acid-base balance (especially in the case of extrarenal losses) and to correct metabolic acidosis; anabolizing hormones are prescribed in order to shorten the catabolic phase.

When diuresis falls below 500 ml/24 h (under conditions of correct water, acid-base and ionic balance) osmotic diuresis may be performed with hypertonic glucose solutions (10—20%) or mannitol (20%) rapidly administered so as not to cause the rebound phenomenon. In the last instance, extrarenal artificial clearance is resorted to.

The shock kidney in burns. In severe burns, intense tissular proteolysis, release of toxic protein products, hyperkaliaemia and hypovolaemia are factors that either alone or combined contribute to the onset of shock kidney.

Early lesions develop within the first 72 hours and consist of tubular degeneration, bilateral cortical necrosis, tubular haemoglobin casts. Acute interstitial nephritis, abscesses and hyperplasia of the glomerular endothelia have been described among the late lesions.

Acute interstitial nephritis does not develop before the fourth day and is always conditioned by secondary infections. In patients with burns some authors have found a marked increase in non-protein nitrogen and polypeptides in the blood, and have attributed to them a prognostic value [6]. It has been asserted that when non-protein nitrogen exceeds 100 mg/100 ml plasma the prognosis is fatal [16].

In experimental animals the injection of serum from a patient with burns produces intoxication of the animal; this led to the conclusion that the toxic state in burns is produced by the proteins derived from the burnt area (acid haematin, methaemoglobin, etc.) [15].

The determination of tissular enzymatic activity (LDH, transaminase, alkaline phosphatase, aldolase) showed certain pathologic levels especially high alkaline phosphatase levels). In contrast to haemorrhagic shock, in which alkaline phosphatase rapidly falls, in burns it decreases more gradually, higher values being maintained over a longer period. In burns the intracellular organelles of the urinary tubules are strongly impaired.

In a patient with burns, the prevention of shock kidney is of the first importance in intensive therapy. Control of pain, oxygen therapy, the rational administration of fluids and proteins, antibiotic therapy, osmotic diuresis, early excision of burnt areas are measures that sustain the organism but also prevent shock kidney.

Shock kidney and anaesthesia. The sensitivity of patients with shock kidney to a series of drugs is well known. In emergency cases the anaesthesia of patients with ARF involves a number of risks that depend upon the patient's constitution and the nature and mode of administration of the drugs. The haemodynamic lability of these patients is very high due to the large amounts of non-protein nitrogen that potentiate the effects of the anaesthetics.

Drugs not having a marked myocardial depressor effect and that are not exclusively eliminated through the urine must be given preference. The anaesthetics that can be used by intravenous route, especially those with a myocardial depressor effect, should be injected very slowly in order to avoid cardiac arrest.

Several of the inhalation anaesthetics are proscribed: ether, because it brings about a discharge of catecholamines; cyclopropane, because it causes vasoconstriction of the renal vessels and may also induce cardiac arrest where there is an uraemic background. Penthrane favours metabolic acidosis and upsets the coagulolytic mechanisms by blocking the enzymatic processes that interfere in coagulation; halothane, whose action is still uncertain, may cause collapse of the BP in hypovolaemic patients and, according to some authors (Powers, 1970), reduce the glomerular filtrate. However, in low concentrations it may be administered in cases of ARF.

Particular attention should be paid, among the intravenous drugs to pentothal which depresses both the renal and myocardial function and may cause cardiac arrest in hypovolaemic patients with uraemia. This particular sensitivity is caused by hypoproteinaemia (the fraction dissolved in pentothal increases), acidosis, elevation of non-protein nitrogen and decrease of the lipid reserves. Hydroxydion and propanidid are hydrolysed in the organism and their inactive metabolites eliminated through the kidneys.

Myorelaxants and their antagonists are excreted by renal route. Tubarin is to be found in a proportion of 75% in the urine but in the absence of renal excretion it produces a redistribution of total water, being metabolized finally in the liver. Gallamine is excreted integrally in the urine. The depolarizing myorelaxants (succinylcholine) are ideal in cases of ARF, being completely metabolized in the body.

Of the analgesics, Fentanyl has the least accentuated diuretic effects; morphine diminishes diuresis due to the release of ADH and increases tubular reabsorption. Peptidin is rapidly detoxified by the liver, but causes intra-operative bleeding by vasomotor paralysis brought about by the liberation of histamine.

Of the vagolytics, scopolamine is preferred to atropin because it is totally metabolized in the organism.

Neuroleptics are degraded in the body; dihydrobenzperidol is better tolerated than the other neuroleptics.

In ARF the best anaesthesia for surgery appears to consist of:

- a premedication: Petidin-Prometazine;
- induction: Pentothal (in small doses of 2 to 3 mg/kg body weight) and intubation with Myorelaxin;
- maintenance: Halothane + nitrogen protoxide + oxygen, or neuroleptanalgesia with Scopolamine in preanaesthesia.

Osmotic diuresis. It is known that osmotic overloading of the primary urine with glucose prevents the reabsorption of water by the renal tubuli [14]. In osmotic diuresis with increase in osmotic charge of the filtrate the concentration of the urine decreases.

Glucose hypertonic solutions (10-20%) and mannitol (20-25%) administered rapidly may produce an efficient osmotic diuresis but may dehydrate the tubular cells and renal interstitium [11, 12].

Some clinical cases have been reported in which repeated administration of Dextran-40 caused ARF on approximately the fourth day. Peritoneal dialysis for six days always solved this type of ARF. Renal biopsy revealed only tubular lesions, the cells being the site of monstrous intumescence [5, 11, 12]. The mechanism of ARF may be anaphylactic or only osmotic. Examination in the electron microscope revealed giant lysosomes and digestive tubulocytes both after Dextran-40 and after hypertonic glucose, dextrose or mannitol. All these substances penetrate within the tubulocyte by pinocytosis and result in histologic lesions known as osmotic nephrosis [5, 14].

Today it appears to be clearly established that repeated doses, or an overdose, of Dextran-40 increase urinary viscosity, lower intratubular flow and dehydrate the tubulocytes, without however causing irreversible lesions. Nevertheless, especially when associated with renal hypoperfusion, as in shock, the urinary output has to be followed up, particularly after Dextran-40. If the urinary output shows a tendency to diminish, crystalloid solutions and natriuretics must be quickly administered.

SELECTED BIBLIOGRAPHY

1. Balș, G. M., Căruntu, Fl., Duminică, A., Ștefănescu, O., Emil. T., Tratamentul imediat al infecțiilor urinare într-o secție de boli infecțioase (The Immediate Treatment of Urinary Infections in an Infectious Diseases Department), *Chirurgia*, 1970, *19* (2), 397–402.
2. Bell, G., The Effect of Noradrenaline and Phenoxybenzamine on the Renal Response to Haemorrhage, *Surg. Gynec. Obst.*, 1970, *130*, (5), 813–820.
3. Burghele, Th., Rugendorff, E. W., *Le rein des états de choc*, Masson, Paris, 1966.
4. Crystal, R. G., Effects of Nephrectomy and Uretral Occlusion on Extracellular Fluid Measurements during Shock, *Surg. Gynec. Obst.*, 1970, *131*, (6), 1109–1114.

5. DIOMI, P., MATHESON, N. A., NORMAN, J. N., SHEARED, J. R., The Renal Response to Dextran-40 in Dogs with Renal Artery Constriction, *Surg.*, 1971, *69*, (2), 256–262.
6. EKLUNG, J., Studies on Renal Function in Burns: II. Early Signs of Impaired Renal Function in Lethal Burns, *Acta Chir. Scand.*, 1970, *136*, (8), 735–740.
7. FREIRE, E. C. S., GABRIEL, R., D'ALMEIDA, E., Traumatismos do aparelho urinario, *Folha Med.*, 1970, *60*, (1), 1–35.
8. GILLENWATER, J. J., WESTERVELT, F. B. Jr., Current Concepts in Pathogenesis and Management of Acute Renal Failure, *J. Urol.*, 1969, *101*, (4), 433–437.
9. GUERRIERO, W. G., CARLTON, C. E. Jr., SCOTT, R. Jr., BEALL, A. C. Jr., Renal Pedicle Injuries, *J. Trauma*, 1971, *11*, (1), 53–62.
10. HINSHAW, B., Autoregulation in Normal and Pathological States Including Shock and Ischemia, *Circ. Res.*, 1971, *28*, (1), 46–50.
11. MAILLOUX, L. SWARTZ, C. D., CAPIZZI, R., KIN, K. E., ONESTI, G. L., RAMIREZ, O., BREST, A. N., Acute Renal Failure after Administration of Low Molecular Weight Dextran, *New Engl. J. Med.*, 1967, *277*, (6), 1113–1122.
12. POWERS, R. S., Relation of Acute Tubular Necrosis to Shock and the Effect of Mannitol, *Amer. J. Surg.*, 1965, *110*, (3), 330–332.
13. POWERS, S. R., Jr., The Maintenance of Renal Function Following Massive Trauma, *J. Trauma*, 1970, *10* (4), 554–564.
14. ROHM, G. F., RETIEF, C. P., JOHNSTON, G. S., van ZYL, J. A., van ZYL, J. J. W., De KIERK, J. N., MURPHY, G. P., Renal and Systemic Circulatory and Metabolic Alterations after Intravenous Glucose Loads in Baboons, *J. Urol.*, 1969, *101*, (4), 460–464.
15. STONE, A. M., STAHL, W. M., Renal Effects of Haemorrhage in Normal Man, *Ann. Surg.*, 1970, *172*, (5), 825–836.
16. * * * Studies on Renal Function in Burns. III. Hyperosmolal States in Burned Patients Related to Renal Osmolal Regulation, *Acta Chir. Scand.*, 1970, *136*, (8), 741–751.

10.4 THE LUNG IN SHOCK

The lung is one of the vital organs whose involvement has a decisive role in shock although the attention of investigators was not drawn to its importance until rather late in the study of shock. The first to presume its importance in the mechanisms of shock were the clinicians and pathologists.

Cahill was among the first to focus attention on the decrease in pulmonary compliance and increase in pulmonary resistance in bacteriemic shock [14, 15]. The most detailed studies of the shock lung were carried out by Blaisdell, Lichtmann, McLaughlin, Moss and Siegel [9, 30, 34, 40, 55].

Today, much is certainly known about the treatment of direct injuries of the chest and lungs, although statistics show that in more of the cases the diagnosis is mistaken or delayed more than 48 hours and the therapeutical interventions are inadequate [8, 30]. It is common knowledge that when one lung only is injured directly, the other lung develops œdema, haemorrhage and, especially, atelectatic shunts and a decrease in P_aO_2 (Berman, 1970; Buckberg, 1970).

But the importance attributed today to the lung in shock is based upon the lesions that develop even when the primary disturbances do not affect lung directly [2, 29, 33, 34, 37].

Approximately one-third of patients who die after major surgery or massive trauma do so from a peculiar and highly lethal form of respiratory failure often referred so as *shock lung*.

In addition to shock lung other eponyms have been used to refer to similar process: wet lung, post-traumatic lung, the respiratory distress syndrome of trauma, oxygen toxicity lung, respiratory lung, pump lung, congestive atelectasis, progressive pulmonary insufficiency, the Melrose lung, the Da Nang lung, progressive stiff lung, postperfusion lung, pulmonary fat embolism syndrome, low-flow lung syndrome, etc.

The shock lung syndrome is a non-specific reaction of the lung, as the same pathologic end result in the pulmonary parenchyma can be obtained from a variety of insults. Sepsis, overhydration, disseminated intravascular coregulation, O_2 toxicity, debris from massive transfusions, fat emboli, immunologic factors, aspiration, central nervous system hypoxia, and adrenergic excesses, either alone, but usually in combination — as they act in shock — affect the lung.

There are countless clinical groups of shocked patients (the so-called 'critical cases' with haemorrhagic shock, burns, infections, embolism, major operations, etc.) who exhibit the triad:

(i) increase in pulmonary resistance (pulmonary barrier);
(ii) pulmonary interstitial œdema (wet lung);
(iii) insufficient oxygenation of the blood fluid (right-left shunt effect).

Some of the common clinical conditions with pulmonary distress include: pulmonary contusion, pulmonary embolus, atelectasis caused by bronchial obstruction, mechanical compression of the lung, hypoventilation from any cause and, very often, *the respiratory distress syndrome of trauma*, a term at present more widely used than *shock lung*.

The lung is considered to be the *target organ*, especially in anaphylactic shock, since it plays such an important role in the activation of fibrinolysis and responds electively to extrapulmonary histamine [6] (*see also* Section 12.5).

A decrease in the ECBV and rheologic disturbances in the course of shock always appear to be drastically affected in the pulmonary space, triggering off progressive distress that ends in respiratory failure and death [18, 29, 50].

It should, however, be mentioned that there are still some authors who question the vulnerability of the lung, at least in primates (Silberschmidt, 1973).

There is a long series of recent articles showing that haemorrhagic shock alone, even severe, does not, actually, produce any measurable damage to the lung, particularly when viewed from the functional standpoint.

Studies of the relation between hemorrhagic shock and postresuscitative pulmonary insufficiency have yielded no clear results.

Collins et al. (1973) studied lung weight, lung water content and arterial oxygen tension over 3 days in bled rats treated by ten fluid regimens. A total of 400 rats were used in the study; treatment regimens including red blood cells produced the lowest mortality and the driest lungs.

Many clinical observations correlate the progressive changes in the clotting mechanism and the development of significant pulmonary dysfunction. This does not answer the question as to whether one causes the other (Milligan, 1974).

The rather devastating effect of sepsis on pulmonary function as well as on myocardial performance has been confirmed (Derks, 1973). A combination of shock and fat embolus is much more damaging to the lungs than is shock alone.

Most clinicians agree, however, that one-third of all shocked patients, regardless of the type of shock, present a picture — virulent and rapidly progressing towards death — of *shock lung* [9, 13, 27, 57, 59].

The sensitivity of the lung in shock is brought about by the following pathologic events:

(i) the onset of a DIC episode in the pulmonary microcirculation;
(ii) platelet and lipid thromboembolism;
(iii) dysfunction of the surfactant;
(iv) crystalloid accumulation in the narrow interstitial space of the lung — wet lung (Figure 10.3).

These disturbances have the following effects:

(i) perfusion-ventilation discrepancy with a right-left shunt effect;
(ii) low pulmonary compliance;
(iii) decrease of $P_a(O_2)$ and increase of $P_a(CO_2)$;
(iv) increased pulmonary resistance;
(v) the onset of congestive atelectatic areas, interstitial and intra-alveolar haemorrhage;
(vi) pulmonary hypertension accompanied by right heart failure and increase in the central venous pressure [3, 18, 30].

DIC in the lung favours:

(i) obstruction of the pulmonary microcirculation, resulting in a haematosis shunting effect and interstitial and alveolar haemorrhage followed by atelectasis;

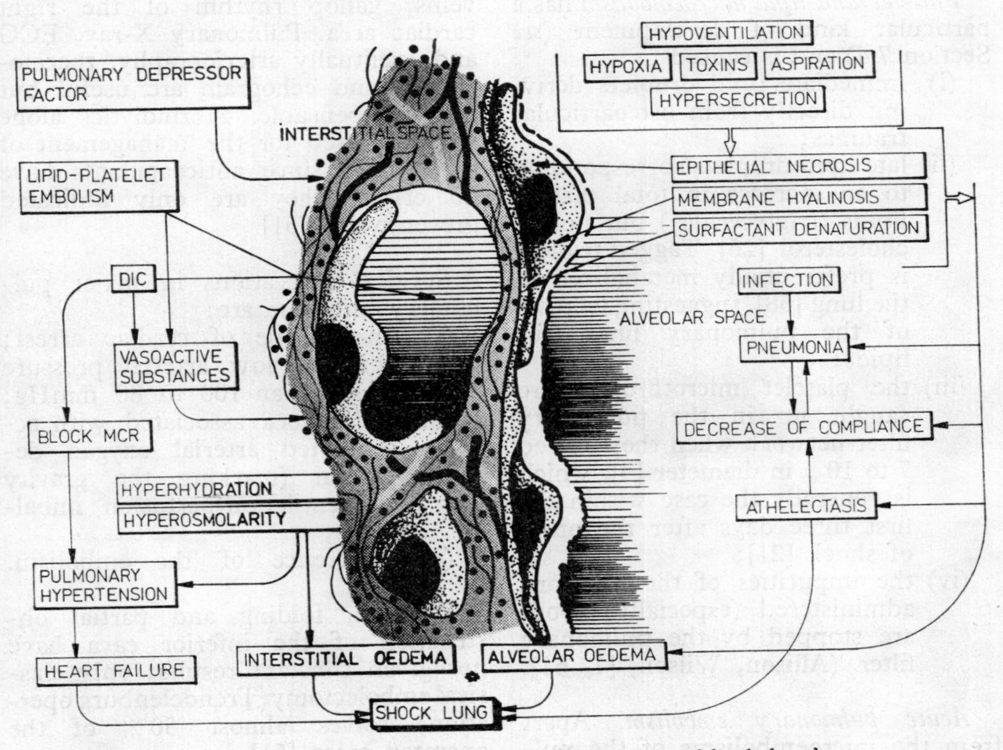

Figure 10.3. The complexity of the factors causing 'shock lung'.

(ii) hyalinosis of the alveolar membranes with a decrease in compliance likewise takes place;
(iii) the release of tissular vasoactive substances (histamine, serotonin, bradykinin, etc.), producing congestive interstitial œdema. There is likewise a release of pulmonary E_2 prostaglandin which acts in the same way as histamine, producing bronchiolar spasm and alveolar atelectasis [7, 32]. Four hours after inducing experimental shock fibrin deposits were found in the pulmonary microcirculation and alveoles; the alveolar walls were also probably damaged by excess FFA [26].

Platelet and lipid microembolism has a particular kind of development (*see* Section 7.4) and is caused by:
(i) immediate lipid droplets deriving directly from osteoarticular traumas;
(ii) late lipid droplets corresponding to an increase in total plasma lipids (triglycerides) but not in cholesterol [26]. Tagged triolein is preferentially metabolized in the lung [58], suggesting tropism of the pulmonary filter for lipids;
(iii) the platelet microthrombi are caught up in the pulmonary filter network when they exceed 7 to 10 μ in diameter [5], which is generally the case within the first three days after the onset of shock [21];
(iv) the impurities of the solutions administered (especially blood) are stopped by the pulmonary filter (Allison, Wilson) [1, 61].

Acute pulmonary embolism. Apart from the microembolisms of the pulmonary, microcirculation is the outstanding drama of a major acute pulmonary embolism. It is difficult to establish its incidence because of its rapid onset which in many instances does not allow for a clinical diagnosis. The incidence at necropsy is impressive: 10% of the causes of death (based on serial histologic sections, some statistics give an incidence of 11%, 52% and even 64%) [1]. The clinical progression of pulmonary embolism to death is hard to foresee: of 100 cases 41 died after 10 minutes, 3 after 15 minutes, 22 after 2 hours, and 34 after 2 weeks [47].

The characteristic clinical signs are: abrupt hypertension, rapid weak pulse, cold skin, increased central venous pressure, turgescence of the jugular veins, gallop rhythm of the right cardiac area. Pulmonary X-ray, ECG and eventually arteriography, thermography and echogram are useful but not indispensable. Fibrinolytics alone must be used for the management of these cases since anticoagulants have no effect (they are only indicated preventively) [51].

Surgical indications in major pulmonary embolism are:
(i) an episode of cardiac arrest;
(ii) constant low arterial pressure of less than 100 to 80 mmHg;
(iii) tachypnœa associated with accentuated arterial oxygen depletion (marking the gravity of ventilation-perfusion imbalance);
(iv) recurrence of the embolism.

Ligature, folding and partial obstruction of the inferior cava have not given constant results; only classical embolectomy (Trendelenburg operation) solves almost 50% of the operated cases [61].

Denaturation of the surfactant. The tensioactive factor that forms the cell-free endoalveolar film (each alveole being similar to a spherule 100 μ in diameter) is a complex of proteins hydrocarbons and phospholipids (including dipalmitoillecithin, sphingomyelin, phosphatydyl dimethylethanolamide, etc.) [48].

A 'factor depressing the surfactant' was found in the plasma collected from the splanchnic pool of an animal in shock [28]. Its origin may be the liver cells. In states of shock it appears that the production and quality of the surfactant diminishes, lowering alveolar ventilation (*see* Section 11.2.2).

The wet lung. Increase in the weight of the lung in shock is a fact commonly observed at necropsy. (The 'wet liver' was also a fact observed at necropsy).

The lung appears to be the target-organ for the accumulation of fluid, although its interstitial space is very small [31]. Normally, sodium ions do not cross the pulmonary capillary membrane; in shock, however, sodium ions accumulate in the interstitium, increasing its osmotic pressure. Whether it passes with the RBC or with platelets, or crosses alone the capillary and/or alveolar wall, whose basement membranes are damaged in places (as observed in shock in the electron microscope), the sodium ion will also accumulate a proportional amount of fluid osmotically in the interstitium. It has been determined isogravimetrically that capillary permeability in the lung shows a twofold increase which would account for the accumulation of fluid in the pulmonary interstitium in the course of shock [43].

Moreover, as long ago as 1958 Von Oettingen communicated a list of 59 chemical substances that may produce states of shock only by affecting directly the permeability of the alveolar and capillary endothelium [46].

Some authors consider it exaggerated to attribute to fluids perfused in excess an essential role in interstitial pulmonary accumulation of fluids [11, 39, 42, 44, 55]. Increase in the osmotic pressure of the blood and reduction of blood volume would thus become the main therapeutical objectives. In point of fact, only the regional ionic imbalance of the pulmonary interstitium must be corrected, the wet lung not implicitly reflecting overall hyperhydration (Siegel, Weedn) [55, 60].

High local concentrations of sodium may produce local osmotic gradients resulting in transudation of plasma water into the interstitial zone and the subsequent saturation of collagen water-binding sites. The use of colloid alone in the treatment of intractable hypovolaemic shock should be seriously questioned. Dextran and mannitol administration has been seen to have no consistent effect on interstitial pulmonary œdema.

It should be pointed out that the usual cause of pulmonary œdema is not increased hydrostatic pressures, but rather capillary damage. Pulmonary œdema does not follow large infusions of saline, given partly as a basic infusion and partly to replace blood loss (Brown, Johansen) [10, 24].

Posthemorrhagic pulmonary insufficiency due to overzealous infusion of Ringer's lactate solution is thought by some to be avoidable if 'overtransfusion' is eliminated, but 'shock lung' appears to develop independently of intravenous therapy (Rocchio) [52].

Saline resuscitation cannot be implicated as a principal cause of post-resuscitative pulmonary insufficiency.

Changes in shock include interstitial œdema, increased endothelial pinocytosis and a dramatic increase in interstitial sodium. Sodium increases in areas of collagen fibres. Acidosis accompanying haemorrhagic shock may perhaps lead to the accumulation of sodium and water in areas occupied by the collagen fibres of the alveolar capillary membrane. Haemorrhagic shock leads to diffuse interstitial œdema in the primate lung, primarily located in the interstitial space of the alveolar capillary membrane; the endothelium and epithelium appear to be undamaged.

The first histochemical study of sodium distribution in the lung (Moss and Das Gupta) [40, 41] showed that pinocytosis appears to be a bulk-carrier of sodium. In haemorrhagic hypotension pools of sodium-rich œdema fluid develop in the interstitium. Collagen appears to possess a basal affinity for sodium.

Likewise worthy of note is the direct toxic effect of oxygen administered for an interval longer than the critical time upon the alveolar endothelium and even upon the intracapillary RBC, which become lysed by peroxidation. This toxic effect of oxygen has been denied by some authors [39, 62].

Sodium lactate appears to offer a real alveolar protection against oxygen toxicity [4, 20]. Actually, administered alone and in large amounts, electrolytes may have a negative influence upon the wet lung, as have colloids administered in excess [36, 60]. On the other hand, crystalloids combined with colloids equilibrate the pulmonary interstitium. The most useful combination seems to be that of Ringer's lactate solution and Dextran [48, 60].

Paraclinical investigations of the shock lung are difficult and pretentious; however, the following may be useful:

(i) arteriogram (may detect atelectasis, embolisms, etc.);
(ii) thermogram and ultrasonoechogram;
(iii) scintigram with ^{133}Xe or ^{99}Tc (detecting atelectasis);
(iv) determination of the alveolar gas pressure.

X-ray, bronchogram and respiratory tests do not furnish sufficient information (Gump, Powers) [22, 50]. Using ^{133}Xe as tracer one can determine, by means of an electronic computer, both the gas flow and the pulmonary blood flow, obtaining rapidly, and without risk in the severest cases, a tridimensional aspect of the pulmonary function, which can be followed up in time [25]. After recording the data concerning respiratory mechanics, gas exchanges and the ventilation-perfusion ratio, the general function of the lung can be calculated with a digital electronic computer, and in terms of the results, the degree of pulmonary involvement and the efficiency of the treatment [49, 50].

Therapeutical principles:

(i) The first object is to obtain freedom of the air pathways (eventually with an bronchoscope); cultures and an antibiogram are immediately performed.
(ii) Controlled ventilation at an intermittent positive pressure (tracheostomy when $P_aO_2 = 60$ mmHg).
(iii) The administration of fluid should be restricted to strict requirements and major antibiotics are administered.
(iv) Vasodilators of the pulmonary microcirculation are given (Dibenzylin or Nosinan) [18, 23].
(v) An adequate proportion of crystalloids and colloids is cal-

culated for the perfusion fluid; alkalinisation is favourable [45].

(vi) Heparin, Dextran-40 and cortisone are on the obligatory list in cases of wet lung, as well as rapid-acting intravenous diuretics; Ringer's lactate solution is unsurpassed for electrolytic re-equilibration of the wet lung.

(vii) Early excision of the damaged pulmonary segment or lobe has likewise been recommended (in irreversible contusion or atelectasis) [54].

Recently, in the clinic the *respiratory index* (RI) has been used as an indicator of the patient's respiratory state. The RI is calculated from Siegel and Farell's equation in which the numerator reflects the alveolo-arterial oxygen difference. An increase of the alveolo-arterial oxygen difference may be an index of the decrease in P_aO_2 and permits control of arterial oxygenation in the clinic.

This equation is:

$$\frac{[(P_B - P_{H_2O}T)\ F_{IO_2} - P_aCO_2] - P_aO_2}{P_aO_2} = \frac{P(A_aDO_2)}{P_aO_2}$$

where:

P_B is the barometric pressure;

$P_{H_2O}T$ is the alveolar water vapour pressure at the patient's temperature (T) (approximately 47 mmHg);

P_aCO_2 is the arterial partial pressure of carbon dioxide assumed to be equal to the alveolar partial pressure of carbon dioxide;

P_aO_2 is the arterial partial pressure of oxygen;

F_{IO_2} is the fractional concentration of oxygen in inspired gas;

$P(A_aDO_2)$ is the alveolar-arterial oxygen difference.

$P(A_aDO_2)$ was divided by P_aO_2 to derive a quotient that reflects more accurately the clinical state

A value of RI of 0.1 to 0.37 is normal. Patients with an RI of 2 or greater were intubated. Those patients who reached an RI of 6 or more had an associated probability of survival of only 12 per cent (Goldfarb, 1975) [19].

SELECTED BIBLIOGRAPHY

1. ALLISON, P. R., Pulmonary Embolism and Thrombophlebitis, *Brit. J. Surg.*, 1967, *5*, Lister Centenary Number, 466–468.
2. ASHBAUGH, D. G., Effect of Ventilatory Methods and Patterns on Physiologic Shunt, *Surg.*, 1970, *68*, (1), 99–104.
3. AYRES, S., M., MUELLER, H., GIANNELLI, Jr. S., The Lung in Shock; Alveolar Capillary Gas Exchange in the Shock Syndrome, *Amer. J. Cardiol.*, 1970, *25*, (6), 588–594.
4. BARBER, R. E., LEE, J., HAMILTON, W. K., Oxygen Toxicity in Man. A prospective Study in Patients with Irreversible Brain Damage, *New Engl. J. Med.*, 1970, *283*, (27), 1478–1484.
5. BERMAN, I. R., SMULSON, M. E., PATTENGALE, R., SCHOENBACH, S. F., Microembolisation pulmonaire après traumatisme des parties molles chez les primates, *Surg.*, 1970, *70*, (2), 246–253.
6. BERNAUER, W., HAHN, F., BECK, E., KURY, H., The Role of Histamine and Catecholamines in Anaphylatoxin-Shock as Compared to the Anaphylactic Shock, *Naunyn-Schmied. Arch.*, 1970, *266*, (3), 208–222.
7. BERRY, E. M., EDMONDS, J. F., WYLLIE, J. H., Release of Prostaglandyn E_2 and Unidentified

Factors from Ventilated Lungs, *Brit. J. Surg.*, 1971, *58*, (3), 189–192.
8. Blair, E., Topozlu, C., Davis, J. H., Delayed or Missed Diagnosis in Blunt Chest Trauma, *J. Trauma*, 1971, *11*, (2), 129–145.
9. Blaisdell, F. W., Lim, Jr. R. C., Stallone, R. J., The Mechanism of Pulmonary Damage Following Traumatic Shock, *Surg. Gynec. Obst.*, 1970, *130*, (1), 15–22.
10. Brown, P. P., McLurdy, J. E., Elkins, R. C., Greenfield, J. L., Effects of Albumin and Urea on Hydrostatic Edema in the Perfused Lung, *J. Surg. Res.*, 1973, *14*, (4), 359–366.
11. Bryant, L. R., Trinkle, J. K., Dubilier, L., Acute Respiratory Pathophysiology after Haemorrhagic Shock, *Surg.*, 1970, *68*, (3), 512–519.
12. Buckberg, G. D., Lipman, C. A., Hahn, J. A., Pulmonary Changes Following Haemorrhagic Shock and Resuscitation in Baboons, *J. Thorac. Cardiovasc. Surg.*, 1970, *59*, (3), 450–460.
13. Budow, J., Kroop, I. G., Pulmonary Blood Volume in Early Shock Following Experimental Myocardial Infarction, *Amer. J. Med. Sc.*, 1967, *254*, (5), 675–678.
14. Cahill, M. J., Jouasset-Strieder, D., Byrne, J. J., Lung Function in Shock, *Amer. J. Surg.*, 1965, *110*, (3), 324–329.
15. Cahill, M. J., Byrne, J. J., Intravenous Gram Negative Endotoxin: its Effect on Pulmonary Diffusing Capacity and Capillary Blood Volume in Dogs, *Clin. Res.*, 1965, *13*, (6), 552–558.
16. Collins, J. A., Braitberg, A. Alexandra, Butcher, H. R. Jr., Changes in Lung and Body Weight and Lung Water Content in Rats Treated for Hemorrhage with Various Fluids, *Surg.*, 1973, *73*, (3), 401–444.
17. Derks, C. M., Peters, R. M., Role of Shock and Fat Embolus in Leakage from Pulmonary Capillaries, *Surg. Gynecol. Obstet.*, 1973, *137* (6), 945–948.
18. Geiger, J. P., Gielchinsky, I., Acute Pulmonary Insufficiency Treatment in Vietnam Casualties, *Arch. Surg.*, 1971, *102*, (4), 400–405.
19. Goldfarb, M. A., Ciurej, T. F., McAslan, T. C., Sacco, W. J., Weinstein, M. A., Cowley, R. A., Tracking Respiratory Therapy in the Trauma Patient, *Amer. J. Surg.*, 1975, *129*, (3), 255–258.
20. Greenfield, L. J., Mc Curdy, W. C., Coalson, J. J., Pulmonary Oxygen Toxicity in Experimental Haemorrhagic Shock, *Surg.*, 1970, *68*, (4), 662–675.
21. Gruner, O. P. N., Plaquettes, fibrinogène et embolie graisseuse après fracture du fémur chez le rat, *Acta Chir. Scand.*, 1971, *137* (5), 407–414.
22. Gump, F. E., Kuney, Y. M., Simultaneous Use of Three Indicators to Evaluate Pulmonary Capillary Damage in Man, *Surg.*, 1971, *70*, (2), 262–270.
23. Hironaka, K., Experimental Study on Fluid Replacement in Puppies with Special Reference to Pulmonary Arterial and Central Venous Pressures, *Arch. Jap. Chir.*, 1970, *39*, (3), 97–103.
24. Johansen, S. H., Bech-Hansen, P., Beck, O., Alveolar-Arterial Oxygen Tension Gradients during Peroperative Replacement of Fluid Loss by Physiologic Saline, *Acta Anaesthesiol. Scand.*, 1972, *16*, (2), 127–131.
25. Jones, R. H., Coulan, C. M., Goodrich, J. K., Sabiston, D. G., Radionuclide Quantitation of Lung Function in Patients with Pulmonary Disorders, *Surg.*, 1971, *70*, (6), 891–903.
26. King, P., Fat Embolism Syndrome, *Med. J. Austr.*, 1970, *57*, (25), 1190–1193.
27. Kuk, M., Coalson, J. J., Massion, W. H., Guenter, C. A., Pulmonary Effects of *E. Coli* Endotoxin; Role of Leukocytes and Platelets, *Ann. Surg.*, 1972, *175*, (1), 26–36.
28. Levitsky, S., Annable, C. A., Park, B. S, Depletion of Alveolar Surface Active Material by Transbronchial Plasma Irrigation of the Lung, *Ann. Surg.*, 1971, *173*, (1), 107–115.
29. Lewin, I., Weil, M. H., Shubin, H., Sherwin, R., Pulmonary Failure Associated with Clinical Shock State, *J. Trauma*, 1971, *11*, (1), 22–35.
30. Lichtmann, M. W., The Problem of Contused Lung, *J. Trauma*, 1970, *10*, (9), 731–739.
31. Lindsay, W. G., Humphrey, E. W., The Sodium Space in Pressure Induced Pulmonary Edema, *Curr. Top. Surg. Res.*, 1970, *2*, (2), 351–358.
32. Lindsey, H. E., Release of Prostaglandins from Embolized Lungs, *Brit. J. Surg.*, 1970, *57*, (10), 738–741.
33. McLaughlin, J. S., Leellacer, R., Attar, S., Cowley, R. A., The Lung in Trauma, *Amer. J. Surg.*, 1970, *36*, (3), 157–162.
34. McLauchlin, J. S., Physiologic Consideration of Hypoxemia in Shock and Trauma, *Ann. Surg.*, 1971, *173*, (5), 667–679.
35. Milligan, G. F., Mac Donald, J. A. E., Mellon, Anne, Ledingham. I. McA., Pulmonary and Hematologic Disturbances during Septic Shock, *Surg. Gynecol. Obstet.*, 1974, *138*, (1), 43–49.
36. Mizumura, S., Experimental Studies on the Effect of Fluid Therapy for Pulmonary

Circulatory Disturbance, *J. Jap. Ass. Thor. Surg.*, 1969, *17*, (5), 82–97.
37. MOSELEY, R. V., DOTY, D. B., Hypoxemia, during the First Twelve Hours after Battle Injury, *Surg.*, 1970, *67*, (5), 765–772.
38. MOSELEY, R. V., VERNICK, J. J., DOTY, D. B., Response to Blunt Chest Injury: A new Experimental Model, *J. Trauma*, 1970, *10*, (8), 673–683.
39. MOSS, G. S., SIEGEL, D. C., COCHIN, A., FRESQUEZ, V., Effects of Saline and Colloid Solutions on Pulmonary Function in Haemorrhagic Shock, *Surg. Gynec. Obst.*, 1971, *133*, (1), 53–58.
40. MOSS, G. S., DAS GUPTA, T. K., NEWSON, B., NYHUS, L. M., Effect of Haemorrhagic Shock on Pulmonary Interstitial Sodium Distribution in the Primate Lung, *Ann. Surg.*, 1973, *177*, (2), 211–221.
41. MOSS, G. S., DAS GUPTA, T. K., NEWSON, B., NYHUS, L. M., Effect of Salin Solution Resuscitation on Pulmonary Sodium and Water Distribution, *Surg. Gynecol. Obstet.*, 1973, *136*, (6), 934–940.
42. MORGAN, W. W., DELEMOS, R., WOLFSDORF, H., NACHMAN, R., BLOCK J., LEIBY, G., WILKINSON, H. A., ALLEN, T., AVERY, M., HALLER, J. A., Jr., A Quantitative and Pathologic Assessment of Lung Injury from Oxygen, with and without Assisted Ventilation, and Air with Assisted Ventilation in Lambs, *Surg. Forum*, 1968, *19*, (2), 265–272.
43. MOTSAY, G. J., ALHO, A. V., SCHULTUZ, L. S., Pulmonary Capillary Permeability in the Post Traumatic Pulmonary Insufficiency Syndrome: Comparison of Isogravimetric Capillary Pressure, *Ann. Surg.*, 1971, *173*, (2), 244–248.
44. NEELY, W. A., ROBINSON, W. T., McMULLAN, M. H., Postoperative Respiratory Insufficiency: Physiological Studies with Therapeutic Implication, *Ann. Surg.*, 1970, *171*, (5), 679–685.
45. O'DRISCOLL, M., Postoperative Pulmonary Atelectasis and Collapse and its Prophylaxis with Intravenous Bicarbonate, *Brit. Med. J.*, 1970, *4*, (5276), 26–28.
46. OETTINGEN, von, W. F., *Poisoning: A Guide to Clinical Diagnosis and Treatment*, Saunders, Philadelphia, 1958.
47. PANETH, M., The Treatment of Pulmonary Embolism, *Brit. J. Surg.*, 1967, *54*, Lister Centenary Number, 468–470.
48. PARIENTE, R., BROUET, G., Le surfactan pulmonaire. Notions physiologiques et physiopathologiques actuelles, *Rev. Prat.*, 1968, *18*, (3), 325–332.
49. PETERS, R. M., HILBERMAN, M., Respiratory Insufficiency; Diagnosis and Control of Therapy, *Surg.*, 1971, *70*, (2), 280–287.
50. POWERS, Jr. S. R., GUMP, F. E., BENDIXEN, H. H., Trauma Workshop Report: The Lung, *J. Trauma*, 1970, *10*, (11), 1047–1049.
51. RICCITELLI, M. L., Pulmonary Embolism: Modern Concepts and Diagnostic Techniques, *J. Amer. Geriat. Soc.*, 1970, *18*, (9), 175–728.
52. ROCCHIO, M. A., DILOLA, V., RAUDALL, H. T., Role of Electrolyte Solutions in Treatment of Hemorrhagic Shock, *Am. J. Surg.*, 1973, *125*, (4), 488–495.
53. ROSEN, A. J. Shock Lung: Fact or Fancy?, *Surg. Clin. N. Amer.*, 1975, 55, (3), 613–620.
54. RUTHERFORD, R. B., VALENTA, J., An Experimental Study of Traumatic Wet Lung, *J. Trauma*, *11*, (2), 146–166.
55. SIEGEL, D. C., MOSS, G. S., COCHIN, A., DAS GUPTA, T. K., Pulmonary Changes Following Treatment for Haemorrhagic Shock: Saline Versus Colloid Infusion, *Surg. Forum*, 1970, *21*, (1), 17–19.
56. SILBERSCHMID, N., SZCZEPANSKI, K. P., LUND, C., Normal Lung Function during Experimental Hemorrhagic Shock, *Eur. Surg. Res.*, 1973, *5*, (suppl. 1), 1–2.
57. SKORNIK, A., DRESSLER, D. P., Lung Bacterial Clearance in Burned Rat, *Ann. Surg.*, 1970, *172*, (5), 837–843.
58. VANDOR, E., ANDA, E., Pulmonary Lipid Metabolism in Experimental Fat Embolism, *Med. Acad. Sc. Hung.*, 1970, *27*, (2), 173–182.
59. WAHRENBROCK, E. A., CARRICO, C. J., AMUNDSEN, D. A., Increased Atelectatic Pulmonary Shunt during Haemorrhagic Shock in Dogs, *J. Appl. Physiol.*, 1970, *29*, (5), 615–621.
60. WEEDN, R. J., COOK, J. A., McELREATH, R. L., GREENFIELD, L. J., Wet Lung Syndrome after Crystalloid and Colloid Volume Repletion, *Curr. Top. Surg. Res.*, 1970, (2), 335–350.
61. WILSON, J. E., PIERCE, A. K., JOHNSON, R. L. Jr., Hypoxemia in Pulmonary Embolism, a Clinical Study, *J. Clin. Invest.*, 1971, *50*, (3), 481–491.
62. WRIGHT, F., STANLEY, L. K., Oxygen Toxicity in Man. A Prospective Study, in Patients after Open Heart Surgery, *New Engl. J. Med.*, 1970, *283*, (27), 1473–1478.

11
ELEMENTS OF THERAPY

The best cure for shock is prevention.

GEORGE, W. CRILE, 1918

11.1 MANAGEMENT OF SHOCK-INDUCING LESIONS
- 11.1.1 Notions of organization, management and treatment
- 11.1.2 The role of surgery

11.2 EMERGENCY TREATMENT
- 11.2.1 Cardiorespiratory resuscitation
- 11.2.2 Artificial ventilation and oxygen therapy

11.3 BLOOD VOLUME SUBSTITUTES
- 11.3.1 The blood
- 11.3.2 Plasma substitutes
- 11.3.3 Crystalloid solutions

11.4 RHEODYNAMIC MEDICATION
- 11.4.1 Medication of the α- and β-receptors
- 11.4.2 Cardiotonics and diuretics
- 11.4.3 Coagulolytic equilibrium medication

11.5 CELLULAR RE-EQUILIBRATION
- 11.5.1 Metabolic medication
- 11.5.2 Corticosteroid therapy

11.1 MANAGEMENT OF SHOCK-INDUCING LESIONS

Today, patients with severe organic and/or functional lesions who require special care in intensive therapy units, and whose lives are in danger, are sometimes referred to as *critical patients;* rheodynamically, and especially metabolically, most of them are almost always in a *state of shock.* There is an 'epidemic' of traumatic injuries today and world statistics show that during the last few years one in every eight hospital beds is occupied by an injured patient, and that the morbidity rate from trauma is much greater than that from cardiac diseases and four times that for cancer. In addition to traumatic injuries another source of critical patients are the coronary diseases and major operations, particularly those performed for cancer.

Regardless of the cause, the 'critical state' of a patient is actually a 'state of shock' and in the management of such cases the treatment must be in harmony with the human body's natural reaction, it must overcome the imbalance that has developed, and it must compensate for the insufficiencies.

To treat a shocked organism means to help and control its reactions when the body struggles in disorder. Otherwise stated, the treatment of shock is an exogenous pharmacotherapeutic monitoring.

The totality of the pathologic events that take place in the human body in shock is simply an intimate, oscillating interweaving of the lesional with the reactionary trends. However correct and well adapted the treatment is to the after shock reactions, failure will be certain if the initial lesion is neglected, for instance an ulcerous vascular fistula or an undrained purulent collection. At necropsy, in the patients who died from 'irreversible shock', pulmonary embolism, mesenteric infarct, biliary ileus, suppurated pancreatitis, peritonitis, massive internal hemorrhages, etc. have been discovered [32].

Surgery performed in time is the only means of removing the cause of 'irreversibility' in severe shock. Emergency surgery is not a cure *per se*, but only part of resuscitation.

On the other hand, one should never overlook the fact that shock does not imply haemorrhage or sepsis only; surgery of the lesions, blood, and antibiotics may often be ineffective and the patient dies from hydroionic cellular imbalance, if it is not promptly corrected. Therefore, for the treatment to be successful one must permanently have in view the influence it will exercise at cellular level and the degree of improvement it will bring about.

It is necessary to sustain alveolar ventilation and the circulating fluids in order to supply the cell with oxygen and organic substances as well as to cleanse the periphery of its acidotic waste; medication of the microcirculation receptors improves tissular perfusion and pharmacologic hormonal doses are meant to readjust the intracellular enzymatic and membrane systems. A wide spectrum of active drugs available today act upon the enzymatic chains and are able to induce 'resuscitation of the shock cell'.

In shock, part of the treatment attempts to correct the lesions generated by the initial traumatic agent and/or those developing in the course of shock. This includes the whole morbid condition, with its anatomic, clinical and therapeutical aspects, as well as emergency surgical and medical pathology covering all the lesions that accompany the picture of shock (*see* Section 11.1.2).

Treatment of the injuries caused in the progress of shock has as its object the making up of deficiencies in the respiratory, myocardial and renal functions (*see* Section 11.2.1). Recovery of the intravascular fluid volume (*see* Section 11.3) and maintenance of the integrity of the cellular subsystems (*see* Section 11.5) also aim at correcting the lesional trend of the endogenous reaction to shock.

How are the after effects of the shock, which nevertheless tend to save the organism, controlled therapeutically? The answer is simple: by any means that may speedily adapt the conditions to the vital interest of the whole macrosystem.

In general, the reactional syndrome of shock is dysharmonious and, in primates, has numerous dangerous imperfections. Hence, it is useful to block the excessive warning that the network of sensory fibres convey through the central neuro-endocrine space which cause an exaggerated response. Moreover, it is useful also to temper the released catecholamines, reducing particularly α-stimulation of the microcirculation and β-metabolic stimulation of glycogenolysis and glycolysis (*see* Section 11.5.1). But the initial increase in glucocorticoids may be accentuated by exogenous doses in order to ensure a better cellular protection (*see* Section 11.5.2). Neuroplegics and reducing substances also arrest the extracellular exodus of electrolytes and enzymatic hypercatabolism.

The judicious management of the two opposed trends in shock may balance the cellular supply and export. Sometimes, when the initial lesion cannot be corrected because of deep cellular affection, controlled hypothermia may for some time lower cellular requirements and the production of protons. This interval should be used to the utmost in order to repair the damage of the shocked organism.

11.1.1 NOTIONS OF ORGANIZATION, MANAGEMENT AND TREATMENT

During the last few years transport and intensive therapy have saved from death an increasing number of 'critical patients'. Intensive therapy implies the general management of severely wounded and shocked patients. It includes cardio-respiratory resuscitation, and active drug therapy with electronic monitoring in intensive therapy units. In some units special wards for the intensive therapy of shocked patients are equipped for all types of resuscitation. In severe injuries functional imbalance of all the vital functions have to be treated simultaneously.

During recent wars the mortality rate on the battlefield was reduced because of early application of cardio-respiratory resuscitation and blood volume re-equilibration near to the front line. Resuscitation during transport from the battlefield continued the treatment up to the moment when the patient was admitted to hospital. Evacuation by air (in particular by helicopter) reduced the interval between the infliction of the wound and the application of intensive therapy necessary for survival by ten to fifteen times that of other transport and proved much superior to earlier methods of evacuation by stages (Kennedy, 1971; Safar, 1971) [68, 123].

The increased speed with which the wounded are transported to special medical units has led to the ideal of 'one to two-hour severe wounded'; it is the casualty with zero blood pressure, increased lactacidaemia, and *p*H varying between severe acidosis and severe alkalosis.

The interval within which a casualty reaches hospital fell to 16 hours in the second World War, to 6 hours in Korea, and from 1 to 2 hours in Vietnam. Increase in the number of vascular lesions dealt with surgically is sufficient proof of the rapidity of modern transport. Similarly, renal lesions have been detected and treated in patients with multiple injuries in a proportion of 10% as against 0.14-4% in the second World War.

In peacetime, the 'critical patient' is as a rule met with on the large highways, towns with crowded traffic, factories, or even in homes. Traffic accidents give a high mortality rate, due particularly to injuries to the abdomen, chest, head and limbs.

A central emergency hospital with many-sided intensive therapy units is ideal for saving a critical patient who is transported quickly. Recent statistics have shown that 54.2% of critical shocked patients conveyed to intensive therapy units necessitated tracheostomy and 52.5% artificial pulmonary ventilation for more than 6 hours [69, 73, 123].

Today, the need for adequate means of transport and a resuscitation team able to start treatment at the site of the accident (blood substitutes, control of acidosis and pain, etc.) has been clearly established.

A practical grouping of traumatic situations, applicable under peace and wartime conditions, has been drawn up by Shires (1970):

The first category includes violent lesions of the vital organs. The patient is brought directly to the operation theatre where laparotomy, thoracotomy, etc. is performed and at the same time the first signs of shock are controlled. The management of shock is continued postoperatively.

The second category includes the lesions that permit an intervention after 1 to 2 hours, during which interval blood can be collected and the patient examined radiographically. Shock is treated preoperatively and continued during and after the operation.

The third category includes all the lesions that do not call for immediate surgery. In the event of shock, adequate treatment is given.

The treatment of potential shock must start at the site of the accident. When necessary, cardiac and respiratory arrest must be dealt with, followed by fluid compensation, the control of pain, and prevention of acidosis by sustaining perfusion of the micro-circulation.

Restoration of fluids, analgesics, immobilisation of fractures, etc. prevent shock and render the subsequent course far milder. First aid to critical patients is therefore of paramount importance [142].

11.1.2 THE ROLE OF SURGERY

Surgeons have always shown a particular interest in the shock that frequently accompanies operation on a *critical surgical patient*, who, because of certain complex clinical, pathophysiologic and therapeutical problems, constantly needs re-evaluation [13, 19, 41, 57, 99, 139, 143].

The moment at which surgery is performed still remains a critical decision that cannot be avoided. Surgical experience during the last decade has brought the operation closer to the onset of shock; the term 'immediate intervention' is more and more often used for many 'critical states'. Resuscitation has always been ineffective where there is bleeding into the visceral cavities and/or serous cavities, or where the focus of infection is not under control. In many instances the subsequent well-being of a severely injured patient depends not on *hours* but on *minutes*. Here, surgery forms an essential part of the complex management of shock.

Surgery is also recommended when shock has proved refractory to all treatment; for by means of an exploratory operation it is possible to detect a lesion in the abdomen or chest that often may be surprisingly superficial, yet quite unobserved.

Surgery may therefore be considered as an element in the aetiologic treatment of the initial shock-inducing lesion, and of pathogenic treatment for arresting the irreversible course of shock [32, 127, 142].

Of particular interest are some of the surgical questions that arise in connection with the patient in shock.

Traumatic coma often dominates the clinical picture, masking the visceral lesions [39, 61]. Foremost in the management of post-traumatic coma is restoration of the respiratory deficiences. It is a known fact that there is often no connection between the state of consciousness and the intensity of the cerebral lesions; of importance in such cases are the clinical pupillary signs, hemiparesis, incontinence, temperature, etc. The echoencephalogram may be useful, and surgery should be performed immediately in the cases of inbending fractures of the skull. Cerebral œdema and agitation must be controlled. Examination of the patient in coma must comply with the indications of a complete scheme allowing for appraisal of all the lesions.

For thoracic injuries the following are imperative:

- recovery of the patency of the tracheobronchial tree;
- repair of the anatomic and functional integrity of the chest wall;
- functional conservation of the pleural vacuum.

Thoracotomy and repair of fractures in a flail chest must be performed as soon as possible [38, 82] in order to reduce haemorrhage and preserve an airway.

Sharp increase in intra-abdominal and/or intrathoracic pressure often result in rupture of the intracavitary organs [106]. Intra-abdominal puncture is still of great value today, especially in acute surgical abdomen with uncertain data. Abdominal puncture will help to establish the diagnosis in 75-85% of the cases and may be repeated if necessary. By this means the problem of intraperitoneal haemorrhage, that has always been an outstanding clinical difficulty, can be solved [134].

Present experience has shown that it is better in traumatic perforations of the colon to resect and solve it completely in a first operative stage, thus removing the danger of postoperative infection [153].

Severe digestive haemorrhage (haematocrit less than 30% and BP under 70 mmHg, according to the well-known criteria of Edelman) demands a surgical intervention with or without a conclusive X-ray examination. When the conventional methods of diagnosis do not prove sufficient in cases of massive gastric haemorrhage selective angiography is recommended, which reveals the topographic diagnosis in 50-60% of the cases (provided the rate of flow of the haemorrhage is more than 0.5 to 0.6 ml/minute) [24, 119].

However severe the patient's general condition a refractory digestive tract haemorrhage imposes surgery, which will often be limited to suture *in situ* of the ulcerous vascular fistula, pyloroplasty and vagotomy, but only when the patient's state permits it, gastric resection *(see* Section 10.1.1).

If bleeding occurs through oesophageal varices the surgeon may restrict the operation to local haemostasis or may from the beginning perform a venous shunt [109].

Vascular surgery has changed the prognosis of arterial lesions [9, 10, 40, 45, 148]. Today successful operations are performed in obstruction of the carotids, rupture of the axillary artery in scapulo-humeral luxation, traumatic thrombosis of the popliteal artery, mesenteric infarct and the vascular lesion that frequently accompany fractures of the bones and muscular attrition (Figure 11.1). It is very difficult to differentiate between rupture of the intima with thrombosis and arterial spasm, the diagnosis necessitating arteriography. Ischaemic muscular contractions may be considered as an emergency for vascular exploration. The collateral circulation may be explored by plethysmography and ultrasonography. Statistics based on 1100 cases of truncular vascular lesions underlined the importance of the collaterals and muscular masses in supplanting the principal arterial pathway [80].

DeBakey and Simeone reported in 1946 that they had found only 3 ruptures of the aorta in 2471 cases of vascular injuries during the second World War, but Billy reported only 138 rup-

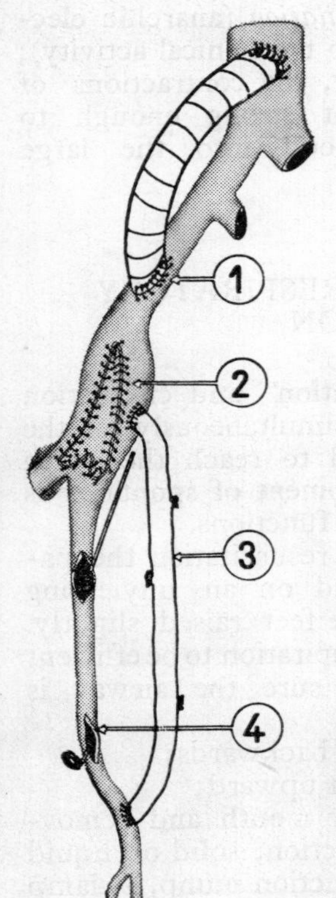

Figure 11.1. Practical therapeutical solutions of vascular lesions:

1— synthetic bypass; *2*— venous patch; *3* — venous autograft; *4* — thrombendarteriectomy.

tures of the aorta (86% thoracic locations), and was able to suture 39 of the cases, with 27 recoveries [14] in Vietnam, owing to the rapid transport available for the casualties.

The technical progress made in extracorporeal circulation and conservation of the cardiopulmonary block has led to an increase in 'surgical incisiveness', so that not only can cardiac wounds and contusions be treated but also necrosis after acute coronary occlusion [34, 144]. Early surgery (within less than 2 weeks) of myocardial infarct gave favourable results in 80% of the cases and lowered the mortality rate from cardiogenic shock to less than 60%. Emergency surgery is indicated for rupture of the septum or chordae tendinae and severe left ventricular failure with cardiogenic shock refractory to treatment. Revascularisation by aorto-coronary by-pass is performed with internal saphena autograft and with or without infarctectomy (*see* Section 12.6).

Surgery is not always indicated for a patient in shock; an operation can be undertaken only if its benefits may be considered to counterbalance the risks (*see* Section 12.2.3). Intraoperative bleeding and the duration of an operation must be taken into account. It is likewise of importance to bear in mind that the patient with multiple injuries, operated on or not is exposed to early and severe complications [72, 25, 134] that have to be recognized immediately and treated, such as:

- the detachment of eschars from the damaged intestinal loop or a damaged vessel (equivalent to the disconnection of an intestinal or vascular suture);
- rupture in two stages of the spleen, liver, kidney, etc.;
- diaphragmatic hernia or strangulated intestinal loop impacted within mesenteral gap;
- stress ulcer with its variants, especially in burns (Curling) and head injuries (Cushing);
- acute renal failure with its serious prognosis;
- acute pancreatitis or pancreatic suppuration with the rapid formation of a false pancreatic cyst;
- thromboembolic accidents alternating with fibrinolytic ones;

- infection with its multiple roots in a subphrenic abscess or infected haematoma, etc.;
- late necrosis of an intestinal loop with a damaged mesentery;
- repeated digestive tract haemorrhage in an operated patient, which usually points to an ignored ulcer of the post-bulbar duodenum;
- prolonged fever may mean insufficient drainage or may indicate the need to interrupt treatment with antibiotics.

However long the list is, it cannot cover all the aspects that the clinical course of shock may take on.

11.2 EMERGENCY TREATMENT

The onset of artificial respiration and circulation within 3 to 5 minutes, before the onset of death is the foremost condition for the survival of a critical patient with cardiorespiratory arrest *(see* Section 4.5). In a patient with at leat partly conserved reserves cardiorespiratory resuscitation may be life-saving but when cardiac arrest appears in a final stage and is the result of metabolic coma or of an irreducible heart failure it fails.

More than 90% of cases of cardiac arrest in the course of an operation recover, whereas outside the operating theatre the mortality rate from cardiac arrest is about 90% [60, 67, 151].

Within the minimal time available, the diagnosis of cardiorespiratory arrest has to be established, based upon the absence of pulsations in the large arteries and of respiration, the presence of fixed mydriasis and cyanosis. The diagnosis must then be of *terminal* or *accidental* cardiac arrest.

According to Safar, accidental cardiac arrest is defined as 'the clinical picture of arrest of the circulation in a patient whose death is not imminent' [123].

The ECG examination may establish one of the following variants:

- *cardiac arrest proper* (no electric or mechanical activity);
- *ventricular fibrillation* (anarchic electric activity; no mechanical activity);
- *inefficient heart*, the contractions of which are not strong enough to pump the blood into the large vessels.

11.2.1 CARDIORESPIRATORY RESUSCITATION

Artificial respiration and circulation must be started simultaneously for the oxygenated blood to reach the nerve cell up to the moment of spontaneous recovery of both functions.

Before starting resuscitation the patient must be laid on an unyielding surface, with the feet raised slightly.

For artificial respiration to be efficient one must make sure the airway is clear by:

- tilting the head backwards;
- pulling the chin upward;
- looking into the mouth and removing any obstruction, solid or liquid with a simple suction pump, a clamp or the index finger covered with gauze;
- by introducing a pharyngeal pipe in order to prevent the patient from swallowing his tongue or obstructing the laryngeal orifice.

Artificial respiration. In emergencies the mouth-to-mouth artificial respiration method is employed; it is more efficient than the classical Silvester, Holger-Nielsen or Schäffer methods which do not supply satisfactory alveolar ventilation [111, 114].

Artificial respiration should be started as soon as possible since an apnoea of 15 seconds lowers oxyhaemoglobin to 85%.

In the mouth-to-mouth method the operator takes up position on the patient's right side, pinches the patient's nostrils, leaning over him and expels air from his own lungs into the patient's mouth, with enough force to make the patient's chest rise. He then moves his mouth and listens for the sound of air being exhaled from the patient and watches for any movement of the chest. This is repeated until the exercise is successful. The rate of respiration must be about 18-22 per minute, as with slight hyperventilation the operator's expired air contains a higher proportion of oxygen (16-17%) and a lower proportion of carbon dioxide (2-3%). When simple kits are available (the Ambu kit for instance) breathing is carried out from mouth-to-mask or after previous endotracheal intubation. Unidirectional valves may also be used or a Safar tube (two S-shaped Guédel pipes).

External massage of the heart is started at the same time as artificial respiration, the proportion of forced breathing to the pressure exercised on the chest being 1:4.

Cardiac massage. Cardiac massage when successful produces the artificial circulation necessary for survival of the nerve cells up to the moment in which spontaneous heart beats appear [60, 144]. For success the massage must be started not more than three or four minutes after the onset of cardiac arrest.

Closed-chest cardiac massage. The technique of closed-chest cardiac massage has been described by Kouvenhoven-Knikerboker. The patient lies on a hard surface (a floor) in dorsal decubitus, with the legs raised 30 to 35 degrees. The operator kneels on one side and applies rhythmic pressure with his hands upward and inward against the lower part of the sternum, using the whole weight of his body so that the sternum is depressed 3 to 5 cm (in the adult) [111]. The arms are then relaxed without taking the hands off the sternum and the chest returns to its normal position. The rate of compressions must be 60 to 80 per minute. The success of the massage is gauged by reappearance of the pulse in the large vessels, synchronous to the rate of compression on the sternum, by disappearance of mydriasis, and by colour returning to the skin.

In some cases, even if the limit of four minutes has been exceeded, the heart may take up its spontaneous contractions, but mydryasis will always persist and the EEG recording will indicate death of the nerve cell *(see* Section 4.5).

Internal cardiac massage. Until recently it was considered that if the heart failed to begin to beat within 'a few minutes' after starting external massage, thoracotomy for internal massage is indicated [123]. Most authors nowadays agree that internal massage is often indicated from the beginning, if the external massage is counter-indicated because of pneumothorax, haemopneumothorax, the presence of flail sternocostal or left costal chest, advanced scleroemphysema, cardiac tamponade [67]. Thoracotomy is performed in the 4th left intercostal space, introducing if possible a self-retaining retractor. Cardiac massage may be done either transpericardially, or after longitudinal incision of the pericardium in front of the phrenic by direct myocardial contact. The right hand grasps the left ventricle with the thumb and thenar portion over the right ventricle and squeezes the heart strongly, in order to close the tricuspid and mitral valves firmly and eject the contents of the ventricular cavities. For a good flow towards the coronaries and cerebral vessels the descending aorta is clamped intermittently.

Certain complications may develop in the course of internal cardiac massage: chondrocostal fractures, especially in patients with a rigid thorax; ruptures of the myocardium or of the liver. The most frequently encountered are ruptures of the heart, haemopericardium, subepicardiac lesions.

The incidence of intraoperative cardiac arrest ranges between 1: 500 and 1: 1200 cases [67, 140].

The result of cardiac massage depends upon the reversibility of the patient's disease, the interval after the heart has stopped, correct application of the technique and the equipment of the operation theatre or resuscitation unit. The efficiency of cardiopulmonary resuscitation decreases in the following order: in the operation theatre, intensive therapy units, wards and outside hospital.

The chemotherapy of cardiac arrest
Metabolic acidosis that develops following arrest of the circulation has a negative influence on the efficiency of cardiac massage, of various drugs and electric defibrillation. Sodium bicarbonate is administered in an 8.4% molar solution, calculated according to the Gibston formula:

$$\text{mEq NaHCO}_2 = \frac{GT}{10}$$

where G is the body weight (in kg), and T the number of minutes elapsing after the heart has stopped beating.

Safar administers in all cases of cardiac arrest 44 mEq sodium bicarbonate every 5 minutes throughout the entire duration of resuscitation.

It is known that THAM (trihydroxymethylaminomethane) buffers cellular acidity and is preferred for this reason. Some authors also consider that it increases the force of myocardial contraction and of the coronary flow and exercises an antiarrhythmic effect. THAM is used in doses of 1 mEq/kg body weight, combined with sodium bicarbonate [29, 151].

Medication stimulating cardiac activity (see also section 11.4.1).

Epinephrine is one of the most efficient drugs for use in cardiac arrest. It stimulates β_1-receptors, increasing the contraction force of the myocardium and cardiac output, but may cause arrhythmias or even ventricular fibrillation, which probably are more frequent in the presence of hypoxia and metabolic acidosis [44, 130]. Epinephrine may be administered by intracardiac route in doses of 1 mg diluted in 10 ml normal saline.

Norepinephrine has a vasodilator effect upon the coronaries, but repeated administration may cause tachyphylaxis. In an acid medium, norepinephrine has a diminished effect and must be associated with bicarbonate.

Metacaraminol and *Neosinephrine* have less drastic α and β effects and can be administered either by intravenous or intramuscular route. Metacaraminol has a positive inotropic cardiac action, increases the cardiac output and causes reflex bradycardia [44].

Isopropylnorepinephrine (Isuprel). The β-stimulating efficiency of this drug is equal to that of epinephrine, increasing the cardiac output and lowering telediastolic pressure. The major indication for this drug is when the heart stops in the course of the Adams-Stokes syndrome. β-adrenergic stimulators are also assumed to have a coronary dilatory effect, after which there is an increase in the supply and utilization of oxygen in the myocardium. By arteriolar dilatation isopropylnorepinephrine lowers peripheral resistance and the work of the heart [60]. The conventional doses are 0.1 mg by the intracardiac route in case of cardiac arrest, or 0.2 mg when administerd in a 5% glucose solution when the contractions start again with bradycardia and hypotension. *Isuprel* does not stimulate the supraventricular ectopic foci but permits cardio-stimulation with a pacemaker [27, 44, 46, 85].

Calcium chloride is administered by the intracardiac route in cardiac arrest or the intravenous route whenever the contractions are inefficient, in a single 10 ml dose (10% solution); the dose may be repeated but it must be borne in mind that large doses produce bradycardia, extrasystolic arrhythmias and even fibrillation [139].

Energodynamic medication. The energodynamic medication consists in hypertonic glucose and insulin; combined with insulin and potassium it influences intramitochondrial metabolism in the myocardial cell, favouring the synthesis of ATP.

Antiarrhythmic medication. An antiarrhythmic medication is used after spontaneous contractions have been taken up again in cases of hypoxic distress of the heart with foci of ectopic excitation. *Inderal* and *Eraldin* (β-blocking agents) protect the heart by competition with catecholamines at the level of the β_1-receptors. They are indicated in arrhythmias maintained by the sympathetic hypertonia of shock. The dose is 1 mg, repeated if necessary under clinical or ECG control. As β-blocking agents bring about depression of the myocardium and decrease of the BP, precautions must taken when administered to hypotensive patients. Procainamide *(Pronestyl)* is an antiarrhythmic but to the same extent toxic for the central nervous system. It is administered intravenously in doses of 100-300 mg in ventricular rhythm disturbances.

Electric defibrillation. Electric defibrillation is applied in cases of ventricular defibrillation, but only after efficient ventilation and cardiac massage with any type of external electric defibrillator. The external electrodes are placed so that the electric current intercepts the ventricle in fibrillation, and defibrillation is started with an electric stimulus of 200 watts; if defibrillation is not obtained the intensity of the current is increased to 400 watts. In internal defibrillation only 50 watts are used. Electric defibrillation is performed under ECG control. Artificial ventilation and heart massage are interrupted only for a few seconds in order to place the electrodes and discharge the electric shock.

When defibrillation is inefficient the acid-base imbalance must be corrected, administering 80-100 mEq sodium bicarbonate and then again attempting defibrillation. If the ventricular fibrillation waves have a low amplitude, 0.05 mg *Isuprel* may be administered, repeating the electric stimulation when the amplitude of the fibrillation waves is high enough. If, on the contrary, ventricular fibrillation is characterized by high amplitude ventricular complexes, procainamide may be administered in repeated doses of 100 mg (the maximum dose is 2 g). When the amplitude of the fibrillation waves falls, electric stimulation is again attempted. Instead of procainamide, xylocaine can be used in doses of 100-200 mg.

If cardiac massage is ineffective a last possibility is that of electric stimulation with an external or internal pacemaker. In the latter case an electrode catheter is introduced through the jugular or basilic vein up to right atrium.

Follow-up of the resuscitated patient. The resuscitated heart may at any moment relapse into ventricular fibrillation or asystolia, demanding monitoring of its electric activity. Similarly, early spontaneous ventilation may be ineffective, and mechanical artificial respirators have to be used [111, 115].

In following up a resuscitated patient particular attention must be paid to maintenance of the respiration and circulation, the appraisal and treatment of eventual neurologic sequelae, antibiotherapy and osmotic diuresis. When cardiac massage is applied after thoracotomy, active aspiration of the pleural cavity under daily X-ray control of the expansion of the lung is necessary. Maintenance of a sufficient microcirculation for perfusion of the vital organs is the factor upon which the patient's prognosis depends.

If α-mimetics have been used they should be suppressed as quickly as possible. When the BP is low and the central venous pressure high, intravenous digitalis may be given, but of first importance is the administration of fluids up to hydric balance and improved myocardial perfusion *(see* Section 11.3). When the BP falls below 70 mmHg, drugs with a minimal vasoconstrictor action on the microcirculation, for instance *Neosinephrine* and *Aramine*, can be used.

11.2.2 ARTIFICIAL VENTILATION AND OXYGEN THERAPY

Regardless of the nature of the patient's condition 80% of the severe cases suffer from various degrees of respiratory failure or anomalies in the transport of gases to the tissues *(see* Section 7.2.1). The same statistics mention that mechanical ventilation lowers the mortality rate of critical patients two to three fold.

During the last decade the number of patients needing respiratory therapy has increased fourfold, and to the same extent the number of patients requiring mechanical respirators. About 20% of cases of acute respiratory failure were introduced in an artificial lung for 12 to 24 hours; 60% needed ventilation for 2 to 7 days and 20% for more than 7 days. Most of the patients requiring mechanical ventilation belonged to the older age-groups (60 to 70 years) [115].

It is, however, also in the last decade that progressive accentuation of pulmonary failure has been reported among patients undergoing prolonged artificial ventilation, without any apparent connection with the cause of the disease. In these patients, the vital capacity and compliance decrease with the onset of variable degrees of hypoxia, phenomena that bear the generic name of 'ventilator lung syndrome'. Microscopic study of the lungs of these patients reveals congestion, œdema, abundant macrophage cells and haemorrhage, as well as intra-alveolar fibrin exudate that forms hyaline membranes *(see* Section 10.4).

The syndrome appears both in patients ventilated with manometric relaxers and in those ventilated with volumetric relaxers (Pontoppidan et al., 1970).

An increasing number of authors have reported on the *wet lung* produced by hydric retention, especially at the level of the lungs, following upon prolonged mechanical ventilation. X-ray examination may diagnose interstitial pulmonary œdema without any clinically evident cardiac disturbances. As remission of these disorders was obtained following the institution of a negative water balance the cause of wet lung was assumed to be excess water retention [107, 108, 136]. During spontaneous ventilation these respiratory disturbances do not appear in patients hydrated and hyperhydrated by parenteral route with Ringer's solution, as in the case of mechanical ventilation.

A continuous positive pressure breathing anaesthesia may have a useful place in treating the patient with compromised cardiovascular function, or in the obese or hypovolemic patient with severe compromised pulmonary function (Wyche, 1973). It has therefore been postulated that positive intermittent pressure of longer duration against a background of slight hyperhydration may generate the wet lung syndrome.

Injury to the alveolar capillary membrane particularly is assumed to cause spread of interstitial pulmonary œdema [129]. In the onset of this pulmonary syndrome an increase in the secretion of the antidiuretic hormone and subclinical heart failure may also be incriminated. Repeated determination of the respiratory parameters may detect in time a pulmonary affection and improve the situation by a negative water balance.

Importance of the alveolar surfactant. Geometrical alveolar stability and a normal balance of the filtration and absorption of gaseous fluids in the lung depend upon the presence of the alveolar surfactant *(see* Section 10.4). The symptoms of degradation and disappearance of the surfactant showed 18-24 hours after it was no longer produced. The activity of the surfactant lowers surface tension and maintains the alveoli open to a low transpulmonary pressure. When there is a paucity of surfactant or it disappears, the alveolar surface tension increases and collapse of the alveoli occurs. There is a consequent right-left shunt and aggravation of hypoxia. Reduction of the alveolar collapse necessitates higher inspiratory pressures, spontaneous ventilation no longer being sufficient.

Anomalies in the production and quality of the surfactant are encountered in multiple clinical situations associated with respiratory failure: alveolar hyperperfusion in haemorrhagic and bacteriemic shock, the respiratory syndrome of burned patients, prolonged mechanical ventilation, fatty embolism, etc.

The early alveolar lesions in bacteriemic shock and the toxicity of oxygen affect the pulmonary microcirculation and favour the intra-alveolar penetration of proteins and fibrinogen. Intracellular proteins are transformed into hyaline membranes, characteristic of the advanced forms of acute respiratory failure [82, 84].

The early manifestations of the surfactant dysfunction include haemorrhagic atelectasis, pulmonary œdema accompanied by right-left shunt, increase in the alveolar capillary oxygenation gradient and reduction of pulmonary compliance. The late manifestation is a respiratory failure necessitating prolonged artificial respiration with high oxygen concentrations and low expiratory pressures.

Although it is obvious that continuous alveolar hyperinsufflation may have a negative influence on the synthesis of surfactant or may alter its quality, there is no reason to believe that periodical hyperinsufflation has the same harmful effect.

Pulmonary toxicity of oxygen. Prolonged therapy with oxygen concentrations diminishes the surfactant both in animals breathing spontaneously and in those with artificial respiration [136]. Histologic study of the lungs of cases undergoing prolonged artificial ventilation before dying, revealed alveolar œdema, intra-alveolar haemorrhage, fibrin exudate with the formation of hyaline membranes, without inflammatory processes. The morphologic alterations were correlated not only to prolonged artificial ventilation but also to the high oxygen concentrations used. Although no relationship has actually been established between the morphologic alterations and oxygen therapy, it is certain that the administration of oxygen in high concentrations is harmful. Thus, it has been demonstrated experimentally in monkeys that spontaneous respiration in an atmosphere saturated with oxygen 100% for 12 days produces the same pulmonary lesions as observed in clinical cases following mechanical ventilation. The same experiment in dogs caused their death within four days after exposure to 100% oxygen. However, the inhalation of oxygen under pressures lower than 300 to 400 mmHg is better tolerated by the human body, over longer periods, without producing significant alterations, but respiratory failure rapidly appears when oxygen pressure is lowered, developing at 1 atmosphere within 24 to 30 hours and at 2 atmospheres within 6 to 12 hours [136].

11.3 BLOOD VOLUME SUBSTITUTES

In order to survive, a patient who has lost part of his circulating fluid needs to refill his vascular system with an adequate amount of fluid (initially irrespective of the nature of the fluid used for repletion).

In practice, the treatment with fluids — subject to the available supply, with stored blood, plasma substituted and crystalloids—is performed at the beginning under control of the central venous pressure, by introducing a catheter into a central vein [52, 53]. The volume and rate of administration depend upon the ECBV, CVP, BP, urinary output and pulse. It is considered that in severe haemorrhagic shock 4 to 5 litres of blood may be administered within a short interval; in some cases a record amount of 40 litres of blood have been administered in 24 hours [4, 17, 23, 59, 140].

Of particular importance, to prevent the cells from suffocating, is maintaining the fluidity of the blood. The control of acidosis and maintenance of the metabolism of cellular organelles is, of course, always useful in shock, but the essential emergency management of such cases, without which the cell cannot survive, is the control of capillary stasis by massive repletion with fluids and the permeability of the microcirculation. Hence, blood substitutes and α- and β-receptor medication are the first to be acted upon.

In states of shock the intravascular fluid has the following three 'aggregation states':

(i) *fluid state* in which it is generally shunted (the actual ECBV);
(ii) *hyperviscous state*, represented by acid blood that sludges as in pooling of the microcirculation and venous system;
(iii) *jellified state*, in disseminated micro and/or macrothromboses (DIC).

As a result, therefore, in the shocked patient a great part of the intravascular fluid is 'fixed', and to this may be added the hypertonic extravascular fluid pooled by the interstitial collagen sponge (*see* Chapter 6).

Certain conclusions may be drawn from the state of aggregation of the circulating fluid:

(i) Isotopic measurement of the ECBV is hardly useful in practice since it expresses only a fraction of the intravascular blood fluid.

(ii) The amount of fluid introduced must of necessity exceed the actual losses since they must also compensate the amounts of fluids 'fixed' in one way or another within the organism.

(iii) It is difficult to estimate the volume of 'fixed' fluids in contrast to the volume actually lost which is comparatively easy to calculate in the clinic [17, 60].

(iv) As the large amount of 'fixed' fluid must also be compensated the total volume of repletion fluids, as appraised by most authors, must be up to 10 000 ml/24 hours [59, 91, 103 128]. However, the amount actually administered is almost always lower (according to the possibility of mobilizing the 'fixed fluid').

(v) The perfusion of huge amounts of fluid implies compulsory monitoring of the CVP, the most important parameter that can serve as a guide in the clinic for assessing the efficiency of blood volume replacement and recovery of myocardial performance. (Useful monitoring is also that of the urinary hourly excretion that will point to the onset of DIC in the renal microcirculation).

Part of the 'fixed' fluid must be released by α-lytics, antisludge medication, anticoagulants, fibrinolytics, alkalines, osmotic substances and reducing agents.

The importance of using crystalloids again in the management of shock is based upon a closer understanding of the retention of fluids. In the initial phases of shock, crystalloids allow for the administration of large amounts of fluids, which then rapidly leave the body, removing the additional amount of fluid (Moss, 1972; Nash, 1967; Shires, 1970).

The interval within which intravascular fluid is lost is of extreme importance for the onset of shock, so is the refilling time far more important than the nature of the fluid used.

It is of greater usefulness to replace the volume lost and promote flow in the microcirculation (in order to wash away the metabolic wastes) than it is to build up again red blood corpuscles by the obligatory administration of blood. (Up to 6 to 5 g% Hb the organism does not suffer if there is a normal circulating volume) [59].

Endogenous transcapillary refilling in response to haemorrhage is very important and many of the phases involved have been recently studied experimentally (Carey, 1973) [22].

It should not be forgotten that stored blood is actually an isotonic colloid suspension which, besides its other unquestionable qualities, also has many disadvantages for the re-equilibration of shock: it is not efficient as an antisludge fluid, it has no appraisable buffering alkalizing effects, and it does not promote osmotic autoperfusion. (Rheomacrodex, Ringer's lactate solution or THAM and glucose are superior to blood, but only from this point of view) (Shoemaker, 1971).

The administration of α-lytics relaxes the microcirculation (renal, pulmonary, etc.) and lowers the BP; hence, α-lytics are administered concomitantly to blood volume replacement (at least compensatory), new amounts of fluid being introduced subsequently.

The following rules are to be complied with for the efficient administration of blood subtitutes:

(i) the fluids, in large amounts, are administered under control of the CVP, BP, urinary output and pulse;

(ii) α-lytic drugs are indicated only after beginning the administration of fluids and are then followed by a greater volume of fluids;
(iii) only the strictly necessary amount of blood should be administered for replacing the erythrocytic mass or plasma, but non-erythrocytic fluids can be used in large amounts: plasma substitutes, crystalloids.

Blood substitutes are fluids which, after infusion into the vascular bed, imitate the functions of blood. Their purpose is to maintain the effective circulating blood volume, providing tissue perfusion. Investigations on 'ideal' blood substitutes which can transport and exchange oxygen are fairly advanced. The maintenance of electrolyte concentrations or buffering capacity are other functions of blood that can be replaced only to a very limited extent.

Blood subtitutes may be separated into: colloids (blood, plasma and plasma substitutes) and crystalloids, or according to a more complete classification:

- erythrocytic fluids
  - blood
  - erythrocytic mass
- non-erythrocytic fluids
  - colloids
    - plasma and its derivatives
    - plasma substitutes (macromolecular solutions)
  - crystalloids
    - polyelectrolytes solutions
    - micromolecular solutions

To these fluids may also be added the 'alimentary fluids' (protein hydrolysates and lipid emulsions).

Of particular importance is the micromolecular isotonic glucose solution that is always present in any repletion association since it supplies energy and fixes potassium. Numerous mixed products, e.g. amino acid and fructose lipid emulsions, are now being prepared.

The practice of the last few years has supplied outstanding information concerning blood volume subtitution. With blood, the constant hazard exists of inoculating viral hepatitis, an effect that cannot be annulled but can be avoided by using a pasteurized plasma protein and albumin [55].

Although suspected, of late, to have untoward renal effects *(see* Section 10.3) dextrans are still irreplaceable for their anti-sludge effect. A new anti-sludge agent is Pluronic F-68 used in experimental haemorrhagic shock [50].

The non-erythrocytic fluids are of use at least in the initial stages of shock. Ringer's lactate solution or gelatins help to maintain a sufficient oxygen pressure up to losses of 40% of the blood volume. Up to losses of 500-1000 ml blood gelatins are excellent [55].

However, Ringer's lactate solution, at a *p*H ranging between 6.4 and 8.4 — the alkaline solutions being more indicated — is now being successfully used in almost all associations in clinical practice [35].

Today there is a tendency to establish the formula of an optimal combination of blood substitutes since the electrolyte deficit particularly is not sufficiently corrected in all cases of shock. The sodium deficit may range between 2.5 and 11 mEq/kg body weight, i.e. 6-25% of the total sodium content of the human body [35].

Shires recommends 1000 ml crystalloids per hour during surgery (up to at most 3 to 4 litres) [129]. As, in shock, the liver is still competent to perform the Cori cycle, sodium may be given in the form of sodium lactate; if the hourly urinary output is maintained within normal limits then isotonic solutions can be administered since the eventual increase in potassium and urea may often point to prerenal causes (increased protein supply, catabolism, etc.).

The optimal ratio of crystalloids to colloids appears to oscillate between 2 to 1 and 3 to 1.

Dillon prefers, for the initial perfusion blood with 1.8% Ringer's lactate solution ($pH = 8.4$) + normal saline, because he considers that, in shock, aldosterone retention of sodium only begins after 150 minutes; therefore within the first hours of shock there is no actual danger of sodium retention [35].

Resuscitation from haemorrhagic shock with a balanced salt solution (especially Ringer's lactate) in conjunction with red blood cell replacement effectively restores normal renal tubular function. Resuscitation with 5% albumin and red blood cells is less effective in restoring renal tubular function (Siegel, 1973) [133].

For a fluid loss of up to 20 ml/kg body weight the best combination (even better than blood) seems to be Dextran-40 + Ringer's lactate solution in amounts two to three times that of the blood lost [21].

Another solution that may lower blood requirements is Dextran-40 + 1.4% sodium bicarbonate + 0.9% NaCl. Also useful is Ringer's solution + 5% albumin + crystalloids (Moss, 1972; Shires, 1970).

Determination of the plasma expander value for Haemaccel, Soldextrin, Dextran-70 and Dextran-40 by double isotopic labelling showed that the last-named is the most efficient and rapid osmocaptor [48]. Because of this property Dextran-40 may even become dangerous, especially during its elimination through the renal tubes, where it may dehydrate the tubulocytes (osmotic nephrosis) (*see* Section 10.3).

Crystalloids likewise have adverse effects, an excess accentuating, although not for certain, interstitial œdema in the wet lung [84, 103, 150]. Ringer's lactate solution with α-lytics and/or 5% albumin has a favourable effect on 'discharge' of the wet lung.

Moreover, α-lytics (according to some authors also β-lytics) always release the pulmonary microcirculation, diminishing the excess fluids in the lungs [63, 132] (*see* Section 10.4).

11.3.1 THE BLOOD

When the shocked patient has lost more than one third of his intravascular fluid, blood becomes the ideal substitute colloidal solution because its red blood cells transport oxygen with its antihypoxic properties.

Loss within a short interval of 30% of the circulating blood volume may bring about circulatory collapse and cardiac arrest; on the other hand, a loss of 50% of the circulating mass within a long interval does not endanger life because the circulating volume is replaced following endogenous mobilization of the interstitial and cellular fluids.

In a healthy adult, the loss of up to 1000 ml blood does not necessitate transfusion. The decrease in haemoglobin below 7 to 5 g% and of haematocrit below 30% reveals a loss greater than 1000 ml of the circulating blood volume and justifies, but is not always an absolute necessity, the administration of blood (Gruber, 1968; Rapin, 1970; Trinkle, 1970) [52, 119, 144].

When the ECBV is sufficient but haematocrit and haemoglobin are reduced, a massive administration of erythrocytes is more rational; it is in the case of haemorrhagic patients initially treated with plasma substitutes whose circulating volume is normalized but who have remained with a marked erythrocytic deficit.

In severe haemorrhagic shock, with alarming CVP and BP values, massive transfusion is compulsory (involving both the rate of administration and the volume administered).

The rapid administration of a large amount of blood (not previously heated to body temperature) in a central vein may produce generalized hypothermia; by vasoconstriction, peripheral blocking of the blood and a fall in the cardiac output, hypothermia may give rise to errors on estimating the fluid volume that must be administered. The first organ to cool down is the heart whose temperature falls to critical values (29-30°—the threshold of ventricular fibrillation). Hypothermia disturbs the bioelectrical mechanisms of the membrane equilibrium, facilitates the release of the potassium cation from the cell, and increases kaliaemia. Hypothermia and hyperkaliaemia together increase the probability of ventricular fibrillation. Previous heating of the blood eliminates these risks, so dangerous for the patient in a critical state of shock [4, 25, 53].

In some resuscitation units, the blood volume is completed, especially in haemorrhagic shock, by equal amounts of blood, 5% glucose and saline solutions. In blood losses not exceeding 1000 ml, albumin (5%) in Ringer's solution can be used with good results; it is now increasingly used in haemorrhagic shock of medium intensity [128, 147].

In cases of injuries accompanied by massive blood depletion it is considered that only 20% of the circulating volume lost must be replaced by blood; the remaining 80% may be substituted by 10% glucose solutions, electrolytes and Ringer's lactate + albumin (5%). Hypertonic glucose solution, with a temporary blood addition has, on the other hand, a constant diuretic and metabolic role.

Frozen blood. In 1953 Lovelock postulated that in the case of frozen blood the drawbacks of storage derive from the concentration of electrolytes and not from the ice crystals. This concept has now been adopted by cryobiologists who consider that haematic lysis, caused by an 80% saline solution can be prevented by freezing the blood [64, 65]. Blood banking by freezing (with glycerol) and storage at a temperature of —85°C is used today in many centres all over the world and has many advantages over the classical methods of conservation at +4°C. This method also appears to solve the question of viral hepatitis, and the storage time has been prolonged from three weeks to several years. It also makes possible the stocking of large amounts of blood (especially of the rarer groups) and permits more perfusions (Huggins, 1966) [64, 65].

The frozen and washed RBC are re-suspended in 250 ml normal saline, then re-concentraed by centrifugation before administration. In 20 000 cases of transfusions with frozen blood there was no case of hepatitis [58, 65].

Fluorocarbon compounds are neutral chemical substances in which hydrogen ions have been replaced by fluorine. For blood replacement fluorocarbon are used in the form of emulsions: FX_{80}, FC_{75}, FC_{43}, FC_{47} (perfluorotetrahydrofuran, perfluorotributylamine, perfluorodecaline, perfluoromethyldecaline). The emulsifiers used include bovine albumin, lipids and polyols as it is Pluronic F68.

The final preparation contains 15-20% fluorocarbon and 2.5-10% Pluronic, together with electrolytes, glucose, buffering systems, and agents improving blood flow, such as hydroxyethyl starch. 'Artificial blood' produced in this way has been used for experimental animals with good results.

Stroma-free haemoglobin solutions are colloidal solutions to transport and exchange oxygen. These solutions have an anticoagulant activity and may become in future an effective blood substitute.

Complications. Apart from viral hepatitis and hypothermia blood transfusions may also produce complications of another nature.

Citrate intoxication only develops following massive, rapid administration of blood, when the quantity of citrate exceeds the conjugation and elimination abilities of the body. The normal plasma citrate concentration is 1 mg%; if it increases to 50-60 mg%, following untimely administration or due to an organic failure to neutralize the citrate, a decrease occurs in the cardiac output, with hypotension, increase in the CVP, ECG signs of hypocalcaemia and dyspnoea caused by increase in the pulmonary artery pressure. It should be recalled that β-blocking agents sensitize the organism to citrate. The administration of calcium extends these phenomena and, moreover, improves cardiac performance.

Blood group or subgroup incompatibility is seldom encountered today because of the preventive measures taken. The symptoms are similar to those of the acute defibrination syndrome. The laboratory tests reveal: reduction up to disappearance of fibrinogen, decrease of factors I, V and VIII and of prothrombin. As blood transfusion is equal to 'fluid tissue transplant' various immunologic aspects have been revealed and new problems have appeared linked to repeated and massive transfusion of the 'dangerous universal donor', of group A genotypes, etc. The number of compulsory compatibility tests has increased as the only means of reducing the risk to a minimum.

Hyperpotassaemia. At the end of the conservation limit (21 days) kaliaemia is 15-20 mEq/litre. Therefore, by using cold and long-stored blood in large amounts kaliaemia may increase to alarming levels, detectable by ECG monitoring. Hyperkaliaemia in shock may be controlled by administering Ca^{++} and glucose solutions.

Metabolic acidosis. After three weeks storage the *p*H of the blood is 6.4-6.5. Massive administration within a short interval adds to the specific acidosis of shock the acidosis of transfused blood. It has been proposed to neutralize the acidity of bank blood by adding 15 mEq $NaHCO_3$ to each flask. Today, however, this is considered contra-indicated since metabolic alkalosis may develop following the metabolism of citrate and liberation of Na^+ [17, 52, 53, 140].

Post-transfusional pulmonary micro-embolisms may be due to microagglutinations (found especially in blood conserved for a long time) which are initially stopped by the pulmonary filter; pulmonary microembolisms accentuate respiratory failure, whose gravity may at times be correlated with the amount of blood transfused (*see* Section 10.4).

Fine screen filtration of massive blood transfusions is beneficial and the side effects negligible. Prevention of posttraumatic pulmonary insufficiency by this means in critically injured patients is impressive (Reul, 1973) [120].

Blood is administered in terms of three parameters: length of storage, amount required, and the patient's characteristics. In the aged with cardiopulmonary failure, blood older than five days should not be administered; blood stored for ten days may be used for youths and in smaller amounts [141]. After ten days 2,3-diphosphoglycerate decreases in the erythrocytes to 5% of the normal quantity (*see* Section 7.2.1). Therefore, the transfusion of blood stored for more than ten days may bring about in the recipient a shift to the left of the oxyhaemoglobin dissociation curve [141].

11.3.2 PLASMA SUBSTITUTES

As plasma substitutes have not the property of transporting oxygen they cannot be considered as integral blood substitutes and can only be used to complete the circulating blood volume [51, 52, 53, 88]. To this end, artificial colloids must have the following properties:

- a sufficiently high molecular weight (between 20 000 and 120 000);
- osmotic pressure and viscosity equal to that of the plasma;
- must not be eliminated from the organism before producing its therapeutical effect;
- must not be toxic, pyrogenic, antigenic and must not produce allergic effects;
- must be stable so that they can be stored for long intervals at different temperatures (Grönwall, 1957; Litwin, 1970) [51, 86].

To these properties the following requirements may be added:

- plasma expander effect for at least 6 to 12 hours, regardless of whether the molecular weight of the colloids is 40 000 or 70 000;
- viscosity low enough not to stress the heart, and that can be administered intravenously;
- not to interfere with haemostasis or coagulation, not to produce agglutination or haematic lysis, and not to modify the blood groups;
- their metabolism and elimination should not alter the function of various organs after repeated administration.

Included in the group of plasma expanders are: dextrans, gelatin derivatives, polyvinylpyrolidon derivatives and starch derivatives.

Dextrans are obtained from sucrose media following the action of dextransucrase, an enzyme belonging to the leukonostock bacteria; sucrose is converted into dextran by an irreversible reaction [1, 51]. In terms of the molecular weight two forms are produced: one with a molecular weight of 70 000 (Dextraven, Intradex, Macrodex, Dextran-70) and another with a weight of 40 000 (Rheomacrodex, Dextran-40). Dextran is supplied either in a saline solution or in a 5% glucose solution. Both colloids have a *p*H of between 4.4 and 5.7 and a buffering capacity of less than 2 mEq [51, 53].

Dextran solutions are the most effective plasma expanders known as yet. They are also the best studied class of colloids.

The Dextran-70 6% solution exercises a greater osmotic pressure than blood (2.5% Dextran-70 and 3.5%

Dextran-40 are approximately isotonic with the blood). One gram of Dextran-70 binds with 20-25 ml water, producing haemodilution but with a normal plasma osmotic pressure. The osmotic pressure exercised by a 10% Dextran-40 solution corresponds *in vitro* to that of a 17% albumin solution. Most of Dextran-70 is eliminated through the urine: 30% in the first 6 hours and 40% within 24 hours; Dextran-40 60% within the first 6 hours and 10% in 24 hours; the rest is temporarily deposited in the liver, kidneys and spleen after which it is split into H_2O and CO_2 at a rate of 70 mg/kg body weight/24 hours [16, 44, 50, 81, 97, 152].

In the microcirculation, Dextrans, especially Dextran-40, improve the capillary flow, preventing haematic agglutination and the sludge phenomenon. Dextrans have antithrombotic properties, following adsorption on the thrombocyte surface and thus reducing their clumping. The antithrombotic effect of Dextran-70 appears to be still more efficient [86].

Some authors have recently assumed Dextran-40 to be the cause of oligouria [92]; however, this effect seems to be caused indirectly by the associated factors with an anurigenic potential (anaesthesia, shock, vasopressors) that produce acute renal failure [36, 37]. As a matter of fact, in 1961, when Dextran-40 was introduced in the clinic more than 3 million litres were used in the whole world yet there were only 34 cases of oligoanuria published and in only 4 of these cases can Dextran-40 be considered directly responsible [97].

On administering Dextran-40 the following should be taken into account:

- being a hypertonic solution it should be administered parallel with correction of the ionic and hydric balance; the prompt administration of major natriuretics (etacrinic acid and furosemide) is indicated whenever the hourly urinary output demands it;
- acute renal failure is a formal contra-indication;
- severe haemorrhage (when the losses exceed 20-30% of the circulating volume) cannot be exclusively compensated for by Dextran-40. In shock, the perfusion with Dextrans should not exceed 20 ml/kg body weight or 1 500 ml in 24 hours [88].

In shock, Dextrans produce the following haemodynamic effects: increase in the cardiac output, in the venous return, CVP and BP; they lower peripheral resistance and improve the flow in the microcirculation.

Gelatin derivatives. This group includes oxypolygelatin (Gelifundol) discovered by Campbell in 1951, and modified fluid gelatin (Plasmagel, Gelofusin, Physiogel, Haemaccel).

Gelatin solutions were the earliest blood substitutes, the first trial infusions being reported by Hogan in 1915.

The molecular weight of gelatins varies between 10 000 and 100 000 (60% of the gelatins have a molecular weight of 20-60 000). Gelatins have numerous qualities that justify their use as plasma blood substitutes:

- no side effects, not even after the infusion of several litres;
- they do not affect haemostasis and have no antigenic properties;
- they are pharmacologically inert;
- they are more rapidly eliminated by renal route than dextrans owing to their smaller molecular weight (overloading is not a real problem).

In view of the fact that they are very well tolerated by the human body even after the administration of several litres, gelatins are now increasingly being used in the clinic [55, 88].

Gelifundol (Biotest) is prepared in normal saline at a colloidosmotic pressure of 39 mmHg. Plasmagel is delivered in 3% concentration and contains 120 mEq/Na^+/l, 27 mEq Ca^{++}/l and 147 mEq Cl^-/l; the colloidosmotic pressure is of 28.5 mmHg.

Physiogel and Plasmagel (Roger-Bellon and Braun) with a concentration of 4.2% contain 120 mEq/Na^+/l, 24 mEq/Ca^{++}/l, 125 mEq Cl^-/l and a colloidosmotic pressure of 27 mmHg.

Haemaccel (Hoechst) with a concentration of 3.5% includes 145 mEq Na^+/l, 5.1 mEq/K^+/l, 12.5 mEq Ca^{++}/l and 162.6 mEq Cl^-/l, with a colloidosmotic pressure of 350-390 mm H_2O (about 27—29 mmHg) and *p*H 7.3.

Gelatins produce a more efficient plasma expander effect in severe shock than in hypovolaemias. Some authors have noted a slight antigenicity of Oxypolygelatin, Plasmagel and Haemaccel [53]. *In vitro*, gelatins increase the viscosity of the blood slightly. Blood pseudoagglutination may occur but a few drops of glycine may prevent any difficulties arising in the determination of the blood group. Haemaccel does not alter the prothrombin and antithrombin time, and may be kept in stock for three years (Physiogel for four years and Gelifundol 150 day) [55].

Polyvinylpyrrolidon (PVP). As PVP is not broken down in the human body its use is very restricted. It cannot be perfused in amounts greater than 1000 ml; at a concentration of 4% it has a colloidosomotic pressure of 670 mm H_2O and *p*H 6. The fraction with a molecular weight of 25 000 is eliminated by urinary route. The volumetric effect depends upon the size of the molecules. Molecules with a weight of 25 000 leave the circulation immediately and have no plasma expander effect; 18% of the PVP molecules administered are to be found in the urine before the perfusion flask is empty. At three hours, 48% of the amount administered has already been eliminated [88].

Used as a 4% solution in normal saline PVP was not superior to dextran and gelatin preparations.

On the other hand, the fractions with a molecular weight greater than 100 000 cannot be eliminated; they are stocked in the RES, the tubular epithelium and other parts.

Starch derivatives, colloid substitutes with great future expectations are now being tested. The only product at present is Hydroxyethyl Starch (HES), rarely used because it is too quickly broken down by serum amylase. Its use as plasma expander in experimental medicine has given approximately the same effects as Dextran-70 [53].

HES was introduced by Wiedersheim in 1957 and prepared by Thompson in 1962 for clinical use. The haemodynamic effects of HES resemble those of Dextran-70 preparations but HES is easier to prepare and less expensive. HES remains for a long time in the organism, and its metabolism has not yet been completely elucidated; this is at present the principal cause limiting its clinical application.

11.3.3 CRYSTALLOID SOLUTIONS

Crystalloid solutions contain substances specific of the organism; they are not toxic and do not cause allergic or immunologic reactions. Although they have the drawback of passing rapidly into the interstitium, in severe injuries with shock the infusion of polyelectrolytic solutions did not precipitate cerebral or pulmonary œdema or cardiac affection as long as the renal function was satisfactory [17, 25, 128].

Crystalloids produce rapid electrolytic equilibration of the interstitium and intravascular fluid without causing a dangerous volumetric loading.

In normal blood donors one litre of normal saline did not increase the circulating blood volume two hours after the end of administration. In hypovolaemic subjects the administration of 500 ml Ringer-lactate solution brought about a temporary increase in the blood volume [128].

As only a quarter of the perfused solution remains in the circulation, normal saline solutions administered to hypovolaemic subjects must be calculated to be three times greater than the losses [17, 21, 25, 52, 147].

Although hypoalbuminaemia has been reported after crystalloid it is temporary and not dangerous, albumin returning to the blood by lymphatic route at an efficient rate of 4 g/hour. Moore likewise stresses that only 3.5-fold the amount of lost blood is able to maintain a normal circulating blood volume. In the absence of colloids the same amount of crystalloids should be administered after 6 hours (*see* Section 6.1.2).

Isotonic solutions. Although some authors have not observed any effect after the perfusion of glucose or polyelectrolytic solutions in the advanced stages of shock, an increasing number assert that they have reduced the mortality rate from 804 to 20% by the massive administration of crystalloids, especially Ringer's lactate solution [21, 35, 37, 48, 52, 53, 60, 83, 86, 122].

Using a potentially reversible experimental shock model, Shires obtained a 20% survival rate after blood transfusion, 30% after blood + plasma, and 70% after blood + large amounts of Ringer's lactate [128]. Hartmann recommends a Ringer solution containing 132 mEq Na^+/l, 4 mEq K^+/l, 4 mEq Ca^{++}/l and 110 mEq Cl^-/l.

It cannot be asserted that therapy with crystalloids is the only one indicated or that it completely replaces blood or plasma substitutes. The most reasonable approach is to combine colloid and crystalloid solutions, well balanced and adapted to the type of shock. Far better results are obtained with crystalloid solutions in haemorrhagic and traumatic shock than in bacteriemic shock [41, 43, 102, 103, 126, 128, 147].

At present, balanced electrolytic solutions are used in amounts two to three times greater than the blood loss, particularly useful in moderate haemorrhage or in severe haemorrhage where the association of colloids is obligatory. Ringer's lactate is used at *p*H 8.5 or Rush's buffered solution containing per 1000 ml: 148 mEq Na^+, 4 mEq K^+, 3 mEq Ca^{++}, 28 mEq lactate and 18 mEq bicarbonate, *p*H 7.4, 300 milliosmols, administered in amounts of 120 ml/kg body weight [122]. Recently a mixture of Ringer's lactate and 5% human albumin was used with promising results, especially in haemorrhagic shock.

One of the most useful practical applications of the modern concept of the pathophysiology of shock probably is the administration of crystalloids in the treatment of hydroelectrolytic imbalance. Today, the great drawback of crystalloids, i.e. that they are not retained in the vascular tree, is considered a 'great quality' since it involves electrolytic equilibration of the intestine and therefore of the cell. As in the first instance the vascular content does not count for resuscitation of the shock cell but for the composition of the interstitial space, electrolytic solutions alone and not colloids may be useful because they manage to penetrate within the intimacy of the tissues, passing the vascular wall frontier. At present it is a question of the general filling of the extracellular space and not of the intravascular one; therefore the administration of crystalloids gives excellent results when administered from the beginning and, of necessity, in large amounts (Moss, 1972).

Crystalloids must also always be combined with colloids as they spoliate the water and electrolytes of the interstitial space, aggravating cellular distress. Hence, the administration of blood or plasma has to be 'buffered' by sufficient amounts of crystalloids.

When dextrans, which violently 'squeeze' the interstitium owing to the superior osmotic pressure of the plasma, are used crystalloids in amounts twice those of the blood or plasma have to be added [96].

Ringer's lactate solution seems to bring about a good re-equilibration of the extracellular space; it does not increase lactic acidosis (and even renders Cori's cycle permeable again in the liver) and does not favour hydration of the lungs, especially when associated with 5% albumin. Shires and Moss recommend the initial administration of large quantities in order to avoid the use of blood even in cases of severe shock and, obviously, the untoward post-transfusional reactions. Ringer's lactate significantly improves colloidal effects when combined with blood, plasma or dextrans [7, 42, 104].

Hypertonic solutions. Brooks compared the effect of two hypertonic solutions (1.8% sodium chloride and 2.74% sodium bicarbonate) with normal saline and 10% glucose. The shocked animals treated with the latter two isotonic solutions, and the control animals, died, whereas those to which were given an osmolarity twice that of the blood survived. Brooks interpreted the results as a proof that the hypertonic solutions augment the extracellular space and diminish the acid-base imbalance [53]. Other authors, however, consider that the hypertonic solutions do not give appreciably better results (35). At any rate, irritating effects and the danger of osmotic nephrosis constantly accompany the administration of hypertonic solutions. The use of normal solutions is considered advisable in practice for correcting the electrolytic imbalance according to the ionogram (Kragelund, 1971).

11.4 RHEODYNAMIC MEDICATION

Rheodynamic medication implies a complex treatment addressed to the receptors of the microcirculation and cardiac pump (fluid and blood import and export systems).

11.4.1 MEDICATION OF THE α- AND β-RECEPTORS

The term 'vasopressor' is often misunderstood, since it is believed to be a drug that increases the blood pressure; yet the blood pressure can be elevated either by arterial vasoconstriction, by increase of cardiac ionotropism, or by increase of the blood flow.

The complex action of drugs with haemodynamic effects have a triple effect:

- increase myocardial rhythm and contractions (*see* Section 11.2.1);
- constrict the arterioles and arteries (the resistent blood vessels);
- constrict the venules and veins (the capacity blood vessels).

Experimental haemodynamic analysis of a wide range of vasoactive drugs has separated them into large groups according to their predominant pharmacodynamic action (Figure 11.2).

The first group of α-mimetics, includes vasopressor drugs with a predominant vasoconstriction effect upon the resistant vessels. Drugs of this group raise the blood pressure by increasing peripheral resistance; increase in the peripheral pressure implicitly results in a corresponding increase in the work of the left ventricle. Moreover, α-mimetics lower the cardiac output and reduce the systemic blood flow.

The second group includes β-mimetic drugs with an inotropic cardiac action and positive chronotropic effect. Their administration increases myocardial efficiency, venous return and the cardiac output by lysis of the precapillary spasm in the microcirculation and by vasodilatation.

Among the vasopressor agents used in the clinic, norepinephrine and aramine exercise a twofold effect as they are both α- and β-mimetics. When used in small amounts the primary action is β-mimetic, and in large doses the α-mimetic effect is predominant. A new vasoconstrictor similar to epinephrine has been recently reported: Ornithinvasopressin [90].

However, the use of vasopressors for their effect upon cardiac performance by a transitory positive inotropic action and improvement of the cardiac output

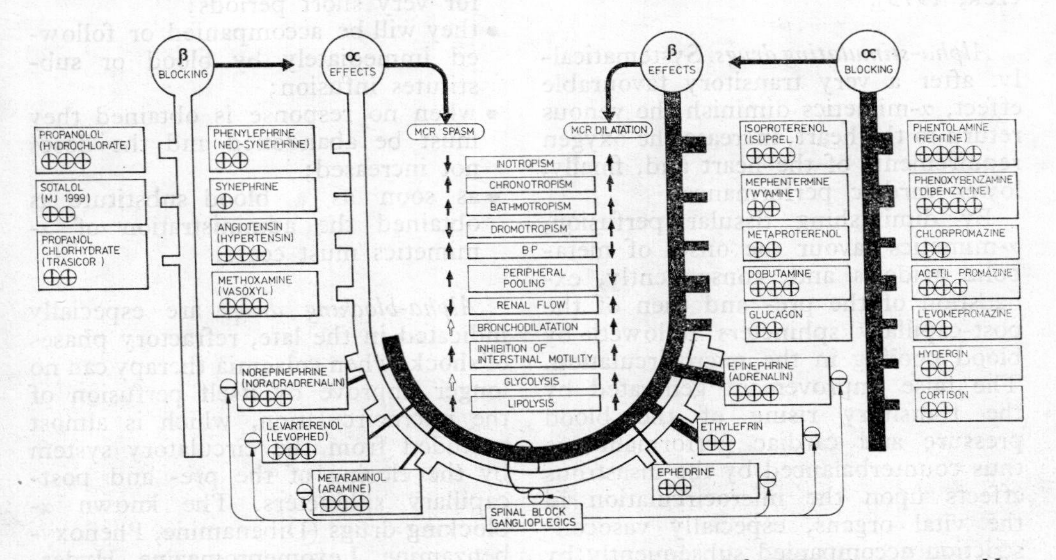

Figure 11.2. The effects of drugs stimulating and blocking α- and β-adrenergic receptors. Note that the effects of β-blocking agents are, in general, of the α type, and of α-blocking agents of the β type. In the lower part are several substances that have a mixed α and β effect, which may be globally blocked by ganglioplegics.

are dearly paid for by diminution of the peripheral blood output, especially by a decrease of the microcirculation space [4, 8, 17, 25, 85, 131].

The peripheral haemodynamic action of β-mimetic drugs is similar to that of α-adrenergic blocking drugs (Phenoxybenzamin, Regitin) (Fitts, 1970; Skinner, 1967).

The reaction of the microcirculation to pharmacodynamic stimulation occurs as follows:

The arterioles react by constriction to stimulation of the α-adrenergic receptors and by vasodilatation to stimulation of the β-receptors. The meta-arterioles, precapillary sphincters and capillary pericytes react in the same way to α- and β-stimulation. As the postcapillary sphincters do not possess α-receptors they do not react to β-stimuli. Arteriovenous anastomoses appear to be both α- and β-sensitive (Hardaway, 1968; Litarczek, 1973).

Alpha-stimulating drugs. Systematically, after a very transitory favourable effect, α-mimetics diminish the venous return to the heart, increase the oxygen requirements of the heart and, finally, lower cardiac performance.

By diminishing tissular perfusion, α-mimetics favour the onset of metabolic acidosis and, consequently, exhaustion of the pre- and then of the post-capillary sphincters followed by blood pooling in the microcirculation. The false improvement generated by the transitory rising of the blood pressure and cardiac performance is thus counterbalanced by the disastrous effects upon the microcirculation in the vital organs, especially vasoconstriction accompanied subsequently by oligoanuria.

Bearing these effects in mind the use of α-mimetics has increasingly been restricted and even excluded by some practitioners in the treatment of shock [135, 155, 157].

Certain vasogenic circulatory failures (hypotension after spinal anaesthesia, after an overdosage of inhalatory drugs or after anaphylactic collapse) may be controlled by the administration of α-mimetics, but only by as much as is necessary to increase the blood pressure to almost normal values (100-110 mmHg); α-stimulation may also be used at times for its antiphlogistic effect in bacteriemic shock (Wosornu, 1970).

For very short periods — and only in the absence of perfusion solutions — α-mimetics may be administered to injured subjects in a state of collapse who have to be rapidly evacuated.

When used in exceptional situations, the following rules must be complied with:

- they should be administered only for very short periods;
- they will be accompanied or followed immediately by blood or substitutes infusion;
- when no response is obtained they must be abandoned and the doses not increased;
- as soon as a blood substitute is obtained the administration of α-mimetics must cease.

Alpha-blocking drugs are especially indicated in the late, refractory phases of shock, when volaemia therapy can no longer improve by itself perfusion of the microcirculation, which is almost excluded from the circulatory system by the closing of the pre- and post-capillary sphincters. The known α-blocking drugs (Dibenamine, Phenoxybenzamine, Levomepromazine, Hydergin) open up the dams closing the microcirculation and improve the tissular circulation, facilitating the exchange of nutrients and cellular metabolism. Alpha-blocking drugs can be administered only with large amounts of fluids that fill the vessels and improve the flow in the microcirculation.

The fall in blood pressure that occurs after α-lytics has to be rapidly corrected by increase in the amount of macromolecules until a normal pressure is re-established. Together with colloids, α-blocking agents improve the flow in the microcirculation by abolishing vasoconstriction, increasing venous return and the cardiac output, and indirectly improving the work of the heart.

Apart from long-acting α-blocking agents (Dibenamine, Phenoxybenzamine), short-term α-blocking agents such as Hydergin and Levomepromazin (in 0.3 mg doses and respectively 2.5 mg, repeated until the expected effect is obtained) have given spectacular clinical results, comparable to the action of Dibenzilin (10 mg increasing up to 70-100 mg doses) [87, 104, 157].

In practice, when shocked patients do not respond to the substitution of fluids they may be treated with fractional, diluted doses of Levomepromazin (2-5 mg) administered by the i.v. route; an immediate effect is obtained: increase of the vascular bed and improvement of tissular perfusion. When administration of the α-blocking agent is not combined with concomitant correction of ECBV, an additional fall in the blood pressure will occur. The volume of colloidal solutions that have to be administered together with α-lytics sometimes exceeds 25 ml/kg body weight [112]. *To this a two- or threefold volume of crystalloids must be added.*

The concomitant administration of α-blocking drugs and macromolecular circulating mass appreciably improves the patient's general condition: his extremities become warmer, pallor and cyanosis disappear, as well as tachycardia and tachypnoea, and diuresis increases improving glomerular filtration. If shock is not too advanced and has not damaged the histofunctional integrity of the cell, the organelles take up their metabolic functions again consequent to the improved tissular perfusion and removal of the accumulated acid detritus.

Beta-stimulating drugs are widely used in shock for both their cardiac and their peripheral effects. The principal pharmacodynamic stimulators of β-adrenergic receptors are: isopropylnorepinephrine (Isuprel, Aleudrin) and orciprenaline (Alupent). On the heart, β-stimulants exercise a positive chronotropic, inotropic, bathmotropic and dromotropic effect; in the periphery they produce vasodilatation of the microcirculation, effectively improving the tissular flow and intracellular enzymatic activity.

Intravenous administration of Isuprel lowers peripheral vascular resistance and increases the heart output by 50% [8, 46, 157]. At the same time an increase takes place in the circulating fluid following modification of the blood volume stocked in the viscera or at the periphery. After Isuprel the cardiac rate increases up to 120-130 beats/min. When combined with polyelectrolytic crystalloid solutions, Isuprel also improves renal flow and filtration. In cardiogenic shock the therapeutic effect of Isuprel is conspicuous: an increase in blood pressure, decrease in the CVP, and elevation of the urinary output. Spectacular results are also obtained with Isuprel in total cardiovascular block with Adams-Stokes syndrome or in severe bradycardia. It does not induce cardiac necrosis, as asserted by some authors [5, 137]. As a rule, the following doses are administered by i.v. route: 0.2 mg Isuprel or 0.5-1 mg Alupent, doses that may be repeated. The former can be administered only when the cardiac rate is under 120 beats/min; it also to a certain extent reduces the effects of bacteriemic shock; its efficiency is due to the blocking of vasoconstriction induced by the endotoxin. Bearing in mind the improved work of the heart the evident clinical improvement obtained with Isuprel can readily be understood.

As Isuprel brings about an increase in the vascular bed — the same as α-blocking agents — a corresponding amount of fluids has to be administered. In contrast to α-blocking agents Isuprel augments the venous return and the cardiac tonus and rhythm, with a favourable effect in shock.

The effect of this drug upon the blood pressure depends upon the doses used. Small doses have a minimal systolic effect but substantially lower diastolic pressure. In hypotension with the decrease of cardiac output it brings about an increase in both systolic and diastolic pressures. After the administration of Isuprel the urinary output increases, and heart failure, consequent to the action of toxins, improves. To conclude, Isuprel increases systolic pressure, lowers diastolic pressure, reduces peripheral resistance, raises the cardiac rate and urinary output, but does not give results in acute pulmonary œdema and acute renal failure [157].

It may be favourably combined with a catecholamine precursor (Dopamine) in haemorrhagic and bacteriemic shock. By its positive inotropic effect synergic with that of Dopamine, Isuprel produces splanchnic vasodilatation and Dopamine elective vasodilatation of the renal microcirculation [6].

Beta-blocking drugs. Berck's pathogenic theory on excessive β-excitation in shock (with a shunting effect on pulmonary haematosis) promoted the use of β-blocking agents, especially in bacteriemic shock. Among the known β-blocking agents today are: dichlorisoproterenol, pronetalol, oxyprenol hydrochlorate (Trasicor and propanolol-Inderal), the last-named being very much used in the clinic.

Beta-blocking substances have a high specificity; they prevent the positive inotropic and chronotropic effects of epinephrine and the positive chronotropic effect of Isuprel. Their action upon the heart is due to specific sympatholytic effects [6, 27, 66, 124]. Inderal administered by i.v. route produces bradycardia and lowers the oxygen consumption of the myocardium. The usual doses are 1 to 2 mg, repeated if necessary under control of the pulse and BP. Some authors recommend 5 mg Inderal every 6 to 12 hours [44]. Inderal is indicated in the cardiogenic shock of myocardial infarct and the advanced stages of haemorrhagic shock, combined with atropin, glucose, digitalis and alkalinisers [28].

11.4.2 CARDIOTONICS AND DIURETICS

Cardiotonics. When in the course of shock the heart gives signs of congestive failure (with a CVP within normal limits) the i.v. administration of a digitalis product is recommended. Diminution in cardiac performance may often be the result of hypovolaemia, decrease in the coronary output, myocardial hypoxia and metabolic acidosis.

In such instances, blood volume replacement, oxygenotherapy and the administration of sodium bicarbonate must precede the digitalis treatment. Therefore, correction of hypovolaemia will lower the CVP following improvement of the cardiac performance. After initially administering 0.4 mg Lanatosid C, the drug is re-injected at 4 to 6 hour intervals; the dose is subsequently reduced to 0.2-0.4 mg. In severe myocardial failure the initial dose may be increased to 0.6 or 0.8 mg; the effects — decrease of the CVP, increase in the BP, and signs of an improved microcirculation — should appear immediately. Elective coronary dilators may be useful.

Digitalization in prolonged or severe endotoxemia might well be beneficial to the heart (Geocaris, 1973) [47].

Not only the change in *p*H alone, but also the change in osmolality and almost certainly the associated volume increase are in fact responsible for the improvement in cardiovascular function (Rowe, 1973) [121].

Diuretics. Diuretics are used in shock whenever there is an imminent acute renal failure or blood volume overloading, with or without acute pulmonary œdema. A 10% mannitol solution may be rapidly administered and in comparatively large amounts (500 ml in 30 minutes). This not only produces osmotic diuresis and lowers the viscosity of the intratubular fluid, the reason for which it is combined with Dextran-40, but it is also a plasma expander, lowering haematocrit but not natraemia [117]. For the control of the *antidiuretic state* (aldosterone and ADH secretion) that accompanies shock, it is recommended Dextran-40 combined with mannitol (the two solutions, however, potentiate the negative effect of osmotic nephrosis induction (*see* Section 10.3). Subsequently, in order to counteract hyperaldosteronism, anti-aldosteronic drugs are used. Diuretics with a rapid action (frusemide, etacrinic acid) have proved in clinical practice to be superior to mannitol [18, 138]. Moreover, mention should be made of the saline diuretics that adjust the electrolytic gradient of arteriolar walls to normal values, especially Na^+, lowering the reactivity of the intrinsic tonus of the microcirculation.

11.4.3 COAGULOLYTIC EQUILIBRIUM MEDICATION
(*see also* Section 12.1)

In shock, the DIC episode is always present and, aggravating (*see* Section 7.4.2). Maintenance of the normal fluidity of the blood is a means of preventing and of treating DIC, and its effects: consumption coagulopathy and fibrinolysis, in its broad proteolytic meaning. For the treatment of coagulolytic equilibrium disturbances there are no certain tests, only orientative ones. The thrombo-elastogram offers data only on the quantitative aspects of the coagulolytic process.

Of practical use are the coagulogram, thrombocytaemia and determination of fibrinogen which, when repeated, may detect the onset of DIC and fibrinolysis. In addition, one should always calculate the dilution effect of colloids and even of crystalloids. The classical dilemma has not yet been solved: should ε-ACA be administered with heparin, in order to block the stimulating action of the latter on fibrinolysis, or avoid the risk of hastening the onset of DIC by an antifibrinolytic? At any rate, the shift from hyper- to hypocoagulability, observed by repeated determination of the coagulation time, is a certain indication of a DIC episode.

The treatment of DIC is first of all pathogenic: the microcirculation flow is improved by administering fluids and α-blocking agents. Rheomacrodex, favouring the supply of negative electric charges also acts by its anti-sludge effect.

Heparin administered in large doses in the early stages of experimental shock reduces the mortality rate; in the late stages, when DIC has set in, heparin no longer influences the course of shock [59]. The use of heparin in active doses should therefore be instituted early. The risks of haemorrhage from a heparin overdose are too great; if bleeding is caused by heparin then an efficient antidote is protamine sulphate.

Heparin acts by the following mechanisms:

- antithrombinic — inhibiting the action of thrombin in transformation of fibrinogen in fibrin;
- antiprothrombinic — preventing the transformation of prothrombin into thrombin (and as it arrests both actions, heparin also has an antithromboplastin effect);
- activates the fibrinolytic system.

When inactivated by acidosis heparin cannot prevent agglutination and disintegration of the platelets, therefore it cannot stop the connecting link with the onset of DIC, especially in bacteriemic shock (see Section 12.3).

Administered by i.v. route in doses of 1 mg/kg body weight heparin becomes active in the plasma within 15 to 30 minutes and maintains its efficiency for 4 hours, after which the dose is repeated (6 mg/kg body weight/24 h). The first injection is 75 mg because part of the initial amount fixes on proteins and platelets. The effect of heparin may be checked by a twofold increase in the Howel or Lee-White time, as well as the thrombin or prothrombin time. After the first i.v. injection perfusion is started with 100 to 150 mg heparin per 500 ml glucose solution (10 drops/min). At 15 to 30 minutes 5 g fibrinogen is administered and then fresh blood or plasma if the prolonged coagulation time suggests that DIC has occurred. As at *p*H 7.2 DIC develops even in the presence of heparin [59] acidosis must be controlled in order to prevent DIC by improving the blood flow and by alkalinization.

Due to calcium consumption in the course of DIC it is advisable to prescribe small, repeated calcium doses only, especially when there is an increase in the BP and CVP.

Accidents with anticoagulants have become increasingly frequent (1-10%). They may be non-haemorrhagic (gastric intolerance, eruptions, alopecia, arthralgia, exfoliative dermatitis, etc.) or haemorrhagic, which in turn, may be slight (ecchymoses, nose bleeding, gingivitis, haematuria) or major (haematemesis and melena, serous intraparietal haemorrhage in the gastrointestinal tract, brain, retroperitoneum, adrenals or hypophysis) [16, 89]. A new means for preventing overdosage of K anti-vitamins is the determination of prothrombin since that of the Quick time has many drawbacks [113].

Fibrinolytics are not easily handled in shock as not enough clinical experience has been gained. Sporadic use of heparin with streptokinase or urokinase has been reported.

The treatment of fibrinolysis consists in fresh blood transfusions or with fibrinogen (Cohn fraction I) in doses of 3 to 6 g (even up to 8 to 20 g) associated with cortisone haemisuccinate (300 up to 1000 mg) and plasmin inhibitors, obtained from the pancreas or parotid, or synthetic inhibitors.

Parotid and pancreatic inhibitors neutralise plasmin molecules one by one. There are two products with a major antiplasmin effect:

- the inhibitor of Kunitz-Northrop, extracted from the pancreas (Iniprol)
- the inhibitor of Frey-Werle, extracted from the parotid gland (Zymofren, Trasylol).

Iniprol is administered in 5 to 10 million units peptidase inhibitors and Zymofren in 500 000 to 750 000 units kallikrein inhibitors, if fibrinolysis is confirmed by laboratory tests. These inhibitors have a broad action, inactivating trypsin, kallikrein, chymotrypsin and plasmin. Having a very short-term effect they are administered by continuous perfusion [4, 59, 103].

Synthetic inhibitors. Epsilon-aminocaproic acid (ε-ACA) inhibits the transformation of plasminogen into plasmin and opposes the activation of plasminogen by activators of tissular (urokinase) or bacterial origin (streptokinase). This acid may be used preventively in pulmonary, prostatic and genital surgery. In supra-acute fibrinolysis i.v. doses of 8 to 12 g/24 hours (0.20-0.50 g/kg body weight), and in the subclinical forms 4 to 8 g [59].

11.5 CELLULAR RE-EQUILIBRATION

Cellular re-equilibration starts from a knowledge of the metabolic oscillations of the enzymatic lines in the shock cell (*see* Chapter 8 and Section 9.3). Biophysical interpretation of the dynamics of the shock cell suggested the existence of therapeutical inflexions which imply a radical change or even inversion of the pharmacotherapeutical attitude (*see* Section 9.6). In the previous chapters frequent reference has been made to drugs with an intracellular action.

11.5.1 METABOLIC MEDICATION

Metabolic medication tends to re-establish the balance of the enzymatic chains. The French School, the initiator of this *intracellular therapy*, communicated the first clinical results [75, 76, 77, 78]. Administration of dihydroxyacetone, to resuscitate glycolysis, acetazolamide, to build up again the quantity of tissular carbonic gas, and alkalinisers, combined with reducing substances and antibiotics sterilising the intestinal flora resulted in survival of 50 to 60% of otherwise fatal cases. But this treatment is addressed to the late stages of shock and is justified by an actual inversion of the situation: glycolysis at first endogenously exaggerated has finally to be sustained therapeutically.

Therapeutical paradoxes actually involve the chemically alert messengers of shock: the catecholamines. The fact that the initial high liberation of catecholamines helps the human body to pass over the immediate critical moments of shock cannot be questioned. But this response is exaggerated both in the microcirculation (by α-stimulation) and along metabolic lines (by α-stimulation). In the stage of early reversible shock hypercatecholaminaemia is moderate due to its interference with ß-lytic and even with β-lytic effects. Large cortisone doses likewise help in this 'attenuation'. The course of shock is rapidly modified by events, however, and the initial therapy is no longer helpful. Hyperglycolysis stagnates and the microcirculation is dammed by clots. In the late reversible and refractory stages of shock it is no longer necessary to buffer glycolysis or to overcome only the precapillary spasm behind which the clots persist; it now becomes necessary to reactivate glycolysis, to stimulate the Krebs cycle and lysis of the clots.

Dihydroxyacetone may help glycolysis, and increase in the concentration of carbon gas may stimulate the Krebs cycle. Control of acidosis and fibrinolytics may refluidify the circulating medium and cortisone, of particular utility in this phase, and stabilises especially the membrane functions upon whose quality depends the entire cellular laboratory.

Moreover, in the late stages of shock the great paradox interferes: catecholamines themselves become necessary, but not in excess and not for their vascular effects, but especially for their energising metabolic efficiency, as the source of endogenous entropy tends to fall out in cellular metabolism.

In this stage, the following therapeutical measures must be taken simultaneously: the administration of catecholamine synthesis precursors (Dopa, Dopamine and especially L-thyrosine), monoaminoxidase inhibitors, in order to stop catecholamine catabolism, and ganglioplegics or neuroplegics to interrupt the adrenergic effect upon the smooth muscular fibre (*see* Section 8.3.6).

Biochemical correction of the *p*H is a secondary treatment; one might call it symptomatic. Repermeabilisation of the microcirculation is a true pathogenic treatment.

An efficient buffer is THAM (0.3 *M* solution); 36 mg in 1000 ml distilled water gives a *p*H of 8.5. It is administered at 120-180 minutes, and monitoring of the arterial and venous *p*H, P_aCO_2, glycaemia and electrolytaemia is obligatory. THAM is 30% ionizable, thus controlling intracellular acidosis. It produces hypoglycaemia, hypopotassaemia and respiratory depression and is an osmotic diuretic. If its administration cannot be closely followed up sodium bicarbonate should be used instead.

Byrne's formula for the administration of THAM is:

THAM mEq = 0.3 × body weight in kg × mEq bicarbonate deficit.

The sodium bicarbonate solution is isotonic (1.4%) or hypertonic (8.4%). The latter is normal (1 ml 8.4% $NaHCO_3$ = 1 mEq sodium or bicarbonate) and may be administered under control according to the ionogram formula: mEq bicarbonate deficit × extracellular hydric space of the body in litres = ml 8.4% $NaHCO_3$ solution to be administered.

11.5.2 CORTICOSTEROID THERAPY

In small and medium doses (100-500 mg daily) cortisone exercises its effect by activation of carbohydrate metabolism (gluconeogenesis and the favouring of phosphorylation) which allows the vascular smooth musculature to sustain its contractions. These doses may reactivate the arterioles and precapillary sphincters to catecholamines.

In large doses (3 to 5 g daily) cortisone has an α-blocking effect, producing lysis of the microcirculation spasm (Lillehei, 1964). Due to its α-blocking effect cortisone is very much used in the severe states of shock that evolve with vasoconstriction of the microcirculation, lowering the cardiac output and CVP [83, 84, 116]. Administration of cortisone in massive doses demands the use of large amounts of fluids; these large doses are not harmful when used for only 47 to 72 hours (*see* Section 4.10.1).

Large steroid hormonal doses (30 to 50 mg/kg body weight) penetrate intimately within the cell, and their efficiency is greater than the opening up of the microcirculation [125]. Steroids not only introduce amino acids in gluconeogenesis reactions but also favour gluconeogenesis from lactate in the liver cell, contributing to the coupling of phosphorylation with dehydrogenation and the building up of ATP again, opposing the glycolytic and lipolytic action of epinephrine. Apart from repermeabilisation of certain anabolic pathways and moderation of some catabolic ones, corticoids also play a fundamental stabilising role: *they remedy the selectivity and polarity of the cellular membranes* [23].

Stability of the lysosomal sacs is one of the main arguments lending support to the use of steroid hormones in large amounts in almost all states of shock. Checked experimentally for a long time, cortisone megadoses have proved effective in the clinic, increasing the proportion of survivals by 20 to 25% in cases of shock considered as refractory to treatment [77, 109, 116, 133].

SELECTED BIBLIOGRAPHY

1. ATIK, M., Dextran 40 and Dextran 70, *Arch. Surg.*, 1970, *94*, (5), 664–672.
2. BĂNDILĂ, T., NOVATSEK, A., TEODORESCU-EXARCU, LUIZA, NIȚESCU, P., Probleme majore de reanimare în insuficiența renală acută la o bolnavă cu septicemie gravă post-abortum (Major Resuscitation Problems in Acute Renal Failure in a Patient with Severe Postabortum Septicaemia), *Chirurgia*, 1970, *XIX*, (8), 741–749.
3. BĂNDILĂ, T., NOVATSEK, A., TEODORESCU-EXARCU, LUIZA, VERDEȘ, A., Insuficiența renală acută după administrarea de Pobilan în scop radiodiagnostic, tratată și vindecată prin hemodializă (Acute Renal Failure After the Administration of Pobilan for Radiodiagnostic Purposes, Treated by Haemodialysis. Recovery), *Chirurgia*, 1971, *XX*, (6), 549–553.
4. BĂNDILĂ, T., NOVATSEK, A., TEODORESCU-EXARCU, LUIZA, Aspecte moderne ale tratamentului în șoc (Modern Aspects in the Treatment of Shock). *Revista sanitară militară,* 1972, *LXXV*, (3), 277–288.
5. BARNER, H. B., JELLINEK, H., KAISER, G. C., Effects of Isoproterenol Infusion on Myocardial Structure and Composition, *Amer. Heart. J.*, 1970, *79*, (2), 237–243.
6. BAUE, E. A., JONESCU, F. E., PARKINS, M. W., The Effects of Beta-adrenergic Receptor Stimulation on Blood Flow, Oxidative Metabolism and Survival in Haemorrhagic Shock, *Ann. Surg.*, 1968, *167*, (3), 403–412.
7. BAXTER, R. C., CANIZARO, C. P., CARRICO, J. C., SHIRES, T. G., Fuid Resuscitation of Haemorrhagic Shock, *Postgrad. Med.*, 1970, *68*, (3), 95–99.
8. BELL, G., The Effect of Noradrenaline and Phenoxybenzamine on the Renal Response to Haemorrhage, *Surg. Gynec. Obst.*, 1970, *130*, (5), 813–820.
9. BERGAN, F., Traumatiske karskader, *T.Norske Laegeforen*, 1970, *90*, (9b), 938–944.
10. BERGER, R. L., Pulmonary Embolectomy for Massive Embolization, *Amer. J. Surg.*, 1971, *121*, (4), 437–441.
11. BERK, J. L., Monitoring the Patient in Shock. What, When and How?, *Surg. Clin. N. Amer.*, 1975, 55 (3), 713–720.
12. BERK, J. L., Use of Vasoactive Drugs in the Treatment of Shock. *Surg. Clin. N. Amer.*, 1975, *55*, (3), 721–728.
13. BERKUTOV, A. I., *Voenno-polevaia Hirurgia* (War Surgery), 2nd edition, Izdatelstvo Meditsina, Moskva, 1968.
14. BILLY, L. J., AMATO, J. J., RICH, N. M., Aortic Injuries in Vietnam, *Surg.*, 1971, *70*, (3), 385–391.
15. BLAIR, E., OLLODART, R., ATTAR, S., COWLEY, A. R., Hyperbaric Oxygenation in the Treatment of Experimental Shock, *Amer. J. Surg.*, 1965, *110*, (3), 348–354.
16. BLANC, M., MICHIELS, R., BEAUFILS, J. P., JUSTRABO, E., Les hématomes bilatéraux des glandes surrénales. Complication du traitement anti-coagulant à propos d'une nouvelle

observation anatomoclinique, *Sem. Hôp.*, 1968, *44*, (5), 302–307.
17. BRÂNZEU, P., Tratamentul stărilor de şoc (The Treatment of Shock States), I.M.F. Timişoara, 1971.
18. BONSOM, R., BOURDAIS, A., THIRE, P., Les dangers des perfusions itératives de mannitol chez l'insuffisant rénal, *Anesth. Anal. Réan.*, 1970, *27*, (5), 843–854.
19. BROWN, R. F., BINNS, J. H., Missile Injuries in Aden, 1974–1967, *Injury*, 1970, *1*, (4), 293–302.
20. CAFRIŢĂ, A., BĂNDILĂ, T., GIURGIU, T., STRÎMBEANU, I., SAFTA, T., SURDULESCU, S., PLEŞCA, M., Ganglioplegicele în şoc (Ganglioplegics in Shock), *Revista sanitară militară*, 1965, *LXI*, (special number), 84.
21. CAREY, J. S., SCHARDSCHMIDT, B. F., CULLIFORD, A. T. Hemodynamic, Effectiveness of Colloid and Electrolyte Solutions for Replacement of Simulated Operative Blood Loss, *Surg. Gynec. Obst.*, 1970, *131*, (4), 679–686.
22. CAREY, J. S., Physiologic Hemodilution: Interrelationships between Hemodynamics and Blood Volume after Acute Blood Loss, *Ann. Surg.*, 1973, *178*, (1), 87–94.
23. CARTER, W. J., THOMAS, S. C. Jr., Circulatory Response to Pharmacologic Levels of Hydrocortisone, *J. Surg. Res.*, 1970, *10*, (3), 437–442.
24. CHAVES, P. C., Hemorragia digestiva, *Rev. Med. Estado Da Guanabara*, 1970, *37*, (2), 165–189.
25. CHIOTAN, N., CRISTEA, I., *Şocul (Shock)*, Editura medicală, Bucureşti, 1968.
26. CIESIELSKI, L., TROJANOWSKI, W., ZIELINSKI, A., Bleeding Following Abdominal Surgery, *Pol. Przegl. Chir.*, 1970, *42*, (12), 1797–1802.
27. CLAUPF, R., Applications thérapeutiques des agents bétabloquants du sympathique, *Presse Méd.*, 1967, *75*, (32), 1651–1653.
28. COHN, J. N., Treatment of Shock Following Myocardial Infarction, *Wisconsin Med. J.*, 1970, *69*, (2), 118–120.
29. COLLINS, J. A., SIMMONS, R. L., JAMES, P. M., Acid Base Status of Seriously Wounded Combat Casualties: II. Resuscitation with Stored Blood, *Ann, Surg.*, 1971, *173*, (1), 6–18.
30. CURELARU, I., Interesul miniventilatoarelor pentru Anestezie (The Interest of Miniventilators for Anaesthesia), *Viaţa medicală*, 1971, *18*, (2), 71–76.
31. DANIELSON, R. A., Differential Diagnosis and Treatment of Oliguria in Post-Traumatic and Postoperative Patients, *Surg. Clin. N. Amer.*, 1975, *65*, (3), 697–711.
32. DAVIS, H., Pathology of Intractable Shock in Man, *Arch. Surg.*, 1967, *95*, (1), 44–48.
33. DEZOTEUX, H., VINCENEUX, P., Les corticoïdes en médecine d'urgence, *Maroc Méd.*, 1969, *49*, (527), 519–521.
34. DIETHRICH, E. B., LIDDICOAT, J. E., DEBAKEY, M. E., Cardiac Preservation, *Bull. Soc. Int. Chir.*, 1970, *29*, (4), 257–260.
35. DILLON, J., LYNCH, L. J., MYERS, R., BUTCHER, H. R., MOYER, C. A., A Bioassay of Treatment of Haemorrhagic Shock. I. The Roles of Blood, Ringer's Solution with Lactate and Macromolecules (Dextran and Hydroxyethyl Starch) in the Treatment of Haemorrhagic Shock in the Anesthetized Dog, *Arch. Surg.*, 1966, *93*, (4), 537–555.
36. DIOMI, P., ERICSSON, L. E. J., MATHESON, A. N., Effects of Dextran 40 on Urine Flow and Composition during Renal Hypoperfusion in Dog with Osmotic Nephrosis, *Ann. Surg.*, 1970, *172*, (5), 813–824.
37. DIOMI, P., MATHESON, A. N., NORMAN, J. N., SHEARER, J. R., The Renal Response to Dextran 40 in Dogs with Renal Artery Constriction, *Surg.*, 1971, *69*, (2), 256–262.
38. DOR, V., NOICLERC, M., CUAVIN, G., MERMET, B., KREITMANN, P., LEONARDELLI, M., AMOROS, J. F., Les traumatismes graves du thorax. Place de l'ostéosynthèse dans leur traitement, *La Nouv. Presse Méd.*, 1972, *1*, (8), 519–524.
39. DUCLUZEAU, R., *Les intoxications barbituriques aiguës. A propos de 1012 observations du service de toxicologie clinique de Lyon*, Masson, Paris, 1971.
40. EASTCOTT, H. G.G., *Arterial Surgery*, Pitman Medical, London, 1969.
41. EBERT, P.A.. *Shock and Haemorrhage*, Butterworth, London, 1970.
42. FILIPESCU, Z., CURELARU, I., *Echilibrarea funcţională a bolnavilor în urgenţă* (Functional Equilibration of Emergency Cases), Editura medicală, Bucureşti, 1963.
43. FLORENT, C., COUSIN., M. T., VOURC'H, G., Réanimation cardiorespiratoire d'urgence. Etude de 69 cas, *Anesth. Anal. Réan.*, 1970, *27*, (4), 541–592.
44. FITTS, T. C. Vasoactive Drugs in Treatment of Shock, *Postgrad. Med.*, 1970, *48*, (3), 105–109.
45. FONTAINE, R., FONTAINE, J. L., La thrombectomie dans les thromboses veineuses iliofémorales aiguës, *J. Méd. Strasbourg*, 1970, *1*, (5), 369–379.
46. GAVRILESCU, S., STREJAN, C., POP, T., COTOI, S., VOICULESCU, V., Utilizarea unor stimulatori β-adrenergici în tratamentul stărilor critice ale aparatului cardiovascular (The Utilization of β-adrenergic Stimulators in the Treatment of Critical States of the Cardiovascular System), *Medicina internă*, 1970, *22*, (10), 1201–1215.

47. GEOCARIS, T. V., QUEBBEMAN, E., DEWOSKIN, R., MOSS. G. S., Effects of Gram-Negative Eudotoxemia on Myocardial Contractility in the Awake Primate, *Ann. Surg.*, 1973, *178*, (6), 715–720.
48. GIEBEL, O., HORATZ, K., Behaviour of Blood Volume and its Components after Replacement with Dextran and Gelatin Plasma Substitutes Following Bleeding in the Healthy Young Male, *Bil. Haemat.*, 1969, *33*, (2), 171–183.
49. GREENE, M. N., *Physiology of Spinal Anesthesia*, Williams an Wilkins Co., Baltimore 1969.
50. GROVER, F. L., NEWMAN, M. M., PATON, B. C., Beneficial Effect of Pluronic F 68 on the Microcirculation in Experimental Hemorrhagic Shock, *Surg. Forum*, 1970, *21*, (1), 30–32.
51. GRÖNWALL, A., *Dextran and its Use in Colloidal Infusion, Solution*, Almquist and Wiksell, Uppsala, 1957.
52. GRUBER, U. F., ALLÖGWER, M., Infusion Therapy in Surgery, *Triangle*, 1968, *8*, (7), 273–283.
53. GRUBER, U. F., *Blood Replacement*, Springer-Verlag, Berlin-Heidelberg-New York, 1969.
54. GRUBER, U. F., Dextran and the Prevention of Postoperative Thromboembolic Complications, *Surg. Clin. N. Amer.*, 975, *55*, (3), 679–685.
55. HABIF, D. V., DEBBAS, E., MIGNE, J., NAHAS, G. G., A Balanced Fluid Gelatin for the Treatment of Haemorrhage, *Ann. Surg.*, 1971, *173*, (1), 85–90.
56. HAGBERG, S., HALJAMÄE, H., RÖCKERT, H., Schock Reactions in Skeletal Muscle: IV. Effect of Hypothermic Treatment on Cellular Electrolyte Responses to Haemorrhagic Schock, *Acta Chir. Scand.*, 1970, *136*, (1), 23–28.
57. HALLER, Jr. J. A., TALBERT, J. L., Trauma Workshop Report: Trauma in Children, *J. Trauma*, 1970, *10*, (11), 1052–1054.
58. HANDIN, R. I., VALERI, C. R., Improved Viability of Frozen Platelets, Program of the 24-th Annual Meeting Chicago, September, 1971.
59. HARDAWAY, R. M., *Syndromes of Disseminated Intravascular Coagulation*, Charles C. Thomas, Springfield-Illinois, 1966.
60. HARDAWAY, R. M., *Clinical Management of Shock*, Charles C. Thomas, Springfield-Illinois, 1968.
61. HARRIS, P., Current Concepts in the Management of Brain Trauma, *J. Roy. Coll. Surg. Edin.* 1970, *15*, (5), 268–282.
62. HARTFORD, C. E. ZIFFREN, S. E., Electrical Injury, *J. Trauma*, 1971, *11*, (4), 331–336.
63. HIRONARTA, K., Experimental Study on Fluid Replacement in Puppies, with Special Reference to Pulmonary Arterial and Central Venous Pressures, *Arch. Jap. Chir.*, 1970, *39*, (3), 97–103.
64. HUGGINS, C. E., Frozen Blood: Principles of Practical Preservation, *Monographs in the Surg. Science*, 1966, *3*, (3), 133–173.
65. HUGGINS, C. E., Frozen Blood, *Europ. Surg. Res.*, 1969, (1), 3–12.
66. JENKINS, K. C., *General Anesthesia and the Central Nervous System*, Williams and Wilkins Co., Baltimore, 1969.
67. JUDE, J. R., BOLOOKI, H., NAGEL, E., Cardiac Resuscitation in the Operating Room: Current Status, *Ann. Surg.*, 1970, *171*, (6), 948–955.
68. KENNEDY, J. H., Criteria for Selection of Patients for Mechanical Circulatory Support, *Ann. J. Cardiol.*, 1971, *27*, (1), 33–40.
69. KINNEY, J. M., Intensive Care of the Critically Ill: A Foundation for Research, J. *Trauma*, 1970, *10*, (11), 949–957.
70. KLEBANOFF, G., PHILLIPS, J., EVANS, W., Use of a Disposable Autotransfusion Unit under Varying Conditions of Contamination. Preliminary Report, *Amer. J. Surg.*, 1970, *120*, (3), 351–354.
71. KNIRSCH, A. K., GRALLA, E. J., Abnormal Serum Transaminase Levels after Parenteral Ampicillin and Carbenicillin Administration, *New Engl. J. Med.*, 1970, (19), *282*, 1081–1082.
72. KRAGELUND, E., Loss of Fluid and Blood to the Peritoneal Cavity during Abdominal Surgery, *Surg.*, 1971, *69*, (2), 284–287.
73. KRAMER, S. G., LIPSON, C. S., Intensive Care Unit Flow Sheet, *Surg.*, 1970, *67*, (4), 590–592.
74. KUBAN, D. J., MATEICKA, W. E., BATAYIAS, G. E., *Autotransfusion for Open Heart Surgery*, Program of the 24th Annual Meeting, Chicago, September, 1971.
75. LABORIT, H., BARON, C., WEBER, B., Traitement du choc hémorragique expérimental dit 'irréversible'. Rôle des groupes SH et de la restauration des réserves intracorticulaires en catécholamines. I. Vue d'ensemble, *Agressol.*, 1969, 10, (3), 189–198.
76. LABORIT, H., BARON, C., WEBER, B., Traitement du choch hémorragique expérimental dit 'irréversible'. Rôle des groupes SH et de la restauration des réserves intracorticulaires en catécholamines. II. Etude expérimentale, *Aggressol.*, 1969, *10* (3), 199–204.
77. LABORIT, H., BARON, C., WEBER, B., Traitement du choc hémorragique expérimental dit 'irréversible'. Rôle des groupes SH et de la restauration des réserves intracorticulaires en catécholamines. III. Etude stéréo-

taxique des stimulations cérébrales, *Agressol.*, 1969, *10*, (3), 205—215.

78. Laborit, H., Baron, C., Effets de la monosemicarbazone de la β-naphtoquinone dans le choc hémorragique expérimental, *Agressol.*, 1971, *12*, (1), 25—30.
79. Lang, T. W., Efect of Venoarterial Pulsatile Partial Bypass on the Renal and Mesenteric Circulations in Cardiogenic Shock, *Ann. J. Cardiol.*, 1971, *27*, (1), 41—45.
80. Levin, P. M., Rich, N. M., Hutton, Jr. J. E., Collateral Circulation in Arterial Injuries, *Arch. Surg.*, 1971, *102*, (4), 392—399.
81. Lewis, F. J., Monitoring of Patients in Intensive Care Units, *Surg. Clin. N. Amer.*, 1971, *51*, (1), 15—23.
82. Lichtmann, M. W., The Problem of Contused Lung, *J. Trauma*, 1970, *10*, (9), 731—739.
83. Lillehei, R. C., Longerbeam, J. K., Bloch, J. H., Manax, W. G., The Nature of Irreversible Shock, *Ann. Surg.*, 1964, *160*, (4), 682—710.
84. Lindsay, W. G., Humphrey, E. W. The Sodium Space in Pressure-Induced Pulmonary Edema, *Curr. Top. Surg. Res.*, 1970, *2*, (2), 351—358.
85. Litarczek, G. Medicaţia vasoactivă şi cardiotonică în tratamentul tulburărilor circulatorii în şoc (Vasoactive and Cardiotonic Medication in the Treatment of Circulatory Disturbances in Shock), *Revista sanitară militară*, 1973, *LXXVI*, (3) 217—222.
86. Litwin, M. S., Chapman, K., Comparison of Effects of Dextran 70 and Dextran 40 Infused into Postoperative Animals, *Europ. Surg. Res.*, 1970, (2), 125—126.
87. Loeb, H. S., Pietras, R. J., Ninos, N. Hemodynamic Responses to Chlorpromazine in Patients in Shock, *Arch. Intern. Med.*, 1969, *124*, (3) 354—358.
88. Lundsgaard-Hansen, P., Hassig, A. Nitschmann, H. S., *Modified Gelatins as Plasma Substitutes*, S. Karger, Basel-New York, 1969.
89. Macon, W. L., Morton, J. H., Adams, J. T., Significant Complications of Anticoagulant Therapy, *Surg.*, 1970, *68*, (3), 571—582.
90. McCaffrey, J. Orinithine Vasopressin (POR 8) as a Cutaneous Vasoconstrictor, *Aust. N.T.J. Surg.*, 1970, *40*, (2), 198—199.
91. McPeak, D. W., Camp, F. R., Conte, N. F., *Blood Component Logistics*, Program of the 24-th Annual Meeting, Chicago, September, 1971.
92. Mailloux, L., Swartz, C. D., Capizzi, R., Kin, K. E., Onesti, G., Ramirez, O., Brest, A., N. Acute Renal Failure after Administration of Low Molecular Weight Dextran, *New Eng. J. Med.*, 1967, *277*, (6), 1113—1122.
93. Mareş, E. Principiile şi tratamentul şocului la etapele de evacuare (The Principles and Treatment of Shock in the Evacuation Stages), *Revista sanitară militară*, 1965, *LXI* (special number), 9—28.
94. Mareş, E., Văideanu, C., Tratamentul şocului pe timpul transportului (The Treatment of Shock during Transport), *Revista sanitară militară*, 1965, *LXI* (special number), 28—32.
95. Markley, K., Protection against Burn, Tourniquet and Endotoxin Shock by Purine Compounds, *J. Trauma*, 1970, *10*, (7), 598—607.
96. Marty, T. A., Zweifach, W. B., High Oncotic Pressure Effects of Dextrans, *Arch. Surg.*, 1970, *101*, (3), 421—424.
97. Matheson, A. N., Diomi, P., Renal Failure after Administration of Dextran 40, *Surg. Gynec. Obst.*, 1970, *131*, (4), 661—668.
98. Mays, E. T., Complex Penetrating Hepatic Wounds, *Ann. Surg.*, 1971, *173*, (3), 421—428.
99. Menegaux, G., *Manuel de pathologie chirurgicale*, tome II, Masson, Paris, 1971.
100. Messmer, K., Hemodilution, *Surg. Clin. N. Amer.*, 1975, *55*, (3), 659—677.
101. Moore, D. C., *Regional Block*, Charles C. Thomas, Springfield-Illinois, 1971.
102. Moss, G. S., A Comparison of Asanguineous Fluids and Whole Blood in the Treatment of Haemorrhagic Shock, *Surg.*, 1969, *129*, (6), 1247—1257.
103. Moss, G. S., Siegel, D. C., Cochin, A., Fresquez, V., Effects of Saline and Colloid Solutions on Pulmonary Function in Haemorrhagic Shock, *Surg. Gynec. Obst.*, 1971, *133*, (1), 53—58.
104. Moss, G. S., An Argument in Favor of Electrolyte Solution for Early Resuscitation, *Surg. Clin. N. Amer.*, 1972, *52*, (1), 3—17.
105. Morin, J. E., Goldenberg, I. S., An Experimental Study of Replacement Solutions in Hypovolemic Shock, *Laval Méd.*, 1967, *38*, (8), 744—749.
106. Muller, W. H., Die traumatische Zwerchfellruptur, *Der Chirurg*, 1970, *41*, (7), 315—320.
107. Nash, G., Blennerhassett, J. B., Pontoppidan, H., Pulmonary Lesions Associated with Oxygen Therapy and Artificial Ventilation, *New Engl. J. Med.*, 1967, *276*, (2), 368—374.
108. Niculescu, G., Cafriţă, A., Baciu, D., Filip, I., Anestezia şi reanimarea în fracturile deschise ale membrului pelvin (Anaesthesia and resuscitation in open fractures

of the leg) *Rev. sanitară militară*, 1965, *LXI* (special number), 66–67.

109. PAPAHAGI, E., POPOVICI, Z., CIUREL, M., Rezultate imediate pe 70 de derivaţii portocave pentru hemoragii digestive, (The Immediate Results of 70 Portocaval Shunts for Digestive Hemorrhage), *Chirurgia*, 1970, *XIX*, 1, 45–52.
110. PETER, K., KLOSE, R., LUTZ, H., Ketanest zur Narkoseeinleitung beim Shock, *Z. Prakt. Anästh.*, 1970, *5*, (6), 296–401.
111. PETERS, R. M., HILBERMAN, M., Respiratory Insufficiency: Diagnosis and Control of Therapy, *Surg.*, 1971, *70*, (2), 280–287.
112. PHILLIPS, S. J., VICK, J. A., The Pretreatment of E. coli Endotoxin Shock with WR 2823: A New Alpha-adrenergic Blocking Agent, *Surg.*, 1971, *69*, (4), 510–514.
113. PLOUVIER, S. R., ARNOLD, M., Importance du dosage du facteur 11 au cours des traitements par les inhibiteurs de la vitamine K, *Agressol.*, 1971, *12*, (1), 41–48.
114. POCIDALO, J. J., *Réanimation et choc*, Flammarion, Paris, 1968.
115. PONTOPPIDAN, H., LAVER, M. B., CEFFIN, B., Acute Respiratory Failure in the Surgical Patient. In *Advances in Surg.*, Chicago, Year Book Medical Publishers, 1970, (4), 163–254.
116. POISVERT, M., DEBRAS, C., Le choc irréversible: quelques réflexions à propos d'un traitement simple que nous utilisons depuis plus de deux ans, *Laval Méd.*, 1970, *41*, (11), 1112–1116.
117. POWERS, S. R. Jr., Relation of Acute Tubular Necrosis to Shock and the Effect of Mannitol, *Amer J. Surg.*, 1965, *110*, (3), 330–332.
118. POWERS, S. R. Jr., MANNAL, R., NECLERIO, M., ENGLISH, M., MARR, C., LEATHER, R., UEDA, H., WILLIAMS, G., CUSTEAD, W., DUTTON, R., Physiologic Consequences of Positive End-Expiratory Pressure (PEEP) Ventilation, *Ann. Surg.*, 1973, *178*, (3), 265–272.
119. RAPIN, M., GROSBIUS, S., KERNBAUM, S., GOULON, M., Digestive Haemorrhage during Acute Circulatory Failure, *Ann. Intern. Med.*, 1970, *121*, (8–9), 673–679.
120. REUL, G. J. Jr., GREENBERG, S. D., LEFRAK, E. A., McCOLLUM, W. B., BEALL, A. C. Jr., JORDAN, G. L. Jr., Prevention of Posttraumatic Pulmonary Insufficiency: Fine Screen Filtration of Blood, *Arch. Surg.*, 1973, *106*, (4), 386–394.
121. ROWE, M. I., ARANGO, A., Cardiovascular Response of Acidotic Newborn to Sodium Bicarbonate, *Arch. Surg.*, 1973, *106*, (3), 327–332.
122. RUSH, B. K. J., Treatment of Experimental Shock: Comparison of the Effects of Norepinephrine, Dibenzyline, Dextran, whole Blood, and Balanced Saline Solutions, *Surg.*, 1967, *61*, (6), 938–944.
123. SAFAR, P., GRENVIK, A., Critical Care Medicine Organizing and Staffing Intensive Care Units, *Chest.*, 1971, *59*, (5), 535–547.
124. SANTAMARINA, B. A. G., KLEIN, T. A., Treatment of Septic Abortion and Septic Shock, *Med. Treatment*, 1970, *7*, (4), 779–788.
125. SCHUMER, W., NYHUS, L. M., Corticosteroid Effect on Biochemical Parameters of Human Oligemic Shock, *Arch. Surg.*, 1970, *100*, (4), 405–408.
126. SCHUMER, W., Evolution of the Modern Therapy of Shock: Science Versus Empiricism, *Surg. Clin. N. Amer.*, 1971, *51*, (1), 3–13.
127. SEVITT, S., Reflections on some Problems in the Pathology of Trauma, *J. Trauma*, 1970, *10*, (11), 962–973.
128. SHIRES, G. T. Role of Sodium-containing Solutions in Treatment of Oligemic Shock, *Surg. Clin. N. Amer.*, 1965, *45*, (5), 365–376.
129. SHIRES, G. T., Initial Management of the Severely Injured Patient, *J.A.M.A.*, 1970, *213*, (11), 1873.
130. SHOEMAKER, W.C., *Shock: Chemistry, Physiology and Therapy*, Charles C. Thomas, Springfield-Illinois, 1967.
131. SHOEMAKER, W.C., The Dilemma of Vasopressors and Vasodilators in the Therapy of Shock, *Surg. Gynec. Obst.*, 1971, *132*, (1), 51–57.
132. SIEGEL, D. C., MOSS, G. S., COCHIN, A., DAS GUPTA, T. K., Pulmonary Changes Following Treatment for Haemorrhagic Shock: Saline Versus Colloid Infusion, *Surg. Forum*, 1970, *21*, (1), 17–19.
133. SIEGEL, D. C., COCHIN, A., GEOCARIS, T., MOSS, G. S., Effects of Saline and Colloid Resuscitation on Renal Function, *Ann. Surg.*, 1973, *177* (1), 51–57.
134. SIIMES, A., VARTAINEN, E., WIDHOLM, O., Post Operative Wound Complications, *Acta Obst. Gynec. Scand.*, 1970, **49**, (3), 255–258.
135. SKINNER, D. B., CAMP, T. F., AUSTREN, W. G., Use of Vasopressor Agents to Increase Somatic Blood Flow, *Arch. Surg.*, 1967, *94*, (5), 610–618.
136. SLADEN, A., LAVER, M. B., PONTOPPIDAN, H., Pulmonary Complications and Water Retention in Prolonged Mechanical Ventilation, *New Engl. J. Med.*, 1968, **279**, (2), 448–453.

137. SMITH, G., A comparison of Some Cardio-vascular Effects of Tubocurarine and Pancuronium in Dogs, *Brit. J. Anesth.*, 1970, *42*, (7), 923–928.
138. STAHL, W. M., STONE, A. M., Prophylactic Diuresis with Ethacrynic Acid for Prevention of Postoperative Renal Failure, *Ann. Surg.*, 1970, *172*, (3), 361–369.
139. STRING, T., ROBINSON, A. J., BLAISDELL, F. W., Massive Trauma Effect of Intravascular Coagulation on Prognosis, *Arch. Surg.*, 1971, *102*, (4), 406–411.
140. STUART, F. P., MOORE, F. D., Effects of Single, Repeated and Massive Manitol Infusion in the Lap, *Ann. Surg.*, 1970, *172*, (2), 190.
141. SUGERMAN, J. H., DAVIDSON, T. D., VIBUL, S., DELIVORIA-PAPADOPOULOS, M., MILLER, D. L., OSKI, A. F., Basis of Defective Oxygen Delivery from Stored Blood, *Surg. Gynec. Obst.*, 1970, *131*, (4), 733–741.
142. ŞUTEU, I., CAFRIŢĂ, A., BĂRBOI, D., Primul ajutor în şoc (First Aid in Shock), *Revista sanitară militară*, 1970, *LXXII*, (3), 359–362.
143. ŞUTEU, I., CAFRIŢĂ, A., BUCUR, I. A., Progrese recente in fiziopatologia şocului (Recent Progress in the Pathophysiology of Shock), *Revista sanitară militară*, 1972, *LXXV*, (3), 267–276.
144. TRINKLE, J. K., Mechanical Support of the Circulation: a New Approach, *Arch. Surg.*, 1970, *101*, (6), 740–743.
145. ŢURAI, I., Patogenia şi tratamentul şocului (Pathogenesis and Treatment of Shock), *Chirurgia*, 1970, *IX*, (4), 205–216.
146. VOLLMAR, J., GRUSS, J. D., LAUBACH, K., Technique de la thrombendartériectomie (Désoblitération hélicoïdale au ring-stripper), *J. Chir.*, 1970, *100*, (1–2), 67–82.
147. VOURC'H, G., de la BASTAIE, R. P., TRINER, L., DEBBAS, E., HABIF, D. V., NAHAS, G. G., La place des solutés sodés équilibrés dans le traitement de l'hémorragie, *Anesth. Anal. Réan.*, 1970, *27*, (5), 737–757.
148. WANGENSTEEN, S. L., DEHOLL, J. D., KIECHEL, S. F., MARTIN, J., LEFER, A. M., Influence of Hemodialysis on a Myocardial Depressant Factor in Haemorrhagic Shock, *Surg.*, 1970, *67*, (6), 935–943.
149. WARDLE, L. N., Septic Shock, *Brit. Med. J.*, 1970, *5698*, (1), 236–239.
150. WEEDN, R. J., COOK, J. A., MCELREATH, R. L., GREENFIELD, L. J., Wet Lung Syndrome after Crystalloid and Colloid Volume Repletion, *Curr. Top. Surg. Res.*, 1970, (2), 335–350.
151. WEIL, M. H., SCUBIN, H., *Diagnosis and Treatment of Shock*, Williams and Wilkins Co., Baltimore, 1967.
152. WEIS, K. H., BRACKEBUCH, H. D., On the Cardio-vascular Effect of Propanolol during Halothane Anaesthesia in Normovolaemic and Hypovolaemic Dogs, *Brit. J. Anesth.*, 1970, *42*, (2), 272–279.
153. WHELAN, C. S., FURCINITTI, J. F., LAVARREDA, C., Surgical Management of Perforated Lesions of the Colon with Diffusing Peritonitis, *Amer. J. Surg.*, 1971, *121*, (4), 374–378.
154. WHELTON, M. J., Paracentesis Abdominis, *Brit. J. Hosp. Med.*, 1970, *4*, (6), 795–799.
155. WIENER, S. N., WEISS, P. H., Radionuclide Imaging in the Care of the Critically Ill Patient, *Surg. Clin. N. Amer.*, 1975, *55*, (3), 729–736.
156. WILSON, R. S., RIE, M. A., Management of Mechanical Ventilation, *Surg. Clin. N. Amer.*, 1975, *55*, (3), 591–601.
157. WOSORNU, J. L., EASMON, C. O., Intravenous Isoprenaline in Treatment of Septic Shock in Man, *Brit. Med. J.*, 1970, *1*, (5698), 723–726.
158. WYCHE, M. Q. Jr., TEICHNER, R. L., KALLOS, T., MARSHALL, B. E., SMITH, T. C., Effects of Continuous Positive-Pressure Breathing on Functional Residual Capacity and Arterial Oxygenation during Intra-abdominal Operations: Studies in Man during Nitrous Oxide and D-Tubocurarine Anesthesia, *Anesthesiology*, 1973, *38*, (1), 68–74.
159. ZAPOL, W. M., Membrane Lung Perfusion for Acute Respiratory Failure, *Surg. Clin. N. Amer.*, 1975, *55*, (3), 603–612.

12 TYPES OF SHOCK

It is a case of forcing too many feet to fit the same shoe.

PROVERB

12.1 HAEMORRHAGIC SHOCK
12.2 TRAUMATIC SHOCK AND ITS VARIANTS
12.2.1 General
12.2.2 Traumatic shock from firearm wounds
12.2.3 The Crush Syndrome
12.2.4 Operative and anaesthetic shock
12.2.5 Blast shock
12.3 BACTERIEMIC SHOCK
12.4 BURN SHOCK
12.5 ANAPHYLACTIC SHOCK
12.6 CARDIOGENIC SHOCK

12.1 HAEMORRHAGIC SHOCK

Haemorrhage is always impressive by the gravity of its clinical picture. Haemorrhagic shock by itself is seldom encountered because in most cases it is accompanied by elements of traumatic, bacteriemic shock, etc. Rupture of an ectopic pregnancy, of oesophageal varices, vascular fistula caused by a perforating ulcer and direct vascular wounds are some of the clinical circumstances that can bring about haemorrhagic shock [13, 17, 21, 66, 69]. When the loss of blood is caused by rupture of a parenchymatous organ or a blood vessel, with bleeding into a preformed cavity, haemorrhagic shock is more difficult to diagnose. In such cases, when the clinical signs are uncertain, an abdominal or pleural puncture will provide the materials needed for a diagnosis (Figure 12.1).

The human body has an efficient mechanism of autoperfusion and draws upon the large reserves of interstitial and even cellular fluid. After two to three hours of haemorrhage a slight decrease occurs in serum protein concentrations, with a fall in haematocrit, because the fluid rapidly taken up by the capillaries lacks proteins. However, proteinaemia is quickly re-established by the release into the circulation of fresh albumins from the hepatic reserves, and by synthesis; thus, remission of the osmotic pressure is obtained by passage of the interstitial fluid into the blood vessels by osmosis. Moore demonstrated that plasma osmotic pressure varies very little during spontaneous blood volume re-equilibration owing to the albumin mobilized from the extravascular hepatosplanchnic pool, a process that starts within the first hour of bleeding. The rate of penetration of albumin into the intravascular space is about 5 g/hour and is uninfluenced by the injection of electrolytic solutions.

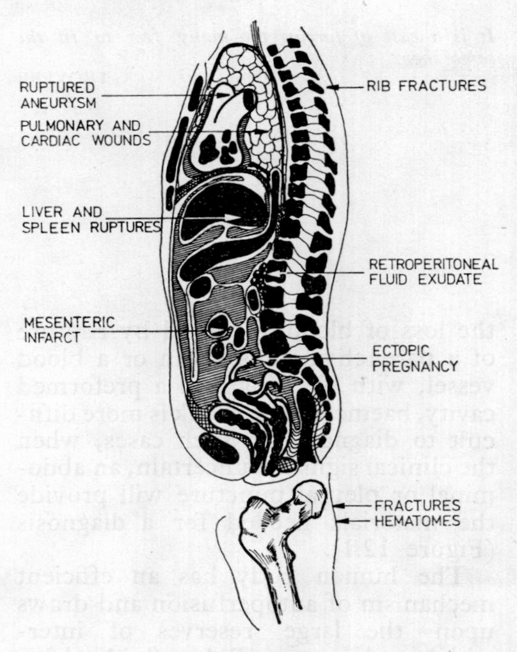

Figure 12.1. The possible non-exterior loss of fluid (internal trapping).

The mobilisation of albumins from the extravascular supply as well as their renewed synthesis is inhibited by colloids (*see* Section 6.2).

In losses of 25% of the blood mass, the circulating blood volume is recovered within 36 hours (at a rate of 33 ml/hour); the actual haematocrit (Ht) value is only reflected after this interval. At the end of the 36 hours 25% of the circulating proteins are newly synthesized and penetrate into the circulation after the onset of bleeding. The loss of blood is proportional to the decrease in haemoglobin (Hb) and Ht only in the haemodilution phase. The erythrocytic mass is, however, very difficult to replace: after a loss of 500 ml blood 40 to 50 days are required for Ht and Hb to reach normal values.

Selective, peripheral vasoconstriction lowers the venous return to the heart, cardiac output and minute volume; the heart increases its rate in response to a decrease of the venous return and cardiac output. If bleeding continues, the compensatory mechanisms are exceeded; the cardiac output falls below critical levels and the blood pressure collapses. The initial vasoconstrictor response to haemorrhage maintains the systolic BP at almost normal levels.

In this stage an attempt to raise the patient may precipitate circulatory failure followed by cardiac arrest. Owing to vasoconstriction the BP measured is not conclusive of the general haemodynamics. This was demonstrated experimentally by the fact that to a 25% fall in the cardiac output there may be a corresponding reduction of the BP of only 7% [27]. A 20% loss of the circulating volume will cause a 45% decrease of the cardiac output and only 15% of the blood pressure.

Haemodynamic effects differ according to the duration of bleeding and to the amount of blood lost within a given interval. It should be emphasised that blood volume depletion is more important than erythrocytic depletion since it may precipitate an abrupt decrease of the ECBV, with the onset of cardiac arrest. However, more important still than the absolute volume lost is the rate at which bleeding occurs and the early institution of therapy. Thus, renewed transfusion of experimental animals after 2 hours results in a survival of 80% whereas after 4 hours almost all die [69].

As haemorrhage is the cause of this type of shock it is of importance to mention the following classification which appears to be of the greatest interest from the clinical point of view.

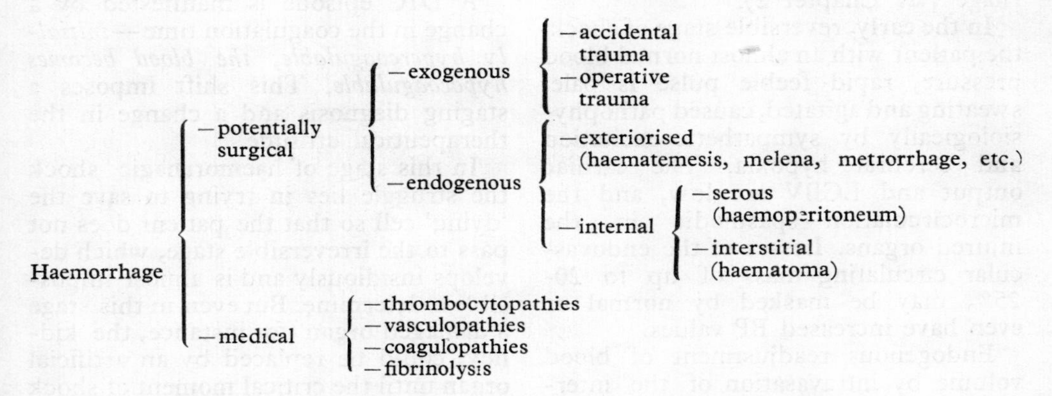

The intensity, rate and duration of bleeding transforms a haemorrhage into haemorrhagic shock.

Recent investigations have clearly shown the dependence of the clinical gravity of shock upon depletion of the circulating blood volume. This makes it possible to assess the blood volume lost and the amount necessary for resuscitation on the basis of clinical signs alone.

Table 12.1 gives the correlation between the blood lost and the fundamental clinical signs.

Haemorrhagic shock certainly represents the prototype of the evolutive stages of all types of shock, as estab-

Table 12.1 CORRELATIONS BETWEEN BLOOD LOSSES AND CLINICAL SIGNS *(modified after Weil and Shubin)*

Severity of haemorrhagic shock	Clinical signs	Proportion of blood losses
No shock (e.g. blood donor)	No clinical signs	up to 10% (500 ml)
Mild shock	Moderate tachycardia; slight increase in BP; slight peripheral vasoconstriction; cold, pale extremities.	10–25% (500–1250 ml)
Moderate shock	Tachycardia 110–120 beats/min; low pulse and systolic pressure; BP 90–100 mmHg; abundant sweating, pallor, oliguria.	(1250–1750 ml) 25–35%
Severe shock	Tachycardia > 120 beats/min; systolic pressure < 60 mmHg (often undetectable); cyanotic, cold extremities.	more than 35% (> 1750 ml)

lished by experiments that have now become classical (Wiggers, Lillehei, Fine, Werle, Simeone et al.) and especially as observed daily in the intensive care ward in cases of severe haemorrhage *(see* Chapter 2).

In the early, reversible stage of shock, the patient with an almost normal blood pressure, rapid feeble pulse is pale, sweating and agitated, caused pathophysiologically by sympathetic excitation and cerebral hypoxia. The cardiac output and ECBV are low, and the microcirculation spasmodic in the injured organs. Losses of the endovascular circulating mass of up to 20-25% may be masked by normal or even have increased BP values.

Endogenous readjustment of blood volume by intravasation of the interstitial fluid depends to a large extent upon the previous state of hydration of the body. If fluids are not administered the microcirculation will become permeable again, not by flow but by tissular acidosis and histamines massively released by the hypoxic mast cells and fibrocytes which will bring about lysis of the precapillary sphincter spasm, opening up the microcirculation like a pouch, since the postcapillary venular spasm persists. The pooling phase specific of late, reversible haemorrhagic shock is well known. In this stage large amounts of fluid have to be administered because expansion of the microcirculation space increases double or even threefold the fluid requirements. At the same time, the venous return diminishes and in some of the visceral veins the *p*H falls below 7 [78].

It is considered that a purely haemorrhagic shock is reversible (by the administration of fluids) up to absolute systemic BP values of 20 mmHg. It should be recalled that against the background of pooled, acid blood in the microcirculation the slightest tissular injury, infection or an erythrothrombocytolytic episode will trigger DIC and the patient passes into the stage of refractory reversible haemorrhagic shock.

A DIC episode is manifested by a change in the coagulation time—*initially hypercoagulable, the blood becomes hypocoagulable.* This shift imposes a staging diagnosis and a change in the therapeutical attitude.

In this stage of haemorrhagic shock the struggle lies in trying to save the 'dying' cell so that the patient does not pass to the irreversible stage, which develops insidiously and is almost impossible todetermine. But even in this stage a damaged organ (for instance, the kidney) could be replaced by an artificial organ until the critical moment of shock is over [13, 65].

Therapeutical features. Haemostasis is essential (and sometimes extremely difficult) in the treatment of lesions that have caused haemorrhagic shock. In external bleeding, compression is followed by clamping, ligature, or vascular repair surgery. In internal bleeding the operative risks and the optimal moment for intervention must be assessed. If, after transfusion, bleeding continues and exceeds 500 ml/8 hours, then surgery is certainly necessary.

Transfusion. The best treatment in haemorrhagic shock is the transfusion of iso-group, iso-Rh blood. For rapid transfusions venous catheters may be introduced into two veins and the blood administered under pressure. In acute haemorrhage, when the BP falls to zero, transfusion precedes only by a single step, anaesthesia and operation. In the early reversible phase the course of a haemorrhagic shock can be arrested only by blood transfusion.

In some instances, fairly large amounts of blood are necessary: up to 6-8 litres. Blood transfusion must not be put off since shock will continue to run its course towards the refractory phase in which transfusion is no longer sufficient to maintain it within reversible limits.

The efficiency parameter of transfusion is in no case the BP (except when it falls below 6 mmHg) but the CVP, whose values of 12 cm of water will confirm normal blood volume. When the performance of the myocardium is normal, lower values may indicate hypo- and respectively hypervolaemia.

Administration of blood substitutes. In medium and severe haemorrhagic shock it is essential to administer blood, and only when not available can one resort to colloidal non-erythrocytic solutions. On administering colloidal solutions (especially dextrans) in amounts greater than 1000 ml errors may appear in the determination of the blood group, disturbances in the coagulation mechanisms and an alarming RBC decrease.

Dextrans with a small molecular weight are more often used since, apart from the properties of the substitutes, they also reduce the viscosity of the blood, with anti-sludge effects. Some authors criticize the use of Rheomacrodex, however, in haemorrhagic shock because it produces an exaggerated dilution of the blood, superposed upon the endogenous one, drastically lowering the coagulation factors and aggravating the haemorrhage. At present, the best volume substitute is Ringer's lactate solution, which gives better results than those obtained by the exclusive administration of colloids. Associated to blood dextrans, or still better to plasma, Ringer's solution will correct the electrolytic imbalance induced by colloids [16] (*see* Section 11.3).

Vasodilator drugs. Experiments and clinical practice have demonstrated that tissular perfusion is more efficient under conditions of systemic hypotension, induced pharmacodynamically but with a capillary bed dilated by the administration of drugs, since it provides almost normal gas exchanges, better than a normal systemic pressure but with a peripheral vascular bed blocked by vasoconstriction. At present there is a general consensus regarding the therapeutical role of α-blocking drugs of the Dibenamine, Phenoxybenzamine, Hydergin and Levomepromazine type. By increasing the vascular bed α-blocking drugs demands the concomitant administration of macromolecular substitutes to fill the vacated space.

Therefore, the administration of α-blocking drugs without blood or a macromolecular solution may induce an abrupt, dangerous hypotension.

In order to remove the peripheral vasoconstrictor barrier and recover the dammed up blood, ganglioplegics may be used (*see* Section 2.8). Cortisone, by its milder and longer acting anti-α-lytic and anti-sludge effect, is useful when given in large doses.

Vasopressor drugs. It is obvious that loss of blood demands replacement. To administer vasopressor drugs only accentuates the post-shock syndrome, already affected by vasoconstriction.

Increased blood pressure following the administration of vasopressors is misleading; it increases only in the large vessels and not in the microcirculation, which is totally excluded from the normal circuit of the blood.

Coagulation defects. Sometimes, owing to coagulation defects, diffuse bleeding may develop in gynaecologic, urologic, pulmonary or cardiovascular surgery.

The 'key' to the known paradox 'haemorrhage generates haemorrhage' is the onset of DIC (*see* Section 7.4).

In abnormal bleeding the thrombocyte count, and still better the thromboelastogram, must be performed. The Quick time, when prolonged, is often due to excessive administration of an anticlotting medication in the hypercoagulation stage of shock. A hepatic failure or, more seldom, a vitamin K deficiency (associated with biliary obstruction) may be involved. The blood samples should always be collected in three test-tubes.

One test-tube is maintained in the thermostat at 37°C. If the clot develops in time, the beginning of lysis should be followed up every hour. To the second test-tube a drop of bovine thrombine is added; if a clot does not form it might suggest the absence of fibrinogen or the presence of a heparin-like anticoagulant. When the absence of clotting is due to heparin, the antidote — protamine sulphate — is added to the test-tube with unclotted blood. Prompt coagulation after the administration of thrombin excludes the absence of fibrinogen or the presence of substances with a heparin-like activity. Coagulation is tested in the third test-tube; if a clot does not form within 15 minutes it is due to a deficit of one or several of the factors acting in the first phas of coagulation.

Patients with excess circulating heparin may be given 50 to 200 mg protamine sulphate, administered very slowly, and eventually combined with epsilon aminocaproic acid to counteract the stimulating effect of heparin on fibrinolysis. When not possible because of the high heparinaemia values and/or severe proteolysis, then 6 to 8 g human fibrinogen should be administered, eventually protected by heparin against a permanent consumption in DIC. Anticoagulants of the coumarin type are efficiently neutralized by vitamin K. If primary or secondary fibrinolysis is suspected then epsilonn aminocaproic acid must be added to the treatment (4 to 8 g i.v.).

The control of acidosis. Lack of response to the treatment may be attributed to hypoxic metabolic acidosis. If to this ventilation acidosis is added, the situation is extremely dangerous. Actually, metabolic acidosis is treated only by re-establishing a normal peripheral circulation and not merely by 'neutralization' with alkaline substances. Acidosis may be controlled chemically with extracellular (sodium bicarbonate) or intracellular alkaline solutions (THAM). Ventilation acidosis is controlled by artificial respiration, frequent aspirations, and even tracheostomy. For the shock kidney to take up its excretion and for acidosis compensation to function again, renal filtration must be improved by an adequate blood volume and the administration of α-lytics. Practically, buffering of the transfused blood with alkalies is not absolutely necessary [16], but buffering of the acidosis generated by the metabolic deficiences of shock is essential (eventually by the cautious administration of calcium). Simple buffering of acidaemia is insufficient; it must be done by perfusion and the washing away of the peripheral acid waste.

12.2 TRAUMATIC SHOCK AND ITS VARIANTS

12.2.1 GENERAL

From the very beginning of the medical practice traumatic shock has been described as a separate entity, apart from haemorrhagic shock, by the fact that there is no important bleeding. Considered for a long time as an 'isovolaemic shock', shock due to a maldistribution of the blood, with a major corticosubcortical component — and consequently treated with drugs rather than with fluids — the idea has gradually changed until today it is considered to have a typical pathogenic origin encountered in almost all types of shock.

Traumatic shock starts by abrupt stimulation of a vast receptor field, following the action of physical agents. In terms of the causal agent the primary lesional syndrome may consist of damaged tissues, a comminuted fracture, a visceral wound, etc. Blast shock, tourniquet shock, crush shock, surgical shock, etc. are all related to traumatic shock and form a true family with common pathogenic traits.

Traumatic shock has been a constant subject of study for many investigators (Shoemaker, Mareş-Şuteu-Văideanu-Cafriţă) [38, 63]. It may be considered the most 'cybernetic' of shocks because it develops by disturbances of reception at the start, followed by neuro-endocrine generalization of the reactional syndrome.

Traumatic shock is mistakenly considered to be *isovolaemic;* it is in point of fact *dysvolaemic* with endovascular hypovolaemia (decrease in ECBV), since, when we speak of blood, we refer to the circulating blood and not to total blood (*see* Section 1.5.1). The central features of traumatic shock are actually those discussed in chapters 3, 4, 5 and 7.

The rheodynamic disturbances of traumatic shock have led to the general classification of stages of shock, the cellular metabolic disturbances being typical (*see* Sections 8 and 9).

As already mentioned, haemorrhagic shock in itself is reversible up to BP values of 20 mmHg due to the absence of factors that might trigger DIC in the microcirculation (where the pooled acid blood 'waits' a favourable moment to jellify). In traumatic shock, DIC always develops, as even small amounts of humors from the crushed tissular areas, lysed platelets or red blood cells are sufficient to initiate acute microthromboses. An association difficult to treat thereupon develops: trauma, which following the onset of the DIC episode, generates haemorrhage. This happens in the typical traumatic shock due to massive crushing of tissues. Bywaters describes a state of shock, with acute renal failure, in people who have been buried for several hours under buildings (ruins) that have collapsed. At necropsy, the stigmata of DIC were always found in muscular necrosis and in the distal tubulocytes. Bywaters compared these lesions with those that occur after transfusion with incompatible blood. The erythrocytes contain a very active thromboplastin, which normally has no visible effects, but in shock may become manifest. According to Hardaway, in an injured patient in shock, 20 ml of haemolysed blood are sufficient to trigger a DIC episode. It is well known that patients with crushed soft tissues tolerate haemorrhage very badly, even when minimal, because they are added to the haemorrhagic consequences of DIC.

In *tourniquet shock* the events are very similar. On releasing the tourniquet the body is flooded with a collection of toxic factors foremost among which are the tissular activators of DIC (*see* Section 7.4). Mixed protease activation (plasmatic and tissular) appears to take place in the limb encircled by a tourniquet, which may be inhibited by large Trasylol doses (experimentally 50 000 U/kg body weight) only when administered before removing the tournique [18].

Bacteriemic shock, which by its frequency and severity, is one of the most feared factors in traumatic shock, is not due to the collapse of antibody genesis as initially believed. In mice with tourniquet shock, antigenic stimulation has revealed that the 7S and 19S immunoglobulin titre shows no change in comparison to the controls (39); this means that diminution in the clearance activity of the RES, inhibition of phagocyte mobilization and decrease of complement factors and of the plasma bactericide titre may be accounted for by metabolic failure of the antibacterial defence cells generated by shock (*see* Section 6 and 8).

12.2.2 TRAUMATIC SHOCK FROM FIREARM WOUNDS

Among the multitude of causal agents of shock (*see* Section 1.3) are firearm wounds.

The features of firearm wounds are well known because of their frequency in cases of major shock. Wounds caused by missiles such as bullets may be grouped as follows:

Laceration and *crushing* when the bullet comes in contact with hard tlssues at lower speed; the gravity of these lesions lies in the damage to vital structures (the spinal column) or vessels crushed against a bone (aorta, cava veins);

Shock waves of short duration (15-25 milliseconds) but great strength, their impact causing muscular attrition and lesions of the cavitary viscera;

Temporary cavitation produced by bullets entering at high speed when rupture of the small vessels and comminuted bone fractures extend over wide areas. Sometimes expansion of the cavity is excessive, exceeding the elasticity of the tissues; thus, in the abdomen, the intraperitoneal pressure falls so suddenly that the stomach and intestine may collapse without being directly affected by the bullet;

Infection. Contrary to older doctrines (Ogilvie) bullets with the chromogen germ *Serratia marcescens* form rusty colonies in gelatin blocks [74]. A bullet is always considered to be a source of infection, infection being produced by the cavitational suction effect which draws in germs from both ends of the wound [28].

Oscillographic, cinematographic recordings and X-rays have made it possible to establish two large lesional effects of firearm wounds: the role of temporary pulsatile cavities in the tissues, and the possibility of infection due to a bullet and/or by aspiration from the cavity (Hopkinson, 1967).

The formula of the energy lost by the tissues is:

$$T_t = \frac{mV_1^2}{2}(1 - e^{-2\alpha x}),$$

where T_t is the energy released to the tissues;

m — the mass of the bullet;
V_1 — the speed of the bullet on impact with the tissues;
α — the slowing down coefficient;
x — the distance travelled by the bullet through the tissues.

The formula of the slowing coefficient likewise helps us to reach a better understanding of the factors upon which the energy ceded by the tissues depends:

$$\alpha = \frac{AC_d}{2m}\rho,$$

where A is the area of the bullet;
C — the drag coefficient
ρ — the density of the medium through which the bullet travels.

The complex dependence of the multiple factors renders the clinical anatomy aspect of firearm wounds often unforeseeable and generates very severe states of shock.

12.2.3 THE CRUSH SYNDROME

At first called 'crush injuries syndrome' it was described by Bywaters in 1941. Five years later Lucke reported on the 'inferior nephron nephrosis syndrome' in burned patients and in transfusion accidents. Then Mallory communicated cases of "haemoglobinuric nephrosis' in traumatic shock. The nephron is considered today as reflecting various aetiologies that will all rapidly lead to the onset of acute renal failure.

The crush syndrome, similar to tourniquet shock, is one of the aetiologic circumstances of the onset of shock; in turn, both represent variants of traumatic shock. The crush syndrome is fairly frequently encountered in natural cataclysms, especially earthquakes, and its incidence varies within narrow limits; for instance, after the earthquake at Agadîr, 118 of 1550 casualties developed the crush syndrome (7.6%) and at Skoplje, 34 of 618 (5.5%). After the bombardment of Hiroshima, about 20% of the casualties developed the symptomatology of the crush syndrome. A similar proportion was recorded in the earthquakes of Peru, Jordan and East Pakistan more recently.

Clinical picture. The syndrome develops following prolonged compression, with destruction of the muscles. After removing the patient from under the weight crushing him, marked œdema develops in the area of the crushed muscles and within 12 to 24 hours a state of hypovolaemic shock develops and is interpreted by the extent of the œdema. With the onset of shock, free haemoglobin and myoglobin casts appear in the urine, and in 79% of the more severe case haemoglobinuric nephropathy sets in. Toxaemia caused by the overflow into the systemic circulation of catabolism products, histamine and kinins aggravates circulatory rheodynamic disturbances.

When first removed from under rubble or ruins the traumatised case has a normal aspect. Within the first hour a white œdematous exudate appears in the area that was crushed, then gradually hypotension sets in and the arterial radial pulse disappears. The patient soon becomes pale, cyanotic and sweats profusely. Urinary catheterism reveals a dark urine with a brownish sediment.

Evolution. Even when the patient recovers from the initial shock after two to three days, acute renal failure, which is the stigma of the crush syndrome, may develop. If diuresis gradually decreases the prognosis is severe. In the absence of haemodialysis, malignant nephropathy evolves to death. There are also, however, benign forms of myoglobinuria, where the renal tubules are not blocked and in which diuresis can be maintained at a satisfactory level.

In terms of the degree of renal involvement three types of nephropathy may be differentiated:

- simple myoglobinuria (without renal involvement);
- myoglobinuria with the presence of myoglobinuric casts;
- malignant nephropathy with acute renal failure.

If early treatment is instituted to sustain systemic haemodynamics the first two nephropathies may run a benign course. Malignant nephropathy is of extreme gravity and requires haemodialysis in addition to the complex antishock treatment.

In the crush syndrome mention must also be made of the myocardial disturbances with ectopic rhythms, particularly when the plasma potassium exceeds 6 to 7 mEq/l.

The acid-base balance is deeply affected, the Astrup tests usually showing a fixed metabolic acidosis. Non-protein nitrogen increases sharply up to 6 to 7 g in three to four days. The digestive phenomena are dominated by intractable vomiting which produces an accentuated hydroionic imbalance.

Preventive treatment. A major priority in the treatment of these cases is to prevent them becoming irreversible. This involves the remission of haemodynamics, initially realised with macromolecules then with crystalloid solutions, particularly slightly hypertonic solutions which also have an osmodiuretic effect when rapidly administered.

To prevent catecholamine vasoconstriction and to maintain an adequate flow, α-blocking agents with a diuretic effect are used to increase the glomerular flow. Nozinan also has an antihistaminic effect, opposing the serous extravasation characteristic of the syndrome.

The onset of bacteriemic shock is prevented by administering wide-spectrum antibiotics.

Alkalinotherapy, initially proposed by Bywaters, has a double role: to control progressive metabolic acidosis and to alkalinize the urine in order to prevent the precipitation of myoglobin in the urinary tubules.

The curative treatment may be surgical or pharmacotherapeutical.

Surgery includes decompressive aponeurotomies to release the œdematous muscles. There is a risk of infection in the devitalised muscles, hence rigorous asepsis of the wounds is of the utmost importance. The necrosed and ischaemic zones must be treated by exercising the devitalized muscular masses or, when the limb is jeopardized, by amputation.

For slow, progressive promotion of the circulation an elastic band should be placed at the root of the limb and gradually released. Local refrigeration gives good results but is not always applicable.

The blood flow in the microcirculation must be re-established as soon as possible by means of blood perfusions, colloid and crystalloid solutions, α-blocking agents, cortisone in large doses, magnesium and antienzymes [18].

The treatment of acute renal failure. An attempt should be made to maintain an increased urinary flow by hydration, osmotherapy and diuretics. Daily control of the plasma and urinary ionogram will indicate the necessary ionic corrections.

Once renal failure has set in haemodialysis must be instituted, since it is the only procedure that can save the patient. When diuresis is resumed the prognosis will soon become favourable. The absence of secretion after four to five haemodialysis treatments shows that the tubular lesions are irreversible, and the prognosis poor.

12.2.4 OPERATIVE AND ANAESTHETIC SHOCK

Operative trauma is a daily potential cause of shock. Although remedial in intention, operative trauma, which the surgeon tries to attenuate and the anaesthetist to correct, maintains its lesional character.

Anaesthesia may induce shock *per se* where there is an overdosage of the anaesthetics used or unsuitable substances that disturb the fragile homeostasis. Operative shock and anaesthesia are in point of fact only other aspects of traumatic shock.

Determination with isotopes of intra-operative depletion has shown that plasma losses are greater than erythrocytic losses, probably due to the capillary venular plasma skimming phenomenon, that lowers haematocrit in this compartment [30].

The parameters of operative shock depend upon surgery and the patient in a relationship that is difficult to graduate systematically. In radical operations for cancer the effect is always added to the patient's diseased constitution, so that in such cases it is almost always a question of 'chronic shock'.

Numerous well-known schemes have been drawn up to establish the operative and anaesthetic risk in terms of emergency, age, associated diseases, etc.; actually, they only underline the fact that every day critical cases are treated in the surgical and intensive care wards in an attempt to prevent or solve shock states.

When a shocked patient is operated on and shock persists after the operation, the most frequent causes are:

- the initial loss of fluids has not been sufficiently compensated for or continues due to the same cause or to new causes brought about by surgery;
- the onset of severe infection (peritonitis, febrile obstruction, abscess, etc.);
- the persistence of a marked ionic imbalance;
- acute coronary obstruction, cardiac tamponade or pulmonary embolism;
- the precipitation of acute pancreatitis;
- the onset of cerebral hypoxic lesions;
- the onset of pneumo- or haemothorax;
- relapsing biliary ileus with hepatorenal insufficiency;
- bleeding of the raw area of the transected part, stress ulcer, hepatic insufficiency or Zollinger-Ellison syndrome may account for the recurrence of bleeding by upper digestive haemorrhage;
- bleeding due to coagulopathy or systemic or loco-regional fibrinolysis.

Preoperative appraisal of the pulmonary and renal functions especially as regards the acid-base balance, are of utmost importance [53, 72, 74].

Hypoxia due to hypoventilation accompanies abdominal surgery just as frequently as thoracic surgery but with an intensity running parallel to the duration of the operation [72].

Postoperatively, there is an immediate critical period of cardiac and respiratory arrest, together with hypomagnesiaemia in some cases [59].

When the neuronal control mechanisms of arterial and venous resistance and of cardiac performance are interrupted pharmacodynamically, hypotension may be very severe. Re-equilibration can be rapidly obtained by the anaesthetist if noted in time; when ventilation dysfunction, acid-base imbalance and coagulation disturbances are added, re-equilibration becomes very difficult. Many general volatile or gas anaesthetics (except nitrogen protoxide) are inotropically negative. Ether is autocompensated by simultaneous adrenergic stimulation. Cyclopropane increases plasma catecholamines but at the same time has a parasympathetic effect, lowering the heart rate. When atropin is used, cyclopropane becomes hypotensive. Halothane produces hypotension both by its myocardial depressor effect and by its peripheral papaverin effect. Rapid anaesthesia with barbiturates lowers the blood pressure owing to their myocardiodepressor effect.

In regional and local anaesthesia hypotension is the result of pre and/or postganglionic neuronal blocking. Hypotension after local anaesthesia is not due to a hypersensitivity to the anaesthetic agent as much as to overdosage. Procaine accidents (convulsions, coma) may be controlled by the intravenous administration of barbiturates, by myorelaxants and oxygenation.

Hypotension is frequent after spinal anaesthesia. High spinal anaesthesia in pregnancy is aggravated by compression of the pregnant uterus upon the cava veins, preventing venous return. Peridural anaesthesia is less frequently accompanied by hypotension, and is electively indicated in patients with a poor constitution. Neurolepanalgesia type II, with or without Pentazocin confers increased safety, as the drugs do not alter the performance of the heart or of the microcirculation.

It is certain that the incidence and intensity of anaesthetic shock depends to a great extent upon the patient's pathologic condition. Adrenocortical insufficiency (especially after frequent cortisol treatment), porphyria, diabetes, alcoholism, neuropathies, etc. should be known and corrected in time to prevent serious accidents.

12.2.5 BLAST SHOCK

Blast is a violent wave of increased atmospheric pressure followed by a wave of decreased pressure displaced at high speed. It causes varied lesions to the human body, in terms of the region upon which the shock wave has exercised its compression and decompression with maximum force [6, 11, 14, 41, 57]. The mild forms of blast shock are not serious, the disorders being of neurovegetative order. However, casualties with blast shock are extremely delicate from the haemodynamic viewpoint, and a simple postural change may cause cardio-respiratory arrest.

According to the impact surface of the blast blow several types of lesions may be differentiated:

- lesions of the pulmonary type;
- lesions of the abdominal type;
- lesions of the nervous type.

Pulmonary lesions. The clinical course takes on two aspects:

The immediately lethal forms in which the apparent lesions are minimal and histologic examination reveals only small pleuropulmonary haemorrhagic foci. Reflex death is produced by the blast consequent to abrupt neuroautonomic inhibition.

Attenuated forms in which at the gross examination the lung has a 'drowned lung' aspect. The lesions are bilateral, somewhat more accentuated in the right lung, and consist of haemorrhage and sometimes alveolar and bronchial rupture. In blast shock the predominant aspect is that of pulmonary failure. The patient is dyspnoeic, with haemoptoic sputum or even haemoptysis, and bleeding of the mucosa. Traumatopnoea is intense. The symptomatology may develop immediately or after a 48 hours symptomless interval. Acute pulmonary œdema develops after two to four days, when a supra-added infection may also appear.

The treatment consists in adequate ventilation by endotracheal intubation or tracheotomy with the introduction of a tube connected to a resuscitator. Aspiration, aerosol therapy with antifoam drugs and antibiotics should be done as often as possible. The central venous pressure is monitored in order to prevent hyperhydration and overloading of the lesser circulation. Massive antibiotherapy is necessary for preventing a supra-added bacteriemic shock. Rupture of the larger bronchi can only be solved surgically (aerostasis). Pneumothorax and haemothorax always necessitate pleurotomy and continuous aspiration respectively.

Abdominal lesions. In general, the lesions are mixed thoraco-abdominal but there are cases in which the abdomen alone is hit by the blast. The violence of the shock may produce mesenteric disinsertion and ruptures of the parenchymatous or cavitary organs. The patient often looks as if he was suffering from severe haemorrhagic shock: the skin and mucosa are pale, the extremities cold, and agitation due to cerebral hypoxia develops. Internal haemorrhage may be checked by intra-abdominal puncture. In case of rupture of a cavitary organ a characteristic pneumoperitoneum appears on the radiographic screen. Abdominal contraction and pain depend upon the organ damaged.

Treatment. Once the diagnosis of internal haemorrhage or peritonitis is established the patient must be operated on after brief resuscitation. Preoperatively large amounts of blood or substitutes will be administered under control of the CVP. In mixed thoraco-abdominal lesions artificial respiration should be applied to help recovery of the pulmonary lesions, until spontaneous breathing is satisfactory.

Cerebrospinal lesions will bring about immediate death or comas of various degrees.

Failure of treatment is due to certain features characteristic of this type of blast shock:

- the violent character and excessive extent of the anatomic lesions (mediastinal emphysaema, suffocating haemopneumothorax, haemopericardium, ruptures of the spleen, liver, cavitary organs);
- particular visceral effects (myocardial ischaemia, arrhythmia, hyposystoles);
- the high incidence of supra-added bacteriemic shock.

12.3 BACTERIEMIC SHOCK

Experimental endotoxin shock. It is assumed that endotoxins penetrating into the circulatory stream react explosively with the figured and plasmatic elements of the blood; the reaction is similar to an anaphylactic one. Similar responses are produced in animals by hydatid fluid, peptones, ascarid fragments or histamine. In the dog, endotoxin shock is believed to be caused by pooling of the blood in the splanchnic venous system, by vasoconstriction of the suprahepatic veins following the release of histamine into the systemic circulation, but this type of pooling has not been observed in the monkey and in humans (Kux, 1971; Hinshow, 1970).

The term of 'endotoxic shock' has been restricted to the laboratory [32, 77]. At present 'bacteriemic shock' is used in order to designate the clinical situations in which infection, without the apparent loss of a mass of blood, is the direct cause of shock. For a better understanding the causal agent ought to be determined, for instance peritonitis with bacteriemic shock produced by *E. coli*. In the clinic, however, the term toxicoseptic shock or simply septic shock is widely used but is less precise than bacteriemic shock. It should likewise be emphasized that apart from the typical bacteriemic shock there is a *pathogenic bacterial component* in almost all the other types of shock.

Although hypotension has been described from time to time in association with bacterial infection, it was Waisbren who, in 1951, first indicated that a precise shock-like state could be seen in some patients with Gram-negative septicaemia.

The development in the human body of a major septic foci may significantly alter homeostasis. When the imbalance reaches the threshold at which metabolic disturbances block cellular respiration, shock accompanied by hypotension or not, develops.

The pattern of bacteriemic shock differs from that in which there is an actual loss of fluid; *due to intercompartmental dysvolaemia bacteriemic shock has a direct tissular insertion being considered the most cellular type of shock* [7, 25, 27, 37, 63].

After following up 9 haemodynamic and biochemical parameters Siegel noted that the gravity of bacteriemic shock is always correlated with the biochemical parameters and seldom with haemodynamic ones [64].

By itself, or accompanying the late stages of other types of shock, bacteriemic shock is the severe complication of a septicaemic state which, not until long ago was almost always lethal; today survival of up to 50-80% of the cases is reported [5, 9, 12, 44, 63].

By their previous lysis Gram-negative germs release endotoxins; Gram-positive cocci release exotoxins and are less often the cause of bacteriemic shock, the same as rickettsia, spirochetae, viruses and fungi. When the Gram-negative bacteria penetrate into the circulation, as a rule from the gastrointestinal tract, they trigger a characteristic shock. Bacteriemic shock is more frequently encountered in older patients who, therefore, have diminished general resistance to the bacterium and a low properdin titre.

Haemorrhagic or traumatic shock may be complicated by infections with Gram-positive or Gram-negative germs; staphylococci cause local œdema whose amplitude is sometimes sufficient to induce a marked deficit of the circulating blood volume. In supra-infections caused by Gram-positive bacteria shock is generated by extravasation of the fluid from the dilated microcirculation. In shock caused by Gram-negative bacteria the initial hypotension is due neither to vasodilatation nor to the loss of fluid, but to spasm of the microcirculation accompanied by the uptake of fluid in the microclots of the DIC episode and its pooling in the interstitium. The clinical course of a shock complicated by meningococcal infection is particularly severe, death occurring within a few hours. Fine [19] found that at the origin of peripheral disturbances in any type of shock there is a neurotropic toxin deriving from Gram-negative bacteria that invade the organism in the phase of late or refractory reversible shock. Endotoxin is a lipopolysaccharide to be found in the wall of the bacterial cell. Endotoxin shock is, in the essence, a sympathicomimetic shock with a double origin; initially, endotoxins increase catecholamine concentrations, then combine with certain blood elements (probably leukocytes or platelets) and form a sympathicomimetic substance with a supplementary action (serotonin is the most likely).

It has been sustained (Fine, 1962) that the absence of the response to the therapy with blood masses in any type of shock is caused by endotoxins, assumed to be the 'factor of irreversibility'. This is based upon the following experimental findings:

- animals that become resistant to endotoxins survive haemorrhagic shock;

- antibiotherapy with an intestinal bacterial flora spectrum visibly lower the mortality rate in all types of shock.

By its characteristic vasoconstriction any prolonged shock alters the detoxifying function of the RES (*see* Chapter 6). In advanced shock, the Gram-negative bacteria as such or the bacterial endotoxins are absorbed in the systemic circulation. As the RFS loses its detoxifying function and the properdin titre falls to zero, the general immunobiologic index collapses and the endotoxins play havoc, the site of election being the microcirculation system. Endotoxins aggravate the vasoconstriction caused by hypovolaemia and accentuate the peripheral tissular perfusion deficit. Hypoproteinemia, leukopenia, and the arrested synthesis of antibodies are factors that contribute to exacerbation of the saprophyte flora.

Bacteriemic shock proper is a false isovolaemic shock, since although the total blood volume is not modified the effective circulating volume is lowered by pooling of the blood in the peripheral circulation. The decrease in the ECBV is brought about by diminution of the venous return to the heart, of the cardiac output and minute-volume. Slowing down of the circulation in the arterlo-capillary system is accounted for by increase in the viscosity of the blood and sludging. Increase in the viscosity is a major pathogenic coefficient. Today it is considered that peripheral damming up of the blood is due especially to diminution in blood fluidity rather than to vasoconstriction (*see* Section 7.4).

In the microcirculation, hypoxic stasis brings about the release of platelet thromboplastin which, together with agglutination of the thrombocytes, triggers the intrinsic mechanism of coagulation and the onset of DIC. Thrombotic blocking of the vital areas then aggravates the state of shock by the onset of a defibrination syndrome.

The genesis of bacteriemic shock has not yet been fully elucidated; the direct action of endotoxins on the myocardium or vascular endothelium, the indirect action of complement mediated by the platelets or by neuronal tropism, or its direct insertion upon the enzymatic chains, results in shock. Therefore endotoxin brings about an anaphyactic response, generating an elective vasospasm in the structures provided with α-adrenergic receptors. Spasm is produced either directly upon the endothelial cell through platelet serotonin or fibrinopeptides A and B, or indirectly as en effect of DIC. At any rate the pathogenic key to bacteriemic shock is the onset of DIC. The tropism of endotoxins for platelets is remarkable (within 8 minutes 97% of the amount injected is to be found in the platelets). Agglutination of the platelets starts after 5 minutes and the RES takes up the endotoxin within 10 minutes of the injection. In the dog, the microcirculation is affected in several of the target organs — intestine, liver, kidney — in which frequent foci of haemorrhagic necrosis appear. The release of biogenic vasoactive substances is manifested by a storm in the intestinal villi which ends by their contraction and rupture. In man it is especially the lung and kidney that suffer; in the severe forms the intestine is constantly affected.

The actual existence of the 'myocardial depressor factor' has been recently demonstrated and confirmed by its isolation especially in cases of bacteriemic shock (*see* Section 7.4.5).

In the pathogeny of bacteriemic shock the CNS has also been incriminated. Reilly proved that endotoxin shock appears more rapidly when the endotoxin is injected into the splanchnic nerves than when administered intravenously. Fever has been attributed to the direct action of endotoxins upon the posterior hypothalamus. It was demonstrated in dogs that ulcerous lesions of the gastrointestinal tract develop after introducing small amounts of endotoxin into the 3rd cerebral ventricle.

An important stage in the irreversibility of endotoxin shock is the release of hydrolases and lysosomal digestion. An intense lysosomal activity appears to take place in the intestine and spleen, where the release of hydrolases and digestion is maximum, the liver representing the target organ. Rapid increase in the lysosomal enzymes lends support to a direct action of the endotoxins upon the lysosomal membrane [80].

The clinical picture of bacteriemic shock consists essentially in:

- an abrupt rise in temperature (39°-40°C);
- tachycardia with a filiform pulse [130/150 beats/min);
- hypotension (less than 80 mmHg with narrowing of the difference);
- chills, pains in the muscles;
- warm and dry skin in the early stages; later it becomes cold, pale and moist, slightly or intensely icteric;
- neuropsychiatric disturbances;
- vomiting, diarrhoea;
- oligoanuria.

Septicaemia usually occurs within 12 hours of the 'initiating' episode and death may supervene 2-3 days after the onset.

The increase in peripheral resistance proves that hypotension is not brought about by 'vasomotor collapse', as initially assumed. The CVP is normal, demonstrating that bacteriemic shock is not the result of congestive failure of the heart. The total plasma volume is normal if vomiting and diarrhoea do not set, in leading to water depletion.

It is considered that approximately 25—30% of the patiensts with Gram-negative bacteria develop shock. The following species, in their order of frequency, have been incriminated in the production of bacteriemic shock: *Escherichia coli*, *Aerobacter-Klebsiella*, *Proteus*, *Pseudomonas pyocyanea*, *Salmonella* and, dating more recently, *Bacteroides genera* and *Serratia marcescens*. The highest mortality rate is reported in bacteriemic shock due to Proteus. Bacteriemic shock is seldom observed in patients under the age of 40 except postabortum, postpartum or due to a chorioamniotitis that may appear at any moment in the course of pregnancy. Diabetes and chronic hepatic insufficiency predispose to bacteriemic shock. The origin of this shock must frequently be looked for in the genito-urinary tract. Many of the cases of bacteriemic shock are associated with purulent cholecystitis, cholangitis, pneumonia, peritonitis, meningococcaemia, infantile diarroea *(E. coli* 0-111-B_2), etc.

An ample clinical study of 244 patients with positive haemocultures(Gram-negative) was recently published by Neely [48]; 152 of these patients developed bacteriemic shock. The patients previously treated with immunosuppressives or suffering from cancer had a higher mortality rate. Of the 152 cases, 101 died; of the 14 patients with jaundice 6 had positive Pseudomonas cultures. Five of the patients presented pancreatitis with amylasaemia greater than 1200 uW. Acute respiratory failure was present in 33 cases of which 28 died. In the 101 cases of septic shock that died, the time elapsing between the diagnosis and death was, on average, 95.2 hours, approximately 4 days.

A diagnosis of bacteriemic shock has to be established as early as possible. In a septicaemic patient there is a striking alternation of pallor and cyanosis, tachycardia and oliguria of less than 25 ml/hour. The fall in BP should not be waited for and the additional sign of jaundice appears much later. Leucocytosis with thrombopenia and fever show that the haemoculture must be rapidly performed.

In point of fact, shock-inducing Gram-positive bacteria are coagulo-positive; they act by their thrombinic effect inducing a DIC episode. Staphylococcal or streptococcal exotoxins also act through the platelets. Pneumococcal infections may give typical pictures of the Waterhouse-Friederichsen syndrome. Table 12.2 gives the differential diagnosis of bacteriemic shock in terms of the Gram-negative or Gram-positive aetiology.

Table 12.2 THE DIFFERENTIAL DIAGNOSIS OF BACTERIEMIC SHOCK IN TERMS OF THE ITS GRAM-NEGATIVE OR GRAM-POSITIVE AETIOLOGY

Bacteriemic shock with Gram-negative germs	Bacteriemici shock with Gram-postive germs
very low BP	moderate BP
cold skin	temperature of the skin and diuresis less affected
low urinary flow	
vomiting and diarrhoea	absence of vomiting and diarrhoea
severe neuropsychiatric disturbances	minor neuropsychiatric disturbances
vasoconstriction of the microcirculation	vasodilatation of the microcirculation
low cardiac output	normal cardiac output
low CVP	normal CVP
increased peripheral resistance	normal peripheral resistance
severe acidosis	absent or moderate acidosis
60−90% mortality rate	30−40% mortality rate

There are four clinical types of bacteriemic shock:

Type 1, the common type of bacteriemic shock. The toxins affect the myocardium directly. The onset is drastic, with a marked decrease of the CO and BP, increase in the CVP, severe oliguria and acidosis. Inotropism may be aided by digitalis and β-stimulating drugs.

Type 2A. The toxins produce vasodilatation in the microcirculation. There is no acidosis and oliguria, the skin is warm, the BP higher than 70 mmHg (added hypovolaemia against this background rapidly transforms it into type 1 shock). Type 2A bacteriemic shock frequently has a Gram-positive aetiology.

Type 2B is similar to type 2A but with acidosis, and a cyanosed, cold skin. The correction of acidosis is of first importance. Cellular metabolic affection is severe.

Type 3. To bacteriemic shock a loss of serous fluid rich in proteins is added (for instance peritonitis, febrile occlusion, etc.). Apart from alkalinisation plasma must be administered. *This is the type of surgical bacteriemic shock.*

Type 4 is the bacteriemic shock which is refractory to treatment and arrived at following the evolution of the other three previous types (due to ineffectiveness of the treatment). This type is severe, associated with jaundice, heart failure and increased CVP values. The treatment includes cortisone, haemodialysis and oxygen administered at higher pressures.

Therapeutical principles. In spite of its clinical and physiologic features bacteriemic shock evolves according to the same general scheme at the origin of which are rheodynamic alterations and cellular damage. The therapy should be ordered as follows: correction of the fluid volume deficit, administration of α-blocking drugs, eventually β-stimulating drugs [49], cortisone, antibiotics.

It is considered that in 50% of the cases of bacteriemic shock less than 48 hours elapse between onset and death; hence, only a treatment instituted within the first hours will save the patient.

The efficiency of the treatment is followed up by the temperature, aspect of the skin, hourly diuresis, CVP, BP, CO and repeated humoral cultures. The following may be used as a scheme for fluid repletion:

At a CVP of 10 cm crystalloids are administered; at a Ht under 30% blood and Rheomacrodex up to 500-600 ml, to which a double amount of crystalloids is added. Cortisone (30-50 mg/kg body weight) in doses repeated every two to four hours, are combined with β-stimulants (Isuprel) and especially β-blocking drugs (thymoxamine or phenoxybenzamine; in practice Dibenzilin 1 mg/kg body weight or Levomepromazine in 3 to 4 mg repeated doses).

In spite of its inconclusive results Isuprel may be administered when the heart rate is less than 120 beats/min, with obligatory ECG monitoring [68]. According to the excessive β-stimulating theory β-blocking drugs have sometimes given favourable results in bacteriemic shock [5]. It is assumed that ß-blocking drugs protect especially the pulmonary 'sponge' (preventing right-left shunts).

The treatment should always be preceded by surgical analysis of each case, since the first cause of the high mortality rate and irreversibility of bacteriemic shock are lesions that can be treated surgically. Visceral perforations, subphrenic urinary or intravisceral abscesses, suture dehiscence, obstruction, necrotic pancreatitis, cholangiocholecystitis, etc. are some of the pathoanatomic 'surprises' in bacteriemic shock. Surgery must be performed in time, that is, immediately after the presumptive diagnosis has been established, because the body can fight against a limited dose of germs but not against a persistent attack starting from an untreated source. Incision and drainage, performed in time lower the mortality rate from 100% to 50% [61, 62].

Antibiotics therapy. An efficient antibiotic therapy is bactericide, synergic and massive. Associations must cover both the Gram-positive and the Gram-negative spectra (ampicillin or Cephalotin, and respectively Colimycin, kanamycin or gentamycin) [56].

The therapy with antibiotics starts as soon as shock has been diagnosed clinically and only after the blood, urine or other infected fluids have been collected for cultures.

However, one of the greatest mistakes is to wait for the results of the cultures. The therapy must be instituted immediately, therefore at first empirically, tripling the chances of survival.

It is recommended after establishing the diagnosis, collecting and seeding, to administer 500 mg tetracyclin and 4 million units penicillin (i.v., diluted in 500-10 ml normal saline) every 6 hours. Kanamycin in 7.5 mg/kg body weight doses, by intramuscular route, may be administered as the initial dose; subsequently the daily dose may be lowered to 1.5 mg/kg body weight (in 3 doses). When the urine falls below 800 ml/24 hours the tetracyclin and kanamycin doses should be reduced proportionally. In bacteriemic shock caused by Pseudomonas a daily intramuscular dose (separated into equal parts) of 2.5 mg polymixin B/kg body weight may be given. Recently, in shock produced by Proteus and Pseudomonas good results have been obtained with Pyopen (20-30 g/day) and gentamycin (1 mg/kg body weight); however as they are toxic for the kidneys and sciatic nerve they should only be administered for short intervals. Carbenicilin combined with gentamycin is useful in burns covering 80—85% of the body surface and accompanied by bacteriemic shock with Pyocyaneus [34].

In the treatment of Gram-negative bacteriaemia nalidixic acid, Cephalorin and ampicillin may be useful. Cephalothin has proved particularly active against Klebsiella, Proteus and staphylococci, colistin against Pyocyaneus, *E. coli* and Proteus (McLean, 1970; Motsay, 1970; Shoemaker, 1970).

Antibiotics may, however, precipitate the onset of shock, abruptly releasing into the systemic circulation endotoxins derived from the dead bacteria. In order to prevent this, antibiotics have to be combined with pharmacological doses of corticosteroids.

In bacteriemic shock, apart from the medication aimed at the β-receptors, active drugs must also be administered for the α-receptors. In bacteriemic shock, α-stimulants are certainly counterindicated because they produce excessive vasoconstriction. Nevertheless, α-stimulants have been used in bacteriemic shock because they stimulate cardiac performance. The precursors of catecholamines (Dopamine) have been tested with certain success [24]. The α-blocking drugs are better indicated, good results being obtained with levomepromazine, hydergin and dibenzylin, always combined with massive blood volume substitution.

Corticotherapy has imposed itself in the therapy of shock, entering paradoxically by the 'door' of bacteriemic shock. Large doses, 3 to 4 g/24 hours, are administered only in actual shock not in cases of bacteriemia. Of the rich spectrum of corticoids two effects are particularly useful in bacteriemic shock: their short and long term α-lytic action and their stabilising effect on the cellular membrane. Some authors [12, 13, 48, 49, 56, 62] report that by administering cortisone in doses of 30 mg/kg body weight, repeated at 2 to 4 hours (i.e. 4 g in the 24 decisive hours of treatment in bacteriemic shock) the mortality rate is reduced by half. At any rate to obtain a real effect at least 500 mg cortisone must be administered in 24 hours.

Medication of the coagulolytic system. The dynamics of coagulolytic disturbances in themselves are a reason for bacteriemic shock, as severe as it is difficult to treat. The administration of heparin at the right moment might avoid the DIC episode with its disastrous effects. Although in the clinic the results are not always conclusive, heparin appears to be the only means of avoiding necrosis of vital organs. Linoleic acid would be useful to avoid the agglutination of platelets but no practical means of determining the endotoxinaemia dose exists and the best moment for its administration is almost always missed.

Once the DIC episode has passed, therapeutical fibrinolysis should be taken up. Fibrinolysins (streptokinase, urokinase, etc.) are not used, however, although a shift towards hypocoagulability has been noted, because it is not possible to determine whether a reactive hyperfibrinolysis does or does not exist.

It is indicated to administer antiplasmins and antienzymes because secondary fibrinolysis almost always occurs. When it has been established that there is no reactive fibrinolysis the haemorrhagic effects of consumption coagulopathy may be corrected with fibrinogen protected by heparin. Together with fresh blood (containing plasmin and its precursors) antiplasmin must be given.

12.4 BURN SHOCK

Burns should be viewed as a complex dynamic process aggravated by the vicious metabolic circles it gives rise to. The process continues after the deleterious agent has ceased its action, and the patient appears to continue 'to burn slowly' (Lorthior). The phenomenîn is due to proteolysis on the surface of the burnt areas, aggravating toxaemia and plasmexodia.

Today, for a precise and rapid appraisal of the degree of infection, of the prognosis, and the listing of the case in the modern anatomoclinical classification, biopsy is recommended, with the taking of samples from the edge of the burn, within the first 24 hours after the accident [20]. Histology may detect bacteria, and also viruses and fungi, and may help to assess the degree of reversibility of the lesion. In second degree burns the endocrine glands, the pilose-sebaceous apparatus and dermal collagen are involved; infection may affect the entire thickness of the skin and transform the second degree burn into a third degree one, in which coagulation necrosis of the derm down to the lax subcutaneous tissue and muscles is specific. Burns of the fourth degree exhibit whitish or brown eschars.

Along the hyperaemic border of burns (the site of election for biopsy) the degree of colonisation with germs and the limits of surgical excision may be determined. The local lesional dynamics of the burn in burns covering more than 15% of the body surface and the effect upon the systemic system are of shock intensity.

There are two periods when shock may appear: within the first 48 hours and after two to three weeks. In the first instance shock is caused by hypovolaemia, which responds to a repletion treatment, and in the second, by supra-added bacteriemic shock that frequently occurs in burns, especially in third degree burns.

However, shock may also develop in burns in other pathophysiologic ways. The fall in cardiac output may precede the effect of plasma losses due to myocardiotoxins released by the burnt tissues, which are known to contain thrombin-like substances and, in addition, the microcirculation in the neighbouring tissues is always thrombosed; Curling ulcers have also been attributed to some thromboses in the microcirculation.

Therapeutical principles. On admission, the cases of severe burns closel resemble those of severe shock. It is considered that when a burn covers more than 15% of the body surface (50% of the burn representing deep lesions) it always generates shock. The intense pain, the loss of fluid in the burned areas, oliguria, etc. are signs of severe shock. Resuscitation of burned patients is an extreme emergency and should be instituted within the first three hours of the accident, when haemodynamic disturbances are still minimal and have not yet affected renal function.

The immediate treatment includes analgesics of the petidin group or combined with Levomepromazine (analgesic and α-blocking agent) in order to assuage the pain, sustenance of the systemic haemodynamics, insuring an efficient diuresis and antibiotic therapy.

The volume of fluids that must be administered is, however, still subject to discussion. Classically, the fluid necessary in case of burns is assessed in terms of the body weight and extent of the burned surface.

Most authors recommend 1 ml macromolecules and 1 to 2 ml crystalloid solution per kg body weight and per proportion of burned surface every 24 hours (50% of the total amount is administered within the first 6 hours).

On taking into account only the body weight a fluid volume equivalent to 10% of the body weight is administered within the first 48 hours. The criterion applied for evaluating the efficiency of vascular refilling is the return of diuresis, increase in the CVP over 10 cm of water, and increase in the BP.

Intensive haemolysis in the cutaneous necrosis foci demands the administration of blood in spite of the haemoconcentration that appears after 24 hours [29]. In burns, the increase in viscosity appears to be caused by increased Ht only to a slight extent [81]. However, in the late phases of burn shock, blood perfusions are not easily tolerated.

Blood should only represent about a quarter of the fluids perfused in the course of the first days.

Initially, vascular refilling is done with a plasma expander (Dextran) and 10% glucose (with insulin). The intermittent administration of Levomepromasin by its effect upon the ARAS and on the microcirculation insures autonomic deafferentation and an α-blocking effect, favouring tissular perfusion and renal filtration. Some authors [42] maintain that Ringer's solution offers results superior to those obtained with human serum albumin.

For general anaesthesia in burns only the drugs that do not produce myocardial depression, hypotension or hypoventilation can be used; analgesia must be powerful and hypnosis moderate. Drugs with an antidiuretic effect are counterindicated. In general, neuroleptanalgesia or ataralgesia which do not affect the vital functions are preferable.

Nitrate hypercatabolism is characteristic of the burned patient; in spite of the supply of glucose solutions, urinary nitrogen rapidly reaches high concentrations, and the nitrogen balance remains negative, regardless of the protein supply. Glycosuria is constant even under conditions of a minimal carbohydrate supply. Serum pseudocholinesterase decreases in proportion to the gravity of shock [54], and hyponatraemia and hypokaliaemia appear (in case of an intact kidney). After some time hypoproteinaemia develops with inversion of the serin/globulin ratio. When the serum albumins fall below 3 g the prognosis becomes uncertain and any eventual grafting procedures should be put off.

The acid-base balance oscillates; after an initial phase of ventilation acidosis (as a result of hyperventilation) metabolic acidosis develops and is revealed by the excessive reduction of bases, standard bicarbonate and total bicarbonates.

The prolonged catabolic phase due to delayed surgery, ineffective treatment or septic complications results in the onset of chronic shock which has a trailing evolution and is difficult to re-equilibrate.

Complications. Complications are frequent in burns, especially thrombotic complications. In the first weeks hypercoagulability with or without clinical manifestations is constant. The burn in itself is a typical acute microangiopathy, and perilesionally masses of amorphous matter, DIC and sludging appear.

The digestive disturbances vary: digestive haemorrhage (due to ulcer of the Curling type) are fairly frequent; in some instances intestinal fragility prevents alimentation and delays taking up of the transit.

Pneumonia may appear in two forms: a miliary form (forming part of the septicaemia) and a peribronchic aerogenic form with Gram-negative germs.

Of particular importance in the treatment of burns appears to be the therapeutical tandem: early grafting and supra-alimentation.

An early graft favours haemodynamic and metabolic therapy by suppressing the uncontrolled sources of protein and fluid losses. Parenteral and peroral supra-alimentation helps the body to fight against infection and to regenerate the plastic protein background. When the face and mouth are burned the patient is fed through an œsogastric tube or by gastrotomy. Through the gastric tube fluid or semifluid food can be introduced, representing up to 3000 to 4000 calories per day (milk, butter, cocoa, sugar, broth, tea, compot, eggs, etc.).

The first graft should be applied only after a satisfactory energy equilibrium is obtained. Proteinaemia should not be below 6 g and albumins not less than 3 g, otherwise grafting is counterindicated.

In general, the bacilli of infected wounds are resistant to antibiotics and for this reason many authors do not recommend antibiotic therapy by parenteral route if the nature of the wound does not absolutely require it. This implies, however, a very strict asepsis on dressing the wounds and in the wards: masks, air filters, sterile gowns. The local treatment (fibrinolytics, irrigation with Dakin's solution) frequently suppresses local suppurations. The avascular nature of burns permits, however, the administration of gentamycin, sulfonamides and silver nitrate, that reduce the number of germs from 100 000 to 10 000 per gram tissue [43]. When systematically given, the antibiogram should first be performed; in burned patients haemoculture may detect in time a bacteriemia.

Vitamin biocatalyzers should not be neglected: vitamins B_1, B_6, B_{12}. Vitamin C is administered in large doses of 3 to 4 g/day. Hydrocortisone haemisuccinate is given, starting from the 3rd day (not more than 50—70 mg/day). In case of ventilation insufficiency an Assistor resuscitator may be used and disconnected only when spontaneous breathing is satisfactory.

After total grafting of the burned surfaces the hypercaloric alimentation may be gradually reduced, but calorie values must still remain high for a long time. When burns involve the upper respiratory tract tracheostomy and aerosols with antibiotics (using an ultrasonic atomizer) are recommended. Dry plasma or human albumin should never be omitted in the treatment of burns. Even if we admit that part of the proteins administered are transformed from plastic substances into substances with an energy and calorie-producing role, albumin by its multiple attributes will help to maintain a satisfactory biologic status. The principal aim in the therapy of burn shock is not to transform the patient's condition into chronic shock, difficult to control and with a high potential of suprainfection and the hazards of a severe bacteriemic shock.

12.5 ANAPHYLACTIC SHOCK

Allergy represents an abnormal response to a given antigen, whereas *hypersensitivity* is the exaggeration of a physiologic response to any substance with an antigenic character.

Hypersensitivity may be:

- Hypersensitivity
  - immediate (circulating antibodies)
    - anaphylaxis and atopia
    - Arthus reaction
    - serum disease
  - late (cellular antibodies)
    - infectious
    - not infectious

Anaphylactic shock belongs to the first type of immediate hypersensitivity; it is a drastic and often dramatic event that may appear after introduction into the body of any substance, but especially proteins. Although apparently a supra-acute clinical manifestation, the roots of anaphylactic shock are to be found in 'immunologic shock' produced by previous contact with the antigen. By its pathologic physiology, anaphylactic shock is a true shock which may include a supra-acute DIC episode that explains many of the clinical signs.

Supra-acute blocking of the microcirculation by DIC has also been checked by the injection of serum albumin which is always accompanied by the appearance of fibrin thrombi in the pulmonary network. Anaphylaxis is constantly accompanied by purpura, thrombocytopenia, fibrinolytic activation and bleeding of the integument and/or intravisceral mucosa, which are the direct effects of DIC. The antigen-antibody complex certainly has thrombogenic properties, demonstrated some time ago by the Arthus phenomenon.

Initially described as sensitization to heterogenous proteins, anaphylactic shock actually occurs from a very wide range of causes: from the bite of an insect, because of intolerance to certain foods or drugs, and to allergies that accompany various pathologic states. Among drugs with a protein, glucidic or lipopolysaccharide structure are sera, protein hormones, enzymes (antienzymes), heparin and dextrans, protein hydrolysates, and among the micromolecular substances penicillin and its metabolites, streptomycin, chloramphenicol, local anaesthetics, salicylates and iodate contrast media.

The interval between the first and second administration of an allergen is called the risk interval, and may range between three days and several years; as a rule it is considered to be 10 to 20 days up to a few months. Antibodies circulate freely but may be placed within the cell, as they are specific immunoglobulins (IgG with a 7 S sedimentation constant).

The antigen-antibody complexes are precipitating macromolecules which, deposited in the endotheliocytes may give rise to DIC and subsequent collapse of the microcirculation (the Arthus phenomenon is the foremost argument). But these complexes may remain for a long time in colloidal suspension in the body fluids. It is true that complement is consumed in this conflict, but anaphylaxis also appears in animals in which the complement complex has been totally removed.

The actual pathogenic link of anaphylactic shock must be looked for in the suite of intermediate vasoactive substances released during the antigen-antibody conflict. The smooth muscle cells of the microcirculation, bronchioles and intestine are target cells, effectors that perform the messages of the violently released active amines.

The genesis of anaphylactic shock includes an *immunologic stage* (conflict of the antigen with the preformed antibody), a *biochemical stage* (avalanche of vasoactive substances that are immediately activated by the antigen-antibody conflict) and a *visceral stage* manifested in the clinic [51].

Introduced into the organism with which it had already been in contact (once or several times) the antigen abruptly joins the antibodies (circulating and/or cellular), a coupling that degranulates the mast cells, breaks down the platelets and triggers the activating system of the coagulostat (Figure 12.2).

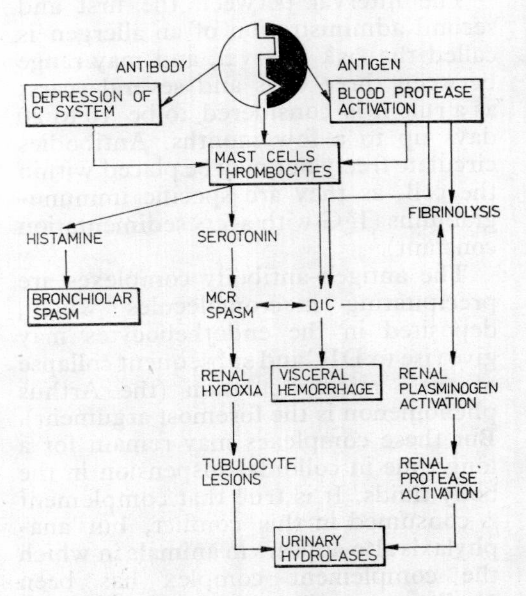

Figure 12.2. The antigen-antibody conflict generates blood activation of proteases, followed by the appearance of hydrolases in the urine.

Histamine, serotonin, bradykinin, acetylcholine, heparin, prostaglandins, enigmatically slow-reacting substances, adenylates, potassium, etc. are explosively produced.

In transfusion shock (anaphylactic) the antigen-antibody conflict also releases a thromboplastic lipoprotein (erythrocytin) that lysis the RBC.

Clinical picture. The impressive symptomatology of anaphylactic shock consists in an extensive reaction of the peripheral circulation, accompanied by a sharp, variable fall in the BP, bronchiolar obstruction with asthmatic attacks, laryngeal œdema, urticaria, Quincke œdema, neurologic and gastrointestinal phenomena and even allergic myocarditis (which in itself may produce a secondary state of shock). Therefore, several clinical forms of anaphylactic shock may be differentiated: respiratory, cutaneous, cardiac, digestive, neurologic, haemorrhagic, the vascular and haemorheodynamic manifestations being however foremost in all cases.

A particular form of anaphylactic shock has been described: hypovolaemic permeabilization, consisting in a slow leaking of the intravascular fluid into the interstitium.

The clinical symptomatology of anaphylactic shock is very rich: premonitory signs, a supra-acute form culminating in collapse, a major classical form and a minor form, monosymptomatic or episodic. Anaphylactic shock evolves rapidly, the end occurring within several minutes by cardiorespiratory arrest, or in 10-15 minutes by asphyxia. The medium and minor forms evolve for 15-20 minutes or are prolonged and attenuated for several hours.

After the primary administration of protein substances (vaccines, milk), colloids (dextrans, gelatins, blood, enzymes, polyvinylpyrolidon, etc.) or following the ingestion of histamine-forming substances, a clinical picture may sometimes develop very similar to that of anaphylactic shock. Actually, in these cases not only is the clinical picture of anaphylactic shock encountered but also its pathophysiologic aspects, without an antigen-antibody conflict at the onset however. These pathologic situations originate directly from the *biochemical link* of vasoactive substances generating anaphylactic-like shock that have been given various names: colloidoclastic shock, protein, histaminic, haematotransfusional shock, etc. All these types of shock may be called by the general term of *anaphylactoid shock* and their therapy is, obviously, the same as for anaphylactic shock proper.

The reaction period in anaphylactic shock is of less than 30 minutes (in the very severe forms 10-15 minutes) but one or several hours may likewise elapse. Although the portal of entry is also of importance in anaphylactic shock (parenteral or oral), as well as the amount introduced, the interval after the previous allergisation, the genetic structure of the individual, etc.

Therapeutical principles. Both diminution of the amount of preformed antibodies and prevention of the antigen-antibody reaction are questionable. Actually, in practice the only possibility is to combat the group of biochemical substances produced by the antigen-antibody conflict and their effect upon the target cells. In the supra-acute forms catecholamines may be used directly, whereas in the subacute forms the therapeutical scheme of a typical shock is applied, hypovolaemia becoming the main objective of the treatment. Cortisone is useful in both instances (in doses of at least 10 mg/kg body weight). Theophylline and methysergid do not give satisfactory results in primates [23]. On the other hand, the inhibitors of histidindecarboxylase and of 5-hydroxytryptophandecarboxylase appear to be efficient (tritoqualine, trade mark Hypostamine and Inhibostamine). To block fibrinolysis antiplasmin may be administered and antienzymes to control protease activation.

The therapeutical attitude that must constantly be kept in view is the clinical course of the patient's condition. It may be applied as follows:
(i) With the first signs of anaphylactic shock a tourniquet is applied for at most 20-30 minutes at the root of the injected limb (injections that may potentially trigger anaphylactic shock are done in the extremities of the limbs).
(ii) 0.5 ml epinephrine 0.1% is then injected by subcutaneous, intramuscular or intravenous route. In the severe crises with cardiac arrest epinephrine may be introduced by the intracardiac route. ECG monitoring is recommended; the dose may be repeated if there is no improvement. Some authors [51] recommend the addition of 0.5 ml epinephrine 0.1% to the injection that might trigger the shock.
(iii) Perfusion is started (eventually in the central veins by catheterisation) with hydrocortisone hemisuccinate in the perfusion fluid.
(iv) Orotracheal aspiration is performed (by tracheostomy or intubation) and oxygen administered.

12.6 CARDIOGENIC SHOCK

The history of cardiogenic shock almost always starts in the hospital medical wards. The subject will therefore be treated only from the 'surgical viewpoint'. The reader may refer to chapter 5.2 in which the myocardium in shock is discussed.

Cardiogenic shock reflects systemic rheodynamic and metabolic disturbances caused by a major defect of the cardiac pump.

The heart function may suddenly become deficient due to myocardial or pericardial, coronary or vascular lesions. The frequent cause of cardiogenic shock is acute coronary occlusion. An increasingly encountered aetiology is myocardial contusion within the framework of chest injuries.

Not all myocardial infarcts cause shock; it is a mistake to diagnose as cardiogenic shock a myocardial infarct established by ECG and whose clinical manifestation is a rhythm disorder and transitory hypotension.

Cardiogenic shock must be understood as a shock secondary to affection of the myocardium, which subsequently evolves according to the pathogenic rules of the classical shock passing through the sequence of peripheral rheologic disturbances [22, 58, 60, 79].

Cardiogenic shock may also have a longer onset following a protracted perfusion due to hyposystolia. It is considered that approximately 12-15% of the patients with myocardial infarct appear to be in a state of cardiogenic shock; on average, shock develops 12 to 18 hours after the onset of coronary occlusion.

When myocardial infarct is complicated by cardiogenic shock the mortality rate soars to 80-85%; about one third of the deaths from myocardial infarct are caused by cardiogenic shock. In this type of shock the cardiac output is always low and vascular resistance and atrial pressure high; some patients exhibit signs of congestive heart failure. When the latter appears early, an increase in the CVP is observed about an hour after the onset of shock. In patients who survive there is a progressive increase in the cardiac output and a return to normal of peripheral resistance.

The symptomatology of the cardiac vascular accident is easy to recognize in the clinic: hypotension with values ranging between 60-80 mmHg a depressed arrhythmic pulse are characteristic. Tachycardia may sometimes replace bradycardia, reflecting a hypervagotonia. The following signs point to an unfavourable prognosis: frequent ventricular extrasystoles, complete atrioventricular block, atrial fibrillation or flutter, ventricular tachycardia (more 130/min) (McFarlane, 1969).

With diminished cardiac output rheodynamic disturbances in the microcirculation and cellular distress develop, with concomitant reduction of renal filtration. An increase takes place in serum transaminases, GOT and GPT, and in LDH; they represent the criteria of differentiation between cardiogenic shock and other forms of shock.

Therapeutical elements. Patients with cardiogenic shock must be treated in specialized units with the necessary equipment for determining the various parameters of the blood circulation (continuous ECG monitoring is essential). The most frequently used drugs in cardiogenic shock are:

(i) Norepinephrine, in moderate doses, sustains the heart and produces medium vasoconstriction.

(ii) Metaraminol (aramine) has a milder action than norepinephrine; it may be administered subcutaneously; its effect lasts 4 hours.

(iii) Isopropylnorepinephrine is particularly effective, especially in cardiogenic shock that evolves with bradycardia; it increases the coronary flow, heart rate, cardiac output and minute-volume. Due to the improved glomerular flow, renal filtration and diuresis increases [31, 71]. In atrioventricular block the best drug is Isuprel.

(iv) α-blocking drugs are less efficient than Isuprel in cardiogenic shock. These drugs should be administered slowly because they may bring about a fall in the B.P. Fluid repletion is compulsory.

Oxygenotherapy increases the alveolar capillary gradient. A mixture of equal parts of oxygen and air increases the oxygen pressure to 300-400 mm Hg and the alveolar pressure to 100 mm Hg, normalizing the P_aO_2 values. Oxygen is administered by aerosols or by humidification.

Hyperoxybarotherapy is the future treatment of cardiogenic shock. Administered at pressures of 3 atmospheres oxygen increases in the plasma to 5 vol %; there is likewise an increase in P_aO_2, in the capillary interstitial pressure gradient, facilitating diffusion and improving cellular hypoxia.

Fibrinolytic therapy. Theoretically, fibrinolysis ought to improve coronary obstruction, prevent eventual thromboembolisms and repair the viability of the ischaemic myocardium. The agents currently used are streptokinase and fibrinolysin; however, fibrinolytic treatment is still in an experimental stage.

Xylocaine has given good results in arrhythmia, especially in tachyarrhythmia. The dose is of 1 mg/kg body weight.

Pethidines are recommended in fractional doses of 25-30 mg. They are administered very slowly to avoid bringing about myocardial depression. Fridberg contra-indicates the intravenous use of digitalis products, especially in the presence of extrasystoles. In the absence of ectopic foci some authors recommend lanatosid C in an initial dose of 0.4-0.8 mg and then every 6 hours 0.4 mg.

Surgery. In spite of the active medication the mortality rate from cardiogenic shock has still not fallen below 85%, suggesting the need for surgical intervention. The first attempts to bypass the central circulation were made with venoarterial and arterio-arterial pumps. In 1960, Bevegard introduced transseptal cannulation of the left atrium, and Senning and Denis, in 1962, developed a closed bypass of the left heart, used at present with fair results [36, 79]. Hypothermia and perfusion of the coronaries with magnesium solutions protect the stopped heart for up to 90 minutes or more [46, 67].

Recently, the benefits of intraaortic balloon pulsation for postcardiotomy myocardial failure were reported, the benefits outweighing the hazard especially in less critically ill patients (Berger, 1973).

Surgical intervention in cases of cardiac trauma or infarct are fairly frequent. Ruptures of the left atrium, crushing against the spinal column in thoracic contusions and penetrating cardiac wounds have been successfully treated [25, 47]. In cardiac wounds from firearms and crushing, simple pericardiocentesis gives good results in 30% of the cases. Emergency operations must be performed in all cardiac wounds caused by firearms, in cases of intractable bleeding, tamponade relapse after aspiration by pericardiocentesis, and wounds that have generated cardiac arrest.

Experimentally, in the dog, an area of 14 cm^2 has been successfully resected from the infarcted myocardium, as demonstrated by the coronographic controls performed five months later. In the clinic, in certain cases of myocardial infarct, aorto-coronary bypass has been carried out in an emergency operation [8, 15, 45, 46, 47]. In practice, it is of the utmost importance to establish the optimal moment for the operation: when the BP is below 90-80 mm Hg, the cardiac index below 2000 ml and oliguria less than 20 ml/hour in spite of an active inotropic treatment, then mechanical circulatory assistance must be given. When no improvement is observed, then the cardiogenic shock must be considered as *refractory to the treatment* and a remedial surgical intervention performed for left ventricular failure. According to Mundth's [47] statistics, surgery (in cases in which about 50% of the contractile area o the left ventricle is infarcted) was performed between 6 hours and 4 days after the onset of cardiogenic shock (on average 57 hours). The intervention consisted in double aorto-coronary bypass with an internal saphena autograft, with or without a coronary endarterectomy. The mortality rate after this operation is not known precisely but has been estimated at less than 50%, which evidently suggests the surgical intervention.

SELECTED BIBLIOGRAPHY

1. ADAM, E., *Şocul traumatic* (Traumatic shock), I.M.F., Timişoara, 1970.
2. BERKUTOV, A. I., *Voenno-polevaia hirurghia* (War Surgery), izd. 2-e, Izdatelstvo Meditsina, Moskva, 1968
3. BERGER, R. L., NOVOGRADAC, W. E., BYRNE, J.J., Surgical and Chemical Denervation of Abdominal Viscera in Irreversible Haemorrhagic Shock, *Ann. Surg.*, 1965, *162*, (2), 191–186.
4. BERGER, R. L., SAINI, V.K., RYAN, T.J., SOKOL, D.M., KEEFE, J.F., Intraaortic Balloon Assist for Postcardiotomy Cardiogenic Shock, *J. Thorac. Cardiovasc. Surg.*, 1973, *66*, (6), 906–915.
5. BERK, J.L., The Role of Adrenergic Blockade in the Treatment of Septic Shock, *Surg. Gynec. Obst.*, 1970, *130*, (6), 1026–1034.
6. BERMAN, J.R., MOSELEY, R.V., DOTY, D.B., GUTIERREZ, V.S., Posttraumatic Alkalosis in Young Men with Combat Injuries, *Surg. Gynec. Obst.*, 1971, *133*, (1), 11–15.
7. BLAIR, E., Hypocapnia and Gram-negative Bacteriemic Shock, *Amer. J. Surg.*, 1970, *119*, (3), 433–439.
8. BUCKLEY, M.J., MUNDTH, E.D., DAGGETT, W.M., DESANCTIS, R.W., SANDERS, C.A. AUSTEN, W.C., Surgical Therapy for Early Complications of Myocardial Infarction, *Surg.*, 1971, *70*, (6), 814–820.
9. BLUCKBERG, G., COHN, J., DARLING, C., *Escherichia Coli* Bacteriemic Shock in Conscious Baboons, *Ann. Surg.*, 1971, *173*, (1), 122–130.
10. CALOGHERA, C., *Şocul hemoragic* (Haemorrhagic Shock), I.M.F., Timişoara, 1970.
11. CANDOLE, C.A., Blast Injuries, *Canad. Med. Ass. J.*, 1967, *96*, (2), 207.
12. CIUCHTA, H.P., Serratia Marcescens Endotoxin Shock in the Dog, *Milit. Med.*, 1970, *135*, (5), 479–482.
13. CLERMONT, H.G., ADAMS, J.T., WILLIAMS, J.S., Source of a Lysosomal Enzymes Acid Phosphatase in Hemorrhagic Shock, *Ann. Surg.*, 1972, *175*, (1), 19–25.
14. CLOUTIER, C.T., Fat Embolism in Vietnam Battle Casualties in Haemorrhagic Shock, *Milit. Med.*, 1970, *135*, (4), 369–373.
15. COHN, L.H., FOGARTY, T.J., DAILY, P.O., SHUMWAY, N.E., Emergency Coronary Artery Bypass, *Surg.*, 1971, *70*, (6), 821–829.
16. COLLINS, J.A., BROITBERG, A., MARGRAFF, H.W., BUTCHER, H.R., Haemorrhagic Shock in Rats, *Arch. Surg.*, 1969, 66, (4), 484–488.
17. DREICHLINGHER, V., DREICHLINGER, T., Infecţiile la bolnavi cu şoc hemoragic (Infections in Patients with Haemorrhagic Shock), *Chirurgia*, 1971, *XX*, (7), 653–658.
18. EIGLER, R.W., KRISTEN, H., STOCK, W., HOFER, I., Der Proteinase Inhibitor Trasylol als Therapeutisches Prinzip beim Tourniquet-Syndrom, *Bul. Soc. Int. Chir.*, 1970, *XXIX*, (4), 197–203.
19. FINE, J., *Ciba Symposium Shock*, Springer, Stockholm-Berlin-Göttingen-Heidelberg, 1962.
20. FOLEY, F.D., Pathology of Cutaneous Burns, *Surg. Clin. N. Amer.*, 1970, *50*, (6), 1201–1210.
21. FROM, A.H.L., GEWURZ, H., GRUNINGER, R.P., Complement in Endotoxin Shock. Effect of Complement Depletion on the Early Hypotensive Phase, *Infection Immunity*, 1970, *2*, (1), 38–41.
22. GAVRILESCU, S., *Şocul Cardiogen* (Cardiogenic Shock), I.M.F., Timişoara, 1970.
23. GIERTZ, H., MITZE, U., Die Wirkung von Theophyllin, Isoprenalin, Orciprenalin und Dexamethason auf den Protrahierten Anaphylaktischen Schock des Meerschweinchens, *Arzneimittel-Forsch.*, 1970, *20*, (10), 1551–1554.
24. GUENTER, C.A., HINSHAW, I.B., Haemodynamic and Respiratory Effects of Dopamine on Septic Shock in the Monkey, *Amer. J. Physiol.*, 1970, *219*, (2), 335–339.
25. HENRY, J.N., GELFAND, E.W., HINCHEY, E.J., Gram Negative Endotoxin Shock due to Serratia Marcescens, *Canad. Med. Ass. J.*, 1970, *102*, (1), 45–48.
26. HEWITT, R.L., SMITH, Jr. A.D., WEICHERT, R.F., DRAPANAS, T., Penetrating

Cardiac Injuries Current Trends in Management, *Arch. Surg.*, 1970, *101*, (6), 683–688.
27. HINSHAW, I.B., MATHIAS, M.C., NANETO, J.A., HOLMES, D.D., Recovery Patterns and Lethal Manifestations of Live *E. coli* Organism Shock, *J. Trauma*, 1970, *10*, (9), 787–794.
28. HOPKINSON, D.A.W., MARSHALL, T.K., Firearm Injuries, *Brit. J. Surg.* 1967, *54*, (5), 344–353.
29. IONESCU, A., RĂDULESCU, V., VASILIU, A., *Arsurile, clinică, fiziopatologie, tratament* (Burns, Clinics, Pathophysiology, Treatment), Editura Medicală, Bucureşti, 1970.
30. KRAGELUND, E., Loss of Fluid and Blood to the Peritoneal Cavity during Abdominal Surgery, *Surg.*, 1971, *69*, (2), 284–287.
31. KUHN, L.A., Shock in Myocardial Infarction. Medical Treatment, *Amer. J. Cardiol.*, 1970, *26*, (6), 578–587.
32. KUX, M., HOLMES, D.D., HINSHAW, I.B., MASION, W.H., Effects of Injection of Live *E. Coli* Organism in Dogs after Denervation of the Abdominal Viscera, *Surg.*, 1971, *69*, (3) 392–398.
33. LEDINGHAM, I. McA., Septic Shock, *Brit. J. Surg.*, 1975, *62*, (10), 777 – 780.
34. LILJEDAHL, S.O., Braunskadebehandling, *Opusc. Med.*, 1970, *15*, (6), 179–194.
35. LITTON, A., Gram-negative Septicaemia in Surgical Practice, *Brit. J. Surg.*, 1975, *62*, (10), 773–776.
36. MCFARLANE, J.K., SCOTT, H.J., CRONIN, R.F.P., Effects of Left Heart Bypass in Experimental Cardiogenic Shock, *J. Thorac. Cardiovasc. Surg.*, 1969, *57*, (2), 214–224.
37. MCLEAN, L.D., MCLEAN, A.P.H., DUFF, J.H., Hemodynamic and Metabolic Abnormalities in Septic Shock, *Postgrad. Med.*, 1970, *48*, (3), 114–122.
38. MAREŞ, E., NESTORESCU, N., BONA, C., Imunologia bolii arşilor (The Immunology of Burnt Patients), *Revista sanitară militară*, 1965, *LXI*, (special number), 85–92.
39. MARKLEY, K., SMALLMAN, E., Protection Against Burn. Tourniquet and Endotoxin Shock by Purine Compounds, *J. Trauma*, 1970, *10*, (7), 598–607.
40. MILSTEIN, B.B., Acute Circulatory Arrest, *Brit. J. Surg.*, 1967, *54*, Lister Centenary Number, 471–473.
41. MOFFAT, W.C., Recent Experience in the Management of Battle Casualties, *J. Roy. Army. Med. Corps*, 1967, *113*, (1), 25–32.
42. MONAFO, W.W., The Treatment of Burn Shock by the Intravenous and Oral Administration of Hypertonic Lactate Saline Solution, *J. Trauma*, 1970, *10*, (7), 575–586.
43. MONCRIEF, J., *Use of Antimicrobial Agents in Burns, in Antimicrobial Therapy*, W.B. Saunders Co., Philadelphia, 1970.
44. MOTSAY, G.J., DIETZMAN, R.H., ERSEK, R.A., LILLEHEI, R.C., Hemodynamic Alterations and Results of Treatment in Patients with Gram-negative Septic Shock, *Surg.*, 1970, *67*, (4), 577–583.
45. MUNDTH, E.D., Circulatory Assistance and Emergency Direct Coronary-Artery Surgery for Shock Complicating Acute Myocardial Infarction, *New Engl. J. Med.*, 1970, *283*, (9), 1382–1384.
46. MUNDTH, E.D., SOKOL, D.M., LEVINE, F.H., AUSTEN, W.G., Evaluation of Methods for Myocardial Protection during Extended Periods of Aortic Cross Clamping and Hypoxic Cardiac Arrest, *Bull. Soc. Int. Chir.*, 1970, *29*, (4), 227–235.
47. MUNDTH, E.D., BUCKLEY, M.J., LEINBACH, R.C., DESANTIS, R.W., SANDERS, C.A., KAUTROWITZ, A., AUSTEN, W.G., Myocardial Revascularization for the Treatment of Cardiogenic Shock Complicating Acute Myocardial Infarction, *Surg.*, 1971, *70*, (1), 78–87.
48. NEELY, W.A., BERRY, D.W., RUSHTON, F.W., HARDY, J.D., Septic Shock: Clinical, Physiological and Pathological Survey of 244 Patients, *Surg.*, 1971, *173*, (4), 657–666.
49. NICKERSON, M., Treatment of Shock Associated with Sepsis, *Int. J. Gynec. Obst.*, 1970, *8*, (4), 636–641.
50. NOON, G.P., BOULAFENDIS, D., BEALL, Jr. A.C., Rupture of the Heart Secondary to Blunt Trauma, *J. Trauma*, 1971, *11*, (2), 122–128.
51. PĂUNESCU-PODEANU, A., *Şocul anafilactic*, (Anaphylactic Shock), I.M.F., Timişoara 1970.
52. PERRY, M.O., THAL, E.R., SHIRES, G.T., Management of Arterial Injuries, *Ann. Surg.*, 1971, *173*, (3), 403–408.
53. POLK, Jr. H.C., VARGAS, A., The Prevention of Postoperative Renal Failure, *Sth. Med. J.*, 1970, *63*, (9), 1068–1071.
54. PRICE, W.R., WOOD, M., COOK, F., GARODIMORE, B., Enzyme Depletion in Major Thermal Burns, *Amer. J. Surg.*, 1970, *120*, (5), 671–675.
55. PROCTOR, H.J., LENTZ, T.R., JOHNSON, Jr. G., HILL, C., Alterations in Baboon Erythrocyte 2,3-Disphosphoglycerate Concentration Associated with Haemorrhagic Shock and Resuscitation, *Ann. Surg.*, 1971, *174*, (6), 923–831.
56. RAPIN, M., DIAMANT, B.F., Traitement du choc septique, *Rev. Prat.*, 1968, *18*, (31 bis), 158–165.

57. ROBERTSON, I., Ballet Injuries, *Proc. Roy. Soc. Med.*, 1967, *60*, (5), 530–538.
58. SANDOE, E., Kardiogent Shock, *Nord. Med.*, 1970, *83*, (22), 681–689.
59. SAWYER, R.B., DREW, M.A., GESIUK, M.H., Postoperative Magnesium Metabolism, *Arch. Surg.*, 1970, *100*, (4), 343–348.
60. SCHEIDT, S., ASCHEIM, R., KILLIP, T., Shock after Acute Myocardial Infarction. A Clinical and Hemodynamic Profile, *Amer. J. Cardiol.*, 1970, *26*, (6), 556–564.
61. SCHMOOKLER, A., Septic Shock. Complications Encountered by the Anesthesiologist, *Minn. Med.*, 1970, *53*, (3), 399–401.
62. SCHWIMMER, D., Septic Shock, Management of Intra-Abdominal Abscesses, *Minn. Med.*, 1970, *53*, (3) 389–391.
63. SHOEMAKER, W.C., MOHR, A., PRINTEN, K.J., Use of Sequential Physiologic. Measurements for Evaluation and Therapy of Uncomplicated Septic Shock, *Surg. Gynec. Obst.*, 1970, *131*, (2), 245–254.
64. SIEGEL, J.H., GOLDWYN, R.M., FRIEDMAN, H.P., Pattern and Process in the Evolution of Human Septic Shock, *Surg.*, 1971, *70*, (2), 232–245.
65. SIEGEL, D.C., REED, P.C., FRESQUEZ, V., COCHIN, A., MOSS, G.S., Renal Effects of Massive Infusions of Adenine during Resuscitation from Haemorrhagic Shock in the Baboon, *Ann. Surg.*, 1971, *174*, (6), 932–938.
66. SKILMAN, J.J., HEDELY-WHYTE, J., PALLOTTA, A.J., Cardiorespiratory Metabolic and Endocrine Changes after Haemorrhage in Man, *Ann. Surg.*, 1971, *174*, (6), 911–922.
67. SKINNER, R.B., RACITI, A., SABATIER, Jr. H.S., Surgery for Acute Myocardial Infarction: Coronary Flow and Heart Work during Total Circulatory Support, *Surg.*, 1970, *68*, (1), 128–135.
68. STARZECKI, P., Effect of Isoproterenol on Survival in Canine Endotoxin Shock, *Ann. Surg.*, 1968, *167*, (1), 33.
69. STRING, T., ROBINSON, A.J., BLAISDELL, F.W., Massive Trauma Effect of Intravascular Coagulation on Prognosis, *Arch. Surg.*, 1971, *102*, (4), 406–411.
70. TEODORESCU-EXARCU, I., *Agresologie chirurgicală generală (General Surgical Stress Science)*, Editura Medicală, Bucureşti, 1968.
71. TERES, D., STRACK, P.R., REGAN T.J., Enhanced Ventricular Function after Beta Blockade in Haemorrhagic Shock, *Arch. Int. Pharmacodyn.*, 1970, *186*, (2), 243–254.
72. THOMSON, D.S., EASON, C.N., Hypoxemia Immediately after Operation, *Amer. J. Surg.*, 1970, *120*, (5), 649–651.
73. THORESBY, M.F.P., DARLOW, H.M., The Mechanism Primary Infection of Bullet Wounds, *Brit. J. Surg.*, 1967, *54*, (5), 359–361.
74. TODD, A.C., Pulmonary Function and Surgery, *J. Amer. Osteopath. Ass.*, 1970, *69*, (9), 856–875.
75. TRINKLE, J.K., Mechanical Support of the Circulation: a New Approach, *Arch. Surg.*, 1970, *101*, (6), 740–743.
76. VIDNE, B., LEVINSON, S., LEVY, M.J., Venous Pressure Monitor, *Surg.*, 1970, *67*, (2), 279–280.
77. WEIL, M.H., SCHUBIN, H., *Diagnosis and Treatment of Shock*, Williams and Wilkins Co., Baltimore, 1967.
78. WEIL, M.H., Experimental and Clinical Studies on Lactate and Pyruvate as Indicators of the Severity of Acute Circulatory Failure (Shock), *Circulation*, 1970, *41*, (8), 989–1001.
79. WILLIAMS, T.B., WORMAN, R.K., JACOBS R.R., CHENK, W.C. Jr., An Electromagnetic Probe for the Quantitation of Intracardiac Flows, *Surg.*, 1970, *67*, (6), 1031–1034.
80. WHITE, IV, R.R., MELA, L., MILLER, L.D., BERWICK, L., Effect of *E. Coli* Endotoxin on Mitochondrial Form and Function, *Ann. Surg.*, 1971, *174*, (6), 983–990.
81. ZINGG, W., SULEV, J.C., MORGAN C.D., Relationship between Viscosity and Hematocrit in Blood of Normal Persons and Burn Patients, *Canad. J. Phyol.*, 1970, *48*, (3), 202–205.

AUTHORS INDEX

Adams-Stokes, 384, 401
Addison, 7, 162, 164, 165
Ahlquist, 9
Allgöwer, 237, 352
Allison, 370
Altura, 218
Ambu, 383
Andersen-Engel, 76
Aristotle, 129
Arthus, 435
Attar, 268

Baglioni, 3
Bainbridge, 8, 186, 356
Ballinger, 1, 5, 12, 14
Barankay, 228
Barker-Sumerson, 76
Baue, 327, 340
Bayliss, 8, 19
Berger, 439
Bergström, 222
Berk, 402
Berman, 269, 367
Bernard, 144, 300, 302, 315, 325, 356
Bernoulli, 53, 252, 253
Beufay, 114
Bevegard, 439
Bigelow, 54
Billy, 381
Bingham, 252
Blalock, 9, 35
Blaisdell, 367
Bland, 237
Bloodgood, 7, 19
Bohr, 248, 286
Bounous, 350, 353
Brandt, 275
Brînzeu, 18, 21, 41
Broido, 213, 272
Brooks, 14, 398
Brown, 371
Brown-Séquard, 7, 19
Buckberg, 367
Bucur, 3
Burghele, 11, 26, 360
Burlui, 359
Burnet, 168
Bușu, 198
Byrne, 36, 37, 101, 406
Bywaters, 9, 419, 421, 422

Cafriță, 56, 259, 419
Cahill, 11, 367
Caloghera, 355
Campbel, 395
Canizaro, 12
Cannon, 8, 9, 19, 106
Carey, 389
Carrel, 212, 313
Carrico, 12
Celsus, 6, 19
Chambers, 10, 258, 275
Chatelier, le, 333
Chiotan, 11, 18, 42
Chiotan-Tabacu-Cristea, 198
Chiricuță, 18, 43
Christmas, 260
Ciurel, 359
Clausius, 331
Cohn, 260
Collins, 368
Conn, 166
Cope, 11
Cori, 299, 303, 357
Couffignal, 6, 115
Crabtree, 3, 284, 285, 293, 309, 327,
Crile, 8, 19, 281, 376
Cristea, 11, 18, 42
Curling, 381, 432, 434
Cushing, 347, 381

Dale, 8, 19, 220
Darrow, 32
Das Gupta, 372
Davis, 21, 43

DeBakey, 381
DeDuve, 114, 328
DePalma, 327
DeTakats, 177, 198
Delbet, 8, 19
DelGuercio, 12
Deloyers, 1
Denis, 439
Derks, 368
Diacenko, 10
Dillon, 5, 18, 113, 391
Diomi, 362
Dobzhansky, 3
Dogru, 12, 188
Donnan 242, 245, 246
Dorfman, 145
Douglas, 8
Dragstedt, 9
Dran, Le, 7
Drummond, 149
DuCailar, 197
Dunphy, 16
Duval, 8

Eccles, 130
Edelman, 380
Ehrlich, 168
Eliot, 10
Embden-Meyerhof, 303, 304, 318
Essiet, 340
Euler, 133

Fabricius, 168
Fahreus-Lindquist, 254
Farrell, 157
Fattorusso, 237
Fick, 361
Fischer, 7, 19
Finch, 159
Fine, 10, 17, 51, 58, 59, 60, 416, 426
Fishberg, 32
Fitts, 400
Fleisher, 237, 244
Forscher, 12
François-Franck, 7
Franck-Starling, 186
Freeman, 9
Frey-Werle, 272
Friedman, 76
Frohlich, 244
Fromm, 349
Fulton, J.F., 12
Fulton, R.L., 113, 213

Gallagher, 158
Gazzaniga, 329
Gecse, 220
Gedda, 119
Gelin, 10, 17, 251, 269
Geocaris, 403
Gibston, 384
Giurgiu, 56
Glenn, 228
Golden, 192, 194
Goldfarb, 373
Golgi, 216
Golz, 7, 19
Goormaghtigh, 159
Grant, 9, 25
Graves, 165
Grigg, 6
Gross, 18
Grönwall, 394
Gruber, 237, 391
Gudden, 132
Guédel, 383
Gump, 372
Gutelius, 113
Guthrie, 7

Hageman, 260, 262, 263, 273
Hagen-Poiseuille, 253, 254
Haldane, 248, 286
Halmagyi, 154, 198
Hardaway, 4, 5, 11, 17, 36, 45, 112, 258, 259, 263, 400, 419
Harison, 32
Harrison, 54
Hartmann, 397
Hayasaka, 11
Heimbach, 187
Hench, 9
Henderson, 8, 287
Henle, 205, 225
Hermann, 115
Hershey, 11, 12, 204, 218
Hiebert, 154
Hift, 327,355
Hinshow, 425
Hogan, 395
Holger-Nielsen, 382
Hollander, 10
Hopkinson, 420
Hortolomei, 198
Howard, 11
Huckabee, 10
Huggins, 392
Hunt, 11

Ingram, 12
Ingvar, 138
Iida, 269
Ionescu, B., 121, 148, 150, 151, 152, 162
Ionescu, P., 56

Jane, 194
Jenkins, 130
Jerne, 168
Johansen, 371
Johnson, 36, 340
Jones, R. G. 359
Josse, 272
Jouvet, 132

Karli, 131
Katchalsky, 335
Kaulla, von, 268
Kendall, 9
Kennedy, 378
Kerr, 196
King, 270
Kirklin, 196
Kinney, 282
Kirimli, 194
Klebanoff, 244
Knight, 12
Knisely, 9, 10, 181, 205, 258, 268
Kobald, 19,
Kouvenhoven-Knikerboker, 383
Kragelund, 398
Kulowski, 10
Kultschitzky-Masson, 221
Kunitz-Northrop, 272
Kupffer, 205, 218, 283
Kussmaul, 286
Kux, 425
Kwaan, 359

Laborit, 11, 18, 22, 26, 38, 39, 53, 99, 104, 175, 216, 217, 259, 283, 306, 314
Lacroix, 203
Lancisi, 132
Laplace, 255
Larcan, 247, 252, 268
Lareng, 203
Larrey, 7, 19
Latta, 7
Lázár, 218
Ledingham, 12
Leena, Mela, 340
Lefer, 228, 275
Leger, 198
Lewis, 10
Lichtmann, 367
Lieberkühn, 221
Lillehei, 4, 10, 12, 44, 51, 181, 350, 352, 356, 406, 416
Linder, 348
Lister, 8, 11, 31,
Litarczek, 18, 24, 110, 400
Litwin, 11, 252, 394
Lorthioir, 432
Lovelock, 392
Lovett, 275
Lucke, 421
Luft, 85

Magnus, 254
Magri, 338
Mahfouz, 206
Malerne, 8
Mallet-Guy, Y., 329
Mallory, 421
Mandache, 11, 18, 26, 42
Mann, 8
Mapothei, 7, 8
Mareş, 18, 22, 41, 43, 419
Markwardt, 268
Matsumoto, 166, 194
Mautner-Pick, 205
Mazor, 4
Mazur, 275
Mays, 355
McAllister, 12
McFarlane, 438
McKay, 11, 348
McLaughlin, 367
McLean, 431
Messmer, 12, 18
Michaelis, 309
Milcu, 121, 123, 148, 151
Milligan, 368
Mills, 11
Mitchinson, 247
Mollaret-Goulon, 142
Monod, 3
Moon, 9, 19, 20
Moore, 10, 11, 16, 17, 22, 23, 38, 40, 295, 306, 413
Monro-Kellie, 138
Morel, 149
Mori, 143
Morris, 7
Moss, 367, 372, 389, 391, 398
Motsay, 431
Moyer, 11
Mundeleer, 286
Mundth, 439
Müller, 158

Nash, 389
Natof, 10
Nauta, 132
Neely, 428
Newton, 247
Nilsson, 268
Noble-Collip, 51
Nunes, 355
Nyhus, 12

Oettingen, von, 371
Ogilvie, 420
Ohno, 119
Olsen, 194, 203
Onsager, 332, 333, 332
Ornellas, 306
O'Shaughnessy, 287
Osler, 345
Osteen, 244

Page, le, 294
Paget, 85
Palade, 85
Palmeiro, 11
Papez, 132
Park, 8, 24
Pascal, 235
Pasteur, 284, 285, 309
Pavlov, 19, 20
Pearson, 53
Petrovici, 85
Petrov, 42
Peyer, 168
Pfiffmer, 9
Phemister, 9
Pirogov, 7, 19
Plouvier, 247
Poiseuille, 235
Poisson, 53
Polonovski, 282, 300
Pontoppidan, 386
Popper, 205
Porter, 12
Postnicov, 42
Powers, 362, 365, 372
Proctor, 248
Pullman, 241

Quénu, 8, 19
Quincke, 436

Randle, 298
Rapin, 391
Raven, 1,
Raynaud, 268
Reeves, 9, 25
Reilly, 427
Reissner, 145
Reitman 51
Reitman-Frenkel, 85
Reul, 394
Reynolds, 253
Richards, 8, 220
Rhoads, 11
Rhodes, 349
Richardson, 7, 19, 349
Richet, 8
Ricker, 10, 20
Riordan, 197
Rittenhouse, 338
Robison, 301
Rocchio, 371
Rosato, 252
Rosenthal, 260
Rosoff, 269
Rouget-Eberth, 199
Rowe, 403
Rush, 397
Rusher, 10
Rutherford, 12, 14

Sadove, 10
Saegesser, 42, 43
Safar, 378. 382, 383, 384
Sardesai, 272
Schäffer, 382
Schloerb, 282
Schumer, 114, 282, 289, 318
Schwartz, 12
Schwartzman-Sanarelli, 164
Schweitzer, 95
Scott-Blair, 254
Scudder, 9
Selye, 9, 13, 18, 22, 26, 27, 38, 104, 167
Senning, 439
Sevitt, 14, 16, 38
Sheehan, 29, 163
Shepro, 12
Shires, 12, 16, 113, 282, 295, 379, 389, 390, 391, 397, 398
Shoemaker, 11, 12, 16, 269, 355, 389, 419, 431
Shorr, 10, 58, 275
Shubin, 11, 239, 415
Shumer, 12
Siegel, D.C., 367, 371, 373, 391
Siegel, J.H., 20, 426
Silberschmidt, 368
Silvester, 382
Simeone, 10, 16, 143, 381, 416
Simmonds, 162
Simon, 194, 203
Singer, 329
Skinner, 400
Slyke, van, 287
Spencer, 12
Spink, 10
Stahl, 340
Staples, 294
Starling, 180, 225, 228
Starling-Bargeton, 180
Stokes, 251
Stoltz, 268
Stone, 362
Storer, 12
Strawitz, 327, 355

Strîmbeanu, 329
Stuart-Prower, 260
Stubbs, 12
Student, 53
Sunder-Plassmann, 12, 18
Sutherland, 10, 123
Swingle, 8

Șuteu, 3, 4, 5, 18, 22, 41, 43, 56, 196, 259, 281, 419

Teodorescu-Exarcu, 2, 12, 15, 22, 32, 38, 39, 104, 226
Thal, 11, 12, 274
Thompson, 396
Travers, 7
Trendelenburg, 8, 21, 370
Trevan, 8
Trinkle, 391
Trunkey, 340

Țurai, 10, 18, 28, 31, 43, 359

Ungar, 130

Van't Hoff, 244
Văideanu, 419
Verney, 161,
Vernon, 18
Vicq d'Azyr, 132
Voigt-Kelvin, 247

Waisbren, 426
Wald-Serra, 53
Walters, 109
Wangensteen, 12, 255, 275
Warburg, 355
Warren, 18
Waterhouse-Friederichsen, 104, 164, 429
Weber, 306
Weedn, 371
Weil, 10, 11, 18, 24, 239, 415
Weissbluth, 241
Werle, 9, 10, 416
Werner, 222
Westerfield, 9
Westphal, 149
White IV, 326, 340
Wiedersheim, 396
Wiggers, 9, 51, 68, 84, 198, 318, 416
Wiggers-Johnson, 51
Williams, 145
Wilson, 370
Wrisberg, 187
Wosornu, 400
Wyche, 387

Yates, 152
Young, 247
Young-Shrago-Lardy, 301

Zederfeldt, 252
Zollinger-Ellison, 166, 347, 423
Zuidema, 12, 14
Zucker, 272
Zweifach, 10, 11, 58